ANNUAL REVIEW OF NUCLEAR AND PARTICLE SCIENCE

ANNUAL REVIEW OF NUCLEAR AND PARTICLE SCIENCE

J. D. JACKSON, *Editor*
University of California, Berkeley

HARRY E. GOVE, *Associate Editor*
University of Rochester

ROY F. SCHWITTERS, *Associate Editor*
Harvard University

VOLUME 29

1979

ANNUAL REVIEWS INC.   4139 EL CAMINO WAY   PALO ALTO, CALIFORNIA 94306

ANNUAL REVIEWS INC.
Palo Alto, California, USA

REPRINTS   The conspicuous number aligned in the margin with the title of
each article in this volume is a key for use in ordering reprints. Available
reprints are priced at the uniform rate of $1.00 each postpaid. The minimum
acceptable reprint order is 5 reprints and/or $5.00 prepaid. A quantity discount
is available.

International Standard Serial Number: 0163-8998
International Standard Book Number: 0-8243-1529-4
Library of Congress Card Number: 53-995

Annual Reviews Inc. and the Editors of its publications assume no
responsibility for the statements expressed by the contributors to this Review.

FILMSET BY TYPESETTING SERVICES, LTD, GLASGOW, SCOTLAND
PRINTED AND BOUND IN THE UNITED STATES OF AMERICA

SOME RELATED ARTICLES IN OTHER
ANNUAL REVIEWS

From the *Annual Review of Astronomy and Astrophysics*, Volume 17 (1979):

Computer Image Processing, Ronald N. Bracewell
Digital Imaging Techniques, W. Kent Ford, Jr.
Physics of Neutron Stars, Gordon Baym and Christopher Pethick

From the *Annual Review of Earth and Planetary Sciences*, Volume 7 (1979):

Geochemical and Cosmochemical Applications of Nd Isotope Analysis, R. K. O'Nions,
    S. R. Carter, N. M. Evensen, and P. J. Hamilton
The Magnetic Fields of Mercury, Mars, and Moon, Norman F. Ness
*The Magnetic Field of Jupiter: A Comparison of Radio Astronomy and Spacecraft
    Observations*, Edward J. Smith and Samuel Gulkis

From the *Annual Review of Energy*, Volume 4 (1979):

United States Energy Alternatives to 2010: The CONAES Study, Harvey Brooks
    and Jack M. Hollander
Assessment of Global Uranium Resources, DeVerle P. Harris
Canadian Energy Policy, John F. Helliwell
Supply Assurance in the Nuclear Fuel Cycle, Henry D. Jacoby and Thomas Neff

From the *Annual Review of Materials Science*, Volume 9 (1979):

From Solid State Research to Semiconductor Electronics, Heinrich J. Welker
Radiation Effects in Structural Materials, J. O. Stiegler and L. K. Mansur

From the *Annual Review of Physical Chemistry*, Volume 30 (1979):

*Developments in Extended X-ray Absorption Fine Structure Applied to Chemical
    Systems*, Donald R. Sandstrom and Farrel W. Lytle
Molecular Beam Studies of Collisional Ionization and Ion-Pair Formation,
    S. Wexler and E. K. Parks
Molecular Spectroscopy, E. Bright Wilson

Annual Review of Nuclear and Particle Science
Volume 29, 1979

CONTENTS

Ann. Rev. Nucl. Part. Sci. 1979. 29 : 1–31

RECENT DEVELOPMENTS IN COMPOUND-NUCLEUS THEORY

✻560

Claude Mahaux

Institut de Physique, Université de Liège, 4000 Liège, Belgium

Hans A. Weidenmüller

Max-Planck-Institut für Kernphysik, 69 Heidelberg, Germany

CONTENTS

1

0163-8998/79/1201-0001$01.00

1   INTRODUCTION

In a celebrated address delivered before the Royal Danish Academy on January 27, 1936, Niels Bohr launched the compound-nucleus model with the sentence: "The phenomena of neutron capture force us to assume that a collision between a high-speed neutron and a heavy nucleus will in the first place result in the formation of a compound system of remarkable stability; the later breaking up of this intermediate system . . . must in fact be considered as a separate process which has no immediate connection with the first stage of the encounter" (Bohr 1936). The evaporation model (Weisskopf 1937) is a mathematical formulation of Niels Bohr's physical picture of the independence of formation and decay of the compound nucleus (CN). It neared its final form in the work of Wolfenstein (1951) and of Hauser & Feshbach (1952). These authors predicted that the angular distribution of the emitted particles is symmetric about 90°. Does this model work? How can it be improved? What are its microscopic foundations? These questions have captured the imagination of many physicists ever since the first pioneering papers were published. Only in the last few years have satisfactory answers finally emerged.

Compound-nucleus theory is devoted to the foundations and to the modifications of the evaporation model. This field is intellectually challenging: it lies at a crossroad of scattering theory, many-body physics, statistical mechanics, and nuclear theory. It is also of considerable practical use: recent theoretical developments have already been incorporated into computer codes that are used daily by applied physicists (Prince 1978). The field recently reached maturity but remains lively, and is steadily expanding, because of a close feedback interaction between theoretical and experimental investigations.

1.1   *The Problems*

The evolution of the evaporation model was promoted by the discovery of new features of CN reactions.

1. The advent of the optical model (Barschall 1952, Feshbach, Porter & Weisskopf 1954) revolutionized the time-honored belief, based on the CN model, that the mean free path of a nucleon inside the nucleus is short in comparison with the nuclear radius. In spite of this discovery, the evaporation model was found to yield good fits to the data.

2. The early theoretical justifications of the evaporation model (Bethe 1937) emphasized the domain of isolated resonances ($D \gg \Gamma$). But the model also works well at excitation energies (several hundred keV above neutron threshold or more) where the average level spacing D is small in comparison with the average total width Γ, or even in comparison with

the partial width in a given channel (strong absorption). This challenged theorists to justify the model for $\Gamma \gtrsim D$.

3. The observation of angular distributions for inelastic scattering that were not symmetric about 90° led to the development of direct interaction (DI) theories and raised the problem of combining the apparently conflicting CN and DI models (Sections 5.1, 5.2, and 6).

4. The discovery of precompound reactions (Sidorov 1962, Holbrow & Barschall 1963) and their subsequent theoretical analysis (Griffin 1966, 1967) somewhat bridged the conceptual gap between CN and DI processes. For $\Gamma \gg D$, the evaporation model describes the behavior of an equilibrated system. This is why the system loses memory of the incident channel (save for the conserved quantum numbers) and, by the same token, why formation and decay of the CN are independent processes. The equilibration of the compound system requires some time. As the compound system develops toward equilibrium, configurations with increasing complexity are populated from the incident channel (Weisskopf 1961, Feshbach, Kerman & Lemmer 1967, Feshbach 1974); precompound reactions consist in the emission of particles from configurations of intermediate complexity, before equilibrium is reached. Their microscopic study considerably enlarged the scope of CN theory (Sections 5.2 and 8).

5. Even at low energy, near neutron threshold, there exist hundreds of thousands of CN resonances per MeV in intermediate and heavy nuclei. The density of these resonances rises roughly exponentially with excitation energy (Huizenga & Moretto 1972). An individual description of the resonances is therefore hopeless, and all derivations of the evaporation model are based on statistical arguments. These show that the evaporation model only describes cross sections averaged over an energy interval of length $I \gg \Gamma, D$. Such a description implies the existence of statistical fluctuations (Ericson 1960, 1963, Brink & Stephen 1963) in the actual (nonaveraged) cross sections. The observation of these fluctuations generated a flurry of theoretical activity in the early 1960s (Ericson & Mayer-Kuckuk 1966), but several questions remained open. Only recently was the theory put on firm grounds (Section 5.3).

6. In recent years, a new range (10^{-14} to 10^{-18} s) of CN lifetimes has become available through the crystal blocking technique (Gibson 1975), which provides a promising way of investigating the CN.

7. CN theory must also encompass the intermediate structure phenomenon (Kerman, Rodberg & Young 1963, Feshbach, Kerman & Lemmer 1967). In particular, the discovery of isolated isobaric analogue resonances located upon a dense background of CN resonances (Fox, Moore & Robson 1964) posed a challenging problem (Section 6.2).

8. With increasing excitation energy, the spacing of isobaric analogue

resonances decreases. Eventually, one encounters two types of overlapping resonances that can be characterized by different isospin quantum numbers. These two classes of states are mixed by Coulomb forces. The effect of isospin mixing on CN reactions has been the subject of several recent experimental and theoretical investigations (Section 7).

1.2   *Main Theoretical Approaches*

The ultimate goal of CN theory is to give a unified description of all the above-mentioned phenomena on the basis of dynamical properties of the nuclear Hamiltonian. Since the theory necessarily involves statistical assumptions, it should also establish a connection with the statistical theory of nuclear spectra (Brody et al 1978). The latter successfully describes the local fluctuations and the global mean values of the distribution of eigenvalues and of eigenvectors of the nuclear Hamiltonian. In particular, statistical spectroscopy has met with spectacular success in explaining the statistical properties of isolated resonances (Porter 1965). CN theory can be viewed as an extension of this theory to yet higher excitation energies, and to the calculation of the averages of the cross sections and of their fluctuations.

In the 1960s, most efforts (Moldauer 1964c, also Section 3) to formulate CN theory were based on a description of CN resonances in terms of poles of the collision matrix (Humblet & Rosenfeld 1961). The distribution of the pole parameters was hoped to obey simple laws, but, unfortunately, numerical calculations showed it to be complicated (Moldauer 1964b, 1968, Sections 3.4 and 4.3). It was later shown analytically (Engelbrecht & Weidenmüller 1973) that for $\Gamma \gtrsim D$ flux conservation (unitarity of the collision matrix) forbids the use of simple assumptions on the distribution of the pole parameters. In retrospect, it is therefore not surprising that CN theory made little progress in the 1960s.

The field received considerable new impetus from the work of Kawai, Kerman & McVoy (1973, Section 5.1). For $\Gamma \gg D$, these authors were the first to give a microscopic support to a conjecture (Equation 2.8) of Vager (1971) concerning the effect of DI on the average CN cross section. Their argument does not fully take into account the unitarity condition 2.1 (Moldauer 1975a). Besides generating intriguing questions on the stringency of the unitarity requirement (Kerman & Sevgen 1976, Section 5.1), this stimulated the development of other theoretical approaches, which fully obey the unitarity condition 2.1.

In these approaches, the statistical assumptions bear on the distribution of the matrix elements of the nuclear Hamiltonian rather than on the distribution of the pole parameters; these assumptions can be linked to the ones made in the statistical theory of nuclear spectra. The difficulty lies in the evaluation of the CN cross sections and of other observables. Indeed, these quantities involve the square (or higher powers) of matrix elements of the operator $(E + i\varepsilon - H)^{-1}$, the statistical distribution of which is not known explicitly. The unitary approaches can be grouped into two main categories:

1. The K-matrix parametrization 4.1 of the S-matrix is used as input, with a distribution of K-matrix parameters as given by the statistical theory of spectra. One performs *numerical* calculations of average CN cross sections and of other observables. This procedure can be used for any value of Γ/D. The calculations of Hofmann et al (1975a, Section 4.2) and of Moldauer (1975a, Section 4.4) have yielded convenient and quite accurate fit formulas. In the absence of DI, the results essentially confirm the validity of the basic physical assumption (Equation 2.6) of Bohr's CN model. The effect of DI can be taken into account analytically by making use of a transformation due to Engelbrecht & Weidenmüller (1973, Section 6.1).

2. Formulas for average CN cross sections and other observables are derived *analytically* from the statistical distribution of the matrix elements of the nuclear Hamiltonian. Aside from the straightforward case of isolated resonances ($\Gamma \ll D$; Lane & Lynn 1957, Section 3.2) this has at present been possible only for strongly overlapping resonances, $\Gamma \gg D$. Systematic rules for expanding the observables in powers of D/Γ were found by Agassi, Weidenmüller & Mantzouranis (1975, Section 5.2). These rules are sufficiently general to yield analytic results not only for average CN cross sections including DI, but also for Ericson fluctuations (Section 5.3), for isospin mixing (Section 7), and for precompound processes. Significant progress is currently being made in the latter field, discussed briefly in Section 8.

2 FACTORIZATION ASSUMPTIONS FOR THE AVERAGE CN CROSS SECTION

The evaporation model assumes that the average CN cross section (Section 2.1) can be factorized (Equations 2.6 and 2.8). The subsequent sections describe to what extent this conjecture has been confirmed or modified by microscopic approaches.

2.1 *Average CN Cross Section: Transmission Coefficients*

A reaction channel is specified by a set of quantum numbers that we generically denote by small indices $a, b, \ldots$. The collision (or S-) matrix element S_{ab} is the ratio of the amplitude of the outgoing wave in channel b to that of the incoming wave in the entrance channel a (Lane & Thomas

1958). The S-matrix is symmetric (time-reversal invariance) and unitary (conservation of flux):

$$S_{ab} = S_{ba}, \qquad (SS^\dagger)_{ab} = \sum_{c=1}^{N} S_{ac}S_{bc}^* = \delta_{ab}, \qquad\qquad 2.1$$

where N is the number of open channels. Henceforth, we denote by $\langle f(E) \rangle$ the average of a quantity $f(E)$ over an interval of size I centered on E; angle brackets will also be used for an average over resonance indices (e.g. Equation 3.2) and for ensemble averages (Section 4.2).

The optical model yields the diagonal elements of the average S-matrix, which is given by $\langle S(E) \rangle = S(E + iI)$ (Brown 1959). DI theory (Austern 1970) aims at finding the off-diagonal elements of $\langle S(E) \rangle$. The fluctuating part of S_{ab} is defined by

$$S_{ab}^{FL}(E) = S_{ab}(E) - \langle S_{ab}(E) \rangle. \qquad\qquad 2.2$$

Except for trivial factors, the integrated cross section is given by $\sigma_{ab} = |S_{ab} - \delta_{ab}|^2$. Its average is the sum of a DI part and the average CN cross section

$$\langle \sigma_{ab} \rangle = \sigma_{ab}^{DI} + \langle \sigma_{ab}^{CN} \rangle, \qquad\qquad 2.3$$

$$\sigma_{ab}^{DI} = |\langle S_{ab} - \delta_{ab} \rangle|^2, \qquad \langle \sigma_{ab}^{CN} \rangle = \langle |S_{ab}^{FL}|^2 \rangle. \qquad\qquad 2.4$$

The unitarity relation 2.1, yields

$$\sum_c \langle S_{ac}^{FL} S_{cb}^{FL*} \rangle = \delta_{ab} - \sum_d \langle S_{ad} \rangle \langle S_{db}^* \rangle = T_{ab}. \qquad\qquad 2.5$$

The last equation defines the generalized transmission coefficients T_{ab} (Satchler 1963). They measure the unitarity deficit of the average S-matrix. When $\langle S \rangle$ is diagonal, $T_{ab} = T_a \delta_{ab}$ where $T_a = 1 - |\langle S_{aa} \rangle|^2$ is the usual transmission coefficient in channel a.

2.2   Hauser-Feshbach Formula

The CN model implies that the formation and the decay of the CN are independent processes. This suggests that the average CN cross section can be factorized:

$$\langle \sigma_{ab}^{CN} \rangle = \mathcal{V}_a \mathcal{V}_b. \qquad\qquad 2.6$$

When DI are negligible, $\langle S \rangle$ and T are diagonal, and Equations 2.4–2.6 yield the "Hauser-Feshbach formula"

$$\langle \sigma_{ab}^{CN} \rangle = T_a T_b \left[\sum_c T_c \right]^{-1}. \qquad\qquad 2.7$$

The factorization assumption 2.6 for $\langle \sigma_{ab}^{CN} \rangle$ along with simple prescriptions for evaluating $\mathscr{V}_a$ formed the evaporation model (Landau 1937, Weisskopf 1937, Weisskopf & Ewing 1940, Blatt & Weisskopf 1952). Its "modern form" (Vogt 1968) added the following two main ingredients. (a) Wolfenstein (1951) and Hauser & Feshbach (1952) took into account the effect of the conservation of parity, and also of the total angular momentum and its projection along the quantization axis. They also used the statistical assumption that the contributions of different orbital angular momenta do not interfere. This enabled them to derive expressions for the angular distributions, which turn out to be symmetric about 90°. (b) The advent of the optical model (Feshbach, Porter & Weisskopf 1954) provided a reliable way of evaluating T_a, in particular of including in T_a the effect of the size resonances discovered by Barschall (1952).

2.3   Direct Interactions

Satchler (1963) suggested that the effect of DI is to replace the assumption 2.6 by $\langle S_{ab}^{FL} S_{cd}^{FL*} \rangle = X_{bd} X_{ac}$. Vager (1971) pointed out that this latter assumption is not in keeping with the symmetry of $\mathbf{S}^{FL}$ and proposed instead

$$\langle S_{ab}^{FL} S_{cd}^{FL*} \rangle = X_{bd} X_{ac} + X_{bc} X_{ad}. \qquad 2.8$$

The matrix $\mathbf{X}$ is related to the generalized transmission matrix by

$$\mathbf{T} = \mathbf{X} \, (\text{trace } \mathbf{X}) + (\mathbf{X})^2 ; \qquad 2.9$$

this follows from Equations 2.5 and 2.8.

In the strong absorption limit, one finds $\mathbf{X} = \mathbf{T} \, (\text{trace } \mathbf{T})^{-1/2}$. If in addition DI are negligible, one finds

$$\langle \sigma_{ab}^{CN} \rangle = (1 + \delta_{ab}) \, T_a T_b \left(\sum_c T_c \right)^{-1} . \qquad 2.10$$

In Vager's model, the appearance of the "elastic enhancement" factor $(1 + \delta_{ab})$ is intimately related to the symmetry of the S-matrix.

3   EARLY ATTEMPTS BASED ON THE POLE PARAMETRIZATION

3.1   The Pole Parametrization

Resonances can be identified with poles of the S-matrix (Humblet & Rosenfeld 1961). Causality requires that these poles be located in the lower half of the complex energy plane. If the poles are simple, and if branch point singularities are disregarded, one can write

$$S_{ab}(E) = S_{ab}^{(0)} - i \sum_{\mu=1}^{\Lambda} \frac{g_{\mu a} g_{\mu b}}{E - \varepsilon_\mu + \frac{1}{2} i \Gamma_\mu},$$ 3.1

where the number Λ of resonances can be infinitely large. Henceforth, the matrix $S^{(0)}$ will be approximated by a constant.

The pole parameters $g_{\mu a}$ (complex), ε_μ, and Γ_μ (real) have a physical meaning only in the case of isolated resonances. Nevertheless, many early papers on CN theory started from the pole parametrization, Equation 3.1 (Moldauer 1978), because $\langle \sigma_{ab}^{CN} \rangle$ can be expressed in terms of a few combinations of the pole parameters (Section 3.3). However, the evaluation of these expressions is possible only if the statistical distribution of the pole parameters is known. Except for the case of isolated resonances (Section 3.2), this is not the case (Section 4.3). Hence, the derivation of a reliable approximation (Section 4.4) for $\langle \sigma_{ab}^{CN} \rangle$ from the pole parametrization encounters severe difficulties (Section 3.4).

3.2   Isolated Resonances

The domain of isolated resonances extends to at most several hundred keV above neutron threshold. There, one easily finds that $\Gamma_\mu = \sum_a |g_{\mu a}|^2 = \sum_a \Gamma_{\mu a}$ and

$$\langle \sigma_{ab}^{CN} \rangle = 2\pi D^{-1} \left\langle \Gamma_{\mu a} \Gamma_{\mu b} \Big/ \sum_c \Gamma_{\mu c} \right\rangle = T_a T_b \left(\sum_c T_c \right)^{-1} W_{ab}.$$ 3.2

Here, D is the average distance between neighboring ε_μ, while $T_a = 2\pi D^{-1} \langle \Gamma_{\mu a} \rangle$. The symbol W_{ab} denotes the width fluctuation correction (WFC)

$$W_{ab} = \frac{\langle \Gamma_\mu \rangle \langle \Gamma_{\mu a} \Gamma_{\mu b} / \Gamma_\mu \rangle}{\langle \Gamma_{\mu a} \rangle \langle \Gamma_{\mu b} \rangle}.$$ 3.3

Various methods for calculating the WFC were recently reviewed by Gruppelaar & Reffo (1977). The WFC can be written as a one-dimensional integral if one assumes that the partial widths $\Gamma_{\mu a}$ are distributed according to a χ^2 law with ν_a degrees of freedom (Lane & Lynn 1957, Dresner 1957). In the domain of isolated resonances, $\nu_a = 1$ (Porter & Thomas 1956), and W_{aa} reaches the value 3 (Thomas 1956, Moldauer 1961) when several channels contribute ($W_{aa} = 1$ if $N = 1$). The unitarity sum rule, Equation 2.5, then shows that for $a \neq b$ W_{ab} is on the average somewhat smaller than unity; one can, however, have $W_{ab} > 1$ for individual channels $a \neq b$. This happens, for instance, if T_a and T_b are both small (Moldauer 1976, Reffo, Fabbri & Gruppelaar 1976).

3.3   *Average CN Cross Section in Terms of Pole Parameters*

As a means to obtain a uniform distribution of pole parameters, Moldauer (1964a,c) introduced the "statistical S-matrix." It is defined by the right-hand side of Equation 3.1, with the specification that the statistical properties of the pole parameters are the same outside the averaging energy interval as within it. We note that the statistical S-matrix involves an infinite number of resonances. Therefore, sum rules between the pole parameters of a model S-matrix for which the number Λ of resonances is finite are not always applicable to the statistical S-matrix. This was emphasized by Moldauer (1969a, 1971) and by Feshbach & Mello (1972). Moldauer (1975a) also pointed out that while $S^{(0)}$ is unitary if Λ is finite (Weidenmüller 1974b), this is not so for the background term $S^{(0)}$ of the statistical S-matrix.

The average CN cross section is given by (Moldauer 1964a,c)

$$\langle \sigma_{ab}^{CN} \rangle = \Phi_{ab} - M_{ab}, \qquad\qquad 3.4$$

where

$$M_{ab} = 2\pi^2 D^{-2} \left| \langle g_{\mu a} g_{\mu b} \rangle \right|^2 - 2i\pi D^{-1} \left\langle \frac{g_{\mu a} g_{\mu b} g_{\nu a}^* g_{\nu b}^*}{(\varepsilon_\mu - \varepsilon_\nu) + \frac{1}{2}i(\Gamma_\mu + \Gamma_\nu)} \right\rangle_{\mu \neq \nu} \qquad 3.5$$

$$\Phi_{ab} = 2\pi D^{-1} \langle |g_{\mu a}|^2 |g_{\mu b}|^2 / \Gamma_\mu \rangle = \langle \theta_{\mu a} \theta_{\mu b} / \theta_\mu \rangle, \qquad 3.6$$

$$\theta_{\mu a} = 2\pi D^{-1} N_\mu |g_{\mu a}|^2, \qquad \theta_\mu = \sum_c \theta_{\mu c}, \qquad N_\mu = \sum_c |g_{\mu c}|^2 / \Gamma_\mu. \qquad 3.7$$

The evaluation of Φ_{ab} and of M_{ab} requires knowledge of the statistical properties of the pole parameters.

3.4   *Difficulties with the Pole Parametrization*

The Hauser-Feshbach formula with WFC (Equation 3.2) agrees remarkably with experimental data even in the case of overlapping resonances. The formal similarity between Equations 3.2 and 3.6 is striking and suggests as a possible explanation that M_{ab} is negligible. In the case of strong absorption, this surmise is given additional weight by the observation (Weidenmüller 1972) that it leads to Vager's factorization 2.8 if one assumes that the complex quantities $g_{\mu a}$ have a Gaussian distribution, and if the fluctuations of the total widths Γ_μ are neglected. Nevertheless, the neglect of M_{ab} is erroneous unless all $T_a \ll 1$, as we now proceed to show.

Moldauer (1967c) extended the unitarity property 2.1 to the complex energy plane. For $\mathscr{E} = E + iI$, this analytic unitary condition $S(\mathscr{E})S^*(\mathscr{E}^*) = 1$ yields

$$2\pi D^{-1} \langle g_{\mu a} g_{\mu b} \rangle = (\langle S^* \rangle^{-1})_{ab} - \langle S_{ab} \rangle. \qquad 3.8$$

In the case of the statistical S-matrix, one has furthermore

$$S_{ab}^{(0)} = \tfrac{1}{2}(\langle S_{ab} \rangle + \langle S^* \rangle_{ab}^{-1}). \qquad 3.9$$

If DI are negligible, $\langle S \rangle$ is diagonal and one can write

$$\langle S_{ab} \rangle = \delta_{ab} \exp (2i\xi_a)(1 - T_a)^{1/2}. \qquad 3.10$$

Equation 3.8 then gives (Moldauer 1967c)

$$2\pi D^{-1} \langle g_{\mu a} g_{\mu b} \rangle = \delta_{ab} \exp (2i\xi_a) T_a (1 - T_a)^{-1/2}. \qquad 3.11$$

This result implies that, in the case of strong absorption, M_{ab} cannot be neglected in Equation 3.4 (Engelbrecht & Weidenmüller 1973, Weidenmüller 1974a,b). Indeed, Equations 3.6, 3.4, and 2.5 give, in the absence of DI,

$$
\begin{aligned}
T_a &= 2\pi D^{-1} \left\langle \frac{\left| g_{\mu a} \right|^2 \left(\sum_b \left| g_{\mu b} \right|^2 \right)}{\Gamma_\mu} \right\rangle - \sum_b M_{ab} \\
&= 2\pi D^{-1} \langle N_\mu | g_{\mu a} |^2 \rangle - \sum_b M_{ab}.
\end{aligned}
\qquad 3.12
$$

Since $N_\mu \geq 1$ (Lane & Thomas 1958, Weidenmüller 1964, Glöckle 1966), Equations 3.11 and 3.12 imply:

$$T_a \geq 2\pi D^{-1} \langle | g_{\mu a} |^2 \rangle - \sum_b M_{ab} \geq T_a (1 - T_a)^{-1/2} - \sum_b M_{ab}. \qquad 3.13$$

This shows that in the case of strong absorption, $\sum_b M_{ab}$ (and therefore $\langle \theta_{\mu a} \rangle$) diverges. Hence, the evaluation of $\langle \sigma_{ab}^{CN} \rangle$ from Equation 3.4 is not simple.

3.5   Discussion

The unitarity condition 2.1 plays an important role both for weak and for strong absorption. For $\Gamma \ll D$, it implies $\Gamma_\mu = \sum_a \Gamma_{\mu a}$. In the absence of DI, this is the main origin of the WFC for $a \neq b$. In the opposite limit, the unitarity condition leads to the difficulty exhibited in Section 3.4. Related to this is the fact that the pole parameters have no physical meaning for $\Gamma \gg D$ and that their distribution is not simple. Therefore, we believe that the pole parametrization 3.1 is not a convenient tool for an analytic derivation of CN theory. Nevertheless, the pole expansion has played an important role in the historical development of the field. It underlies much of the work of Moldauer, whose relentless efforts were vital in maintaining interest and faith in CN theory and eventually led him to a useful expression for $\langle \sigma_{ab}^{CN} \rangle$ (Section 4.4).

4 NUMERICAL RESULTS FROM THE K-MATRIX PARAMETRIZATION

The difficulties with the pole expansion derive from the unitarity condition. Therefore, it seems appropriate to start from a manifestly unitary parametrization of S.

4.1 The K-Matrix

The unitarity property of the S-matrix can be explicitly exhibited by writing

$$S = (1 + i\mathbf{K})(1 - i\mathbf{K})^{-1}, \qquad\qquad 4.1$$

$$K_{ab} = K_{ab}^{(0)} + \sum_{\mu=1}^{\Lambda} \frac{\gamma_{\mu a}\gamma_{\mu b}}{E_\mu - E}. \qquad\qquad 4.2$$

Here, all parameters are real and $K_{ab}^{(0)}$ is symmetric and independent of energy. This parametrization is encountered, for instance, in R-matrix theory (Lane & Thomas 1958) and in the shell-model approach to nuclear reactions (Mahaux & Weidenmüller 1969).

For isolated resonances ($E_\mu \simeq \varepsilon_\mu$ and $\gamma_{\mu a}^2 \simeq \frac{1}{2}\Gamma_{\mu a}$) the partial width amplitudes $\gamma_{\mu a}$ are random variables normally distributed with zero mean (Porter & Thomas 1956), and the E_μ display the level repulsion predicted by Wigner (1957). These statements are supported by analyses of experimental data on isolated resonances (Liou 1972). For overlapping resonances, theoretical arguments based on random-matrix models for the nuclear Hamiltonian show that the statistical distribution of the K-matrix parameters is the same as for isolated resonances. An approach based on this distribution is thus equivalent to a dynamical approach. In both cases, the input is a statistical model for the underlying Hamiltonian.

For overlapping resonances, it is possible to invert the matrix $(1 - i\mathbf{K})$ analytically only if N is small (Thomas 1955) or in the case of simplified "picket-fence" models. Such models may be used in suggesting generalizations (Moldauer 1967b, Bertram 1978a,b, Mello & McVoy 1978), but they have been unable so far to include fully the statistical distribution of the K-matrix parameters. The papers reviewed in Sections 4.2–4.5 are therefore based on numerical calculations.

4.2   Ensemble Average of the CN Cross Section

Tepel et al (1974) and Hofmann et al (1975a,b) computed $\langle \sigma_{ab}^{CN} \rangle$ directly from Equations 4.1 and 4.2, thus sidestepping the pole expansion. Rather than calculating the energy average of σ_{ab}^{CN}, these authors computed its ensemble average; this reduces the error due to the finite value of the number of resonances. They numerically verified that these two averages are equal. A theoretical proof of the latter "ergodicity" property in the limit $D/I \to 0$ was found by Richert & Weidenmüller (1977), who started from the parametrization 4.1. French, Mello & Pandey (1978) recently gave an elegant proof of the ergodicity property by using its relation with the asymptotic behavior of the autocovariance function. For the latter, they used the expression recently derived by Agassi, Weidenmüller & Mantzouranis (1975) in the limit $D \to 0$ (Section 5.2), to which this second proof of the ergodicity is thus also restricted.

We now list the main results obtained by Hofmann et al (1975a) in the case when DI are negligible (DI are discussed in Section 6). Results 1 and 2 below could be proved analytically in the limit $\Lambda \gg N$, and have been found numerically to be accurate for Λ as small as 100 and for $N \leqslant 10$. Results 3, 4, and 5 have only numerical support and should be deemed "usually accurate" rather than "exact."

1. $\langle S_{ab}^{FL} S_{cd}^{FL*} \rangle$ vanishes unless two pairs of indices coincide.
2. $\langle \sigma_{ab}^{CN} \rangle$ and $\exp(-2i\xi_a + 2i\xi_b) \langle S_{aa}^{FL} S_{bb}^{FL*} \rangle$ do not depend on the phases ξ_c of the elements of the (diagonal) $\langle S \rangle$-matrix (Equation 3.10).
3. One can factorize $\langle \sigma_{ab}^{CN} \rangle$ for $a \neq b$:

$$\langle |S_{ab}^{FL}|^2 \rangle = V_a V_b \left(\sum_c V_c \right)^{-1} [1 + \delta_{ab}(W_{aa} - 1)]. \qquad 4.3$$

The average (Equation 2.5) of the unitarity property then yields the relation between the Vs and the transmission coefficient T_a:

$$T_a = V_a + V_a^2 \left(\sum_c V_c \right)^{-1} (W_{aa} - 1). \qquad 4.4$$

This equation can be solved by iteration (Kawai, Kerman & McVoy 1973, Hofman et al 1975a).

4. The elastic enhancement factor can be parametrized as follows

$$W_{aa} = 1 + 2[1 + T_a^{0.3 + 1.5 T_a/(\Sigma_c T_c)}]^{-1}$$

$$+ 2 \left[\left(T_a - N^{-1} \sum_d T_d \right) \Big/ \left(\sum_c T_c \right) \right]^2. \qquad 4.5$$

A simpler but less accurate expression is $W_{aa} = 1 + 2/(1 + T_a^{1/2})$ (Tepel et al 1974). Note that $W_{aa} = 3$ for isolated resonances as it should (Section 3.2), while $W_{aa} = 2$ for strong absorption (Section 2.3).

5. For $a \neq b$, one can write

$$\langle S_{aa}^{FL} S_{bb}^{FL*} \rangle = \exp(2i\xi_a - 2i\xi_b)\, \Omega_a \Omega_b. \qquad 4.6$$

This quantity is needed for the calculation of $\langle \sigma_{ab}^{CN} \rangle$ in the presence of DI (see Equation 6.3). An accurate parametric expression for the factor Ω_a is

$$\Omega_a = y_a(1 - y_a + 4y_a^2);$$

$$y_a = V_a \left(\sum_c V_c \right)^{-1/2} (1 - T_a)^{1/2} \left(1 + 0.15 \sum_c T_c \right)^{-1}. \qquad 4.7$$

4.3   Properties of the Distribution of the Pole Parameters

The calculation of $\langle \sigma_{ab}^{CN} \rangle$ from the pole expansion (Equations 3.4–3.7) requires knowledge of the distribution of the pole parameters. This led Moldauer (1964b, 1967a, 1968, 1972, 1975a) to study some properties of the distribution of the pole parameters $g_{\mu a}$, ε_μ, and Γ_μ by numerically inverting the matrix $(\mathbf{1} - i\mathbf{K})$ (Equation 4.1). The main outcome of these and of other investigations is the following.

1. The Wigner repulsion among the ε_μ progressively disappears as $\langle \Gamma_\mu \rangle / D$ increases; the level spacing distribution approaches an exponential.

2. There exists a correlation between the values of N_μ belonging to neighboring ε_μ. One can show formally that $N_\mu \gtrsim 1$ (Lane & Thomas 1958, Weidenmüller 1964). The lower limit is reached for isolated resonances (Humblet 1964, Glöckle 1966).

3. The distribution of total widths is very skewed. It is peaked near the value

$$\Gamma_{\text{corr}} = (2\pi)^{-1} D \sum_a T_a. \qquad 4.8$$

Figure 1 illustrates this property, and also features 4 and 5 below. Its upper part suggests that Γ_{corr} can be identified with the correlation width, i.e. that h/Γ_{corr} gives the mean lifetime of the compound nucleus (Section 5.3).

4. Feature 3 implies that $\langle \Gamma_\mu \rangle \gg \Gamma_{\text{corr}}$, unless all $T_a \ll 1$. This statement can be made more quantitative by comparing definition 4.8 with the relation 4.10 derived by Simonius (1974). This author first showed that

$$|\det \langle \mathbf{S} \rangle| = \exp(-\pi \langle \Gamma_\mu \rangle / D), \qquad 4.9$$

where det $\langle S \rangle$ denotes the determinant of $\langle S \rangle$. This result holds even in the presence of DI. If these are negligible, $\langle S \rangle$ is diagonal and Equations 2.5 and 4.9 yield

$$2\pi D^{-1} \langle \Gamma_\mu \rangle = 2\pi D^{-1} \sum_a \langle \Gamma_{\mu a} \rangle = - \sum_a \ln (1 - T_a), \qquad 4.10$$

where $\Gamma_{\mu a} = N_\mu^{-1} |g_{\mu a}|^2$. In the one-channel case, this result (Equation 4.10) had been demonstrated by Moldauer (1969b), who improved an earlier derivation of Ullah (1968). A physical interpretation of Equation 4.10 has been given by Gibbs (1969) in the case of weak absorption.

5. Large values of Γ_μ are not infrequent and are preferentially associated with large values of $|g_{\mu a}|^2$. This may give rise to intermediate-structure-like peaks in the average CN cross section (Moldauer 1967a, Barnard et al 1974). The occurrence of large values of $|g_{\mu a}|^2$ can be understood from Equation 3.11.

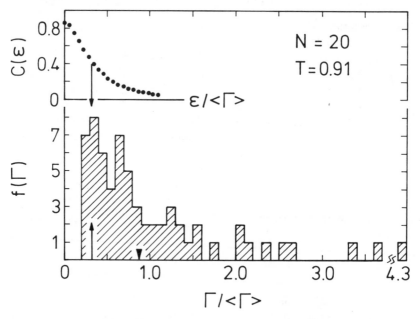

Figure 1 Adapted from Moldauer (1975a). The lower histogram shows the distribution of the total widths computed from Equations 4.1 and 4.2 with 65 resonances and 20 open channels, all with equal transmission coefficients $T = 0.91$. The black triangle shows the value of $\langle \Gamma_\mu \rangle$ deduced from Equation 4.10. The upward-pointing arrow indicates the value of $\Gamma_{\text{corr}}/\langle \Gamma_\mu \rangle$. In keeping with Equation 5.5, it nearly coincides with the half-width of the autocorrelation function $C(\varepsilon)$ (Equation 5.6), see downward-pointing arrow. The values of $C(\varepsilon)$ correspond to the ensemble average of 190 independent autocorrelation functions of the statistically computed elastic and nonelastic cross sections.

6. Level-level and channel-channel correlations exist between the quantities $\theta_{\mu a}$ and θ_{vb}: $\langle\theta_{\mu a}\theta_{\mu b}\rangle \neq 0$ for $a \neq b$ and $\langle\theta_{\mu a}\theta_{va}\rangle \neq 0$ for $\mu \neq v$. Sevgen (1974, 1976) showed that $\langle N_\mu\rangle > 1$ is a necessary and sufficient condition for the existence of level-level correlations between $g_{\mu a}$ and g_{va}.

Unfortunately, these results are insufficient for an analytic evaluation of $\langle\sigma_{ab}^{CN}\rangle$ from the pole expansion. This led Moldauer (1975a, Section 4.4) to compute $\langle\sigma_{ab}^{CN}\rangle$ from his numerical values of the pole parameters. His proposed parametric expression fits his results and can be formulated as a rather general "M-cancellation" scheme.

4.4 *M-Cancellation Scheme in the Pole Parametrization*

In the case of medium or strong absorption, the success (Equation 4.3) of the factorization assumption 2.6 requires that in Equation 3.4 the quantities $\langle\theta_{\mu a}\theta_{\mu b}/\theta_\mu\rangle$ and M_{ab} largely cancel. The numerical investigations of Moldauer (1975a) demonstrate this cancellation. His results for $\langle\sigma_{ab}^{CN}\rangle$ are in good quantitative agreement with the following M-cancellation approximation, which is inspired by Equation 3.2. In the absence of DI (see Section 6) one sets

$$T_a = \langle |t_{\mu a}|^2 \rangle. \qquad 4.11$$

The formula

$$\langle\sigma_{ab}^{CN}\rangle = \left\langle |t_{\mu a}|^2 \, |t_{\mu b}|^2 \Big/ \sum_c |t_{\mu c}|^2 \right\rangle = T_a T_b \left(\sum_c T_c \right)^{-1} W_{ab} \qquad 4.12$$

is a good approximation, if one assumes that the quantities $|t_{\mu a}|^2$ and $|t_{\mu b}|^2$ are uncorrelated random variables distributed according to a χ^2 distribution law with a suitably chosen "effective" number of degrees of freedom, v_a and v_b. Moldauer (1975a) plotted fitted values of v_a versus $\sum_c T_c$, for a few values of T_a. The effective number of degrees of freedom and the elastic enhancement factor W_{aa} are close to 2 if $\sum_c T_c$ is large compared to unity. This is in good agreement with Vager's surmise 2.10 and with the results described in Section 4.2.

The M-cancellation scheme replaces an earlier proposal by Moldauer (1964a,c) to neglect M_{ab} while introducing an effective transmission coefficient that involved an "overlap parameter," Q. It can be used to evaluate quantities other than the average CN cross section; it yields, for instance (Moldauer 1975b)

$$\langle S_{aa}^{FL} S_{bb}^{FL*}\rangle = \langle\sigma_{ab}^{CN}\rangle \left[\left(\frac{2}{v_a} - 1\right)\left(\frac{2}{v_b} - 1\right) \right]^{1/2}. \qquad 4.13$$

Frenkel (1978) showed that the latter relation is exact when $T_a = T_b$, if one assumes that its left-hand side only depends on **T**.

4.5   *Discussion*

In Sections 4.2 and 4.4, we described two ways of parametrizing $\langle \sigma_{ab}^{CN} \rangle$. They are both based on the numerical study of the *K*-matrix parametriza-

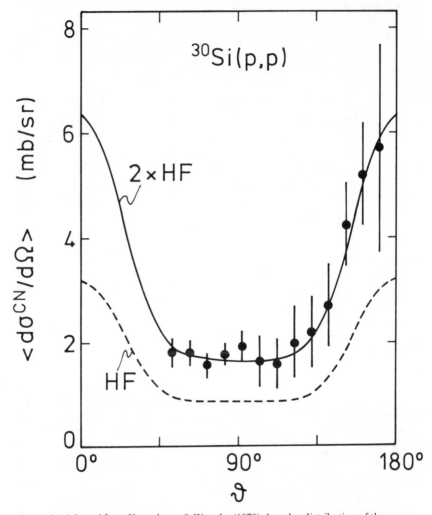

Figure 2   Adapted from Kretschmer & Wangler (1978). Angular distribution of the average CN cross section for the elastic scattering ^{30}Si(p,p)^{30}Si, for $8.5 < E_p < 10.7$ MeV. The dots represent the measured values; the direct elastic contribution has been evaluated from the analyzing power, and has been subtracted. The dashed curve shows the Hauser-Feshbach expression $T_p^2 / \sum_c T_c$ (suitably normalized), and the full curve represents twice this value (Equation 2.10). A least-square fit yields the result 2.09 ± 0.14 for the elastic enhancement factor.

tion 4.1, which is largely equivalent to a dynamical starting point. The agreement between the two parametrizations is excellent (Moldauer 1975a,b), except for $T_a, T_b \ll 1 \ll \sum_c T_c$ (Moldauer 1976). In the latter case, the M-cancellation scheme is more accurate; however, $\langle \sigma_{ab}^{CN} \rangle$ is then quite small so that this case is somewhat academic. The relative merits of the two parametrizations for practical purposes have been discussed by Moldauer (1975a,b), Gruppelaar & Reffo (1977), and others. The M-cancellation approximation has the advantage of joining smoothly to the case of isolated resonances (Section 3.2). Its use in numerical codes (Prince 1978) presents a difficulty: one must interpolate the values of the "effective" number of degrees of freedom v_a, which were given by Moldauer (1975a) only for three values of T_a (0.0975, 0.75, and 0.91).

Both approximations fail to reproduce accurately some cross-section ratios in cases with fewer than about 10 channels. These ratios are better fitted by another parametrization for $\langle \sigma_{ab}^{CN} \rangle$, recently proposed by Bertram (1979) and suggested by the investigation of a picket-fence model (Bertram 1978).

The value $W_{aa} = 2$ for the elastic enhancement factor in the strong absorption limit had already been conjectured in the 1960s (Thomas 1956, Satchler 1963, Moldauer 1964c). It is in excellent agreement with the experimental value $W_{aa} = 2.09 \pm 0.14$ obtained recently by Kretschmer & Wangler (1978) from a simultaneous fluctuation analysis of the cross section and the analyzing power of the reaction $^{30}Si(\vec{p},p)^{30}Si$ (Figure 2).

5   ANALYTICAL RESULTS IN THE STRONG ABSORPTION LIMIT

The expressions for the S-matrix upon which the theories surveyed in Sections 5.1 and 5.2 are based appear in the microscopic approaches to nuclear reactions of Feshbach (1962) and of Mahaux & Weidenmüller (1969), respectively. These theories both include DI and show analytically that Vager's conjecture 2.8 is correct in the strong absorption limit, to which they are restricted. The theory of Kawai, Kerman & McVoy (1973, Section 5.1) only takes into account the average of the analytic unitarity condition 2.1. The theory of Agassi, Weidenmüller & Mantzouranis (1975, Section 5.2) is based on a systematic expansion of the observables in powers of (D/Γ). It encompasses Ericson fluctuations (Section 5.3), isospin mixing (Section 7), and precompound processes (Section 8).

5.1   Theories Based on Average Unitarity

Kawai, Kerman & McVoy (1973) were the first authors to include the effect of DI in a dynamical CN theory. In order to circumvent the difficulty that DI introduce correlations among the quantities $g_{\mu a}$ and $g_{\mu b}$ ($\langle g_{\mu a} g_{\mu b} \rangle \neq 0$ for $a \neq b$), they wrote the S-matrix in the form

$$S_{ab}(E) = \langle S_{ab} \rangle - i \sum_\mu \frac{\hat{g}_{\mu a}(E)\hat{g}_{\mu b}(E)}{E - \hat{\varepsilon}_\mu + \frac{1}{2}i\hat{\Gamma}_\mu}, \qquad 5.1$$

which implies $\langle \hat{g}_{\mu a}(E)\, \hat{g}_{\mu b}(E)\rangle = 0$. An expression for the energy-dependent quantities $\hat{g}_{\mu a}(E)$ can be found by dividing the Hamiltonian into an off-diagonal optical-model part (DI) plus the rest, and by using the projection operator technique of Feshbach (1962).

In the limit of strong absorption, Kawai et al (1973) omitted the fluctuations of $\hat{\Gamma}_\mu$, neglected part of the energy dependence of $\hat{g}_{\mu a}$, and argued on dynamical grounds that the complex quantities $\hat{g}_{\mu a}$ are normally distributed, uncorrelated, random variables with the same dispersion for their real and imaginary parts. As shown later by Weidenmüller (1972), the latter properties automatically yield Vager's conjecture 2.8. The fact that the energy dependence of $\hat{g}_{\mu a}$ is not fully taken into account requires that nonunitary forms of the S-matrix appear in intermediate stages of the formal manipulations (Moldauer 1974, 1975a). This is the main reason why Kawai et al did not stumble across the difficulties discussed in Section 3.4. Since they nevertheless obtained the correct result (Equation 2.8) in the case $\Gamma \gg D$, it appears that the analytic unitarity condition 2.1 need not always be strictly enforced. The *average* unitary condition 2.5 was taken into account by Kawai et al in order to derive the relation 2.9, which connects the matrices **T** and **X**. Their theory was recently extended to the case of transfer reactions populating dense-lying resonances (Kerman & McVoy 1979, private communication).

Kerman & Sevgen (1976) investigated the possible usefulness of relaxing somewhat the requirement of analytic unitarity. They pointed out that any parametric form for the S-matrix is a simplification. Therefore, they argue that it is perfectly tolerable that such a parametrization be in slight disagreement with the unitarity condition 2.1. Defining a unitary defect matrix **B**(E) by $\mathbf{S}(E)\mathbf{S}^\dagger(E) = \mathbf{1} - \mathbf{B}(E)$, they accordingly consider as realistic those expressions for the S-matrix that have (for E real) the properties

$$|B_{ab}(E)| \ll 1, \qquad \langle B(E) \rangle = 0. \qquad 5.2$$

The latter relation expresses the fact that the average unitarity condition 2.5 should be fulfilled.

Note that $\langle \mathbf{B}(E) \rangle \neq \mathbf{B}(E+iI)$, because $\mathbf{B}(\mathcal{E})$ has poles in the upper as well as in the lower half of the complex $\mathcal{E}$-plane. The inequality in Expression 5.2 does not imply that $|B_{ab}(\mathcal{E})|$ is small for $\mathcal{E}$ complex. In particular, $B_{ab}(E+iI)$ may be comparable to unity. This is the case in the work of Kawai, Kerman & McVoy (1973) in which $\mathbf{B}(E+iI) = \mathbf{T}(E)$.

This corresponds to a sizeable deviation from analytic unitarity in the case of strong absorption:

$$S(E + iI)\, S^{\dagger}(E - iI) = 1 - B(E + iI) = 1 - T(E). \qquad 5.3$$

Since this approach nevertheless yields the correct expression for the average CN cross section, it is plausible that it fulfills the inequality in Expression 5.2, although this has not yet been shown.

Kerman & Sevgen (1976) discussed the attractive possibility that a slight modification of the pole parametrization 3.1 might be sufficient to make M_{ab} negligible in Equation 3.4. They proposed one such model, which interpolates smoothly between the correct weak and strong absorption limits. Relations 3.5 and 3.11 suggest that the large size of $\sum_b M_{ab}$ probably arises from the poles associated with large values of $|g_{\mu a}|^2/D$ and of Γ_μ (Section 4.3 and Figure 1). It would be of interest to investigate whether the omission of these poles renders $\sum_b M_{ab}$ negligible while leaving the average CN cross section practically unchanged.

The additional flexibility gained by allowing $B(\mathscr{E})$ to differ slightly from zero for $\mathscr{E}$ real while letting it be large for $\mathscr{E} = E + iI$ must be handled with some caution. This warning is obvious in the case of the S-matrix. Small modifications of $S(E)$ for E real may lead to large changes of $\langle S(E) \rangle = S(E + iI)$; however, physically acceptable models should all yield the value of $\langle S(E) \rangle$ given by direct reaction theory. The same holds true for the models that fulfill relations 5.2: they are physical only provided they yield practically the same average CN cross sections as the unitary models reviewed in Section 4.

5.2   Equilibrium Limit of a Theory for Preequilibrium Processes

The formation of the CN proceeds through a succession of interactions that populate a chain of quasibound configurations. These can be classified by their degree of "complexity," which we call m. For instance, m may denote the number of particles and holes with respect to a closed-shell reference state ("exciton number"). From class m, a further collision between nucleons can either lead to another quasibound configuration n or to the escape of a particle from the system. In the latter case, one has "preequilibrium" or "precompound" decay (Section 8) unless the complexity of class m is such that one has reached the equilibrium characteristic of the CN. Hence, a satisfactory theory of preequilibrium decay should yield as a limiting case (no precompound emission) expressions for CN cross sections. This was recently achieved (Agassi & Weidenmüller 1975, Agassi et al 1975). These authors considered only strongly overlapping resonances, i.e. the limit $D \to 0$. In order to evaluate the ensemble averages of the observables in this limit, they first expanded the S-matrix in powers of the strength of the residual interaction V. Inserting this expansion into the expressions for the observables, they gave rules for finding the dominant contribution of each term of the resulting multiple Born series in the limit $D \rightarrow 0$. As basic input they assumed (Brody et al 1978) that the matrix elements of V are random variables with a Gaussian distribution. Finally, they resummed the series of dominant contributions.

In the absence of DI and to order D/Γ, these authors found

$$\langle S_{ab}^{FL} S_{cd}^{FL*} \rangle = (\delta_{ac}\delta_{bd} + \delta_{ad}\delta_{bc}) \sum_{m,n} T_{am} \Pi_{mn} T_{bn}, \qquad 5.4$$

where T_{am} is the transmission coefficient for class m and where Π_{mn} gives the relative probability of population of configurations with complexity n if one starts from class m. The matrix Π obeys a probability balance equation that is equivalent to the master-equation approach (Blann 1975) to preequilibrium decay. If the lifetime of the CN is large compared to the internal equilibration time of the system (or, more formally, if only one eigenvalue of Π^{-1} is close to zero), Equation 5.4 yields relations 2.10. The authors furthermore calculated the contributions to $\langle \sigma_{ab}^{CN} \rangle$ ($a \neq b$) that are of order $(D/\Gamma)^2 \propto (\sum_c T_c)^{-2}$. They found that these contributions are in agreement with the formulas 4.3 and 4.4 with $W_{aa} = 2$. In the presence of DI, and to order D/Γ, Agassi et al (1975) obtain Vager's conjecture 2.8.

5.3   Ericson Fluctuations: The Lifetime of the Compound Nucleus

The standard theory of Ericson fluctuations (Ericson & Mayer-Kuckuk 1966, Richter 1974) applies to the case of strong absorption and many open channels. It is based on the following three assumptions: (a) All widths Γ_μ in the pole expansion 3.1 are equal. (b) The elements of S^{FL} are not correlated with each other. (c) They have a bivariate Gaussian distribution. The results described in the preceding sections show that these assumptions are questionable. Figure 1 shows that assumption (a) is incorrect, in particular that the average width $\langle \Gamma_\mu \rangle$ is much larger than the correlation width Γ_{corr} (Equation 4.8) near which the distribution of the Γ_μ is peaked. Assumption (b) is violated when DI are important. Finally, assumption (c) is suspect in view of the correlations implied by analytic unitarity. A reinvestigation of the theory of Ericson fluctuations was thus necessary. This was performed in the framework of the dynamical approach discussed in Section 5.2 (Agassi et al 1975, Richert et al 1975).

In the limit $D/\Gamma \rightarrow 0$, these authors proved that assumption (c) is correct. For larger values of D/Γ and in the absence of DI, numerical studies of the K-matrix parametrization 4.1 (Tepel 1975, Richert et al 1975) have shown that the distribution of S^{FL}_{ab} can significantly deviate from a bivariate Gaussian. This is partly due (Krieger 1967) to the cut-off $|S^{FL}_{ab}| \leqq 1$ implied for $a \neq b$ by the unitarity condition. This cut-off is irrelevant in the strong absorption limit since one has then $\langle|S^{FL}_{ab}|\rangle \ll 1$.

In the limit $D \to 0$, Agassi et al (1975) established the following relation

$$\langle S^{FL}_{ab}(E)\, S^{FL*}_{cd}(E+\varepsilon)\rangle = [T_{ac}T_{bd} + T_{ad}T_{bc}]\, [\text{trace } \mathbf{T} + i2\pi D^{-1}\varepsilon]^{-1}. \qquad 5.5$$

This result includes the effect of DI. It demonstrates that the auto-correlation function

$$C_{ab}(\varepsilon) = \frac{\langle\sigma^{CN}_{ab}(E+\varepsilon)\,\sigma^{CN}_{ab}(E)\rangle - \langle\sigma^{CN}_{ab}(E)\rangle^2}{\langle[\sigma^{CN}_{ab}(E)]^2\rangle - \langle\sigma^{CN}_{ab}(E)\rangle^2} \qquad 5.6$$

has the standard Lorentzian form

$$C_{ab}(\varepsilon) = C(0)\,[1+(\varepsilon/\Delta)^2]^{-1}, \qquad 5.7$$

and that the correlation width can be identified with $(2\pi)^{-1}D \times$ trace $\mathbf{T}$ as suggested by Figure 1. The average density matrix decays like $\exp(-t/\tau)$, where $\tau = h/\Gamma_{\text{corr}}$ is the mean life-time of the CN. This quantity may be directly measurable by the crystal blocking technique (Gibson 1975), which gives experimental access to the "time development function." The latter has been theoretically investigated by Yoshida & Yazaki (1975) in the framework of the M-cancellation approximation (Section 4.4), and by Matsuzaki (1978), who performed a numerical study based on the unitary random matrix model described in Section 4.2. Present measurements (Kanter et al 1979) are limited to the case of weak or medium absorption. In a recent paper, Fernandez Diaz & Sirotkin (1978) illustrate the fact that isolated doorway states can modify the shape of the auto-correlation function.

6   DIRECT INTERACTIONS AND INTERMEDIATE STRUCTURE

The presence of DI reveals itself in the fact that the average S-matrix is not diagonal. Section 5 discusses dynamical proofs that Vager's conjecture 2.8 gives the correct way of including the effect of DI in the case $\Gamma \gg D$. However, we have not yet shown how to take DI into account for arbitrary values of Γ/D. This is the purpose of Section 6.1, where we describe how the change of channel representation introduced by Engelbrecht & Weidenmüller (1973) brings one back to the case of vanishing DI.

Isolated doorway states can also give rise to nondiagonal values of

$\langle \mathbf{S} \rangle$ (Feshbach, Kerman & Lemmer 1967). Their effect can usually be treated in the same way as that of DI. Isolated isobaric analogue resonances appear to provide the best experimental tests of the validity of the theory (Section 6.2).

6.1   Transformation to a Diagonal Average Collision Matrix

When $\langle \mathbf{S} \rangle$ is not diagonal, Equation 3.8 shows that $\langle g_{\mu a} g_{\mu b} \rangle \neq 0$ for $a \neq b$. Correspondingly, one then has $\langle \gamma_{\mu a} \gamma_{\mu b} \rangle \neq 0$ for the residues of the K-matrix (Moldauer 1964b, Hofmann et al 1975a). This is why the expressions derived from the numerical calculations described in Section 4 are not valid when DI are present. Uncorrelated partial width amplitudes can be introduced by the change of representation of Engelbrecht & Weidenmüller (1973). This transformation defines quantities $\tilde{\gamma}_{\mu a}$ and $\tilde{g}_{\mu a}$ with $\langle \tilde{\gamma}_{\mu a} \tilde{\gamma}_{\mu b} \rangle = 0 = \langle \tilde{g}_{\mu a} \tilde{g}_{\mu b} \rangle$ if $a \neq b$. The new parameters are independent of energy, so that reliable results are obtained for all values of the ratio Γ/D. Therefore, this method goes beyond those reviewed in Sections 5.1 and 5.2, which apply only for $\Gamma \gg D$.

The transmission matrix $\mathbf{T}$ in Equation 2.5 is Hermitian and positive semidefinite. It can therefore be diagonalized by a unitary matrix

$$(\mathbf{U}\mathbf{T}\mathbf{U}^{\dagger})_{ab} = \delta_{ab}\tilde{T}_a, \qquad 0 \leq \tilde{T}_a \leq 1. \qquad\qquad 6.1$$

This relation implies that the symmetric matrix $\langle \tilde{\mathbf{S}} \rangle = \mathbf{U}\langle \mathbf{S} \rangle \mathbf{U}^T$ commutes with its Hermitian adjoint. By a theorem on such matrices, it can therefore be diagonalized. Equation 6.1 shows that $\langle \tilde{\mathbf{S}} \rangle \langle \tilde{\mathbf{S}}^{\dagger} \rangle$ is already diagonal; this implies that $\langle \tilde{\mathbf{S}} \rangle$ and $\langle \tilde{\mathbf{S}}^{\dagger} \rangle$ are diagonal too:

$$\langle \tilde{S}_{ab} \rangle = (\mathbf{U}\langle \mathbf{S} \rangle \mathbf{U}^T)_{ab} = \delta_{ab} \exp(2i\tilde{\xi}_a)(1 - \tilde{T}_a)^{1/2}. \qquad\qquad 6.2$$

Multiplying relation 3.8 from the left by $\mathbf{U}$ and from the right by $\mathbf{U}^T$, one finds furthermore that $\langle \tilde{g}_{\mu a} \tilde{g}_{\mu b} \rangle = 0$ for $a \neq b$, with $\tilde{g}_{\mu a} = \sum_b U_{ab} g_{\mu b}$.

These results suggest that the symmetric and unitary matrix $\tilde{\mathbf{S}} = \mathbf{U}\mathbf{S}\mathbf{U}^T$ has the same properties as the S-matrix in the absence of DI. In the case of weak absorption, this suggestion was proved to be valid by Engelbrecht & Weidenmüller (1973); an analytical proof that holds for arbitrary absorption strength was found by Hofmann et al (1975a) in the limit $\Lambda \to \infty$, N fixed. It has been verified numerically (Hofmann et al 1975a, Moldauer 1975b) that this suggestion remains accurate for Λ as small as 100, $N \leq 10$.

All quantities of physical interest (average CN cross section, autocorrelation functions) can thus be obtained by expressing $\mathbf{S}$ in terms of $\tilde{\mathbf{S}}$ and by applying to $\tilde{\mathbf{S}}$ the formulas given in Section 4.2 or in Section 4.4. For instance, one has

$$\langle S_{ab}^{FL} S_{cd}^{FL*} \rangle = \sum_{efgh} U_{ea}^* U_{fb}^* U_{gc} U_{hd} \langle \tilde{S}_{ef}^{FL} \tilde{S}_{gh}^{FL*} \rangle. \qquad 6.3$$

In the case of strong absorption in all channels ($\tilde{T}_a \simeq 1$ for all a), it can be checked that only the term with $e = g$ and $f = h$ survives. One then retrieves Vager's surmise 2.8, with $X_{ab} = T_{ab}$ (trace $\mathbf{T}$)$^{-1/2}$, or equivalently the results of Kawai, Kerman & McVoy (1973) (who in fact used the U-transformation in this limit), of Agassi, Weidenmüller & Mantzouranis (1975), and of Moldauer (1975b) who used the M-cancellation scheme to evaluate $\langle \tilde{S}_{ef}^{FL} \tilde{S}_{gh}^{FL*} \rangle$.

In the presence of DI, expression 6.3 does not in general vanish, even if the indices a,b,c,d are all different. This implies for instance that the fluctuating part of the S-matrix can contribute to the polarization of the emitted particles. Since this fluctuating part is identified with the CN contribution, this shows that DI force the CN to keep a certain memory of the way it was formed: Bohr's CN model is then invalidated.

The U-transformation yields the content of the CN memory regardless of its dynamical origin, which must lie in the interplay of statistical and nonstatistical properties of the nuclear Hamiltonian. Since $\mathbf{U}$ is determined by $\langle \mathbf{S} \rangle$, finding this dynamical origin amounts to constructing a model for the off-diagonal elements of $\langle \mathbf{S} \rangle$, i.e. in practice to deriving the distorted wave Born approximation (DWBA). This problem has been investigated by Bloch (1957), Hüfner, Mahaux & Weidenmüller (1967), Ratcliff & Austern (1967), Lane (1973), and Agassi, Weidenmüller & Mantzouranis (1975). The model used by these authors essentially assumes that the channels are coupled directly in a nonstatistical fashion by matrix elements of the residual interaction, whereas the CN levels are coupled to the channels in a statistical fashion, i.e. without correlations. Essentially the same model underlies the theoretical analysis (Lane 1971, Cugnon & Mahaux 1975) of the correlations observed in neutron radiative capture between the partial widths for neutron decay on the one hand and for gamma decay on the other hand.

6.2   Intermediate Structure

Although it is not purely academic (Coope et al 1977, Smith et al 1978), the application of Equation 2.8 or 6.3 to DI data is difficult because of the many contributing angular momenta. Experimental checks of the formalism are simpler when the channel-channel correlations can be ascribed to the effect of an isolated doorway state, mixed with a dense background of overlapping resonances of the same spin and parity (Mahaux 1973). The approach of the preceding section applies to this case if $\Gamma_d \gg I \gg \Gamma$, where Γ_d is the total width of the intermediate structure and Γ the average width of the background resonances. Formulas have been given by Agassi, Weidenmüller & Mantzouranis (1975) and by Richert, Simbel & Weidenmüller (1975).

Isobaric analogue resonances provide the best-understood example of isolated doorway states (MacDonald 1978, Lane et al 1978, Mitchell et al 1979). They are thus well suited for an experimental check of the theory. The requirement of maximizing the CN cross section in the channels coupled through the isobaric analogue resonance is met by nuclei in the vicinity of mass 90, for which comparatively few neutron channels are open. Several experimental groups have measured the angular distribution, analyzing power, spin-flip probability, spin-flip asymmetry, and cross-correlation functions for the outgoing protons (Kretschmer & Graw 1971, Albrecht et al 1973, Graw et al 1974, Blanke et al 1975, Davis et al 1975, Clement et al 1977, Haynes et al 1976, Feist et al 1977, King & Slater 1977). Typically, the cross section is written in the form 2.3, and two different fits to the data are performed: one is based on the expression for $\langle \sigma_{ab}^{CN} \rangle$ where DI are neglected, and one is based on Equation 6.3. While some cases are inconclusive, the majority of the cases shows that much better fits are obtained when the U-matrix approach 6.3 is used.

7   ISOSPIN MIXING IN CN REACTIONS

In the absence of electromagnetic interactions, CN theory must take into account the fact that isospin (τ) is a conserved quantum number: its influence on the autocorrelation function was studied by Robson, Richter & Harney (1973). In reality, Coulomb forces mix states with different isospins. Studies of isospin mixing in CN reactions can thus indicate the size of the Coulomb matrix elements. In a typical experiment, a $\tau = 0$ target is bombarded with alpha particles and one compares the alpha yields to two low-lying states having $\tau = 0$ and $\tau = 1$, respectively.

In the following, we are concerned with a situation where only two classes of levels exist, one with isospin $\tau_<$ and one with isospin $\tau_> = \tau_< + 1$, and where the resonances belonging to each class overlap strongly: $\Gamma_> \gg D_>$ and $\Gamma_< \gg D_<$. Let us call τ_A and τ_p the isospins of the target and of the projectile, respectively. In general, the channel $a = A + p$ can feed resonances with both isospins $\tau_>$ and $\tau_<$. We suppress the z-components and abbreviate the Clebsch-Gordan coefficients by symbols like $C_< = C(\tau_A, \tau_p; \tau_<)$. One can define isospin-dependent transmission coefficients, for instance $T_{a<} = C_<^2 T_a$. Grimes et al (1972) suggested the following expression for the average CN cross section in the presence of isospin mixing:

$$\langle \sigma_{ab}^{CN} \rangle = (1-f) \frac{T_{a>} T_{b>}}{\sum_c T_{c>}} + \frac{(T_{a<} + f T_{a>})(T_{b<} + f T_{b>})}{\sum_c (T_{c<} + f T_{c>})}, \qquad 7.1$$

where f denotes the fraction of the width of the $\tau_>$ states leading to the population of $\tau_<$ states. These authors showed that this expression leads to a better understanding of various proton-induced reactions [they also considered Ericson fluctuations; see, however, Robson, Richter & Harney (1973)]. The expression 7.1 was subsequently applied to other cases (Porile et al 1974, Porile & Grimes 1975); additional terms were included by Lux, Porile & Grimes (1977); alternative forms were also proposed by Shikazono & Terasawa (1975) and Robson (1975). Isospin mixing in precompound reactions was considered by Feinstein (1977).

The arguments put forward in the references above are largely intuitive. A more rigorous derivation of such prescriptions, and an extension of the theory, was given by Harney et al (1977) and by Weidenmüller et al (1978). These authors included the two mechanisms for isospin mixing first introduced by Robson (1965) in his theory of isobaric analogue resonances, namely "internal" and "external" mixing. They used the formalism developed by Agassi & Weidenmüller (1975) (see Section 5.2) and derived formulas for the cross-section correlation functions and for the average CN cross section. The latter is identical in form to Equation 7.1, as pointed out by Lane (1978); it contains an explicit expression for f in terms of the strengths of internal and external mixings. An experimental distinction between these two causes of mixing is thus impossible, since only f is accessible. Harney et al (1977) showed that the contribution of external mixing may be important: its theoretical upper limit is close to the empirical value of f. Weidenmüller et al (1978) and Simpson & Wilson (1979) pointed out that particular correlations exist between pairs of "mirror" channels; these have recently been investigated by Simpson et al (1978).

Lane (1978) also derived an expression similar to 7.1 in the case of isolated $\tau_>$ states ($\Gamma_> \ll D_>$, $\Gamma_< \gg D_<$), with an adapted expression for f. He then suggested a formula that interpolates between the results for $\Gamma_> \gg D_>$ and $\Gamma_> \ll D_>$. It would be of interest to investigate the accuracy of this suggestion through computer simulation of cross sections using a random-number generator (Sections 4.2 and 4.4).

8 PRECOMPOUND REACTIONS

In Sections 1 and 5.2, we mentioned that precompound reactions consist of the emission of particles from configurations of intermediate complexity, populated while the system is on its way towards equilibrium. We cannot here survey the recent progress in this attractive field, which bridges the gap between DI and CN processes and thus provides a severe testing ground for any "unified" theory. The reader is referred to the following: Semiphenomenological models have been reviewed by Blann (1975). A more recent survey by Bunakov (1978) includes a discussion of some microscopic theories, whose aim is in part to understand previous models in terms of statistical properties for the matrix elements of the nuclear Hamiltonian and in part to extend these models to the description of angular distributions.

The latter have been fairly well (Bisplinghoff et al 1976) described by Irie et al (1975, 1976) and by Mantzouranis and collaborators (Mantzouranis et al 1975, 1976, Mantzouranis 1976a,b). According to the basic model used by the latter authors, the incident beam of "fast" particles is degraded both in energy and in angle through a series of two-body collisions. If the "fast" particle is reemitted before full equilibrium has been attained, the angular distribution will be forward peaked. Recently, this model was extended to the case of complex clusters (Machner 1978, private communication). A qualitatively similar model has been put forward by Feshbach, Kerman, and Koonin (in preparation); its essential features have been described by Feshbach (1973, 1976, 1978). In this approach the distinction between the fast particle and the other nucleons is replaced by the distinction between (a) shell-model configurations built from bound single-particle orbitals; (b) configurations containing at least one continuum orbital. Transitions between states of different complexities in group (a) are labeled "multistep compound" and those between states in group (b) "multistep direct." For the former, one assumes rapid equilibration within each class of given complexity (Section 5.2). These configurations yield an angular distribution symmetric about 90°. The multistep direct processes yield an asymmetric angular distribution, with an expression very similar in form to that given by Mantzouranis et al. Only preliminary numerical applications of this approach are now available (Feshbach 1973). The degree of complexity of the dominant configurations is still a matter of debate (compare references Mantzouranis 1976a, Tamura et al 1977, Tsai & Bertsch 1978).

What is the theoretical basis for these models? Their outstanding feature is the use of occupation *probabilities* of classes of given complexity, whereas the Schrödinger equation describes the time-evolution of *amplitudes*. A microscopic derivation of models for precompound reactions must therefore be based upon the existence of a mechanism that destroys phase correlations between different amplitudes or on a loss of information about such correlations. Agassi, Weidenmüller & Mantzouranis (1975) tackled this problem by using a shell-model basis. They group the configurations according to their degree of complexity m (number of "excitons," Section 5.2). They assume that these classes are populated in a chain of increasing complexity and that the matrix elements (of the nuclear two-body interaction) that connect the states of different complexity have stochastic properties. The equations thus obtained correspond closely with those of the semiphenomenological "exciton model" (Blann 1975). The theory of Agassi et al contains the equilibrated state of the evaporation model as a limit (Section 5.2). Since they considered different values of the total angular momentum separately, they did not obtain expressions for the angular distributions directly from this approach. In principle, angular distributions are included in the theory being developed by Feshbach, Kerman & Koonin (Feshbach 1976, 1978). These authors also use a random-matrix model and a chaining hypothesis. As outlined above, they distinguish between bound and continuum shell-model configurations. The statistical assumptions bear on the matrix elements connecting all these classes. It appears convenient to adopt a plane wave representation for the nucleons in the continuum. The theory has apparently not yet been shown to lead to the evaporation model as a limit.

In the approach of Bunakov et al (Bunakov & Nesterov 1976, Bunakov 1977, 1978), the use of statistical assumptions in a stationary theory is replaced by that of sufficiently short wave packets in a time-dependent theory, and by a statistical average over the initial conditions.

The work of Rumyantsev & Kheifets (1975) goes beyond all these theoretical approaches. It relates directly to the nuclear many-body Hamiltonian. In principle it includes the possibility that fast particles can be "splashed out" from the nucleus by the time dependence induced in the mean field by the incident particle (Belyaev & Rumyantsev 1974). Kinetic equations of the type used in preequilibrium models are derived with the help of a hypothesis first introduced by Bogolyubov in quantum statistical mechanics. The applicability of this hypothesis in the nuclear context has not yet been investigated. The works of Mantzouranis & Pauli (1977), Orland & Schaeffer (1979), and Wong & Tang (1978) have a similar starting point.

9   SUMMARY AND OUTLOOK

CN theory in its modern form is based upon a random-matrix model for the nuclear Hamiltonian. It shares this basis with the statistical theory of nuclear spectra, and a unified theory of the statistical properties of nuclei is gradually emerging.

Right above neutron threshold, the CN resonances are isolated ($\Gamma \ll D$); CN theory has been satisfactory in this domain for about two decades. Most recent developments concern higher excitation energies where $\Gamma \simeq D$ or $\Gamma \gg D$, and where DI or precompound processes play a significant role.

In the intermediate domain ($\Gamma \simeq D$) observables have been accurately parametrized. These parametrizations are mainly based on fits of large-scale numerical calculations that use a random-matrix model as input; an analytical method has been found to include the effect of DI. It is desirable to find a wholly analytical treatment of this domain $\Gamma \simeq D$. We believe that this will require new approaches. Entropy has, for instance, recently been suggested as a useful concept in CN theory (Mello 1979, Nemes & Seligman 1978, private communication).

In the domain $\Gamma \gg D$, the recurrence time $2\pi h/D$ is very much larger than all other characteristic times of the nuclear system. This corresponds to the thermodynamic limit, within which Bohr's CN model works best; it follows from the statistical equilibration of the nuclear system. Analytical methods have been developed to show this microscopically, and to calculate the modifications of the CN model due to DI, precompound processes, intermediate structure, and isospin mixing.

CN theory has been very much consolidated over the past five years. Major developments may nevertheless still be expected, both on its borders, and in its core. Can the angular distribution of precompound particles be derived in the framework of the general theory? Can the precompound emission of deuterons, tritons, and alpha particles be understood? Is it true (as postulated in the statistical approach) that the phases of the individual nucleon-nucleon scattering amplitudes are sufficiently random *ab initio*, allowing the replacement of amplitudes by probabilities? At what energy, and how, does the transition take place to a semiclassical cascade model with spatially localized nucleon-nucleon collisions? Can the fusion of heavy ions be incorporated into CN theory? Are the statistical methods applicable to the analysis of deeply inelastic collisions between heavy ions? In these collisions, the nuclear system partially equilibrates. The fragments separate without ever reaching the CN. Collective degrees of freedom not encountered in preequilibrium processes enrich the observed phenomena and enlarge the theoretical description (Weidenmüller 1979).

Perhaps the most fundamental problem concerns the foundations of both CN theory and the statistical theory of nuclear spectra: Which properties of the nucleon-nucleon interaction justify the use of random-matrix models for the nuclear Hamiltonian? This difficult question has barely been investigated (Bloch 1969). Its answer might provide a link with the quantum-statistical description of other physical systems.

Literature Cited

Agassi, D., Weidenmüller, H. A. 1975. *Phys. Lett. B* 56: 305
Agassi, D., Weidenmüller, H. A., Mantzouranis, G. 1975. *Phys. Lett. C* 22: 145
Albrecht, R., Mudersbach, K., Wurm, J. P., Zoran, V. 1973. *Proc. Int. Conf. Nucl. Phys., Munich 1973*, ed. J. de Boer, H. J. Mang, p. 577. Amsterdam: North-Holland
Austern, N. 1970. *Direct Nuclear Reaction Theories.* New York: Wiley. 390 pp.
Barnard, E., deVilliers, J., Reitman, D., Moldauer, P., Smith, A., Whalen, J. 1974. *Nucl. Phys. A* 229: 189
Barschall, H. H. 1952. *Phys. Rev.* 86: 431
Belyaev, S. T., Rumyantsev, B. A. 1974. *Phys. Lett. B* 53: 6
Bertram, W. K. 1978. *Aust. J. Phys.* 31: 151
Bertram, W. K. 1979. *Aust. J. Phys.* In press
Bethe, H. A. 1937. *Rev. Mod. Phys.* 9: 69
Bisplinghoff, J., Ernst, J., Löhr, R., Mayer-Kuckuk, T., Meyer, P. 1976. *Nucl. Phys. A* 269: 147
Blanke, E., Genz, H., Richter, A., Schrieder, G. 1975. *Phys. Lett. B* 58: 289
Blann, M. 1975. *Ann. Rev. Nucl. Sci.* 25: 123–66
Blatt, J., Weisskopf, V. F. 1952. *Theoretical Nuclear Physics.* New York: Wiley. 864 pp.
Bloch, C. 1957. *Nucl. Phys.* 4: 503
Bloch, C. 1969. *Proc. Int. Conf. Statistical Mechanics. J. Phys. Soc. Jpn.* 26: 57–60
Bohr, N. 1936. *Nature* 137: 344
Brink, D. M., Stephen, R. O. 1963. *Phys. Lett.* 5: 77
Brody, T. A., Flores, J., Mello, P. A., French, J. B., Wong, S. S., M. 1978. *Notas Fis.*, Vol. 1, No. 3
Brown, G. E. 1959. *Rev. Mod. Phys.* 31: 893
Bunakov, V. E., Nesterov, M. M. 1976. *Phys. Lett. B* 60: 417
Bunakov, V. E. 1977. *Sov. J. Nucl. Phys.* 25: 271
Bunakov, V. E. 1979. *Proc. Trieste Course on Nuclear Theory for Applications.* Vienna: IAEA. In press
Clement, H., Graw, G., Zenger, R., Zöllner, G. 1977. *Nucl. Phys. A* 285: 109
Coope, D. F., Tripathi, S. N., Schell, M. C., Weil, J. L., McEllistrem, M. T. 1977. *Phys. Rev. C* 16: 2223
Cugnon, J., Mahaux, C. 1975. *Ann. Phys. NY* 94: 128
Davis, S., Glashausser, C., Robbins, A. B., Bissinger, G., Albrecht, R., Wurm, J. P. 1975. *Phys. Rev. Lett.* 34: 215

Dresner, L. 1957. *Proc. Int. Conf. on Neutron Interactions with the Nucleus, Columbia University, 1957.* Report CU-175, p. 71. New York: Columbia Univ.
Engelbrecht, C. A., Weidenmüller, H. A. 1973. *Phys. Rev. C* 8: 859
Ericson, T. 1960. *Phys. Rev. Lett.* 5: 430
Ericson, T. 1963. *Ann. Phys. NY* 23: 390
Ericson, T., Mayer-Kuckuk, T. 1966. *Ann. Rev. Nucl. Sci.* 16: 183–206
Feinstein, R. L. 1977. *Ann. Phys. NY* 107: 222
Feist, J. H., Kretschmer, W., Pröschel, P., Graw, G. 1977. *Nucl. Phys. A* 290: 141
Fernandez Diaz, J. R., Sirotkin, V. K. 1978. *Nucl. Phys. A* 312: 17
Feshbach, H. 1962. *Ann. Phys. NY* 19: 287
Feshbach, H. 1973. See Albrecht et al 1973, p. 631
Feshbach, H. 1974. *Rev. Mod. Phys.* 46: 1
Feshbach, H. 1976. *Proc. Int. Conf. on the Interaction of Neutrons with Nuclei, Lowell 1976*, ed. E. Sheldon. Natl. Tech. Inf. Serv. US Dept. Commerce, Springfield, Va. 1645 pp.
Feshbach, H. 1978. *Notas Fis.* 1: 111
Feshbach, H., Mello, P. A. 1972. *Phys. Lett. B* 39: 461
Feshbach, H., Kerman, A. K., Lemmer, R. H. 1967. *Ann. Phys. NY* 41: 230
Feshbach, H., Porter, C. E., Weisskopf, V. F. 1954. *Phys. Rev.* 96: 448
Fox, J. D., Moore, C. F., Robson, D. 1964. *Phys. Rev. Lett.* 12: 198
French, J. B., Mello, P. A., Pandey, A. 1978. *Phys. Lett. B* 80: 17
Frenkel, A. 1978. *Phys. Rev. C* 17: 418
Gibbs, W. R. 1969. *Phys. Rev.* 181: 1414
Gibson, W. M. 1975. *Ann. Rev. Nucl. Sci.* 25: 465–508
Glöckle, W. 1966. *Z. Phys.* 190: 391
Graw, G., Clement, H., Feist, J. H., Kretschmer, W., Pröschel, P. 1974. *Phys. Rev. C* 10: 2340
Griffin, J. J. 1966. *Phys. Rev. Lett.* 17: 478
Griffin, J. J. 1967. *Phys. Lett. B* 24: 5
Grimes, S. M., Anderson, J. D., Kerman, A. K., Wong, C. 1972. *Phys. Rev. C* 5: 85
Gruppelaar, H., Reffo, G. 1977. Reprint Technical Notes, *Nucl. Sci. Eng.* 62: 756
Harney, H. L., Richter, A., Weidenmüller, H. A. 1977. *Phys. Rev. C* 16: 1774
Hauser, W., Feshbach, H. 1952. *Phys. Rev.* 87: 366
Haynes, C. F., Davis, S., Glashausser, C., Robbins, A. B. 1976. *Nucl. Phys. A* 270: 269

Hofmann, H. M., Richert, J., Tepel, J. W., Weidenmüller, H. A. 1975a. *Ann. Phys. NY* 90:403

Hofmann, H. M., Richert, J., Tepel, J. W. 1975b. *Ann. Phys. NY* 90:391

Holbrow, C. H., Barschall, H. H. 1963. *Nucl. Phys.* 42:264

Hüfner, J., Mahaux, C., Weidenmüller, H. A. 1967. *Nucl. Phys. A* 105:489

Huizenga, J. R., Moretto, L. G. 1972. *Ann. Rev. Nucl. Sci.* 22:427–64

Humblet, J., Rosenfeld, L. 1961. *Nucl. Phys.* 26:529

Humblet, J. 1964. *Nucl. Phys.* 57:386

Irie, Y., Hyakutake, M., Matoba, M., Sonoda, M. 1975. *J. Phys. Soc. Jpn.* 39:537

Irie, Y., Hyakutake, M., Matoba, M., Sonoda, M. 1976. *Phys. Lett. B* 62:9

Kanter, E. P., Kollewe, K., Komaki, K., Leuca, I., Temmer, G. M., Gibson, W. M. 1979. *Nucl. Phys. A* 299:230

Kawai, M., Kerman, A. K., McVoy, K. W. 1973. *Ann. Phys. NY* 75:156

Kerman, A. K., Rodberg, L. S., Young, J. E. 1963. *Phys. Rev. Lett.* 11:422

Kerman, A. K., Sevgen, A. 1976. *Ann. Phys. NY* 102:570

King, H. T., Slater, D. C. 1977. *Nucl. Phys. A* 283:365

Kretschmer, W., Graw, G. 1971. *Phys. Rev. Lett.* 27:1294

Kretschmer, W., Wangler, M. 1978. *Phys. Rev. Lett.* 41:1224

Krieger, T. J. 1967. *Ann. Phys. NY* 42:375

Landau, L. D. 1937. *Phys. Z. Sowjetunion* 11:556

Lane, A. M. 1971. *Ann. Phys. NY* 63:171

Lane, A. M. 1973. In *Proc. Int. Conf. on Photonuclear Reactions and Applications, Asilomar 1973*, ed. B. L. Berman, Vol. 2, p. 803. Lawrence Livermore Laboratory, Calif.

Lane, A. M. 1978. *Phys. Rev. C* 18:1525

Lane, A. M., Dittrich, T. R., Mitchell, G. E., Bilpuch, G. E. 1978. *Phys. Rev. Lett.* 41:454

Lane, A. M., Lynn, J. E. 1957. *Proc. Phys. Soc. London LXX* 8-A:557

Lane, A. M., Thomas, R. G. 1958. *Rev. Mod. Phys.* 30:257

Liou, H. I. 1972. *Phys. Rev. C* 5:974

Lux, C. R., Porile, N. T., Grimes, S. M. 1977. *Phys. Rev. C* 15:1308

MacDonald, W. M. 1978. *Phys. Rev. Lett.* 40:1066

Mahaux, C. 1973. *Ann. Rev. Nucl. Sci.* 23:193–218

Mahaux, C., Weidenmüller, H. A. 1969. *Shell-Model Approach to Nuclear Reactions*. Amsterdam: North-Holland. 347 pp.

Mantzouranis, G. 1976a. *Phys. Lett. B* 63:25

Mantzouranis, G. 1976b. *Phys. Rev. C* 14:2018

Mantzouranis, G., Agassi, D., Weidenmüller, H. A. 1975. *Phys. Lett. B* 57:220

Mantzouranis, G., Pauli, H. C. 1977. *Z. Phys. A* 281:165

Mantzouranis, G., Weidenmüller, H. A., Agassi, D. 1976. *Z. Phys. A* 276:145

Matsuzaki, H. 1978. *Nucl. Phys. A* 308:95

Mello, P. A., McVoy, K. W. 1978. *Notas Fis* 1:172

Mello, P. A. 1979. *Phys. Lett. B* 81:103

Mitchell, G. E., Dittrich, T. R., Bilpuch, E. G. 1979. *Z. Phys. A* 289:211

Moldauer, P. A. 1961. *Phys. Rev.* 123:968

Moldauer, P. A. 1964a. *Phys. Rev.* 135:B642

Moldauer, P. A. 1964b. *Phys. Rev.* 136:B947

Moldauer, P. A. 1964c. *Rev. Mod. Phys.* 36:1079

Moldauer, P. A. 1967a. *Phys. Rev. Lett.* 18:249

Moldauer, P. A. 1967b. *Phys. Rev.* 157:907

Moldauer, P. A. 1967c. *Phys. Rev. Lett.* 19:1047

Moldauer, P. A. 1968. *Phys. Rev.* 171:1164

Moldauer, P. A. 1969a. *Phys. Rev. Lett.* 23:708

Moldauer, P. A. 1969b. *Phys. Rev.* 177:1841

Moldauer, P. A. 1971. *Phys. Rev. C* 3:948

Moldauer, P. A. 1972. In *Statistical Properties of Nuclei*, ed. J. B. Garg, p. 335. New York: Plenum

Moldauer, P. A. 1974. In *Proc. Int. Symp. on Correlations in Nuclei, Hungary 1973*, p. 319. Budapest: Hungarian Physical Society

Moldauer, P. A. 1975a. *Phys. Rev. C* 11:426

Moldauer, P. A. 1975b. *Phys. Rev. C* 12:744

Moldauer, P. A. 1976. *Phys. Rev. C* 14:764

Moldauer, P. A. 1978. See Bunakov 1978

Orland, H., Schaeffer, R. 1979. *Z. Phys. A* 290:191

Porile, N. T., Grimes, S. M. 1975. *Phys. Rev. C* 11:1567

Porile, N. T., Pacer, J. C., Wiley, J., Lux, C. R. 1974. *Phys. Rev. C* 9:2171

Porter, C. E. 1965. *Statistical Theories of Spectra: Fluctuations*. New York: Academic. 576 pp.

Porter, C. E., Thomas, R. G. 1956. *Phys. Rev.* 104:483

Prince, A. 1978. See Bunakov, 1978

Ratcliff, K. F., Austern, N. 1967. *Ann. Phys. NY* 42:185

Reffo, G., Fabbri, F., Gruppelaar, H. 1976. *Lett. Nuovo Cimento* 17:1

Richert, J., Simbel, M. H., Weidenmüller, H. A. 1975. *Z. Phys. A* 273:195

Richert, J., Weidenmüller, H. A. 1977. *Phys. Rev. C* 16:1309

Richter, A. 1974. *Nuclear Spectroscopy and Reactions*, ed. J. Cerny, Part B, p. 343.

New York: Academic
Robson, D. 1965. *Phys. Rev.* 137: B535
Robson, D. 1975. See Richter 1974, Part D, p. 245
Robson, D., Richter, A., Harney, H. L. 1973. *Phys. Rev. C* 8:163; 1975. *Phys. Rev.* erratum 11:1867
Rumyantsev, B. A., Kheifets, S. A. 1975. *Sov. J. Nucl. Phys.* 21:267
Satchler, G. R. 1963. *Phys. Lett.* 7:55
Sevgen, A. 1974. *Phys. Lett. B* 52:306
Sevgen, A. 1976. *Phys. Lett. B* 63:139
Shikazono, N., Terasawa, T. 1975. *Nucl. Phys. A* 250:260
Sidorov, V. A. 1962. *Nucl. Phys.* 35:253
Simonius, M. 1974. *Phys. Lett. B* 52:279
Simpson, J. J., Wilson, S. J., Dixon, W. R., Storey, R. S., Kuehner, J. A. 1978. *Phys. Rev. Lett.* 40:154
Simpson, J. J., Wilson, S. J. 1979. *Phys. Rev. C* 19:272
Smith, A., Guerther, P., Moldauer, P., Whalen, J. 1978. *Nucl. Phys. A* 307:224
Tamura, T., Udagawa, T., Feng, D. H., Kan, K. K. 1977. *Phys. Lett. B* 66:109
Tepel, J. W., Hofmann, H. M., Weidenmüller, H. A. 1974. *Phys. Lett. B* 49:1
Tepel, J. W. 1975. *Z. Phys. A* 273:59
Thomas, R. G. 1955. *Phys. Rev.* 97:224
Thomas, R. G. 1956. See Porter 1965, p. 561

Tsai, S., Bertsch, G. F. 1978. *Phys. Lett. B* 73:247
Ullah, N. 1968. *Phys. Rev. Lett.* 20:1510
Vager, Z. 1971. *Phys. Lett. B* 36:269
Vogt, E. 1968. *Advances in Nuclear Physics* 1:261. New York: Plenum
Weidenmüller, H. A. 1964. *Ann. Phys. NY* 29:378
Weidenmüller, H. A. 1972. *Phys. Lett. B* 42:304
Weidenmüller, H. A. 1974a. *Nukleonika* 19:387
Weidenmüller, H. A. 1974b. *Phys. Rev. C* 9:1202
Weidenmüller, H. A. 1979. *Progress in Particle and Nuclear Physics*, ed. D. H. Wilkinson. Oxford: Pergamon. In press
Weidenmüller, H. A., Richter, A., Harney, H. L. 1978. *Phys. Rev. C* 18:1953
Weisskopf, V. F. 1937. *Phys. Rev.* 52:295
Weisskopf, V. F. 1961. *Phys. Today* 14(7):18
Weisskopf, V. F., Ewing, D. H. 1940. *Phys. Rev.* 57:472
Wigner, E. P. 1957. See Porter 1965, pp. 199, 223
Wolfenstein, L. 1951. *Phys. Rev.* 82:690
Wong, C. Y., Tang, H. H. K. 1978. *Phys. Rev. Lett.* 40:1070
Yoshida, S., Yazaki, K. 1975. *Nucl. Phys. A* 255:173

Ann. Rev. Nucl. Part. Sci. 1979. 29: 33–68
Copyright © 1979 by Annual Reviews Inc. All rights reserved

NUCLEAR PHYSICS WITH POLARIZED BEAMS   ✳5602

Charles Glashausser[1]

Physics Department, Rutgers University, Piscataway, NJ 08854

CONTENTS

1   INTRODUCTION

Polarized beams are useful for solving problems in nuclear structure and nuclear reactions. Many of these problems are, of course, closely related to spin. But polarization measurements can be decisive, for example, in establishing the wave function of a nuclear state; they can also be crucial in deciding whether a nuclear reaction proceeds to the final state directly or through an intermediate process. In fact, in many of the most interesting experiments, spin is of surprisingly little relevance to the basic problem; although polarization obviously depends on spin, the fundamental physics at issue often does not.

[1] Supported in part by the National Science Foundation.

33

0163-8998/79/1201-0033$01.00

The relationship between low energy polarization experiments and more general nuclear physics is often lost in translation. The details of coordinate systems and notation have become so complicated that an outsider at a conference on nuclear physics with polarized beams needs an interpreter; even an expert on polarization in high energy physics would find himself ill prepared for a typical low energy polarization talk. The aim of this review is to consider the achievements of polarization experiments from a general nuclear physics standpoint. In so doing, I gloss over the details of the polarization experiments and the notation, and concentrate in this limited space on explicating instead the details of the nuclear physics.

A small vocabulary of common expressions is sufficient to describe the most important features of most experiments (Haeberli 1974). For spin-one-half particles, the most frequently measured quantity is the analyzing power, A_y. This is a left-right or spin up–spin down asymmetry measurement with a beam polarized transverse to the reaction plane. An A_y measurement in a nuclear reaction is denoted $A(\vec{a}, b)B$, where the arrow indicates that the incident particle has a measured polarization. Closely related, but not necessarily identical except in special cases, is the polarization, p. This is the transverse polarization of outgoing particles if the incident beam is unpolarized, and is denoted $A(a, \vec{b})B$. To measure p normally requires a second scattering from a target with a known analyzing power. Differences between p and A_y in inelastic scattering can arise if the transverse component of the spin of the incident particle is flipped during the scattering. The spin-flip probability S can be measured in various ways. The most straightforward method conceptually is to measure the transverse polarization of the outgoing inelastically scattered particles when the incoming beam is transversely polarized. This is a particular example of a very general type of experiment called polarization transfer measurements, all of which involve determining the polarization state of the products of a reaction initiated by a polarized beam.

When particles of spin greater than one-half are involved, the notation becomes more complicated. Because a deuteron from a polarized ion source, for example, may have three possible orientations relative to its momentum, it is not enough to state the difference in the number of particles with parallel and antiparallel spins (i.e. the vector polarization); you must also state how many have projections of zero. If you rotate to a different z axis, you now require many parameters to describe the alignment or tensor polarization. Symmetry requirements reduce the total number of possible independent analyzing powers to four. The vector analyzing power is commonly labeled iT_{11}; this corresponds to A_y for spin-one-half particles. There are three tensor analyzing powers generally but not universally labeled T_{20}, T_{21}, and T_{22}. For higher spins, correspondingly higher rank tensors are necessary for the description.

All these experimentally determined quantities can be defined in terms of combinations of scattering-matrix elements. For elastic scattering of spin-one-half particles from a spin zero nucleus, for example, the scattering amplitude at a particular angle can be written as a sum of a spin-independent amplitude g and a spin-dependent amplitude h. Whereas the differential cross section σ is then $|g|^2 + |h|^2$, the analyzing power A_y is Im $g^*h/[|g|^2 + |h|^2]$. This illustrates a general feature of polarization measurements: They depend on interference of amplitudes (Simon & Welton 1953). In most situations, h is small and so an A_y measurement is important to determine it precisely. Even if h is not an interesting quantity, however, the A_y measurements also yield information on the phase of g that is not derivable from σ.

A good illustration of the power of polarization measurements to determine quantities essentially unrelated to spin is the inelastic scattering of polarized protons at low energies, say below 10 MeV, where compound-nucleus and direct processes often interfere. This work is described in detail in Section 2.2.2, but a few words here are useful. The scattering matrix $S_{cc'}$ can then be conveniently written as a sum of a term slowly varying with energy and one that fluctuates rapidly:

$$S_{cc'} = \langle S_{cc'} \rangle + S_{cc'}^{\text{fl}}, \text{ with } \langle S_{cc'}^{\text{fl}} \rangle \equiv 0. \qquad 1.$$

The index c denotes the elastic channel; the index c' denotes a particular partial wave in the inelastic channel leading to a particular final state. The energy-averaged compound-nucleus cross section is then

$$\langle \sigma \rangle^{\text{CN}} \propto \sum_{c'c''} \langle S_{cc'}^{\text{fl}} S_{cc''}^{\text{fl}*} \rangle. \qquad 2.$$

The compound-nucleus polarization cross section is

$$\langle \sigma p \rangle^{\text{CN}} \propto \text{Im} \sum_{c' \neq c''} \langle S_{cc'}^{\text{fl}} S_{cc''}^{\text{fl}*} \rangle. \qquad 3.$$

If the matrices for scattering from the initial channel c to different final channels c' and c'' are uncorrelated, i.e. if $\langle S_{cc'}^{\text{fl}} S_{cc''}^{\text{fl}*} \rangle = 0$ for $c' \neq c''$, then $\langle \sigma p \rangle^{\text{CN}} = 0$, but $\langle \sigma \rangle^{\text{CN}} \neq 0$. In other words, a nonzero polarization implies channel-channel correlations that play a significant role in contemporary compound-nucleus theories. These correlations contribute also to σ, but they are generally very difficult to identify there because of the large background from the sum of terms of the form $|S_{cc'}|^2$.

Since the sensitivity to interference in polarization measurements is generally less familiar than their utility in measurements of spin, the former aspect is emphasized here. In the ideal polarization experiment, the extra sensitivity is used to advantage to learn something that could not be discovered otherwise. Examples of this are fairly numerous in low energy compound-nucleus reactions. It is very important also in $(\vec{p}, \gamma)$ reactions in the region of the giant resonances. On the other hand, direct $(\vec{p}, p')$ reactions to the giant dipole resonance have yielded little of value, probably because A_y is *too* sensitive to many unknown parameters.

These topics are explored in detail in the following sections. We first talk about reactions that proceed via the compound nucleus. Almost all the interesting work has been concerned with the inelastic scattering of protons. The most elegant results are not widely known and have not been previously reviewed. In the section on direct reactions, we consider a much wider class of processes with considerable diversity in the fruitfulness of polarization measurements. Many polarization results in this area are more widely known (Glashausser 1974, Haeberli 1977), so we confine our attention to an overview of the field and to a discussion of some new results that illustrate the comments of this introduction. Section 4 describes closely related results of polarization work in photonuclear reactions.

This review is not exhaustive. Some areas, notably the few-nucleon problem, tests of fundamental conservation laws, and model-independent spin and parity measurements, are entirely omitted, mostly for reasons of space. In the areas that I do consider, my discussion is quite selective; though I have tried to include a wider range of papers in the bibliography, this is by no means a complete list of interesting polarization papers.

2   COMPOUND-NUCLEUS REACTIONS

2.1   *Wave Functions of Isobaric Analog States*

Ever since they were first observed, isobaric analog resonances (IAR) were promised as a vehicle for determining components of core-excited states comparable to single-particle stripping and pickup reactions for simple states. The wave function of an excited $\frac{3}{2}^+$ state in ^{89}Sr, for example, may be written in terms of the weak-coupling model as

$$\psi(\tfrac{3}{2}^+) = a\,|0^+, \tfrac{3}{2}^+\rangle + b\,|2_1^+, \tfrac{1}{2}^+\rangle + c\,|2_1^+, \tfrac{3}{2}^+\rangle$$
$$+ d\,|2_1^+, \tfrac{5}{2}^+\rangle + e\,|2_1^+, \tfrac{7}{2}^+\rangle + \cdots$$

The IAR corresponding to this at high excitation in the compound system ^{89}Y decays both to the ground state of ^{88}Sr, with a $d_{3/2}$ partial decay width proportional to a, and to the 2_1^+ state in ^{88}Sr with partial widths proportional to $b, c, d, \ldots$ in the channels $s_{1/2}, d_{3/2}, d_{5/2}, \ldots$ respectively. A neutron-stripping experiment on a ^{88}Sr target can determine the amplitude a; in the same way, proton elastic scattering from ^{88}Sr at the energy of the corresponding IAR determines the elastic width, which is also a measure of a. Whereas the simple stripping experiment is not sensitive in first order to the remaining terms, inelastic scattering at the IAR to the 2_1^+ state of ^{88}Sr, on the contrary, can in principle measure both their magnitudes and their phases. Since there are so many parameters, a simple σ measurement is not sufficient. The method that has now been applied to several nuclei involves measuring also an angular distribution of A_y, as well as, in favorable cases, angular distributions of S with a polarized beam. The latter measurement also yields the difference ΔS between the spin-flip probabilities for incoming spin-up and spin-down polarized beams.

The results of such a measurement are shown in Figure 1. Inelastic measurements were carried out at 7.08 MeV on a ^{88}Sr target; this energy corresponds to a $\frac{3}{2}^+$ IAR in ^{89}Y. At this energy, the direct reaction (DR) amplitude is small but nonnegligible. In fact, if the DR amplitude were zero, the values of A_y would all be zero (Harney 1968). Thus the scattering matrix for the solid curves in this figure must include both DR amplitudes, via the distorted-wave Born approximation (DWBA), and IAR amplitudes. The parameters in the calculation at hand were the signs and magnitudes of the inelastic partial widths for $s_{1/2}$, $d_{3/2}$, and $d_{5/2}$ decay. These were then translated into the wave-function amplitudes shown in Table 1 using renormalized single-particle widths. The error bars, which estimate errors due to imperfect fits and do not include, for example, possible errors in the DWBA or single-particle widths, are

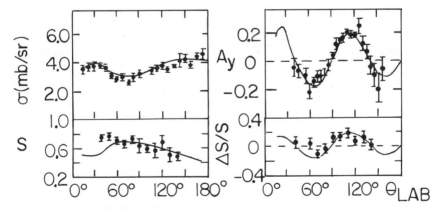

Figure 1 Cross-section, analyzing-power, spin-flip probability, and spin-flip asymmetry measurements at the 7.08-MeV ($\frac{3}{2}^+$) isobaric analog resonance in ^{88}Sr + p. The curves were calculated by varying the resonance parameters in a code that included both the resonance and a DWBA direct reaction background (Haynes et al 1976).

impressively small. They indicate that the method determines even very small components of the wave function with high precision. A comparison with the intermediate-coupling wave functions of Spencer et al (1971) shows good agreement.

A measure of the accuracy with which this method can be applied (but not of its intrinsic accuracy) is the comparison of the results obtained for the same states by two different groups. Angular distributions of σ and A_y at low-lying resonances in ^{138}Ba + p were measured at Rutgers (Davis et al 1976); at Erlangen, excitation functions at several back angles were measured over the same region (Clement & Graw 1977). Spin-flip probabilities for the first few states are small, so these data are not included. Values of the amplitudes for two states derived by the two groups are listed in Table 2; they are also compared with intermediate-coupling wave functions of Vanden Berghe et al (1971). The good agreement between all three sets of values tends to confirm both the method and the theoretical wave functions, although there are some exceptions. The sign of the $f_{5/2}$ amplitude in the wave function for the 10.0-MeV $\frac{5}{2}^-$ state, for example, is opposite in the two experimental determinations, and the theoretical value is midway between the two experimental ones.

Given the apparent wide applicability of this technique, it has thus far been applied in relatively few cases. In addition to the Sr and Ba data cited, the Erlangen group carried out an extensive study of the $N = 82$ isotones through Sm, and some work has been done on ^{86}Sr (King & Slater 1977), ^{90}Zr (Feist et al 1977), ^{92}Mo (Davis et al 1975), ^{124}Sn (Boyd et al 1974), and ^{126}Te (Van Eeghem & Verhaar 1976) isotopes. The basic reason for this is the general shift in nuclear physics interest away from light ions at low energies. But some intrinsic problems deserve further study. The primary problem is the lack of a good calibration, in the sense that ^{208}Pb is a good calibration for single-particle

Table 1 Comparison of normalized experimental amplitudes with theoretical wave functions for the 2.00-MeV $\frac{3}{2}^+$ level in ^{89}Sr, the parent of the 7.08-MeV analog state in ^{89}Y

Amplitude	Experiment[a]	Theory[b]
$\|0^+ \times d_{3/2}\rangle$	$+0.67$	$+0.69$
$\|2^+ \times s_{1/2}\rangle$	-0.06 ± 0.06	-0.01
$\|2^+ \times d_{3/2}\rangle$	0.00 ± 0.06	$+0.28$
$\|2^+ \times d_{5/2}\rangle$	-0.74 ± 0.02	-0.65
$\|2^+ \times g_{7/2}\rangle$	—	-0.16

[a] Haynes et al (1976).
[b] Spencer et al (1971).

Table 2   Experimental and theoretical wave functions for states in ^{139}Ba

Amplitude	Experiment[a]	Experiment[b]	Theory[c]
A. $E_p = 10.00$ MeV ($\frac{7}{2}^-$ IAR)			
$\|0^+ \times 2f_{7/2}\rangle$	0.96 ± 0.10	0.89	0.94
$\|2^+ \times 3p_{3/2}\rangle$	-0.22 ± 0.05	-0.17	-0.17
$\|2^+ \times 2f_{5/2}\rangle$	$-0.25{}^{+0.25}_{-0.11}$	0.29	-0.04
$\|2^+ \times 2f_{7/2}\rangle$	$-0.26{}^{+0.14}_{-0.09}$	-0.15	-0.27
B. $E_p = 10.63$ MeV ($\frac{3}{2}^-$ IAR)			
$\|0^+ \times 3p_{3/2}\rangle$	0.79 ± 0.08	0.70	0.75
$\|2^+ \times 3p_{1/2}\rangle$	-0.28 ± 0.11	-0.30	-0.16
$\|2^+ \times 3p_{3/2}\rangle$	-0.24 ± 0.08	-0.13	-0.20
$\|2^+ \times 2f_{5/2}\rangle$	$0.06{}^{+0.17}_{-0.11}$	0.22	-0.09
$\|2^+ \times 2f_{7/2}\rangle$	-0.61 ± 0.09	-0.43	-0.57

[a] Davis et al (1976).
[b] Clement & Graw (1977).
[c] Vanden Berghe et al (1971).

stripping. Second, in heavier nuclei, resonances of high spin often overlap so the number of parameters to vary simultaneously becomes very large. In addition, there are the usual problems apparent even in elastic scattering at IAR, e.g. "standard" single-particle widths are not really standard and are often not reliable. These intrinsic problems can be overcome with some effort, which will be worthwhile insofar as the wave functions studied are important. The competing method of higher order excitation of this same type of state with heavy ions seems intrinsically much less reliable. The reliability of the present method is also subject to con-firmation by comparing theory and experiment for general $(\vec{p}, p'\gamma)$ corre-lations.

2.2   Compound-Nucleus Reaction Mechanisms

2.2.1   ELASTIC SCATTERING   Only recently has the fundamental basis of compound-nucleus theory been understood in depth. Data have generally been analyzed by means of Hauser-Feshbach theory, a statistical theory that can be simply derived only by making assumptions about un-correlated amplitudes that are not always true. Yet Hauser-Feshbach theory works in a wide variety of situations where the sophisticated observer might well expect it to fail. Why it works has been the subject of considerable recent discussion, much of it technical and somewhat tedious, often hinging on statistical ideas unfamiliar to most practitioners.

Yet the story is ultimately very elegant and very interesting; it has attracted some of the best theorists of the day, and it is clearly summarized in the article by Mahaux & Weidenmüller in this volume. The new theories discussed there are most important when compound and direct amplitudes mix. Polarized beams have permitted important tests of several aspects of these theories. It is useful to discuss these experiments here both because they may suggest other such tests and also because they are a good illustration of the power of polarization measurements.

These experiments rely on the fact noted in the Introduction that the quantities $\langle \sigma p \rangle$ and $\langle \sigma A_y \rangle$ depend on a coherent sum of amplitudes, and so average to zero when the amplitudes involved are uncorrelated. For elastic scattering, this has been used to show that the quantities $\langle \sigma p \rangle^{CN}$ and $\langle \sigma A_y \rangle^{CN}$ must be zero provided the average is taken over a large enough energy region (Thompson 1967). This is clear from Equation 3, since $c' = c$ only. (The notation CN on these quantities implies that they depend only on S^{fl} in Equation 1. A compound-nuclear process is *defined* as one that depends on the fluctuating term. The definition is reasonable in terms of the traditional method of distinguishing compound-nuclear processes from direct processes by their time scale.)

Measured values of $\langle \sigma A_y \rangle$ in the presence of both CN and DR amplitudes thus depend only on $\langle S \rangle$ in Equation 1. This fact was first applied by Kretschmer & Graw (1971) to the determination of σ^{CN} in the vicinity of an IAR. Excitation functions of both σ and σA_y were measured. Resonance parameters were chosen to provide a good fit to $\langle \sigma A_y \rangle$ with a small optical model background included. The same parameters were then used to generate cross sections that we can label σ^{DR}; the difference between the measured σ and σ^{DR} is σ^{CN}. This procedure is model independent insofar as the optical model parameters play no role. The work is important also because it provides a "rather model-independent" way to obtain IAR parameters in the presence of compound elastic background.

More recently, Kretschmer & Wangler (1978) applied a variant of the same technique to elastic scattering from ^{30}Si in the region of Ericson fluctuations. Contemporary Hauser-Feshbach theories seem to agree that in this many-channel, strong absorption limit the ratio σ^{CN}/σ^{HF} should be two. Here σ^{HF} is the ordinary Hauser-Feshbach cross section, without width fluctuation corrections. This ratio, called the compound elastic enhancement factor $W_{\alpha\alpha}$, had previously been measured only with large error bars.

The crucial quantity σ^{CN} was again fixed by a measurement of σ and σA_y as a function of energy. Now, however, σ^{DR} was determined using a formula derived by Henneck & Graw (1977), who showed that the percentage y_D of direct reaction (σ^{DR}/σ) is given by

$$[R(\sigma) + R(\sigma A)]\langle \sigma \rangle^{-2} = 1 - y_D^2,$$

with $R(x) = \langle x^2 \rangle - \langle x \rangle^2$. The quantity σ^{DR} determined this way is shown with error bars at the top of Figure 2. The quantity $\sigma - \sigma^{DR}$ is again a reasonably model-independent determination of σ^{CN}; the model dependence now arises from the Ericson fluctuation model itself.

To confirm their result, Kretschmer & Wangler then used just the method described above for IAR (Kretschmer & Graw 1971). They fitted $\langle \sigma A_y \rangle$ with only the optical model and used the parameters to obtain a second determination of σ^{DR}. These results are shown as the solid curves in Figure 2; the agreement is reasonable but certainly not perfect. The

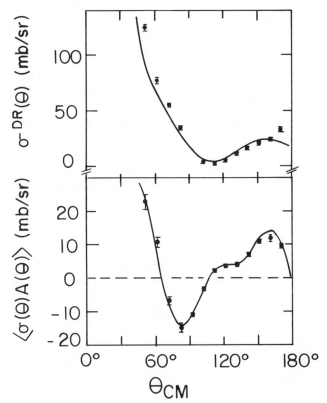

Figure 2   Angular distributions of the direct reaction cross section σ^{DR} and the averaged analyzing-power cross section $\langle \sigma A_y \rangle$ for elastic proton scattering from ^{30}Si at a mean energy of 9.84 MeV. The σ^{DR} data points were determined using the Henneck & Graw (1977) method. The solid curves are optical model calculations with parameters adjusted slightly to fit $\langle \sigma A_y \rangle$ (Kretschmer & Wangler 1978).

quantity σ^{HF} was then calculated using the same optical parameters. The ratio $W_{\alpha\alpha}$ so determined turns out to be 2.09 ± 0.14, in good agreement with theory. Although this number is claimed to be model independent, the necessity of using the optical model to determine σ^{HF} coupled with the imperfect fit to σ^{DR} indicates that this statement is not precise. In addition, it should be noted that the σ^{CN} determined this way is small, smaller than the discrepancy between the two determinations of σ^{DR} at both forward and backward angles. Thus the values of σ^{CN} determined from $\langle \sigma A_y \rangle$ alone, i.e. from the method of Kretschmer & Graw, would be very different from the more accurate values determined from the fluctuation analysis. This problem was not present in the original IAR analysis of Kretschmer & Graw because there the optical parameters had little effect, and the values of σ^{CN} were much larger. However, it does suggest that it might be useful to compare the σ^{CN} determined this way with the σ^{CN} determined by another method that claims to be model independent, e.g. crystal-blocking lifetimes (Kanter 1977).

2.2.2 INELASTIC SCATTERING   Whereas Equation 1 shows that the polarization cross section $\langle \sigma p \rangle$ depends on interference between different partial waves in the exit channel, the corresponding expression for $\langle \sigma A_y \rangle$ demands interference between different partial waves in the entrance channel (Harney 1968). For elastic scattering this has the consequence that $\langle \sigma p \rangle^{CN} = \langle \sigma A_y \rangle^{CN} = 0$, as we just noted. For inelastic scattering, the possibilities are more numerous. Consider inelastic proton scattering on a spin-zero nucleus, for example, at the energy of a single isolated IAR without background of any kind. In this unrealizable situation, A_y would necessarily be zero because only one entrance channel is important, but p would generally not be zero if the IAR decayed by several partial waves. Of course, in the practical situation, there is either compound or direct background or both. Even a single resonance can interfere with a DR background to yield nonzero A_y (and p). But interference with compound background will not give rise to additional A_y or p, provided the compound states have truly random phases and a sufficient energy average is taken to average these phases to zero.

One of the important consequences of the new developments in compound-nucleus reaction theory in the past few years has been the realization that compound-nucleus phases cannot be assumed to be random in the presence of direct reactions. Limitations imposed principally by the unitarity of the S matrix require that decay amplitudes from the compound states into different channels, for example, partial waves to a given final state, be correlated. The new theories also suggest methods for treating these correlations in certain circumstances.

It has been difficult to find evidence for these correlations, possibly because for most direct reactions many different values of the total spin and parity J^π are involved in the reaction. Since no correlations are postulated between amplitudes for different J^π, the net effect of correlations that might be present for a single J^π tends to be small in typical reactions. Relative to the energy widths of typical compound states near an analog state in a heavy nucleus, however, the IAR itself is very wide. Reactions proceeding through the IAR itself can then be considered direct, i.e. described by $\langle S \rangle$ in Equation 1; other, DWBA-type, contributions to $\langle S \rangle$ are negligible at low enough energy. Reactions that proceed through the dense narrow compound states under the resonance should be described by S^{fl}. Since the IAR strongly selects a given J^π, such reactions might be expected to show more easily the correlation effects noted above. The way to see the correlations is to measure a nonzero value of the polarization p^{CN} of the particles scattered by the term S^{fl} in the S matrix.

Experimentally, however, a single value of p is measured at a given angle and energy, and there is no model-independent way to separate the relative contributions of $\langle S \rangle$ and S^{fl} to the actual value of p. The quantity p^{IAR} is determined by the parameters of the resonance; if the wave function of the IAR were known, p^{IAR} (and σ^{IAR}) would be known also. If there are no channel-channel correlations, the contribution of S^{fl} to p, i.e. p^{CN}, is zero and the following relations hold:

$$\sigma = \sigma^{CN} + \sigma^{IAR} \qquad\qquad 4.$$

$$p = p^{IAR}(\sigma^{IAR}/\sigma). \qquad\qquad 5.$$

In the new compound-nucleus theories, p contains an additional contribution also from p^{CN}, which in simple cases can be shown to equal p^{IAR} (Graw et al 1974).

Experiments that support the new theories have now been carried out at several laboratories. A new method of measuring p was developed that permitted a relatively rapid determination of p without double scattering (Boyd et al 1971). Angular distributions of p and σ for inelastic scattering to the 2_1^+ state of ^{92}Mo were measured by Davis et al (1975); the incoming energy, 6.55 MeV, corresponds to a $\frac{3}{2}^+$ IAR in ^{93}Tc. All attempts to fit the data using Equations 4 and 5 above, i.e. without correlations, yielded poor agreement. When correlations were included by setting $p^{CN} = p^{IAR}$, resonance parameters could be found that produced excellent fits. The result is thus convincing evidence for the necessity of including the channel-channel correlations in the vicinity of an IAR.

It was possible to achieve this result because the measured values of p at this resonance are relatively large (about 0.30). The values of course do not approach 1.0 simply because of the large *uncorrelated* compound-nucleus background in channels with J^π different from the J^π of the resonance. The IAR induces correlations in S^{fl} only in its own channel, and off resonance no polarization is predicted.

Further work in this area has concentrated on applying the full apparatus of the new compound-nucleus theories to the analysis of data similar to that of Davis et al (1975). The simple result that $p^{CN} = p^{IAR}$ in that case agrees with the data, but a somewhat smaller value of p^{CN} could not be eliminated by the fit. In fact $p^{CN} = p^{IAR}$ only if *all* excitation of the IAR via the S^{fl} term takes place because of the mixing of the $T_>$ IAR into the $T_<$ CN states; the wave function of the IAR then imposes its own phase relationships on each of the many normally disordered CN states. But at least some cross section in the channel with the J^π of the IAR would be present even in the absence of the IAR; the nonzero off-resonance cross section shows this. The new CN theories provide a method for calculating it.

The only extensive analysis was carried out by Feist et al (1977), who measured σ and $p\sigma$ for several $\frac{3}{2}^+$ resonances in $^{90}Zr + p$ near the (p, n) threshold. Data for three states are shown in Figure 3. The solid and dashed curves are, respectively, best fits with and without correlations. The quantity σ_b is the calculated cross section in channels other than $\frac{3}{2}^+$. Note that σA_y is significantly nonzero at the two higher resonances, which indicates that DWBA-type direct reaction background cannot be

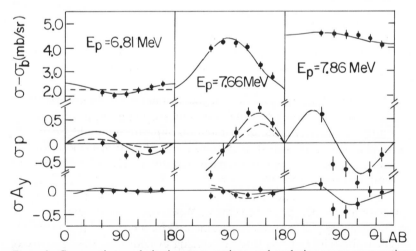

Figure 3 Cross sections, polarization cross sections, and analyzing-power cross sections for inelastic proton scattering from ^{90}Zr to the first 2^+ state. The incident energies correspond to $\frac{3}{2}^+$ analog resonances. The solid and dashed lines correspond to best fits with and without channel-channel correlations, respectively (Feist et al 1977).

neglected. As the energy increases, the importance of correlations decreases, and nearly identical results are obtained for the 7.86-MeV resonance with and without them. This occurs because the CN contribution to σ decreases as the energy of the resonances increases. Feist's calculation, with the almost rigorous application of the new theory prescribed by Hofmann et al (1975), yields results significantly different in detail from results obtained previously by application of the simple formulation described above for ^{92}Mo. The differences arise in part from previous neglect of the transmission at off-resonant energies, but also in part from a different prescription for the width fluctuation correction. However, the basic physical notion that the polarization of the compound contribution and the polarization of the direct contribution to the resonance are about the same remains approximately valid.

2.3   Ericson Fluctuations and Intermediate Structure

Fluctuations in measured cross sections as a function of energy were first predicted by Ericson (1960); developments since then were reviewed by Richter (1974). The fluctuations arise when the widths, Γ, of states in the compound nucleus are larger than the separation, D, between them. Autocorrelation analyses then yield coherence widths, Γ^c, as well as the ratio $(1 - y_D^2)/N_{eff}$, where y_D is the percentage direct reaction and N_{eff} is the effective number of spin channels taking part in the reaction. If N_{eff} can be reliably estimated, the compound and direct scattering can be separated.

While such analyses were once very popular because they seemed a rich source of information about the compound nucleus, interest waned as the uncertainties of the analysis became apparent. Problems with trend reduction, finite range of the data, bias errors, and the like cast doubt on the results obtained. In addition, the new compound-nucleus work has shown that Γ^c in many cases is actually much smaller than the average width of states in the compound nucleus (Moldauer 1975).

While interest in these randomly fluctuating cross sections was high, a corresponding interest in apparent nonrandom behavior developed. Isolated large "bumps" amid the fluctuating cross sections came to be labeled "intermediate structure," because they appeared to have properties suggestive of the simple modes of excitation in the nucleus predicted by Kerman et al (1963). Distinguishing between a large random fluctuation and a nonrandom peak whose origin lies in a significantly different dynamics is difficult, and the number of claimed intermediate structures has diminished rapidly as the statistical analyses have become more sophisticated.

In recent years some work in this area has been done with polarized beams. The formula cited in Section 2.2 for y_D (Henneck & Graw 1977), for example, uses A_y to eliminate the dependence on N_{eff}. In the search for a nonstatistical peak as evidence of an interesting structure at high excitation in the nucleus, it is much better to have two independent measurements, σ and A_y, that reveal the anomaly rather than σ alone. In addition, A_y is expected to be more sensitive than σ to the presence of such a structure because it depends on interference with the DR background. Thus inelastic proton scattering has been measured for ^{26}Mg and ^{27}Al in the region of compound-nucleus excitation corresponding to the giant resonances seen as final states in direct reactions (Glashausser et al 1975, Kaita 1978). Some of the data observed for A_y in the excitation of the 2_2^+ state in ^{26}Mg are shown in Figure 4.

The A_y fluctuates for the same reason that σ fluctuates. In fact, the compound contribution to σA_y at any particular energy is nonzero for this same reason: The summation of random phases from many overlapping compound states yields zero only on the average. On the other hand, it is clear from inspection of this figure that the average of σA_y taken over, say, 5 MeV does not yield zero. This can be attributed either to direct reactions, intermediate structure, or both. With data only up to 9.5 MeV, the large bumps were interpreted as intermediate structure (Glashausser et al 1975), mostly on the basis of statistical tests devised by Baudinet-Robinet & Mahaux (1974). Evidence for intermediate structure in A_y for ^{27}Al+p was also noted, but there were no anomalous bumps in the excitation function of σ for either reaction.

These results would be of clear interest if they signalled the presence of widespread intermediate structure at excitations of 15 MeV or so in

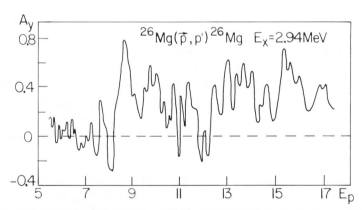

Figure 4   Excitation function of the analyzing power at 140° for inelastic proton scattering from ^{26}Mg (Kaita 1978). The curve has been drawn through data points measured in 50-keV steps (approximately the coherence width) and then averaged over 150 keV.

s-d shell nuclei. Subsequent developments, however, give little reason for optimism. First, the statistical methods employed in the ^{26}Mg analysis were questioned by Haglund et al (1976), who applied similar methods to the analysis of computer-generated data. Thorough investigation of this criticism (Kaita 1978) showed that the statistical tests originally used do, in fact, yield proper results, in the sense that they do not indicate spurious intermediate structure in statistical "data." However, actual data contain nonstatistical long-range trends that must be eliminated before the tests can be applied; a definitely reliable method for doing this has not yet been devised. With reasonable assumptions about the trends in the complete data of Figure 4, the statistical tests do indicate nonrandom structure, but the actual trend may be significantly different.

The point of such discussion is to suggest that intermediate structure must be unmistakable before it can be accepted. To be useful, it must be seen in several channels with large cross correlations, and, even better, in several nuclei. Investigations of $(\vec{p}, p')$ reactions on ^{19}F (Eng et al 1975), ^{23}Na (C. Glashausser et al, unpublished data), ^{28}Si (Hurd et al 1977), and ^{39}K (M. Sosnowski et al, unpublished data), as well as $(\vec{p}, \alpha)$ reactions on the first two of these, have not revealed any bumps comparable to those seen in ^{26}Mg + p, although the behavior cannot be entirely described by statistical models. Thus further progress in this area is certainly possible, but polarized beams apparently have not provided a crucial breakthrough.

In the one example where a polarized beam was used to unambiguously determine a supposed intermediate structure, the result was negative. This was one of the very few apparently well-established cases, namely, a $\frac{1}{2}^{+}$ IAR in ^{70}Ge + p that was split into several smaller peaks apparently of the same spin and parity (Temmer et al 1971). A measurement of A_y across this resonance revealed nonzero values at several of the peaks (Terry et al 1978). This shows these states must have spin greater than one half, and thus casts considerable doubt on the intermediate structure assignment.

This experiment is one of the many examples of the spin sensitivity of polarized beams. Since these are generally widely known, I prefer to concentrate here on the sensitivity of polarized beams to interference and thus to reaction mechanisms. Topics worthy of discussion in this context but omitted here for reasons of space include the observation of systematic A dependence in the widths of isospin-forbidden resonances (Ikossi et al 1976), and the interpretation of threshold anomalies in (d, p) reactions (Graw & Hategan 1971). Other technical aspects of IAR have been fruitfully investigated with polarized beams as well. The measurement of A_y in preequilibrium decays is only now beginning. And the investigation of other aspects of the new compound-nucleus theories, in a region of a more general type of direct background than an IAR, has not yet been attempted.

3 DIRECT REACTIONS

Simple direct reactions of light ions at low energy are reasonably well understood. Although detailed fits to data often cannot be achieved, the gross features are sufficiently well explained by the DWBA so that most are interesting primarily as instruments of spectroscopy. Reactions that proceed via several steps, however, do continue to evoke enthusiasm. In addition, in many situations it is difficult to decide a priori whether an apparently simple reaction proceeds by a direct or a circuitous route, particularly if two or more particles are transferred.

The nuclear physics motivation for using polarized beams in direct reactions is then rather unambiguous. They are widely used in spectroscopy to determine spins, generally using the dependence of A_y on the total angular momentum transfer J. Our concern with the sensitivity of polarization measurements to interference is stimulated more, however, by reactions that take place via multiple paths. Analyzing powers are generally affected much more than σ by a contribution to the reaction amplitude that is not a single step. In such reactions, an A_y measurement may be absolutely necessary if any definite results at all are to be derived from the experiment. In addition to these motivations, the explicit dependence of the reaction on spin-dependent interactions may be of general importance to nuclear physics. Such is the case with proton inelastic scattering at moderate energies, where the validity of microscopic models can more readily be decided for states excited only by spin-dependent interactions because the core polarization effects are generally smaller. The D-state amplitudes in the deuteron, triton, and ^{3}He wave functions are fundamental quantities that may be susceptible to measurement via transfer reactions. And, of course, the spin-orbit term in the optical potential can be significant in the explanation of details of cross-section angular distributions.

The amount of data available on polarization in direct reactions has been growing dramatically. Except for the most exotic multiparticle pickup reactions, some results have appeared for essentially all reactions initiated by protons and deuterons, and many with ^{3}He and tritons as well. A polarized ^{6}Li beam at Heidelberg has yielded some interesting data, and isolated results with heavier ions have been published. On the other hand, almost all the experiments have been measurements of A_y; few double-scattering measurements and no angular correlation measurements have been carried out with polarized beams.

3.1   *Elastic and Inelastic Scattering*

3.1.1 PROTONS Inelastic proton scattering is useful to nuclear physics most obviously because it is a measure of the deformation of the nucleus. A number, call it β, is determined by comparing the magnitude of σ with a DWBA calculation; β is generally considered to be the deformation of the central potential. Interesting recent work has focused on such questions as whether it is indeed β itself or a closely related quantity like the potential moment that is in fact determined from scattering data (Hamilton & Mackintosh 1978). In such a context, the measurement of A_y in inelastic proton scattering is relatively unimportant, except as a measure of the overall reliability of the reaction theory. Unless higher order effects contribute, the value of A_y does not depend on β, but it does depend on the ratio β_{so}/β, where β_{so} is the deformation of the spin-orbit potential. This ratio should be approximately 1.0 if a collective model of the nuclear states is appropriate; variations up to about 1.5 are probably attributable to different radii for the spin-orbit and central potentials.

Much higher values of this ratio have been observed on occasion, and these values change with the energy of the incident particle. A ratio of about 3.0 gives the best fit to the ^{90}Zr 2_1^+ state at 20 MeV, decreasing to about 1.0 at 40 MeV (de Swiniarski et al 1978). Equally large energy variations have been observed for ^{54}Fe (Van Hall et al 1977). Such a phenomenon would be more interesting if it could be explained. The very large values were at first attributed to a stronger two-body spin-orbit interaction between two protons than between a proton and a neutron, since the two transitions in ^{90}Zr and ^{54}Fe have a larger proton component in lowest order (Raynal 1971). However, the observed energy dependence makes this interpretation unlikely, and suggests instead an interpretation in terms of a resonance-like phenomenon. Geramb and his collaborators interpreted many excitation functions for inelastic scattering, particularly to weak states at large angles, in terms of intermediate excitation of giant resonances (Perrin et al 1977). While this reaction model allows considerable freedom in adjusting parameters and is certainly not thoroughly confirmed, the evidence in its favor is strong. Thus it is tempting to consider that such a mechanism might play a role in the energy variation of the ratio β_{so}/β. A calculation of this type has been performed for ^{54}Fe (Amos & Smith 1974); the authors are unable to achieve a good fit but they do not consider their failure definitive.

Attempts to use A_y measurements to advantage in direct excitation of giant resonances have also been unsuccessful. At a bombarding energy of 60 MeV, Kocher and co-workers (1976) were unable to confirm a suggested 2^+ assignment for a peak in the resonance region in ^{58}Ni($\vec{p}$, p′)

with an A_y measurement. The situation is not unusual: σ is fitted reasonably well and A_y is fitted less well, just because A_y is generally a sensitive fault-finder and the theory is not perfect. A similar attempt to distinguish between different spins in the comparable resonance in ^{90}Zr was made by the group at Grenoble (Martin et al 1979). While evidence for $L = 2$ and $L = 4$ admixtures is observed, the result is not convincing, particularly because A_y is fitted poorly.

The simple A_y measurement can be of value in distinguishing between one-step and two-step reaction mechanisms. This is discussed in detail in Section 3.2.2, but here it is useful to point out a recent $(\vec{p}, p')$ measurement that includes a (p, d)(d, p) channel as an alternate to the direct route (Aoki et al 1978). This is a measurement of the 1^+ to 0^+ 2.31-MeV transition in ^{14}N at an incident energy of 21 MeV. The transition is weak because it requires spin flip and isospin flip of the incident proton and because the wave functions produce large cancellations of amplitudes. The individual contributions from the two-step process and from single-step processes involving only the tensor term or only the central terms in the two-body interaction give a poor description of the data. When these are coherently added, it becomes evident that the complex path is dominant at large angles and the direct path at forward angles, but each

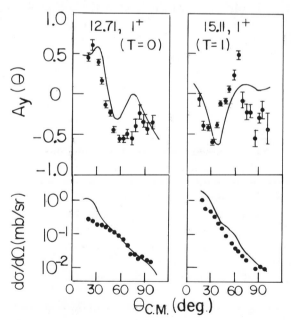

Figure 5   Analyzing powers and cross sections for inelastic proton scattering from ^{12}C at 65 MeV. The solid curves are microscopic DWBA calculations (Hosono et al 1978).

is significant in both domains. The result is reasonably good agreement with the data, but uncertainties remain.

It would be appropriate here to require satisfactory prediction of further experimental quantities, such as S and ΔS. The same comment can be made for the work shown in Figure 5, done at the University of Osaka (Hosono et al 1978). Cross sections and analyzing powers at 65 MeV for excitation of the isospin pair of states in ^{12}C at 12.71 MeV (1^+, $T = 0$) and 15.11 MeV (1^+, $T = 1$) are illustrated, together with sophisticated microscopic model predictions. Both states have essentially the same particle-hole components, with $p_{3/2}$-$p_{1/2}$ transitions overwhelmingly dominant. Yet the differential cross sections are significantly different from each other, and the analyzing powers dramatically so. Interestingly, the gross differences in the shapes of the A_y data are described by the theory (they result mostly from exchange contributions) but the more subtle differences between the cross-section data are hardly suggested.

The ambiguities in such a calculation are large. The effective microscopic two-body interaction is not well known and includes spin-orbit and tensor terms as well as a spin-spin term; each is different for p–p and p–n interactions. The other large uncertainty is the importance of higher order processes, such as the (p, d)(d, p) transfer discussed above. Since the wave functions are relatively simple, however, the core polarization expected to be small, and the two states so well matched, it is a beautiful test case for a microscopic theory. It also shows a clear necessity for a polarized beam and not just to measure A_y.

Joel Moss and his collaborators at Texas A & M University made a major effort in this direction, with measurements of the spin-flip probability S for transitions in light nuclei such as those just discussed (Moss et al 1977a, 1977b). The transitions are all unnatural parity transitions from or to a 0^+ state, so the selection rules require a spin transfer Δs of one unit for a single-step transition. Large spin-flip probabilities are expected from $\Delta s = 1$ transfers according to DWBA calculations, and Moss et al obtain good agreement with data for ^{6}Li and ^{14}N at about 35 MeV. On the other hand, multistep processes can take place with $\Delta s = 0$ and thus are likely to lead to small values of S. In their earlier work, Moss et al neglected the possibility of multistep contributions and attributed discrepancies between small measured values of S and large predicted values to problems in the two-body force, particularly in its tensor components. Examples are the ^{12}C(p, p')^{12}C and ^{11}B(p, p')^{11}B reactions.

Beginning with their most recent work on the ^{16}O 2^- state at 8.88 MeV, they have a different interpretation (Moss et al 1978). They now attribute the particularly dramatic discrepancy between theory ($S > 0.8$)

and experiment ($S < 0.25$) to multistep contributions (though no coupled-channels calculations are performed). Even when S is greatly affected, however, the spin-flip cross section, σS, should have no contribution from multistep processes in the limit that such processes are pure $\Delta s = 0$. It is σS, then, that they suggest comparing with microscopic theories to determine problems with wave functions or two-body interactions. The σS for ^{12}C are fitted well with conventional interactions, which indicates that the effective force is appropriate. However, the σS for ^{16}O is still almost a factor of five too low, and this is now interpreted as a problem with the random phase approximation wave functions for ^{16}O.

This is a significant statement, that the isoscalar M2 strength in ^{16}O is retarded. While the basic aim of microscopic theories of inelastic scattering has always been to show that a microscopic description is possible, a related goal has been to determine spectroscopic information as well. Even at this stage in the development of low energy microscopic theories, such spectroscopic quantities cannot be presumed to be reliable in the sense that a $B(E2)$ measurement from electron scattering is. With the assimilation of sufficient data of the sort just described, however, the microscopic theories may receive a strong confirmation.

3.1.2 DEUTERONS   The scattering of polarized deuterons, rich in potential, has thus far yielded relatively little information of general physics interest. The problems and the possibilities arise from the high spin of the deuteron. As noted in the Introduction, four polarization measurements are now possible with a polarized beam. Additional spin-dependent potentials appear in the optical model description of the scattering (Satchler 1960). These are a T_r term [dependent on $(\mathbf{s} \cdot \mathbf{r})^2$], a T_p term [dependent on $(\mathbf{s} \cdot \mathbf{p})^2$], and a T_L term [dependent on $(\mathbf{L} \cdot \mathbf{s})^2$]. While theoretical estimates of the magnitudes and shapes of these terms have been made, in practice data are fitted phenomenologically assuming some tractable form.

With a wealth of parameters to adjust, you might expect good agreement between the model and elastic-scattering experiments. For σ and iT_{11}, this is correct, but the extra parameters are not needed since σ and iT_{11} depend very little on the tensor potentials. For the tensor analyzing powers, however, fits are generally only fair, and some essential element of physics is clearly missing. At 5.5 MeV, around the Coulomb barrier, good fits have been achieved for all five observables, but even at moderately higher energies, the fits become noticeably inferior (Knutson & Haeberli 1975). The results do suggest that a T_r term is useful in fitting the tensor data, particularly T_{21} and a weighted combination of T_{20} and T_{22} that is particularly sensitive to the tensor and not the vector potentials. The magnitude of T_r varies but often is about 1.0 MeV; with a second-derivative shape approximately the same as folding-model predictions, the radius parameter seems to be larger than the radius of the nucleus. Little work has been done with T_p at these energies; Goddard (1977) showed that its effects are almost indistinguishable from the effects of T_r unless the bombarding energy is much larger. The potential T_L is expected theoretically to be very small, and the few analyses are consistent with this expectation.

Interesting physics is the basis of work on the spin dependence of the deuteron potential, but present results are probably too phenomenological to be of general significance in nuclear physics. For example, the breakup of the deuteron penetrating the nucleus is an interesting problem in itself, and it may be responsible for an imaginary component of the vector spin-orbit potential (Rawitscher & Mukherjee 1978). Recently Goddard & Haeberli (1978) found evidence for such a complex $\mathbf{L \cdot s}$ potential in their analysis of complete elastic-scattering data for ten nuclei at 10 and 15 MeV. The imaginary term is consistently found to be about half the magnitude of the real term, even though adding this term yields only a small improvement in χ^2. Since the final χ^2 in most of the examples is much larger than one, and since nonnegligible effects like coupled channels were neglected, this result must be considered tentative. A complex T_r potential was also used in this analysis, so that the number of parameters was indeed large.

Our description of proton scattering focused almost entirely on inelastic scattering, since the vector spin-orbit potential for protons is well known. For deuterons there is little inelastic scattering to report, since the elastic scattering is not sufficiently well understood. Several interesting results, however, involve coupled-channels problems. One is the anomaly in elastic-scattering iT_{11} measurements for Se isotopes noted by Nurzynski et al (1977): The magnitude of iT_{11} at corresponding peaks and valleys increases almost linearly with mass number. This has been related to an effective change in the imaginary spin-independent potential, which, in turn, can be explained by a coupled-channels calculation (Ohnishi et al 1978). Inelastic data on Fe and Ni isotopes at 12 MeV reveal a wide variety of shapes for iT_{11} that also change in a regular way with mass number (Brown et al 1973). These results have been interpreted as indicative of changes between vibrational and rotational nuclei; however, a similar analysis at 15 MeV disagrees (Baker et al 1975). Finally, it is encouraging to be able to note here the solution of the long-standing J-dependence puzzle for inelastic scattering (Love & Baker 1975). Cross-section and iT_{11} data for elastic and inelastic deuteron scattering are shown in Figure 6. The data for the $\frac{5}{2}^-$ and $\frac{7}{2}^-$ states are almost identical to the corresponding data for the 2_1^+ state in ^{62}Ni (dashed lines); the $\frac{1}{2}^-$ data appear anomalous. Such an effect was first observed with proton cross sections, but the differences between the $\frac{1}{2}^-$ and the other states in A_y for protons were very small. All these facts are explained when the effects of the intrinsic quadrupole moment of the ground state of ^{63}Cu are included in a coupled-channels calculation. The solid lines in Figure 6 are the result of such a calculation.

3.1.3 ^{3}He, t, AND ^{6}Li   Some elastic-scattering results have appeared for heavier ions at single energies corresponding to the unique polarized source for each ion. The measured A_y are surprisingly large if one's intuition is guided by simple models. For the triton, for example, the spin-orbit potential V_{so} has a phenomenological strength of about 6 MeV, about three times larger than expected (Hardekopf et al 1975). It is also noteworthy that the diffuseness a_{so} of the spin-orbit potential for ^{3}He ions is small, about 0.3 fm, compared to the normal values, about 0.75 fm, found for the triton (Burcham et al 1975). The values of a_{so} appear

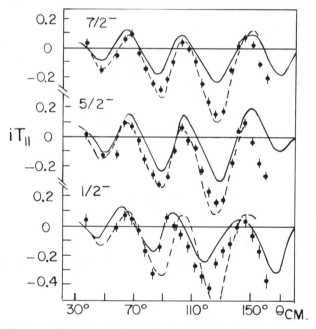

Figure 6 Vector analyzing power for inelastic scattering of deuterons from ^{63}Cu at 15 MeV (Love & Baker 1975). The dashed curve is the same in each data set and corresponds to the vector analyzing power for the 2_1^+ state in ^{62}Ni. The solid curves are the result of a DWBA calculation that includes the quadrupole moment of the ground state of ^{63}Cu.

well defined and the difference between them is unexplained. The values of V_{so}, on the other hand, are subject to considerable uncertainty because they depend critically on the parameters of the central potential. In fact, sophisticated single (Weiss et al 1976, Amakawa & Kubo 1976) and double (Petrovich et al 1978) folding-model calculations yield good agreement with ^{6}Li scattering data. These calculations indicate that for ions much heavier than Li, the effects of the spin-orbit potential should be small.

3.2 *Transfer Reactions*

3.2.1 J DEPENDENCE   Interest in the J dependence of cross-section angular distributions was originally focused equally on its reaction mechanism and nuclear structure aspects. The differences in σ for $J = l + \frac{1}{2}$ and $J = l - \frac{1}{2}$ transfer are often substantial and might be useful in assigning unknown spins in (d, p) and (p, d) reactions, for example. In fact, however, very few new spins have been determined this way despite considerable effort to establish the systematics of the effect and explain apparent discrepancies. On the other hand, such studies provoked generally unanswerable questions about the importance of various mechanisms in producing the J dependence. While DWBA calculations occasionally agree quite well with experimental data, it can be reasonably said that no satisfactory general qualitative explanation of J dependence in σ for (p, d) and (d, p) reactions has appeared, and all the calculations have yielded little physical insight. A recent study of the energy dependence of (d, p) reactions on ^{56}Fe clearly reveals the present unsatisfactory situation (Kishida & Ohnuma 1978). Even the forward-angle J dependence of $l = 3$ transfer cannot be explained. The effect of the D state of the deuteron, once thought to be the basic reason for the differences there, has been shown to be small in exact finite-range calculations.

For A_y measurements the picture is just reversed. Qualitative explanations of the effect are reasonably successful; these are based on the semi-classical model of Newns (1953) or simple extensions (Verhaar 1969), or they may be specifically for sub-Coulomb stripping (Vigdor et al 1973). The J dependence is generally unmistakable; often the sign of A_y is reversed while the magnitudes remain closely constant in going from $J = l + \frac{1}{2}$ to $J = l - \frac{1}{2}$ transitions. Unless higher order effects are important, the DWBA calculations have little trouble yielding the correct gross features of the J dependence. While detailed agreement is often lacking, it is generally not worthwhile to pursue the reasons for the discrepancies. The J dependence in A_y is thus an ideal instrument of spectroscopy; the signature is clear and the theoretical basis is understood. After establishing the existence of the J dependence in a particular reaction, then, continued measurements are useful only to answer a specific nuclear structure question. Initial observations and some recent samples from the extensive literature are cited below.

The basic effect appears to be universal in transfer reactions initiated by light polarized ions, although the magnitude of the differences in a particular reaction depends of course on the bombarding energy, the l transfer, and the reaction Q value. The J dependence has now been observed in the vector analyzing power for the following single-nucleon transfer reactions: $(\vec{p}, d)$ (Mayer et al 1971), $(\vec{d}, p)$ (Yule & Haeberli 1967), $(\vec{d}, n)$ (Hilscher et al 1971), $(\vec{d}, t)$ (Liers et al 1971), $(\vec{d}, {}^{3}He)$ (Mayer et al 1974), $(\vec{t}, \alpha)$ (Flynn et al 1976), $({}^{3}\vec{He}, d)$ (Roman et al 1977a), and $({}^{3}\vec{He}, \alpha)$ (Karban et al 1976) reactions. A clear J dependence has been established in $(\vec{d}, \alpha)$ (Ludwig et al 1978) and $(\vec{p}, \alpha)$ (Tagishi et al 1978, Van Hall et al 1978) reactions. Evidence for J dependence in the tensor analyzing powers, particularly T_{22}, in $(\vec{d}, p)$ reactions has also appeared (Johnson et al 1973, Knutson et al 1973, Basak et al 1978) and been explained (Santos 1976), but the effect is not always definitive. The J dependence in a given reaction has generally been observed over a rather small energy region, but recent work with the cyclotrons at Karlsruhe, Osaka, and Texas A & M is beginning to extend this range considerably.

A good example of the use of J dependence in the solution of a pure nuclear structure problem is the recent study of the $(\vec{d}, {}^{3}He)$ reaction at 52 MeV by Bechtold et al (1977). The analyzing powers iT_{11} were measured for several 1p and 1d proton-hole states in ${}^{15}N$, ${}^{27}Al$ and ${}^{39}K$ to determine empirical spin-orbit splittings. For the ${}^{40}Ca$ target, A_{y} for the known $\frac{3}{2}^{+}$ state is almost the inverse of A_{y} for the states assigned $\frac{5}{2}^{+}$. While tentative $\frac{5}{2}^{+}$ assignments had been made previously, questions were raised and this experiment nicely confirms the tentative spins. The differences between the $\frac{1}{2}^{-}$ and $\frac{3}{2}^{-}$ states observed in this work are not so marked, but firm spin determinations were still possible. The separation energy between the centroids of the $p_{1/2}$ and $p_{3/2}$ orbits was thus reliably determined to be 6.8 MeV in ${}^{15}N$, somewhat larger than the traditional value of 6.2 MeV.

The rather exotic $(\vec{t}, \alpha)$ reaction has been one of the most prolific in terms of new interesting spin assignments based on J dependence. After the J dependence in this reaction was first observed in a good spherical nucleus, the Los Alamos group, in collaboration with a group from McMaster, applied it to the study of the systematics of single quasi-particle states in heavy deformed nuclei. Fifteen targets between samarium and osmium have now been observed with an energy resolution that is usually about 20 keV or less. Typical results for the ${}^{190}Os(\vec{t}, \alpha){}^{189}Re$

reaction are illustrated in Figure 7, along with single-step DWBA calculations (Hirning et al 1977). The calculations are not necessary to establish the spins, since an empirical calibration from the same reaction on ^{188}Os is superior. However, it is interesting to note that the direct calculation agrees quite well with the data for essentially all reasonably strong states, and also for many that are weak. The curves for the doublet at 0.481 and 0.501 MeV show the power of the polarized beam in deciding spins even for overlapping levels.

For odd-A targets, it is always possible to have at least two competing J values in a (d, p) reaction unless the final state has spin zero. Quin et al (1976) showed that even when two l values and three J values are important, it is possible to make a meaningful comment about a theoretical prediction. In their study of the ^{21}Ne($\vec{d}$, p)^{22}Ne reaction to the 4.46-MeV 2^+ state, they were able to disentangle approximately equal contributions from $\frac{1}{2}^+$, $\frac{3}{2}^+$, and $\frac{5}{2}^+$ transfer using empirical analyzing powers from neighboring even-even nuclei. The results agreed well with shell-model calculations. While the assignments appear convincing, they could usefully be checked with appropriate measurements of tensor analyzing powers.

3.2.2 MULTISTEP PROCESSES As computing power has advanced, so has interest in determining the importance of the many possible alternate routes by which an apparently simple transfer reaction might proceed.

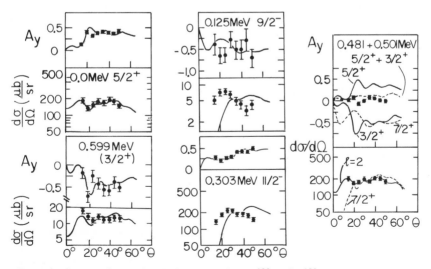

Figure 7 Cross sections and analyzing powers for the ^{190}Os($\vec{t}$, α)^{189}Re reaction at 15 MeV. The solid curves are the result of DWBA calculations (Hirning et al 1977).

Coupled-channels calculations for inelastic scattering have certainly proved to be worthwhile. The significance of complicated multistep methods for transfer reactions has been more controversial, although there is no doubt that many reactions proceed by circuitous paths. Once one indirect path is permitted, many are allowed and there are few intuitive principles to discriminate between them. For a two-particle transfer reaction like (p, t), even strong states may be excited primarily by either inelastic scattering plus simultaneous transfer of two nucleons or by sequential transfer through an intermediate state such as (p, d)(d, t). While in principle the strength of such intermediate paths should often be fixed by deformation parameters, spectroscopic factors, and normalization parameters for a given reaction, in practice there is usually so much arbitrariness that the necessity for including a particular process is seldom unambiguous. Measurements with a polarized beam may provide confirmation; since they are generally very sensitive to the relative phases of competing mechanisms, they could be crucial. Analyzing-power measurements are the simplest of these, but more difficult polarization transfer experiments might be necessary. Some very elegant examples of A_y experiments are described below.

The first of these is a study of the ^{40}Ca($\vec{d}$, p)^{41}Ca reaction at 11.0 MeV by Boyd et al (1978); the aim was to test structure calculations in ^{41}Ca. The states of interest have small cross sections and high spins. Careful analysis of inelastic alpha scattering on ^{41}Ca (deVoigt et al 1974) had previously indicated that the $\frac{11}{2}^+$ state is essentially a pure $|f_{7/2} \times 3^-\rangle$ configuration and the $\frac{15}{2}^+$ state essentially a pure $|f_{7/2} \times 5^-\rangle$ configuration. Single-step DWBA predictions for the (d, p) reaction yielded fair agreement with σ data, but poor fits to A_y indicated that this naive

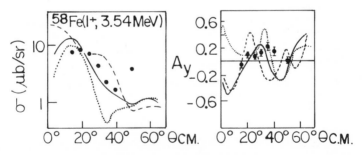

Figure 8   Cross section and analyzing power for the ^{56}Fe($\vec{t}$, p)^{58}Fe reaction to the 3.54-MeV state at 15 MeV. The curves correspond to multistep calculations. The dashed curve is the result for a pure sequential (p, d)(d, t) process. The dotted curve includes inelastic scattering preceded or followed by direct transfer of two neutrons. The solid curve is the best fit obtained by coherently adding these two processes (Boyd et al 1977).

assumption about the reaction mechanism was invalid. Use of the coupled-channels Born approximation, in which the reaction proceeds via excitation of the core with prior or subsequent transfer of an $f_{7/2}$ particle, produced much better fits to both σ and A_y, good enough that one is reasonably confident about the proposed structure.

Two-nucleon transfer reactions are rich in potential complexity, and the need for information beyond the cross section is more obvious. A particularly good example is the investigation of the (t, p) reaction leading to unnatural parity states in ^{58}Fe and ^{60}Ni (Boyd et al 1977). These states are forbidden in the first-order zero-range DWBA, but the observed σ are often substantial. The σ and A_y data for excitation of the 3.54-MeV level in ^{58}Fe are shown in Figure 8. The solid curve is the best fit that could be achieved by combining a sequential transfer amplitude with an inelastic scattering plus two-nucleon transfer amplitude. The dashed curve is the result for a pure sequential process, while the dotted curve is the very different prediction for a pure inelastic plus transfer process.

In such an analysis it is clear that the A_y data are necessary, but even with the additional data, some ambiguity remains. Nagarajan et al (1977) analyzed ^{208}Pb(p, t)^{206}Pb data leading to the 3^+ state using a pure single-step reaction model and a finite-range DWBA code in which the transition is permitted via two-neutron transfer with $\Delta s = 1$. This is the same state for which previous calculations suggested that sequential transfer was appropriate. No A_y data are available here, but they are certainly necessary. However, once this direct path is allowed to add coherently to the several possible indirect paths, the parameter space may become unmanageably large. Whether this same direct path is also important in the (t, p) work cited above is not known.

Probably the most beautiful example of the sensitivity of A_y measurements to interference between several reaction mechanisms is the $(\vec{p}, t)$ work of Yagi et al (1978). They measured angular distributions of σ and A_y for seven nuclei in the $N = 50$–82 shell. The critical observation is shown by the A_y data in Figure 9 for transitions to the 2_1^+ states in ^{102}Pd, with 56 neutrons, and ^{126}Te, with 74 neutrons: The phase of A_y toward the end of the shell is almost opposite the phase at the beginning of the shell. This change occurs gradually through the shell, and is clearly noticeable although the magnitudes of A_y are quite small. The cross sections for both the 0^+ and 2^+ transitions change very little, and A_y for the 0^+ transition is also almost the same at the beginning and end of the shell. All of these facts have been explained in detail by Yagi and co-workers with reaction calculations based on a microscopic analysis of the structure of the states. Both the direct transition (1) and the two-step process (2) involving inelastic excitation in both the entrance and exit channels are important. Each of these mechanisms individually gives rather similar A_y for ^{102}Pd and ^{126}Te, as the figure shows. It is only when the two amplitudes are coherently added $(1+2)$ that the dramatic differences between the two nuclei are predicted, comparable to the differences observed. The interference term is thus responsible. The authors proved from their microscopic structure calculations that the sign of this term should change to give constructive interference at the beginning of the shell and destructive interference at the end of the shell for the 0^+ to 2^+ transition. This statement depends on the nuclear force used in the calculation and indicates that A_y is then a sensitive test of the theory. Although the final fits are not perfect, so many details of σ and A_y for 0^+ and 2^+ transitions in the two nuclei are predicted that the authors' conclusions are very convincing.

3.2.3 TENSOR ANALYZING POWERS AND THE D STATE   General nuclear physics interest in the measurement of tensor analyzing powers in single-nucleon transfer reactions initiated by deuterons concerns mostly the D state. In $(\vec{d}, p)$ reactions, it is the D state of the deuteron that is important, whereas in $(\vec{d}, t)$ and $(\vec{d}, {}^3He)$ reactions, it is the D-state com-

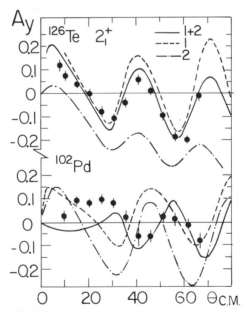

Figure 9 Analyzing power for (p, t) transitions to the first excited 2^+ states in ^{126}Te and ^{102}Pd at 22 MeV. The curves labeled 1 and 2 correspond to one-step and two-step processes, respectively. The $1+2$ curve is the prediction when both processes are coherently added (Yagi et al 1978).

ponent of the relative motion between the transferred particle and the residual deuteron. In each case the quantity measured is D_2, the ratio of integrated D-state (u_2) and S-state (u_0) wave functions heavily weighted toward large values of the separation parameter r:

$$D_2 = \frac{1}{15} \int_0^\infty u_2(r)\, r^4 \, dr \left/ \int_0^\infty u_0(r)\, r^2 \, dr. \right.$$

With some approximations within the standard DWBA, the tensor analyzing powers scale almost linearly with D_2. For the deuteron, D_2 is largely determined by the deuteron quadrupole moment Q, but wave functions that yield the proper Q have a variation of about 10% in D_2. The D-state probability also does not uniquely fix D_2. Thus a proper measurement of D_2 is important in determining the asymptotic deuteron wave function. The relation between the values of D_2 for both the triton and ^{3}He and other observable parameters is less clear, but an accurate determination for either particle should be significant.

Only sub-Coulomb stripping appears capable of producing accurate values for D_2 for the deuteron; the best data are for ^{208}Pb (Knutson 1977). Excellent fits (which have been very beautifully explained semiclassically) yielded a value for D_2 of 0.432 ± 0.032 fm^2, about two standard deviations smaller than the values calculated from existing deuteron wave functions. Recent calculation of corrections to the standard DWBA indicate now that this value should be increased (R. C. Johnson, private communication). The D_2 is then probably consistent with the standard wave functions, but some uncertainty remains about the importance of corrections due to deuteron breakup. Whether the experiment and analysis will ever be reliable enough to discriminate between deuteron wave functions is still uncertain.

In $(\vec{d}, t)$ and $(\vec{d}, {}^3\text{He})$ reactions, the errors in both the theoretically expected and the experimentally determined values of D_2 are less well known. Tensor analyzing powers for the $(\vec{d}, t)$ reaction on ^{118}Sn at 12.0 MeV are illustrated in Figure 10 (Knutson et al 1975); similar data exist now for other nuclei (Brandan & Haeberli 1977). The dashed curves in these figures are DWBA calculations in which the D state was neglected; the solid curves include the D state with $D_2 = -0.24$ fm^2. The D state is clearly the overwhelming contribution to the tensor analyzing power, and the fits are quite good for both states. The values of D_2 extracted from $(\vec{d}, t)$ data range between -0.20 and -0.30 fm^2; in $(\vec{d}, {}^3\text{He})$, the range is about -0.22 to -0.37 fm^2 (Roman et al 1977b, Brandan & Haeberli 1977). Theoretical values appear to be slightly smaller than this (Santos et al 1979).

4   GIANT RESONANCES AND THE ($\vec{p}$, γ) REACTION

Polarized beams have considerably increased the detailed nuclear structure information that can be extracted from proton capture experiments leading to the giant resonance region. I do not review the impact of polarized beams on special problems of photonuclear reactions, but instead graphically illustrate the fact that polarized beams are useful here for the same reasons they are useful in other reactions.

Most of the work on ($\vec{p}$, γ) reactions has been done either at Stanford or at TUNL. The most impressive results from the large number of mostly light nuclei studied at Stanford come from the investigation of the giant resonance region in ^{16}O via the ^{15}N($\vec{p}$, γ) reaction (Hanna et al 1974). Angular distributions of σ and A_y are measured at a number of energies. These are fitted at each energy with Legendre polynomials or associated Legendre polynomials, respectively:

$$\sigma(\theta) = A_0(E)\left[1 + \sum_{k=1} a_k(E) P_k(\cos\theta)\right]$$

$$A_y(\theta)\sigma(\theta) = A_0(E)\sum_k b_k(E) P_k^1(\cos\theta).$$

The allowed values of k are constrained by selection rules; in practice for ^{16}O, k assumes all values up to four when E1, possibly M1, and E2 radiation are included. For E1 radiation in ^{16}O, proton capture is limited to $s_{1/2}$ and $d_{3/2}$ partial waves, while for E2 the allowed partial waves are $p_{3/2}$ and $f_{5/2}$. It is instructive to write down a typical Legendre

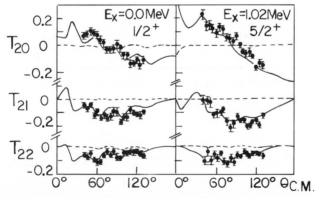

Figure 10   Tensor analyzing powers for the ^{118}Sn($\vec{d}$,t)^{117}Sn reaction to two states at 12.0 MeV. The solid curves and the dashed curves are DWBA predictions with and without the D state, respectively (Knutson et al 1975).

expansion coefficient for σ and A_y in terms of the partial wave amplitudes and phases (Glavish 1975):

$$A_0 a_2 = 1.06\, s_{1/2} d_{3/2} \cos(s, d) - 0.38\, d_{3/2}^2$$
$$+ 0.63 p_{3/2}^2 + 0.71 f_{5/2}^2 - 0.44 p_{3/2} f_{5/2} \cos(p, f)$$
$$A_0 b_2 = -0.53\, s_{1/2} d_{3/2} \sin(s, d)$$
$$+ 0.36\, p_{3/2} f_{5/2} \sin(p, f).$$

Note that this expression for b_2, like that for all b_k, depends entirely on the interference between partial wave amplitudes; the cross-section coefficients do not require interference although such terms do contribute. In proton capture reactions on an even-even nucleus, the distinction is even more striking and is similar to the rules for proton inelastic scattering at an IAR discussed earlier. As Weller et al (1974a) pointed out for proton capture on ^{14}C, A_y will be zero unless there are two interfering states of different J^π or two interfering multipoles.

With the measurement of both σ and A_y in ^{15}N$(\vec{p}, \gamma)^{16}$O, the Stanford group had sufficient data in the form of a_k and b_k values to determine the amplitudes and relative phases of all contributing partial waves in a model-independent way. This certainly could not have been done with σ measurements alone. Because of the squared terms occurring in the a_k, two solutions are allowed; model-dependent methods are necessary to distinguish between them. These two solutions, shown in Figure 11, are the basic nuclear physics information from the experiments.

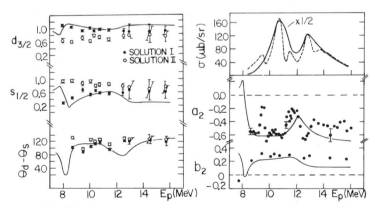

Figure 11 Experimental data and theoretical predictions for the ^{15}N$(\vec{p}, \gamma_0)^{16}$O reaction. The data are the cross section (broken line in upper figure), the expansion coefficients a_2 and b_2 (points with some error bars), and partial wave amplitudes and relative phases (points with error bars corresponding to two phase-shift solutions). The solid curves are theoretical predictions (Mavis 1977).

In the IAR scattering experiments, it is possible to extract a (model-dependent) wave function of the IAR. In (p, γ) reactions, the link to a wave function is less direct because the states are more complicated. In fact, the two dominant peaks in σ, at excitations of 22.3 and 24.45 MeV, correspond to highly collective states whose dominant components are $d_{5/2} p_{3/2}^{-1}$ and $d_{3/2} p_{3/2}^{-1}$ respectively; these components cannot be directly observed in the (p, γ) reaction from the $\frac{1}{2}^-$ ground state of ^{15}N. Calculations by Mavis (1977), however, shown by the solid curves in the figure, have been able to reproduce the main features of σ and the extracted amplitudes as a function of energy. These calculations, based on the simple shell model and including two 1p–1h doorway states, indicate a clear preference for phase shift solution I. Later detailed A_y measurements and analysis showed that the discrepancy apparent in the a_2 coefficient around 9.5 MeV could be remedied by including a secondary 4p–4h doorway at around 21 MeV, which mixes with a primary doorway whose resonance energy is 22.15 MeV (Calarco et al 1977).

The group at Stanford has also looked at this same region of excitation in ^{16}O through elastic and inelastic scattering of polarized protons on ^{15}N, but preliminary results do not reveal any anomalous behavior indicative of excitation of giant resonances as compound states (La Canna et al 1977). On the other hand, Weller and various collaborators found evidence for resonances in elastic scattering on ^{14}C (Weller et al 1974b), ^{56}Fe (Weller et al 1976b), and ^{13}C (Weller et al 1978) with quantum numbers consistent with the giant dipole resonance (GDR) built on the ground state of the compound nucleus. They measured σ and A_y at many angles as a function of incident energy. Generally σ and A_y vary smoothly with energy, although often some wide structure appears, particularly in the A_y excitation function. All the data can be fitted reasonably well with slowly varying optical parameters. The apparent resonances are detected only when a detailed phase shift analysis is carried out. Difficulties are encountered in this analysis chiefly because of the many parameters, but for each nucleus it has been possible to determine a set of phase shifts (not necessarily unique) that do not fluctuate wildly from one energy to the next. In fact, it usually turns out that most of the phase shifts are essentially constant or at least monotonic with energy, but anomalous behavior suggestive of a resonance is found in one or two partial waves. In ^{14}C + p, for example, an apparent $\frac{3}{2}^+$ resonance was observed, while in ^{56}Fe + p, the spin was $\frac{5}{2}^+$. In ^{15}N, the GDR is predominantly $\frac{3}{2}^+$ (Weller et al 1976a). In ^{57}Co, the (p, γ) work gave two possible solutions, predominantly $\frac{3}{2}^+$ or predominantly $\frac{9}{2}^+$ for the GDR (Cameron et al 1976). It is tempting to choose the $\frac{5}{2}^+$

solution on the basis of the elastic-scattering work, and, indeed, there is other (model-dependent) evidence for choosing the $\frac{5}{2}^+$ solution.

Are these elastic-scattering resonances the giant dipole states? Are they resonances? While the favorable evidence seems strong, Haeberli (1976) suggested that at least for ^{57}Co, the anomalous behavior of the phase shift does not indicate a resonance. Also, while the positions and total widths of the supposed resonances appear reasonable, the ratio of Γ_p/Γ is too large. In support of the proposed very interesting connection Weller et al seem to observe between proton scattering and capture reactions, it should be pointed out that the work of Geramb and collaborators discussed briefly in Section 3.1.1 is a similar example in a somewhat different context (Perrin et al 1977a). In addition, unusual energy dependence of A_y in the region of the giant resonance has been observed in other nuclei, but no comparable phase shift analysis has been carried out (Roy et al 1977). Finally, the apparent intermediate structure discussed in Section 2.3 occurs in the same region of excitation energy. Considerable evidence thus suggests that some of the same states are excited in both reactions, so it would be very useful to have confirming experiments, perhaps by polarization transfer measurements with a polarized beam.

5 CONCLUSION

A polarized beam is one of the standard resources at the disposal of an experimentalist in the solution of general nuclear physics problems. With polarized ion sources widely available and straightforward to operate, use of a polarized beam is no longer exotic and demands no unusual expertise. But it may be inappropriate. This review has illustrated the kinds of nuclear physics questions that a polarized beam may be helpful in studying.

Obviously, polarization measurements can establish spin assignments; often the sign of A_y is sufficient and no calculation is necessary. More sophisticated investigations may also be based essentially on this kind of yes-or-no measurement. The energy average of σA_y eliminates compound-nucleus contributions (Section 2.2.1); a nonzero value of p requires correlations (Section 2.2.2).

The magnitude and shape of A_y may also be informative. This is true if the number of free parameters is limited but too large to fix unambiguously using σ alone. The determinations of amplitudes and phases at analog resonances (Section 2.1) and giant dipole resonances (Section 4) are good examples. Deciding whether higher order processes contribute significantly also falls into this category. The $(\vec{p}, t)$ investigation of Yagi et al is superb (Section 3.2.2); the A_y measurements allowed interesting physics to be extracted from what often becomes a computational exercise. Analyzing-power data are useful in such cases as long as a perfect fit is not demanded, for A_y may be too delicate an instrument, beyond the capabilities of an imperfect reaction theory. This appears to be the difficulty in direct inelastic scattering, especially for microscopic analyses (Section 3.1.1). Here the polarized beam seems more appropriately confined to its traditional use in determining the spin-dependent terms in the effective interaction, and this by spin-flip measurements.

The focus of light ion nuclear physics interest is shifting toward higher energy; polarized beams are an integral part of most intermediate energy machines. The lessons from the low energy experience can be usefully applied in the new regime.

Literature Cited

Amakawa, H., Kubo, K. I. 1976. *Nucl. Phys. A* 266:521
Amos, K., Smith, R. 1974. *Nucl. Phys. A* 226:519
Aoki, Y., Nagano, K., Kunori, S., Yagi, K. 1978. *Proc. Symp. Nuclear Direct Reaction Mechanisms*, Fukuoka, Japan, 1978
Baker, F. T., Glashausser, C., Robbins, A. B., Ventura, E. 1975. *Nucl. Phys. A* 253:461
Basak, A. K., Griffith, J. A. R., Karban, O., Nelson, J. M., Roman, S. 1978. *Nucl. Phys. A* 295:111
Baudinet-Robinet, Y., Mahaux, C. 1974. *Phys. Rev. C* 9:723
Bechtold, V. et al. 1977. *Phys. Lett. B* 72:169
Boyd, R., Davis, S., Glashausser, C., Haynes, C. F. 1971. *Phys. Rev. Lett.* 27:1590
Boyd, R. N. et al. 1974. *Nucl. Phys. A* 228:253
Boyd, R. N., Alford, W. P., Flynn, E. R.; Hardekopf, R. A. 1977. *Phys. Rev. C* 15:1160
Boyd, R. N. et al. 1978. *Phys. Rev. C* 14:946
Brandan, M. E., Haeberli, W. 1977. *Nucl. Phys. A* 287:213
Brown, R. C. et al. 1973. *Nucl. Phys. A* 208:589
Burcham, W. E. et al. 1975. *Nucl. Phys. A* 246:269; cf *Univ. Birmingham Prog. Rep.*, 1978
Calarco, J. R., Wissink, S. W., Sasao, M., Weinhard, K., Hanna, S. S. 1977. *Phys. Rev. Lett.* 39:925
Cameron, C. P. et al. 1976. *Phys. Rev. C* 14:553
Clement, H., Graw, G. 1977. *Nucl. Phys. A* 285:109
Davis, S. et al. 1975. *Phys. Rev. Lett.* 34:215
Davis, S., Glashausser, C., Robbins, A. B., Bissinger, G. 1976. *Nucl. Phys. A* 270:285
de Swiniarski, R., Pham, D.-L., Bagieu, G. 1978. *Phys. Lett. B* 79:47
de Voigt, M. J. A., Cline, D., Horoshko, R. N. 1974. *Phys. Rev. C* 10:1798
Eng, J. et al. 1975. *Bull. Am. Phys. Soc.* 20:694
Ericson, T. 1960. *Adv. Phys.* 9:425
Feist, J. H., Kretschmer, W., Pröschel, P., Graw, G. 1977. *Nucl. Phys. A* 290:141
Flynn, E. R., Hardekopf, R. A., Sherman, J. D., Sunier, J. W., Coffin, J. P. 1976. *Phys. Rev. Lett.* 36:79
Glashausser, C. 1974. In *Nuclear Spectroscopy and Reactions B*, ed. J. Cerny, pp. 197–231. New York: Academic
Glashausser, C. et al. 1975. *Phys. Rev. Lett.* 35:494
Glavish, H. 1975. *Proc. Fourth Symp. Polarization Phenomena, Zürich*, p. 317
Goddard, R. P. 1977. *Nucl. Phys.* 291:13
Goddard, R. P., Haeberli, W. 1978. *Phys. Rev. Lett.* 40:701
Graw, G., Clement, H., Feist, J. H., Kretschmer, W., Pröschel, P. 1974. *Phys. Rev. C* 10:2340
Graw, G., Hategan, C. 1971. *Phys. Lett. B* 37:41
Haeberli, W. 1974. In *Nuclear Spectroscopy and Reactions A*, ed. J. Cerny, pp. 151–91. New York: Academic
Haeberli, W. 1976. *Phys. Rev. C* 14:2322
Haeberli, W. 1977. *Proc. Int. Conf. Nucl. Structure*, Tokyo, p. 435

Haglund, R. F. Jr., Bowen, J. M., Thompson, W. J. 1976. *Phys. Rev. Lett.* 37: 553

Hamilton, J. K., Mackintosh, R. S. 1978. *J. Phys. G* 4: 557

Hanna, S. S. et al. 1974. *Phys. Rev. Lett.* 32: 114

Hardekopf, R. A., Veeser, L. R., Keaton, P. W. Jr. 1975. *Phys. Rev. Lett.* 35: 1623

Harney, H. L. 1968. *Phys. Lett. B* 28: 249

Haynes, C. F., Davis, S., Glashausser, C., Robbins, A. B. 1976. *Nucl. Phys. A* 270: 269

Henneck, R., Graw, G. 1977. *Nucl. Phys. A* 281: 261

Hilscher, D., Davis, J. C., Quin, P. A. 1971. *Nucl. Phys. A* 174: 417

Hirning, C. R. et al. 1977. *Nucl. Phys. A* 287: 24

Hofmann, H. M., Richert, J., Tepel, F. W., Weidenmüller, H. A. 1975. *Ann. Phys.* 90: 403

Hosono, K. et al. 1978. *Phys. Rev. Lett.* 41: 621

Hurd, J. et al. 1977. *Bull. Am. Phys. Soc.* 22: 587

Ikossi, P. G., Thompson, W. J., Clegg, T. B., Jacobs, W. W., Ludwig, E. J. 1976. *Phys. Rev. Lett.* 36: 1357

Johnson, R. C. et al. 1973. *Nucl. Phys. A* 208: 221

Kaita, R. 1978. *A study of nonstatistical structure in the compound nucleus using polarized proton beams.* PhD thesis. Rutgers Univ., New Brunswick, NJ

Kanter, E. 1977. *An investigation of compound nuclear lifetimes in the presence of direct reactions.* PhD thesis. Rutgers Univ., New Brunswick, NJ

Karban, O. et al. 1976. *Nucl. Phys. A* 269: 312

Kerman, A. K., Rodberg, L. S., Young, J. E. 1963. *Phys. Rev. Lett.* 11: 422

King, H. T., Slater, D. C. 1977. *Nucl. Phys. A* 283: 365

Kishida, N., Ohnuma, H. 1978. See Aoki et al, 1978

Knutson, L. D., Stephenson, E. J., Rohrig, N., Haeberli, W. 1973. *Phys. Rev. Lett.* 31: 392

Knutson, L. D., Haeberli, W. 1975. *Phys. Rev. C* 12: 1469

Knutson, L. D. et al. 1975. *Phys. Rev. Lett.* 35: 1570

Knutson, L. D. 1977. *Ann. Phys.* 106: 1

Kocher, D. C., Bertrand, F. E., Gross, E. E., Newman, E. 1976. *Phys. Rev. C* 14: 1392

Kretschmer, W., Graw, G. 1971. *Phys. Rev. Lett.* 27: 1294

Kretschmer, W., Wangler, M. 1978. *Phys. Rev. Lett.* 41: 1224

La Canna, R., Glavish, H. F., Calarco, J. R., Hanna, S. S. 1977. *Stanford Univ. Prog.*

Rep. 1977, p. 27

Liers, H. S., Rathmell, R. D., Vigdor, S. E., Haeberli, W. 1971. *Phys. Rev. Lett.* 26: 261

Love, W. G., Baker, F. T. 1975. *Phys. Rev. Lett.* 35: 1219

Ludwig, E. J., Clegg, T. B., Jacobs, W. W., Tonsfeldt, S. A. 1978. *Phys. Rev. Lett.* 40: 441

Martin, P. et al. 1979. *Nucl. Phys. A* 315: 291

Mavis, D. G. 1977. *Polarized capture studies of the giant resonances in ^{16}O and ^{12}C.* PhD thesis. Stanford Univ., Palo Alto, Calif.

Mayer, B. et al. 1974. *Phys. Rev. Lett.* 32: 1452

Mayer, B., Gosset, J., Escudié, J. L., Kamitsubo, H. 1971. *Nucl. Phys. A* 177: 205

Moldauer, P. A. 1975. *Phys. Rev. C* 11: 426

Moss, J. M., Cornelius, W. D., Brown, D. R. 1977a. *Phys. Lett. B* 69: 154

Moss, J. M., Cornelius, W. D., Brown, D. R. 1977b. *Phys. Lett. B* 71: 87

Moss, J. M., Cornelius, W. D., Brown, D. R. 1978. *Phys. Rev. Lett.* 41: 930

Nagarajan, M. A., Strayer, M. R., Werby, M. F. 1977. *Phys. Lett. B* 68: 421

Newns, H. C. 1953. *Proc. Phys. Soc. London Ser. A* 66: 477

Nurzynski, J. et al. 1977. *Phys. Lett. B* 67: 23

Ohnishi, H., Noya, H., Tanifuji, M. 1978. *Phys. Lett. B* 76: 256

Perrin, G. et al. 1977a. *Phys. Lett. B* 68: 55 and references therein

Petrovich, F., Stanley, D., Parks, L. A., Nagel, P. 1978. *Phys. Rev. C* 17: 1642

Quin, P. A., Thomson, J. A., Babcock, R. W. 1976. *Phys. Lett. B* 60: 448

Rawitscher, G. H., Mukherjee, S. 1978. *Phys. Rev. Lett.* 40: 1486

Raynal, J. 1971. *CEN-SACLAY Rep. D.Ph./t* 71/1. 68 pp.

Richter, A. 1974. See Glashausser 1974, pp. 343–91

Roman, S. et al. 1977a. *Nucl. Phys. A* 284: 365

Roman, S. et al. 1977b. *Nucl. Phys. A* 289: 269

Roy, R. et al. 1977. See Haeberli 1977, p. 492

Santos, F. D. 1976. *Phys. Rev.* 13: 1145

Santos, F. D., Eiró, A. M., Barroso, A. 1979. *Phys. Rev. C* 19: 238

Satchler, G. R. 1960. *Nucl. Phys.* 21: 116

Simon, A., Welton, T. A. 1953. *Phys. Rev.* 90: 1036

Spencer, J. E., Cosman, E. R., Enge, H. A., Kerman, A. K. 1971. *Phys. Rev. C* 3: 1179

Tagishi, Y. et al. 1978. *Phys. Rev. Lett.* 41: 16

Temmer, G. M., Maruyama, M., Mingay, D. W., Petrascu, M., Van Bree, R. 1971. *Phys. Rev. Lett.* 26:1341

Terry, G. H., Hausman, H. J., Donoghue, T. R., Suiter, H. R., Wallace, P. H. 1978. *Phys. Rev. Lett.* 41:934

Thompson, W. J. 1967. *Phys. Lett. B* 25:454

Vanden Berghe, G., Heyde, K., Waroquier, M. 1971. *Nucl. Phys. A* 165:662

Van Eeghem, W. P. Th. M., Verhaar, B. J. 1976. *Nucl. Phys. A* 258:70

Van Hall, P. J. et al. 1977. *Nucl. Phys. A* 291:63

Van Hall, P. J., Melssen, J. P. M. G., Poppema, O. J., Smits, J. W. 1978. *Phys. Lett. B* 74:42

Verhaar, B. J. 1969. *Phys. Rev. Lett.* 22:609

Vigdor, S. E., Rathmell, R. D., Haeberli, W. 1973. *Nucl. Phys. A* 210:93

Weiss, W. et al. 1976. *Phys. Lett. B* 61:237

Weller, H. R. et al. 1974a. *Phys. Rev. Lett.* 32:177

Weller, H. R., Roberson, N. R., Rickel, D., Tilley, D. R. 1974b. *Phys. Rev. Lett.* 33:657

Weller, H. R. et al. 1976a. *Phys. Rev. C* 13:922

Weller, H. R., Szücs, J., Kuehner, J. A., Jones, G. D., Petty, D. T. 1976b. *Phys. Rev. C* 13:1055

Weller, H. R. et al. 1978. *Phys. Rev. C* 18:1120

Yagi, K. et al. 1978. *Phys. Rev. Lett.* 40:161; see Aoki et al 1978

Yule, T. J., Haeberli, W. 1967. *Phys. Rev. Lett.* 19:756

Ann. Rev. Nucl. Part. Sci. 1979. 29:69–119

NUCLEI FAR AWAY FROM   ✳5603
THE LINE OF BETA STABILITY:
Studies by On-Line Mass Separation

P. G. Hansen

Institute of Physics, University of Aarhus, DK-8000 Denmark,[1]
and CERN, CH-1211 Geneva, Switzerland

CONTENTS

[1] Present address.

69

0163-8998/79/1201-0069$01.00

1 INTRODUCTION

The last decade has seen great progress in the experimental methods for producing intense and pure beams of mass-separated radioactivity and also in the sophistication with which these beams are used. The main aim of this research has been the study of structure and properties of nuclei far from the line of beta-stability; this aspect was the subject of three major conferences (Lysekil Conference 1966, Leysin Conference 1970, Cargèse Conference 1976), the proceedings of which detail the field's development. The application of mass separation to nuclear reaction studies was reviewed by Klapisch (1969).

The study of nuclear properties under conditions that allow N and Z to vary over a wide range provides new insight into nuclear stability and structure, and several surprises have been encountered. At the same time, new types of experiments become possible, because the far unstable nuclei have decay modes and quantum numbers different from those found near stability, e.g. decays by particle emission, strength-function behavior of the beta-decay process, and heavier nuclei with low isospin. Finally, the ability to predict the properties of exotic nuclei has important applications, especially in astrophysics and to the nuclear chain reaction.

The present paper approaches the field from an experimental point of view. A relatively detailed discussion of the experimental methods for on-line mass separation is followed by examples of the physics applications. These include a brief mention of some applications to nuclei near stability, for which the high intensities now available have permitted new experiments. The format here does not permit a discussion of some of the older techniques: in-beam gamma-ray spectrometry (Diamond 1970), helium-jet transport (Macfarlane & Griffioen 1963, Wollnik 1976), and fast chemistry (Herrmann & Denschlag 1969), all of which have been extremely important for the study of far unstable nuclei. The essential point is that compared with on-line mass separation these techniques are not highly selective, and that therefore they are most useful if the production cross sections are large, or if the products provide a characteristic experimental signature (for example, spontaneous fission).

2   EXPERIMENTAL TECHNIQUES FOR MASS AND ELEMENT SEPARATION

It is instructive to begin with an example that clearly demonstrates the needs for mass and element separation and also some of the technical problems. Figure 1 shows the yield curve of the reaction $^{93}Nb(p,5pXn)^{89-x}Rb$, which has produced (D'Auria et al 1977) the most neutron-deficient rubidium isotopes observed to date, including the self-conjugate $^{74}_{37}Rb$ with a half-life of 64.9 ± 0.5 ms. Because this experiment covers approximately 10 orders of magnitude in intensity, it is evident that with chemical separation alone the study of beta and gamma radiations from the rare lighter products would have been very difficult. Mass separation alone would have been only marginally better, as the high yields of isobars near stability again would have obscured the products of interest.

Figure 1 also illustrates that the delay between production and collection of the product may seriously affect the yields of short-lived species. The systems based on stopped reaction fragments, which are subsequently liberated, ionized, and accelerated, are inherently slow, although important progress has been made in the last few years, as can be seen from the inset in Figure 1. Other and much faster methods, discussed in Section 2.2, are based on the separation of unslowed fragments. Both techniques are discussed in a recent review by H. Wollnik (to be published).

2.1   On-Line Isotope Separation

The first experiment of this type was reported by Kofoed-Hansen & Nielsen (1951a,b). An emanating target of uranium, i.e. one that releases noble gases, was irradiated with neutrons from the Copenhagen cyclotron, and noble gases from fission were swept into the ion source of an electromagnetic isotope separator. The 10-s isotope ^{91}Kr represented the most neutron-rich product studied, although radioactivity was also observable at the heavier masses. A personal account of this pioneering work was given by Kofoed-Hansen (1976). The experiments, however, were soon interrupted "because of some rearrangements of the equipment at the Institute" and were never resumed.

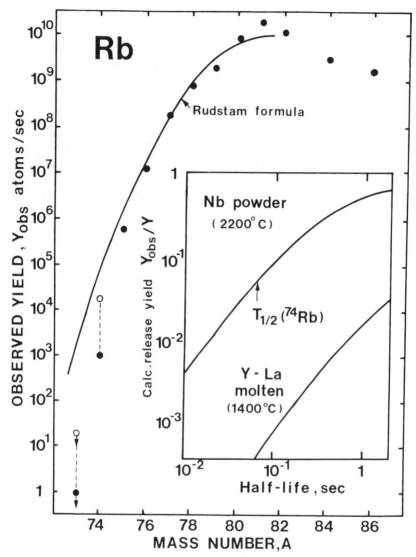

Figure 1 Observed yields (in atoms s^{-1}) of rubidium isotopes and (*inset*) loss factors due to radioactive decay in an isotope-separation experiment (D'Auria et al 1977). The black circles are the experimental yields at the collector corresponding to a 1-μA beam of 600-MeV protons impinging on a 50 g cm^{-2} target of niobium powder (grain diameter 20 μ) at a temperature of 2200°C. The curve shows, in an arbitrary normalization, the spallation cross sections calculated from Rudstam's semiempirical formula (1966). For short half-lives, the yields are reduced by decay losses. The release yield as a function of half-life is shown in the inset; the niobium powder target provides a 100 times higher yield of 65-ms ^{74}Rb than the molten-metal target (Y-La alloy at 1400°C) used in older work (Ravn et al 1972), in which this isotope remained undetected. The open circles for ^{73,74}Rb show the production yields corrected for decay losses.

In the mid-1960s experiments based on on-line mass separation again began to emerge, primarily because new advances in nuclear theory pointed to a number of interesting phenomena to be expected for far unstable nuclei (Lysekil Conference 1966). But the advent of more powerful techniques in nuclear instrumentation [Ge(Li) detectors, small computers] also played a role. Modern techniques of on-line isotope separation at particle accelerators and nuclear reactors were reviewed at several conferences (Talbert 1970, Andersson & Holmén 1973, Amiel & Engler 1976), but the reader is referred especially to a forthcoming paper by Ravn (1979).

2.1.1 THE ANATOMY OF AN ON-LINE ISOTOPE SEPARATOR   As an example of an on-line installation, Figure 2 shows the largest of the kind, CERN's ISOLDE-2 Facility, which went into operation in December 1974 (Sundell et al 1973, Ravn et al 1976, Kugler 1973). The forerunner, ISOLDE-1, was described by Kjelberg & Rudstam (1970). A characteristic feature of the new ISOLDE is the extended use of parallel operation to serve several experiments simultaneously. A switchyard that allows an elegant solution to this problem has been suggested by Andersson (1973).

The details of an on-line setup will depend very much on the bombarding particle:

High-energy protons   Owing to the long range of high-energy protons, it is advantageous to use very thick targets. At 600 MeV, a target thickness of 150 g cm^{-2} is near the optimum, and, as the incoming beam is lost essentially through nuclear reactions, the production rates are very high. A 5-μA proton beam will yield of the order of 3×10^{13} reactions per second, and a corresponding background of fast neutrons. After the irradiation, the radioactivity in the target and surrounding construction parts will be at the kilocurie level. Experiments with intense proton beams at high energy therefore require elaborate safety precautions, but also provide intensities that over a wide mass range cannot be matched by other techniques.

Experiments using relativistic projectiles are being carried out at CERN with 600-MeV protons, 910-MeV ^{3}He, and 1-GeV ^{12}C (in preparation); at Orsay with 156-MeV protons and 210-MeV ^{3}He (Foucher et al 1973, Paris et al 1976); and in Leningrad (Berlovich 1973, 1976) with 1-GeV protons. A feature of special interest in Orsay's ISOCELE installation is that it is based on a medium-current isotope separator designed for handling milliampere ion beams at 40–50 keV. This is important (Putaux et al 1974) because it allows the target to be operated at higher temperatures than with a low-current separator: The operational temperature will normally be limited by the onset of massive evaporation from the

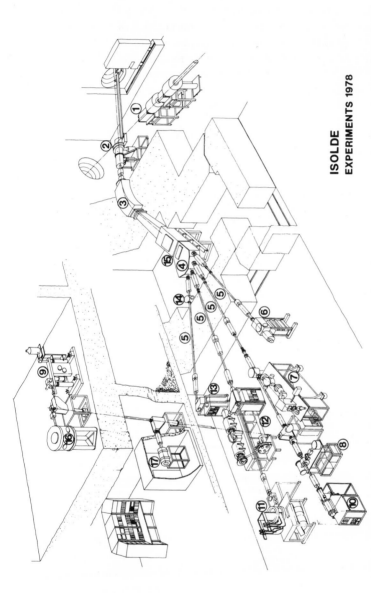

**ISOLDE
EXPERIMENTS 1978**

Figure 2 The on-line isotope separator ISOLDE-II with the experiments that were on the floor in 1978. Not shown are electronics, power supplies, on-line computers, lasers, etc. The 600-MeV proton beam (1) is focused on the target and ion-source unit (2), and the 60-keV ions are mass-analyzed in the magnet (3). Individual masses are selected in the electrostatic switchyard (4) and distributed through the external beam lines (5) to the experiments. These comprise alpha and proton spectroscopy (6), high-resolution mass spectrometry (7), beta-gamma spectrometry (8 and 9), range measurements of ions in gases (10), optical-pumping and laser spectroscopy on mercury (11), atomic beam magnetic resonance (12), collection of radioactive sources for off-line work [hyperfine interactions in solids, determination of shifts in the energies of K x rays, targets for nuclear reaction studies (13, 14, 15)], beta-decay Q values measured by coincidences with a magnetic "orange" spectrometer (16), and spectroscopy of beta-delayed neutrons (17).

target itself; this interferes with the operation of the ion-source and quenches the beam of interest.

A unique application of projectiles at very high energies is the production of extremely neutron-rich light isotopes through fragmentation reactions. The 24-GeV proton beam of the CERN Proton Synchrotron permitted Klapisch and collaborators (Klapisch et al 1969, Thibault et al 1975) to study sodium isotopes with masses up to 35.

Heavy ions   This category also includes experiments using slow light ions. These experiments are characterized by thin targets and low overall production rates, but for far unstable products this may be compensated by very high cross sections. Many combinations of target and projectile are possible, so that it becomes easier to find conditions for producing a given element. Many such experiments have been performed (Äystö et al 1976, Bogdanov et al 1976, Burkard et al 1976, Dumont et al 1978, Fransson et al 1973, Karnaukhov et al 1974, Spejewski et al 1973, Yoshizawa et al 1976). A special advantage with heavy ions is that the nuclear reaction and the evaporation step may be separated by stopping the recoils from the reaction in a catcher at high temperature (see, for example, Karnaukhov et al 1974, Mlekodaj et al 1976, Kirchner & Roeckl 1976). An interesting alternative is to use a helium jet to transport the recoil into the ion source; the RAMA system (Cerny et al 1977) was recently used for the separation of elements that would have been hard to obtain by other techniques.

Reactor neutrons   A recent summary (Wohn 1977) lists 11 experimental programs that use reactor neutrons to produce fission products in connection with an on-line separator. One approach to this problem is to use an external neutron beam, so that the geometry of the target and ion-source system resembles that of experiments based on accelerators. This arrangement was used in the SOLIS system of Amiel (1966) and in several other experiments, including the planned TRISTAN-II (Wohn 1977) at the Brookhaven high-flux reactor. It has the advantage of leaving the vital parts of the experiments accessible, but it does not fully use the large neutron intensities inside the reactor. A more radical approach was adopted by Rudstam and collaborators (Rudstam 1976, Borg et al 1971), who placed an integrated target and ion-source unit near the reactor core in a flux of up to 4×10^{11} n cm^{-2} s^{-1}. To operate the most delicate parts of the equipment in high radiation fields clearly requires a very careful design, especially as unforeseen maintenance operations may be extremely time consuming, but the intensities obtained compare favorably with beam experiments at the most powerful reactors.

2.1.2 TARGET AND ION SOURCE   The essential problem is to find a configuration that will produce a beam of the element of interest only. This has been discussed extensively at conferences (Andersson & Holmén 1973, Amiel & Engler 1976), and the reader is referred to a series of comprehensive papers by Ravn and collaborators (Ravn et al 1975 and 1979; see also Carraz et al 1978a,b). In the following only a few examples are given.

The early work naturally concentrated on elements that could be separated rather simply. The noble gases emerge in a clean form from emanating targets at room temperature (Kofoed-Hansen & Nielsen 1951a,b, Patzelt 1970) and can be ionized in a plasma ion source. Likewise, zinc, cadmium, and mercury are the only volatile elements emerging from targets of molten germanium, tin, and lead, respectively (Hagebø & Sundell 1970, Hagebø et al 1970). Surface ionization, on the other hand, selects the alkali elements from the complex mixture emerging from a hot uranium-graphite target (Klapisch 1969).

Speed and versatility may in certain cases be more important than chemical selectivity. The Rudstam group (Borg et al 1971) showed that a number of fission products could be obtained from a plasma ion source lined with uranium and graphite. Similar results for the elements around lead were obtained (F. Hansen et al 1973) by proton irradiation of a thorium ceramic at temperatures up to 2000°C. An important step forward was the development of a new plasma ion source by Kirchner & Roeckl (1976). When this source is operated in the FEBIAD mode (forced electron beam induced arc discharge) it is possible to use gas pressures very much below the threshold pressure of a normal arc-discharge ion source. The authors quote yields as high as 50% (for xenon) and point out that the low gas pressure leads to cleaner spectra, better line shapes, and a long lifetime for the source.

Recent work has greatly extended the use of molten-metal targets (Ravn et al 1975), thus the yields of mercury from molten lead at 800°C can be increased by vibrating the melt, a method that breaks up an oxide film. Liquid-metal targets can also be adapted to heavy-ion work (Dumont et al 1978). Further developments (Carraz et al 1978a,b) concern the extended use of surface ionization, in particular to earth alkalies and rare earths, and the use of a wide range of refractory target materials including metals (V, Nb, Mo, Hf, Ta, W), carbides (TiC, VC, CeC, ThC, UC), and certain other compounds ($BaZrO_3$, CeS). Very recently, tests indicated the possibility of producing the halogens pure and in good yields as negative ions (B. Vosicki, personal communication).

According to the review by H. L. Ravn (1979), a total of forty-four elements can now be separated, not including elements available as daughter products from beta and alpha decay. For about 20 of these, the targets are close to ideal: good yields of one element only. For products with short half-lives, however, an additional important parameter is the time delay between formation of a radioactive atom and its release from the ion source.

2.1.3 RELEASE RATES   The discussion given here is largely based on a review article by H. L. Ravn (1979), but considerable additional information can be found in Andersson & Holmén (1973), Ravn et al (1975), Amiel & Engler (1976), and Carraz et al (1978a,b). The comments that follow apply mainly to the bulky targets used with high-energy projectiles and reactor neutrons; it is usually much easier to obtain high release rates in heavy-ion work.

So far, the fastest systems developed are based on fine-grained solids at high temperature. If the grains are characterized by an average grain radius R and diffusion constant D, application of Fick's second diffusion law gives the probability $p(t)$ per unit time that an atom produced at time zero crosses the grain surface at time t:

$$p(t) = (6D/R^2) \sum_{n=1}^{\infty} \exp \left[-n^2\pi^2 Dt/R^2 \right]. \hspace{2cm} 1.$$

An expression of this type and reasonable assumptions for D and R give a good fit to release data for many elements (Carraz 1978a). For a product with mean life T_m much shorter than the average delay time, the yield is

$$y = 3(DT_m)^{1/2}/R, \hspace{4cm} 2.$$

where the fact that the yield decreases with the square root only of the mean life implies that high-temperature powder targets are extremely favorable for the production of short-lived isotopes. An example of this is shown in Figure 1, where the upper curve in the inset was calculated from Equation 1 and, for small values of the half-life, is well represented by Equation 2.

The release from liquid-metal targets is characterized by a single exponential (Hagebø & Sundell 1970)

$$p(t) = \mu \exp(-\mu t), \hspace{4cm} 3.$$

where the average delay time μ^{-1} is typically 30–100 s. The release yield for a mean life T_m is then

$$y = \mu/(\mu + T_m^{-1}) \simeq \mu T_m \quad \text{(for small } T_m), \hspace{2cm} 4.$$

which is the expression shown as the lower curve in the inset of Figure 1. The kinetic mechanism underlying Equation 3 is interpreted as either a surface evaporation step or a diffusion-controlled step through a (Nernst) diffusion layer. The second interpretation now appears to agree best with the experimental data and with present knowledge about diffusion and heat transfer in liquid metals.

2.1.4 COMPARISON OF DIFFERENT PRODUCTION METHODS   It is probably useful at this point to make a few comments on the merits of different methods for producing secondary beams of radioactivity. As an example, Figure 3 compares production yields of cesium (an element easily acces-

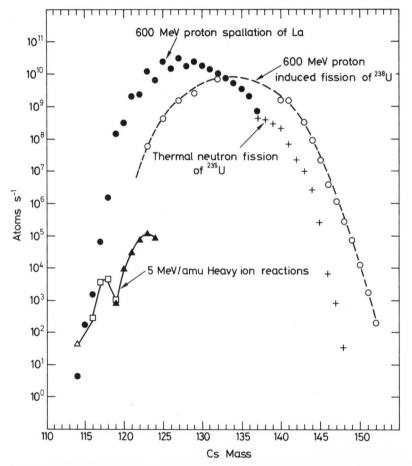

Figure 3 Production (in atoms s^{-1} arriving at the collector plate) of cesium isotopes by various techniques: (a) spallation of molten lanthanum with 600-MeV protons, (b) fission of uranium carbide with 600-MeV protons, (c) heavy-ion reactions with the GSI Unilac, and (d) with reactor neutrons at the TRIGA reactor in Mainz. (From Ravn 1979.)

sible by several techniques) produced with heavy ions, high-energy protons, and reactor neutrons. High-energy protons offer larger yields for almost all products except for the most neutron-deficient ones, where heavy ions now show a lead (E. Roeckl, to be published). The advantage to the user in having both neutron-rich and neutron-deficient products available is, of course, also considerable.

Heavy ions, on the other hand, offer the only possibility for studying isomers with very high spin and the additional advantage of small yields near stability, so that the unavoidable cross contamination of the rare products by more abundant ones will be very much smaller than it would be in an experiment with high-energy protons.

It is also striking that more neutron-rich radioactivity is presently obtained from fission with high-energy protons than with reactor neutrons. However, accelerator time is much more precious than reactor time, so that a reactor experiment will likely have much more running time available. For experiments not intensity limited, total running time is a better figure of merit than beam delivered.

Decay losses begin to be appreciable for half-lives of 0.1–1 s, and the practical limit for on-line isotope separation now seems to be 1 ms. Even shorter half-lives will require recoil separators; these, discussed in the following, allow experiments on half-lives as short as 0.1 μs.

2.2 *Separation of Fast Reaction Products*

Heavy-ion reactions and fission create products with high charge and high velocity. Several schemes exist for performing mass, element, or velocity separation directly on the unslowed fragments. These techniques were recently reviewed by Armbruster (1976) and Sistemich (1976). Recoil separation offers a number of advantages in comparison with on-line mass separation: (*a*) measurements of products with lifetimes down to 0.1 μs; (*b*) efficiency independent of atomic number so that all elements can be separated; (*c*) many possibilities for detailed reaction studies involving energy and yield measurements on primary reaction products.

2.2.1 THE PARABOLA SPECTROMETER   The largest parabola spectrometer, LOHENGRIN (Armbruster et al 1976), is installed at the ILL high-flux reactor in Grenoble. Combined magnetic and electric deflections separate the fragments according to the charge/mass ratio, so that different velocities correspond to different points along the parabolic collector. For a product with an independent fission yield of 2%, the intensity on the 72-cm long collector and for the thickest target (0.4 mg cm^{-2}) is 1.6×10^4 atoms s^{-1}. The mass assignment is unique, and charge resolution for the isobars may be obtained by measuring the energy loss in solids or by time-of-flight techniques. The large collector area presents some technical problems, and a gas jet has been used to concentrate the radioactivity; this, however, introduces some element discrimination.

2.2.2   THE GAS-FILLED SEPARATOR   A fast heavy ion moving in a gas has an average charge proportional to its velocity, so that the deflection in a magnetic field becomes approximately independent of velocity and proportional to the product of mass and a function of Z. The gas-filled separator uses this fact to provide both mass and element resolution (Cohen & Fulmer 1958). Figure 4 shows the separator JOSEF placed at the FRJ-2 reactor of the Kernforschungsanlage Jülich. JOSEF, the most powerful gas-filled spectrometer so far developed, provides a mass resolution of 79 and an atomic number resolution of 28, which means that

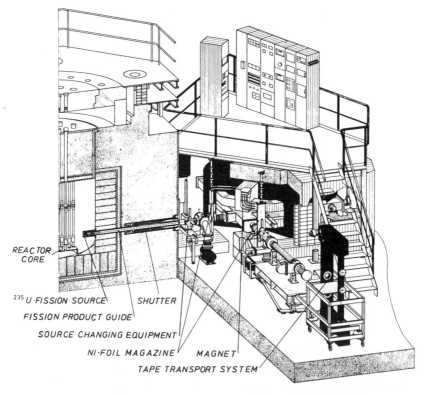

Figure 4   Overall view of JOSEF. Fission fragments emerging from a ^{235}U source near the reactor core are transported by an electrostatic particle guide to a 312° analyzing magnet filled with gas at a pressure of a few torr. The separated fragments are collected on a tape transport system or concentrated by a gas transport system (Lawin et al 1976).

interference from other masses and elements remains a problem—in part, however, resolvable by varying the gas (N_2/He) and its pressure. The total rate on the 200-cm^2 collector for a fission product with a yield of 2% is 7×10^6 atoms s^{-1} (Armbruster 1976). This high intensity gives favorable conditions for nuclear spectroscopy.

2.2.3 THE KINEMATIC SEPARATOR In a heavy-ion reaction involving a projectile with mass A_1 and velocity v_1, the fusion product with a target of mass A_2 will have velocity $v = v_1 A_1 (A_1 + A_2)^{-1}$, and will move nearly along the beam direction. The heavy-ion separator SHIP (Ewald et al 1976; for a detailed description see also Faust 1978) at GSI in Darmstadt consists of two separated velocity filters, which deflect the primary beam but allow fusion products to reach an exit slit. In a narrow sense this instrument falls outside the topic of the present paper, since it provides no mass resolution. It is, however, interesting in that it provides a very high efficiency (5–50%, comparable to that of an on-line isotope separator) and a reduction of the primary particles in the beam by an impressive factor of 10^9–10^{12}.

3 THE PROPERTIES OF NUCLEAR GROUND STATES

This and the following sections give examples of applications of on-line mass separation, starting with the characteristics of the nuclear ground state: mass, spin, magnetic and electric moments, and charge radii. The studies of ground-state properties as a function of nuclear composition are especially important, even though spectroscopists are both able and courageous enough to attack the excited states in a wholesale way, dealing with hundreds and probably soon thousands of excited levels in a single nucleus. The advantage of the atomic techniques for measuring ground-state properties is that they are impartial and model-independent; consequently, each systematic scan over a wide mass range may reveal surprises.

3.1 Nuclear Masses

The binding energies of nuclear ground states have great fundamental and practical significance. They are obtained experimentally as mass differences, either from mass spectroscopy or from Q values of nuclear reactions and decays. As a pair of nuclides will usually be connected through several independent measurements, it is necessary to carry out a data reduction in order to arrive at a single mass table. The most recent mass table was prepared by Wapstra & Bos (1977), who also surveyed input data and discussed the problems associated with arriving at a set of "best" values; a somewhat more detailed and very readable description of this work is contained in the dissertation of Bos (1977).

The general theoretical problem in the understanding of nuclear masses is that of incorporating the effects of nuclear structure. A "structureless" theory, such as von Weizsäcker's liquid-drop model or its modern "droplet" version (Myers 1977), can very accurately account for the nuclear binding energies, but higher precision requires an understanding of the effects of the shell structure and of nuclear deformations. Much of the progress here, as well as in the closely related question of the shape of the fission barrier, is due to insight provided by Swiatecki (Swiatecki 1966, Myers & Swiatecki 1966a,b) and Strutinsky (1966, 1967). We return to the question of shapes and shells in the discussion of nuclear spectroscopy (Sections 4 and 5), but refer now to the recent review by Ragnarsson et al (1978).

It would be outside the scope of the present work to discuss the various theoretical mass formulas, their foundations and performance. Much of this information is provided in a review article edited by Maripuu (1976), who has brought together tables with data from eight current mass formulas. The role of experiments on nuclear masses, however, goes beyond providing additional input for coming mass adjustments and providing a scoreboard for existing theories. Since the masses depend on structure, systematic experiments in new regions will perhaps reveal unexpected structure.

3.1.1 DIRECT MASS DETERMINATION The possibility of using mass spectroscopy for determining the masses of short-lived radioactive isotopes was discussed at the Leysin Conference by W. H. Johnson (1970), but the first program of this kind (Thibault et al 1975, Epherre et al 1979) is more recent.

In the first series of experiments (Klapisch et al 1973, Thibault et al 1975) the masses of ^{11}Li and $^{26-32}$Na were determined with a magnetic sector spectrometer of relatively low resolution. The most striking finding was that the heaviest sodium isotopes ^{31,32}Na, produced in bombardments of uranium with 24-GeV protons from the CERN PS, were considerably more bound than expected from theory; Campi et al (1975) and Beiner et al (1975) interpreted this as the onset of a new deformed region, so that $N = 20$ is no longer a magic number for the sodium isotopes.

The second series of experiments (Epherre et al 1979) again studied two alkali elements: rubidium and cesium, produced as mass-separated beams at the ISOLDE facility. In order to obtain sufficient precision it was necessary to introduce a second mass-separation stage in the form of

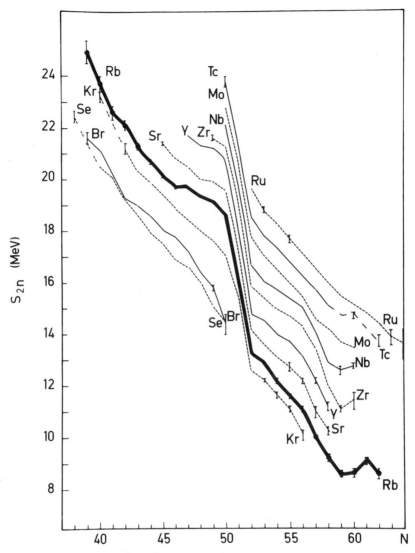

Figure 5 Experimental two-neutron separation energies S_{2n} as a function of the neutron number of the heaviest of the pair N, $N-2$. The heavy line represents the results for rubidium obtained by Epherre et al (1979), while values from neighboring elements are from the 1977 mass adjustment (Wapstra & Bos 1977). In the rubidium data the following effects are noted: (*a*) the increased binding at the lighter masses, largely due to the Wigner term, linear in the absolute value $|T_Z|$ of the isospin projection, but maybe also containing a contribution from the existence of a deformed region near ^{80}Zr; (*b*) a strong drop at $N = 50$ representing the well-known major neutron shell; (*c*) a smaller drop at $N = 56$, presumably representing the closure of the $d_{5/2}$ subshell; and finally, (*d*) the bump at $N \geqq 60$ seems a clear indication of the onset of a region of deformed nuclei (see Section 4.2.2 and Figure 10).

a Mattauch-Herzog spectrometer consisting of a spherical electrostatic deflector followed by a homogeneous magnetic sector. This spectrometer was normally operated at a resolving power of 5000. The incoming ISOLDE beam of 60 keV was stopped, reionized, and accelerated to 9 keV, and the transmitted ions were counted with an electron multiplier. The magnetic field is kept constant, while all electric potentials (for acceleration and deflection) have constant ratios but vary in absolute values (V_A, V_B, V_C) in a way that makes ions of different masses (M_A, M_B, M_C) follow the same trajectory. From precise measurements of the potentials, the masses of one unknown and two references (A, B, C) are now related by $M_A V_A = M_B V_B = M_C V_C$. The precision varied from ± 24 keV for ^{90}Rb to ± 380 keV for the rare, self-conjugate ^{74}Rb (see Figures 1 and 5).

3.1.2 Q VALUES The determination of the energy balance in nuclear reactions and radioactive decay processes also provides mass differences.

Nuclear reactions In the light nuclei, multinucleon transfer reactions provide a powerful tool for studying ground-state masses (and excited levels) of nuclei far from stability. These techniques were discussed by

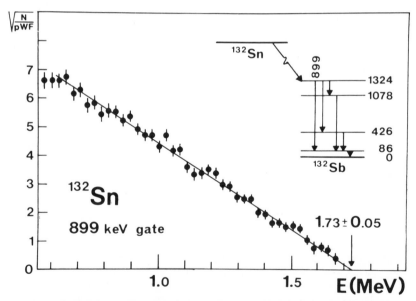

Figure 6 The binding energy of the doubly magic nucleus ^{132}Sn is of special importance. It was determined by Aleklett, Lund & Rudstam (1976) in experiments on ^{132}Sn and ^{132}Sb. The figure shows the Fermi-Kurie plot of the ^{132}Sn beta spectrum gated on the 899-keV gamma ray (see the decay scheme shown in the inset).

Cerny (1976), Benenson et al (1976), and Pardo et al (1978), who describe the use of reactions such as (^{3}He, ^{6}He), (^{7}Li, ^{8}B), (^{4}He, ^{8}He), (^{3}He, ^{8}Li), and even (^{7}Li, ^{2}He), leading to an unbound proton pair. A special advantage of reaction spectroscopy is that it allows the observation of particle-unbound exotic nuclei as final states in reactions such as ^{8}C, ^{12}O, ^{16}Ne, and ^{15}F (Kekelis et al 1978). Another interesting possibility is the use of double charge exchange reactions with pions for making exotic nuclei; recently the reaction ^{18}O$(\pi^-, \pi^+)^{18}$C was used to determine the mass of ^{18}C (Seth et al 1978).

Beta decay   The determination of mass differences from the end points of continuous $\beta^\pm$ spectra becomes very difficult for far unstable nuclei because of their complex decay schemes. Some knowledge of the decay scheme is indispensable, and it is usually necessary to measure the beta particles in coincidence with gamma rays. The beta spectra have been detected with scintillation counters, solid-state counters, and magnetic spectrometers. Experiments of this type have been carried out at ISOLDE (Westgaard et al 1972, 1975), OSIRIS (Aleklett et al 1976), UNISOR (Kern et al 1976), LOHENGRIN (Stippler et al 1978a,b, Keyser et al 1978, 1979), and elsewhere. As an example, Figure 6 shows data from an experiment, which determined the ^{132}Sn mass excess to -76.59 ± 0.08 MeV. This value differs by about 3 MeV from the theoretical predictions of modern Hartree-Fock calculations.

Alpha decay   Gauvin et al (1975) prepared a compilation of data on alpha decay, but new results keep emerging in all of the three known regions of alpha emitters.

The region heavier than lead ($Z = 82$) is classical for alpha-decay studies. A number of new neutron-deficient isotopes have been added for the elements Pb-Ra. Thus, a recent experiment used the velocity filter SHIP (Schmidt et al 1979) to identify the new radioactivities ^{215,217m,218}Pa with half-lives of 14, 1.6, and 0.12 ms, respectively.

About 130 alpha emitters are now known in the region from samarium ($Z = 62$) to lead. The systematics of the Q_α values is shown in Figure 7, in which one especially notes a gap above Gd ($Z = 64$) first discussed by Rasmussen et al (1953). This gap presumably is due to a spherical subshell closure between $d_{5/2}$, $g_{7/2}$, and $h_{11/2}$, which leads to the "magic" properties of ^{146}Gd discussed below in Section 4.2.1.

A third region of alpha emitters characterized by $50 < N < 82$ is now emerging near the tellurium isotopes ^{108,109}Te found by Macfarlane & Siivola (1964, Macfarlane 1967). In a series of experiments with the on-line isotope separator at the Darmstadt UNILAC, Kirchner et al (1977) and Roeckl et al (1978) observed several new alpha emitters in "over-shoot" reactions such as ^{58}Ni + ^{58}Ni, and they established a Q-value systematics.

The Q values in the α chains provide a rich set of data for determining masses of far unstable nuclei. Unfortunately, the masses of the daughter products that terminate the chains are usually unknown, and an effort to determine these "key" masses by beta measurements would likely be highly rewarding.

Beta-delayed particles   The EC, β^+ decay of extremely neutron-deficient nuclei often will populate excited states that decay by proton or alpha emission. The Q value for the (EC, particle) reaction can be determined by counting positrons (or annihilation radiation) in coincidence with the particle; the EC/β^+ ratio is strongly energy dependent and hence gives a precise determination of the experimental window energy $Q - S_p$, where S_p is the proton separation energy. Jonson et al (1976) gave a table of the quantity $Q - S_p$ and Bos et al (1974) discussed the systematics.

It is also possible to determine $Q - S_p$ from the end point of the particle

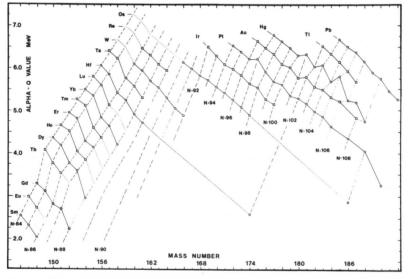

Figure 7   The systematics of alpha-decay Q values in the rare-earth region ($82 \leq N \leq 126$, $62 \leq Z \leq 82$) as presented by S. Mattsson (private communication and to be published). The energies are presented as the sum of the kinetic energy to the ground state of the daughter, the recoil energy and the correction for orbital electron screening. Note especially the gap between Gd ($Z = 64$) and Tb ($Z = 65$), indicative of a subshell closure between $d_{5/2}$, $g_{7/2}$ on one side, and $h_{11/2}$ on the other side (see also Section 4.2.1.).

spectrum (Hardy 1976, see also Section 5.2.3 for a discussion of the average properties of delayed-proton spectra); this method has led to errors and should only be attempted for spectra with good counting statistics.

Beta-delayed gamma rays   Hornshøj et al (1977b) proposed an interesting new method for measuring Q values. They note that the gamma-ray spectrum following electron-capture beta decay becomes extremely complex for high Q_{EC} values. Thus for the case of ^{145}Gd they estimate that at 4.4 MeV the gamma spectrum will have one line per keV, so that fluctuations (Section 5.2) are expected to be small. Assuming that the strength functions vary slowly with energy, they extrapolate the square root of the intensity to the end point $Q - B_K$, where B_K is the K-electron binding energy. The value obtained for ^{145}Gd was $Q - B_K = 4925 \pm 70$ keV. This method should be applicable in many other cases.

3.2 Nuclear Spins, Moments, and Radii

The hyperfine structure (HFS) of atomic spectra is the classical source of information on static nuclear properties. With the advent of modern on-line mass separation as well as of a number of new techniques in atomic physics, studies of the HFS of radioactive atoms have become possible on a large scale, and have led to a number of very striking results. This intersection between atomic and nuclear physics, probably representing the most important single use of on-line separators, was reviewed by Otten (1976) and by Klapisch & Jacquinot (1979). Much of the recent progress in the area stems from applications of laser spectroscopy reviewed by Murnick & Feld (1979) in a companion paper to the present one.

3.2.1 THE MERCURY ISOTOPES   The most complete series of studies of HFS is for mercury ($Z = 80$), for which data have been taken from mass 181 to mass 206. The on-line experiments have all worked with the 2537-Å line $[^3P_1(6s6p)\text{-}^1S_0(6s^2)]$, which has the advantage that the magnetic HFS and the isotope shifts are large compared with the Doppler width; hence special techniques for suppressing the Doppler broadening are not necessary.

The first experiments on the light mercury isotopes were carried out (Bonn et al 1971, 1972) by radiation-detected optical pumping (RADOP). Vapor of neutral radioactive mercury atoms was irradiated with circularly polarized light from a mercury lamp, and via the hyperfine coupling the polarization was transferred to the nucleus. An asymmetry in the beta-decay rate was obtained when the frequency of the lamp matched that of the radioactive atom. The frequency of the lamp was controlled by placing it in a variable magnetic field. The nuclear g factor could be determined by applying nuclear magnetic resonance at a matching point. A detailed description of the optical-pumping experiments on mercury was given by Huber et al (1976a) and by Bonn et al (1976), who also cited the extensive theoretical literature resulting from these experiments. The data, shown in Figure 8, are briefly discussed below and in Section 4.2.2; the crucial question at the completion of the optical-pumping experiments was the isotope shifts of the light, even mercury isotopes, which have spin zero and therefore can not be studied by polarization techniques.

Purely optical techniques were therefore required (Kühl et al 1977, Dabkiewicz et al 1978, 1979). A tunable dye laser was pumped by a

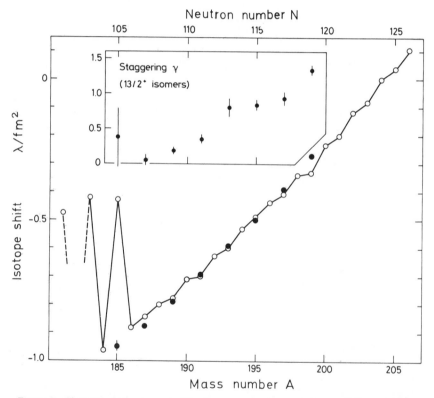

Figure 8 Changes of the charge radii of mercury isotopes relative to ^{204}Hg. (Adapted from Dabkiewicz et al 1979 and Stroke et al 1979). The size parameter is defined as $\lambda = \delta\langle r^2 \rangle - 1.1 \times 10^{-3}\delta\langle r^4 \rangle + $ higher terms. Open circles indicate ground states and solid circles isomers. With the exception of ^{185m}Hg the experimental error is smaller than the diameter of the circle. The line connects ground states of neighboring isotopes. The inset shows the even-odd staggering parameter for the $\frac{13}{2}^{+}$ isomers, defined as $\gamma = 2(\langle r^2 \rangle_{N+1} - \langle r^2 \rangle_N)/(\langle r^2 \rangle_{N+2} - \langle r^2 \rangle_N)$, for even neutron number N.

400-kW pulsed nitrogen laser, and the ultraviolet (2537 Å) laser beam was generated by frequency doubling in an ammonium dihydrogen phosphate crystal. The beam traversed a resonance cell (containing short-lived mercury radioactivity delivered from the ISOLDE facility) and a reference cell with a stable mercury isotope. Resonance-scattered radiation was detected in photomultipliers. The laser method provided isotope-shift data for the isotopes ^{184,186,188}Hg (Kühl et al 1977) as well as for several odd-mass isotopes and isomers (Duke et al 1977, Dabkiewicz et al 1979).

The systematics of the mercury radii is given in Figure 8, in which the absolute scale has been derived (Bonn et al 1976) from atomic Hartree-Fock calculations. The measured x-ray isotope shifts (Lee et al 1978) confirm this scale; still, if used directly, they would increase the λ values of Figure 8 by 23%. Some of the main conclusions are as follows.

The overall slope   The general slope in $\langle r^2 \rangle$ as a function of A is a factor of two less steep than expected if the nuclear charge radius within a sequence of isotopes is proportional to $A^{1/3}$. This is a well-known phenomenon, discussed, for example, in Stacey (1966). It can be understood in terms of the dependence of the proton potential on the neutron excess (see, for example, Bohr & Mottelson 1969).

The deformed isotopes $^{181,183,185}Hg$ The optical-pumping experiments showed that the lightest odd isotopes of mercury have the same charge radii as isotopes with 13 neutrons more. Bonn et al (1972) immediately interpreted this as the onset of strong permanent quadrupole deformations. All subsequent work has strengthened this notion and the three ground states of ^{181,183,185}Hg presumably have the Nilsson quantum numbers $\frac{1}{2}^-[521]$.

The even isotopes $^{184,186}Hg$ The even mercury isotopes down to ^{184}Hg all fall on the same line. There is no indication that ^{184,186}Hg are "soft": their radii do not show any tendency to approach those of ^{181,183,185}Hg. A 0^+ state that may represent the deformed state was discovered at 375 keV in ^{184}Hg by Cole et al (1976). From the measured lifetime of 0.9 ns, and using the fact that the EO transition probability represents an off-diagonal matrix element of r^2, Kühl et al (1977) estimate the mixing between the ground state and the (assumed) deformed state at 375 keV to be only 0.7%. As an example of "shape coexistence," ^{184}Hg (and ^{185}Hg, see below) may thus be second only to the fission isomers.

The isomer pair $^{185,185m}Hg$ Dabkiewicz et al (1979) measured the isotope shift of an isomer ^{185m}Hg by laser spectroscopy and the spin ($I = \frac{13}{2}$) by the Hanle effect. The isomer shift between the two ^{185}Hg states is

0.52 ± 0.02 fm^2, by far the largest value ever recorded for any nucleus. The only comparable case is situated right at the transition between the spherical and deformed rare earths; here the isomer shift between the low-lying levels of ^{153}Eu as measured from muonic x rays corresponds to approximately 0.1 fm^2 (Boehm 1974).

The even-odd staggering Stroke et al (1979) presented an extremely interesting systematics of the even-odd staggering of the isotope shifts of the $\frac{13}{2}^+$ isomers. They find that the staggering parameter γ increases systematically with increasing neutron number as shown in the inset of Figure 8.

3.2.2 ATOMIC BEAM MAGNETIC RESONANCE   This classical method for measuring spins and magnetic moments remains valid and useful. It is easily adapted to on-line use, it can be applied to many elements, and the counting times are short. An experimental setup especially adapted for on-line use was described by Ekström et al (1978a). In brief outline, the general technique uses a magnet with an inhomogeneous field to select certain magnetic sublevels of the atomic HFS from an atomic beam. In a subsequent homogeneous field, a radio-frequency (RF) transition is induced to a sublevel whose atomic magnetic moment has the opposite sign. The beam is subsequently analyzed in a second inhomogeneous field, and only atoms that have undergone an RF transition ("resonance flip-in") are transmitted. For spin measurements, the RF needs only to be set to the values corresponding to the possible spins, and transmission through the spectrometer gives a yes-or-no answer. Measurements of magnetic moments require scanning of the RF and take longer.

Since 1958 the Gothenburg-Uppsala Group has measured approximately 200 spins (C. Ekström, personal communication). Their recent experiments using mass-separated beams from ISOLDE have dealt with the elements In, Rb, Cs, Tm, Au, and Fr. The data for the neutron-deficient rubidium and cesium isotopes are evidence for deformation effects (Ekström et al 1977a,b, 1978b,c, 1979).

3.2.3 ATOMIC BEAM LASER SPECTROSCOPY   A new technique for optical spectroscopy combines laser-induced optical transitions in an atomic beam with a six-pole analyzing magnet, and uses as the transmission detector a mass spectrometer. This combination offers Doppler-free spectroscopy, which is essential in the lighter elements, and it avoids the chemical problems often encountered when tracer amounts of radioactivity must be contained in an optical cell. In the initial experiments on-line at the Orsay synchrocyclotron, the sodium isotopes ^{21,22,24,25}Na were studied (Huber et al 1975, 1976a,b). In a second series of experiments, the 24-GeV proton beam of the CERN PS was used to create extremely neutron-rich isotopes of sodium, and optical measurements were performed to mass 31 (G. Huber et al, personal communication and to be published). In a subsequent experiment the same technique was applied to rubidium and cesium produced at the ISOLDE on-line isotope separator. The experimental setup involved a double mass separation; data on isotope shifts for the $6s_{1/2}$-$7p_{1/2}$ were taken down to mass 123 (Huber et al 1978). In the most recent work a laser working in the infrared was used to study the $6s_{1/2}$-$6p_{1/2}$ transition with greatly improved resolution and stability (R. Klapisch et al, personal communication and to be published).

3.2.4 OTHER OPTICAL TECHNIQUES FOR ON-LINE EXPERIMENTS   Following early in-beam work on alkalies performed in Heidelberg and Karlsruhe, optical pumping of rubidium and cesium radioactivity has been performed using mass separation, both at CERN's ISOLDE and at the Mainz reactor (Bonn et al 1978, Fischer et al 1978).

A new concept with far-reaching possibilities, also due to the Mainz group (Anton et al 1978a,b, Schinzler et al 1978), is collinear spectroscopy. The accelerated ion beam is deflected into the beam of a single-frequency dye laser. Subsequently it is neutralized in a charge-exchange cell with cesium vapor. The resonance scattered photons from about 20 cm of the path are detected in a photomultiplier tube. The essential point is that the velocity distribution is narrowed by a factor of $\frac{1}{2}(kT/eU)$, where T is the temperature of the ion source and U is the acceleration voltage. This reduction amounts to typically 10^{-3} and leads to essentially Doppler-free spectroscopy. The neutralization of the beam is essential because the resonance lines of ions in general are in the UV region inaccessible with present single-mode dye lasers, but applications to ions in excited states also seem attractive (E. W. Otten, personal communication). Results by the collinear technique for $^{138-142}$Cs were recently reported by Bonn et al (1979).

4 NUCLEAR SPECTROSCOPY

An important inspiration for spectroscopic studies of far unstable nuclei has been the expectation of finding new regions with "simple" properties: new "magic" regions and new deformed regions. To the author's knowledge, the opening guns in this campaign were fired by B. R. Mottelson at a small conference in Copenhagen in 1961; he pointed out that certain regions, for example $50 < Z, N < 82$, ought to show well-developed rotational spectra similar to those then known in the ^{25}Mg, "rare earth," and "actinide" regions. The first evidence for a new deformed region comprising the light barium isotopes was obtained soon after (Sheline et al 1961). Since then, many experimental and theoretical results have deepened our knowledge of the interplay between shells and deformations; a review of this large field would be impossible here. Instead, the reader is referred to the survey given by Ragnarsson et al (1978), and in a compact form by Sheline (1976). The mention of nuclear stability in new regions would not be complete without evocation of the superheavy nuclei, near $(N, Z) = (184, 114)$, which to date have escaped detection, and for which at least some calculations fail to make $Z = 114$ a magic number (see, for example, Lombard 1976).

Another region of great fundamental importance is the $Z = N$ line, because of its relation to nuclear isospin. The proton-rich nuclei, in particular, offer exciting possibilities for performing unique experiments on beta decay, as was pointed out at the Lysekil Conference by Damgaard (1966) and Sorensen (1966). Experimentally, this field is a very difficult one, and only in the last few years has on-line mass separation become fast and efficient enough to contribute in parallel with conventional techniques.

4.1   Light Nuclei

4.1.1   THE PROTON-RICH NUCLEI   The difficulty encountered in producing proton-rich nuclei is that the cross-sections drop very rapidly with decreasing A, as seen in Figures 1 and 3. A good measure of the difficulty in producing nuclei along the $N = Z$ line can therefore be obtained simply by considering which are the heaviest radioactivities produced in a sequence with a given isospin projection $T_Z = \frac{1}{2}(N - Z)$.

For $T_Z = \frac{1}{2}$ the heaviest member observed is $^{97}_{48}$Cd found at ISOLDE (Elmroth et al 1978). Lighter members of this series (^{69}Se, ^{73}Kr) played an important role in the development of a new technique for measurements of short lifetimes, discussed in Section 5.2.3.

In the family of self-conjugate nuclei, the heaviest known member is $^{74}_{37}$Rb, which has isospin $T = 1$ and a half-life of 64.9 ± 0.5 ms (D'Auria et al 1977). As shown in Figure 1, this isotope can be produced in copious amounts. In the even-Z family with $T_Z = 0$ the last proton will be more bound and the beta half-lives will be much larger. It should therefore not be too hard to go beyond $^{72}_{36}$Kr with a half-life of 16.7 ± 0.6 s (Schmeing et al 1973), but there seems to be a general feeling that the doubly magic $^{100}_{50}$Sn will be nearly impossible to reach with present techniques.

The mirror nuclei with $T_Z = -\frac{1}{2}$ decay predominantly by a superallowed beta transition to the ground state of the $T_Z = +\frac{1}{2}$ partner. The isotope $^{55}_{28}$Ni with a half-life of 183 ± 5 ms was produced in a (^{3}He, 2n) reaction on ^{54}Fe and detected by direct beta counting of the target (Hornshøj et al 1976, 1977a, Edmiston et al 1976). Recently ^{71}Kr with a half-life of 102 ± 9 ms was produced by on-line isotope separation (G. T. Ewan et al, to be published).

In the $T_Z = -1$ family the odd members are beta-delayed proton and alpha emitters with 90-ms $^{44}_{23}$V as the heaviest known case (Cerny et al 1971). The heaviest even member is 0.26-s $^{46}_{24}$Cr (Zioni et al 1972).

The series of nuclei with $T_Z = -\frac{3}{2}$ is known from $^9_6$C to $^{61}_{32}$Ge, which has a half-life of 40 ms (Cerny 1976, Vieira et al 1979). Their decay is characterized by numerous groups of beta-delayed protons. Their properties were discussed by Hardy (1976) and Cerny (1976).

Many attempts have been made to observe radioactivity with $T_Z = -2$, most recently by Robertson et al (1976). The first member of the series $^{32}_{18}$Ar with a half-life of $75 \pm ^{70}_{30}$ms was found in experiments at ISOLDE by Hagberg et al (1977b). As can be seen from Figure 9, only the proton decay of the isobaric analogue state was detectable. Two more cases, the isotopes $^{20}_{12}$Mg and $^{24}_{14}$Si with half-lives near 100 ms, were found by

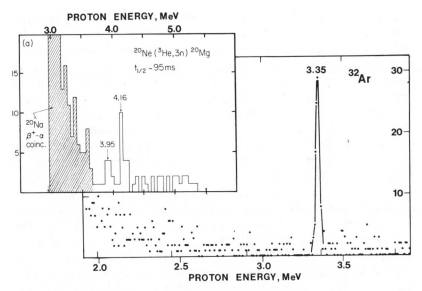

Figure 9   Beta-delayed proton spectra from the decay of the $(T, T_Z) = (2, -2)$ nuclei ^{32}Ar and ^{20}Mg [from Hagberg et al (1977b) and Moltz et al (1979)]. The proton peaks arise from the decay of the $T, T_Z = 2, -1$ isobaric analogue states, estimated to be populated in beta decay with an intensity of $\sim17\%$ (^{32}Ar) and 3% (^{20}Mg). The large background at low energy in the ^{20}Mg spectrum comes from ^{20}Na, which is produced in large quantities, and which is transmitted because the RAMA system has no chemical selectivity.

Moltz et al (1979) and Äystö et al (1979) using the RAMA system (Cerny et al 1977). Again only the isobaric analogue state was detectable (Figure 9).

The experiments on the $T_Z = -2$ family primarily provide a very exacting check on the isobaric multiplet equation. It would, however, be extremely interesting to use the beta decay of the $T_Z = -2$ nuclei to investigate other parts of the beta strength function, which is expected to be dominated by the Gamow-Teller giant resonance. Such experiments would require greatly improved intensities, but do not appear to be excluded.

Finally, the advantages of on-line isotope separation are not restricted to the study of rare isotopes. In many precision experiments, especially those dealing with weak branches, the high purity of an element- and mass-separated sample may be essential. This, of course, holds true in all regions of the nuclear chart, but seems especially to be the case for certain precision experiments on the beta decay of light nuclei. An example is the recent study (Hagberg et al 1979b) of the apparent anomaly of the Cabbibo angle in the beta decay of ^{35}Ar.

4.1.2 THE NEUTRON-RICH NUCLEI Unlike their proton-rich counterparts, the neutron-rich nuclei are not distinguished by special properties and selection rules; their spectroscopy is qualitatively the same as for all heavier nuclei with $N > Z$. From an experimental point of view, it is, however, interesting to note that only in the region of the lightest nuclei are the neutron "drip lines" accessible. The limit of proton instability has been reached in many places, very likely for cases as heavy as ^{73}Rb and ^{113}Cs (D'Auria et al 1977, 1978). On the neutron-rich side, the drip line lies much further from stability; it has been reached in $^{11}_{3}$Li, bound against the emission of two neutrons by 170 ± 80 keV (Thibault et al 1975), but not in heavier elements.

The use of the fragmentation reactions with high-energy protons (Section 2.1.1) allows the production of very neutron-rich light nuclei. The effort has until now been concentrated on sodium for which a long series of experiments (Klapisch et al 1969, 1972, 1973, Thibault et al 1975, Détraz et al 1979) gradually has taken the known isotopes out to $^{35}_{11}$Na. The heavier sodium isotopes have complex beta-decay schemes, in which the final states in many cases are unstable against neutron emission (Roeckl et al 1974).

As discussed in Section 3.1.1, the masses of the heavy sodium isotopes indicate a new deformed region near $N = 20$. This interpretation has been further examined by Détraz et al (1979), who observed the gamma rays following the beta decay of ^{32}Na. An intense gamma ray of 0.886 MeV was interpreted as evidence for rotational structure in ^{32}Mg. There are calculations by Barranco & Lombard (1979) indicating that the 3^- level will also come low in the $N = 20$ region and that even permanent octupole deformations may occur. Further spectroscopic studies in this region will clearly be of appreciable interest.

4.2   Heavy Nuclei

4.2.1   DOUBLY MAGIC SPHERICAL NUCLEI   The tin isotope ^{132}Sn with 50 protons and 82 neutrons represents a close analogy to ^{208}Pb, but has a half-life of only 40 s (Kerek et al 1972). It is produced in thermal fission, but as the yield is relatively low (0.14%) it was not until the advent of on-line mass separation that more detailed studies of the region around it could be undertaken. It became possible to study excited levels in ^{132}Sn when Kerek et al (1973) found the mother activity 0.12-s ^{132}In; they interpreted a 4.041-MeV gamma ray as representing the first excited state in ^{132}Sn. This result was confirmed in experiments at JOSEF by Lauppe et al (1977), who isolated a 1.7-μs isomer ^{132m}Sn at 4.847 keV. The decay of this level populates three lower-lying levels: 4.041, 4.415, and 4.714 MeV. The 4.041-MeV level can be interpreted either as a 2^+ or as a 3^- state (Dehesa et al 1978), which are expected to be relatively close to each other; further experiments are clearly needed in order to establish the quantum numbers of this and other ^{132}Sn levels.

The radioactive decay of ^{132}Sn further provides a unique opportunity for studying the beta decay of a doubly magic nucleus. In most nuclei the pair-occupation coefficients (u^2 and v^2) are not well enough known to allow a precise discussion of the Gamow-Teller decay rates, and in particular of the role of spin-isospin polarization effects (for example, see Żylicz et al 1966). The pairing corrections, however, are expected to be unity in the decay of ^{132}Sn to a 1^+ level at 1.324 MeV in ^{132}Sb (Kerek et al 1972). This level must represent the state $[\pi 2d_{5/2}, v2d_{3/2}^{-1}]_{1^+}$ for which the pure shell model gives a calculated log ft of 2.63; the measured value of 3.95 ± 0.05 [recalculated using the Q value measured by Aleklett et al (1976), Figure 6] is a measure of the core-polarization effect, which thus amounts to a factor of 20 (J. Blomqvist, cited by Kerek et al 1972). A more detailed theoretical analysis would require precise knowledge of the single-particle energies in the ^{132}Sn region. Spectroscopic studies of this problem are in progress at several places. Recently, Sistemich et al (1978) found evidence for the three lowest single-proton states in $^{133}_{51}$Sb: $g_{7/2}$(g.s.), $d_{5/2}$(0.963 MeV), and $h_{11/2}$(2.792 MeV). The large gap between $d_{5/2}$ and $h_{11/2}$ makes $Z = 64$ (corresponding to filled $g_{7/2}$ and $d_{5/2}$ shells) a near-magic number for spherical shape. The most direct spectroscopic evidence for this subshell closure probably comes from the Q_α systematics (see Section 3.1.2 and Figure 7). The near-magic properties of the gadolinium isotope $^{146}_{64}$Gd were exploited in a series of interesting experiments by Kleinheinz and collaborators (Kleinheinz et al 1978a,b, Ogawa et al 1978). They find that the first excited state in ^{146}Gd is 3^-, a property that this nucleus shares only with ^{208}Pb and possibly with ^{132}Sn. The 1579-keV 3^- state has a lifetime of 1.06 ± 0.13 ns corresponding to a $B(E3)$ of 37 single-particle units, the same enhancement as in the ^{208}Pb case (Kleinheinz et al 1978b). The first 2^+ level is situated at 1971 keV (Ogawa et al 1978). Just as in the ^{208}Pb region, the (sub-) shell closure is associated with isomerism (Broda et al 1978), and an alternative explanation of these isomers in terms of oblate "yrast traps" seems less likely (Pedersen et al 1977).

4.2.2 DEFORMED AND TRANSITIONAL NUCLEI   In addition to the three classical regions of deformed nuclei, four additional ones have been found to date. The main interest in these lies not in what they can teach us about the structure of strongly deformed nuclei; the "old" regions already provide an abundant number of cases, so that the rotational nuclei and their intrinsic states are well understood. The interest lies rather in what can be inferred from the observation of nuclear spectra as a function of Z and N en route to the new regions and, of course, in the mere fact that these regions exist, thus providing a test of our theoretical notions of a nuclear energy surface determined by shapes and shells (Ragnarsson et al 1978). Off the stability line, the spectroscopy of deformed nuclei thus merges with that of transitional nuclei.

The systematics of nuclear collective motion is not discussed here; the subject was recently treated by Scharff-Goldhaber et al (1976) in the light of the variable moment of inertia (VMI) model. There appears to be a smooth transition from the spectra of near-magic nuclei to those of strongly deformed nuclei, which is illustrated by the relation between the moment of inertia and the transition quadrupole moment that together define a "main series" (Goldhaber & Scharff-Goldhaber 1978).

One of the new regions of deformation, the one near ^{32}Mg, was discussed in Sections 3.1.1 and 4.1.2. The remaining three are illustrated in Figure 10, which shows the systematics of the first 2^+ level in even-even nuclei of the elements strontium, zirconium, and barium. The regions of deformation are characterized by low-lying 2^+ levels; for the elements in question, the 2^+ energies fall off on either side of the magic neutron number. The light barium isotopes formed the first new deformed region to be detected (Sheline et al 1961), and the knowledge of this region has recently been greatly extended by the atomic physics methods discussed in Sections 3.2.2 and 3.2.3. The transition to low 2^+ energies in the heavy barium isotopes experimentally is a recent addition, but simply represents the entry into well-known "rare earth" region.

The proton number 40 has a chameleon character (Sheline et al 1972) as it represents a subshell gap for spherical nuclei and at the same time a shell gap for a 2:1 prolate deformation. This character is clearly confirmed by the systematics in Figure 10, which shows spherical shape (high 2^+ energies) for the magic neutron number 50 (and for the subshell at $N = 56$), while on the wings the neutron numbers 40 and 60, which like $Z = 40$ favor prolate deformation, lead to two new regions of deformed nuclei. The one near $(Z, N) = (40, 60)$ was found by Cheifetz et al (1970) in measurements of radiations from fission fragments, and recent experiments with the mass separator JOSEF by Sistemich and collabora-

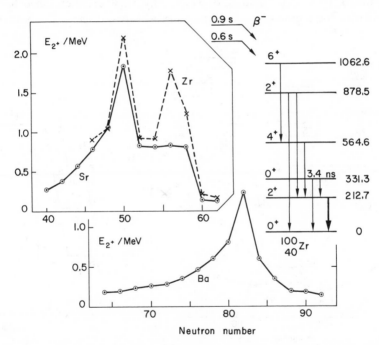

Figure 10 Systematics of the energies of the first 2^+ state for strontium ($Z = 38$), zirconium ($Z = 40$), and barium ($Z = 56$) as a function of the neutron number. Also shown is the level scheme of ^{100}Zr populated in the β^- decay of two isomers of ^{100}Y (Khan et al 1977, 1978). The 2^+ systematics is largely from Sakai & Rester (1977) with some new data points: ^{98}Sr, $E_{2+} = 144.4$ keV (Wollnik et al 1977; G. Nyman, personal communication); ^{100}Sr, $E_{2+} = 129.2$ keV, from ^{100}Rb with $T_{1/2} = 55$ ms (Azuma et al, to be published); ^{146}Ba, $E_{2+} = 181.0$ keV, from ^{148}Cs with $T_{1/2} = 343 \pm 7$ ms (Azuma et al, to be published; J. A. Pinston and F. Schussler, private communication 1978; Lund & Rudstam 1976); ^{148}Ba, $E_{2+} = 141.7$ keV, from ^{148}Cs with $T_{1/2} = 170 \pm 7$ ms (Azuma et al, to be published).

tors (Sistemich et al 1977, 1979, Khan et al 1977, 1978) have greatly extended our knowledge of this region. As an example, the inset in Figure 10 shows a simplified version of the ^{100}Y decay schemes, in which the most striking feature is the 331.3-keV 0^+ level with its large E2 and E0 transition probabilities to the ground-state band. An experiment at ISOLDE (Azuma et al, to be published) detected the 2^+ level in ^{100}Sr at 129.2 keV. Scaled with the usual $A^{-5/3}$ law for the inertial parameter, this represents the lowest 2^+ first excited state detected in any nucleus: in ^{158}Gd it would correspond to a 2^+ at 60 keV (experimental value 79.5 keV), and in ^{240}Pu to 30 keV (experimental value 42.9 keV). Only for shape isomers are lower 2^+ values found; thus the fission isomer in ^{240}Pu has its 2^+ at 22 keV (Specht et al 1972).

Many spectroscopic studies, of which only a few are mentioned here, dealt with the region of the light mercury isotopes, discussed in Section

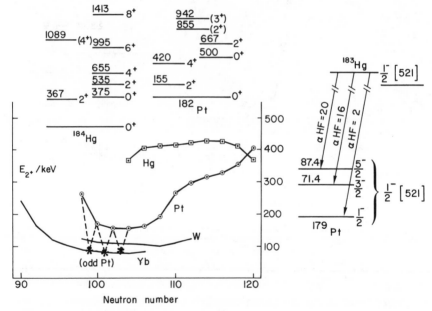

Figure 11 Systematics of the energies of the first 2^+ state for ytterbium ($Z = 70$), tungsten ($Z = 74$), platinum ($Z = 78$), mercury ($Z = 80$) as a function of the neutron number (Sakai & Rester 1977). For ^{176}Pt and ^{178}Pt, the correct 2^+ energies are 263 keV and 170.7 keV (Hagberg et al 1978b). Also shown are the level schemes of ^{184}Hg (Cole et al 1976, Rud et al 1973), ^{182}Pt (Cailliau et al 1974a), and the alpha-decay scheme of ^{183}Hg (Hagberg et al 1978b, 1979a), where the alpha transitions to the ground-state rotational band in ^{179}Pt are characterized by their hindrance factors. The inertial parameters for the $\frac{1}{2}^-[521]$ bands in ^{177,179,181}Pt were inferred from the alpha fine structure; in the 2^+ systematics they are represented by the quantity $6h^2/2\mathscr{I}$ (asterisks).

3.2.1. This region is of special interest because the proximity of the proton number to the magic value 82 favors a spherical shape, but on the other hand, the proximity of the neutron number to the midshell value $\frac{1}{2}(82 + 126) = 104$ favors deformed shape (Frauendorf & Paskevich 1975). The net result over most of the range is a transitional behavior, which is well illustrated by the behavior of the 2^+ energies of platinum shown in Figure 11 and also by the 0^+ energies, which drop low exactly at the middle of the shell, indicative of the onset of permanent prolate deformations (Finger et al 1972, Cailliau et al 1974a,b, Husson et al 1977, and papers cited therein). This behavior is typical of "soft" or beta-unstable nuclei (Sheline 1960). In the mercury isotopes, on the other hand, the 2^+ states remain at about 400 keV; this observation made by using heavy ions for the cases ^{184,186}Hg provided the first indication that the anomalous isotope shifts (Figure 8) were restricted to odd-mass nuclei. The heavy-ion experiments (Proetel et al 1973, 1974, Rud et al 1973) also gave evidence that the yrast band was developing a rotational character very low in the spectra of ^{184,186}Hg (compare the decay schemes of ^{182}Pt, ^{184}Hg in Figure 11). The low-lying 0^+ levels observed in these nuclei (Béraud et al 1977, Cole et al 1976) suggest deformed shape isomers with little mixing. Recent evidence concerning the odd-platinum isotopes $^{177-183}$Pt (Hagberg et al 1978b, 1979b, Visvanathan et al 1979) indicates that these have strongly deformed shapes that can be described in terms of the Nilsson model. The favored alpha decays of $^{181-185}$Hg (see the example in Figure 11) clearly identify the rotational bands built on the $\frac{1}{2}^-$[521] states of the daughters (Hagberg et al 1978b). As the moments of inertia in this band are not expected to be governed by Coriolis effects alone, the low values of $6\hbar^2/2\mathcal{I}$ shown in Figure 11 may indicate a shape staggering similar to that of the mercury mothers.

4.3 The Systematics of Alpha-Decay Widths

The role of alpha-decay studies in spectroscopy was illustrated in the preceding section; the use of alpha-decay Q values for determining nuclear masses was discussed in Section 3.1.2. This section deals with some systematic features of alpha-decay widths that have emerged in the study of alpha radioactivity in the cis-lead region. Very many alpha emitters are now known in the region $84 \leqq N \leqq 126$, and with the recent addition of five alpha-radioactive isotopes of tantalum (Hofmann et al 1979), all elements above samarium show alpha radioactivity. In the next lower region in N, the alpha radioactivity is concentrated in an island at $N = 56-59$; a total of eight cases are known there (Roeckl et al 1978).

The systematics of s-wave alpha widths is shown in Figure 12. These are high just above the closed shells $N = 50, 82, 126$, and the decrease

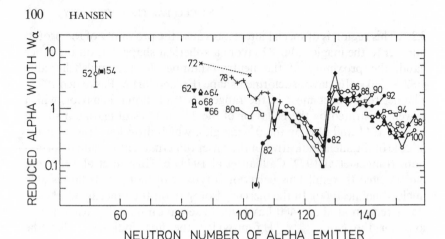

Figure 12 The systematics of reduced alpha widths W for s-wave alpha decay of even-even nuclei as a function of the neutron number N_M of the emitters (Hornshøj et al 1974, Hagberg et al 1977a, Roeckl et al 1978). The unit is normalized so that $W_\alpha(0^+)$ for ^{212}Po is unity.

between the shells has until now been observed only for $84 \leqq N_M \leqq 127$ and for $N_M \geqq 130$, where N_M is the neutron number of the alpha emitter. Only lead ($Z = 82$) seems to fall outside the systematics. Hornshøj et al (1974) observed that the d-wave widths are very low for mercury ($Z = 80$) but seem to be rising again strongly for platinum ($Z = 78$). The d-wave widths in the decay of the odd-mass deformed mercury nuclei ^{181,183,185}Hg are larger by a factor 3–10 than those of the neighboring even nuclei (Hagberg et al 1978b, 1979a).

5 AVERAGE AND STATISTICAL PROPERTIES IN THE BETA DECAY OF FAR UNSTABLE NUCLEI

The beta decays of exotic nuclei have high Q values and can populate a large number of excited levels. An instructive example is the isotope ^{116}Cs with a calculated Q_{EC} of 11 MeV. The estimated density of levels with spin I at the top of the excitation spectrum of the ^{116}Xe daughter is $2 \times 10^4 \times (2I + 1)$ MeV^{-1}, so that for a mother spin of 2, the average spacing will be about 3 eV. Therefore it is not surprising that the beta-delayed particle spectra associated with the ^{116}Cs decay (Figure 13) show no detectable fine structure.

In view of this complexity, the states populated in the beta decay as well as their subsequent decay by gamma rays and particles can best be described statistically in terms of average quantities, such as level densities and strength functions, and in terms of the probabilities of fluctuations around the average values. This description of nuclear states has been developed especially in connection with the theory of neutron resonance reactions (Lynn 1968), and its application to the beta decay of far unstable nuclei has been reviewed by Hansen (1973). A number of developments since then were covered at the Cargèse Conference (1976); see, in particular, the papers by Hardy and by Jonson et al.

The main motive for studying gross effects in beta decay is the possibility of learning about nuclei at 3–12-MeV excitation energy, their structure, strength functions, level densities, etc. At the same time, this research

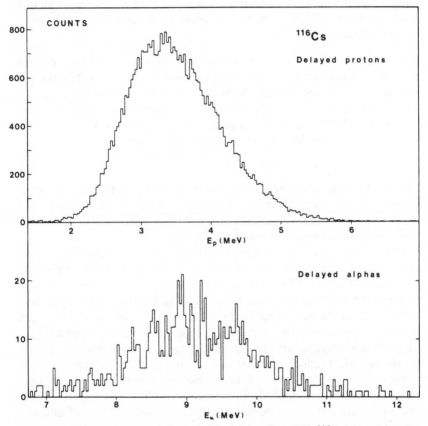

Figure 13 Delayed-proton and delayed-alpha spectra from 3.5-s ^{116}Cs measured with a surface-barrier counter telescope. The radioactivity was produced at ISOLDE from a target of molten lanthanum (see (Figure 3). The absolute alpha branch is $(8 \pm 2) \times 10^{-5}$ and the proton/alpha ratio is 47 ± 2. (From Hagberg et al 1978a, D'Auria et al 1978.) A 0.7 ± 0.2-s isomer was observed by positron counting, both by Bogdanov et al (1977) and by the ISOLDE group, which, however, was unable to detect the alphas and protons attributed to this radioactivity.

opens up the possibility of predicting parameters important to other fields, especially beta half-lives, total beta spectra, beta-delayed neutron spectra, and level densities. Such parameters are of importance for the r-process of nucleosynthesis (see, for example, Kodama & Takahashi 1975), for the electron capture in highly evolved stars (Egawa et al 1975), and in nuclear reactor technology.

5.1 General Properties

On-line mass separation has furnished the majority of all data on strength-function phenomena in beta decay. Before turning to the experimental results in Section 5.2, it is, however, necessary to introduce briefly some of the basic ideas.

5.1.1 THE BETA STRENGTH FUNCTION   A strength function for a nuclear reaction or decay expresses the total reduced transition probability per unit interval of excited states. As beta transition probabilities normally are expressed in terms of the ft value, it is most convenient to define (Duke et al 1970) the beta strength function $S_\beta(E)$ as a reciprocal ft value calculated per MeV of final levels in the daughter nucleus:

$$S_\beta(E) = b(E)[f(Z, Q-E)T_{1/2}]^{-1},$$ 5.

where $b(E)$ is the absolute beta intensity per MeV of final levels in the daughter with atomic number Z, Q the total energy available, and f the usual statistical rate function. The concept of a strength function is most useful if the variations with energy are slow.

Theoretically, the beta strength function can be calculated in a single-particle model as a sum of transition probabilities between initial and final nucleon states. The blocking, due to the Pauli principle, and also pairing corrections play an essential role in this picture. This approach, however, from the beginning has been known to be less straightforward than it seems because of the existence of strong correlation effects that lead to important renormalizations of the coupling constants. For allowed decays this effect reduces the Fermi transition probability typically by a factor of 10^6 and the Gamow-Teller transition probability typically by a factor of 10, for forbidden decays typically by factors of 3–10. [For a discussion, see the recent review by Ejiri & Fujita (1978) of effective coupling constants in medium and heavy nuclei.]

Full microscopic calculations of β^+ strength functions have been carried out for allowed Gamow-Teller transitions by Ivanova et al (1976), who used the random-phase approximation. Their results are in good agreement with experiments on light barium and xenon isotopes (Bogdanov et al 1978, Hornshøj et al 1972a).

The "gross theory" developed by Yamada and collaborators (Yamada 1965, Takahashi & Yamada 1969, Takahashi 1971, 1972, Kodama & Takahashi 1975) takes a more sweeping approach. Starting from a Fermi gas picture, it includes in a semiempirical fashion the collective states, which are responsible for the strong renormalizations. The strength to low-lying levels is not neglected, but is represented by the "tails" of the giant resonance states. This theory gives a good overall picture of beta strength functions (see Figure 15), and it is sufficiently simple to allow a global systematics. It has been used to calculate beta half-lives (Takahashi et al 1973). The most recent version (Kondoh & Yamada 1976) incorporates effects of nuclear shell structure.

5.1.2 FLUCTUATIONS The experimentally measured strength function is expected to fluctuate because of the finite number of levels contributing to the average. The study of these fluctuations is of interest for two reasons: (a) they give rise to a "noise level," which may make the observation of real structural effects difficult, and (b) they provide a new way of measuring the nuclear level density. Fluctuation phenomena in beta-decay processes were discussed previously by Hansen (1973) and by Jonson et al (1976); the following outlines only a few of the main features.

It is easily seen (Egelstaff 1958) that fluctuations in the level spacings are much less important than fluctuations in the transition probabilities. The latter are governed by the Porter-Thomas law, which for a single reaction channel can be written

$$p(x) = (2\pi x)^{-1/2} \exp(-x/2), \qquad\qquad 6.$$

where $p(x)$ is the probability density for observing a reduced width x. Here, x is measured in units of its average, so that $\langle x \rangle \equiv 1$. The variance on x is two. To understand how fluctuations appear, it is important to note that this distribution is strongly asymmetric: a sample will contain many values close to zero and a few large values.

From the Porter-Thomas law it is possible to derive distributions governing many other experimental situations (Jonson et al 1976); here two frequently occurring situations are considered. Assume that the experiment observes a weak particle branch with width y from a state populated in beta decay with width x. If the *total* width of the intermediate states is approximately constant, as is often the case, the observed intensities of y will be proportional to the product $v = xy$. The variable v obeys the product distribution law

$$p(v) = \pi^{-1} v^{-1/2} K_0(v^{1/2}), \qquad\qquad 7.$$

where K_0 denotes the modified Bessel function. This distribution has a mean value of one and a variance of eight. Finally, consider the case of two competing particle channels x and y, which are Porter-Thomas distributed. The branching ratio for particle x can be expressed in terms of the average widths $\langle \Gamma_x \rangle$ and $\langle \Gamma_y \rangle$:

$$\left\langle \frac{\Gamma_x}{(\Gamma_x + \Gamma_y)} \right\rangle = \langle \Gamma_x \rangle^{1/2} (\langle \Gamma_x \rangle^{1/2} + \langle \Gamma_y \rangle^{1/2})^{-1}, \qquad 8.$$

which strongly favors the weak branch. If the decay of a group of states is followed as a function of time, the decay law is no longer exponential. Let particle x be the one observed, and the intensity as a function of time is on the average

$$\langle I_x(t) \rangle = \frac{\langle \Gamma_x \rangle}{4\hbar} \left(\frac{\langle \Gamma_x \rangle t}{\hbar} + \frac{1}{2} \right)^{-3/2} \left(\frac{\langle \Gamma_y \rangle t}{\hbar} + \frac{1}{2} \right)^{-1/2}, \qquad 9.$$

corresponding to a t^{-2} decay law for large values of t. An expression similar to Equation 9 was derived by Malaguti et al (1971) for use in blocking experiments.

The arguments outlined above are, of course, valid irrespective of whether the individual states can be resolved or not, but the analysis poses additional problems in the latter situation. If σ denotes the Gaussian resolution parameter of the detector, D the average level spacing, and Γ the average natural width of the levels in question, the situation most often encountered is $\sigma \gg D \gg \Gamma$. This type of fluctuation was first discussed by Egelstaff (1958); an appropriate name for the phenomenon would be Porter-Thomas fluctuations. [The case for $\Gamma \gtrsim \sigma \gg D$ is known as Ericson fluctuations (Ericson 1963).]

Mathematical techniques for analyzing fluctuations in unresolved nuclear spectra can be adapted from those used for interpreting noise in electrical circuits (Rice 1944–1945). For the simple case of a spectrum $g(x)$ with one intermediate spin, constant level spacing D, and experimental resolution parameter σ ($\gg D$), one finds the autocorrelation function

$$\psi_g(\tau) = \langle g(x)g(x+\tau) \rangle$$

$$= 1 + \frac{\alpha D}{2\pi^{1/2}\sigma} \exp\left(-\frac{\tau^2}{4\sigma^2} \right), \qquad 10.$$

where α is the variance of the intensity of a line with average intensity unity. This expression permits the determination of the quantity $\alpha D/\sigma$ and (independently) of the experimental resolution. A related result is that for a detector with resolution $W_{1/2}$ (full width at half maximum) the average spacing between peaks will be $2.2\ W_{1/2}$ (Jonson et al 1976).

Several recent investigations have resorted to Monte Carlo techniques in order to calculate the fine structure of delayed-particle spectra (Gjøtterud et al 1978, Hardy et al 1977, 1978). The primary aim of these calculations has been pedagogical: even if the initial strength functions are structureless (Hardy et al name this structureless element "pandemonium"), a considerable fine structure may appear, as illustrated in Figure 14. Therefore, claims that a "prominent line structure" is evidence for a "selective beta population" are unjustified. Hardy et al (1978) point out that nuclear structure effects extracted from real data must be obtained along lines that would give the correct (structureless) picture of pandemonium from its "data." A second and probably more important applica-

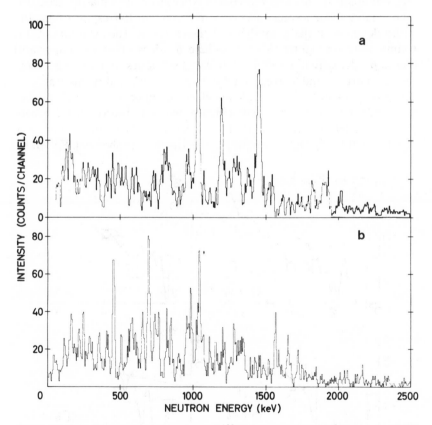

Figure 14   Beta-delayed neutron spectrum of ^{135}Sb from experiment (*a*) (Kratz et al 1979) and from a "pandemonium" calculation (*b*) (Hardy et al 1978). The three strong lines in the experimental spectrum have been placed as single transitions between excited levels in ^{135}Te and the 1.28-MeV 2^+ level of ^{134}Te. An "autopsy" of the three strongest lines in the pandemonium spectrum shows that these are exceedingly complex: they have contributions from about 200 single transitions!

tion of pandemonium-type calculations is deriving level-density parameters in the cases in which suitable analytical expressions are not readily
available (Gjøtterud et al 1978). Increased precision could probably be
obtained by comparing autocorrelation functions of data and pseudodata.

5.2   *Experimental Results*

5.2.1   HALF-LIVES AND TOTAL-ABSORPTION GAMMA SPECTROMETRY   The
beta feed to excited levels populated in electron capture may be measured
directly in a large 4π NaI spectrometer (Duke et al 1970). With a spectrometer of modest dimensions (two 6-in. diameter × 4-in. crystals face-to-
face were used in this and subsequent work) some gamma-ray escape is
unavoidable and it becomes necessary to correct for the response function,
which depends on the multiplicity of the cascades. The extension of this
technique to the lighter elements (where β^+ decay becomes important)
and to β^- decay requires more elaborate techniques to avoid registering
the beta particle, and to avoid annihilation radiation and bremsstrahlung.
For positron emitters a 4π anticoincidence counter was used (Hornshøj
et al 1975a), while for β^- a coincidence demand ensured that the beta
particle had left the central part of the spectrometer (Johansen et al 1973,
Aleklett et al 1975). Data from the β^- experiments are shown in Figure 15.

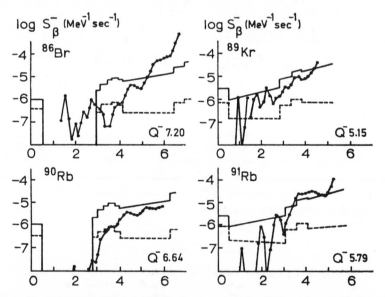

Figure 15   Calculated β^- strength functions compared with experiment (Kodama &
Takahashi 1975). The dots are experimental points (Johansen et al 1973, see also Aleklett
et al 1975), the lines show the theoretical total (*solid*) and first forbidden (*dashed*)
contributions.

The clear conclusion emering from the total-absorption experiments is that β^+, EC, and β^- strength functions have a very different energy dependence. While the former vary slowly with energy, the latter show a very pronounced increase. This asymmetry originates in the neutron excess of the heavier nuclei, and it can be understood on the basis of a single-particle picture with correlations.

The beta half-lives alone provide information about the beta strength. A demonstration of this was given by Takahashi et al (1973), who calculated the beta-decay half-lives for $3 \leq Z \leq 100$ from the gross theory. It is striking that the trend with mass and even the absolute values are well represented by a theory with essentially one free parameter. It is also possible to represent the electron-capture half-lives in terms of an average beta strength function $\overline{S_\beta}$ (Duke et al 1970, Hansen 1973), which in the range $95 < N < 126$ is seen to drop regularly with increasing neutron number. This drop represents two effects: the closing up of the allowed channels as the neutrons and the protons come one shell out of phase, and the effects of neutron-proton correlations.

5.2.2 DELAYED NEUTRONS   Pappas & Sverdrup (1972) were the first to conclude from an analysis of delayed-neutron emission probabilities that the β^- strength function must increase strongly with energy, but that the (traditional) assumption of proportionality to level density was an over-estimate. Later work has strengthened these conclusions.

The advent of ^{3}He-filled neutron spectrometers has greatly increased our knowledge of the beta spectra of beta-delayed neutrons (Shalev & Rudstam 1974, Rudstam et al 1974, Franz et al 1974, Kratz et al 1976, 1979). The neutron spectra, in general, show a pronounced fine structure (Figure 14), which has tempted the interpretation as a selection-rule effect (see, for example, Klapdor 1976). The interpretation as Porter-Thomas fluctuations (Hansen 1973) is supported by pandemonium calculations (Hardy et al 1978), which demonstrate that "neutron decay schemes" based on energy fits are likely to be wrong. For this reason, the identification of low-lying bumps in the β^- strength function cannot yet be regarded as certain.

Calculations of delayed-neutron spectra under different assumptions have been reported in several of the papers quoted above and also by Takahashi (1972). Delayed-neutron emission plays an important role in the final stages of the astrophysical r-process and leads to an important smoothing of the abundance curve of the nuclides (Kodama & Takahashi 1973).

5.2.3 DELAYED PROTONS   Delayed-proton radioactivity was reviewed by Cerny & Hardy (1977). Both experimentally and theoretically this process seems to be on a firmer footing than the corresponding process for neutrons. From pγ coincidence measurements the branching ratios to final states are known in many cases and furnish a valuable check on the theoretical calculations (Hornshϕj et al 1972a,b, Jonson et al 1976).

Evidence for effects of the shell structure   The delayed-proton spectra of tellurium, xenon, and barium isotopes show a pronounced "bump," which must represent the $\pi g_{9/2} \rightarrow \nu g_{7/2}$ beta transition (Bogdanov et al 1973, 1978, Hornshϕj et al 1972b). Calculations of these structures have been performed by Ivanova et al (1976). It is not surprising that this pheno- menon is more easily detectable in the neutron-deficient nuclei, in which only a few allowed channels remain open.

Measurements of level densities   The density of intermediate states popu- lated in beta decay can be determined through fluctuation analysis as outlined in Section 5.1.2. As an example, Figure 16 shows the measured fine structure of the ^{99}Cd delayed-proton spectrum. To determine the fluctuation amplitudes from such a spectrum introduces the delicate question that these, in principle, should be measured from the "true" value, which is, of course, both unknown and operationally undefined. This problem can be solved by a mathematical filtering technique, to which an electrical analogue would be the determination of a high-

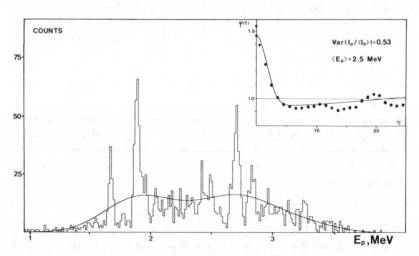

Figure 16   The measured proton energy spectrum of 16-s ^{99}Cd and a smoothed spectrum obtained with a folding function of second order. The inset shows the autocorrelation function $\psi(\tau)$ for a spectrum formed as the quotient of the two spectra. The solid curve shows a theoretical fit to the autocorrelation function, from which a level density parameter $a = 8.5$ was obtained (Elmroth et al 1978).

frequency noise component in the presence of a background of fluctuations of lower frequency in the dc level (Jonson et al 1976, Elmroth et al 1978).

The systematics of the level-density parameter a for the tin region is shown in Figure 17, which in a striking way demonstrates the strong influence of the nearby doubly magic ^{100}Sn on the point derived from the ^{99}Cd proton spectrum (Figure 16). The success of the semiempirical formula of Truran et al (1970) is reassuring.

Lifetimes of excited states in the 10^{-16}-s range   In the process of beta-delayed proton emission, a K vacancy is created simultaneously with the population of a proton-unstable excited state. If the x ray is emitted before the proton, its energy will correspond to that of the beta-decay daughter; if it is emitted after, it will correspond to the element one unit lower in Z.

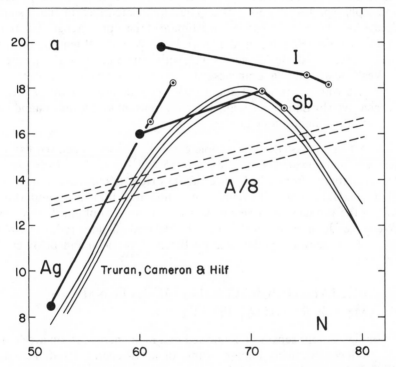

Figure 17   The level density parameter a (in MeV^{-1}) as a function of the neutron number N for the silver ($Z = 47$), antimony ($Z = 51$), and iodine ($Z = 53$). Experimental points are connected by heavy lines. The large solid circles are from fluctuation analysis of delayed-proton spectra (Jonson et al 1976, Elmroth et al 1978), whereas the smaller open circles, corresponding to nuclei near stability, are from neutron resonance data compiled by Erba et al (1961). The thin curves correspond to the semiempirical formula given by Truran et al (1970), and the dashed lines to the usual estimate $a = A/8$.

Consequently, the intensity ratio of the two x rays will provide a measure of the nuclear lifetime, as the atomic one is accurately known. It is important to take the nonexponential decay law (Equation 9) into account. This technique has been applied to the isotopes ^{69}Se and ^{73}Kr by Hardy and collaborators (Hardy et al 1976, Macdonald et al 1977, Asboe-Hansen et al 1978). The lifetimes are in good agreement with statistical-model estimates.

5.2.4 DELAYED ALPHAS   Beta-delayed alpha emission has been noted for a long time in ^{212,214}Bi ("long-range alphas") and in the light nuclei, but the observation of this process in the EC decay of very neutron-deficient nuclei such as ^{181}Hg and ^{114,116,118,120}Cs is more recent (Hornshøj et al 1975b,c, Hagberg et al 1978a). Figure 13, which gives references to some of the newer work, shows as an example the proton and alpha spectra of 3.5-s ^{116}Cs. The alpha intensities in the cesium decays have been used to derive an estimate of the alpha strength function at low energies (Hagberg et al 1978a). In units of the Wigner limit, the result of 0.05–0.10 MeV^{-1} agrees well with simple theoretical estimates.

Weak delayed-alpha branches have been observed in the isotopes ^{113,115}Xe with p/α ratios of about 10^3. The enhancement of the weak branch given by Equation 8 is apparently important for the understanding of these alpha intensities (Hagberg 1978).

5.2.5 DELAYED FISSION   For completeness the process of delayed fission is briefly mentioned. The beta decay of 55-s ^{232}Am leads to fission with 1% probability, from which an estimate of the fission barrier may be obtained (Habs et al 1978). Beta-delayed fission might play an important role in the production of heavy elements in the astrophysical r-process (Wene & Johansson 1974). The β^--delayed fission radioactivities ^{236,238}Pa were recently identified by Batist et al (1977) and Belov et al (1976).

6   SPECIAL APPLICATIONS OF INTENSE BEAMS OF RADIOACTIVITY

An on-line isotope separator connected to a high-energy proton accelerator produces extremely intense beams of mass-separated radioactivity (Figures 1 and 3). Not surprisingly, these beams also have applications outside the field of far unstable nuclei.

6.1   Radioactive Targets

An ion beam of 10^{11} atoms s^{-1} collecting on a 4-mm^2 surface for one day corresponds to a target thickness of the order of 50 μg cm^{-2}. This is sufficient for certain reaction spectroscopy experiments with low-energy accelerators, such as Van de Graaffs.

Nevertheless, the first experiments of this type studied (n, p) and (n, α) reactions following thermal-neutron capture. Long-lived radio-activities produced at ISOLDE were placed in an external neutron beam at the ILL high-flux reactor in Grenoble. Among the reactions observed were 34.5-d ^{84}Rb(n, p) ^{84}Kr(0^+ and first 2^+) (Andersson et al 1976), a 16.1-h ^{76}Br(n, p) ^{76}Se(0^+ and three 2^+) (Andersson et al 1978), and the (n, p) and (n, α) reactions on 34.8-d ^{37}Ar (Asghar et al 1978).

6.2   Optical Spectroscopy of Francium

A combination of atomic beam and laser techniques of the type described in Section 3.2.3 has recently permitted the first observation of an optical transition in francium ($Z = 87$), the D_2 line at a wavelength of 717.7 nm (Liberman et al 1978 and to be published). This element until then remained the only one with $Z < 100$ for which no optical transition had been reported; the difficulties were due to the short half-lives (the longest-lived francium isotope has $T_{1/2} = 22$ min) and low production rates. Spallation of uranium or thorium targets at ISOLDE provided 10^8 atoms s^{-1} of mass-separated radioactivity, enough for the optical experiments, which have now opened the way for a study of francium nuclei through their hyperfine structure.

6.3   Shifts in the Energies of K X Rays

The intense sources of electron-capture radioactivity available at ISOLDE permit a study of shifts in the energies of K x rays with very high precision. The measurements are made with a crystal-diffraction spectrometer in the DuMond geometry; an important new feature is a special double-source arrangement (G. L. Borchert et al, to be published), which permits a comparison of the energies of two sources with an error as low as 2×10^{-4} of the natural K line width. The experiments have provided the first observation of shifts due to the 1s hyperfine structure, an effect that accounts entirely for the measured energy difference of 0.112 ± 0.011 eV between Xe $K_{\alpha 1}$ x rays from sources of ^{131}Cs and ^{132}Cs. The hyperfine shifts in electron-capture sources of K x rays are brought about by a new selection rule in the atomic F quantum number (Borchert et al 1977). Further experiments give evidence for contributions to the energies from shake-off effects in the photoionization process (Borchert et al 1978a), and for an effect that reflects the difference in electronic structure between the elements that occur as mother and daughter in the beta-decay process (Borchert et al 1978a,b).

6.4   *Implantation of Radioactivity*

Although high intensities usually are not essential in implantation experiments, this type of research probably constitutes a major and, until recently, unexploited use of on-line installations. One possibility that goes beyond what can be done off-line is to implant activities that subsequently through beta-decay processes give rise to the element or isotope of interest. Experiments at ISOLDE have until now made use of implanted radioactivity for low-temperature experiments, perturbed $\gamma\gamma$ angular correlations for the study of quadrupole interactions in solids, Mössbauer spectroscopy, and for a study of ranges of slow heavy ions in gases (I. Berkès, H. Haas, G. Weyer, G. Sidenius, personal communication).

7   CONCLUDING REMARKS

This article summarizes the progress in many fields linked by a common experimental technique, and it is hardly possible to offer a summary of the summary. Instead a brief comment on future prospects may be in order.

Although the modern phase of research with on-line mass separators began about a decade ago, some of the most striking developments have come within the last few years. For the moment, the field is changing very rapidly, and experiments are now probing deeply into unknown regions of the nuclear chart, occasionally with surprising results. Very many experiments are possible and as the number of groups involved is small, it is necessary to be extremely selective in the choice of problem. This is probably an advantage, although the interest in a problem in nuclear physics often seems proportional to the number of accelerators that can tackle it. On the side of the experimental techniques, the limited scale of the effort may pose a serious drawback because it is difficult to devote sufficient time to the development of new techniques. To be convinced that this is essential to the whole field the reader needs only look back at Section 2 to be reminded of how much ingenuity was invested in the development of the present methods for producing beams of radioactive ions. In the future as well, progress will be determined by a mere handful of innovative experimenters.

ACKNOWLEDGMENTS

The author is indebted to CERN and to Aarhus University for their untiring and wholehearted support of his work. He is grateful to all participants in the ISOLDE Collaboration for the enthusiasm and team spirit that has always characterized this undertaking. Finally, he takes the opportunity to express special gratitude to Björn Jonson and Helge Ravn for a particularly close and friendly companionship; the present paper in many ways reflects their viewpoints and contributions. Comments on the manuscript by Kohji Takahashi have been appreciated.

Literature Cited[a]

Aleklett, K., Lund, E., Rudstam, G. 1976. *Nucl. Phys. A* 281:213–20 [3.1.2, 4.2.1]

Aleklett, K., Nyman, G., Rudstam, G. 1975. *Nucl. Phys. A* 246:425–44 [5.2.1]

Amiel, S. 1966. *Ark. Fys.* 36:71–76 [2.1.1]

Amiel, S., Engler, G., eds. 1976. *Proc. Int. Conf. on Electromagnetic Isotope Separators and Related Ion Accelerators, 4th, 1976, Nucl. Instrum. Methods,* 139:363 pp. [2.1, 2.1.2, 2.1.3]

Andersson, G. 1973. See Andersson & Holmén 1973, pp. 356–58 [2.1.1]

Andersson, G., Asghar, M., Emsallem, A., Hagberg. E., Jonson, B. 1976. *Phys. Lett. B* 61:234–36 [6.1]

Andersson, G., Asghar, M., Emsallem, A., Hagberg, E., Jonson, B., Tidemand-Petersson, P. 1978. *Phys. Scr.* 18:165–66 [6.1]

Andersson, G., Holmén, G., eds. 1973. *Proc. Int. EMIS Conf. on Low-Energy Ion Accelerators and Mass Separators, 8th, Skövde, Sweden, 1973.* Gothenburg, Sweden: Chalmers Univ. Technol. 519 pp. [2.1, 2.1.2, 2.1.3]

Anton, K. R., Kaufman, S. L., Klempt, W., Moruzzi, G., Neugart, R., Otten, E.-W., Schinzler, B. 1978a. *Phys. Rev. Lett.* 40:642–45 [3.2.4]

Anton, K. R., Kaufman, S. L., Klempt, W., Neugart, R., Otten, E.-W., Schinzler, B. 1978b. *Hyperfine Interactions* 4:87–90 [3.2.4]

Armbruster, P. 1976. See Cargèse Conference 1976, pp. 3–14 [2.2, 2.2.2]

Armbruster, P., Asghar, M., Bocquet, J. P., Decker, R., Ewald, H., Greif, J., Moll, E., Pfeiffer, B., Schrader, H., Schussler, F., Siegert, G., Wollnik, H. 1976. *Nucl. Instrum. Methods* 139:213–22 [2.2.1]

Asboe-Hansen, P., Hagberg, E., Hansen, P. G., Hardy, J. C., Hornshøj, P., Jonson, B., Mattsson, S., Tidemand-Petersson, P. 1978. *Phys. Lett. B* 77:363–66 [5.2.3]

Asghar, M., Emsallem, A., Hagberg, E., Jonson, B., Tidemand-Petersson P. 1978. *Z. Phys. A.* 288:45–48 [6.1]

Äystö, J., Moltz, D. M., Cable, M. D., von Dincklage, R. D., Parry, R. F., Wouters,

J. M., Cerny, J. 1979. *Phys. Lett. B* 82:43–46 [4.1.1]

Äystö, J., Rantala, V., Valli, K., Hillebrand, S., Kortelahti, M., Eskola, K., Raunemaa, T. 1976. *Nucl. Instrum. Methods* 139:325–29 [2.1.1]

Barranco, M., Lombard, R. 1979. *Phys. Lett. B.* In press [4.1.2]

Batist, L. Kh., Berlovich, E. Ye., Gavrilov, V. V., Novikov, Yu. N., Orlov, S. Yu., Tikhonov, V. I. 1977. *Leningrad Institute for Nuclear Physics (AN USSR) Rep. 363* [5.2.5]

Beiner, M., Lombard, R. J., Mas, D. 1975. *Nucl. Phys. A* 249:1–28 [3.1.1]

Belov, A. G., Gangrsky, Yu. P., Kucher, A. M., Marinescu, G. M., Miller, M. B., Kharisov, I. F. 1976. *Dubna Rep. JINR P15-9795* [5.2.5]

Benenson, W., Kashy, E., Mueller, D., Nann, H. 1976. See Cargèse Conference 1976, pp. 235–45 [3.1.2]

Béraud, R., Meyer, M., Desthuilliers, M. G., Bourgeois, C., Kilcher, P., Letessier, J. 1977. *Nucl. Phys. A* 284:221–38 [4.2.2]

Berlovich, E. Ye., Batist, L. Kh., Blinnikov, Yu. S., Bondarenko, V. A., Gavrilov, V. V., Elkin, Yu. V., Lemeshko, G. G., Mesilev, K. A., Mironov, Yu. T., Moros, F. V., Novikov, Yu. N., Orlov, S. Yu., Panteleev, V. N., Polyakov, A. G., Sergienko, V. A., Smolskii, C. L., Tarasov, V. K., Trikhonov, V. I., Tschigolev, N. D. 1976. *Izv. Akad. Nauk. SSSR (Ser. Fiz.)* 40:2036–40 [2.1.1]

Berlovich, E. Ye., Ignatenko, E. I., Novikov, Yu. N. 1973. See Andersson & Holmén 1973 [2.1.1]

Boehm, F., ed. 1974. *Atomic Data Nucl. Data Tables* 14:509–98 [3.2.1]

Bogdanov, D. D., Demyanov, A. V., Karnaukhov, V. A., Petrov, L. A., Voboŕil, J. 1978. *Nucl. Phys. A* 303:145–53 [5.1.1, 5.2.3]

Bogdanov, D. D., Karnaukhov, V. A., Petrov, L. A. 1973. *Yadernaya Fiz.* 18:3–11 [5.2.3]

Bogdanov, D. D., Voboŕil, J., Demyanov, A. V., Karnaukhov, V. A., Petrov, L. A.

[a] Square brackets indicate the section in which the reference is quoted.

1976. *Nucl. Instrum. Methods* 136:433–35 [2.1.1]

Bogdanov, D. D., Vobořil, J., Demyanov, A. V., Petrov, L. A. 1977. *Phys. Lett. B* 71:67–70 [5]

Bohr, Aa., Mottelson, B. R. 1969. *Nuclear Structure*, Vol. I, p. 163. New York: Benjamin. 471 pp. [3.2.1]

Bonn, J., Buchinger, F., Dabkiewicz, P., Fischer, H., Kaufman, S. L., Kluge, H.-J., Kremmling, H., Kugler, L., Neugart, R., Otten, E.-W., von Reisky, L., Rodriguez-Giles, J. M., Steinacher, H.-J., Spath, K. P. C. 1978. *Hyperfine Interactions* 4:174–78 [3.2.4]

Bonn, J., Huber, G., Kluge, H.-J., Köpf, U., Kugler, L., Otten, E.-W. 1971. *Phys. Lett. B* 36:41–43 [3.2.1]

Bonn, J., Huber, G., Kluge, H.-J., Kugler, L., Otten, E.-W. 1972. *Phys. Lett. B* 38:308–11 [3.2.1]

Bonn, J., Huber, G., Kluge, H.-J., Otten, E.-W. 1976. *Z. Phys. A* 276:203–17 [3.2.1]

Bonn, J., Klempt, W., Neugart, R., Otten, E.-W., Schinzler, B. 1979. *Z. Phys. A* 289:227–28 [3.2.4]

Borchert, G. L., Hansen, P. G., Jonson, B., Lindgren, I., Ravn, H. L., Schult, O. W. B., Tidemand-Petersson, P. 1978a. *Phys. Lett. A* 65:297–300 [6.3]

Borchert, G. L., Hansen, P. G., Jonson, B., Lindgren, I., Ravn, H. L., Schult, O. W. B., Tidemand-Petersson, P. 1978b. *Phys. Lett. A* 66:374–76 [6.3]

Borchert, G. L., Hansen, P. G., Jonson, B., Ravn, H. L., Schult, O. W. B., Tidemand-Petersson, P. 1977. *Phys. Lett. A* 63:15–18 [6.3]

Borg, S., Bergström, I., Holm, G. B., Rydberg, B., de Geer, L.-E., Rudstam, G., Grapengiesser, B., Lund, E., Westgaard, L. 1971. *Nucl. Instrum. Methods* 91:109–16 [2.1.1, 2.1.2]

Bos, K. 1977. *Determination of atomic masses from experimental data.* Dissertation, Faculty of Science, Univ. Amsterdam. 91 pp. [3.1]

Bos, K., Gove, N. B., Wapstra, A. H. 1974. *Z. Phys.* 271:115–27 [3.1.2]

Broda, R., Ogawa, M., Lunardi, S., Maier, M. R., Daly, P. J., Kleinheinz, P. 1978. *Z. Phys. A* 285:423–24 [4.2.1]

Burkard, K. H., Dumanski, W., Kirchner, R., Klepper, O., Roeckl, E. 1976. *Nucl. Instrum. Methods* 139:275–80 [2.1.1]

Cailliau, M., Foucher, R., Husson, J. P., Letessier, J. 1974a. *J. Phys.* 35:L233–36 [4.2.2]

Cailliau, M., Foucher, R., Husson, J. P., Letessier, J. 1974b. *J. Phys.* 35:469–82 [4.2.2]

Campi, X., Flocard, H., Kerman, A. K., Koonin, S. 1975. *Nucl. Phys. A* 251:193–205 [3.1.1]

Cargèse Conference. 1976. *Proc. Int. Conf. on Nuclei Far From Stability, Cargèse, 1976, 3rd, Rep. CERN 76-13.* 608 pp. [1, 5]

Carraz, L. C., Haldorsen, I. R., Ravn, H. L., Skarestad, M., Westgaard, L. 1978a. *Nucl. Instrum. Methods* 148:217–30 [2.1.2, 2.1.3]

Carraz, L. C., Sundell, S., Ravn, H. L., Skarestad, M., Westgaard, L. 1978b. *Nucl. Instrum. Methods.* 158:69–80 [2.1.2, 2.1.3]

Cerny, J. 1976. See Cargèse Conference 1976, p. 225–34 [3.1.2, 4.1.1]

Cerny, J., Hardy, J. C. 1977. *Ann. Rev. Nucl. Sci.* 27:333–51 [5.2.3]

Cerny, J., Goosman, D. R., Alburger, D. E. 1971. *Phys. Lett. B* 37:380–82 [4.1.1]

Cerny, J., Moltz, D. M., Evans, H. C., Vieira, D. J., Parry, R. F., Wouters, J. M., Gough, R. A., Zisman, M. S. 1977. *Rep. LBL-7156,* pp. 1–13. In *Proc. Isotope Separator On-Line Workshop,* ed. R. E. Chrien, *Rep. BNL-50847* [2.1.1, 4.1.1]

Cheifetz, E., Jared, R. C., Thompson, S. G., Wilhelmy, J. B. 1970. *Phys. Rev. Lett.* 25:38–43 [4.2.2]

Cohen, B. L., Fulmer, C. B. 1958. *Nucl. Phys.* 6:547–60 [2.2.2]

Cole, J. D., Hamilton, J. H., Ramayya, A. V., Nettles, W. G., Kawakami, H., Spejewski, E. H., Ijaz, M. A., Toth, K. S., Robinson, E. L., Sastry, K. S. R., Lin, J., Avignone, F. T., Brantley, W. H., Rao, P. V. G. 1976. *Phys. Rev. Lett.* 37:1185–88 [3.2.1, 4.2.2]

Dabkiewicz, P., Buchinger, F., Fischer, H., Kluge, H.-J., Kremmling, H., Kühl, T., Müller, A. C., Schüssler, H. A. 1979. *Phys. Lett. B* 82:199–203 [3.2.1]

Dabkiewicz, P., Duke, C., Fischer, H., Kühl, T., Kluge, H.-J., Kremmling, H., Otten, E.-W, Schüssler, H. A. 1978. *J. Phys. Soc. Jpn. Suppl.* 44:503–8 [3.2.1]

Damgaard, J. 1966. See Lysekil Conference 1966, pp. 651–56 [4]

D'Auria, J. M., Carraz, L. C., Hansen, P. G., Jonson, B., Mattson, S., Ravn, H. L., Skarestad, M., Westgaard, L. 1977. *Phys. Lett. B* 66:233–35 [2, 4.1.1, 4.1.2]

D'Auria, J. M., Grüter, J. W., Hagberg, E., Hansen, P. G., Hardy, J. C., Hornshøj, P., Jonson, B., Mattsson, S., Ravn, H. L., Tidemand-Petersson, P. 1978. *Nucl. Phys. A* 301:397–410 [4.1.2, 5]

Dehesa, J. S., Lauppe, W.-D., Sistemich, K., Speth, J. 1978. *Phys. Lett. B* 74:309–12 [4.2.1]

Détraz, C., Guillemaud, D., Huber, G., Klapisch, R., Langevin, M., Naulin, F.,

Thibault, C., Carraz, L.-C., Touchard, F. 1979. *Phys. Rev C* 19:164–76 [4.1.2]

Diamond, R. M. 1970. See Leysin Conference 1970, pp. 65–107 [1]

Duke, C. L., Hansen, P. G., Nielsen, O. B., Rudstam, G. 1970. *Nucl. Phys. A* 151: 609–33 [5.1.1, 5.2.1]

Duke, C. L., Fischer, H., Kluge, H.-J., Kremmling, H., Kühl, T., Otten, E. W. 1977. *Phys. Lett. B* 66:303–6

Dumont, G., Pattyn, H., Huyse, M., Lhersonneau, G., Verplancke, J., van Klinken, J., de Raedt, J., Sastry, D. L. 1978. *Nucl. Instrum. Methods* 153:81–92 [2.1.1, 2.1.2]

Edmiston, M. D., Warner, R. A., McHarris, W. C., Kelly, W. H. 1976. See Cargèse Conference 1976, pp. 258–61 [4.1.1]

Egawa, Y., Yokoi, K., Yamada, M. 1975. *Prog. Theor. Phys.* 54:1339–55 [5]

Egelstaff, P. A. 1958. *Proc. Phys. Soc.* 71: 910–24 [5.1.2]

Ejiri, H., Fujita, J. I. 1978. *Phys. Rep. C* 38:85–131 [5.1.1]

Ekström, C., Ingelman, S., Wannberg, G. 1978a. *Nucl. Instrum. Methods* 148:17–28 [3.2.2]

Ekström, C., Ingelman, S., Wannberg, G., Skarestad, M. 1977a. *Nucl. Phys. A* 292: 144–64 [3.2.2]

Ekström, C., Ingelman, S., Wannberg, G., Skarestad, M. 1977b. *Hyperfine Interactions* 4:165–69 [3.2.2]

Ekström, C., Ingelman, S., Wannberg, G., Skarestad, M. 1978c. *Phys. Scr.* 18:51–53

Ekström, C., Ingelman, S., Wannberg, G., Skarestad, M. 1979. *Nucl. Phys. A* 311: 269–83 [3.2.2]

Ekström, C., Wannberg, G., Heinemeier, J. 1978b. *Phys. Lett. B* 76:565–68 [3.2.2]

Elmroth, T., Hagberg, E., Hansen, P. G., Hardy, J. C., Jonson, B., Ravn, H. L., Tidemand-Petersson, P. 1978. *Nucl. Phys. A* 304:493–502 [4.1.1, 5.2.3]

Epherre, M., Audi, G., Thibault, C., Klapisch, R., Huber G., Touchard, F., Wollnik, H. 1979. *Phys. Rev. C* 19: 1504–22 [3.1.1]

Erba, E., Facchini, U., Menichella-Saetta, E. 1961. *Nuovo Cimento* 22:1237–60 [5.2.3]

Ericson, T. 1963. *Ann. Phys. NY* 23:390–414 [5.1.2]

Ewald, H., Güttner, K., Münzenberg, G., Armbruster, P., Faust, W., Hofmann, S., Schmidt, K. H., Schneider, W., Valli, K. 1976. *Nucl. Instrum. Methods* 139:223–25 [2.2.3]

Faust, W. 1978. *Das Geschwindigkeitsfilter SHIP—Beschreibung, Ausmessung, Berechnung des Wirkungsgrades und erste Experimente*, PhD thesis, Univ. Giessen [2.2.3]

Finger, M., Foucher, R., Husson, J. P., Jastrzebski, J., Johnson, A., Astner, G., Erdal, B. R., Kjelberg, A., Patzelt, P., Höglund, Å., Malmskog, S. G., Henck, R. 1972. *Nucl. Phys. A* 188:369–408 [4.2.2]

Fischer, H., Dabkiewicz, P., Freilinger, P., Kluge, H.-J., Kremmling, H., Neugart, R., Otten, E.-W. 1978. *Z. Phys. A* 284: 3–8 [3.2.4]

Foucher, R., Paris, P., Sarrouy, J. L. 1973. See Andersson & Holmén 1973, pp. 341–45 [2.1.1]

Fransson, K., Ugglas, M. Af., Engström, A. 1973. *Nucl. Instrum. Methods* 113:157–68 [2.1.1]

Franz, H., Kratz, J.-V., Kratz, K.-L., Rudolph, W., Herrmann, G., Nuh, F. M., Prussin, S. G., Shibab-Eldin, A. A. 1974. *Phys. Rev. Lett.* 33:859–62 [5.2.2]

Frauendorf, S., Paskevich, V. V. 1975. *Phys. Lett. B* 55:365–68 [4.2.2]

Gauvin, H., Le Beyec, Y., Livet, J., Reyss, J. L. 1975. *Ann. Phys.* 9:241–70 [3.1.2]

Gjøtterud, O. K., Hoff, P., Pappas, A. C. 1978. *Nucl. Phys. A* 303:295–312 [5.1.2]

Goldhaber, A. S., Scharff-Goldhaber, G. 1978. *Phys. Rev. C* 17:1171–78 [4.2.2]

Habs, D., Klewe-Nebenius, H., Metag, V., Neumann, B., Specht, H. J. 1978. *Z. Phys. A* 285:53–57 [5.2.5]

Hagberg, E. 1978. Dissertation, Dept. Phys., Chalmers Univ. Technol. Gothenburg, Sweden. 45 pp. [5.2.4]

Hagberg, E., Hansen, P. G., Hardy, J. C., Hornshøj, P., Jonson, B., Mattsson, S., Tidemand-Petersson, P. 1977a. *Nucl. Phys. A* 293:1–9 [4.3]

Hagberg, E., Hansen, P. G., Hardy, J. C., Huck, A., Jonson, B., Mattsson, S., Ravn, H.-L., Tidemand-Petersson, P., Walter, G. 1977b. *Phys. Rev. Lett.* 39: 792–95 [4.1.1]

Hagberg, E., Hansen, P. G., Hornshøj, P., Jonson, B., Mattsson, S., Tidemand-Petersson, P. 1978a. *Phys. Lett. B* 73:139–41 [5, 5.2.4]

Hagberg, E., Hansen, P. G., Hornshøj, P., Jonson, B., Mattsson, S., Tidemand-Petersson, P. 1978b. *Phys. Lett. B* 78: 44–47 [4.2.2, 4.3]

Hagberg, E., Hansen, P. G., Hornshøj, P., Jonson, B., Mattsson, S., Tidemand-Petersson, P. 1979a. *Nucl. Phys. A* 318: 29–44 [4.2.2, 4.3]

Hagberg, E., Hardy, J. C., Jonson, B., Mattsson, S., Tidemand-Petersson, P., 1979b. *Nucl. Phys. A* 313:276–82 [4.1.1]

Hagebø, E., Sundell, S. 1970. See Kjelberg & Rudstam 1970, pp. 65–80 [2.1.2, 2.1.3]

Hagebø, E., Kjelberg, A., Patzelt, P., Rudstam, G., Sundell, S. 1970. See

Kjelberg & Rudstam 1970, pp. 95–107 [2.1.2]

Hansen, F., Lindahl, A., Nielsen, O. B., Sidenius, G. 1973. See Andersson & Holmén, 1973, pp. 426–31 [2.1.2]

Hansen, P. G. 1973. *Adv. Nucl. Phys.* 7:159–227 [5, 5.1.2, 5.2.1]

Hardy, J. C. 1976. See Cargèse Conference 1976, pp. 267–76 [3.1.2, 4.1.1, 5]

Hardy, J. C., Carraz, L. C., Jonson, B., Hansen, P. G. 1977. *Phys. Lett. B* 71: 307–10 [5.1.2]

Hardy, J. C., Jonson, B., Hansen, P. G. 1978. *Nucl. Phys. A.* 305:15–28 [5.1.2, 5.2.2]

Hardy, J. C., Macdonald, J. A., Schmeing, H., Andrews, H. R., Geiger, J. S., Graham, R. L., Faestermann, T., Clifford, E. T. H., Jackson, K. P. 1976. *Phys. Rev. Lett.* 37:133–36 [5.2.3]

Herrmann, G., Denschlag, H. O. 1969. *Ann. Rev. Nucl. Sci.* 19:1–32 [1]

Hofmann, S., Faust, W., Münzenberg, G., Reisdorf, W., Armbruster, P., Güttner, K., Ewald, H. 1979. *Z. Phys. A.* In press [4.3]

Hornshøj, P., Erdal, B. R., Hansen, P. G., Jonson, B., Aleklett, K., Nyman, G. 1975a. *Nucl. Phys. A* 239:15–28 [5.2.1]

Hornshøj, P., Hansen, P. G., Jonson, B., Ravn, H. L., Westgaard, L., Nielsen, O. B. 1974. *Nucl. Phys. A* 230:365–79 [4.3]

Hornshøj, P., Højsholt-Poulsen, L., Rud, N. 1976. See Cargèse Conference 1976, pp. 120–23 [4.1.1]

Hornshøj, P., Højsholt-Poulsen, L., Rud, N. 1977a. *Nucl. Phys. A* 288:429–38 [4.1.1]

Hornshøj, P., Nielsen, H. L., Rud, N. 1977b. *Phys. Rev. Lett.* 39:537–40 [3.1.2]

Hornshøj, P., Tidemand-Petersson, P., Bethoux, R., Caretto, A. A., Grüter, J. W., Hansen, P. G., Jonson, B., Hagberg, E., Mattsson, S. 1975b. *Phys. Lett. B* 57:147–49 [5.2.4]

Hornshøj, P., Wilsky, K., Hansen, P. G., Jonson, B. 1975c. *Phys. Lett. B* 55:53–55 [5.2.4]

Hornshøj, P., Wilsky, K., Hansen, P. G., Jonson, B., Nielsen, O. B. 1972a. *Nucl. Phys. A* 187:599–608 [5.1.1, 5.2.3]

Hornshøj, P., Wilsky, K., Hansen, P. G., Jonson, B., Nielsen, O. B. 1972b. *Nucl. Phys. A* 187:609–23 [5.1.1, 5.2.3]

Huber, G., Bonn, J., Kluge, H.-J., Otten, E.-W. 1976a. *Z. Phys. A* 276:187–202 [3.2.1]

Huber, G., Klapisch, R., Thibault, C., Duong, H. T., Juncar, P., Liberman, S., Pinard, J., Vialle, J.-L., Jacquinot, P. 1967b. *CR Acad. Sci.* 282:119–24 [3.2.3]

Huber, G., Thibault, C., Klapisch, R., Duong, H. T., Vialle, J. L., Pinard, J.,

Juncar, P., Jacquinot, P. 1975. *Phys. Rev. Lett.* 34:1209–11 [3.2.3]

Huber, G., Touchard, F., Büttgenbach, S., Thibault, C., Klapisch, R., Liberman, S., Pinard, J., Duong, H. T., Juncar, P., Vialle, J. L., Jacquinot, P., Pesnelle, A. 1978. *Phys. Rev. Lett.* 41:459–62 [3.2.3]

Husson, J. P., Liang, C. F., Richard-Serre, C. 1977. *J. Phys. Lett.* 38:L245–48 [4.2.2]

Ivanova, S. P., Kuliev, A. A., Salamov, D. I. 1976. *Yadernaya Fiz.* 24:278–85 [5.1.1, 5.2.3]

Johansen, K. H., Bonde Nielsen, K., Rudstam, G. 1973. *Nucl. Phys. A* 203:481–95 [5.2.1]

Johnson, W. H. 1970. See Leysin Conference 1970, pp. 307–19 [3.1.1]

Jonson, B., Hagberg, E., Hansen, P. G., Hornshøj, P., Tidemand-Petersson, P. 1976. See Cargèse Conference 1976, pp. 277–98 [3.1.2, 5, 5.1.2, 5.2.3]

Karnaukhov, V. A., Bogdanov, D. D., Dem'yanov, A. V., Koval, G. I., Petrov, L. A. 1974. *Nucl. Instrum. Methods* 120: 69–76 [2.1.1]

Kekelis, G. J., Zisman, M. S., Scott, D. K., Jahn, R., Vieira, D. J., Cerny, J., Ajzenberg-Selove, F. 1978. *Phys. Rev. C* 17:1929–38 [3.1.2]

Kerek, A., Holm, G. B., Carlé, P., McDonald, J. 1972. *Nucl. Phys. A* 195: 159–76 [4.2.1]

Kerek, A., Holm, G. B., de Geer, L.-E., Borg, S. 1973. *Phys. Lett B* 44:252–54 [4.2.1]

Kern, B. D., Weil, J. L., Hamilton, J. H., Ramayya, A. V., Bingham, C. R., Riedinger, L. L., Zganjar, E. F., Wood, J. L., Gowdy, G. M., Fink, R. W., Spejewski, E. H., Carter, H. K., Mlekodaj, R. L., Lin, J. 1976. In *Atomic Masses and Fundamental Constants*, ed. J. H. Sanders, A. H. Wapstra, 5:81–87. New York: Plenum [3.1.2]

Keyser, U., Berg, H., Münnich, F., Hawerkamp, K., Schrader, H., Pfeiffer, B. 1979. *Z. Phys. A.* 289:407–13 [3.1.2]

Keyser, U., Münnich, F., Weigert, L. J. 1978. *J. Phys. G* 4:L119–21 [3.1.2]

Khan, T. A., Lauppe, W.-D., Sistemich, K., Lawin, H., Sadler, G., Seliĉ, H. A. 1977. *Z. Phys. A* 283:105–20 [4.2.2]

Khan, T. A., Lauppe, W.-D., Sistemich, K., Lawin, H., Seliĉ, H. A. 1978. *Z. Phys. A* 284:313–17 [4.2.2]

Kirchner, R., Klepper, O., Nyman, G., Reisdorf, W., Roeckl., E., Schardt, D., Kaffrell, N., Peuser, P., Schneeweiss, K. 1977. *Phys. Lett. B* 70:150–54 [3.1.2]

Kirchner, R., Roeckl, E. 1976. *Nucl. Instrum. Methods* 139:291–96 [2.1.1, 2.1.2]

Kjelberg, A., Rudstam, G., eds. 1970. *The*

ISOLDE Isotope Separator On-Line Facility at CERN. Rep. CERN 70-3. 140 pp. [2.1.1]

Klapdor, H. V. 1976. *Phys. Lett. B* 65:35–38 [5.2.2]

Klapisch, R. 1969. *Ann. Rev. Nucl. Sci.* 19:33–60 [1, 2.1.1, 2.1.2]

Klapisch, R., Jacquinot, P. 1979. *Rep. Progr. Phys.* In press [3.2]

Klapisch, R., Prieels, R., Thibault, C., Poskanzer, A. M., Rigaud, C., Roeckl, E. 1973. *Phys. Rev. Lett.* 31:118–21 [3.1.1, 4.1.2]

Klapisch, R., Thibault, C., Poskanzer, A. M., Prieels, R., Rigaud, C., Roeckl, E. 1972. *Phys. Rev. Lett.* 29:1254–57 [4.1.2]

Klapisch, R., Thibault-Philippe, C., Détraz, C., Chaumont, J., Bernas, R., Beck, E. 1969. *Phys. Rev. Lett.* 23:652–55 [4.1.2]

Kleinheinz, P., Lunardi, S., Ogawa, M., Maier, M. R. 1978a. *Z. Phys. A* 284:351–52 [4.2.1]

Kleinheinz, P., Ogawa, M., Broda, R., Daly, P. J., Haenni, D., Beuscher, H., Kleinrahm, A. 1978b. *Z. Phys. A* 286: 27–29 [4.2.1]

Kodama, T., Takahashi, K. 1973. *Phys. Lett. B* 43:167–69 [5.2.2]

Kodama, T., Takahashi, K. 1975. *Nucl. Phys. A* 239:489–510 [5, 5.1.1, 5.2.1]

Kofoed-Hansen, O. 1976. See Cargèse Conference 1976, pp. 65–70 [2.1]

Kofoed-Hansen, O., Nielsen, K. O. 1951a. *Phys. Rev.* 82:96–97 [2.1, 2.1.2]

Kofoed-Hansen, O., Nielsen, K. O. 1951b. *K. Dan. Vidensk. Selsk. Mat. Fys. Medd.* 26(7), 16 pp. [2.1, 2.1.2]

Kondoh, T., Yamada, M. 1976. *Prog. Theor. Phys. Suppl.* 60:136–60 [5.1.1]

Kratz, K.-L., Rudolph, W., Ohm, H., Franz, H., Hermann, G., Ristori, C., Crancon, J., Asghar, M., Crawford, G. I., Nuh, F. M., Prussin, S. G. 1976. *Phys. Lett. B* 65:231–34 [5.2.2]

Kratz, K.-L., Rudolph, W., Ohm, H., Franz, H., Zendel, M., Herrmann, G., Prissin, S. G., Nuh, F. M., Shibab-Eldin, A. A., Slaughter, D. R., Halverson, W., Klapdor, H. V. 1979. *Nucl. Phys. A* 317: 335–62 [5.2.2]

Kühl, T., Dabkiewicz, P., Duke, C., Fischer, H., Kluge, H.-J., Kremmling, H., Otten, E.-W. 1977. *Phys. Rev. Lett.* 39:180–83 [3.2.1]

Kugler, E. 1973. See Andersson & Holmén, 1973, pp. 382–87 [2.1.1]

Lauppe, W.-D., Sistemich, K., Khan, T. A., Lawin, H., Sadler, G., Selič, H. A., Schult, O. W. B. 1977. *J. Phys. Soc. Jpn. Suppl.* 44:335–40 [4.2.1]

Lawin, H., Borgs, J. W., Fabbri, R., Grüter,

J. W., Khan, T. A., Lauppe, W.-D., Sadler, G., Selič, H. A., Shaanan, M., Sistemich, K. 1976. *Nucl. Instrum. Methods* 139:227–33 [2.2.2]

Lee, P. L., Boehm, F., Hahn, A. A. 1978. *Phys. Rev. C* 17:1859–61 [3.2.1]

Leysin Conference 1970. *Proc. Int. Conf. on the Properties of Nuclei Far From the Region of β Stability, Leysin, 1970. Rep. CERN 70-30,* Vols. 1–2, 1171 pp. [1]

Liberman, S., Pinard, J., Duong, H. T., Juncar, P., Vialle, J. L., Jacquinot, P., Huber, G., Touchard, F., Büttgenbach, S., Pesnelle, A., Thibault, C., Klapisch, R. 1978. *CR Acad. Sci. Ser. B* 286:253–55 [6.2]

Lombard, R. J. 1976. *Phys. Lett. B* 65:193–95 [4]

Lund, E., Rudstam, G. 1976. *Phys. Rev. C* 13:1544–51 [4.2.2]

Lynn, J. E. 1968. *The Theory of Neutron Resonance Reactions,* Oxford: Clarendon. 503 pp. [5]

Lysekil Conference 1966. *Nuclides Far Off the Stability Line,* ed. W. Forsling, C. J. Herrlander, H. Ryde, Stockholm: Almqvist & Wiksell, 686 pp. Also published in *Ark. Fys.* 36:1–686 [1, 2.1]

Macdonald, J. A., Hardy, J. C., Schmeing, H., Faestermann, T., Andrews, H. R., Geiger, J. S., Graham, R. L. 1977. *Nucl. Phys. A* 288:1–22 [5.2.3]

Macfarlane, R. D. 1967. *Ark. Fys.* 36:431–43 [3.1.2]

Macfarlane, R. D., Griffioen, R. D. 1963. *Nucl. Instrum. Methods* 24:461–64 [1]

Macfarlane, R. D., Siivola, A. T. 1964. *Phys. Rev. Lett.* 14:114–15 [3.1.2]

Malaguti, F., Uguzzoni, A., Verondini, E. 1971. *Lett. Nuovo Cimento* 2:629–34 [5.1.2]

Maripuu, S., ed. 1976. *Atomic Data Nucl. Data Tables* 17:411–608 [3.1]

Mlekodaj, R. L., Spejewski, E. H., Carter, H. K., Schmidt, A. G. 1976. *Nucl. Instrum. Methods* 139:299–303 [2.1.1]

Moltz, D. M., Äystö, J., Cable, M. D., von Dincklage, R. D., Parry, R. F., Wouters, J. M., Cerny, J. 1979. *Phys. Rev. Lett.* 42: 43–46 [4.1.1]

Murnick, D. E., Feld, M. S. 1979. *Ann. Rev. Nucl. Part. Sci.* 29:411–54 [3.2]

Myers, W. D. 1977. *Droplet Model of Atomic Nuclei,* New York: Plenum. 150 pp. [3.1]

Myers, W. D., Swiatecki, W. J. 1966a. *Ark. Fys.* 36:343–52 [3.1]

Myers, W. D., Swiatecki, W. J. 1966b. *Nucl. Phys.* 81:1–60 [3.1]

Ogawa, M., Broda, R., Zell, K., Daly, P. J., Kleinheinz, P. 1978. *Phys. Rev. Lett.* 41: 289–92 [4.2.1]

Otten, E. W. 1976. *Hyperfine Interactions* 2:127–49 [3.2]

Pappas, A. C., Sverdrup, T. 1972. *Nucl. Phys. A* 188:48–64 [5.2.2]

Pardo, R. C., Kashy, E., Benenson, W., Robinson, L. W. 1978. *Phys. Rev. C* 18:1249–53 [3.1.2]

Paris, P., Berg, V., Caruette, A., Obert, J., Putaux, J. C., Sarrouy, J. L. 1976. *Nucl. Instrum. Methods* 139:251–56 [2.1.1]

Patzelt, P. 1970. See Kjelberg & Rudstam 1970, pp. 83–91 [2.1.2]

Pedersen, J., Back, B. B., Bernthal, F. M., Bjørnholm, S., Borggreen, J., Christensen, O., Folkmann, F., Herskind, B., Khoo, T. L., Neiman, M., Pühlhofer, F., Sletten, G. 1977. *Phys. Rev. Lett.* 39:990–93 [4.2.1]

Proetel, D., Diamond, R. M., Kienle, P., Leigh, J. R., Maier, K. H., Stephens, F. S. 1973. *Phys. Rev. Lett.* 31:896–98 [4.2.2]

Proetel, D., Diamond, R. M., Stephens, F. S. 1974. *Phys. Lett. B* 48:102–4 [4.2.2]

Putaux, J. C., Obert, J., Aguer, P. 1974. *Nucl. Instrum. Methods* 121:615–16 [2.1.1]

Ragnarsson, I., Nilsson, S. G., Sheline, R. K. 1978. *Phys. Rep.* 45:1–87 [3.1, 4, 4.2.2]

Rasmussen, J. O., Thompson, S. G., Ghiorso, A. 1953. *Phys. Rev.* 89:33–48 [3.1.2]

Ravn, H. L. 1979. *Phys. Rep.* In press [2.1]

Ravn, H. L., Carraz, L. C., Denimal, J., Kugler, E., Skarestad, M., Sundell, S., Westgaard, L. 1976. *Nucl. Instrum. Methods* 139:267–73 [2.1.1]

Ravn, H. L., Sundell, S., Westgaard, L. 1972. *Phys. Lett. B* 39:337–38 [2]

Ravn, H. L., Sundell, S., Westgaard, L. 1975. *Nucl. Instrum. Methods* 123:131–44 [2.1.2, 2.1.3]

Rice, S. O. 1944–1945. *Bell Syst. Tech. J.* 23,24:162 pp. [5.1.2]

Robertson, R. G. H., Bowles, T., Freedman, S. J. 1976. See Cargèse Conference 1976, pp. 254–57 [4.1.1]

Roeckl, E., Dittner, P. F., Détraz, C., Klapisch, R., Thibault, C., Rigaud, C. 1974. *Phys. Rev. C* 10:1181–88 [4.1.2]

Roeckl, E., Kirchner, R., Klepper, O., Nyman, G., Reisdorf, W., Schardt, D., Wien, K., Fass, R., Mattsson, S. 1978. *Phys. Lett. B* 78:393–96 [3.1.2, 4.3]

Rud, N., Ward, D., Andrews, H. R., Graham, R. L., Geiger, J. S. 1973. *Phys. Rev. Lett.* 31:1421–23 [4.2.2]

Rudstam, G. 1966. *Z. Naturforsch. Teil A* 21:7–12 [2]

Rudstam, G. 1976. *Nucl. Instrum. Methods* 139:239–49 [2.1.1]

Rudstam, G., Shalev, S., Jonsson, O. C.

1974. *Nucl. Instrum. Methods* 120:333–44 [5.2.2]

Sakai, M., Rester, A. C. 1977. *Atomic Data Nucl. Data Tables* 20:441–74 [4.2.2]

Scharff-Goldhaber, G., Dover, C. B., Goodman, A. L. 1976. *Ann. Rev. Nucl. Sci.* 26:239–317 [4.2.2]

Schinzler, B., Klempt, W., Kaufman, S. L., Lochmann, H., Moruzzi, G., Neugart, R., Otten, E.-W., Bonn, J., von Reisky, L., Spath, K. P. C., Steinacher, J., Weskott, D. 1978. *Phys. Lett. B* 79:209–12 [3.2.4]

Schmeing, H., Hardy, J. C., Graham, R. L., Geiger, J. S., Jackson, K. P. 1973. *Phys. Lett. B* 44:449–52 [4.1.1]

Schmidt, K. H., Faust, W., Münzenberg, G., Clerc, H.-C., Lang, W., Pipelenz, K., Vermeulen, D., Wohlfarth, H., Ewald, H., Güttner, K. 1979. *Nucl. Phys. A* 318:253–65 [3.1.2]

Seth, K. K., Nann, H., Iversen, S., Kaletka, M., Hird, J., Thiessen, H. A. 1978. *Phys. Rev. Lett.* 41:1589–92 [3.1.2]

Shalev, S., Rudstam, G. 1974. *Nucl. Phys. A* 230:153–72; 235:397–409 [5.2.2] ˙

Sheline, R. K. 1960. *Rev. Mod. Phys.* 32:1–24 [4.2.2]

Sheline, R. K. 1976. See Cargèse Conference 1976, pp. 351–63 [4].

Sheline, R. K., Ragnarsson, I., Nilsson, S. G. 1972. *Phys. Lett. B* 41:115–21 [4.2.2]

Sheline, R. K., Sikkeland, T., Chanda, R. N. 1961. *Phys. Rev. Lett.* 7:446–49 [4, 4.2.2]

Sistemich, K. 1976. *Nucl. Instrum. Methods* 139:203–12 [2.2]

Sistemich, K., Lauppe, W.-D., Khan, T. A., Lawin, H., Selič, H. A., Bocquet, J. P., Monnand, E., Schussler, F. 1978. *Z. Phys. A* 285:305–13 [4.2.1]

Sistemich, K., Lauppe, W.-D., Lawin, H., Seyfarth, H., Kern, B. D. 1979. *Z. Phys. A* 289:225–26

Sistemich, K., Lauppe, W.-D., Lawin, H., Seyfarth, H., Kern, B. D. 1979. *Z. Phys. A* 289:225–26 [4.2.2]

Sistemich, K., Lauppe, G., Khan, T. A., Lawin, H., Lauppe, W.-D., Selič, H. A., Schussler, F., Blachot, J., Monnand, E., Bocquet, J. P., Pfeiffer, B. 1977. *Z. Phys. A* 281:169–81 [4.2.2]

Sorensen, R. A. 1966. See Lysekil Conference 1966, pp. 657–61 [4].

Specht, H. J., Weber, J., Konecny, E., Heunemann, D. 1972. *Phys. Lett. B* 41:43–46 [4.2.2]

Spejewski, E. H. et al 1973. See Andersson & Holmén 1973, pp. 318–23 [2.1.1]

Stacey, D. N. 1966. *Rep. Prog. Phys.* 29:171–215 [3.2.1]

Stippler, R., Münnich, F., Schrader, H.,

Bocquet, J. P., Asghar, M., Decker, R., Pfeiffer, B., Wollnik, H., Monnand, E., Schussler, F. 1978a. *Z. Phys. A* 284:95–102 [3.1.2]

Stippler, R., Münnich, F., Schrader, H., Hawerkamp, K., Decker, R., Pfeiffer, B., Wollnik, H., Monnand, E., Schussler, F. 1978b. *Z. Phys. A* 285:287–91 [3.1.2]

Stroke, H. H., Proetel, D., Kluge, H. J. 1979. *Phys. Lett. B* 82:204–7 [3.2.1]

Strutinsky, V. M. 1966. *Ark. Fys.* 36:629–32 [3.1]

Strutinsky, V. M. 1967. *Nucl. Phys. A* 95:420–42 [3.1]

Sundell, S., Hansen, P. G., Jonson, B., Kugler, E., Ravn, H. L., Westgaard, L. 1973. See Andersson & Holmén 1973, pp. 335–40 [2.1.1]

Swiatecki, W. J. 1966. *Ark. Fys.* 36:325–27 [3.1]

Takahashi, K. 1971. *Prog. Theor. Phys.* 45:1466–92 [5.1.1]

Takahashi, K. 1972. *Prog. Theor. Phys.* 47:1500–16 [5.1.1, 5.2.2]

Takahashi, K., Yamada, M. 1969. *Prog. Theor. Phys.* 41:1470–1503 [5.1.1]

Takahashi, K., Yamada, M., Kondoh, T. 1973. *Atomic Data Nucl. Data Tables* 12:101–42 [5.1.1, 5.2.1]

Talbert, W. L., Jr. 1970. See Leysin Conference 1970, pp. 109–41 [2.1]

Thibault, C., Klapisch, R., Rigaud, C., Poskanzer, A. M., Prieels, R., Lessard, L., Reisdorf, W. 1975. *Phys. Rev. C* 12:644–57 [2.1.1, 3.1.1, 4.1.2]

Truran, J. W., Cameron, A. G. W., Hilf, E. 1970. See Leysin Conference 1970, pp. 275–306 [5.2.3]

Vieira, D. J., Gough, R. A., Cerny, J. 1979. *Phys. Rev. C* 19:177–87 [4.1.1]

Visvanathan, A., Zganjar, E. F., Wood, J. L., Fink, R. W., Riedinger, L. L., Turner, F. E. 1979. *Phys. Rev. C* 19:282–84 [4.2.2]

Wapstra, A. H., Bos, K. 1977. *Atomic Data Nucl. Data Tables* 19:175–297; 20:1–126 [3.1, 3.1.1]

Wene, C.-O., Johansson, S. A. E. 1974. *Phys. Scr. A* 10:156–62 [5.2.5]

Westgaard, L., Aleklett, K., Nyman, G., Roeckl, E. 1975. *Z. Phys. A* 275:127–44 [3.1.2]

Westgaard, L., Żylicz, J., Nielsen, O. B. 1972. See Kern et al 1976, pp. 94–104 [3.1.2]

Wohn, F. K. 1977. *Rep. IS-4270.* Available from Nat. Tech. Info. Serv., US Dept. Commerce, Springfield, Ill. 63 pp. [2.1.1]

Wollnik, H. 1976. *Nucl. Instrum. Methods* 139:311–18 [1]

Wollnik, H., Wohn, F. K., Wünsch, K. D., Jung, G. 1977. *Nucl. Phys. A* 291:355–64 [4.2.2]

Yamada, M. 1965. *Bull. Sci. Eng. Res. Lab. Waseda Univ.* No. 31/32:146–55 [5.1.1]

Yoshizawa, Y., Noma, H., Horiguchi, T., Katoh, T., Amemiya, S., Itoh, M., Hisatake, K., Sekikawa, M., Chida, K. 1976. *Nucl. Instrum. Methods* 134:93–100 [2.1.1]

Zioni, J., Jaffe, A. A., Friedman, E., Haik, N., Schechtman, R., Nir, D. 1972. *Nucl. Phys. A* 181:465–76 [4.1.1]

Żylicz, J., Hansen, P. G., Nielsen, H. L., Wilsky, K. 1966. See Lysekil Conference 1966 [4.2.1]

Ann. Rev. Nucl. Part. Sci. 1979. 29:121–60

HOPES AND REALITIES FOR ×5604
THE (p,π) REACTION

D. F. Measday

Physics Department, University of British Columbia, Vancouver,
British Columbia, Canada V6T 1W5

G. A. Miller

Department of Physics, University of Washington, Seattle, Washington 98195

CONTENTS

1   INTRODUCTION

1.1   Phenomenology

The study of pion production in proton-nucleus reactions has produced
some concrete advances along with much frustration and some confusion.
At present there is intense activity in this field, both theoretical and
experimental, and it seems worthwhile to review both the accomplish-
ments and problems in order to aid future activity.

   We limit ourselves to a discussion of the (p,π) reaction where the final
state is a coherent nucleus in its ground or excited state. In this way
we ignore the vast majority of proton-induced reactions in which a pion

121

is produced, as normally the nucleus is shattered and several different fragments are emitted. The reaction that we discuss is thus distinguished by the fact that the pion takes away all, or almost all, the available kinetic energy. Although most experiments have studied the (p,π^+) reaction, there are several publications on the (p,π^-) reaction and a few on the (p,π^0) reaction. Similarly there are measurements on related reactions such as (n,π^0), (π^+,p), and (π^-,n), and we shall draw freely on such results and make comparisons where useful.

The pion production reaction has had several applications. The earliest use was in the verification of the spin of the pion by comparing the cross section for the reaction $pp \rightarrow \pi^+ d$ with that for $\pi^+ d \rightarrow pp$ and by using detailed balance to determine $(2s_\pi + 1)$ (Durbin et al 1951, Clark et al 1951). More recently this comparison was used to test detailed balance to the level of 5% (e.g. Rose 1967). Furthermore the observation of the $\pi^- d \rightarrow n + n$ reaction, for pions captured from an atomic s state, implied that the intrinsic parity of the pion was negative (Panofsky et al 1951, Chinowsky & Steinberger 1954).

The next significant advance came with the measurements on the two reactions $p + d \rightarrow {}^3H + \pi^+$ and $p + d \rightarrow {}^3He + \pi^0$. By detecting the triton and helium ions, the ratio of the cross sections was found to be approximately $2:1$ (Harting et al 1960) confirming the value predicted by isospin

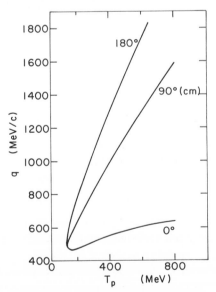

Figure 1 Momentum transfer versus proton (lab) kinetic energy, ${}^{16}O(p,\pi^+){}^{16}O_{g.s.}$.

invariance [when minor corrections caused by the π^+ Coulomb interaction and $\pi^+ - \pi^0$ mass difference are included (Köhler 1960)]. Similarly, searches for forbidden reactions have failed, for example, $d + d \nrightarrow {}^4He + \pi^0$.

We now discuss the physics of the (p,π) reaction on nuclei. There is a simple classical analogy—bremsstrahlung. An electron moving with constant velocity does not radiate. However, application of an electromagnetic field causes an acceleration and photons are produced. Similarly a free proton does not decay into a pion and a neutron, i.e. the reaction $p \rightarrow n + \pi^+$ cannot take place because energy and momentum cannot be simultaneously conserved. Therefore a nucleus or another nucleon is required to supply the necessary strong field, which may be applied to the incoming proton or outgoing pion and neutron.

The strong field applied by a nucleon or nucleus to the incident proton, for example, causes it to accelerate and acquire momenta different from the free space value. In quantum mechanical language we say that the proton is off-shell, i.e. $\mathbf{p}^2 \neq E^2 - m^2$. Now an off-shell proton can emit an on-shell pion and neutron, or an on-shell proton can emit an off-shell pion and become an on-shell neutron. Thus an appropriate question to ask about meson production is which particle or particles are the most off-shell, and why?

The truly unique aspect of pion production or absorption is that the reaction takes place at very high momentum transfer. Figure 1 gives the momentum transfer to the nucleus for the reaction ${}^{16}O(p,\pi^+){}^{17}O_{g.s.}$ for pion angles of $0°$, $90°$, and $180°$ c.m. The minimum momentum transfer in this reaction is about 480 MeV c^{-1}, because the pion mass must be created from the kinetic energy of the proton. This transfer is considerably larger than that of most nuclear reactions, and more important still, it is much larger than typical momenta of single nucleons in the nucleus.

The nuclear Fermi momentum is about 270 MeV c^{-1}; single nucleon wave functions drop rapidly with increasing momentum at large momenta. The change from 270 MeV c^{-1} to 500 MeV c^{-1} gives a decrease of a factor of 100 or more in some typical nuclear wave functions (Elton & Swift 1967, Negele 1970) shown in Figure 2. These wave functions are consistent with the binding energy and root mean square radii, but the short-distance behavior is not constrained by experimental observables. Thus the values of the wave functions at large momentum transfer shown in Figure 2 are essentially guesses; the probability of a nucleon having a momentum of 400 or 500 MeV c^{-1} is generally expected to be very small.

The expectation that the (p,π) reaction will be a rarity is borne out by experiment; for example, at 180 MeV the reaction ${}^{40}Ca(p,\pi^+){}^{41}Ca_{g.s.}$

has a total cross section of 0.7 μb (Pile 1978) compared to the total reaction cross section of 0.5 b; this factor of 10^6 complicates both the theoretical and experimental aspects of the subject.

An early hope was that the (p,π) reaction would produce information about high momentum components of nuclear wave functions. Such information is not obtained in a simple way, however, because there are several reaction mechanisms. On the other hand, one is able to study a system by producing in free space one of the constituents that bind the system. This is in contrast with hadron-hadron collisions at high energies, in which individual gluons cannot be produced. Furthermore, a successful treatment of the (p,π) reaction will ultimately require handling the relativistic aspects of a many-body system of several components and may have a wide area of application.

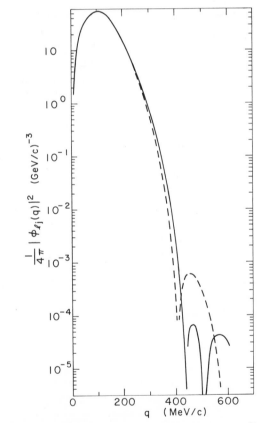

Figure 2   Modulus squared of momentum-space wave functions, $\tilde{\phi}_{lj}(q)$, for the p-shell ($l = 1, j = \frac{3}{2}$) from Negele 1970 (*solid line*) and Elton & Swift 1967 (*dashed curve*).

Most of the recent data has been accumulated at near-threshold energies. For example, pion production experiments with nuclear targets were performed at 140–200 MeV at Indiana (Bent et al 1978, Pile et al 1979), 185 MeV at Uppsala (Dahlgren et al 1971, 1973a,b, 1974, Höistad et al 1978, 1979a,b) and 200 MeV at TRIUMF (Auld et al 1978). These data show a few qualitative trends. (See Figures 8–17.)

The differential cross sections for the (p,π^+) reaction generally vary rapidly with angle (Figures 8–10). The shape of the angular distribution depends on the final state wave function for the transferred neutron (Figure 10) and also changes dramatically with energy (Figure 9). The (p,π^+) reaction excites many final states; often final states of complicated nuclear structure have larger cross sections than final states reached by a simple transfer of one neutron. Another general feature is that for reactions on a given target final states having high angular momentum are preferentially excited (Figure 10). Not much is known about pion production using heavy nuclei as targets, but cross sections tend to decrease from several hundred nb sr^{-1} for the peak in light nuclei to about 50 nb sr^{-1} for ^{90}Zr and ^{208}Pb.

At higher energies there are results from SATURNE, several experiments between 320 and 810 MeV (Beurtey et al 1978b, Aslanides et al 1978), and more recently from LAMPF. One very important feature of the SATURNE data is a direct comparison of the reactions ^{9}Be$(p,\pi^+)^{10}$Be and ^{9}Be$(d,p)^{10}$B at equivalent momentum transfer for which the spectra and angular distributions are similar (Figure 7).

The (p,π^-) reaction has also been observed for a few light nuclear targets (e.g. Couvert et al 1978, Dahlgren 1973a,b). The angular distributions, in contrast with those for π^+ production, are typically flat in shape and of the order of a few tens of nb sr^{-1} (Figure 15). The energy dependence of the (p,π^-) cross section is not yet known. The (p,π^-) reaction could also be used as a purely spectroscopic tool to reach nuclei that are hard to investigate with more traditional probes. Two specific examples, although many others could be quoted, are ^{12}C$(p,\pi^-)^{13}$O and ^{28}Si$(p,\pi^-)^{29}$S. Nothing is currently known about ^{29}S, and the only investigation of the excited states of ^{13}O used the (p,π^-) reaction (Couvert et al 1978). However, many levels are known in their mirror pairs ^{13}B and ^{29}Al.

Recent dramatic advances for the (p,π^+) reaction have come with the introduction of polarized proton beams. For many years strong asymmetry effects in the reaction $p + p \rightarrow \pi^+ + d$ were known, i.e. that for transversely polarized protons more pions are produced on one side of the beam than the other. Such effects can reach 60% (G. Jones 1978; Figure 5) but even larger asymmetries are now observed in nuclear reactions (Auld et al 1978). A strange aspect of these discoveries is that significant changes in the angular variation of the asymmetry from nucleus to nucleus have not yet been observed (Figure 16), and this suggests some universal explanation.

Apart from stripping (d,p) and pick-up (p,d) mentioned already, the (p,γ) and (γ,p) reactions share many similarities with the (p,π) reaction, including high momentum transfer. Knowledge of both photo- and pion production reactions will be increased by comparison of the two. The (γ,p) and (p,γ) (high energy) reactions are much harder to study experimentally, although some data from MIT (Matthews et al 1977), Bonn (Arends et al 1978), and Indiana (Kovash et al 1978) are now available. No direct comparison with (p,π) has yet been made, although some calculations of (p,γ) are now available (Londergan et al 1976, Schoch 1978).

Three of the more recent reviews on (p,π) reactions, two by Höistad (1977, 1978) and one by Aslanides (1976) are excellent phenomenological introductions to the subject. Also relevant is the review of Dieperink & de Forest (1975) on knock-out processes and removal energies.

1.2   Theoretical Approaches

The various theoretical approaches to this reaction are briefly described here. Three approaches have been used to calculate cross sections in the energy domain of $T_p < 300$ MeV, sometimes termed the subthreshold energy region. The first model is called the single-nucleon model (SNM) because the momentum transfer is taken up by a single-nucleon wave function and is displayed in Figure 3a and b. The high momentum components of typical nuclear wave functions are very small and also poorly determined. Most SNM treatments allow the incident proton to become slightly off-shell by virtue of its interaction with the average field (optical potential) of the nucleus. The other two models are the distorted-wave Born approximation (DWBA) and the two-nucleon mechanism (TNM), which both allow the large momentum transfer to occur via the exchange of a virtual pion between the incident proton and a target nucleon. These models may be called meson off-shell models.

In the DWBA the proton emits an off-shell meson that is scattered onto its mass shell by the other nucleons. The scattering probability is estimated by using a pion-nucleus wave function as obtained from an effective pion-nucleus potential called the optical potential. Because of the lack of high momentum content of typical bound-state wave functions (and because of the lack of low momentum content of initial scattering wave functions), pion distortions play a significant and even dominant role in the production process. Pion distortion effects influence the magnitude of cross sections and their angular distributions to an extent that modifies the dynamics of the reaction mechanism from those obtained from a simple plane wave stripping reaction (G. A. Miller & Phatak 1974). Indeed DWBA calculations in which a realistic pion-nucleon interaction is employed to generate the π-nucleus optical potential are much closer in spirit to TNM calculations than those of the SNM.

In the TNM one explicitly calculates a two-nucleon pion-production operator that is inserted between appropriate initial and final nuclear wave functions. In this way, pion and other meson exchange processes may be calculated from an assumed Hamiltonian. Some of the effects that may be included in a TNM calculation are shown in Figure 3. The various diagrams represent pion rescattering (Figure 3c); pion emission from the target (Figure 3d); rho meson exchange (Figure 3e); effects of nucleon-nucleon correlations (Figure 3f); and Mandelstam's (1958) diagram (Figure 3g). Note that the term of Figure 3c is explicitly included in a DWBA calculation if one uses a pion-nucleus optical potential that is derived from a pion-nucleon T-matrix with reasonable off-shell properties. A correct evaluation of the TNM is very difficult

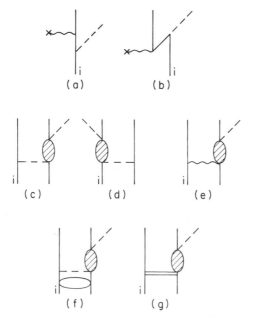

Figure 3 Pion production reaction mechanisms. The pions are represented by dashed lines, nucleons by solid lines and the rho meson by a wavy line. The "x" with a wavy tail represents the nucleon-nucleus off-shell transition matrix. The crosshatched ellipse represents off-shell meson-nucleon scattering. The double bar represents any two-nucleon process leading to a delta-nucleon intermediate state.

whereas DWBA calculations are straightforward (G. A. Miller 1977). The only evaluations of the TNM have used approximations in which the pion production operator is proportional to a delta function in the relative two-nucleon coordinate (Dillig & Huber 1977, Reitan 1971, 1972, Grossman et al 1974).

In the meson off-shell models the terms of Figure $3c-f$ are carefully treated, but those of Figure $3a$ and b are treated crudely (if at all). On the other hand, detailed calculations of the terms of Figure $3a$ and b are made in the nucleon off-shell models, but the other terms of Figure 3 are essentially ignored. Both nucleon and meson off-shell effects are relevant, but there is no present approach in which both aspects are treated consistently and precisely.

An interesting comparison of the three approaches can be made for complex final states. There is much evidence that two-particle one-hole states are strongly excited in pion production reactions (for example, Dahlgren et al 1974, Amann et al 1978). The TNM may be directly applied in this case because the two-nucleon effective operator can convert an incident fast proton into a bound nucleon and also excite a target nucleon. The DWBA treatment must be modified so as to allow the pion-nucleus optical wave function to have components in which the target nucleus has been excited. This can be done without great difficulty (Eisenberg et al 1974, G. A. Miller 1974). In the SNM the process of exciting a nuclear state does not occur, unless the initial proton-nucleus wave function has components in which the target is excited.

Another interesting case is the (p,π^-) reaction that requires the target to acquire two units of charge, while gaining only one nucleon. The TNM may be directly applied through the $p+n \rightarrow p+p+\pi^-$ components of the effective two-body operator. Similarly the DWBA may be generalized so that the final π^- wave function includes components with a virtual π^0. In the SNM one needs the proton to become a neutron via an initial charge-exchange reaction. The neutron then emits the π^-. Another possibility is that the final state consists of a Δ^{++} coupled to the rest of the nucleus. In this case a proton could emit a π^- and become a Δ^{++}.

Although there is much disagreement on details of the theory there is one point on which there is virtually unanimous agreement—the pion optical potential is very important even in TNM calculations. Meson production is extremely sensitive to the short-distance behavior of the pion wave function. Calculations that employ the Kisslinger or Lorentz-Lorenz models (which adequately describe the angular distributions for pion-nucleus elastic scattering) lead to calculated (p,π) cross sections that are too large by a typical factor of about 30 (W. B. Jones & Eisenberg

1970, Keating & Wills 1973, G. A. Miller 1974, Rost & Kunz 1973, LeBornec et al 1976). Considerable improvement is obtained if one constructs the pion optical potential from a pion-nucleon interaction that has a smooth off-shell behavior (G. A. Miller & Phatak 1974).

The use of the (p,π) reaction to provide constraints on the pion-nucleus optical potential may become its most outstanding contribution to knowledge of meson-nucleus dynamics.

2   PION PRODUCTION IN NUCLEON-NUCLEON COLLISIONS

For a proton of 500 MeV, $\lambda = 0.18$ f, which is much smaller than the average internucleon distance in the nucleus (1.9 f). Thus the incoming proton sees the nucleus as distinguishable nucleons, and therefore normal pion production on a nucleus is related to pion production in free nucleon-nucleon collisions. For the (p,π) reaction the recoil nucleus acts as a whole and this adds many complications, yet somewhere the pion must come from a nucleon-nucleon collision even if the other effects mask this simple birth. We therefore first discuss the phenomenology of nucleon-nucleon inelastic scattering.

Seven reactions produce a single pion and are listed in Table 1. Threshold is at about 300 MeV and significant pion production is detectable at 400 MeV. Double-pion production has a threshold at 600 MeV but is negligible up to 1000 MeV so we neglect it here. Also given in Table 1 is the isospin decomposition for the total cross section of each reaction (such simple decomposition is not possible for the differential cross sections). By convention the subscripts in σ_{01} mean that the initial nucleon-nucleon state has $T = 0$ and the final nucleon-nucleon state has $T = 1$.

Table 1   Isospin decomposition for the total cross sections of single-pion production

Reaction	Isospin decomposition
$p+p \rightarrow \pi^+ + d$	$\sigma_{10}(d)$
$p+p \rightarrow \pi^+ + p + n$	$\sigma_{10}(np) + \sigma_{11}$
$p+p \rightarrow \pi^0 + p + p$	σ_{11}
$n+p \rightarrow \pi^0 + d$	$\frac{1}{2}\sigma_{10}(d)$
$n+p \rightarrow \pi^0 + n + p$	$\frac{1}{2}[\sigma_{10}(np) + \sigma_{01}]$
$n+p \rightarrow \pi^+ + n + n$	$\frac{1}{2}[\sigma_{11} + \sigma_{01}]$
$n+p \rightarrow \pi^- + p + p$	$\frac{1}{2}[\sigma_{11} + \sigma_{01}]$

Now the simplest mechanism for pion production is to pass through an intermediate $\Delta(1232)$ $[J = \frac{3}{2}, T = \frac{3}{2}]$, i.e. $N+N \to N+\Delta \to N+N+\pi$. Since the nucleon has $T = \frac{1}{2}$, for this model the intermediate state has to be $T = 1$ and so if isospin is conserved the initial state must also be $T = 1$, thus $\sigma_{01} = 0$. Although there is much experimental confusion, the latest data indicate a very small value for σ_{01}. This means that only three parameters are needed to describe the seven reactions. The values of these three cross sections as a function of energy are given in Lock & Measday (1970). The largest near threshold is $\sigma_{10}(d)$, i.e. the reaction $(p+p \to \pi^+ +d)$, which thus immediately sets it up as a possible archetype for the (p,π^+) reaction, and so we discuss it in more detail.

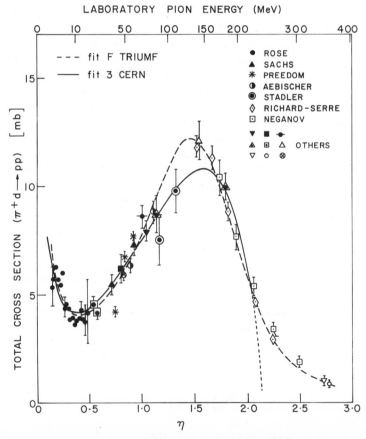

Figure 4 Total cross sections for $\pi^+d \to pp$. The data are from Spuller & Measday (1975) and include data from Aebischer et al (1976), Neganov & Parfenov (1958), Preedom et al (1978), Sachs et al (1958), and Stadler (1954).

The $(p + p \rightarrow \pi^+ + d)$ reaction can be analyzed in terms of transitions between angular momentum states. The initial phenomenological approach was that of Gell-Mann & Watson (1954), who for regions close to threshold hypothesized that only s- and p-wave pions would be produced. This leads to the following expressions:

$$\sigma_{(pp \rightarrow \pi d)} = \alpha\eta + \beta\eta^3$$

$$\frac{4\pi(X + 1/3)}{\beta\eta^3} \frac{d\sigma}{d\Omega} = X + \frac{(X + 1/3)}{\eta^2} \frac{\alpha}{\beta} + \cos^2 \theta,$$

1.

where η = pion momentum in the center of mass in units of $m_\pi c$. The parameters α, β, and X are related to the transition amplitudes, and if the meson wavelength is longer than the critical distance for meson production, these parameters should vary slowly with η. As there were few measurements available at that time, Gell-Mann & Watson (and many subsequent authors) were forced to assume that these parameters were constants.

It took 15 years to accumulate data sufficient to show that X was not a constant (Richard-Serre et al 1970), but no specific energy dependence could be recommended. Then Afnan & Thomas (1974) proposed that α should also have an energy dependence. Their Fadeev calculations showed that a reasonable parametrization was $\alpha = \alpha_0 + \alpha_1\eta$ with various deuteron wave functions giving a similar energy dependence, i.e. $\alpha_1 \approx -0.2$ mb, but that the values of α_0 varied significantly. Later it was shown that the total cross-section data favored $\alpha_0 = 0.25 \pm 0.05$ mb (Spuller & Measday 1975), a value more pleasing than earlier estimates because it also satisfied the low energy pion relations of Brueckner et al (1951) (Rose 1967). The uncertainty of such determinations of α_0 can be seen clearly in Figure 4. Fit F has $\alpha_0 = 0.27 \pm 0.04$ mb, while Fit 3 has $\alpha = 0.175 \pm 0.20$ mb. The earlier determination Fit 3 of CERN assumed that α was constant, as was normal at that time. Note carefully that the region most sensitive to α is $\eta < 1$ where the curves differ very little.

There is also strong evidence for pions in higher partial waves, but the effects can usually be considered as minor corrections to the basic model. The polarization data of Dolnick (1971) demanded d-wave pions at $T_p \approx 400$ MeV, while the cross-section data of Richard-Serre et al (1970) required f waves by $T_p \approx 500$ MeV; however, as more precise data appear these limits approach the threshold at 288 MeV. It has long been expected that measurements using polarized proton beams would reveal small amplitudes with a pion of large angular momentum. If a polarized proton beam is used, there can be an azimuthal dependence in the angular distribution of the produced pions. More pions can be produced on one side of the beam than on the other. Recent data, in which this asymmetry is measured, are now of excellent quality as can be seen in Figure 5. These data of G. Jones (1978) show that d-wave pions are important at $T_p = 350$ MeV, and so the phenomenological model must be expanded. More partial waves require more data to fix the variables. It is now necessary to parametrize the differential cross section for polarized protons as

$$\frac{d\sigma}{d\Omega} = \gamma_0 + \gamma_2 \cos^2 \theta + \gamma_4 \cos^4 \theta$$

$$+ P \sin \theta (\lambda_0 + \lambda_1 \cos \theta + \lambda_2 \cos^2 \theta + \lambda_3 \cos^3 \theta), \qquad 2.$$

in which P is the incident nucleon's degree of polarization in the x direction. (The incident nucleon's momentum is parallel to the positive

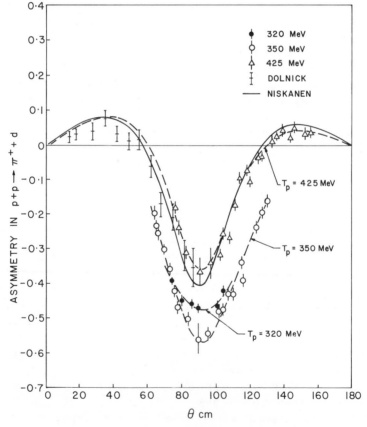

Figure 5   Asymmetry in $p + p \rightarrow \pi^+ d$ (Jones 1978). The solid curve is from Niskanen (1978).

z direction.) If we take $\sigma_L = d\sigma/d\Omega(P = 1)$ and $\sigma_R = d\sigma/d\Omega(P = -1)$, the asymmetry of Figure 5 is given by $(\sigma_L - \sigma_R)/(\sigma_L + \sigma_R)$. The parameter γ_4 is dominated by p-f interference, λ_1 by s-d interference, λ_2 by p-d interference, and λ_3 by d-d effects. The parametrization of Equation 2 was introduced by Mandl & Regge (1955).

Because of these complications and the lack of data to fix such a multitude of parameters, a phenomenological analysis becomes impossible and a model calculation becomes necessary. Almost all calculations have used the well-established importance of the $\Delta(1232)$ in the final state as evidenced by the dominance of the transition $^1D_2 \to (^3S_1)p_2$ (in pp $\to$ dπ) near resonance. The first attempt at a comprehensive description of pion production was that of Mandelstam (1958), who assumed that the production process was given by the Feynman graph of Figure 3g, in which the rectangle represents any interaction between the two nucleons and the cross-hatched oval region at the top represents the π-nucleon interaction that is dominated by the (3,3) resonance.

The important simplification that the contribution of the rectangle does not vary with energy was made, and the entire diagram was taken to be proportional to $\exp[i\delta(q)]\sin\delta(q)/q^2$ where $\delta(q)$ is the (3,3) phase shift and q is the pion-nucleon relative momentum (in the final state). A reasonable representation of the data in the resonance region was obtained by using only three free parameters to represent various wavefunction overlap integrals.

If one is to extract information about the deuteron wave function or use the understanding gained from the pp $\to$ dπ^+ reaction for treating pion production on heavy nuclei, a more detailed treatment is necessary. During the last twenty years much effort has been spent studying the details of Mandelstam's rectangle and in performing the overlap integrals using modern nucleon-nucleon and deuteron wave functions.

Dealing with the rectangle has proved to be a difficult task. At present there is agreement that the data can be explained almost solely by using a treatment in which the rectangle is taken as a single-meson exchange, but there is disagreement on the details. Because of the pseudoscalar nature of the pion, the integral over the momentum (K) of the intermediate pion includes a factor K^2 in the numerator, which tends to make the value of the pion-exchange term too large to fit the experimental data. This problem can be alleviated by using a cutoff on the pion's momentum or by including a rho-exchange graph in addition to the pi-exchange one. In the latter case the destructive interference of the two terms reduces the calculated cross sections.

Schiff & Tran Thanh Van (1968) performed a covariant dispersion relation calculation including pion exchange; good agreement was obtained when the Ferrari & Selleri (1961, 1962, Amaldi 1967) cutoff was used. Goplen et al (1974) performed a one-pion-exchange contribution calculation using a cutoff that was effective at about 300 MeV c^{-1}. Brack et al (1977) included rho exchange together with pion exchange and obtained a reasonable fit to total cross-section data. Green & Niskanen (1976) and Niskanen (1977, 1978) performed pion- and rho-exchange calculations in which the delta is taken as part of an N-Δ component of the initial state. These authors constrain the various coupling constants and form factors by including the effects of the virtual delta in their nucleon-nucleon potential and by adjusting parameters so that the total nucleon-nucleon potential reproduces the phase shifts of the Reid potential at low energies. Niskanen also obtains an adequate representation of the $p\vec{p} \to d\pi^+$ data. A recent paper by Alberg et al (1978) uses a nonrelativistic dispersion relation approach to show that a proper treatment of the off-energy shell behavior of the pion-nucleon scattering amplitude leads to an effective cutoff, which significantly damps the pion momenta at about 500 MeV. These authors obtain cross sections that compare reasonably well with the data, yet no rho exchange is necessary. Keister & Kisslinger (1979) studied the effects of using a Lorentz-covariant off-shell pi-nucleon T-matrix.

Although there is disagreement on the question of the importance of rho-meson exchange terms, it seems that the pp $\to d\pi^+$ reaction is well understood in the $(3,3)$ resonance region.

Most recent theoretical work has concentrated on the resonance region. However, Delacroix & Gross (1977) showed that the effects of including the antinucleonic components of the deuteron wave function on the pp $\to d\pi^+$ reaction near threshold are small but nonnegligible.

In addition to testing models of meson-nucleon dynamics, the reaction np $\to d\pi^0$ has been used to verify isospin conservation. Note (Table 1) that

$$(\text{np} \to \pi^0 \text{d}) = \tfrac{1}{2}\sigma_{10}(\text{d}) = \tfrac{1}{2}\sigma(\text{pp} \to \pi^+\text{d}). \qquad 3.$$

This relation has two consequences. First, of course, the value of the cross section can be verified, but unfortunately neutron-induced reactions are renowned for the difficulty of getting an absolute normalization. However, the relative energy dependence of the two reactions can be successfully checked (Bartlett et al 1970). Easier to check experimentally is the form of the differential cross section. For the reaction pp $\to \pi^+$d the differential cross section has no terms of odd power in cos θ because of the identical nature of the two protons in the initial state. However, for the reaction np $\to d\pi^0$, one might imagine that the cross section has a slight asymmetry about 90°, caused by isospin mixing. The two existing experimental results (Bartlett et al 1970, Wilson et al 1971) show that isospin-violating effects are of the order of 0.5%. Calculations (Cheung et al 1979) indicate that any observable effects will also be about 0.5%.

Because of this confidence in isospin, it is not uncommon to turn the problem around. In experiments on np elastic scattering, the reaction $np \to \pi^0 d$, which is detected at the same time, is used as a means of calibrating the neutron intensity, using the "known" cross section obtained via the reaction $pp \to \pi^+ d$ (Evans et al 1976).

3   FEW-BODY TARGETS

The lightest nuclei have been popular targets for the (p,π) reaction as they offer certain technical advantages. Take, for example, the following: $p+d \to \pi^+ + t$ and $p + {}^3He \to \pi^+ + {}^4He$. In these reactions the separation between the ground state and the unbound continuum is many MeV and there are no bound excited states. Thus a poor resolution spectrometer is quite sufficient to define the reaction. Furthermore, because the recoiling final nucleus is light, it takes away considerable momentum and can be detected, if need be. Another advantage is the relatively large cross section.

Particular favorites have been the fraternal twins $p+d \to \pi^+ + t$ and $p+d \to \pi^0 + {}^3He$. Isospin invariance tells us that the ratio of these two cross sections should be 2:1. Fortunately the recoil triton and helium three can be detected as long as the energy of the incident proton is high enough. A classical experiment done at the CERN SC used 591 MeV protons; the recoil particles were detected at the same laboratory angle and then the rates were compared (Harting et al 1960). Two minor corrections have to be made to the data; first, the conversion to center-of-mass solid angles, and second, the conversion to the same angle in the center of mass.

The naive ratio of 2.00 as given by isospin invariance has to be corrected for electromagnetic effects due to the difference of nuclear wave functions and the $\pi^+ - \pi^0$ mass difference. The corrections were estimated by Köhler (1960), who obtained good agreement with the experimental ratio of 2.26 ± 0.11. Similarly, slightly less precise experiments have also been made at 743 MeV (Booth 1963) and at 800 MeV (Lo 1977).

Experiments have now been performed on the reaction ${}^3He(\pi^-,n)^2H$ so quite interesting comparisons can be made. Unfortunately, at present, results do not exist at the same center-of-mass energy so corrections have to be applied. The two reactions have the same cross section, within the uncertainty in making the necessary corrections. Some earlier pd $\to$ $t\pi^+$ data at 470 MeV suggested a sharp backward peak at 180° (Dollhopf et al 1973), but more recent measurements have not confirmed this effect

(W. C. Olsen and E. G. Auld, private communication). There are also some data for the $p + d \to {}^3\text{He} + \pi^0$ reaction (for example, Banaigs et al 1973, Fredrickson et al 1976).

Another reaction of importance for isospin invariance is $d + d \to \alpha + \pi^0$ in which charge symmetry is tested (Henley 1969). Several experiments searched for this reaction by using deuteron energies between 400 and 800 MeV (Akimov et al 1962, Poirier & Pripstein 1963, Banaigs et al 1974). The dynamics of this reaction are complicated so that analyses are difficult (Greider 1961). Crudely estimated by using a comparison with $dd \to \gamma\alpha$ (Arends et al 1976), the breaking of charge symmetry is less than about one percent.

The reactions ${}^4\text{He}(\pi^-,n){}^3\text{H}$ (Källne et al 1978) and ${}^3\text{He}(p,\pi^+){}^4\text{He}$ (Tatischeff et al 1976) are compared in Figure 6. Because the data are available at similar center-of-mass energies, they can be compared directly using detailed balance. Both the absolute cross section and the shape of the angular distribution are in excellent agreement. We refer the reader to Fearing (1979) for a more detailed discussion of pion production from few-body targets.

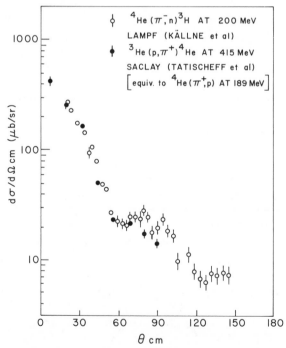

Figure 6 Comparison of ${}^4\text{He}(\pi^-,n){}^3\text{H}$ and ${}^3\text{He}(p,\pi^+){}^4\text{He}$ from Källne et al (1978) and Tatischeff et al (1976).

4   THE (p,π^+) REACTION ON NUCLEI

4.1   *Experimental Methods*

The topic of the (p,π^+) reaction on nuclei at subthreshold energies was initiated by the Uppsala group of Dahlgren, Grafstrom, Höistad, and Åsberg (Dahlgren et al 1971, 1973a,b). This initial work started with a simple 90° spectrometer with a hodoscope of only four channels, which meant that a spectrum at one angle took as long as 24 hours to obtain. Because of the interest generated by their successful efforts, sophisticated spectrometers have been designed specifically to study such reactions. Measurements have now been made at many other laboratories including Indiana, LAMPF, Orsay, Saturne, and TRIUMF.

The first requirement is a good quality proton beam, preferably polarized and variable in energy. Indiana can cover the region from threshold (between 130 and 150 MeV) up to 200 MeV. Both the Orsay and Uppsala synchrocyclotrons are now being modified for variable proton energy up to 200 MeV. TRIUMF can cover 180–520 MeV, Saturne can cover from threshold up to 3 GeV, while LAMPF, in theory, can cover all energies up to 800 MeV but in practice energies lower than 800 MeV are seldom scheduled. At all these laboratories the pion is detected in a magnetic spectrometer with a typical resolution of about 200 keV to distinguish different excited states in the residual nucleus. The spectrometer should have a short flight path, especially for work close to threshold because of the decay of the pions that have a life-time of 26 ns. For example, the Uppsala spectrometer has a flight path of 1.3 m and gives a loss of 20% of the pions at 25 MeV. Several laboratories have one spectrometer for low energy and one for high energy (>100 MeV) pions.

To give an impression of the quality of the data, we show in Figure 7 a spectrum for the reaction $^6\text{Li}(p,\pi^+)^7\text{Li}$ at $T_p = 600$ MeV (Bauer et al 1977) and compare it with a spectrum for the reaction $^6\text{Li}(d,p)^7\text{Li}$ (Beurtey et al 1978a) at $E_d = 698$ MeV for a similar momentum transfer. It is clear that the reactions are giving similar information about the excited states in ^7Li. Several levels can be distinguished and angular distributions can be determined.

Apart from the special case of the reaction $(\pi^+ + d \rightarrow p + p)$, there have been far fewer experiments on the (π^+,p) reaction. The limitation has of course been the low flux and poor energy resolution of pion beams from synchrocyclotrons. With the meson factories now reaching full intensity, one can anticipate a greater interest in this reaction. The first study to separate nuclear levels was that of Amato et al (1974) who used 70-MeV pions from the 184″ SC at Lawrence Berkeley Laboratory (LBL). With targets of Li, Be, C, and O, they achieved an energy resolution of 3 MeV and obtained several points on angular distributions. A similar energy resolution was obtained by Bachelier et al (1977) for the reaction $^{16}O(\pi^+,p)^{15}O$ with $E_\pi = 66$ MeV. Recent experiments at LAMPF achieved better than 2-MeV resolution (Amann et al 1978). Newer

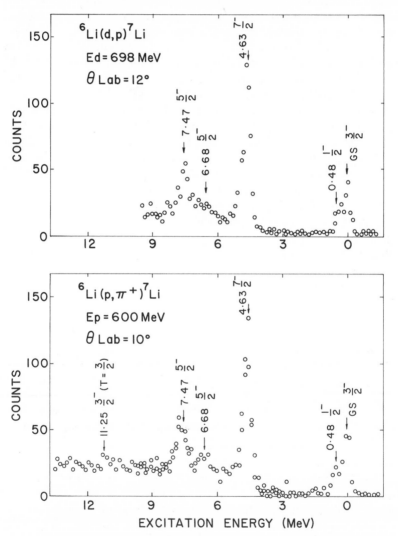

Figure 7   Comparison of the $^6Li(d,p)^7Li$ (Beurtey et al 1978a) and $^6Li(p,\pi^+)^7Li$ (Bauer et al 1977) at a similar momentum transfer (~ 550 MeV c^{-1}).

experiments, being performed on the EPICS channel at LAMPF, will achieve a substantially better resolution. Such experiments can be compared with the (p,d) reaction for high energy protons such as the work at Saturne on $^{12}C(p,d)^{11}C$ with 700-MeV protons (Baker et al 1974). The (p,d) reaction at high energy is complicated by (a) the importance of the deuteron D state; (b) the effects of Δ excitation; and (c) two-step processes (Rost & Shepard 1975, Grabowski et al 1976). The (π^+,p) reaction is also plagued by complications, often different, so the comparison can be fruitful.

One word of warning should be added about absolute normalizations. The Indiana data generally seem to be larger than those of Uppsala by a constant factor of 1.8. However, the data of Indiana and TRIUMF are consistent (Pile et al 1979, Bent et al 1978, Auld et al 1978).

In the past two years the amount of available data has grown rapidly. Prior to this period tests of theories were seriously impeded by a lack of information about either energy or target dependence of the reaction. The presently existing array of data for the (p,π^+) reaction is shown in

Table 2   The (p,π^+) data for heavy nuclei

Target	Energy	Final states resolved	Reference
^{6}Li	600 MeV	lowest 3 states	Bauer et al 1977
^{9}Be	185 MeV	lowest 7 states	Dahlgren et al 1971, 1973a
	410, 605, 809 MeV	lowest 5 states	Aslanides et al 1978
^{10}B	145–198 MeV	ground state	Pile 1978, Pile et al 1979
	154 MeV	ground state	Le Bornec et al 1976
	185 MeV	ground state	Dahlgren et al 1974
	160–185 MeV	ground state	Höistad et al 1979a
	320–606 MeV	lowest 2 states	Beurtey 1978b
^{12}C	185 MeV	lowest 7 states	Dahlgren et al 1971, 1973b
	200 MeV	lowest 4 states	Auld et al 1978
	600 MeV	ground state	Domingo et al 1970
^{13}C	185 MeV	ground state	Dahlgren et al 1973a
	600 MeV	ground state	Domingo et al 1970
^{16}O	185 MeV	lowest 3 states	Dahlgren et al 1974
^{26}Mg	180 MeV	lowest 2 states	Höistad et al 1978
^{28}Si	185 MeV	ground state	Dahlgren et al 1974
^{40}Ca	185 MeV	ground state	Dahlgren et al 1974
	140–200 MeV	ground state	Pile 1978, Pile et al 1979
	160–180 MeV	ground state	Höistad et al 1979a
^{90}Zr	180 MeV	lowest 2 states	Höistad et al 1979b
	160 MeV	ground state	Bent et al 1978
^{208}Pb	180 MeV	lowest 2 states	Höistad et al 1979b
	160 MeV	lowest 3 states	Bent et al 1978

Table 2 and presents a formidable challenge to even the most dedicated theorists. However, there still is a lack of data for targets heavier than ^{40}Ca. This stems from the experimental difficulty in resolving the cross sections to different final states.

4.2   Theory

In order to improve our interpretation of the data, we describe the various theoretical approaches (SNM, DWBA, and TNM) in some detail. Consider first the reaction in which the final state consists of a neutron moving in a single-particle well supplied by the target ground state.

In the single-nucleon model (SNM) the momentum transfer is almost solely taken up by the wave function of the bound final-state neutron. If the pion and proton distortions are ignored, the T-matrix for this reaction, T_{fi}, is given by

$$T_{\text{fi}} = \sqrt{\frac{4\pi}{E}}\frac{f}{\mu} \int \mathrm{d}^3 r \phi^*_{nljm}(\hat{r}, m_{\text{s}})\, e^{-i\mathbf{k}\cdot\mathbf{r}} H_I\, e^{i\mathbf{p}\cdot\mathbf{r}} \chi_{1/2}(m_{\text{i}}), \qquad 4.$$

in which f is the renormalized coupling constant $f^2 = 0.080$, E is the total energy of the pion, and μ is the pion mass. The neutron wave function $\phi_{nljm}(\hat{r}, m_{\text{s}})$ has the orbital and angular momentum quantum numbers $nljm$. The momenta of the incident proton and final pion are $\mathbf{p}$ and $\mathbf{k}$ respectively and the initial spin of the nucleon is m_{i}. The pion-nucleon interaction is a nonrelativistic version of the pseudoscalar one and is given by

$$H_I = \boldsymbol{\sigma}\cdot\left(1+\frac{\lambda\omega_k}{2M}\right)\mathbf{k} - \frac{\lambda\omega_k}{2M}\boldsymbol{\sigma}\cdot\mathbf{p}. \qquad 5.$$

The values $\lambda = 0$ or $\lambda = 1$ correspond to the static and Galilean invariant interactions respectively. The problems associated with the nonrelativistic reduction of relativistic vertex functions are discussed in Ho et al (1975), G. A. Miller (1974), Bolsterli et al (1974), Eisenberg et al (1975), Friar (1975), Bolsterli (1976), and Brack et al (1977).

Equation 4 shows that in the SNM the cross section is proportional to the square of the Fourier transform of the radial neutron wave function, $\tilde{\phi}_{nlj}(q)$ and $q = |\mathbf{p}-\mathbf{k}|$. If the SNM were correct, the observed angular distributions would have rapid variations with angle similar to those of Figure 2. However, because $\phi_{nlj}(q)$ is so small other mechanisms are likely to contribute. Furthermore if one uses the Galilean invariant interaction, the second term of Equation 5 tends to cancel the first term (Grossman et al 1974).

An alternative (RSNM) to the SNM avoids the nonrelativistic reduction of the vertex operator by using relativistic bound wave functions that are solutions of the Dirac equation (L. D. Miller & Weber 1976, 1978, Brockman & Dillig 1977). In this model the small components of the neutron's wave functions are explicitly included. Such components may be as large in magnitude at $q = 500$ MeV c^{-1} as the large ones, and a qualitative representation of the 185-MeV ^{40}Ca$(p,\pi^+)^{41}$Ca$_{g.s.}$ reaction has been obtained with the RSNM (L. D. Miller & Weber 1978). Note that both the SNM and RSNM predict that the polarization asymmetry is zero. Furthermore, both models suffer from the effects of the lack of orthogonality of the initial and final nuclear wave functions (Noble 1978).

One obvious correction to these models is to include the interaction between the incident proton and the target by treating the proton wave function as a wave distorted by a complex optical potential. At medium energies the proton optical potential is typically only about 10% the magnitude of the proton's kinetic energy. Hence proton distortions do not greatly influence the angular distributions, although the magnitude of the cross sections is decreased by the absorptive well (W. B. Jones & Eisenberg 1970). The proton spin-orbit potential does produce a non-zero polarization asymmetry which can be of the order of the data (Weber & Eisenberg 1979).

A more important correction is the inclusion of pion-nucleus interactions, which cause the momentum of the pion in the medium, $\mathbf{K}$, to be very different from $\mathbf{k}$. If $K \gg k$, the effective momentum transfer is $\mathbf{p} - \mathbf{K}$, which can be much smaller in magnitude than q, so that the calculated cross section is expected to increase. One convenient way of including such effects is to distort the pion's wave function in an optical potential. However, typical pion optical potentials fall into two categories: those that allow K to be very different from k and those that do not. Examples of the former are the Kisslinger (1955), Laplacian (Fäldt 1972, Lee & McManus 1971), and separable models (Landau & Tabakin 1972, Londergan et al 1974, Ernst & Miller 1975). Examples of the latter are the Glauber (1959), simple local, and modified Laplacian models (Höistad 1978). If one uses the wave functions from the former set of optical potentials one is doing a meson off-shell calculation. If on the other hand one uses the wave functions of the latter models one is essentially using the SNM. Calculated (p,π) cross sections are extremely sensitive to the optical wave function used.

If the pion's wave function is expected to differ substantially from a plane wave, one does a DWBA calculation in which the transition matrix (using the static approximation to H_I) is given by

$$T_{\mathrm{fi}} = \sqrt{\frac{4\pi}{E}} \frac{f}{\mu} i \int \mathrm{d}^3 r \phi^*_{nljm}(\mathbf{r}, m_{\mathrm{s}}) [\boldsymbol{\sigma} \cdot \nabla \phi_\pi(\mathbf{r})]^* \psi^{(+)}_{\mathrm{p}}(\mathbf{r}, m_{\mathrm{i}}). \qquad 6.$$

Here $\psi_p^{(+)}(\mathbf{r}, m_i)$ represents the incident proton's wave function. The pion's wave function, $\phi_\pi(\mathbf{r})$, is obtained from the solution of the wave equation

$$-\nabla^2\phi_\pi + 2EU_E\phi_\pi + 2EV_c\phi_\pi = k^2\phi_\pi, \qquad\qquad 7.$$

in which U_E represents the pion-nucleus optical potential and V_c the Coulomb potential of the nucleus. The energy E of the pion, neglecting recoil of the nucleus, is $E^2 = k^2 + m_\pi^2$. The three-dimensional integral is performed by making partial wave expansions of ψ_p and ϕ_π, doing the angular integral analytically and the radial integral numerically.

Part of the difficulty in obtaining a reliable optical potential (for use in Equation 6) results from a problem in handling the pion-nucleon (3, 3) resonance. Consider U_E, which to a first approximation is the folding of a pion-nucleon transition matrix with the nuclear density, and contains the effects of the strong angular momentum dependence. In the simplest approach

$$U_E(\mathbf{K}, \mathbf{k}') = t_{\pi N}(E, \mathbf{K}, \mathbf{k}')\rho(|\mathbf{K} - \mathbf{k}'|), \qquad\qquad 8.$$

in which $t_{\pi N}(E, \mathbf{K}, \mathbf{k}')$ is the pion-nucleon transition matrix (effects of the momentum and spin of the struck nucleon are neglected here to simplify our discussion), and $\rho(q)$ is the Fourier transform of the nuclear density.

Kisslinger (1955) introduced the following parametrization for $t_{\pi N}$

$$t_{\pi N}^{(K)}(E, \mathbf{K}, \mathbf{k}') = a_E + b_E\mathbf{K}\cdot\mathbf{k}', \qquad\qquad 9.$$

in which the s- and p-wave amplitudes are made explicit by the use of a p-wave projection operator $\mathbf{K}\cdot\mathbf{k}'$ and in which the spin flip term is neglected.

Substituting Equation 9 into Equation 8 leads to very simple optical potentials (in the coordinate space representation), and the Kisslinger potential has been quite popular. Now ϕ_π depends strongly on the behavior of the pion-nucleon T-matrix for large K, but the right-hand side of Equation 9 diverges as K approaches infinity. This causes the resultant wave functions to have kinks at distances well inside the nuclear radius. The effects of the Kisslinger potential are quite dramatic in (p,π^+), and the cross sections obtained are too large (typically by a factor of 30) to be consistent with the data (W. B. Jones & Eisenberg 1970, Keating & Wills 1973, G. A. Miller 1974).

A popular modification of the Kisslinger potential includes the second-order and higher order terms in $t_{\pi N}$ in the calculation of U_E. In the Lorentz-Lorenz (LL) model, introduced by Ericson & Ericson (1966) the higher order terms cause the optical potential to decrease by about 30%. However, the use of the LL wave functions gives cross sections about twenty times larger than experimental ones (W. B. Jones & Eisenberg 1970, Keating & Wills 1973).

A more realistic treatment uses a separable pion-nucleon T-matrix of the form

$$t_{\pi N}(\mathbf{K}, \mathbf{k}') = \frac{V_0(K)V_0(k')}{V_0^2(k)} a_E + b_E \frac{V_1(K)V_1(k')}{V_1^2(k)} \mathbf{K} \cdot \mathbf{k}'. \qquad 10.$$

The factors $v_{0,1}(K)$ approach zero smoothly for large K so that t_{nN} is finite in the large K limit. These functions can be obtained from separable potential (Landau & Tabakin 1972, Londergan et al 1974) or the Chew-Low model (Chew & Low 1956, Dover et al 1974, Ernst & Johnson 1978).

The use of Equation 10 instead of Equation 9 results in wave functions without kinks. Calculations that use Equation 10 to generate optical potentials have successfully reproduced the $^{12}C(p,\pi)^{13}C$ data (G. A. Miller & Phatak 1974).

The Kisslinger potential has been used successfully in many studies of pion-nuclear elastic and inelastic scattering. However, its use in meson production calculations results in anomalously large cross sections. Furthermore the use of the Lorentz-Lorenz correction does not sufficiently repair the damage. On the other hand, predictions improve considerably if one constructs the pion optical potential from a pion-nucleon interaction with a more realistic off-shell behavior. The use of the (p,π) reaction to provide constraints on the pion-nucleus optical potential may enhance our knowledge of pion-nucleus dynamics.

It is worthwhile to examine the content of the DWBA by examining the term, $T_{fi}^{(1)}$, of first order in $t_{\pi N}$. To isolate this term, use the wave function, $|\Delta\phi_\pi\rangle$, with

$$|\Delta\phi_\pi\rangle = [\nabla^2 + k^2]^{-1} 2EU_E |\mathbf{k}\rangle, \qquad 11.$$

in which $|\mathbf{k}\rangle$ represents the plane wave, instead of ϕ_π in Equation 6. One finds

$$T_{fi}^{(1)} = \frac{4\pi}{E} \frac{f}{\mu} (2E) \int d^3 K \phi_{nljm}^* [(\mathbf{p} - \mathbf{K}), m_s]$$

$$\times \frac{\sigma \cdot \mathbf{K} t_{\pi N}(\mathbf{K}, \mathbf{k})}{-\mathbf{K}^2 + \mathbf{k}^2} \rho(|\mathbf{K} - \mathbf{k}|) \chi_{1/2}(m_i), \qquad 12.$$

where for simplicity we have taken the proton wave function to be a plane wave. A diagrammatic representation of Equation 12 is given in Figure 3b. Note that the bound-state wave function is evaluated at a momentum $\mathbf{p} - \mathbf{K}$, which is smaller in magnitude than q. Thus pion rescattering effects strongly enhance the calculated cross section. The use of Equation 9 in Equation 12 leads to very large values of $T_{fi}^{(1)}$ because of the K^4 factor in the dK integral.

Equation 12 can be used to qualitatively predict the calculated angular distributions if one assumes that the dominant contributions to the integral occur for the smallest values of $|\mathbf{K} - \mathbf{p}|$; if $\hat{K} = \hat{p}$, it is minimized, and the p-wave rescattering term is proportional to $\hat{p} \cdot \mathbf{k}$. Grossman et al first emphasized this tendency of the p-wave rescattering term to lead to minima in the angular distributions at about $90°$. Such minima are also obtained in DWBA calculations that include a p-wave rescattering term (G. A. Miller 1974). Note that if wave-function orthogonality is taken into account the resulting SNM T-matrix is also proportional to $\hat{p} \cdot \mathbf{k}$ (Noble 1978). The nuclear wave functions appearing in Equation 9 give rise to additional angular variations in the cross section.

A more ambitious approach than the DWBA, the two-nucleon model (TNM), explicitly calculates the process by which a pion is produced during the interaction of two nucleons. By introducing specific dynamic models for the production process, such as the graphs of Figure 3, one is able to calculate the effects of the surrounding nuclear medium on the interacting pair as well as to use dynamic models consistent with production data from nucleon-nucleon collisions. In addition to the graph (Figure 3c), which is included in the DWBA calculation, one can include effects of ρ-meson exchange, Pauli correlations, and short-range dynamic correlations (Figure 3d–g). The TNM was introduced by Ruderman (1952) and calculations and refinements were made by Reitan (1971, 1972), Grossman et al (1974), Dillig & Huber (1977), and Fearing (1974, 1975ab, 1977).

Calculations of the TNM are more difficult to pursue than those of the DWBA and a variety of approximations have been used. Dillig & Huber replace the $\boldsymbol{\sigma} \cdot \mathbf{K} t_{n\text{N}}(\mathbf{K}, \mathbf{k})$ term of Equation 12 and other meson exchange terms by $\boldsymbol{\sigma} \cdot \mathbf{K}_e t(\mathbf{K}_e, \mathbf{k})$, where $\mathbf{K}_e = 2\mathbf{p}/3 + \mathbf{k}/3$. Grossman et al approximate their integrands by a term proportional to $\delta(\hat{K} - \hat{p})$. None of the approximations used by Dillig & Huber and Grossman et al have been checked in an exact calculation. Fearing constructs the two-nucleon production operator directly from the $pp \to d\pi^+$ data. However, many approximations are necessary to obtain a direct relationship between the two-nucleon production data and the proton-nucleus amplitude (Green & Maqueda 1979).

The TNM predicts that reactions that excite the target ground state in addition to transferring a single nucleon will also be important. This is because the pion-nucleon interaction can excite a nucleon above the Fermi sea. In that case the final nuclear state has components of single-particle wave functions revolving around excited states of the target.

Because of the inherent similarities between the DWBA and the TNM — that is, the DWBA is included in a proper evaluation of the TNM — it is difficult to distinguish the two models. Furthermore the reaction dynamics of the two models are very similar, namely, the transfer of momentum is aided by the exchange of a virtual pion of relatively high momentum.

It is also true that the differences between the SNM and DWBA or TNM are not profound. This is because in the SNM the interaction of the bound neutron with the other nucleons allows the reaction to proceed. These interactions also arise from meson exchanges so that the distinctions between the various models are blurred.

4.3   *Experimental Data*

In an attempt to extract unambiguous information about the reaction dynamics or nuclear-wave functions, we review the data to see if there are any qualitative systematic features.

Of crucial importance is the shape and energy dependence of the angular distributions. A rapid variation in shape is a signature for the SNM, but the TNM and DWBA can also give rise to a rapid variation both from $\mathbf{k} \cdot \mathbf{p}$ terms in the transition operator and from nuclear wave-function effects. Clearly an analysis of the reaction over a wide range of energies is necessary. Whereas detailed experimental information about the $^{10}\text{B}(\text{p},\pi^+)^{11}\text{B}_{\text{g.s.}}$ and $^{40}\text{Ca}(\text{p},\pi^+)^{41}\text{Ca}_{\text{g.s.}}$ exists, as shown in Figures 8

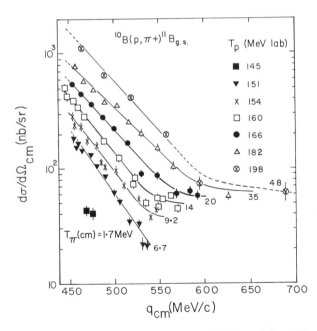

Figure 8   Angular distributions as a function of energy. This figure is from Pile et al (1979).

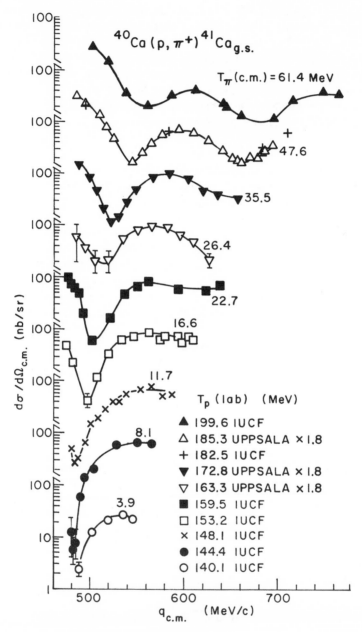

Figure 9 Angular distributions as a function of lab kinetic energy. The data labeled IUCF are from Pile et al (1979) and those labeled Uppsala are from Dahlgren et al (1974) and Höistad et al (1979a). This figure is from Pile et al (1979) and private communication.

and 9, a corresponding theoretical analysis does not. The most striking feature of the data is the energy dependence of the minimum that occurs in the forward direction for the reaction on a ^{40}Ca target, Figure 9. On the other hand, the angular distributions for reactions with ^{10}B are basically monotonic (Figure 8) although this is a rarity. The angular distributions are similar at higher energies (Beurtey et al 1978b). For most of the targets studied at 185 MeV the cross sections seem to have minima in the vicinity of 90°. Examples are reactions on ^{12}C (Figure 10), ^{16}O, ^{28}Si, and ^{40}Ca (Figure 11). For heavier targets there is much less data. However, experiments with ^{90}Zr and ^{208}Pb targets performed by groups at Indiana and Uppsala give some evidence for 90° minima at 185 MeV and for forward-angle minima at lower energies.

The minima may be interpreted as resulting from effects of the $\mathbf{k} \cdot \mathbf{p}$ term, or from zeros in the Fourier transform of the wave function of the bound neutron. For example, it is possible to fit the ^{40}Ca at 185 MeV with SNM calculations (L. D. Miller & Weber 1978) in which the $\mathbf{k} \cdot \mathbf{p}$

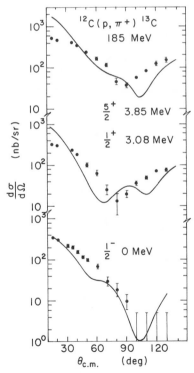

Figure 10   Angular distributions for different final states. The solid curve is from G. A. Miller & Phatak (1974). The data are from Dahlgren et al (1973b).

term is lacking. In this case the minima is close to a minima in $\tilde{\phi}_{nlj}(\mathbf{q})$. Calculations on ^{12}C in both DWBA and TNM require the presence of the $\mathbf{k} \cdot \mathbf{p}$ term in the pi-nucleon scattering matrix to reproduce the position and depth of the observed minima for that target (Grossman et al 1974, G. A. Miller 1974). In ^{40}Ca the final nucleon has an angular momentum

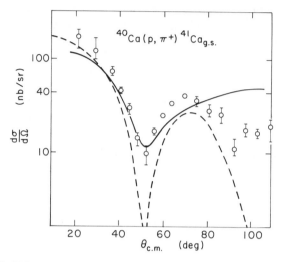

Figure 11   The TNM (*solid curve*; Dillig & Huber 1977) versus the RSNM (*dashed curve*; L. D. Miller & Weber 1978). The data are from Dahlgren et al (1974).

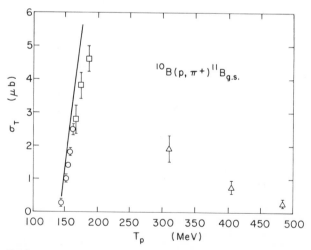

Figure 12   Total cross sections as a function of lab kinetic energy. The data represented by circles are from Pile et al (1979), squares are from Höistad (1977) (scaled up by 1.8), and triangles are from Beurtey et al (1978b).

of $\frac{7}{2}$, a large value that tends to reduce the angular momentum mismatch. Thus in the ^{40}Ca reaction the SNM may be more important than in the ^{12}C reaction.

There is not much information available on the energy dependence of the total (p,π^+) cross section to a single final state in a nucleus heavier than ^{4}He. The ^{10}B$(p,\pi^+)^{11}$B$_{g.s.}$ reaction (Figure 12) is the most studied. The rapid rise of σ_T near threshold follows the energy dependence of the Coulomb penetrabilities. The presumed peak in σ_T results from the influence of the (3, 3) resonance. Studies of $\sigma_T(T_p)$ for other targets could provide new information about the imaginary part of the pion-nucleus optical potential.

There is one aspect of the data not explained by the simplest version of the SNM. The reaction (p,π^+) very often leads to final states consisting of a single neutron orbiting about an excited state of the target (Amann et al 1978, Dahlgren et al 1971, 1973ab). Such a process is forbidden in the SNM and occurs naturally (Grossman et al 1974) in the TNM. Consider, for example, the population of the 6.87-MeV $\frac{5}{2}^+$ state in ^{13}C. As the 3.85-MeV $\frac{5}{2}^+$ state has a spectroscopic factor of 0.9, and the 6.87-MeV state must be orthogonal to the lower-lying state, core-excited components of the second $\frac{5}{2}^+$ state dominate its structure. The cross section to the 6.87-MeV state is as large (Figure 13) as the cross section to the ground state of ^{13}C (Figure 10). The theoretical curves are from Grossman et al (1974). Note, however, that with the use of coupled channels, proton and pion wave functions allow such processes to occur in the SNM and DWBA treatments (G. A. Miller 1974, Eisenberg et al 1974).

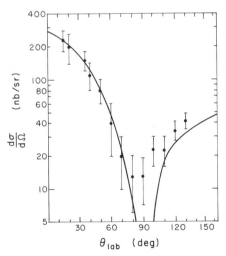

Figure 13   The ^{12}C$(p,\pi^+)^{13}$C$(\frac{5}{2}^+$, 6.9 MeV) reaction. The data are from Dahlgren et al (1973b) and the solid curve is from Grossman et al (1974).

Another example occurs in the strong excitation of the $\frac{1}{2}^- T = \frac{3}{2}$, 12.5-MeV excited state of ^{11}C, which is populated in the ^{12}C$(\pi^+,p)^{11}$C reaction (Amann et al 1978). This state has a large one-particle, two-hole component $1p_{1/2}(1p_{3/2})^{-2}$ based on the ^{12}C core and is strongly populated in the reaction ^{13}C(p,t)^{11}C (Cosper et al 1968). The cross section to excite the 12.5-MeV state is as large as the cross section to excite the ground state, which is dominated by a single $p_{3/2}$ hole configuration.

The prevalence of such processes in which a neutron is put into a nucleus and at least one core nucleon is excited, confirms the multinucleon aspect of the production mechanism.

Another general feature of the data is that for reactions on a given target, final states with high angular momentum are preferentially excited. For example, the $\frac{5}{2}^+$ state at 3.85 MeV is more strongly excited than the $\frac{1}{2}^-$ ground state (Figure 10). A similar phenomenon is observed for reactions involving ^{16}O: the excitation of the $\frac{5}{2}^+$ ground state of ^{17}O (Dahlgren et al 1974) is preferred. A dramatic example is seen in the ^{16}O$(\pi^+,p)^{15}$O data of Bachelier et al (1977), who observed excitation of

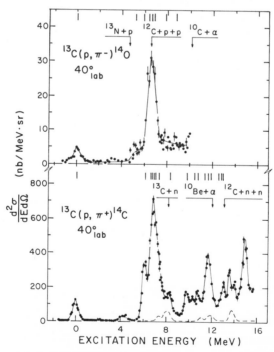

Figure 14   Comparison of the ^{13}C(p,π^-)^{14}O$_{g.s.}$ and ^{13}C(p,π^+)^{14}C$_{g.s.}$ (Dahlgren et al 1973c) spectra.

the $\frac{1}{2}^-$ ground states and $\frac{3}{2}^-$ excited states. The excitation of the excited state is preferred by a factor of about 10. For light nuclei this angular momentum selection rule seems to be generally true, but for heavy nuclei the evidence is less clear.

The preference for excitation of high angular momentum final states is easily understood in terms of angular momentum matching. The wave function of the emerging pion is dominated by components of low angular momentum, for energies in the subthreshold region, but the incident proton mainly has components of large angular momentum. This difference must be compensated by the angular momentum of the final nuclear state, and states of high angular momentum are preferred.

Another quantity of interest is the dependence of the magnitude of the cross sections on the mass of the heavy nucleus. For targets of mass 40 or less the cross sections generally have peak values of several hundred nb sr^{-1}. The observed cross sections for pion production on Zr and Pb targets seem to be smaller and have maximum values of about 50 nb sr^{-1}, at 185 MeV. This target dependence has not yet been explained. So far only data at forward angles have been taken. At backward angles the cross sections are larger at 155 MeV (Bent et al 1978).

5   THE (p,π^-) REACTION ON NUCLEI

The process in which the production of a negative pion in a proton-nucleus collision leads to a bound final nucleus is interesting because only one nucleon is inserted into the target but the total nuclear charge is increased by two. Hence this process must explicitly involve at least two nucleons and the SNM predicts a zero cross section.

Because of the double charge-exchange character of the (p,π^-) reaction, little is known about the final nucleus. As experimental techniques are perfected, this reaction will become a useful spectroscopic tool. An example of this aspect is the experiment by Couvert et al (1978), who used 613-MeV protons from Saturne I to study ^{9}Be(p,π^-)^{10}C and ^{12}C(p,π^-)^{13}O. Some excited states in ^{13}O were identified for the first time.

A mention should be made of the (π^-,p) reaction. As the cross section is expected to be smaller than the (π^+,p) reaction, the experimental difficulties are too great at present, except for the one case of a study at rest. For the reaction ^{12}C(π^-,p)^{11}Be, Coupat et al (1975) found a branching ratio of $(4.5 \pm 0.8) \times 10^{-4}$ for the two bound levels in ^{11}Be, i.e. the $\frac{1}{2}^+$ ground state and the $\frac{1}{2}^-$ excited state at 0.32 MeV.

The integrated cross section for the (p,π^-) reaction is typically a factor of 10 smaller than that for the (p,π^+) reaction, yet the nuclear physics aspects for the two reactions seem similar. Figure 14 illustrates the spectra for the two reactions $^{13}C(p,\pi^-)^{14}O$ and $^{13}C(p,\pi^+)^{14}C$. As the final nuclei are a mirror pair, one is not surprised to observe that the relative excitation of different levels is comparable in the two cases. Here, however, the resemblance ends as the angular distribution of the (p,π^-) reaction is relatively structureless. An example of mirror transitions is given in Figure 15.

The relative size of the (p,π^-) and (p,π^+) cross sections, as well as the different angular distributions for the two reactions, may easily be understood by noting that the (p,π^-) reaction to a definite final state involves the changing of a neutron in an occupied orbit into a proton in a specific unoccupied orbit. This phase space limitation severely hinders the cross section. Furthermore the valence nucleons involved in the (p,π^-) reaction are typically in different locations from the core nucleons that participate in the TNM or DWBA models for the (p,π^+) production process. For example, the $^{13}C(p,\pi^-)^{14}C_{g.s.}$ process must involve a neutron in a $p_{1/2}$ orbital changing into a proton in the same orbital. The form factor of the $p_{1/2}$ orbital is not forward peaked, which is in contrast to the form factor of the entire nucleus, and the resulting angular distributions are flatter for (p,π^-) than for (p,π^+). This valence charge-exchange effect occurs in all of the reaction mechanisms considered below.

Calculations of the (p,π^-) process were performed by Reitan (1971), Grossman et al (1974), Kisslinger & Miller (1975) and Dillig & Huber (1976a,b). Several reaction mechanisms contribute to the (p,π^-) process.

Perhaps the easiest mechanism to visualize is the process in which the

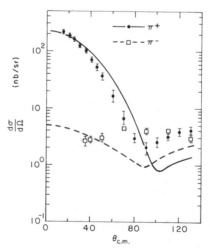

Figure 15   Comparison of the $^{13}C(p,\pi^-)^{14}O_{g.s.}$ and $^{13}C(p,\pi^+)^{14}C_{g.s.}$ angular distributions. The calculations are from Dillig & Huber (1976b).

incoming proton experiences a charge-exchange reaction with the nucleus, becoming a neutron which then emits the final π^-. Calculations of this mechanism for the reaction $^{26}Mg(p,\pi^-)^{27}Si$ were performed by Kisslinger & Miller, who find that this process is very important near threshold but negligible for incident proton energies greater than about 200 MeV. This result depends crucially on the impulse approximation used to estimate the (p,n) charge-exchange matrix element.

In view of the importance of the pion-nucleon rescattering effects in positive pion production, one must consider the process in which the incoming proton emits a π^0 that undergoes a charge-exchange reaction. Coupled-channel calculations of this process for an ^{26}Mg target were also made by Kisslinger & Miller, who found that the pion charge-exchange process is generally smaller than the proton charge-exchange one. However, the predictions of the (π^+,π^0) reaction on nuclei implied by the treatment of Kisslinger & Miller have not been checked against experiment.

The TNM, by virtue of its dependence on the strongly isospin-dependent pion-nucleon scattering, contains predictions of the (p,π^-) process. Relevant contributions of the TNM are shown in Figure 3c–g. For example, the virtual pion in Figure 3c could be a π^0 and the final one a π^-. Grossman, Lenz & Locher (1974) predicted the ratio of the cross sections for the reactions $^{12}C(p,\pi^+)^{13}C_{g.s.}$ to $^{12}C(p,\pi^-)^{13}O_{g.s.}$ at 185 MeV to be 240 at 20°, a number in good agreement with the observed ratio.

Dillig & Huber (1976b) qualitatively explained the $^{13}C(p,\pi^-)^{14}O_{g.s.}$ (Figure 15) and $^9Be(p,\pi^-)^{10}C_{g.s.}$ data at 185 MeV. Only p-shell nucleons are involved in the reaction and the (p,π^-) transition amplitude varies slowly with angle. Note however that the calculations of Dillig & Huber suffer from the approximations mentioned in Section 4.2.

The use of the (p,π^-) reaction on nuclei as a probe to determine Δ^{++} components of nuclear wave functions was suggested by Kisslinger & Miller (1975). If one uses a model in which the Δ^{++} and a nucleon are treated similarly, the reaction may occur when a proton emits a π^- and becomes a Δ^{++}. The probability amplitude to find a Δ^{++} in the nucleus, or Δ^{++} wave function, may be estimated in perturbation theory. Kisslinger & Miller took the shape of their Δ^{++} wave functions from perturbation calculations (Rost, Schaeffer, and Kisslinger, unpublished), but as the spectroscopic factor $S_{\Delta^{++}}$ (total Δ^{++} probability) was unknown, a value of .0025, which obtains for the deuteron (Jena & Kisslinger 1974), was used. However, subsequent work (R. Schaeffer, private communication) showed that $S_{\Delta^{++}}$ should be smaller by a factor of 10^{-3}. Hence Kisslinger & Miller's results for the Δ^{++} should be scaled by this factor. They chose to study the reaction $^{26}\text{Mg}(p,\pi^+)^{27}\text{Si}(\frac{5}{2}^+)_{\text{g.s.}}$ because a Δ^{++} in a $\frac{5}{2}^+$ orbital should have a large orbital momentum of 4 in addition to the standard value of 2. Their calculations, at 185 MeV, even when scaled by 10^{-3}, predict a cross section larger by a factor of 10 than the data of Höistad et al (1978). However, Kisslinger & Miller used the local Laplacian potential to generate the pion distorted waves. This results in a predicted $^{26}\text{Mg}(p,\pi^+)^{27}\text{Mg}_{\text{g.s.}}$ cross section that is much larger than the experimental one. The Δ^{++} transfer should be recalculated using improved pion optical potentials or other mechanisms that would permit a simultaneous explanation of the $(p,\pi^{\pm})$ cross sections.

6   THE (p,π^0) REACTION ON NUCLEI

The (p,π^0) reaction is very similar to the (p,γ) reaction and some data could be very useful. Unfortunately the difficulty involved in π^0 detection has choked experimental investigation. The only (p,π^0) study to bound states for nuclei heavier than ^4He is that at Indiana (Bacher et al 1976) where two lead glass Cerenkov counters were used in coincidence, each being at $90°$ to the beam, facing each other. The data are tantalizing in that the reaction was clearly observed, yet further details are hard to extract. A much more sophisticated detector will be needed, but medium resolution π^0 spectrometers have been used in other studies, so the future holds promise (MacDonald et al 1977, Bowles et al 1978). Also, some attempts at activation experiments on the reaction $^{209}\text{Bi}(p,\pi^0)^{210}\text{Po}$ are beginning to succeed (Friesel et al 1978). The total (p,π^0) cross sections appear much larger than those for (p,π^+), especially near threshold where, of course, the π^0 does not experience a Coulomb barrier.

7   POLARIZATION EFFECTS IN THE $(\vec{p},\pi^+)$ REACTION

It has long been known that polarization effects were important in the reaction $(p+p \rightarrow \pi^+ + d)$, yet everyone was surprised to find that there are even more pronounced asymmetries for the $(\vec{p},\pi^+)$ reaction on nuclei. It is reminiscent of the problem of proton elastic scattering off nuclei, which was confusing for so long because the polarization was observed to be much stronger than for free nucleon-nucleon scattering. However, Bethe (1958) showed that the difference arose from the chance cancellation of certain scattering amplitudes.

The $(\vec{p},\pi^+)$ experiment of Auld et al (1978) was performed at 200 MeV with the polarized proton beam at TRIUMF. A Browne-Buechner spectrograph was placed on one side of the proton beam and the spin of the incident beam was inverted. The results for $^9\text{Be}(\vec{p},\pi^+)^{10}\text{Be}$ are

shown in Figure 16. The asymmetries reach the very high value of 70%
which is higher than that from the pp interaction, and are similar for
both the 0^+ ground state and the 2^+ excited state at 3.37 MeV. This
similarity is even more surprising when one remembers that the angular
distributions to these two levels are very different, that for the 3.37-MeV
level is very much more isotropic (Dahlgren et al 1973a). Not only are
the asymmetries similar for one nucleus, but they are even similar for
different p-shell nuclei; furthermore there is a resemblance to the asymmetries for the reaction $pp \rightarrow \pi^+ d$. Recent Indiana data (P. Pile et al,
private communication), taken at lower energies, show general similarities
in asymmetries obtained using different targets. However, there are some
cases in which the asymmetry has a different angular dependence. These
facts strongly support the TNM or DWBA: In the SNM the asymmetry
can only be produced via the spin-orbit distortion (Noble 1975, 1976).
Weber & Eisenberg (1979), however, obtain large asymmetries from spin-
orbit distortion. Young & Gibbs (1978) obtain large asymmetries from
effects of pion distortion.

Experiments using other targets are needed to see if the asymmetries
are the same for all nuclei. Furthermore, the question of the relative
importance of proton spin-orbit and pion rescattering effects should be
carefully studied.

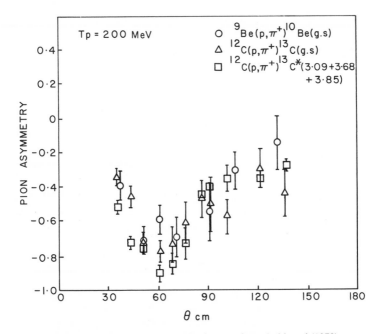

Figure 16 Pion asymmetry. The data are from Auld et al (1978).

8   PION PRODUCTION BY LIGHT IONS

There have been one or two experiments concerning pion production with ions such as d and ^{3}He. The first attempt was that by Wall et al (1974, 1976) who found very small cross sections $[\sim 10^{-36}$ cm^2 (sr · MeV)$^{-1}]$ for the reaction ^{12}C(He,π^0)X and Pb(^{3}He,π^0)X at incident energies of 180 and 200 MeV. A recent successful experiment was reported on the reactions ^{12}C(π^+,d)^{10}C and ^{10}B(π^+,d)^{8}B at $E_\pi = 50$ MeV and a few angles (Amann et al 1978). The cross sections of approximately 1000 nb sr^{-1} are about 10 times smaller than anticipated from a simple plane-wave calculation in which a pion is absorbed on a bound pair of neutrons (Eisenstein & G. A. Miller 1975). However, when the deuteron distortion effects are included (Betz 1977), good agreement with experiment is obtained.

9   CONCLUSIONS

The reaction $(p + p \to \pi^+ + d)$ is already quite well understood and, when results become available from experiments in progress, our knowledge of this fundamental reaction will be excellent. Pion production reactions such as $(n + p \to \pi^0 + d)$ and $(d + d \to {}^4He + \pi^0)$ serve as important tests of isospin in the strong interaction. For the (p,π) reaction on other nuclei, the situation is in a healthy state of flux. The experimental data are improving with the use of variable-energy protons becoming standard, and with polarized proton beams being used at several laboratories. Spectrometers large and small are being built and should provide data at many energies and for many nuclei.

The existing measurements have already helped clarify the interpretation of the reaction mechanism. The copious excitation of two-particle, one-hole states has demonstrated the multinucleon aspects of pion production. Furthermore all calculations are extremely sensitive to the details of the pion-nucleus optical potential. The use of the Kisslinger, local Laplacian, and LLEE potentials in DWBA calculations predicts cross sections that are much larger than experimental ones. Considerable improvement is obtained with more realistic models. The use of the pion production reaction in learning about the pion optical potential may become its most important contribution to our knowledge of meson-nucleus dynamics.

It is also true that the reaction is quite difficult to understand, and it is not clear that we have learned anything about nuclear structure. However, pion production has had an important impact on our ideas of nuclear wave functions. For example, in order to understand the (p,π) reaction it has become necessary to seriously investigate Δ and anti-nucleonic components of wave functions. With further insight, more precise information about such atypical aspects of nuclear structure may be obtained.

The (p,π) reaction has an interesting past, an exciting present and an excellent future.

We thank E. Auld, R. D. Bent, G. Jones, and P. H. Pile for sharing their data with us prior to publication. This work was supported in part by the US Department of Energy. One author (D.F.M.) wishes to acknowledge the hospitality of the Nuclear Physics Division at CEN Saclay where an initial draft of this review was written.

Literature Cited

Aebischer, D., Favier, B., Greeniaus, L. G., Hess, R., Junod, A., Lechanoine, C., Niklès, J.-C., Rapin, D., Werren, D. W. 1976. *Nucl. Phys. B* 106:214–38

Afnan, I. R., Thomas, A. W. 1974. *Phys. Rev. C* 10:109–25

Akimov, Yu. K., Savchenko, V., Soroko, L. M. 1962. *Sov. Phys. JETP* 14:512–23

Alberg, M., Henley, E. M., Miller, G. A., Walker, J. F. 1978. *Nucl. Phys. A* 306:447–67

Amaldi, U. 1967. *Rev. Mod. Phys.* 39:649–56

Amann, J. F., Barnes, P. D., Doss, K. G. R., Dytman, S. A., Eisenstein, R. A., Sherman, J. D., Wharton, W. R. 1978. *Phys. Rev. Lett.* 40:758–61

Amato, J., Burman, R. L., Macek, R., Oostens, J., Shlaer, W., Arthur, E., Sobottka, S., Lam, W. C. 1974. *Phys. Rev. C* 9:501–7

Arends, J., Eyink, J., Hegerath, T., Hartmann, H., Mecking, B., Nöldeke, G., Rost, H. 1976. *Phys. Lett. B* 62:411–12

Arends, J., Eyink, J., Hegerath, T., Hartmann, H., Mecking, B., Nöldeke, G., Rost, H. 1978. *Bonn-HE-78-18*

Aslanides, E. 1976. *Proc. Int. Topical Conf. on Meson Physics*, ed. P. D. Barnes, R. A. Eisenstein, L. S. Kisslinger, pp. 204–20, New York: AIP

Aslanides, E., Bauer, T., Bertini, R., Beurtey, R., Bimbot, L., Bing, O., Boudard, A., Brochard, F., Catz, H., Chaumeaux, A., Couvert, P., Duchazeaubeneix, J. C., Duhm, H., Garreta, D., Gorodetzky, P., Habault, J., Hibou, F., Igo, G., Le Bornec, Y., Lugol, J. C., Matoba, M., Tatischeff, B., Terrien, Y. 1977. *Phys. Rev. Lett.* 39:1654–56

Aslanides, E., Bauer, T., Bertini, R., Beurtey, R., Bimbot, L., Bing, O., Boudard, A., Brochard, F., Bruge, G., Catz, H., Chaumeaux, A., Couvert, P., Duchazeaubeneix, J. C., Duhm, H., Garreta, D., Gorodetzky, P., Habault, J., Hibou, F., Igo, G., Le Bornec, Y., Lugol, J. C., Matoba, M., Tatischeff, B., Terrien, Y. 1978. "Scattering and Reaction Cross-Sections Measured at SPES I," ed. G. Bruge, p. 90. *Rapport Interne DPh-N/ME/78-1, Saclay*

Auld, E. G., Haynes, A., Johnson, R. R., Jones, G., Masterson, T., Mathie, E. L., Ottewell, D., Walden, P., Tatischeff, B. 1978. *Phys. Rev. Lett.* 41:462–65

Bachelier, D., Boyard, J. L., Hennino, T., Jourdain, J. C., Radvanyi, P., Roy-Stéphan, M. 1977. *Phys. Rev. C* 15:2139–42

Bacher, A. D., Debevec, P. T., Emery, G. T., Pickar, M. A., Gotow, K., Jenkins, D. A. 1976. *Bull. Am. Phys. Soc.* 21:983

Baker, S. D., Bertini, R., Beurtey, R., Brochard, F., Bruge, G., Catz, H., Chaumeaux, A., Cvijanovich, G., Durand, J. M., Favre, J. C., Fontaine, J. M., Garreta, D., Hibou, F., Legrand, D., Lugol, J. C., Saundinos, J., Thirion, J. 1974. *Phys. Lett. B* 52:57–59

Banaigs, J., Berger, J., Goldzahl, L., Risser, T., Vu Hai, L., Cottereau, M., Le Brun, C. 1973. *Phys. Lett. B* 45:394–98

Banaigs, J., Berger, J., Goldzahl, L., Vu Hai, L., Cottereau, M., Le Brun, C., Fabbri, F. L., Picozza, P. 1974. *Phys. Lett. B* 53:390–92

Bartlett, D. F., Friedberg, C. E., Goulianos, K., Hammerman, I. S., Hutchinson, D. P. 1970. *Phys. Rev. D* 1:1984–95

Bauer, T., Beurtey, R., Boudard, A., Bruge, G., Chaumeaux, A., Couvert, P., Duhm,

158   MEASDAY & MILLER

H. H., Garreta, D., Matoba, M., Terrien, Y., Aslanides, E., Bertini, R., Brochard, F., Gorodetzky, P., Hibou, F., Bimbot, L., Le Bornec, Y., Tatischeff, B., Dillig, M. 1977. *Phys. Lett. B* 69:433–36

Bent, R. D., Debevec, P. T., Pile, P. H., Pollock, R. E., Marrs, R. E., Green, M. C. 1978. *Phys. Rev. Lett.* 40:495–98

Bethe, H. A. 1958. *Ann. Phys. NY* 3:190–240

Betz, M. 1977. "The (π,d) reaction." PhD thesis. MIT. Cambridge, Mass. 150 pp.

Beurtey, R., Bimbot, L., Boudard, A., Bruge, G., Chaumeaux, A., Couvert, P., Escudié, J. L., Fontaine, J. M., Garçon, M., Le Bornec, Y., Schecter, L., Tabet, J. P., Terrien, Y. 1978a. See Aslanides et al 1978, p. 74

Beurtey, R., Bimbot, L., Bing, O., Boudard, A., Brissaud, I., Bruge, G., Chaumeaux, A., Couvert, P., Escudié, J. L., Garçon, M., Garreta, D., Gorodetzky, P., Le Bornec, Y., Schecter, L., Tabet, J. P., Tatischeff, B., Terrien, Y. 1978b. See Aslanides et al 1978, p. 92

Bolsterli, M. 1976. *Phys. Rev. D* 14:2008–15

Bolsterli, M., Gibbs, W. R., Gibson, B. F., Stephenson, G. J. 1974. *Phys. Rev. C* 10:1225–26

Booth, N. E. 1963. *Phys. Rev.* 132:2305–8

Bowles, T., Geesaman, D. F., Holt, R. J., Jackson, H. E., Laszewski, R. M., Specht, J. R., Rutledge, L. L. Jr., Segel, R. E., Redwine, R. P., Yates-Williams, M. A. 1978. *Phys. Rev. Lett.* 40:97–99

Brack, M., Riska, D. O., Weise, W. 1977. *Nucl. Phys. A* 287:425–50

Brockmann, R., Dillig, M. 1977. *Phys. Rev. C* 15:361–64

Brueckner, K., Serber, R., Watson, K. M. 1951. *Phys. Rev.* 81:575–78

Cheung, C. Y., Henley, E. M., Miller, G. A. 1979. *Phys. Rev. Lett.* In press

Chew, G. F., Low, F. E. 1956. *Phys. Rev.* 101:1570–79

Chinowsky, W., Steinberger, J. 1954. *Phys. Rev.* 95:1561–64

Clark, D. L., Roberts, A., Wilson, R. 1951. *Phys. Rev.* 83:649

Cosper, S. W., McGrath, R. L., Maples, C. C., Goth, G. W., Fleming, D. G. 1968. *Phys. Rev.* 176:1113–19

Coupat, B., Bertin, P. Y., Isabelle, D. B., Vernin, P., Gerard, A., Miller, J., Morgenstern, J., Picard, J., Saghai, B. 1975. *Phys. Lett. B* 55:286–88

Couvert, P., Bruge, G., Beurtey, R., Boudard, A., Chaumeaux, A., Garçon, M., Garreta, D., Gugelot, P. C., Moss, G. A., Platchkov, S., Tabet, J. P., Terrien, Y., Thirion, J., Bimbot, L., Le Bornec, Y., Tatischeff, B. 1978. *Phys. Rev. Lett.* 41:530–33

Dahlgren, S., Grafström, P., Höistad, B., Åsberg, A. 1971. *Phys. Lett. B* 35:219–21

Dahlgren, S., Grafström, P., Höistad, B., Åsberg, A. 1973a. *Nucl. Phys. A* 204:53–64

Dahlgren, S., Grafström, P., Höistad, B., Åsberg, A. 1973b. *Nucl. Phys. A* 211:243–53

Dahlgren, S., Grafström, P., Höistad, B., Åsberg, A. 1973c. *Phys. Lett. B* 47:439–41

Dahlgren, S., Grafström, P., Höistad, B., Åsberg, A. 1974. *Nucl. Phys. A* 227:245–56

Delacroix, E., Gross, F. 1977. *Phys. Lett. B* 66:337–40

Dieperink, A. E. L., de Forest, T. Jr. 1975. *Ann. Rev. Nucl. Sci.* 25:1–26

Dillig, M., Huber, M. G. 1976a. *Nuovo Cimento Lett.* 16:293–98

Dillig, M., Huber, M. G. 1976b. *Nuovo Cimento Lett.* 16:299–303

Dillig, M., Huber, M. G. 1977. *Phys. Lett. B* 69:429–32

Dollhopf, W., Lunke, C., Perdrisat, C. F., Roberts, W. K., Kitching, P., Olsen, W. C., Priest, J. R. 1973. *Nucl. Phys. A* 217:381–99

Dolnick, C. L. 1971. *Nucl. Phys. B* 22:461–77

Domingo, J. J., Allardyce, B. W., Ingram, C. H. Q., Rohlin, S., Tanner, N. W., Rohlin, J., Rimmer, E. M., Jones, G., Girardeau-Montant, J. P. 1970. *Phys. Lett. B* 32:309–12

Dover, C. B., Ernst, D. J., Friedenberg, R. A., Thaler, R. M. 1974. *Phys. Rev. Lett.* 33:728–31

Durbin, R., Loar, H., Steinberger, J. 1951. *Phys. Rev.* 83:646–48

Eisenberg, J. M., Noble, J. V., Weber, H. J. 1974. *Proc. Conf. High Energy Physics and Nuclear Structure, Uppsala, 1973,* ed. G. Tibell. Amsterdam: North-Holland

Eisenberg, J. M., Noble, J. V., Weber, H. J. 1975. *Phys. Rev. C* 11:1048–50

Eisenstein, R. A., Miller, G. A. 1975. *Phys. Rev. C* 11:2001–7

Elton, L. R. B., Swift, A. 1967. *Nucl. Phys. A* 94:52–72

Ericson, M., Ericson, T. E. O. 1966. *Ann. Phys. NY* 36:323–62

Ernst, D. J., Johnson, M. B. 1978. *Phys. Rev. C* 17:247–58

Ernst, D. J., Miller, G. A. 1975. *Phys. Rev. C* 12:1962–67

Evans, M. L., Glass, G., Hiebert, J. C., Jain, M., Kenefick, R. A., Northcliffe, L. C., Bonner, B. E., Simmons, J. E., Bjork, C. W., Riley, P. J., Bryant, M. C., Cassapakis, C. G., Dieterle, B., Leavitt, C. P., Wolfe, D. M., Werren, D. W. 1976. *Phys. Rev. Lett.* 36:497–500

Fäldt, G. 1972. *Phys. Rev. C* 5:400–12

Fearing, H.W. 1974. *Phys. Lett. B* 52:407–10

Fearing, H. W. 1975a. *Phys. Rev. C* 11:1210–26

Fearing, H. W. 1975b. *Phys. Rev. C* 11:1493–96

Fearing, H. W. 1977. *Phys. Rev. C* 16:313–21

Fearing, H. W. 1979. In preparation

Ferrari, E., Selleri, F. 1961. *Nuovo Cimento* 27:1450–83

Ferrari, E., Selleri, F. 1962. *Suppl. Nuovo Cimento* 28:454

Fredrickson, D. H., Carroll, J., Goitein, M., Macdonald, B., Perez-Mendez, V., Stetz, A., Heusch, C. A., Kline, R. V. 1976. *LBL Rep. 4838*

Friar, J. L. 1975. *Phys. Rev. C* 10:955–57

Friesel, D. L., Singh, P. P., Ward, T. E., Doron, A., Yavin, A. 1978. *IUCF Technical and Scientific Report*, pp. 36–37

Gell-Mann, M., Watson, K. M. 1954. *Ann. Rev. Nucl. Sci.* 4:219–70

Glauber, R. J. 1959. *Lectures in Theoretical Physics*, ed. W. E. Brittan, pp. 315–414. New York: Intersciences

Goplen, B., Gibbs, W. R., Lomon, E. L. 1974. *Phys. Rev. Lett.* 32:1012–15

Grabowski, J., Fleming, D. G., Vogt, E. W. 1976. *Can. J. Phys.* 54:870–88

Green, A. M., Maqueda, E. 1979. *Nucl. Phys.* In press

Green, A. M., Niskanen, J. A. 1976. *Nucl. Phys. A* 271:503–24

Greider, K. R. 1961. *Phys. Rev.* 122:1919–20

Grossman, Z., Lenz, F., Locher, M. P. 1974. *Ann. Phys. NY* 84:348–431

Harting, D., Kluyver, J. C., Kusumegi, A., Rigopoulos, R., Sachs, A. M., Tibell, G., Vanderhaeghe, G., Weber, G. 1960. *Phys. Rev.* 119:1716–25

Henley, E. M. 1969. *Isospin in Nuclear Physics*, ed. D. H. Wilkinson, pp. 16–72. Amsterdam: North-Holland

Ho, H. W., Alberg, M., Henley, E. M. 1975. *Phys. Rev. C* 12:217–24

Höistad, B. 1977. *Proc. 7th Int. Conf. on High-Energy Physics and Nuclear Structure*. Ed. M. P. Locher, pp. 215–23. Basel, Stuttgart: Birkhäuser

Höistad, B. 1978. *Advances in Nuclear Physics*, Vol. 11. New York: Plenum

Höistad, B., Johansson, T., Jonsson, O. 1978. *Phys. Lett. B* 73:123

Höistad, B., Dahlgren, S., Johansson, T., Jonsson, O. 1979a. *Nucl. Phys.* In press

Höistad, B., Johansson, T., Jonsson, O. 1979b. *Phys. Lett. B* 79:385–88

Jena, S., Kisslinger, L. S. 1974. *Ann. Phys. NY* 85:251–82

Jones, G. 1978. *Nucleon-Nucleon Interactions—1977 (Vancouver)*, ed. H. Fearing, D. Measday, A. Strathdee, pp. 292–304.

New York: AIP

Jones, W. B., Eisenberg, J. M. 1970. *Nucl. Phys. A* 154:49–64

Källne, J., Thiessen, H. A., Morris, C. L., Verbeck, S. L., Burleson, G. R., Devereaux, M. J., McCarthy, J. S., Bolger, M. E., Moore, C. F., Goulding, C. A. 1978. *Phys. Rev. Lett.* 40:378–81

Keating, M. P., Wills, J. G. 1973. *Phys. Rev. C* 7:1336–40

Keister, B. D., Kisslinger, L. S. 1979. *Nucl. Phys.* In press

Kisslinger, L. S. 1955. *Phys. Rev.* 98:761–65

Kisslinger, L. S., Miller, G. A. 1975. *Nucl. Phys. A* 254:493–512

Köhler, H. S. 1960. *Phys. Rev.* 118:1345–50

Kovash, M. A., Blatt, S. L., Boyd, R. N., Donoghue, T. R., Hausman, H. J., Bacher, A. D. 1978. *Bull. Am. Phys. Soc.* 23:926

Landau, R. H., Tabakin, F. 1972. *Phys. Rev. D* 5:2746–54

Le Bornec, Y., Tatischeff, B., Bimbot, L., Brissaud, I., Holmgren, H. D., Källne, J., Reide, F., Willis, N. 1976. *Phys. Lett. B* 61:47–49

Lee, H. K., McManus, H. 1971. *Nucl. Phys. A* 167:257–70

Lo, J. 1977. *Proc. 7th Int. Conf. on High Energy Physics and Nuclear Structure*, ed. M. P. Locher. Basel, Stuttgart: Birkhäuser

Lock, W. O., Measday, D. F. 1970. *Intermediate Energy Physics*. London: Methuen

Londergan, J. T., McVoy, K. W., Moniz, E. J. 1974. *Ann. Phys. NY* 86:147–77

Londergan, J. T., Nixon, G. D., Walker, G. E. 1976. *Phys. Lett. B* 65:427–31

MacDonald, R., Beder, D. S., Berghofer, D., Hasinoff, M. D., Measday, D. F., Salomon, M., Spuller, J., Suzuki, T., Poutissou, J. M., Poutissou, R., Depommier, P., Lee, J. K. P. 1977. *Phys. Rev. Lett.* 38:746–49

Mandelstam, S. 1958. *Proc. R. Soc. A* 244:491–523

Mandl, F., Regge, T. 1955. *Phys. Rev.* 99:1478–83

Matthews, J. L., Bertozzi, W., Leitch, M. J., Peridier, C. A., Roberts, B. L., Sargent, C. P., Turchinetz, W., Findlay, D. J. S., Owens, R. O. 1977. *Phys. Rev. Lett.* 38:8–10

Miller, G. A. 1974. *Nucl. Phys. A* 224:269–300

Miller, G. A., Phatak, S. C. 1974. *Phys. Lett. B* 51:129–32

Miller, G. A. 1977. *LAMPF Summer School on Nuclear Structure with Pions and Protons*, ed. R. L. Burman, E. F. Gibson, pp. 96–136, LA-6926c, Los Alamos

Miller, L. D., Weber, H. J. 1976. *Phys. Lett. B* 64:279–82

Miller, L. D., Weber, H. J. 1978. *Phys. Rev. C* 17:219–26

Neganov, B. S., Parfenov, L. B. 1958. *Sov. Phys. JETP* 7:528–29

Negele, J. W. 1970. *Phys. Rev. C* 1:1260–1325

Niskanen, J. A. 1977. *Phys. Lett. B* 1978. 71:40–42

Niskanen, J. A. 1978. *Nucl. Phys. A* 298:417–31

Noble, J. V. 1975. *Nucl. Phys. A* 244:526–32

Noble, J. V. 1976. See Aslanides 1976, pp. 221–36

Noble, J. V. 1978. *Phys. Rev. C* 17:2151–58

Panofsky, W. K. H., Aamodt, R. L., Hadley, J. 1951. *Phys. Rev.* 81:565–74

Pile, P. H. 1978. *Near threshold positive pion production by protons on nuclei.* PhD thesis, Indiana Univ., Bloomington, Ind. 144 pp.

Pile, P. H., Bent, R. D., Pollock, R. E., Debevec, P. T., Marrs, R. E., Green, M. C., Sjoreen, T. P., Soga, F. 1979. *Phys. Rev. Lett.* 42:1461–64

Poirier, J. A., Pripstein, M. 1963. *Phys. Rev.* 130:1171–77

Preedom, B. M., Darden, C. W., Edge, R. D., Marks, T., Saltmarsh, M. J., Gabathuler, K., Gross, E. E., Ludemann, C. A., Bertin, P. Y., Blecher, M., Gotow, K., Alster, J., Burman, R. L., Perroud, J. P., Redwine, R. P. 1978. *Phys. Rev. C* 17:1402–8

Reitan, A. 1971. *Nucl. Phys. B* 29:525–28

Reitan, A. 1972. *Nucl. Phys. B* 50:166–93

Richard-Serre, C., Hirt, W., Measday, D. F., Michaelis, E. G., Saltmarsh, M. J., Skarek, P. 1970. *Nucl. Phys. B* 20:413–40

Rose, C. M. 1967. *Phys. Rev.* 154:1305–13

Rost, E., Kunz, P. D. 1973. *Phys. Lett. B* 43:17–19

Rost, E., Shepard, J. R. 1975. *Phys. Lett. B* 59:413–15

Ruderman, M. 1952. *Phys. Rev.* 87:383–84

Sachs, A. M., Winick, H., Wooten, B. A. 1958. *Phys. Rev.* 109:1733–49

Schiff, D., Tran Thanh Van, J. 1968. *Nucl. Phys. B* 5:529–59

Schoch, B. 1978. *Phys. Rev. Lett.* 41:80–82

Spuller, J., Measday, D. F. 1975. *Phys. Rev. D* 12:3550–55

Stadler, H. L. 1954. *Phys. Rev.* 96:496–502

Tatischeff, B., Bimbot, L., Frascaria, R., Le Bornec, Y., Morlet, M., Willis, N., Beurtey, R., Bruge, G., Couvert, P., Garreta, D., Legrand, D., Moss, G., Terrien, Y. 1976. *Phys. Lett. B* 63:158–60

Wall, N. S., Craig, J. N., Berg, R. E., Ezrow, D., Holmgren, H. D. 1974. *Proc. Uppsala Conf. on High Energy Physics and Nuclear Structure.* Amsterdam, North-Holland. pp. 279

Wall, N. S., Craig, J. N., Ezrow, D. 1976. *Nucl. Phys. A* 268:459–68

Weber, H. J., Eisenberg, J. M. 1979. *Nucl. Phys. A* 312:201–06

Wilson, S. S., Longo, M. J., Young, K. K., Haddock, R. P., Helland, J. A., Schrock, B. L., Cheng, D., Perez-Mendez, V. 1971. *Phys. Lett. B* 35:83–86

Young, S. K., Gibbs, W. R. 1978. *Phys. Rev. C* 17:837–41

Ann. Rev. Nucl. Part. Sci. 1979. 29: 161–202
Copyright © 1979 by Annual Reviews Inc. All rights reserved

ALPHA TRANSFER REACTIONS IN LIGHT NUCLEI

✳5605

H. W. Fulbright[1]

Nuclear Structure Research Laboratory, University of Rochester, Rochester, New York 14627

CONTENTS

INTRODUCTION

General Description

This paper discusses recent progress in the study of alpha transfer reactions, focusing principally on the target mass region $A = 16$ to $A = 64$, the sd shell and the lower part of the fp shell. In that region

[1] Work supported by the National Science Foundation.

0163-8998/79/1201-0161$01.00

enough information has accumulated during the past seven or eight years to reveal systematic trends and areas of agreement and disagreement between experimental results and theoretical predictions.

The alpha particle has remarkable properties. Its ground state consists of two protons and two neutrons in the most symmetrical arrangement possible, with spin $J = 0$, parity $\pi = +$, and isospin $T = 0$. It has no excited state below 20 MeV. Its binding energy is high, 28.3 MeV. Removal of one of its nucleons costs about 20 MeV. When a nucleus is formed in a state from which it can decay by single-nucleon emission, the lifetime is very short, roughly 10^{-20} s, governed by the penetrability of a barrier due to the combined effects of nuclear and Coulomb potentials and angular momentum. Lifetimes of natural alpha particle emitters are much longer, in the range from about 3×10^{-7} s to 10^{15} years. Evidently each alpha particle emitted consists of two neutrons and two protons clustered in the form necessary for escape. The tight binding makes alpha particle emission energetically favorable. Gamow's early theory of alpha decay, which assumed the existence of an alpha particle moving inside a potential well and quantum-mechanical tunneling through the barrier, explained quantitatively many of the systematic characteristics of alpha decay, in particular the variation of lifetime with decay energy (Geiger-Nuttall law). Recent theories of alpha decay take into account details of the wave functions of the nuclei involved.

The inverse of alpha decay, alpha particle capture, can in principle be observed in laboratory experiments. Alpha particle capture (or scattering) is a resonant process exhibiting variations in cross section with center-of-mass (c.m.) bombarding energy corresponding to the energy and width Γ of the level involved. The widths of the long-lived ground states of natural emitters are of course very small.

Alpha capture and resonant scattering experiments are feasible in cases where the required alpha particle energy is high enough with respect to the barrier height to give sufficient cross sections. Many such experiments have been made among the lighter elements, but because of the barrier problem few have been made with targets of mass greater than 40. The levels accessible via simple alpha capture are usually limited to an energy range of several MeV.

There is a more generally useful process. If the four nucleons to be captured are carried up to the nuclear surface as part of an energetic heavy-ion projectile, from which they are transferred to the target nucleus, the conservation laws will allow formation of both bound and unbound states. The inverse process, the transfer of four nucleons to the passing projectile, may also occur. Thus we may write $A + a \rightarrow B + b$, where $B = A \pm 4$ and $a = b \pm 4$. Inasmuch as the target in a nuclear reaction is always in its ground state, the only two states of A and B between which both types of experiments can be made are the ground states. Typical four-nucleon transfer reactions are $(^6\text{Li},\text{d})$, $(^7\text{Li},\text{t})$, $(^{13}\text{C},^9\text{Be})$, $(^{16}\text{O},^{12}\text{C})$ and their inverses, and $(^3\text{He},^7\text{Be})$ and $(^4\text{He},^8\text{Be})$. The cross sections for formation of various levels of nucleus B are found from the energy spectra of the outgoing particles seen at various angles and from the conditions of the experiment. The derivation of further information connecting the ground state of the target nucleus and the levels of the residual nucleus to which transitions have occurred is possible only if the reaction mechanism is known and if there exists a detailed theory for it. At the microscopic level indeed we should expect complications because four nucleons are being transferred coherently into shell model orbits and many orbits may be involved. The picture is similar to that for two-nucleon transfer reactions, already extensively studied; a formal connection between two- and four-nucleon transfer reactions is discussed later.

Consider the reaction A(a,b)B. In the classical compound-nucleus (CN) mode, the projectile a coalesces with the nucleus A forming a compound nucleus which, after much nucleon-nucleon interaction, decays by emitting particle b, usually leaving nucleus B in a highly excited state. The CN mode is usually dominant for small bombarding energies and for the higher continuum part of the spectrum. It is usually the principal contributor to the total reaction cross section. The variation of its cross section with beam energy may be rapid, indicating resonances. The relative importance of its contributions to the discrete spectrum of low-lying levels, which are of primary interest to us, varies with the structure of the nuclei involved and with the strength and nature of the competing modes. When a level is excited purely by the CN process the corresponding angular distribution is symmetrical about 90°. A well-known example is the $^{12}\text{C}(^6\text{Li},\text{d})^{16}\text{O}$ reaction to the 8.87-MeV $J^\pi = 2^-$ level, where the selection rules favor the CN mode and the symmetrical angular distribution is well fitted via Hauser-Feshbach theory (Bethge et al 1967).

Reaction modes involving less complete coalescence are also possible. For example, the four nucleons may be transferred sequentially or sequentially in pairs on a much shorter time scale. The projectile or the target nucleus may be raised to an excited state during the course of the reaction. The reaction may proceed through a number of different coherent modes. In most of these cases there is no reaction theory of practical spectroscopic value. Two-step coupled-channel reactions are an exception to be considered later.

Fortunately, the simplest mechanism, direct stripping or pickup, usually predominates when high energy beams are used, for example for $(^6\text{Li},\text{d})$

when $E \gtrsim 25$ MeV, and in every case well above the Coulomb barrier. In the direct reaction theory one assumes that an alpha cluster preexisting in the projectile is simply transferred intact to A where its four nucleons are neatly incorporated to form B, with minimal disturbance of A. The transfer occurs in a single step during the short time when the projectile is passing the target nucleus. Accordingly, in a stripping reaction the angular momentum transferred to the capturing nucleus A is $J = L$, where L is the orbital angular momentum of the captured alpha. The change of parity is $\Delta\pi = (-1)^L$. The change of isospin is $\Delta T = 0$. The evidence for this model comes mainly from the agreement normally found between angular distributions observed and the diffraction-type distributions calculated from distorted-wave Born approximation (DWBA) theory. Other supporting evidence comes from the fact that in cases where the parity-change rule is violated (unnatural parity cases) the strength is weak, usually less than 10% that for normal cases. All the more complicated reaction modes would typically give angular distributions different from those of direct alpha stripping or pickup. A smooth variation of cross sections with bombarding energy in accord with DWBA predictions also can support the alpha transfer model, but this evidence is not equally strong for all types of four-nucleon transfer reaction. See Figure 1.

Figure 2 illustrates schematically the transformation connecting two neutrons and two protons in jj coupling shell model orbits, and the alpha particle that they become in alpha decay or alpha pickup reactions (Harada 1961, Kurath & Towner 1974). The symbol j stands for the

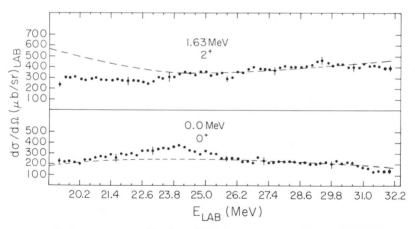

Figure 1 Excitation curves for ^{16}O(^{6}Li,d)^{20}Ne ground state and first excited state at 7.5° laboratory angle. The dashed lines are DWBA predictions (House & Kemper 1978). Curves for the same reaction are given by Gunn et al (1977) and for ^{60}Ni(^{6}Li,d)^{64}Zn by Fulbright & Bennett (1978).

quantum numbers (nlj). The alpha particle wave function is best expressed in LS coupling with L = orbital angular momentum, $S = 0$, and $T = 0$. The four nucleons are first coupled as a proton-proton (di-proton) pair and a neutron-neutron (di-neutron) pair. The pairs are then coupled together under the following restriction: in the alpha particle the protons and neutrons are in spatially symmetric s states with intrinsic spin zero, so that only those components in the original pairs will contribute. A large number of different di-neutron di-proton coupling combinations may be able to contribute, e.g. 70 for $L = 0$ transfers in the fp shell. Calculations based on this scheme were made by Kurath & Towner (1974) to find alpha transfer strengths in the fp shell for various assumed simple configurations of two neutrons and two protons, as is discussed below. The predominance of the direct stripping and pickup modes and the availability of efficient DWBA computer programs make possible comparisons of experimental results with theoretical predictions based on nuclear models. We usually assume that if the experimental angular distribution is well fitted via DWBA calculations, the reaction proceeds entirely by direct alpha transfer. Then the strength of the transition is expressed in the form of an experimental spectroscopic factor defined as

$$S_{exp} = \frac{(d\sigma/d\Omega)_{exp}}{(d\sigma/d\Omega)_{DWBA}} \qquad\qquad 1.$$

calculated as an average (with reasonable weightings) over the measured angular distribution. In principle S_{exp} can be compared directly with a

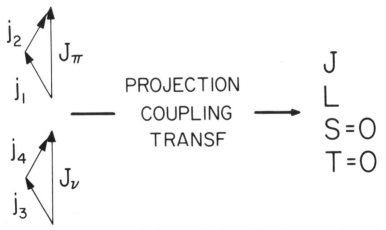

Figure 2 Schematic illustration showing coupling of two shell model protons with quantum numbers j_1 and j_2 to a resultant J_π $[j = (n,l,j)$ etc] and similarity for two neutrons to a resultant J_ν, projecting out of 0s spin-zero parts, coupling and transformation to L-S representation appropriate for an alpha particle cluster.

corresponding theoretical spectroscopic factor S produced by calculations based on nuclear model wave functions. In this way we can test model wave functions of levels that can be connected via alpha transfer. Of course, it is precisely the goal of alpha transfer spectroscopy to determine as much as possible about those wave functions. In practice the accuracy of the result is limited by uncertainties in $(d\sigma/d\Omega)_{DWBA}$ due to the approximate nature of the theory and by ambiguities in the input parameters, as discussed at the end of the next section and in the section on data reduction.

History

In 1970, when Klaus Bethge's article in *Annual Review of Nuclear Science* (Bethge 1970) summarizing the pioneering work appeared, alpha transfer reactions were already known to be a promising tool for nuclear spectroscopy. Most of the experiments done had been confined to target nuclei in the p or sd shell. Use of the (^{6}Li,d) and (^{7}Li,t) reactions had been explored, mostly with beam energies below 20 MeV, and evidence of a strong direct transfer component in each reaction had been seen. Also recognized was the fact that ^{7}Li, viewed as an alpha-triton cluster, is in a p state of relative motion, making the analysis of (^{7}Li,t) reactions more difficult than the corresponding (^{6}Li,d) process in which the ^{6}Li can be viewed as an alpha particle and a deuteron in a relative s state. On the other hand, comparisons of the angular distributions from the two reactions had led to a suspicion that the (^{6}Li,d) reaction mechanism included a substantially larger compound-nucleus component and that (^{6}Li,d) data would therefore tend to be difficult to analyze. At that time heavy-ion accelerators were rapidly appearing. Alpha transfer data had already been obtained from the (^{16}O,^{12}C) reaction, which was shortly to be investigated vigorously at Saclay and elsewhere (Lemaire 1973). Studies had also been made with such reactions as (^{3}He,^{7}Be) and (^{4}He,^{8}Be). Experimental results in many cases did not yet include angular distributions. Nonetheless the strong suspicion existed that direct reaction mechanisms would usually prove dominant, especially at higher beam energies. Alpha transfer cross sections were expected to fall steadily with increasing target mass, on the average, and doubt existed that they would be of practical use with targets heavier than calcium. A multinucleon transfer theory had appeared (Smirnov & Chlebowska 1961, Rotter 1968a,b, 1969), later to be streamlined and extended for alpha transfer reactions by Ichimura et al (1973) and others. Thus by 1970 a vigorous start had already been made both in experiment and theory. The promise for nuclear spectroscopy had been recognized and some of the problems to be explored had been stated.

Since then the field has been active, both in experiment and in theory. Several types of alpha transfer reactions have been shown to give the strongly L-transfer-dependent angular distributions characteristic of direct reactions when the beam energy is high and selection rules are favorable. This has made possible spin and parity assignments. The (^{6}Li,d) reaction and its inverse have proved to be the most generally useful alpha transfer reactions, as is considered later.

New, efficient computer codes have appeared, facilitating DWBA analysis. The zero-range program DWUCK of Kunz (P. D. Kunz, University of Colorado report, unpublished) is widely used. Of equal importance are the "exact" finite-range codes such as LOLA (Perrenoud & DeVries 1971, DeVries 1973a), which are free of the zero-range assumption. They are necessary for proper analysis of heavy-ion stripping and they offer good prospects for absolute cross-section calculations.

A substantial start has been made in connecting alpha transfer data with details of nuclear structure. This involves the magnitudes of cross sections as well as the shapes of angular distributions. Besides the experimental problems, there are problems in reducing the data, even where the reaction mode can be assumed essentially entirely direct. No DWBA procedure has yet been established guaranteed to reduce cross sections to absolute spectroscopic factors S_{exp}, or even to produce correct relative S_{exp} values for data from targets distributed over a range of masses, although progress is being made in these directions. Finally, there are difficulties in calculating theoretical spectroscopic factors for comparison with the experimental values. Working with shell model wave functions is desirable, but technical problems are associated with the large size of the basis sets typically required. To avoid these difficulties certain simplifying methods have been employed. The most successful of these is based on the SU(3) model of Elliott (1958), which depends on a spatial symmetry of shell model wave functions in the mass region $A \leq 28$, where spin-orbit coupling is not too strong. There the simplest SU(3) wave functions often have a large overlap with alpha cluster states (Bayman & Bohr 1958–1959) and ordinary shell model wave functions (Harvey 1968) and evaluation of S-values by SU(3) methods is relatively straightforward (for example, see Draayer 1975). Two other devices that have been employed with fp shell and heavier nuclei are the pairing vibration (PV) (Betts 1975, 1977) and interacting boson approximation (IBA) (Bennett & Fulbright 1978) methods, both of which originated in the study of two-nucleon transfer reactions. They are rough approximations, but each has advantages that are mentioned later. Recent calculations of alpha transfer spectroscopic factors made with detailed shell model wave functions are also discussed later.

Although many calculations involving cluster model wave functions have been made in recent years, often using the resonating group method (Wheeler 1937), their results are not described here. For examples see Goldberg et al (1975) and Van Oers et al (1978) and references they cite. Quartet states (Arima et al 1970) also are not discussed here.

Reviews of alpha transfer reactions are given by Bethge (1970), Becchetti (1978, 1979), Lemaire (1973), and Mallet-Lemaire (1978). Pertinent information can be found in many other places, including the conference proceedings cited above, Kurath (1976), Hodgson (1978), Anyas-Weiss et al (1974), and Siemssen (1978).

EXPERIMENTAL

Types of Reactions

Direct alpha stripping or pickup experiments forming discrete low-lying levels are the main source of spectroscopic information considered here. Reactions leading to continuum levels via compound-nucleus, doorway state, or preequilibrium processes are of little interest for spectroscopy. Angular correlation studies of the alpha particles emitted promptly by decaying unbound levels formed in alpha transfer reactions have led to a number of spin and parity assignments (Artemov et al 1971, 1975, 1977, Sanders 1977, Sanders et al 1977). Results from alpha decay and resonant alpha scattering and capture have been compared with those from transfer reactions and in a number of cases consistency has been seen, as is discussed below.

Experience has shown that when all factors, including experimental convenience, spectroscopic power, and ease of analysis of results, are weighed, among direct alpha transfer reactions (^{6}Li,d) and (d,^{6}Li) are generally the most useful. Since mass-5 nuclei are unstable, ^{6}Li is the lightest projectile available for alpha stripping. The stopping-power problem is less severe with Li than with heavier ions, so self-supporting target foils can often be used without serious loss of resolution. For unique L-transfer values up to 4 or 5 the (^{6}Li,d) and (d,^{6}Li) reactions yield strongly distinctive angular distributions reminiscent of single-nucleon transfer, whereas some of the reactions with heavier ions do not. For example, in the case of (^{16}O,^{12}C), angular distributions seen with a 40-MeV beam show similar broad peaks for all L transfers from 0 to 4 (Lemaire 1973), although at 60 MeV the results are somewhat more favorable, at least for the $L = 0$ transitions (Mallet-Lemaire 1978). The (^{6}Li,d) cross sections, while small, are not prohibitively small, ranging up to about 300 μb/sr in the sd shell and 100 μb/sr in the fp shell. An overall view of the cross-section variation found with targets throughout the periodic table for forming ground states is shown in Figure 3 (Becchetti

1978). Angular distributions from (^{7}Li,t) are more difficult to fit, particularly $L = 0$ cases; finite-range DWBA analysis is required because of the structure of ^{7}Li, for present purposes a ^{4}He-^{3}H cluster in a p state. The (p,pα) measurements of Roos et al (1977) show that the ^{6}Li wave function has a large overlap ($\sim 58\%$) with an (alpha + d) cluster structure in an s state. The fact that the deuteron has no excited state is an advantage in that the simplest spectra possible are obtained. Also many different spin states may be excited in a single experiment, i.e. there is a broad "Q window."

The (^{6}Li,d) reaction has a few practical limitations. The breakup process ^{6}Li → α + d produces a continuum of deuterons that, added to the continuum from unresolved levels, obscures discrete lines in the upper excitation part of the spectrum. With heavy targets, where the reaction Q is negative and cross sections are small, light-element contaminants can interfere significantly. In this respect the (d,^{6}Li) reaction also suffers with light targets, but the relative cross sections are more favorable.

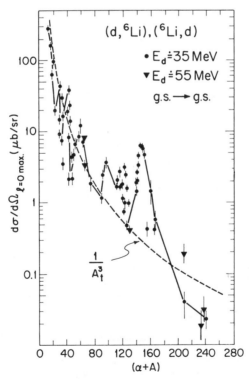

Figure 3 Variation of cross section for ground-state (^{6}Li,d) and (d,^{6}Li) transitions throughout the periodic table. Values are cross sections at first secondary maximum (Becchetti 1978).

This report deals largely with the results of (^{6}Li,d) and (d,^{6}Li) reactions because they have overwhelmingly produced the greatest amount of useful spectroscopic information. Nonetheless, for some purposes other reactions may be especially valuable. For example, one can make use of the two alpha particles into which ^{8}Be decays to identify the (^{4}He,^{8}Be) reaction; thus Sanders et al (1977) were able to study reactions of the type ^{16}O(^{12}C,^{8}Be)^{20}Ne*(α)^{16}O conveniently observing the ^{8}Be at $\theta = 0°$ to simplify the angular correlation analysis.

Here is a partial list of targets used in recent (^{6}Li,d), (d,^{6}Li), and (^{7}Li,t) studies:

(^{6}Li,d), sd shell:
^{16,17,18}O, ^{19}F, ^{20,21,22}Ne, ^{23}Na, ^{24,25,26}Mg, ^{27}Al, ^{28,29,30}Si, ^{31}P, ^{32}S, ^{35}Cl, ^{36,38,40}Ar.

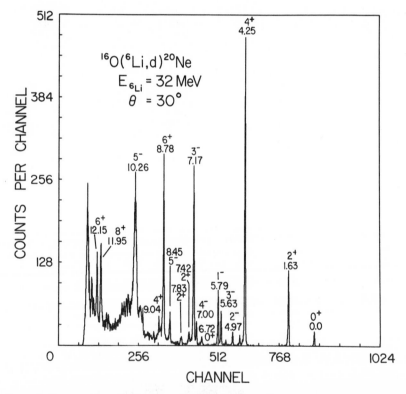

Figure 4   Spectrum from the ^{16}O(^{6}Li,d)^{20}Ne reaction at 32 MeV and 30°. Unnatural parity states are seen at $E_x = 4.97$ MeV (2$^-$) and at 7.00 MeV (4$^-$). The ground-state rotational band and a rotational band based on the 1$^-$ state at 5.79 MeV are strongly excited (Anantaraman et al 1979).

(^{6}Li,d), fp shell:
^{40,42,44,46,48}Ca, ^{46,48,50}Ti, ^{51}V, ^{50,52,54}Cr, ^{54,56,58}Fe, ^{58,60,62,64}Ni, ^{90}Zr.

(d,^{6}Li):
^{14}N, ^{16}O, ^{19}F, ^{20}Ne, ^{24,25,26}Mg, ^{40,42}Ca, ^{46,48}Ti, ^{56}Fe, ^{58}Ni, and many heavier nuclei up to ^{208}Pb.

(^{7}Li,t):
^{9}Be, ^{12,13}C, ^{11}B, ^{14}N, ^{16}O, ^{20}Ne, ^{24}Mg, ^{28}Si, ^{40}Ca, ^{58}Ni.

Throughout this paper all known corrections to published experimental (and theoretical) data will be applied.

Spectra from (^{6}Li,d) and (d,^{6}Li)

Examples of (^{6}Li,d) spectra are seen in Figures 4, 5, and 6. In the lower excitation region discrete lines are seen. The continuum that builds up in the high excitation region (Figure 5) results from the usual excitation

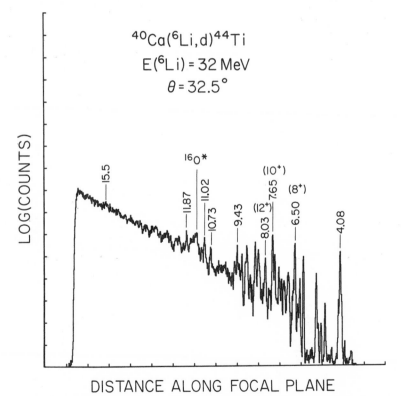

Figure 5 Part of (^{6}Li,d) spectrum from 250-μg/cm^2 ^{40}Ca target on 10-μg/cm^2 carbon backing (Strohbusch et al 1978).

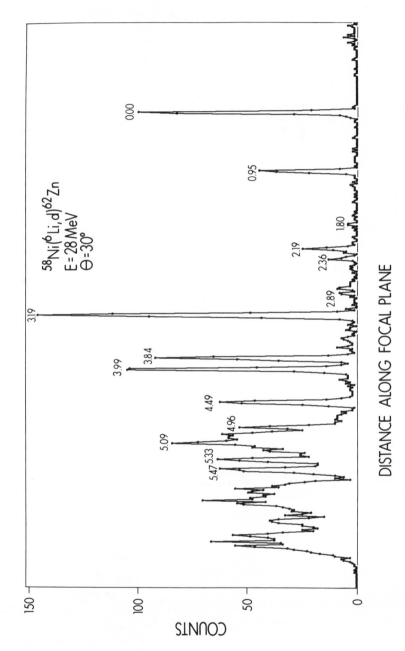

Figure 6 Spectrum from ^{58}Ni(^{6}Li,d)^{62}Zn. A better-than-average spectrum from a clean, self-supporting target (Fulbright et al 1977).

of overlapping levels and to the breakup of ^{6}Li in the vicinity of the target nucleus. Breakup is known to proceed partly via excitation of the 2.185-MeV $J^\pi = 3^+$ state of ^{6}Li and partly via other processes (Scholz et al 1977, Ost et al 1972, 1974, Pfeiffer et al 1973, Castenada et al 1978). The continuum part of the (d,^{6}Li) spectra is usually weaker. Energy resolutions typically range from 30 to 150 keV. Sharp lines are often seen even for formation of unbound levels, as in the ^{58}Ni $\rightarrow$ ^{62}Zn spectrum of Figure 6, where all levels above $E_x = 3.32$ MeV are unstable against alpha particle emission.

In the same spectrum are seen several strong lines from the formation of negative-parity levels, e.g. a $J^\pi = 3^-$ line at 3.19 MeV. Since the ^{58}Ni target has $J^\pi = 0^+$, negative-parity levels cannot be formed by adding four nucleons in the fp shell; another major shell must also be involved. In the simplest picture, one nucleon is probably being added to the $g_{9/2}$ subshell. This shell crossing is common. Theoretically it is expected but inconvenient to handle, partly because of the associated increase in size of the basis space required to represent the states involved. Spurious center-of-mass motion also presents a problem (Ichimura et al 1973).

The (^{6}Li,d) reaction is selective. It favors levels according to the parity selection rule for direct alpha transfer, $\Delta\pi = (-1)^L$, where L is the transferred angular momentum, particularly for high bombarding energies. With fp shell targets no transition violating this rule with a strength greater than a few percent of normal has yet been reported, but few places exist where weak violations could easily be seen (Fulbright et al 1977). With p and sd shell targets, especially at low bombarding energies, weak violations of the parity selection rule are fairly common. Their appearance [e.g. the 2^- and 4^- lines in Figure 4 and examples from ^{24}Mg(^{16}O,^{12}C)^{28}Si in Peng et al (1976)] is evidence that other reaction modes, for example compound nucleus or multistep, are present. The (^{6}Li,d) reaction favors unbound levels known to have large alpha particle widths. It also shows selectivity depending on the structure of the levels populated. For example, in ^{16}O $\rightarrow$ ^{20}Ne (Figure 4) the (sd)^4 ground-state band and the negative-parity (sd)^3(fp) band beginning with the 1^- level at 5.79 MeV are favored. It often favors levels excited via two-nucleon transfer reactions such as (^{3}He,n) and (t,p), as is discussed later. Kurath & Towner (1974) explained the fact that the (^{6}Li,d) reaction frequently favors states also excited strongly by inelastic scattering of alpha particles, e.g. in the case of ^{54}Fe(^{6}Li,d)^{58}Ni and ^{58}Ni(α,α')^{58}Ni: the overlap integrals of (α,α') and α-transfer reaction theory contain common factors such that coherence effects can cause similar enhancements. A corresponding correlation occurs between inelastic scattering and two-nucleon transfer (Broglia et al 1971).

Angular Distributions from ($^6Li,d$) and ($d,^6Li$) Reactions

Figures 7 through 9 show typical angular distributions from (^{6}Li,d) experiments done with sd and fp shell targets. Similar distributions are seen with the (d,^{6}Li) reaction. They have forward-peaked diffraction shapes typical of single-particle transfer. In cases where a single L transfer is expected, the corresponding alpha transfer DWBA calculations usually produce good fits, at least for L values up to 4 or 5. With reactions from a particular target the fits for one or two L values are sometimes inferior to those for other Ls obtained with the same DWBA parameters, as seen in Figure 7. Excellent fits can be seen in Figure 9 for $L = 0, 1, 2, 3,$ and 4 transitions to ^{62}Zn. In some cases of high-lying levels the angular distributions fall smoothly with angle, roughly as the average DWBA trend, but cannot be fitted, e.g. in ^{36}Ar(^{6}Li,d)^{40}Ca, above $E_x = 10$ MeV (C. L. Bennett, J. Tōke, H. W. Fulbright, private communication). In most, but perhaps not all, of these cases the level density is high and the purity of the observation is in doubt. The angular distribution for the entire lower part of the continuum seen in Figure 5 is strongly forward-peaked, probably because of a dominant ^{6}Li breakup contribution indicated by observations on other nuclei (Ollerhead et al 1964, Pfeiffer et al 1973).

Measurements have usually been limited to angles below 70° where the chief spectroscopic value lies, but at least one measurement, made on ^{58}Ni(^{6}Li,d)^{62}Zn at 28 MeV and carried back to 170°, has shown that cross sections continue to fall with angle according to DWBA expectations. Hence this particular measurement failed to yield evidence for a CN contribution (Jundt et al 1975). As would be expected from the alpha transfer model, data from doubly even targets can usually be fitted with calculations for a single value of L transfer, except when unresolved levels are involved. Data from odd-A targets require separation of the expected (incoherent) contributions from several values of L, although a single L is often dominant (for example in ^{21}Ne → ^{25}Mg; Anantaraman et al 1976).

Angular distributions for $L > 4$ become rather flat and characterless; distinguishing between them is difficult. When unnatural parity levels are formed the angular distributions are not like those of stripping. Some have symmetry about $\theta = 90°$ and can be fitted via the Hauser-Feshbach statistical CN theory, for example for ^{12}C(^{6}Li,d)^{16}O, 8.87-MeV 2$^-$ (Bethge et al 1967), various (^{3}He,^{7}Be) reactions (Pisano & Parker 1976), and (^{7}Li,t) processes (Cobern et al 1976). Some angular distributions, not completely symmetrical about 90°, have defied fitting by coupled-channels-Born-approximation (CCBA) calculations; they may arise from a mixture of CN, coupled-channel and multistep processes such as

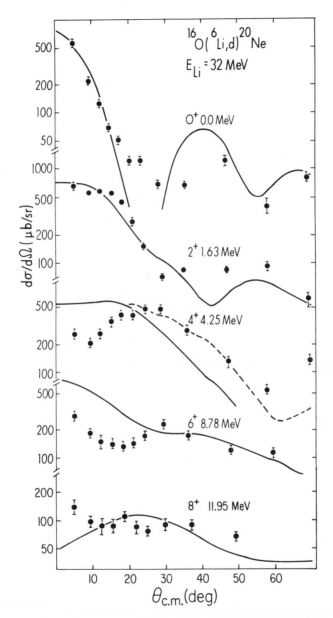

Figure 7 Angular distributions and DWBA curves for members of a ground-state rotational band in ^{20}Ne. For most other states in this experiment fits are better (Anantaraman et al 1979).

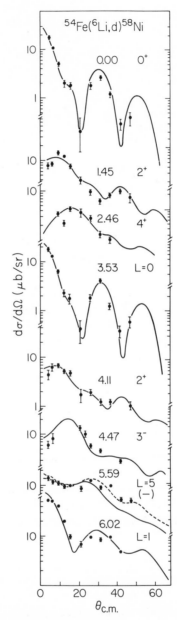

Figure 8 Typical fp shell angular distributions from a 28-MeV experiment on ^{54}Fe (Fulbright et al 1977). DWBA curves averaged over observation angle window.

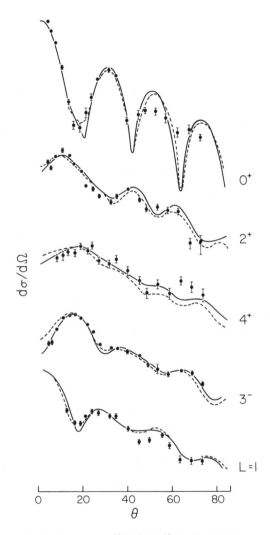

Figure 9 Angular distributions from ^{58}Ni(^{6}Li,d)^{62}Zn. Dashed lines represent DWBA calculations with SBF parameters; solid lines represent DWBA calculations with SBF parameters adjusted via least-squares fitting to the 3^{-} state, and averaged as in Figure 8 (Fulbright et al 1977).

(^{6}Li,α)(α,d) (Gunn et al 1977, Palla et al 1978). Eswaran et al (1979) have analyzed the results of ^{16}O(^{6}Li,dγ)^{20}Ne angular correlation experiments via CCBA theory.

DATA REDUCTION

Use of DWBA Analysis

The simple direct alpha transfer reaction model calls for DWBA analysis according to a model in which the alpha is transferred as a unit. From the analysis come values of S_{exp} (Equation 1) for comparison with theoretical spectroscopic factors S. Many discussions of DWBA theory can be found in the literature, e.g. Satchler (1964), Austern et al (1964), Austern (1970). Many efficient computer codes are available for DWBA calculations, e.g. in zero-range approximation the DWUCK code (P. D. Kunz, University of Colorado report, unpublished), and in full finite-range the LOLA code of DeVries (1973a). The input parameters for the usual zero-range DWBA calculations are (a) numbers characterizing the reaction (L transfer, masses, energies), (b) a set of parameters characterizing the elastic scattering states of the input and output channels between which transitions are calculated (real and imaginary potential well depths, sizes, and shapes), and (c) parameters characterizing the bound-state wave function of the alpha particle after capture, or before pickup (well radius and shape, binding energy, number of nodes N). [Also, the value of a normalization factor D_0^2, stored in the program, can be changed optionally. This parameter absorbs a spectroscopic factor corresponding to the overlap of the projectile a with a ($b+\alpha$) cluster structure. In the Rochester version of DWUCK it normally has the value 10^4 MeV^2 fm^3.] From these the program calculates the required elastic scattering and bound-state wave functions before proceeding to the main calculation. Alternatively, previously calculated wave functions may be introduced.

The number of nodes N specified for the bound-state wave function is obtained from the expression $2N + L = \sum_{i=1}^{4} (2n_i + l_i)$ where n_i and l_i are quantum numbers of harmonic oscillator states into which the four nucleons go (lowest value of n_i is 0).

This gives $2N + L = 8$ for the sd shell, 12 for the fp shell. However, some wave functions include substantial hole components from a lower shell or particle components from a higher shell. For example, the analysis of Fortune et al (1972) indicates the presence of strong fp shell components in the 8.3-MeV $J^\pi = 0^+$ level of ^{20}Ne. Such cases are easy to treat when a clearly dominant simple particle-hole structure is involved. Otherwise, use of a microscopic bound-state wave function calculated beforehand from a known or assumed shell model structure can be considered. For transitions with a change of parity, e.g. when low-lying 1^- or 3^- levels are formed via stripping reactions on doubly even targets, a single nucleon is usually assumed to have occupied a (low-lying) orbit of the next major shell.

In finite-range calculations the wave function of the projectile (in stripping) is required. For ^{6}Li, the wave function of Jain et al (1973) is often used.

The zero-range calculations lead to absolute values of S_{exp} only if a correct overall normalization constant η is available:

$$\frac{d\sigma}{d\Omega}\bigg|_{exp} = S_{exp}\eta \left[D_0^2 \frac{d\sigma}{d\Omega} \right]_{DWBA} \qquad 2.$$

The constant η can be evaluated in a particular case if a trustworthy value of S, known from theory, can be substituted for S_{exp}. In most work in the literature η has simply been set equal to unity, which leads to values of S_{exp} on an arbitrary scale.

In principle exact finite-range DWBA calculations produce absolute cross sections. Unfortunately the results are very sensitive to the values of several input parameters. For example, an increase of 10% in the radius parameter of the bound-state wave function may double the cross section with little change in the angular distribution. As a result calculated cross sections must usually be assumed uncertain by a factor of two or three.

In practice the quality of fits to (^{6}Li,d) data is no better with finite-range than zero-range calculations, and the relative values of S_{exp} found by the two are usually nearly the same, so for reasons of economy most ^{6}Li data have been reduced via zero-range calculations. For proper analysis of the results of heavier-ion reactions such as (^{16}O,^{12}C), where recoil and Coulomb effects are large, finite-range calculations are necessary.

DWBA Parameters for ^{6}Li Reactions

Values of S_{exp} depend critically on the values of the parameters used in the DWBA calculations. Parameters for the elastic scattering waves of the entrance and exit channels are traditionally obtained from optical model analysis of elastic scattering data. For deuterons the parameter set of Newman et al (1967) is often used. In the early 1970s when alpha transfer experiments began in earnest, good ^{6}Li optical model parameters were not known for energies above about 25 MeV. The analysis of ^{6}Li scattering data typically produces ambiguous results, but Chua et al (1976) analyzed 50-MeV elastic scattering data and got sets of parameters that have given excellent fits of (^{6}Li,d) and (d,^{6}Li) angular distributions. Only volume absorption is included. Earlier, a trial-and-error search revealed a parameter set that fitted well a wide range of (^{6}Li,d) data from experiments made in the fp shell at energies near 30 MeV, the SBF set of Strohbusch et al (1975). It includes only surface absorption. Its entrance channel radius includes an allowance for the size of the ^{6}Li ion, i.e. $r \propto (A^{1/3} + 1.9)$ instead of the usual $A^{1/3}$. This is significant, because the cross section is quite sensitive to that parameter: in one case in the Ni region an increase of about 5% in r was enough to reduce the cross section threefold. How the entrance channel radius should be changed with A is important when comparisons are being made across a range of target masses, as is considered below. A common feature of the Chua and SBF parameters is the large real potential well depth of the entrance channel, ~ 250 MeV. A potential set originally employed for ^{40}Ca(^{6}Li,d)^{44}Ti at 32 MeV (Strohbusch et al 1974) has been used throughout the sd shell. Many other sets can be found in the literature. Different sets can all give good angular distribution fits and produce essentially the same relative cross sections for various transitions (from reactions on a single target) whether they involve surface absorption, volume absorption, or both, because the reaction occurs so strongly at the nuclear surface that the interior conditions are relatively unimportant. However, when used to calculate ^{6}Li elastic scattering angular distributions, not all of these sets give good fits.

Unbound Levels

The DWBA programs are written to calculate and use wave functions for bound states of the transferred particle. Unbound levels require special treatment. To estimate S_{exp} for an unbound level transition one can make the DWBA calculation for an artificially small positive value of binding energy, or for a series of small positive values followed by extrapolation. A better procedure (Vincent & Fortune 1970) uses a previously calculated realistic unbound-state wave function in the DWBA calculation. The treatment of unbound-state transitions has also been discussed by Bunakov et al (1971), Arima & Yoshida (1974), Davies et al (1976), and DeVries (1976). Apagyi et al (1976a,b) made detailed calculations of alpha decay amplitudes of various states of ^{16}O.

Reduced Widths

In many cases the same unbound levels seen in alpha transfer reactions have also been studied via resonant scattering or resonant capture of alpha particles or by their alpha decay. Often the reduced widths γ^2_{exp} of R-matrix theory (Lane & Thomas 1958) are known from analyses based on the model of an alpha bound in a quasi-stable orbit (Ajzenberg-Selove 1975, 1976, 1977, 1978, Endt & van der Leun 1978). It is therefore appropriate to express alpha transfer strengths in terms of reduced widths for direct comparison. The relation is

$$\gamma_{exp}^2 = \frac{\hbar^2 r_c}{2\mu} S_{exp} [\phi^{norm}(r_c)]^2, \qquad\qquad 3.$$

where S_{exp} is found in a DWBA analysis using $\phi^{norm}(r_c)$ as the $A + \alpha$ wave function (normalized within a sphere of large, but finite radius), r_c is the channel radius at which the evaluation is made, and μ is the reduced mass of the $A + \alpha$ system (Ichimura et al 1973, Davies et al 1976).

Fortune et al (1973) showed that data from $^{16}O + \alpha$ resonant scattering and $^{16}O(^6Li,d)^{20}Ne$ agreed and that the reduced widths of the 6^+ and 8^+ members of the ground-state rotational band were approximately equal, as was expected theoretically. They apparently assumed $S_{exp} = \Gamma_{exp}/\Gamma_0$, where Γ_0 is the single-particle width. (See Bunakov et al 1971 and Davies et al 1976, but also Arima & Yoshida 1974.) They used single-particle widths calculated for an alpha particle moving in a Woods-Saxon plus Coulomb well with typical radius and diffuseness, and with the well depth adjusted to give the correct resonant energy for a particle of the appropriate L value. In the DWBA calculations they used wave functions found with the same well parameters.

In a related way Davies et al (1976) compared reduced widths from alpha decay of heavy elements with reduced widths derived from $(^{16}O, ^{12}C)$ reactions (on $^{204,207,208}Pb$ and ^{209}Bi targets) at 93 MeV. Identical alpha + target potentials were used in LOLA calculations of the strengths of the two types of reactions, and r_c was taken to be the radius of the last maximum of ϕ^2. In four of five cases good agreement was found in the values of the two sets of reduced widths, which shows an overall consistency in the measurements, the data reduction, and the mechanisms assumed. In a parallel attack DeVries et al (1976) used the results of elastic scattering and total reaction cross-section measurements for alpha particles on ^{208}Pb and ^{209}Pb to limit the range of suitable potential sets, which were then employed to calculate γ^2 for the alpha decay of five heavy elements. The values were in rough agreement with those just described and about 1000 times greater than those predicted by conventional alpha decay theories. However, absolute values of S_{exp} depend on detailed knowledge of the $A + \alpha$ wave function and they remain uncertain.

Becchetti (1978) lists examples of good agreement in directly observed relative alpha decay lifetimes and branching ratios and those inferred from the results of $(d, ^6Li)$ measurements in the rare earth region, but agreement in absolute values was not achieved (J. Jänecke, personal communication).

Systematic Trends of Spectroscopic Factors

We can now look for systematic trends in the variation of S_{exp} through the sd and fp shells. Recognizing that uncertainty exists in choosing the best scheme of DWBA parameter variation to use across this wide range of nuclei, we may nonetheless expect that if the parameter variation is smooth any inherent sharp discontinuity in S_{exp} will be seen. We cannot, however, be sure that inherent slow variations will be correctly recognized, nor that artificial slow variations will not be introduced by the DWBA analysis.

Figure 10 shows relative S values for ground-state transitions derived from (^{6}Li,d), (d,^{6}Li), (^{3}He,^{7}Be), (p,pα), and (α,2α) experiments on even-even sd shell nuclei, normalized at $A_> = 20$. ($A_>$ is the final nucleus in stripping, the target nucleus in pickup and knockout.) Here and elsewhere error bars are omitted to avoid clutter; authors' estimated experimental cross-section errors are usually $\pm 20\%$ to $\pm 50\%$, with additional error

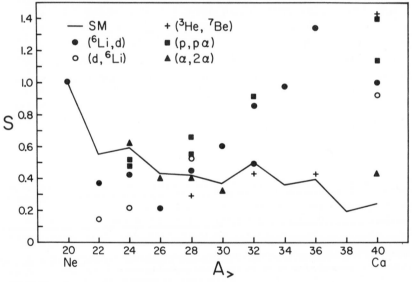

Figure 10 Values of S for alpha transfer from or to even-even nuclei in the sd shell, normalized at $A_> = 20$. The line connects values calculated from shell model wave functions by Chung et al (1978, 1979). The (^{6}Li,d) data are from Anantaraman et al (1974, 1976, 1977a,b, 1978a,b, 1979), Cook (1978), Eswaran et al (1977), and N. Anantaraman, private communication. The (d,^{6}Li) data are from Cossairt et al (1976) and Vander Molen et al (1975). The (^{3}He,^{7}Be) data are from Audi et al (1975). The (p,pα) data are from Bachelier et al (1976), Chant (1978), and N. S. Chant, E. T. Carey, personal communication. The (α,2α) data are from Sherman et al (1976). The lower (^{6}Li,d) point at mass 32 is from 32-MeV data, the upper point from 36-MeV data.

in S_{exp} introduced in the DWBA analysis depending on the quality of the fit. The shell model predictions of Chung et al (1978, 1979) shown for comparison are discussed later.

In general it is hard to guess how much of each of the disagreements in S_{exp} is experimental and how much is analytic. Two (^{6}Li,d) points are shown for mass 32, the lower obtained at 32 MeV, the upper at 36 MeV; here better agreement should be expected. According to Anantaraman (personal communication), the disagreements between the (^{6}Li,d) and (d,^{6}Li) points at $A = 22$ and 24 essentially disappear when the (d,^{6}Li) data are analyzed with the same DWBA reduction as used for the (^{6}Li,d) data (either zero-range or finite-range), but then a disagreement appears at $A = 28$. The striking rise seen in the (^{6}Li,d) points at $A = 34$ and 36 are discussed later. The (^{3}He,^{7}Be) pickup results follow the shell model curve in the mass-32–40 region except at $A = 40$ where a profound departure occurs. The (p,pα) points follow the (^{6}Li,d) trend; they depart from the SM curve in the mass-32–40 region in much the same way, but lie somewhat higher at $A = 40$. The normalized (α,2α) points give the best experimental agreement with the trend of SM predictions.

Figure 11 shows (^{6}Li,d) S values (solid circles) for even and odd targets compared with predictions of Chung et al (1978, 1979).

In Figure 12 are seen experimental (^{6}Li,d) S values for doubly even nuclei from $A = 20$ through $A = 68$. The sd shell values are those of Figure 10. The fp shell values indicated by closed and open circles show data reduced by two different procedures: the closed circles come from reduction and normalization to the sd shell values as described by Anantaraman et al (1975); the open circles come from reduction via zero-range (DWUCK) calculations using SBF parameters and normalization to give average agreement with the closed circles in the mass region 44–48. The values indicated by open squares are those of Hanson et al (1978) with data reduced via DWUCK calculations using ^{6}Li optical parameters of Chua et al (1976) and deuteron parameters of Newman et al (1967) normalized to agree with the solid circle set at ^{44}Ti. [After reduction with SBF parameters, data available from various (d,^{6}Li) experiments at 28 MeV (Martin et al 1973, Bedjidian et al 1972, Fortune et al 1975, Ceballos et al 1973, 1974) and 55 MeV (Medsker et al 1975, Hansen et al 1977), not plotted here, are in fair internal agreement and in spotty agreement with the (^{6}Li,d) results.] The solid circles and squares follow the same average trend, in spite of their different origins and analyses. Hanson et al (1978) pointed out that two interesting steps can be seen in the line connecting the square points, one between ^{54}Fe and ^{56}Fe, the other between ^{62}Ni and ^{62}Zn and that these occur where according to the simplest shell model picture the $p_{3/2}$ subshell is beginning to be filled with neutrons and with neutrons and protons, respectively. The first rise was predicted by Kurath & Towner (1974). These rises are attributed to the fact that microscopic bound-state wave functions containing substantial p-wave components are relatively strong in the surface region where the reaction occurs. Similar effects are seen in two-nucleon transfer, e.g. Bayman & Hintz (1968).

The open circles of the lower curve do not rise with A nearly as strongly as do the upper points. In the Zn region they are only one fourth as high. The most important reason for this is that the SBF entrance channel (^{6}Li + target) radius increases approximately as ($A^{1/3} + 6^{1/3}$), in

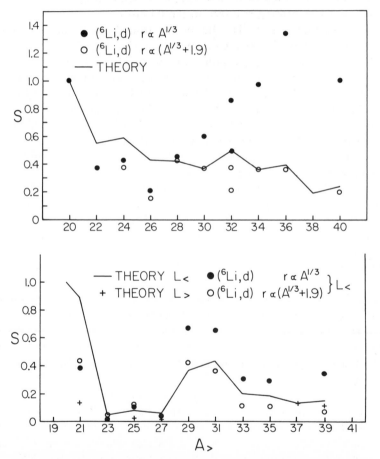

Figure 11   (*Upper*) Solid circles and theory plot are as in Figure 10; open circles are from DWBA reduction with indicated entrance channel radius variation with A. (*Lower*) Similar plot for odd-mass cases (data from references under Figure 10).

contrast to the usual $A^{1/3}$ dependence. The calculated cross section is very sensitive to this radius. At the present stage of development the arbitrariness in choosing parameters and the sensitivity of DWBA analysis to their values (together with the approximations inherent in the DWBA formalism) preclude defining a priori a "correct" reduction scheme giving trustworthy S_{exp} values, absolute or relative, over a range of As; without additional information there is no way of telling which of the DWBA reductions reflected in Figure 12 is best.

Here (p,pα) and (α,2α) results could in principle be useful. Knockout reactions are simpler than stripping reactions in that the inner structures of the protons and alpha particles need not be taken into account in the distorted-wave-impulse-approximation (DWIA) analysis used (Lim & McCarthy 1964a,b, Jackson & Berggren 1965, Chant & Roos 1977). Some of the (p,pα) and (α,2α) values differ from each other by factors of about two, but perhaps it is significant that in the Zn region both fall between the two (^{6}Li,d) curves. A program to find methods of analysis bringing

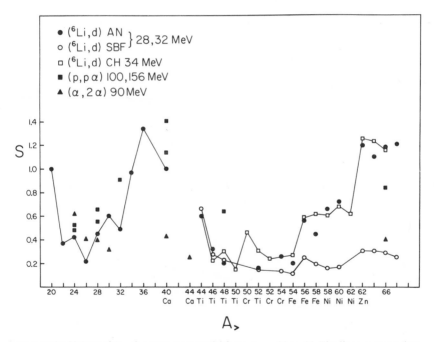

Figure 12 Values of S_{exp} for even-even nuclei from $A_> = 20$ to 66. The lines connect data points. The sd shell data are those of Figure 10. In the fp shell the open circles are values from Fulbright et al (1977) and Bastawros et al (1979), the closed circles are from the same data reduced as in Anantaraman et al (1975), and the squares are the values of Hanson et al (1978).

the knockout and transfer reaction data into harmony might prove valuable.

In the development of single-nucleon transfer studies via direct reactions the relatively simple shell model wave functions for low-lying states of nuclei were used to establish normalization factors for zero-range DWBA calculations. Similar methods could be used to normalize the DWBA formulas for alpha transfer reactions. Shell model calculations to be described below have been used to calculate spectroscopic factors for transitions to positive-parity levels in the sd shell, in ^{44}Ti, and in ^{62,64,66,68}Zn. If those results are taken as absolute standards then DWBA normalization factors can be obtained for the various regions. An attempt to achieve consistency over this range is described below.

Values of S_{exp} for the lowest-lying 1^-, 2^+, 3^-, and 4^+ states of fp shell nuclei are plotted against A in Figure 13 from data reduced with SBF parameters in the zero-range program DWUCK. Several $J^\pi = 1^-$ level assignments are uncertain because of incompleteness of angular distributions and absence of confirmation (Fulbright et al 1977). In the same reference systematic variations of the ratio $S_{exp}(2_1^+)/S_{exp}(g.s.)$ are shown

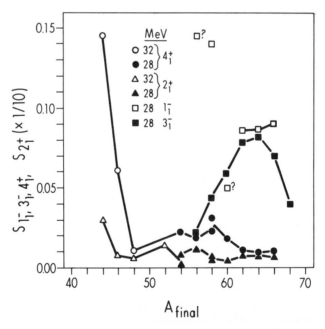

Figure 13 Trends of average values of S_{exp} for the first excited 1^-, 2^+, 3^-, and 4^+ states of fp shell nuclei from ^{44}Ti to ^{64}Zn. Data reduction with DWUCK and SBF parameters (Fulbright et al 1977).

displayed against target Z and A for doubly even fp shell targets. Other systematic variations are discussed in later sections connecting experiment and nuclear structure. Here and throughout the notation 2_1^+ refers to the lowest-lying 2^+ state, 2_2^+ to the next higher 2^+ state, etc.

ALPHA TRANSFER AND NUCLEAR STRUCTURE

Qualitative Results

In many cases simple arguments making use of measured alpha transfer cross sections can shed light on the structure of the levels formed. Thus the selectivity seen in these and other multinucleon transfer reactions can be used to investigate multiparticle-multihole (mp-mh) configurations of levels in nuclei near closed shells.

For example (Betts et al 1977, Fortune et al 1979), consider ^{40}Ca, which according to the model of Gerace & Green (1967) has excited states having mp-mh character with respect to a closed sd shell ($N = Z = 20$) core. Accordingly, the two-particle (2p) transfer reactions ^{38}Ar(^{3}He,n) and ^{42}Ca(p,t) should excite only the 0p-0h and 2p-2h configurations of ^{40}Ca, while the alpha transfer reaction ^{36}Ar(^{6}Li,d) should excite 0p-0h, 2p-2h, and 4p-4h configurations. In experiments the 0^+ level in ^{40}Ca at 3.35 MeV is excited strongly via (^{6}Li,d) and very weakly via (^{3}He,n), a result taken to indicate its predominantly 4p-4h character, in agreement with the prediction of Gerace & Green. The 2^+, 4^+, and 6^+ members of a $K^\pi = 0^+$ rotational band based on that configuration were also identified in the (^{6}Li,d) reaction. On the assumption that the structure of the ground-state band of ^{44}Ti is similar, except for the absence of four holes in its core structure (weak coupling model), the transfer strengths for those levels seen in the ^{40}Ca(^{6}Li,d)^{44}Ti reaction (Strohbusch et al 1974) should be the same, respectively, as those seen for the members of the 4p-4h band of ^{40}Ca. That proved true, except for the 0_2^+ state of ^{40}Ca, which showed only half the strength shown by its analog, the ground state of ^{44}Ti. Betts et al (1977) attributed the difference to mixing of the 3.35-MeV state and the ground state through small 2p-2h components, a mixing presumed also to cause the 3.35-MeV level to lie about 0.5 MeV above its counterpart while all other band members lie at closely corresponding energies (Figure 14).

The fp shell furnishes another example. On ^{50}Cr and ^{54,56}Fe targets the (^{6}Li,d) reaction strongly produces low-lying excited 0^+ levels in ^{54}Fe and ^{58,60}Ni, respectively, but on Ni targets it does not correspondingly excite such levels in Zn. Since the targets in the cases where strong excitation is seen have two-proton or two-neutron holes in the $f_{7/2}$ shell, while in the other cases they do not, the results can be taken to indicate that the levels of interest in the final nuclei also have the same large two-nucleon hole components and that the nucleons of the transferred alpha particles have gone into orbits outside the $f_{7/2}$ shell, a hypothesis supported by the behavior seen in the (^{3}He,n) reaction.

Figure 15 illustrates an interesting correspondence between (p,t) and (d,^{6}Li) pickup reaction yields in ground-state transitions from the same Sn targets (Becchetti & Janecke 1975). The correspondence is taken to indicate that the two protons transferred in the (^{6}Li,d) reaction act primarily as spectators. In each case the yields from ^{117}Sn and ^{119}Sn are about half as great as those from neighboring even-even nuclei. In analyzing the (p,t) results Fleming et al (1971) attributed the blocking seen here to a Pauli principle effect associated with the presence of a single unpaired $s_{1/2}$ neutron in the $J^\pi = \frac{1}{2}^+$ ground state of each of the odd-mass nuclei involved. Thus according to the simplest view neither the two-nucleon nor the four-nucleon direct reaction should be possible, involving that neutron.

Calculation of S from Shell Model Wave Functions

DESCRIPTION   The straightforward calculation of theoretical alpha particle transfer spectroscopic factors S from detailed shell model wave functions, where available, is obviously attractive. A large amount of theoretical work has been done to make that possible (Smirnov & Chlebowski 1961, Rotter 1968a,b, 1969, Ichimura et al 1973). These methods have been applied in numerical calculations made by McGrory

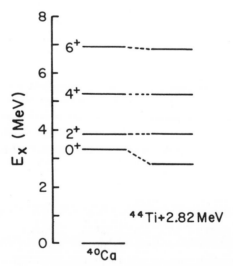

Figure 14   Correspondence of rotational bands in ^{40}Ca and ^{44}Ti.

(1973a,b), Bennett (1976, 1977), Conze (1976), Chung et al (1978), Bennett et al (1978, 1979), and others. Kurath (1973) calculated spectroscopic factors for alpha transfer in the 0p shell by using Cohen-Kurath intermediate coupling wave functions.

USE OF SU(3) WAVE FUNCTIONS   For nuclei in the lower half of the 1s0d shell, up to about mass 28, the SU(3)-SU(4) classification of levels (Elliott 1958, Harvey 1968) provides good approximate wave functions for most low-lying positive-parity levels. Beyond mass 28, spin-orbit forces become too strong and ruin the symmetry underlying the model. Near ^{20}Ne, SU(3) wave functions are known to have a large overlap with shell model wave functions and cluster wave functions. All the members of the $K^\pi = 0^+$ ground-state rotational band in ^{20}Ne are identified with the leading representation $(\lambda,\mu) = (8,0)$, and other bands are identified with other irreducible representations, e.g. (8,2) for a $K^\pi = 2^-$ band beginning at $E_x = 4.97$ MeV (Anantaraman et al 1979). With these wave functions used for the initial and final states of the target and residual nuclei, alpha stripping and pickup strengths have been calculated in the SU(3) basis, assuming the transfer of a $(0s)^4$ cluster of four nucleons with maximum symmetry [4]. The SU(3) basis is in fact a natural one in which to calculate the overlap factors associated with the transfer of an alpha particle.

A number of these pure SU(3)-SU(4) symmetry limit results have been

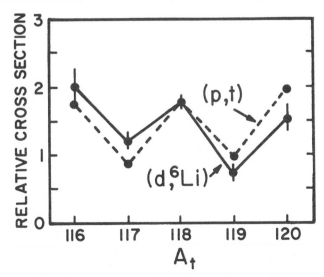

Figure 15 Correspondence between yields from two-nucleon and four-nucleon pickup reactions on even-mass and odd-mass tin nuclei.

compared with results obtained by using wave functions generated in large-scale shell model calculations in the SU(3) basis. There is good agreement between the two calculations, showing that for these nuclei the effect of SU(3) symmetry mixing on alpha strengths is small. By combining the SU(3)/R(3) relative spectroscopic factors tabulated by Draayer (1975) with the SU(6)/SU(3) coefficients of fractional parentage calculated by Hecht & Braunschweig (1975), it is possible to predict strength variations from one nucleus to the next and to calculate the effect of SU(3) representation mixing in the initial and final states.

USE OF LARGE-SCALE jj COUPLING WAVE FUNCTIONS   The computer code of Bennett (1977), used for extensive calculations discussed below, is based on the formalism of Ichimura et al (1973). It calculates the spectroscopic amplitude

$$A_{NL} = S^{1/2} = (B/A)^{N+\frac{1}{2}L} G_{NL} \langle \psi_B \| \psi_\alpha(NL) \| \psi_A \rangle [(2T_B+1)(2J_B+1)]^{-1/2}$$

4.

using ψ_A and ψ_B, the initial and final state wave functions, in the jj representation. Here $\psi_\alpha(NL)$ is the wave function of the alpha transferred, assumed to be the pure SU(3) representation $(\lambda,\mu) = (8,0)$ in the sd shell or (12,0) in the fp shell, but expressed in terms of the jj-coupled SM basis states. The wave functions are in the same jj representation and the calculations make use of the Rochester–Oak Ridge SM code (French et al 1969). The factors G_{NL} are tabulated in Ichimura et al (1973). Equal harmonic oscillator size parameters were assumed for the alpha particle and nuclear wave functions. If that assumption is not made, other SU(3) representations will enter, e.g. (8,2) in addition to (12,0) in the fp shell, and there will be additional terms in the expansion for A_{NL}, for which the $G_{NL}(\lambda,\mu)$ factors are given by Hecht & Sato (1975).

Calculation of S by Schematic Methods

The methods just described are useful in calculating S for transitions where initial and final state wave functions are known. Their predictions of relative strengths of formation of various levels within the same nucleus and absolute strengths for ground-state transitions of nuclei of many masses in the sd and fp shells are compared with experimental results below. However, detailed wave functions are not known in much of the fp shell and beyond and this fact has prompted the development of two schematic models used with doubly even targets; the pairing vibration model, used for $L = 0$ transitions, and the interacting boson model, for transitions to $0_1^+, 0_2^+, 2_1^+$, and 4_1^+ levels.

PAIRING VIBRATION MODEL   The pairing vibration (PV) model that Bohr (1968) and Bohr & Mottelson (1975) originally applied to two-nucleon transfer has been developed for four-nucleon transfer and applied in the fp shell by Betts (1975, 1977) and Stein et al (1977). In this model ^{56}Ni is taken as a basic core and various states of even-even nuclei are pairing vibration excitations of that core. The "quanta," di-neutron and di-proton pairs coupled to zero angular momentum and unit isospin, may either be added to or removed from the core. States are labeled $\psi = |n_a T_a n_r T_r T T_z\rangle$, where n_a and n_r are the numbers of added and removed quanta, T_a and T_r are the corresponding isospins, and T and T_z are the total isospin and its z-component. Usually one has $T_z = T$. From this model simple expressions for S were derived. The two transferred quanta may be alike or different, so there are three possible combinations, aa, ar, rr. Since most removed quanta come from the $f_{7/2}$ shell while most added quanta go into $(f_{5/2}, p_{3/2}, p_{1/2})$ configurations, several different unknown normalization constants are needed. However, trading on the similarity between the PV scheme and the coupling in Figure 2 and the di-neutron di-proton factorization (Equations 12a and 21, in Kurath & Towner 1974), Betts was able to use measured (^{3}He,n) and (p,t) and (t,p) strengths to account for the strengths of a number of $L = 0$ ground-state transitions. Further calculations have recently been made by Broglia et al (1978) and Vitturi et al (1978) using approximate microscopic wave functions in PV analyses of $L = 0$ alpha transfer transitions in the mass regions near 60 and 116. Their results follow a pattern set by the di-neutron and di-proton transfer cross sections.

INTERACTING BOSON MODEL   The interacting boson model of Arima & Iachello (1976, 1977), which has proved valuable in two-nucleon transfer studies, has been adapted to alpha transfer by Bennett & Fulbright (1978) and applied in the fp shell. Here again a ^{56}Ni core is assumed. Now the states involved are assumed to have simple boson structure, i.e. ground states have one s boson, 2_1^+ states have one d boson and the 0_2^+, 2_2^+, and 4_1^+ states have two d bosons. Each boson is constructed from a neutron or proton pair. The theory is completely symmetrical in neutron and proton bosons. Simple expressions for S are derived in terms of the numbers of s and d bosons and of the boson capacities $\Omega = j + \frac{1}{2}$ of the shells involved. Undetermined normalization constants again appear. With arbitrary normalization the results agree well with experiment for g.s. transitions (^{40}Ca $\rightarrow$ ^{44}Ti excepted) and for the ratio $S(2_1^+)/S(0_1^+)$, but not very well for the ratio $S(4_1^+)/S(0_1^+)$. This model takes into account the Pauli principle through the Ωs, but it does not distinguish between neutron and proton bosons.

Numerical Calculation of Spectroscopic Factors

Experimental results for a number of alpha transfer experiments in the
0p shell are compared with SU(3) predictions in Table 1. Numerical S
values based on the use of SU(3) wave functions in the lower part of
the sd shell can be found in the theoretical references cited above, in the
sd shell experimental papers, and in Tables 2, 3, and 4. In Figure 16 they
are compared with S_{exp} for ground-state transitions. For ^{23}Na the

Table 1 S_{exp}(g.s.) in the 1p shell

Transition	(d,^{6}Li)[a]	(d,^{6}Li)[b]	(^{3}He,^{7}Be)[c]	(α,^{8}Be)[d]	(p,pα)[e]	(α,2α)[f]	SU(3)[g]
^{5}He $\leftrightarrow$ ^{9}Be	—	—	—	2.94	—	—	6.12
^{6}Li $\leftrightarrow$ ^{10}B	—	—	—	0.89	—	—	0.015
^{7}Li $\leftrightarrow$ ^{11}B	—	—	—	1.07	—	—	3.58
^{8}Be $\leftrightarrow$ ^{12}C	0.43	3.06[h]	1.54	3.06[h]	—	0.89	3.06
^{9}Be $\leftrightarrow$ ^{13}C	—	—	—	2.05	—	—	2.23
^{10}B $\leftrightarrow$ ^{14}N	—	—	—	2.29	—	—	3.82
^{11}B $\leftrightarrow$ ^{15}N	—	—	—	1.28	—	—	2.23
^{12}C $\leftrightarrow$ ^{16}O	1.14	0.98	0.85	1.35	1.57	1.07	1.29
^{14}C $\leftrightarrow$ ^{18}O	0.62	—	—	—	—	—	0.80
^{15}N $\leftrightarrow$ ^{19}F	—	0.33	—	—	—	—	0.32[i]
^{16}O $\leftrightarrow$ ^{20}Ne	1.00	—	1.00	—	1.00	—	1.00

[a] Vander Molen et al (1975).
[b] Bedjidian et al (1972) normalized at 8 $\leftrightarrow$ 12.
[c] Audi et al (1975).
[d] Wozniak et al (1976).
[e] N. Chant, personal communication.
[f] Sherman et al (1976) normalized to 0.48 for ^{24}Mg(α,2α)^{20}Ne.
[g] Kurath (1973), Vander Molen et al (1975).
[h] Normalized to the SU(3) value.
[i] Shell model, Gutbrod et al (1971).

Table 2 Relative S_α values for the ^{19}F(^{6}Li,d)^{23}Na reaction to states with known structure

				Experiment		Theory	
E_x (MeV)	J^π	K	L	36 MeV[a]	16 MeV[b]	SU(3)[a]	SM(CW)[c]
0.0	$\frac{3}{2}^+$	$\frac{3}{2}$	2	1.00	1.00	1.00	1.00
0.440	$\frac{5}{2}^+$	$\frac{3}{2}$	2	0.26	0.40	3.38	0.78
2.076	$\frac{7}{2}^+$	$\frac{3}{2}$	4	1.61	1.98	2.29	2.81
2.391	$\frac{1}{2}^+$	$\frac{1}{2}$	0	10.8	4.0	6.66	5.02
2.640	$\frac{1}{2}^-$	$\frac{1}{2}$	1	0.09	0.27	—	—
2.704	$\frac{9}{2}^+$	$\frac{3}{2}$	4	0.4	0.66	5.83	1.39
2.982	$\frac{3}{2}^+$	$\frac{1}{2}$	2	0.62	0.85	3.52	0.64
3.678	$\frac{3}{2}^-$	$\frac{1}{2}$	1	0.09	0.50	—	—

[a] Eswaran et al (1977).
[b] Fortune et al (1978).
[c] C. L. Bennett, W. Chung, J. van Hienen, B. H. Wildenthal, 1968, privately circulated tables to be published.

Table 3   Relative spectroscopic strengths for ^{20}Ne(^{6}Li,d)^{24}Mg at 32 MeV

E_x (MeV)	J^π	K	(λ,μ)	$S_{\exp}(^6\text{Li,d})^d$	$S[\text{SU(3)}]^a$	$S(\text{PW})^b$	$S(\text{CW})^c$
0.0	0^+	0	(8,4)	1.00	1.00	1.00	1.0
1.37	2^+	0	(8,4)	0.42	0.37	0.43	0.32
4.12	4^+	0	(8,4)	0.03	0.00	0.03	<0.01
4.24	2^+	2	(8,4)	0.13	0.05	0.04	0.09
6.01	4^+	2	(8,4)	0.28	0.46	0.34	0.46
6.43	0^+	0	(4,6)	3.87	0.86	0.86	0.83
7.35	2^+	0	(4,6)	0.37	0.22	0.05	0.17
8.12	6^+	0	(8,4)	0.09 ± 0.04	0.15	0.00	0.03
8.65	2^+	—	—	0.22	—	—	—

[a] Pure SU(3) limit results using (λ,μ) quantum numbers shown in column 4.
[b] From shell model wave functions in SU(3) representation with Preedom-Wildenthal interaction (Conze 1976).
[c] From shell model wave functions in jj representation with Chung interaction. From C. L. Bennett, W. Chung, J. van Hienen, B. H. Wildenthal, 1968, privately circulated tables to be published.
[d] Anantaraman et al (1977a).

Table 4   Relative spectroscopic strengths for α transfer reactions on ^{24}Mg

			Theory		Experiment		
E_x (MeV)	J^π	(λ,μ)	SU(3)^a	SM^b	$(^6\text{Li,d})^a$	$(^{16}\text{O},^{12}\text{C})^c$	$(^{12}\text{C}^8\text{Be})^d$
0.0	0^+	(0,12)	1.00	1.00	1.00	1.00	1.00
1.78	2^+	(0,12)	0.23	0.17	0.21	0.19	0.19
4.62	4^+	(0,12)	0.10	0.08	0.10	—	0.09
4.98	0^+	—	—	0.13	0.70	0.58	—
6.69	0^+	(12,0)	0.28	0.48	1.00	1.71	—

[a] Draayer et al (1974).
[b] C. L. Bennett, W. Chung, J. van Hienen, B. H. Wildenthal, 1968, privately circulated tables to be published.
[c] Peng et al (1976), 48 MeV.
[d] Mathiak et al (1976).

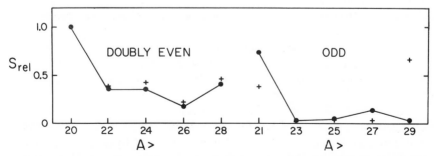

Figure 16   Comparison of values of $S_{\exp}$ with SU(3) predictions, both normalized to unity at $A_> = 20$. Same data as in Figure 11.

agreement between SU(3) and shell model predictions is poor; the experimental values generally follow the SM predictions. For ^{24}Mg shell model predictions with two different interactions agree fairly well with each other and with SU(3) and, except for the 6.43-MeV 0^+ state, they agree with experiment. For ^{28}Si a similar agreement is seen, with the 6.69-MeV 0^+ state yield two or three times greater than predicted.

This pattern of general agreement except for excited 0^+ states is seen for most even-even nuclei in the lower half of the sd shell, but an exception is found in the case of the members of the ground-state band of ^{20}Ne, which, according to SU(3) predictions, should all be formed with equal intensity, whereas the actual values relative to the ground state, are 1, 0.41, 0.22, 0.20, and 0.51 for the 0^+, 2^+, 4^+, 6^+, and 8^+ states, respectively (Anantaraman et al 1979). Another exception is provided by ^{22}Ne, for which the values for the 2^+ and 4^+ members of the ground-state band are 0.32 and 0.36, respectively, compared with 0.77 and 0.36 predicted by SU(3) (Anantaraman et al 1977a). Core polarization may be significant here: the discrepancy disappears when 38-MeV data are analyzed with known particle-hole components in the ^{16}O ground-state wave function taken into account. But with 32-MeV data that is not enough, so other reaction modes may be significant (Anantaraman et al 1979). The large alpha widths of the 8.3- and 6.72-MeV 0^+ levels in ^{20}Ne were explained by Fortune et al (1973) in terms of (fp)^2 and (fp)^4 admixtures. According to Ichimura et al (1973), (sdg) components may also be present.

The strong excitation of 1^- and 3^- levels with Fe and Ni targets provides further strong evidence of configuration mixing. According to the model of Figure 2 and Kurath & Towner (1974), such excitation should be possible through substantial (fp)^3(sdg)^1 components (calculations by R. G. Markham, unpublished Rochester report 1973). For practical reasons the shell model cannot be used to treat these cases. The schematic models are by their nature of no use here.

We now consider (not in chronological order) three sets of calculations made with Bennett's code.

1. The sd shell. In the sd shell Chung et al (1978, 1979) calculated S values for most cases where alpha stripping or pickup experiments are feasible. The wave functions used for the nuclei were those of Chung (1976) in the full (sd)^4 basis. The $(\lambda,\mu) = (8,0)$ representation was used for the alpha particle. Some of their results are shown in Figures 10 and 11 together with values of S_{exp} for various reactions. For even-A targets the predicted variation of S with A is not matched by S_{exp}, except for the $(\alpha,2\alpha)$ results. For odd-A targets only (^{6}Li,d) results are shown and the trends match fairly well.

2. ^{40}Ca → ^{44}Ti. Bennett (1976) calculated S values in the full (fp)^4 shell model space using both Kuo-Brown and modified Kuo-Brown matrix elements (Kuo & Brown 1968). In the latter case his results were identical with those obtained earlier by McGrory (1973a,b). A full comparison with experimental values is impossible because the shell model in this basis

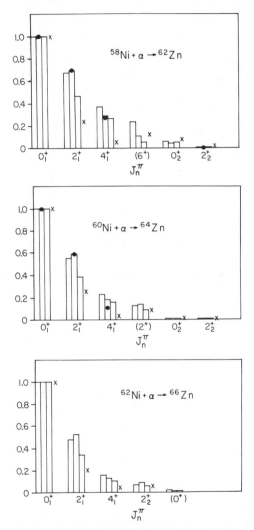

Figure 17   Relative spectroscopic factors for ^{58,60,62}Ni → ^{62,64,66}Zn. Theoretical S values calculated from SM wave functions by Bennett et al (1978, 1979) for these choices of ψ_α(NL): left bar, ^{60}Zn; center bar, (12,0); right bar, (12,0)+(8,2). Dots represent pseudo-SU(3), crosses represent S_{exp} (Fulbright et al 1977).

does not predict all the states seen in nature, but Bennett found that the pattern of observed strengths is about the same as that predicted except for the ground state, which gave $S_{exp} \simeq 4S$. See also DeVries (1973b).

3. ^{58,60,62,64}Ni → Zn. Calculations of S were made by Bennett et al (1978, 1979) with wave functions that van Hienen et al (1976) had obtained by diagonalizing the adjusted surface delta interaction (ASDI) of Koops & Glaudemans (1977) in the full $(f_{5/2}, p_{3/2}, p_{1/2})^n$ space. Three different approximations were used for $\psi_\alpha(NL)$: the pure (12,0) representation of SU(3), the mixed (12,0) and (8,2) representation, and finally the ^{60}Zn wave function (ASDI) in the 0_1^+, 2_1^+, 4_1^+, and 6_1^+ states. The overlap of the ^{60}Zn wave function with the (12,0) representation in this space is large. Figure 17 gives a comparison with the experimental results of Fulbright et al (1977) obtained by zero-range DWBA analysis with the SBF parameter set. Also see Table 5. The predicted relative S values for forming the 2_1^+ and 4_1^+ states are consistently higher than the experimental values. The disagreement disappears, on the average, if the DWBA calculations are made with the 2_1^+ and 4_1^+ bound-state radius parameters reduced 6.5% and 12.5% from the value for the 0_1^+ state, respectively. A similar effect was reported by Arima & Yoshida (1974), who found that consistency between theoretical and experimental alpha decay rates required smaller channel radii for the higher spin states of a number of nuclei.

When the values of the normalization ratio $\eta = S_{exp}/S_{SM}$ are compared, one sees internal consistency in the separate sets for transitions to the 0_1^+, 2_1^+, and 4_1^+ states (Table 6). When the $(\lambda, \mu) = (12,0)$ mean values for the 0_1^+ and 2_1^+ cases from the Zn region are compared with corresponding ^{44}Ti (SBF set) values, the latter are higher by a factor of about 1.65. When

Table 5 Relative strengths for ^{60}Ni(^{6}Li,d)^{64}Zn

E_x (MeV)[a]	J^π	S_{exp}			S	
		28 MeV[b]	34 MeV[a]	36 MeV[b]	SM[c]	MP[a]
0.00	0^+	1.00	1.00	1.00	1.00	1.00
0.99	2^+	0.26	0.22	0.21	0.37	0.22
1.80	2^+	weak	0.01	—	0.009	0.001
1.91	0^+	v. weak	<0.02	—	0.005	0.01
2.30	4^+	0.028	0.02	—	0.16	0.002
2.98	3^-	0.24	0.25	0.37	—	0.25
3.63[d]	(1^-)	0.26	0.25	—	—	0.003
3.83	(5^-)	—	0.07	—	—	0.004

[a] Betts et al (1978) (multipole pairing model).
[b] Fulbright et al (1977).
[c] Bennett et al (1978, 1979), (12,0)+(8,2) alpha operator.
[d] From the 28-MeV work E_x for this state was 3.70 MeV.

Table 6 Average normalization factors for three types of transitions and three α particle operators[a]

Level	^{60}Zn	(12,0)+(8,2)	(12,0)	^{40}Ca → ^{44}Ti (12,0)	R[b]
0_1^+	16.3 ± 0.9	16.6 ± 2.9	20.85 ± 3.5	37.1	1.78 ± 0.3
2_1^+	7.05 ± 1.1	10.54 ± 2.8	8.62 ± 1.9	13.1	1.52 ± 0.3
4_1^+	2.33 ± 1.1	3.36 ± 1.7	3.63 ± 1.6	8.3[c]	2.29 ± 1.0

[a] Values of $N = 10^{-1} \eta$ given in columns 2–4 are averages over transitions to states of ^{62,64,66,68}Zn, except those in the 4_1^+ row, which include no contribution from ^{68}Zn. Errors: rms deviations from the mean. (^{68}Zn from Bastawros et al 1979, others from Fulbright et al 1977.)

[b] Ratio of values in column 5 to those of column 4. Errors reflect only those of column 4.

[c] Probably too large. Includes contribution from two 4^+ states; shell model predicts only one.

the comparison is made with ^{44}Ti values obtained by using SBF parameters but with the normal exit channel radius parameter $r_r = 1.14$ rather than the special value 1.10 specified for Ca isotopes, the latter are higher by a factor of about 1.3.

The pseudo SU(3) method was used by Braunschweig et al (1978) to calculate S values for alpha transfer to Ni targets. It is an approximate method that assumes a doubly closed shell at ^{56}Ni and depends on the fact that the next orbits to be filled ($f_{5/2}, p_{3/2}, p_{1/2}$) have the same j values as the orbits of the sd shell ($d_{5/2}, d_{3/2}, s_{1/2}$). In relative values the results (Figure 17) are in excellent agreement with those from the shell model calculations. In absolute values, however, the agreement is poor; the SM-to-SU(3) ratio is 2 for ^{62}Zn and 4.6 for ^{64}Zn.

Normalization and the Parameters of DWBA

The shell model S values can now be employed in trying to find a DWBA parameter scheme allowing a single normalization factor to be used in all three areas. The theoretical predictions for the ^{44}Ti and ^{62}Zn 0_1^+ transitions are nearly equal ($S = 0.019$ and 0.020, respectively) and about an order of magnitude lower than for ^{20}Ne, so the rise in S_{exp} with A seen in Figure 12 for the fp shell is too large for DWBA parameters using the conventional $A^{1/3}$ entrance channel radius variation and too small for the SBF set. Use of the ($A^{1/3} + 1.9$) variation in the sd shell produces a dramatic effect seen in Figure 11. [DWUCK calculations: ^{6}Li parameters of Chua et al (1976), except for entrance channel radius; deuteron parameters of Newman et al (1967); fits, on the average, slightly inferior.] The previously unexplained rise in S_{exp} to a high peak at $A_> = 36$ disappears and many of the points now fall quite close to the Chung et al (1978, 1979) predictions. Figure 11 also shows that the agreement is improved in the odd-target case. When an ($A^{1/3} + 1.3$) variation is used over the full range of even-A data, agreement between experiment and theory is good at ^{44}Ti, within about 20% in the Zn region, and mostly within 30% to 60% in the ^{24}Mg to ^{40}Ca region, S_{exp} being too high there. (N. Anantaraman and H. W. Fulbright, private communication.) It seems unlikely that this evidence of the importance of the entrance channel radius is entirely fortuitous. A more complete investigation of zero-range DWBA parameterization is needed, with attention also given to other parameters, and similarly for finite-range DWBA, which ultimately may be required for satisfactory agreement between theory and experiment.

The value of this procedure will be limited by weaknesses on the theoretical side. One is truncation of the shell model space; for example, we find that neglect of core polarization (particle-hole components in the core) of the target nucleus can significantly affect the values of S for alpha transfer reactions on ^{16}O (Anantaraman et al 1979, Eswaran et al 1979) and ^{40}Ca (Bennett 1976). Other weaknesses are uncertainty about the residual interaction and the fact that the harmonic oscillator shell model wave functions used in most theoretical calculations of S do not exhibit the correct behavior at the nuclear surface where the reaction mainly occurs. In addition there is the troublesome possibility pointed out by Fliessbach (1975) and also explored by others (Jackson & Rhoades-Brown 1978, Goldflam & Tobocman 1978) that the usual calculation of S is significantly deficient in using a projection operator not normalized to unity. This is connected with the need for overall antisymmetrization. Fliessbach & Mang (1976) proposed to correct the usual theory of alpha decay accordingly in the hope of increasing the calculated decay rates by factors of the order of 10 to 1000 needed to bring them into agreement with experiment (Mang 1964). The importance of the effect for alpha transfer reactions is hard to assess, on the one hand because reducing the theory to numerical predictions involves approximations of uncertain accuracy, and on the other hand because devising a direct experimental test is difficult. Fliessbach & Manakos (1977) give corrected values of S for alpha transfer to ^{16}O, ^{20}Ne, and ^{24}Mg targets, but no clear improvement in agreement with experiment is evident. Fliessbach (1978a) discusses the dependence of the correction on momentum transfer and in a paper on applications (1978b) gives for the ratio $S(^{36}\text{Ar} + \alpha \rightarrow {}^{40}\text{Ca})/S(^{16}\text{O} + \alpha \rightarrow {}^{20}\text{Ne})$ the value unity, which is close to the experimental value found by conventional DWBA analysis, but much larger than the theoretical value 0.24 of Chung et al (1978, 1979).

CONCLUSION

In this paper we have seen that angular distributions from alpha transfer reaction studies can be used with confidence to determine or confirm spins and parities of levels and that a strong start has been made toward using the absolute strengths of transitions for testing wave functions. However, we have seen that deficiencies in the accuracy of measured cross sections should be corrected; $\pm 10\%$ error limits, rather than the previous $\pm 20\%$ to $\pm 50\%$, would now seem an appropriate compromise. Removing deficiencies on the theoretical side mentioned in the previous section will probably happen slowly, but a faster, steady improvement in DWBA procedures is probable. In any event, alpha particle transfer spectroscopy is already a well-established method in the sd and fp shells.

The author is deeply indebted to many collaborators in the field, particularly C. L. Bennett and U. Strohbusch. He is grateful to his colleague N. Anantaraman for providing numerical data and for critically reading the manuscript of this paper.

Literature Cited

Ajzenberg-Selove, F. 1975. *Nucl. Phys. A* 248:1–156

Ajzenberg-Selove, F. 1976. *Nucl. Phys. A* 268:1–204

Ajzenberg-Selove, F. 1977. *Nucl. Phys. A* 281:1–148

Ajzenberg-Selove, F. 1978. *Nucl. Phys. A* 300:1–224

Anantaraman, N., Bennett, C. L., Draayer, J. P., Fulbright, H. W., Gove, H. E., Tōke, J. 1975. *Phys. Rev. Lett.* 35:1131–34

Anantaraman, N., Draayer, J. P., Gove, H. E., Tōke, J., Fortune, H. T. 1978a. *Phys. Rev. C* 18:815–19

Anantaraman, N., Draayer, J. P., Gove, H. E., Trentelman, J. P. 1974. *Phys. Rev. Lett.* 33:846–48

Anantaraman, H., Gove, H. E., Lindgren, R. A., Tōke, J., Trentelman, J. P., Draayer, J. P., Jundt, F. C., Guillaume, G. 1979. *Nucl. Phys. A* 313:445–66

Anantaraman, N., Gove, H. E., Tōke, J., Draayer, J. P. 1976. *Phys. Lett. B* 60:149–52

Anantaraman, N., Gove, H. E., Tōke, J., Draayer, J. P. 1977a. *Nucl. Phys. A* 279:474–92

Anantaraman, N., Gove, H. E., Tōke, J., Fortune, H. T. 1978b. *Phys. Lett. B* 74:199–201

Anantaraman, N., Gove, H. E., Trentelman, J. P., Draayer, J. P., Jundt, F. C. 1977b. *Nucl. Phys. A* 276:119–29

Anyas-Weiss, N., Cornell, J. C., Fisher, P. S., Hudson, P. N., Menchaca-Rocha, A., Millener, D. J., Panagiotou, A. D., Scott, D. K., Strottman, D., Brink, D. M., Ellis, P. J., Engeland, T. 1974. *Phys. Rep. C* 12:201–72

Apagyi, B., Fai, G., Nemeth, J. 1976a. *Nucl.*

Phys. A 272:303–16

Apagyi, B., Fai, G., Nemeth, J. 1976b. *Nucl. Phys. A* 272:317–26

Arima, A., Gillet, V., Ginocchio, J. 1970. *Phys. Rev. Lett.* 25:1043–46

Arima, A., Iachello, F. 1976. *Ann. Phys.* 99:253–317

Arima, A., Iachello, F. 1977. *Phys. Rev. C* 16:2085–89

Arima, A., Yoshida, S. 1974. *Nucl. Phys. A* 219:475–515

Artemov, K. P., Gol'dberg, V. Z., Petrov, I. P., Rudakov, V. P., Serikov, I. N., Timofeev, V. A. 1971. *Phys. Lett. B* 37:61–64

Artemov, K. P., Gol'dberg, V. Z., Petrov, I. P., Rudakov, V. P., Timofeev, V. A., Wolski, R., Szmider, J. 1975. *Yad. Fiz.* 21:1169–77 (Trans. *Sov. J. Nucl. Phys. USA*)

Artemov, K. P., Gol'dberg, V. Z., Petrov, I. P., Rudakov, V. P., Serikov, I. N., Timofeev, V. A., Wolski, R., Szmider, J. 1977. *Yad. Fiz.* 26:9–13 (Trans. *Sov. J. Nucl. Phys. USA*)

Audi, G., Detraz, C., Langevin, M., Pougheon, F. 1975. *Nucl. Phys. A* 237:300–8

Austern, N. 1970. *Direct Nuclear Reaction Theories.* London: Wiley. 390 pp.

Austern, N., Drisko, R. M., Halbert, E. C., Satchler, G. R. 1964. *Phys. Rev. B* 133:3–16

Bachelier, D., Boyard, J. L., Hennino, T., Holmgren, H. D., Jourdain, J. C., Radvanyi, P., Roos, P. G., Roy-Stephen, M. 1976. *Nucl. Phys. A* 268:488–512

Bastavros, A. M., Bennett, C. L., Fulbright, H. W., Markham, R. G. 1979. *Nucl. Phys. A* 315:493–99

Bayman, B. F., Bohr, A. 1958–1959. *Nucl. Phys.* 9:596–99

Bayman, B. F., Hintz, N. M. 1968. *Phys. Rev.* 172:1113–23

Becchetti, F. D. 1978. See Van Oers et al 1978, pp. 308–21

Becchetti, F. D. 1979. Alpha cluster pickup in heavy nuclei. Presented at 2nd Oaxtepec Meet. Nucl. Phys., Mexico, Jan. 1979. *Univ. Mich. Rep.* 23 pp.

Becchetti, F. D., Janecke, J. 1975. *Phys. Rev. Lett.* 35:268–70

Bedjidian, M., Chevallier, M., Grossiord, J. Y., Guichard, A., Gusakow, M., Pizzi, J. R., Ruhla, C. 1972. *Nucl. Phys. A* 189:403–8

Bennett, C. L. 1976. *Alpha particle spectroscopy in ^{44}Ti via the ($^{6}Li,d$) reaction.* PhD thesis. Univ. Rochester, NY. 114 pp.

Bennett, C. L. 1977. *Nucl. Phys. A* 284:301–6

Bennett, C. L., Fulbright, H. W. 1978. *Phys. Rev. C* 17:2225–28

Bennett, C. L., Fulbright, H. W., van Hienen, J. F. A., Chung, W., Wildenthal, B. H. 1978. See Van Oers et al 1978, pp. 702–3

Bennett, C. L., Fulbright, H. W., van Hienen, J. F. A., Chung, W., Wildenthal, B. H. 1979. *Phys. Rev. C* 19:1099–1106

Bethge, K. 1970. *Ann. Rev. Nucl. Sci.* 20:255–88

Bethge, K., Meier-Ewert, K., Pfeiffer, K., Bock, R. 1967. *Phys. Lett. B* 24:663–65

Betts, R. R. 1975. See Goldberg et al 1975, p. 458

Betts, R. R. 1977. *Phys. Rev. C* 16:1617–25

Betts, R. R., Fortune, H. T., Bishop, J. N., Al-Jadir, M. N. I., Middleton, R. 1977. *Nucl. Phys. A* 292:281–87

Betts, R. R., Stein, N., Sunier, J. W., Woods, C. W. 1978. *Phys. Lett. B* 76:47–50

Bohr, A. 1968. *Nucl. Struct. Symp., Dubna*, pp. 179–89. Vienna: IAEA

Bohr, A., Mottelson, B. R. 1975. *Nuclear Structure*, Vol. II. Reading, Mass.: Benjamin. 748 pp.

Braunschweig, D., Hecht, K. T., Draayer, J. P. 1978. *Phys. Lett. B* 76:538–42

Broglia, R. A., Ferreira, L., Kunz, P. D., Sofia, H., Vitturi, A. 1978. *Phys. Lett. B* 79:351–55

Broglia, R. A., Riedel, C., Udagawa, T. 1971. *Nucl. Phys. A* 169:225–38

Bunakov, V. E., Gridnev, K. A., Krasnov, L. V. 1971. *Phys. Lett. B* 34:27–30

Castanada, C. M., Smith, H. A., Singh, P. P., Jastrzebski, J., Karwowski, H. 1978. *Phys. Lett. B* 77:371–75

Ceballos, A. E., Erramuspe, H. J., Ferrero, A. M. J., Sametband, M. J., Testoni, J. E., Bes, D. R., Maqueda, E. E. 1973. *Nucl. Phys. A* 208:617–25

Ceballos, A. E., Erramuspe, H. J., Ferrero, A. M. J., Sametband, M. J., Testoni, J. E.,

Bes, D. R., Maqueda, E. E., Perazzo, R. P. J., Reich, S. L. 1974. *Nucl. Phys. A* 228:216–28

Chant, N. S. 1978. See Van Oers et al 1978, pp. 415–31

Chant, N. S., Roos, P. G. 1977. *Phys. Rev. C* 15:57–68

Chua, L. T., Becchetti, F. D., Janecke, J., Milder, F. L. 1976. *Nucl. Phys. A* 273:243–52

Chung, W. 1976. *Empirical renormalization of shell model hamiltonians and magnetic dipole moments of sd shell nuclei.* PhD thesis. Michigan State Univ. 144 pp.

Chung, W., van Hienen, J., Wildenthal, B. H., Bennett, C. L. 1978. See Van Oers et al 1978, pp. 580–81

Chung, W., van Hienen, J., Wildenthal, B. H., Bennett, C. L. 1979. *Phys. Lett. B* 79:381–84. [The figures in this paper are correct, but the tables of shell model and SU(3) S values are in error. An Erratum will be published (C. L. Bennett, private communication).]

Cobern, M. E., Pisano, D. J., Parker, P. D. 1976. *Phys. Rev. C* 14:491–505

Conze, M. 1976. PhD thesis. Technische Hochschule, Darmstadt; quoted by Anantaraman et al 1977

Cook, R. E. 1978. *Alpha particle spectroscopy in ^{30}Si and ^{34}S.* MS thesis. Univ. Rochester, NY. 57 pp.

Cossairt, J. D., Bent, R. D., Broad, A. S., Becchetti, F. D., Janecke, J. 1976. *Nucl. Phys. A* 261:373–84

Davies, W. G., DeVries, R. M., Ball, G. C., Forster, J. S., McLatchie, W., Shapira, D., Töke, J., Warner, R. E. 1976. *Nucl. Phys. A* 269:477–92

DeVries, R. M. 1973a. *Phys. Rev. C* 8:951–60

DeVries, R. M. 1973b. *Phys. Rev. Lett.* 30:666–68

DeVries, R. M. 1976. *Comput. Phys. Commun.* 11:249–56

DeVries, R. M., Lilley, J. S., Franey, M. A. 1976. *Phys. Rev. Lett.* 37:481–84

Draayer, J. P. 1975. *Nucl. Phys. A* 237:157–81

Draayer, J. P., Gove, H. E., Trentelman, J. P., Anantaraman, N., DeVries, R. M. 1974. *Phys. Lett. B* 53:250–52

Elliott, J. P. 1958. *Proc. R. Soc. Ser. A* 245:128–45; 562–81

Endt, P. M., van der Leun, C. 1978. *Nucl. Phys. A* 310:1–752

Eswaran, M. A., Gove, H. E., Cook, R. E., Draayer, J. P. 1977. *Univ. Rochester Rep. UR-NSRL-183A.* 17 pp.

Eswaran, M. A., Boyd, R. N., Sugarbaker, E., Cook, R. E., Gove, H. E. 1979. *Nucl. Phys. A* 313:467–76

Fleming, D. G., Blann, H. M., Fulbright, H. W. 1971. *Nucl. Phys. A* 163:401–17

Fliessbach, T. 1975. *Z. Phys.* 272:39–46

Fliessbach, T. 1978a. *Z. Phys.* 288:219–26

Fliessbach, T. 1978b. *Z. Phys.* 288:211–17

Fliessbach, T., Manakos, P. 1977. *J. Phys. G* 3:643–56

Fliessbach, T., Mang, H. J. 1976. *Nucl. Phys. A* 263:75–85

Fortune, H. T., Al-Jadir, M. N. I., Betts, R. R., Bishop, J. N., Middleton, R. 1979. *Phys. Rev. C* 19:756–64

Fortune, H. T., Betts, R. R., Bishop, J. N., Al-Jadir, M. N. I., Middleton, R. 1975. *Phys. Lett. B* 55:439–42

Fortune, H. T., Betts, R. R., Garrett, J. D., Middleton, R. 1973. *Phys. Lett. B* 44:65–67

Fortune, H. T., Middleton, R., Betts, R. R. 1972. *Phys. Rev. Lett.* 29:738–40

Fortune, H. T., Powers, J. R., Middleton, R., Bethge, K., Pilt, A. A. 1978. *Phys. Rev. C* 18:255–64

French, J. B., Halbert, E. C., McGrory, J. B., Wong, S. S. M. 1969. *Adv. Nucl. Phys.* 3:193–257

Fulbright, H. W., Bennett, C. L. 1978. *Ann. Rep. Nucl. Struct. Lab. 1977*, pp. 53–54. Univ. Rochester

Fulbright, H. W., Bennett, C. L., Lindgren, R. A., Markham, R. G., McGuire, S. C., Morrison, G. C., Strohbusch, U., Tőke, J. 1977. *Nucl. Phys. A* 284:329–64. [This paper requires corrections: In Table 10, to be consistent with the experimental data, footnote (j) should read $S_{g.s.} = 0.30$ and all (but top) $S/S_{g.s.}$ values in column 4 should accordingly be multiplied by 0.67; In Table 12, the value 0.070 for the 2.83-MeV state should be normalized to read 0.22; In Table 6 and Figure 10, footnote (b) should read $S_{g.s.} = 0.26$ (and the vertical scale of Figure 10 should be tripled).]

Gerace, W. J., Green, A. M. 1967. *Nucl. Phys. A* 93:110–32

Goldberg, D. A., Marion, J. B., Wallace, S. J., eds. 1975. *Proc. 2nd Int. Conf. on Clustering Phenomena in Nuclei.* Univ. Maryland, College Park, Md. 651 pp.

Goldflam, R., Tobocman, W. 1978. *An Alternate Treatment of Exchange Effects in the Theory of Radioactive Decay.* Preprint. Case Western Reserve Univ., Cleveland. 39 pp.

Gunn, G. D., Boyd, R. N., Anantaraman, N., Shapira, D., Tőke, J., Gove, H. E. 1977. *Nucl. Phys. A* 275:524–32

Gutbrod, H. H., Yoshida, H., Bock, R. 1971. *Nucl. Phys. A* 165:240–58

Hansen, O., Maher, J. V., Vermeulen, J. C., Put, L. W., Siemssen, R. H., van der Woude, A. 1977. *Nucl. Phys. A* 292:253–66

Hanson, D. L., Stein, N., Sunier, J. W., Woods, C. W., Betts, R. R. 1978. See Van Oers et al 1978, pp. 716–17

Harada, K. 1961. *Prog. Theor. Phys.* 26:667–79

Harvey, M. 1968. *Adv. Nucl. Phys.* 1:67–151

Hecht, K. T., Braunschweig, D. 1975. *Nucl. Phys. A* 244:365–434

Hecht, K. T., Sato, H. 1975. See Goldberg et al 1975, p. 454

Hodgson, P. E. 1978. *Nuclear Heavy-Ion Reactions.* Oxford: Clarendon. 588 pp.

House, L. J., Kemper, K. W. 1978. *Phys. Rev. C* 17:79–82

Ichimura, M., Arima, A., Halbert, E. C., Terasawa, T. 1973. *Nucl. Phys. A* 204:225–78

Jackson, D. F., Berggren, T. 1965. *Nucl. Phys.* 62:353–72

Jackson, D. F., Rhoades-Brown, M. 1978. *Antisymmetrization in Alpha Decay and Alpha Transfer.* Preprint. Univ. Surrey, Guildford. 18 pp.

Jain, A. K., Grossiord, J. Y., Chevallier, M., Gaillard, P., Guichard, A., Gusakow, M., Pizzi, J. R. 1973. *Nucl. Phys. A* 216:519–40

Jundt, F., Coffin, J. P., Fulbright, H. W., Guillaume, G. 1975. *Phys. Rev. C* 12:1366–67

Koops, J. E., Glaudemans, P. W. M. 1977. *Z. Phys. A* 280:181–209

Kuo, T. T. S., Brown, G. E. 1968. *Nucl. Phys. A* 114:241–79

Kurath, D. 1973. *Phys. Rev. C* 7:1390–95

Kurath, D. 1976. Nuclear structure effects in cluster transfer. *Proc. Int. Sch. "Enrico Fermi," LXII*, ed. H. Faraggi, R. A. Ricci, pp. 58–81. Amsterdam, New York, Oxford: North-Holland. 591 pp.

Kurath, D., Towner, I. S. 1974. *Nucl. Phys. A* 222:1–12

Lane, A. M., Thomas, R. G. 1958. *Rev. Mod. Phys.* 30:257–353

Lemaire, M. C. 1973. *Phys. Rep. C* 7:279–336

Lim, K. L., McCarthy, I. E. 1964a. *Phys. Rev. B* 133:1006–1016

Lim, K. L., McCarthy, I. E. 1964b. *Phys. Rev. Lett.* 13:446–48

Mallet-Lemaire, M. C. 1978. See Van Oers et al 1978, pp. 271–90

Mang, H. J. 1964. *Ann. Rev. Nucl. Sci.* 14:1–28

Martin, P., Viano, J. B., Loiseaux, J. M., Le Chalony, Y. 1973. *Nucl. Phys. A* 212:304–16

Mathiak, E., Eberhard, K. A., Cramer, J. G., Rossner, H. H., Stettmeier, J., Weidinger, A. 1976. *Nucl. Phys. A* 259:129–56

McGrory, J. B. 1973a. *Phys. Lett. B* 47: 481–83

McGrory, J. B. 1973b. *Phys. Rev. C* 8: 693–710

Medsker, L. R., Headley, S. C., Fortune, H. T., Duray, J. R. 1975. *Phys. Rev. C* 11: 100–2

Newman, E., Becker, L. C., Preedom, B. M., Hiebert, J. C. 1967. *Nucl. Phys. A* 100: 225–35

Ollerhead, R. W., Chasman, C., Bromley, D. A. 1964. *Phys. Rev. B* 134: 74–89

Ost, R., Bethge, K., Gemmeke, H., Lassen, L., Scholz, D. 1974. *Z. Phys.* 266: 369–71

Ost, R., Speth, E., Pfeiffer, K. O., Bethge, K. 1972. *Phys. Rev. C* 5: 1835–39

Palla, G., Eppel, D. E., Strohbusch, U. 1978. *Z. Phys. A* 287: 369–72

Peng, J. C., Maher, J. V., Oelert, W., Sink, D. A., Cheng, C. M., Song, H. S. 1976. *Nucl. Phys. A* 264: 312–40

Perrenoud, J. L., DeVries, R. M. 1971. *Phys. Lett. B* 36: 18–20

Pfeiffer, K. O., Speth, E., Bethge, K. 1973. *Nucl. Phys. A* 206: 545–57

Pisano, D. J., Parker, P. D. 1976. *Phys. Rev. C* 14: 475–90

Roos, P. G., Chant, N. S., Cowley, A. A., Goldberg, D. A., Holmgren, H. D., Woody, R. 1977. *Phys. Rev. C* 15: 69–83

Rotter, I. 1968a. *Nucl. Phys. A* 122: 567–76

Rotter, I. 1968b. *Fortschr. Phys.* 16: 195–259

Rotter, I. 1969. *Nucl. Phys. A* 135: 378–94

Sanders, S. J. 1977. *Spectroscopy of high spin states in ^{16}O and ^{20}Ne*. PhD thesis. Yale Univ., New Haven, Conn. 98 pp.

Sanders, S. J., Martz, L. M., Parker, P. D. 1977. *Proc. Int. Conf. on Nuclear Structure, Tokyo, 1977*, pp. 625–26

Satchler, G. R. 1964. *Nucl. Phys.* 55: 1–33

Scholz, D., Gemmeke, H., Lassen, L., Ost, R., Bethge, K. 1977. *Nucl. Phys. A* 288: 351–

64; also Ost, R. 1974. PhD thesis. Univ. Heidelberg, Germany

Sherman, J. D., Hendrie, D. L., Zisman, M. S. 1976. *Phys. Rev. C* 13: 20–34

Siemssen, R. H. 1978. *J. Phys. Soc. Jpn.* 44: 137–53 (Suppl.); *Proc. Int. Conf. Nucl. Structure, Tokyo, 1977*. 896 pp.

Smirnov, Y. F., Chlebowska, D. 1961. *Nucl. Phys.* 26: 306–20

Stein, N., Sunier, J. W., Woods, C. W. 1977. *Phys. Rev. Lett.* 38: 587–91

Strohbusch, U., Bauer, G., Fulbright, H. W. 1975. *Phys. Rev. Lett.* 34: 968–71

Strohbusch, U., Fink, C. L., Zeidman, B., Markham, R. G., Fulbright, H. W., Horoshko, R. N. 1974. *Phys. Rev. C* 9: 965–72

Strohbusch, U., Stwertka, P. M., Fulbright, H. W., Bennett, C. L. 1978. *Ann. Rep. Nucl. Struct. Lab., 1977*. Univ. Rochester, pp. 49–51

Vander Molen, A., Becchetti, F. D., Janecke, J. 1975. See Goldberg et al 1975, p. 413

van Hienen, J. F., Chung, W., Wildenthal, B. H. 1976. *Nucl. Phys. A* 269: 159–88

Van Oers, W. T. H., Svenne, J. P., McKee, J. S. C., Falk, W. R., eds. 1978. *Clustering Aspects of Nuclear Structure and Nuclear Reactions, Winnipeg, 1978*. AIP Conf. Proc. No. 47. New York: Am. Inst. Phys. 792 pp.

Vincent, C. M., Fortune, H. T. 1970. *Phys. Rev. C* 2: 782–92

Vitturi, A., Ferreira, L., Kunz, P. D., Sofia, H. M., Bortignon, P. F., Broglia, R. A. 1978. *Preprint NBI-78-35*. Niels Bohr Inst., Copenhagen. 55 pp.

Wheeler, J. A. 1937. *Phys. Rev.* 52: 1083–1106

Wozniak, G. J., Stahel, D. P., Cerny, J. 1976. *Phys. Rev. C* 14: 815–34

Ann. Rev. Nucl. Part. Sci. 1979. 29: 203–42

HYPERON BEAM PHYSICS   �֍5606

J. Lach
Fermi National Accelerator Laboratory,[1] Batavia, Illinois 60510

L. Pondrom[2]
Department of Physics, University of Wisconsin, Madison, Wisconsin 53706

CONTENTS

[1] Operated by Universities Research Association Inc. under contract with the US Department of Energy.
[2] Supported in part by the US Department of Energy.

1 INTRODUCTION

Rochester & Butler (1947) first observed the decay of a neutral particle in a cloud chamber exposed to cosmic radiation. The mass of the neutral particle was at least 800 m_e, and represented a hitherto unknown phenomenon. Thus the age of strange particles began. Subsequent work by Armenteros et al (1951) showed that one neutral particle was heavier than the proton, with a quoted mass of (2203 ± 12) m_e. In present notation this particle is a strange baryon, the Λ. Charged baryons in this mass range were also discovered. All of these particles exhibited the peculiar property that they were produced by strong interaction, yet decayed by weak interaction into strongly interacting particles. Thus the reaction $\pi^- p \xrightarrow{\text{strong}} \Lambda \xrightarrow{\text{weak}} \pi^- p$ is possible. To account for this effect a new quantum number called "strangeness" was introduced, which was assumed conserved by the strong and electromagnetic interactions. All ordinary strongly interacting particles were given $S = 0$, whereas the new baryons, called hyperons, were given $S = -1$ and $S = -2$. This idea was proposed by Gell-Mann (1953). By the time of the review articles by Dalitz (1957) and Gell-Mann & Rosenfeld (1957), the classification scheme for the spin-parity-$1/2^+$ hyperons (the Λ, an I-spin singlet with $S = -1$, the $\Sigma^+,\Sigma^0,\Sigma^-$, an I-spin triplet also with $S = -1$, and the Ξ^0,Ξ^- I-spin doublet with $S = -2$) was established in notation still currently used. Table 1 shows the physical characteristics of these particles adopted and updated from the compilation of the Particle Data Group (1978).

The next step in classification of the hyperons was incorporation of

Table 1 Properties of the hyperons[a]

Name	Mass (MeV/c^2)	Lifetime (s)	Decay length (cm per GeV/c)	Magnetic moment ($eh/2m_p c$)
Λ	1115.60 ± 0.05	$(2.632 \pm 0.020) \times 10^{-10}$	6.93	-0.6138 ± 0.0047[b]
Σ^+	1189.37 ± 0.06	$(0.802 \pm 0.005) \times 10^{-10}$	2.02	2.95 ± 0.31
Σ^0	1192.47 ± 0.08	$(0.58 \pm 0.13) \times 10^{-19}$	—	—
Σ^-	1197.35 ± 0.06	$(1.483 \pm 0.015) \times 10^{-10}$	3.71	-1.48 ± 0.37
Ξ^0	1314.9 ± 0.6	$(2.90 \pm 0.10) \times 10^{-10}$	6.59	-1.20 ± 0.06[c]
Ξ^-	1321.32 ± 0.13	$(1.654 \pm 0.021) \times 10^{-10}$	3.75	-1.85 ± 0.75
Ω^-	1672.2 ± 0.4	$(0.82 \pm 0.06) \times 10^{-10}$[d]	1.47	—

[a] Particle Data Group (1978) except as otherwise noted; spin parity $1/2^+$ except for Ω^-, which has spin parity $3/2^+$.

[b] Schachinger et al (1978).

[c] Bunce et al (1979).

[d] Bourquin et al (1978).

I-spin and hypercharge $Y = N + S$, where N is the baryon number, into the larger group SU(3). This procedure was proposed independently by Gell-Mann (1961) and Ne'eman (1961), and unified the six strange hyperons and the neutron and proton into an octet of $1/2^+$ baryons, shown schematically in Figure 1. The figure also shows the $3/2^+$ decuplet, in which only the most massive member, the Ω^-, is sufficiently long-lived to be observed in a hyperon beam. The fundamental representations of the SU(3) group, from which the other representations can be formed, are triplets called **3** and **3***. The identification of the **3** with actual particles with remarkable properties was made, again independently, by Gell-Mann (1964) and Zweig (1964). Gell-Mann named the particles in the **3** representation quarks, and those in the **3*** antiquarks. The quark model of strange and nonstrange hadrons constructs all of the particles from products of the form $3 \otimes 3^*$ for mesons and $3 \otimes 3 \otimes 3$ for baryons. The number of quarks must be increased to accommodate the newer charmed particles (Chinowsky 1977).

The SU(3) classification scheme incorporates over 100 hadronic states (both mesons and baryons) into multiplets, gives the mass splittings within each multiplet, and describes decay rates for those particles that decay by strong or electromagnetic interaction (Samios et al 1974). The group has also been used with success to relate strange baryon or meson total cross sections to their nonstrange counterparts, and to predict properties of strange particle elastic and diffraction scattering off nucleons (Quigg & Rosner 1976).

There is a corresponding octet of antibaryons with the same masses and lifetimes but with opposite strangeness and hypercharge. They have all been observed, but only the antiproton has been extensively studied.

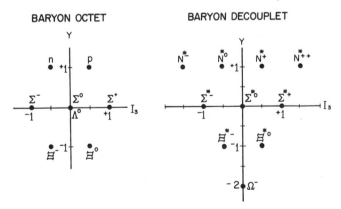

Figure 1   The lowest mass $1/2^+$ octet representation and $3/2^+$ decuplet representation of the baryons.

Application of SU(3) to the production of strange particles in hadronic collisions has not been as successful because high energy particle production is not well understood, and it is difficult to separate phenomena that might be related by group theory from other effects. Reactions of the type $p + p \rightarrow \Lambda + X$, where X is not observed are called "inclusive" and are reviewed by Bøggild & Ferbel (1974). The invariant cross section $E \, d^3\sigma/dp^3$ is usually written as either a function of $(x, p_\perp)$ or $(y, p_\perp)$, where $x = p_{\parallel}^*/p_{\parallel max}^*$, $y = 1/2 \ln [(E + p_{\parallel}^*)/(E - p_{\parallel}^*)]$, $p_{\parallel}^* = $ longitudinal momentum, and $p_\perp = $ transverse momentum of the produced particle in the center of mass. Strange baryons are produced about 10% of the time in pp collisions, and all final-state baryons exhibit the "leading-particle effect" to some degree. If a baryon is incident in the initial state, baryons occurring in the final state are enhanced relative to mesons. The hyperons Λ, $\Sigma^\pm$, Σ^0 show this effect quite distinctly as $x \rightarrow 1$, while the enhancement of Ξ^- and Ξ^0 is less marked. The production of anti-hyperons in pp collisions resembles antiproton production and has no leading-particle effect at all. This leading-particle enhancement has important practical consequences in the design of hyperon beams.

As shown in Table 1 the hyperons, with the exception of the Σ^0, all have lifetimes in the 10^{-10} s range. They all predominantly decay non-leptonically into pion-nucleon for Λ, $\Sigma^\pm$, into pion-Λ for Ξ, and into

Table 2 Nonleptonic decays[a]

Particle	Mode	Branching[b] ratio	Decay parameters[c]	
			α	ϕ
Λ	$p\pi^-$	$(64.2 \pm 0.5)\%$	$+0.647 \pm 0.013$	$(-6.5 \pm 3.5)°$
Λ	$n\pi^0$	$(35.8 \pm 0.5)\%$	$+0.646 \pm 0.044$	—
Σ^+	$p\pi^0$	$(51.6 \pm 0.7)\%$	-0.979 ± 0.016	$(36 \pm 34)°$
Σ^+	$n\pi^+$	$(48.4 \pm 0.7)\%$	$+0.072 \pm 0.015$	$(167 \pm 20)°$
Σ^0	$\Lambda\gamma$[d]	100%	—	—
Σ^-	$n\pi^-$	100%	-0.069 ± 0.008	$(10 \pm 15)°$
Ξ^0	$\Lambda\pi^0$	100%	-0.478 ± 0.035[e]	$(21 \pm 12)°$
Ξ^-	$\Lambda\pi^-$	100%	-0.392 ± 0.021	$(2 \pm 6)°$
Ω^-	ΛK^-	$(67.0 \pm 2.2)\%$[f]	0.06 ± 0.14[f]	—
Ω^-	$\Xi^0\pi^-$	$(2.46 \pm 1.9)\%$[f]	—	—
Ω^-	$\Xi^-\pi^0$	$(8.4 \pm 1.1)\%$[f]	—	—

[a] Particle Data Group (1978).
[b] Ignores rare modes (see Table 3).
[c] The parameter α is the final-state baryon helicity. The decay parameters β and γ are defined by $\beta = \sqrt{1 - \alpha^2} \sin \phi$, $\gamma = \sqrt{1 - \alpha^2} \cos \phi$. See text.
[d] The Σ^0 decays mainly by electromagnetic interaction.
[e] Bunce et al (1978).
[f] Bourquin et al (1978).

ΛK^- or $\Xi\pi$ for the Ω^-. The properties of the nonleptonic decays are shown in Table 2. In each nonleptonic decay the I-spin of the final state must differ from the I-spin of the initial state by half an integer. Amplitudes for which $|\Delta I| = 1/2$ seem to dominate. The phenomenological description of the parameters of the decays, and the sensitivity to various symmetry tests were reviewed by Overseth & Pakvasa (1969). The experimental situation is periodically reviewed by Overseth as a part of the Particle Data Group report (1978).

The nonleptonic two-body decay is often used in a hyperon beam to detect the presence of the parent hyperon. In this case the α parameters, listed in Table 2, are especially important, because $\alpha \neq 0$ leads to an asymmetrical distribution of decay products if the hyperons are polarized. In the general case, a $J = 1/2 \to J = 1/2$ transition without parity conservation can involve both an $l = 0$ amplitude, S, and an $l = 1$ amplitude, P. The three components of the final-state baryon polarization are described in terms of constants α, β, and γ: $\alpha = 2\,\mathrm{Re}\,S^*P/(|S|^2+|P|^2)$, $\beta = 2\,\mathrm{Im}\,S^*P/(|S|^2+|P|^2)$, and $\gamma = (|S|^2-|P|^2)/(|S|^2+|P|^2)$. If the parent hyperon has polarization $\mathbf{P}_Y$, the final-state baryon distribution is of the form $(1+\alpha\,\mathbf{P}_Y\cdot\hat{\mathbf{k}})$, where $\hat{\mathbf{k}}$ is the momentum unit vector of the final-state baryon in the hyperon rest frame. Alternatively if the parent hyperon is unpolarized, the final-state baryon is longitudinally polarized with $\langle\mathbf{P}\rangle = \alpha\hat{\mathbf{k}}$. The polarization of the final-state baryon in the general case also involves the parameters β and γ multiplied by P_Y. The formula was first written down by Lee & Yang (1957).

Table 3 shows the branching ratios and decay parameters for the rarer

Table 3 Semileptonic decays[a]

Particle	Mode	Branching ratio	g_A/g_V	g_V/g_A
Λ	$pe^-\bar{\nu}_e$	$(8.13\pm0.29)\times10^{-4}$	-0.62 ± 0.05	—
	$p\mu^-\bar{\nu}_\mu$	$(1.57\pm0.35)\times10^{-4}$	—	—
Σ^+	$\Lambda e^+\nu_e$	$(2.02\pm0.47)\times10^{-5}$	—	—
	$ne^+\nu_e$	$<0.5\times10^{-5}$	—	—
Σ^-	$ne^-\bar{\nu}_e$	$(1.08\pm0.04)\times10^{-3}$	0.385 ± 0.070	—
	$n\mu^-\bar{\nu}_\mu$	$(0.45\pm0.04)\times10^{-3}$	—	—
	$\Lambda e^-\bar{\nu}_e$	$(0.60\pm0.06)\times10^{-4}$	—	0.24 ± 0.23
Ξ^0	$\Sigma^+e^-\bar{\nu}_e$	$<1.1\times10^{-3}$ [b]	—	—
Ξ^-	$\Lambda e^-\bar{\nu}_e$	$(0.69\pm0.18)\times10^{-3}$	—	—
Ω^-	$\Lambda e^-\bar{\nu}_e$	$\sim10^{-2}$ [c]	—	—

[a] Particle Data Group (1978).
[b] Only upper limits exist for the semileptonic channels of the Ξ^0.
[c] Bourquin et al (1978).

semileptonic decays of the hyperons. For many of the semileptonic final states only the branching ratio is known, and that with large uncertainty. The ratio g_A/g_V given in the table for $\Lambda \to pe^- \bar{\nu}_e$, $\Sigma^- \to ne^- \bar{\nu}_e$, and $\Sigma^- \to \Lambda e^- \bar{\nu}_e$ represents the axial vector to vector coupling constant. The determination of this coupling-constant ratio requires a measurement of the momentum correlation of the hyperon decay products. Cabibbo (1963) proposed a model based on SU(3) that unifies the semileptonic decays of strange and nonstrange baryons. The present data are consistent with this model (Tanenbaum et al 1975). A more detailed discussion of these decays is given below.

2 HYPERON BEAM CONSTRUCTION

The term hyperon beam refers to a secondary beam of particles, usually produced by striking a metal or refractory target with the primary proton beam of an accelerator, which is collimated to a reasonably small solid angle, and which contains a useful flux of hyperons together with a background of the more copiously produced particles. If the beam is negatively charged, pions are the largest background. If it is positively charged, then protons usually are most numerous. In a neutral beam both neutrons and γ rays are present in numbers exceeding the flux of hyperons. Charged beams can be deflected by dipole magnets and focused by quadrupoles like any other secondary beam. In general, a hyperon beam supplies a high flux of short-lived particles that can be used to study their decays and their interactions with ordinary matter in a manner similar to the one traditionally employed for the more common longer-lived particles.

The fourth column of Table 1 illustrates a central design problem; the given constants, when multiplied by the hyperon momentum (in GeV/c), yield the decay lengths of the particles (in cm) in the laboratory. For example, a 10-GeV/c Λ beam is attenuated by $1/e$ in only 69 cm. Since the fractional yield of hyperons at the target remains essentially constant at about 10% as the energy is increased, good hyperon beams clearly become easier to make at higher energy, provided that the length necessary to shield the experiment from the primary production target does not also scale with the energy. This is indeed the case, because adequate shielding for hadronic cascade requires a thickness of material that grows only logarithmically with the primary energy. Thus, as seen below, beam lines designed to operate at 400 GeV are only about a factor of two longer than those designed for 30 GeV, which gives a substantial advantage in signal-to-noise ratio to the higher energy beams.

3 FIRST-GENERATION HYPERON BEAMS

3.1 *Introduction*

Although hyperons were observed earlier in short neutral beams designed for study of the K_s^0-K_L^0 system (Jensen et al 1969), not until the late 1960s was special effort placed on the design of charged and neutral beams specifically for hyperon studies. Two companion beams were constructed nearly simultaneously for charged hyperons, one at the CERN PS and the other at the Brookhaven AGS. Sandweiss (1971) previewed these beams before their operation. In addition, a short neutral beam for Λ and Ξ^0 hyperons was built at the CERN PS. A similar neutral beam was subsequently built at Brookhaven.

Table 4 summarizes the operational characteristics of these beams. The Brookhaven AGS charged-beam layout is shown in Figure 2

Table 4 Characteristics of hyperon beams[a]

Location	E_p (GeV)	I_p (per pulse)	Target	Θ (mrad)	E_Y (GeV)	L (m)	I_Y (per pulse)	I_{tot}
AGS[b] charged	29	1.5×10^{11}	Be	0	17–26	4.4	200 Ξ^- 2 Σ^-	30,000 π^-
PS[c] charged	24	10^{11}	B$_4$C W Al	10	13–20	3	50 Σ^- 1 Ξ^-	25,000 π^-
PS[d] neutral	24	10^9	Pt	175	3–15[e]	2	500 Λ	100,000 n
AGS[f]	30	10^9 [g]	Ir	72	8–16	2	500 Λ	[h]
Fermilab[i] neutral	400	10^8	Cu Be Pb	0–10	70–350	6	5000 Λ	200,000 n 400,000 γ
SPS[j] charged	200	4×10^{10}	BeO	2 − 6 +	70–140	12	4000 Σ^- 46 Σ^+	10^6 π^- 10^6 p

[a] Column headings are: E_p, I_p = primary beam energy and intensity; production target; θ = production angle; E_Y = hyperon beam energy; L = magnetic channel length; I_Y, I_{tot} = typical hyperon and total fluxes.
[b] Data from Hungerbuhler et al (1974b, 1975).
[c] Data from Badier et al (1972a).
[d] Data from Geweniger et al (1974).
[e] For a neutral beam, this is the approximate energy spread after the magnetic channel.
[f] Private communication from D. Jensen (1978).
[g] Proton maximum flux capability 2×10^{11}.
[h] γ rays selectively removed by 10X$_{rad}$ of lead at entrance to collimator.
[i] Data from Skubic et al (1978).
[j] Data from Bourquin et al (1979).

(Hungerbuhler et al 1974a); the corresponding CERN charged-beam layout is shown in Figure 3 (Badier et al 1972a). The design of both these beams was shaped by two requirements; first that the beam be as compact as possible to reduce hyperon decay losses. This set a premium on high magnetic fields and high spatial-resolution detectors used to determine the hyperon trajectories. The second goal was to interact the secondary hadrons produced outside of the beam phase space as far upstream in the beam as practical; before they could decay to muons or generate other backgrounds nearer the experimental apparatus. Care was taken to make the coils of the dipole magnets sufficiently large so that those muons produced in the target region and deflected away from the experimental apparatus by the central magnetic field were not redirected into the apparatus when they passed through the coils into the return flux region. Since the actual hyperon channel region was only a few square centimeters in cross section, high magnetic fields (~ 32 kG) could be achieved by shimming standard dipole magnets.

In the Brookhaven beam the hyperons were detected by their decay

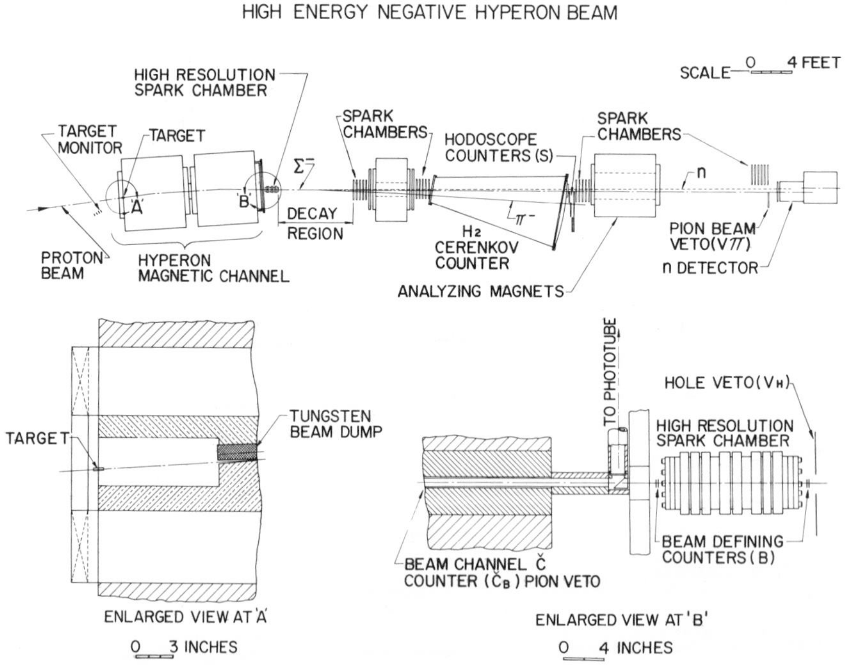

Figure 2 The configuration of the BNL charged-hyperon beam showing in detail the beam dump region and the beam channel Cerenkov counter with high resolution spark chamber.

immediately after a high resolution ($\sigma = 100$ μm), high pressure spark chamber (Willis et al 1971). This counter allowed the hyperon trajectory to be reconstructed and extrapolated back to the production target as a verification of its origin; it also served to measure its momentum. The pion background was suppressed by a threshold gas Cerenkov counter in the downstream end of the magnetic channel and used to tag particles with $m \lesssim m_p$.

The CERN beam embodied a set of superconducting quadrupole magnets in the hyperon channel; this produced a parallel beam and allowed the use of a DISC Cerenkov counter for particle identification. The performance of these high resolution differential gas Cerenkov counters was described by Litt & Meunier (1973). A pressure scan with the DISC counter showed clearly resolved peaks corresponding to Σ^- and Ξ^- as well as the other more copious hadrons. Both beams had useable fluxes of negative hyperons with tolerable backgrounds of pions in the beam and of general background radiation—mostly muons that were not stopped in the shielding. The Σ^+ beam was impractical to use because of the shorter decay length of the hyperons and the high flux of protons. Each group undertook an experimental program to (*a*) study hyperon production yields; (*b*) measure hyperon interaction cross sections; (*c*) investigate the various hyperon decay modes.

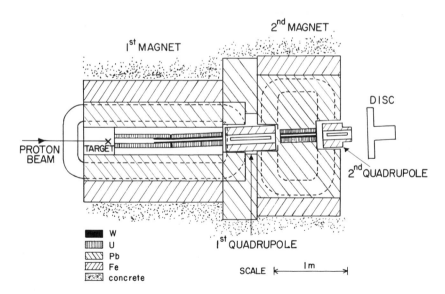

Figure 3 The CERN charged-hyperon beam. Note that the two superconducting quadrupoles fit snugly between the coil of the dipoles and take little extra space.

3.2 Hyperon Production Cross Sections

Badier et al (1972a) obtained ratios Σ^-/π^- and Ξ^-/π^- for various production angles, secondary beam momenta, and targets by comparing the
areas under the peaks of their DISC curves and making suitable corrections for hyperon decays between the production target and the DISC.
Their hyperon cross sections were then obtained by using the normalized
π^- inclusive cross sections of Eichten et al (1972).

Hungerbuhler et al (1975) also measured ratios Σ^-/π^- and Ξ^-/π^- by
detecting the decays $\Sigma^- \to n\pi^-$ and $\Xi^- \to \Lambda\pi^-$ in the downstream
spectrometer shown in Figure 2. These ratios could be extrapolated
back to the hyperon production target by correcting them for detection
losses and hyperon decays in the magnetic channel. Invariant inclusive
cross sections $E\,d^3\sigma/dp^3$ for Σ^- and Ξ^- production on beryllium at 0-
mrad production angle and various secondary momenta between 17 and
26 GeV/c were then obtained by using the π^- invariant cross sections
given in the empirical fit of Wang (1970). Allaby et al (1970) found that
the Wang fit does not reproduce π^- production measurements well near
the kinematic limit. Hence this group used the Wang fit only in the
region away from the kinematic limit where it is in good agreement
with data. A target interaction monitor and the known scaling properties
of the hyperon magnetic channel with momentum allowed them to
compute the hyperon invariant cross section in the full momentum range
of their measurements.

Figure 4 compares the hyperon invariant cross section measurements
as a function of x of Hungerbuhler et al (1975) at 0 mrad and those of
Badier et al (1972a) at 10 mrad. The data of Badier et al have been
converted from a tungsten target to a beryllium target by using the
measured A dependence from Table 2 of their paper. Cross sections for
the more common particles at 12.5 mrad are also shown in the figure
from the work of Allaby et al (1970). Because of uncertainties in the
angle dependence of the cross sections, no attempt was made to convert
the various experiments to the same production angle. Care was taken to
compute x in a manner consistent with conservation of charge, baryon
number, and strangeness, because differences in the maximum allowed
longitudinal momenta, although small, are not negligible at these energies.
Some ambiguity in p_{max} remains because of Fermi motion in the complex
nucleus and because of differences between a neutron and a proton
target. It is interesting to note that above $x \approx 0.75$ the Σ^- cross section
exceeds the π^- cross section, which demonstrates the leading-particle
effect. The Σ^- yield is roughly 1% of the proton yield. The Ξ^- cross

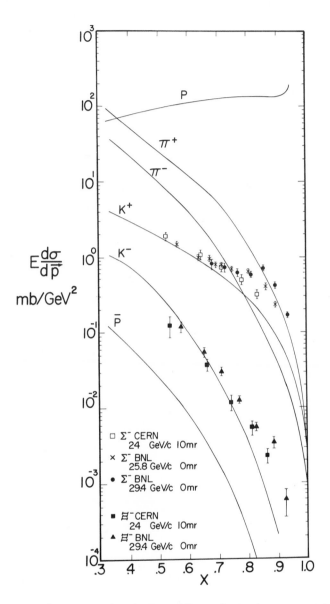

Figure 4   Invariant cross sections for Σ^- and Ξ^- production by protons on beryllium as a function of x. These data are from Hungerbuhler et al (1975) and Badier et al (1972a). Production cross sections of other particles taken from Allaby et al (1970) are shown for comparison. Note the differences in incident momentum and production angle.

section drops off more rapidly with increasing x, more like the $\bar{p}$ cross section and is smaller than the Σ^- cross section by a factor of about $1/25$ at $x = 0.7$.

3.3 Hyperon Interaction Cross Sections

3.3.1 INTRODUCTION These companion negative hyperon beams had sufficient flux to permit practical measurements of various hyperon interaction cross sections in a more or less conventional fashion. Thus precision was considerably improved over previous work, where hyperon interactions were studied in bubble chambers.

The two original charged-hyperon groups studied Σ^- interactions in hydrogen; the CERN group also used deuterium. At CERN Badier et al (1972b) measured the Σ^-p and Σ^-d total cross sections, and Blaising et al (1975) and Blaising (1977) measured Σ^-p elastic scattering, all with cryogenic targets. At Brookhaven Majka et al (1976) and Majka (1974) studied Σ^-p and Ξ^-p elastic scattering with a liquid-hydrogen target. The same group, Hungerbuhler et al (1974b), and Hungerbuhler (1973) studied Σ^- and Ξ^- interactions in a plastic scintillator target. Subsequently Arik et al (1977) and Arik (1976) studied Y^* production in Σ^- nucleus reactions with the Brookhaven charged-hyperon beam on H, Al, Cu, W, and Pb targets. The CERN neutral hyperon beam, listed in Table 4 but not yet described, also played an important role in hyperon interaction studies. Gjesdal et al (1972) obtained Λp, $\bar{\Lambda}p$, Λd, $\bar{\Lambda}d$ total cross sections by $CH_2 - C$ or $D_2O - H_2O$ subtraction, and Dydak et al (1977) measured the Σ^0 lifetime by studying the process $\Lambda + \gamma \to \Sigma^0$, where the "$\gamma$ ray" was supplied by the Coulomb field of a heavy nucleus.

The SU(3) quark model predicts differences between strange and ordinary particle total cross sections on a given target. The relation between hydrogen total cross sections,

$$\sigma_T(\Lambda p) - \sigma_T(pp) = \sigma_T(K^-n) - \sigma_T(\pi^+p) \qquad 1.$$

was first written by Lipkin & Scheck (1966), who used (a) simple quark counting rules to decompose the amplitude, and then (b) the optical theorem to relate the forward amplitude to the total cross section. Other relations of this type are

$$\sigma_T(pp) - \sigma_T(\Sigma^-p) = \sigma_T(\pi^-p) - \sigma_T(K^-p) + 2[\sigma_T(K^+p) - \sigma_T(K^+n)], \qquad 2.$$

$$\sigma_T(pp) - \sigma_T(\Sigma^-n) = \sigma_T(\pi^-p) - \sigma_T(K^-p), \qquad 3.$$

$$\sigma_T(pd) - \sigma_T(\Sigma^-d) = \sigma_T(\pi^-d) - \sigma_T(K^-d), \qquad 4.$$

and

$$\sigma_T(\Sigma^-p) + \sigma_T(\Sigma^-n) = 2\sigma_T(\Lambda p). \qquad 5.$$

The pertinent nucleon-nucleon, pion-nucleon, and kaon-nucleon cross-section data needed to predict the hyperon cross sections via these equations are supplied in the review article by Giacomelli (1976).

Giacomelli also discusses differential elastic scattering at high energies throughout the complete range of momentum transfers. High energy differential elastic cross sections are written $d\sigma/d|t|$, in units mb $(GeV/c)^{-2}$, where the variable t is the square of the 4-momentum transfer. The variables commonly used to describe elastic processes $a + b \rightarrow a + b$, are $s = -(\mathbf{P}_a + \mathbf{P}_b)^2$, $t = -(\mathbf{P}_a - \mathbf{P}'_a)^2$, $u = -(\mathbf{P}_a - \mathbf{P}'_b)^2$, where $\mathbf{P}_a$ is a 4-vector with metric $\mathbf{P}_a \cdot \mathbf{P}_a = -m_a^2$, the prime denotes the final state, and $s + t + u = 2(m_a^2 + m_b^2)$. For particle a incident on target b at rest, the square of the total available energy, is given by $s = m_a^2 + m_b^2 + 2m_b E_a$, and $t = 2m_a^2 - 2E_a E'_a + 2\mathbf{p}_a \cdot \mathbf{p}'_a$. Here $\mathbf{p}$ represents a 3-vector. In terms of the laboratory kinetic energy of the recoiling target particle, the variable $t = -2m_b(E'_b - m_b) = -2m_b T'_b$. The experimental cross section $d\sigma/d|t|$ is often empirically fit by an exponential of the form

$$d\sigma/d|t| = A \exp(bt + ct^2). \qquad 6.$$

For small $|t|$, the cross section is approximately exponential in $|t|$ with slope parameter b. By convention, an effective slope parameter at $|t| = 0.2$ $(GeV/c)^2$ is often used to compare various experiments. At $|t| = 0$, $d\sigma/d|t| = A$, which is called the optical point. If we assume that the forward amplitude is predominantly imaginary, A is directly related to the total cross section through the optical theorem, so that

$$A \, [mb/(GeV/c)^2] = 5.095 \times 10^{-2} \, [\sigma_T(mb)]^2.$$

A very simple geometrical picture of high energy, small $|t|$ diffraction scattering relates the slope parameter b and the total cross section to the radius R of the disc: $\sigma_T \sim R^2$, $b \sim R^2$, or $\sigma_T/b =$ constant (see Giacomelli 1976). Thus larger total cross sections lead to steeper elastic scattering slopes in this model. The logarithmic growth with energy of σ_T is interpreted as a change in the radius of the disc. These concepts are useful in discussing hyperon elastic scattering.

3.3.2 TOTAL CROSS-SECTION MEASUREMENTS The high resolution DISC Cerenkov counter was the hyperon detector for Badier et al (1972b) in their total cross-section measurements. Two such counters were used, one on either side of a cryogenic hydrogen (H) or deuterium (D) target. A vacuum target (V) was used to correct for absorption in the target walls. Multiwire proportional chambers (MWPCs) on either end of the DISCs measured the trajectory of the hyperon before and after the hydrogen target, while the Cerenkov light in the DISCs identified hyperons in the presence of the pion background. The angular accept-ance of the second DISC was made large (± 12 mrad compared to ± 5 mrad for DISC 1) to accept diffractively scattered Σ^-'s as well as beam Σ^-. The beam was tuned to 18.7 GeV/c. The number of Σ's scattered through angles $\theta < \theta_{max}$ per incident beam Σ was measured for each of the three targets H, D, V. To obtain the correct total cross sections $\sigma_T(\Sigma^- p)$ and $\sigma_T(\Sigma^- d)$ as $\theta_{max} \to 0$, it was necessary to take multiple Coulomb scattering into account, and to add the forward elastic $\Sigma^- p$ or $\Sigma^- d$ nuclear scattering contribution, while subtracting the contribu-tions from Coulomb single scattering and nuclear-Coulomb interference. The difference $\sigma_T(\Sigma^- d) - \sigma_T(\Sigma^- p) = \sigma_T(\Sigma^- n) - \delta$, where δ corrects for screening of one nucleon by the other. The theory of this effect has been worked out by Glauber (1960) and Harrington (1964). Badier et al (1972b) obtained the following cross sections at 18.7 GeV/c:

$$\sigma_T(\Sigma^- p) = 34.0 \pm 1.1 \text{ mb} \qquad 7.$$

$$\sigma_T(\Sigma^- d) = 61.3 \pm 1.4 \text{ mb} \qquad 8.$$

$$\sigma_T(\Sigma^- n) = 30.0 \pm 1.8 \text{ mb.} \qquad 9.$$

The short neutral beam constructed at the CERN PS and described by Geweniger et al (1974) was designed to measure Λ and $\bar{\Lambda}$ total cross sections by $CH_2 - C$ or $D_2O - H_2O$ subtraction. The absorption of

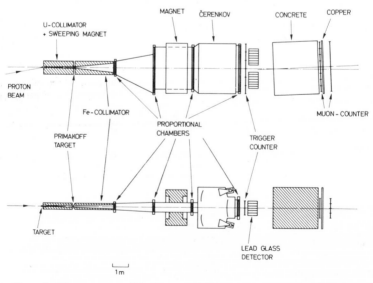

Figure 5   The apparatus of Dydak et al (1977) used to measure the Σ^0 lifetime.

neutral hyperons in the two targets was measured at the same time to eliminate systematic errors due to intensity fluctuations by splitting the target transverse to the beam. Downstream of this target arrangement the decays $\Lambda \to p\pi^-$ and $\bar{\Lambda} \to \bar{p}\pi^+$ were detected in a MWPC magnetic spectrometer. Their apparatus, configured for a different experiment, is shown in Figure 5. Gjesdal et al (1972) noted that, although the simultaneous transmission method eliminates many troubles in obtaining the total cross section, care had to be taken to insure that the decays observed were really $\Lambda \to p\pi^-$ and $\bar{\Lambda} \to \bar{p}\pi^+$, that it was clear which half of the target the hyperon traversed, and that those hyperons which either scattered in the target or were produced there by beam neutrons be eliminated from the transmitted hyperon flux. Cross sections were obtained as a function of hyperon momentum in 15 bins between 6 and 21 GeV/c. Observing no significant variation in cross section with momentum, the authors quote the following average values for 6 GeV/$c \leq p_\Lambda \leq 21$ GeV/c:

$$\sigma_T(\Lambda p) = 34.6 \pm 0.4 \text{ mb} \qquad\qquad 10.$$

$$\sigma_T(\Lambda d) - \sigma_T(\Lambda p) = 31.2 \pm 0.7 \text{ mb} \qquad\qquad 11.$$

$$\sigma_T(\Lambda n) = 34.0 \pm 0.8 \text{ mb}, \qquad\qquad 12.$$

where Equation 12 was obtained from Equation 11 by applying the screening correction discussed above. For $\bar{\Lambda}$ with an average momentum of $p_{\bar{\Lambda}} = (9.2 \pm 2.0)$ GeV/c, Gjesdal et al obtained $\sigma_T(\bar{\Lambda}p) = 56 \pm 11$ mb and $\sigma_T(\bar{\Lambda}n) = 46 \pm 20$ mb. Further work by Eisele et al (1976) in the same beam gave the more accurate result $\sigma_T(\bar{\Lambda}p) = 49.3 \pm 3.7$ mb at $p_{\bar{\Lambda}} = 8.3 \pm 2.7$ GeV/c.

Equation 1 can be solved for $\sigma_T(\Lambda p)$ and predicts $\sigma_T(\Lambda p) = 35.2 \pm 0.6$ mb in this energy range. The difference, $\sigma(\text{predicted}) - \sigma(\text{measured}) = 0.6 \pm 0.7$ mb, is in good agreement with the simple quark model. The equality of Λp and Λn cross sections is expected from charge symmetry, since the Λ has I-spin zero. Use of the charged-hyperon results of Equations 7 and 9, together with Equation 5, gives $\sigma_T(\Lambda p) = 32.0 \pm 1.1$ mb. Here the difference, $\sigma(\text{predicted}) - \sigma(\text{measured}) = -2.6 \pm 1.2$ mb, is not in such good agreement with the quark model. The problem is that the measured $\Sigma^- n$ cross section, Equation 9, is too small. Equation 2 predicts $\sigma_T(\Sigma^- p) = 35.0 \pm 0.9$ mb, in good agreement with Equation 7, whereas Equation 3 predicts $\sigma_T(\Sigma^- n) = 38.5 \pm 0.8$ mb, about four standard deviations larger than the experimental result in Equation 9.

3.3.3 SCATTERING CROSS-SECTION MEASUREMENTS It is helpful to detect the recoil proton as a signature for $\Sigma^- p$ elastic scattering, and both the CERN experiment of Blaising et al (1975) and the Brookhaven experiment of Majka et al (1976) did this. Since a proton of $T_p \leq 50$ MeV does not have sufficient range to leave the target vessel, a practical lower limit was placed on $|t| \geq 0.1$ $(\text{GeV}/c)^2$.

The CERN experiment covered the range $0.2 \leq |t| \leq 0.38$ $(\text{GeV}/c)^2$ at 17.2-GeV/c incident Σ^-, and collected 2826 Σ^-p elastic events. The Brookhaven experiment covered the range $0.1 \leq |t| \leq 0.23$ $(\text{GeV}/c)^2$ at 23-GeV/c incident Σ^-, and collected 6200 Σ^-p elastic events. They also obtained 67 Ξ^-p events in the same data sample. Since negative pions were copiously supplied by both beams, it was convenient to measure the π^-p cross section as well, which was already well known in this $|t|$ range from other experiments. The CERN group normalized their Σ^-p relative to π^-p. The Brookhaven group normalized both cross sections absolutely, and used their π^-p result as a check on the overall technique.

Figure 6 shows the results of these two experiments. The fit of the Σ^- data to the form of Equation 6, shown in Figure 6, did not require a t^2 term in the exponent. The results are summarized in Table 5. The slopes agree with each other, and are in good agreement with the prediction of the diffraction model: $b_{\Sigma^-p} = b_{pp}\sigma_T(\Sigma^-p)/\sigma_T(pp)$, which gives $b_{\Sigma^-p} = 8.7 \pm 0.5$ $(\text{GeV}/c)^{-2}$. The overall normalizations of the two experiments disagree by three standard deviations, assuming that the elastic cross

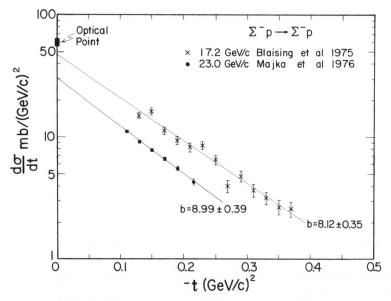

Figure 6 A comparison of the Σ^-p elastic scattering cross sections from the CERN and BNL charged-hyperon beams.

Table 5 Σ^-p elastic scattering $d\sigma/d|t| = Ae^{bt}$

Expt	p_Σ (GeV/c)	$|t|$ range (GeV/c)^2	b (GeV/c)$^{-2}$	A_{meas} mb per (GeV/c)^2	A_{opt}^b mb per (GeV/c)^2
BNL[a]	23	0.1–0.23	8.99 ± 0.39	30 ± 3	59 ± 3
CERN[c]	17.2	0.12–0.38	8.12 ± 0.35	47.4 ± 4.9	59 ± 3

[a] Majka et al (1976).
[b] Derived from $\sigma_T = (34 \pm 1)$ mb of Badier et al (1972b).
[c] Blaising et al (1975).

section changes a negligible amount between 17 and 23 GeV. Both extrapolations to $t = 0$ are below the optical point, the Brookhaven result being 29 ± 4 mb low. A careful measurement of the low $|t|$ region in high energy pp elastic scattering by Barbiellini et al (1972) showed that the slope parameter below $|t| \approx 0.1$ (GeV/c)^2 was distinctly larger than above that $|t|$ value by about 1.5 to 2.0 (GeV/c)$^{-2}$. This slope change at small $|t|$ appears to be a general phenomenon in hadron elastic scattering and has been noticed particularly in experiments performed in the Coulomb-nuclear interference region (Lach 1977). A recent summary of pp and π^-p slope changes as a function of t is given by Burq et al (1978). A slope change of this magnitude increases the cross section extrapolated to $t = 0$ by about 15%, bringing the CERN result up to 54 mb, and the Brookhaven result to 35 mb; the agreement is still poor.

3.3.4 Σ^0 LIFETIME Dydak et al (1977) measured the lifetime of the Σ^0 hyperon in a remarkable experiment in the CERN neutral hyperon beam. The decay rate $\Gamma(\Sigma^0 \to \Lambda\gamma)$ should be proportional to the square of the transition magnetic moment $\mu_{\Sigma\Lambda}$, which was predicted from SU(3) by Coleman & Glashow (1961). The expected Σ^0 lifetime based on these considerations is $\tau_{\Sigma^0} = 0.7 \times 10^{-19}$ s, corresponding to a flight path of 0.2 Å per GeV/c momentum, or a mass width $\Delta m = 9.4$ keV. Neither the flight distance nor the mass width can be measured by existing techniques. The decay rate can be measured indirectly by studying the reaction $\Lambda + Z \to \Sigma^0 + Z$, where the coulomb field of the nucleus Z supplies the "γ" for the inverse process $\Lambda + \gamma \to \Sigma^0$. This mechanism, called the Primakoff effect, was discussed in detail by Dreitlein & Primakoff (1962). The cross section is proportional to Z^2/τ_{Σ^0}; it is sharply peaked in the forward direction within an angle of about 2 mrad for 10-GeV/c lambdas; and it increases logarithmically with increasing energy. Competing strong interactions that convert Λ's into Σ^0's do not have these characteristics.

The experimental procedure then was to insert a thin, high Z target in the neutral beam (U and Ni were used) and to look for $\Lambda\gamma$ downstream of the target by detecting the decay $\Lambda \to p\pi^-$ and converting the γ ray. A very sharp peak was expected for $(\Lambda\gamma)$ events with invariant mass $m_{\Lambda\gamma} \approx m_\Sigma$ at very small angles relative to the (unobserved) incident beam Λ. The apparatus used by Dydak et al to do this is shown in Figure 5. The $\Lambda \to p\pi^-$ was measured in a multiwire proportional chamber spectrometer, and the γ-ray energy and location were recorded in an array of 84 lead glass blocks each 12.7 radiation lengths deep. The $\Lambda \to p\pi^-$ mass resolution had a $\sigma = 1.5$ MeV/c^2. The energy resolution of the lead glass was $\Delta E/E = 0.11/\sqrt{E(\text{GeV})}$ full width at half maximum. By calculating the centroid of the electromagnetic shower energy Dydak et al obtained a position resolution $\sigma = 2.7$ cm. The reconstructed Λ momentum vector intersected the plane of the Primakoff target at a point P, giving the origin of the Σ^0. A line between P and the production target center gave the direction of the incident Λ to within an error $\sigma = 1.2$ mrad.

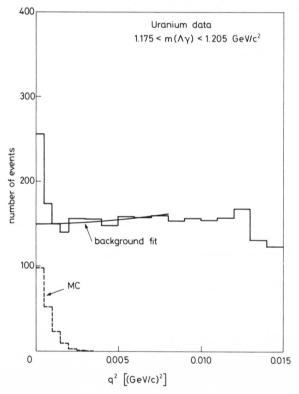

Figure 7 The data of Dydak et al (1977) used to extract the Σ^0 lifetime. The Primakoff effect peak is clearly visible at small q^2.

The uranium data for those $\Lambda\gamma$ events with mass $1.175 \leqq m(\Lambda\gamma) \leqq$ 1.205 GeV/c^2 are shown in Figure 7 plotted as a function of $q^2 = p_\Sigma^2 \theta_\Sigma^2$. The background resulted predominantly from decays of beam Ξ^0's, $\Xi^0 \rightarrow \Lambda\pi^0$, where only one γ ray from the decay $\pi^0 \rightarrow \gamma\gamma$ was detected by the lead glass. The peak at small q^2, was ascribed entirely to the Primakoff effect. Analysis of these data and similar results from nickel gave a lifetime $\tau_{\Sigma^0} = 0.58 \pm 0.13 \times 10^{-19}$ s, in good agreement with the SU(3) prediction.

3.4 Rare Decays

3.4.1 DESCRIPTION OF SEMILEPTONIC DECAYS Hyperon beta decays are very difficult to study. The branching ratios, given in column 3 of Table 3, are all small. In addition, two neutral particles are always involved, a neutrino and a neutral baryon. If the parent baryon is a Λ, or the daughter is a neutron, usually only the baryon direction is measured, not its energy. Thus a process like $\Sigma^- \rightarrow n e^- \bar{\nu}_e$ or $\Lambda \rightarrow p e^- \bar{\nu}_e$ has a zero-constraint kinematic fit. Decays like $\Sigma^- \rightarrow \Lambda e^- \bar{\nu}_e$, or $\Xi^- \rightarrow \Lambda e^- \bar{\nu}_e$, where $\Lambda \rightarrow p\pi^-$ can be observed in the final state, pick up a helpful extra constraint. The Λ decay asymmetry allows the final-state baryon spin direction to be measured as well. But for the other decays the zero-constraint fit leads to a quadratic ambiguity and two solutions for the energy of the neutral baryon. Low rates and weak kinematic constraints make it difficult to accumulate many events, and make the data sample prone to contamination by background, which is often composed of mis-identified nonleptonic decays of the same hyperons.

The semileptonic decays are especially interesting because the theory of such weak interactions is highly developed, and the interpretation of accurate experimental data is clear. In the notation of Marshak, Riazzudin, & Ryan (1969), the Hamiltonian for the current-current coupling of baryons to leptons is written $H = (G/\sqrt{2})(J_\mu^+ j_\mu + \text{h.c.})$. The phenomenological form of the baryon current for a decay like $\Sigma^- \rightarrow n e^- \bar{\nu}_e$ is

$$J_\mu = J_\mu^V + J_\mu^A, \qquad\qquad 13.$$

$$J_\mu^V = \bar{\psi}_n [f_1 \gamma_\mu + (f_2/m_\Sigma)\sigma_{\mu\nu} q_\nu] \psi_\Sigma, \qquad\qquad 14.$$

and

$$J_\mu^A = \bar{\psi}_n [g_1 \gamma_\mu \gamma_5 + (g_2/m_\Sigma)\sigma_{\mu\nu}\gamma_5 q_\nu] \psi_\Sigma, \qquad\qquad 15.$$

where J_μ^V and J_μ^A are the vector and axial vector parts, and $q_\lambda = (p_e + p_\nu)_\lambda = (p_\Sigma - p_n)_\lambda$ is the 4-momentum of the lepton pair. The γ_μ, $\mu = 1, 2, 3, 4$, are 4×4 Hermitian Dirac matrices that anticommute among themselves: $\gamma_\mu\gamma_\nu + \gamma_\nu\gamma_\mu = 2\delta_{\mu\nu}$. The matrix $\gamma_5 = \gamma_1\gamma_2\gamma_3\gamma_4$ anticommutes with all the other γ's. The matrix $\sigma_{\mu\nu}$ is the antisymmetric expression

$$\sigma_{\mu\nu} = \frac{1}{2i}(\gamma_\mu\gamma_\nu - \gamma_\nu\gamma_\mu).$$

The terms f_1, f_2, g_1, g_2 are form factors, and can depend on q^2. The range is $m_e^2 \leqq -q^2 \leqq (m_\Sigma - m_n)^2$. The vector and axial vector form factors f_1 and g_1 are "large" compared to the weak magnetism and induced pseudotensor terms f_2 and g_2, which are multiplied by $\sim E_e/m_\Sigma$. In a $\Delta S = 0$ beta decay like $n \to pe^- \bar{\nu}_e$, the form factor $g_2 \neq 0$ would imply the existence of an axial vector current even under G parity, called "second class" by Weinberg (1958). The G parity operation involves a rotation in I-spin space, which cannot be applied directly to a $\Delta S = 1$ decay. In the limit of exact SU(3) symmetry, however, it is replaced by an equivalent operation that predicts $g_2 = 0$ for $\Delta S = 1$ as well. Since SU(3) symmetry does not hold to an accuracy better than the mass differences within the multiplet, the existence of a g_2 term in a $\Delta S = 1$ decay of the order of $(m_\Sigma - m_\Lambda)/m_\Sigma$ would not imply second-class currents. This point is discussed by Pritchett & Deshpande (1972).

3.4.2 SEMILEPTONIC DECAY EXPERIMENTS   The decay $\Sigma^- \to n e^- \bar{\nu}_e$ was studied in the Brookhaven negative hyperon beam by Tanenbaum et al (1975), and in the CERN negative-hyperon beam by Decamp et al (1977). Both groups obtained $|g_1/f_1|$ by measuring the recoil-neutron kinetic energy distribution. The Brookhaven group collected 3507 events, while the CERN group had 519 events. The Brookhaven group also collected 55 $\Sigma^- \to \Lambda e^- \bar{\nu}_e$ and 11 possible $\Xi^- \to \Lambda e^- \bar{\nu}_e$. A second Brookhaven group using the same hyperon beam obtained 127 $\Sigma^- \to \Lambda e^- \bar{\nu}_e$ and 15 $\Xi^- \to \Lambda e^- \bar{\nu}_e$ candidates (Herbert et al 1978). The results of these experiments combined with previous world averages are shown in Table 3.

The experimental requirements for a high statistics study of $\Sigma^- \to n e^- \bar{\nu}_e$ are (a) electron identification and momentum measurement; and (b) neutron detection. To select and reconstruct beta decay event candidates one must (a) eliminate $\Sigma^- \to n\pi^-$ where a π^- is mistaken for an e^-; and (b) handle the neutron energy twofold kinematic ambiguity.

The apparatus used by Tanenbaum et al is shown in Figure 2. Neutrons were detected by a downstream calorimeter composed of iron plates and multiwire proportional chambers. The neutron interaction point was determined to ± 7 mm in this array. The neutron energy information was not sufficiently precise to provide a useful constraint. The proton counter was used to detect daughter $\Lambda \to p\pi^-$. Events of the form $\Sigma^- \to$ negative particle in the fiducial region were reconstructed as two-body decays $\Sigma^- \to n\pi^-$. The majority of such events were peaked at the Σ^- mass, and were caused by accidental fast muon counts in the Cerenkov counter in time with a beam $\Sigma^- \to n\pi^-$ decay. Time and space correlations between the scintillators that detected the negative particle and the signal from the appropriate Cerenkov counter segment were used to reduce the nonleptonic background. Further reduction was achieved by requiring $m(n\pi^-) < 1165$ MeV/c^2. After final aperture cuts 3507 leptonic decay events remained.

The shape of the neutron energy spectrum is smeared by the kinematic ambiguity in neutron energy. Tanenbaum et al retained both solutions for the neutron energy, and used a maximum likelihood fit to a surface in $(E_n^{\text{upper}}, E_n^{\text{lower}})$ space. Their final result was $|g_1/f_1| = 0.435 \pm 0.035$.

The detection apparatus of Decamp et al at CERN used their DISC Cerenkov counter for Σ^- identification, a 2-m long threshold Cerenkov counter filled with a H_2-CH_4 mixture at atmospheric pressure for π^- rejection and e^- identification; a downstream neutron detector with optical spark chambers and iron plates; and two streamer chambers, one for the $\Sigma^- \to e^-$ vertex, and one after an analyzing magnet for e^- momentum measurement. Event candidates reconstructed as $\Sigma^- \to n\pi^-$ showed a strong peak at the Σ^- mass, again due predominantly to accidentals in the H_2-CH_4 Cerenkov counter. A cut on low pulse height in the counter reduced the nonleptonic background substantially. The sample of beta decays was further purified by requiring $m(n\pi^-) < 1170$ MeV/c^2 and a low value of chi-squared for the hypothesis $\Sigma^- \to ne^-\bar{\nu}_e$ calculated from the eight measured quantities $\mathbf{p}_\Sigma$, $\mathbf{p}_e$, and $\mathbf{p}_n/|\mathbf{p}_n|$. They estimated a residual background of 1.6% $\Sigma^- \to n\pi^-$ in the remaining 519 $\Sigma^- \to ne^-\bar{\nu}_e$ candidates. They weighted the two solutions for E_n equally in the analysis, and obtained $|g_1/f_1| = 0.17 \pm^{0.07}_{0.09}$ from a likelihood fit to the resulting neutron energy distribution. The CERN and Brookhaven results disagree by three standard deviations. This disagreement is reflected in the error assigned to this quantity in Table 3 by the Particle Data Group. The number $|g_1/f_1| = 0.385 \pm 0.070$, is the world average of the two hyperon beam experiments plus several bubble chamber experiments each with less than 100 events.

Two groups at Brookhaven measured $\Sigma^- \to \Lambda e^-\bar{\nu}_e$. Herbert et al (1978) obtained a decay rate for this mode and for $\Xi^- \to \Lambda e^-\bar{\nu}_e$ as well. These rates, combined with previous work, are listed in Table 3. Tanenbaum et al (1975) measured f_1/g_1 for $\Sigma^- \to \Lambda e^-\bar{\nu}_e$ from the Λ energy distribution and angular correlation terms in the Σ^- rest frame for 55 events. The result is $f_1/g_1 = -0.25 \pm 0.35$ assuming $f_2 = 0$ and is quoted as the reciprocal of g_1/f_1 because it is expected that $f_1/g_1 \approx 0$ for this particular decay. This follows from the conserved vector current hypothesis, which relates $\Sigma^- \to \Lambda e^-\bar{\nu}_e$ to the $\Sigma^0 \to \Lambda\gamma$ decay and predicts $f_1/g_1 \ll 1$ (Cabibbo & Gatto 1960).

3.4.3 SEARCH FOR THE RARE MODE $\Xi^0 \to p\pi^-$ Geweniger et al (1975) used the CERN neutral hyperon beam, where the Ξ^0 flux is about 1% of the Λ flux, to search for the $\Delta S = 2$ decay $\Xi^0 \to p\pi^-$. No decay that violates the selection rule $|\Delta S| \leqq 1$ in first-order weak interactions is known to exist, so a search for this process is of fundamental interest. A two-body decay of this type is constrained, and the total momentum vector and mass of the parent particle can be determined from the vectors $\mathbf{p}_p$ and $\mathbf{p}_\pi$. The momentum of either particle in the Ξ^0 rest frame is 300 MeV/c, higher than that for any known decay mode of other particles in the neutral beam. It is an ideal candidate for a search for $\Delta S = 2$ decays, and a neutral hyperon beam is an ideal place to perform the search.

The group collected 10^9 events, corresponding to about 10^8 $\Lambda \to p\pi^-$, and roughly 10^6 beam Ξ^0's available to decay by the $\Delta S = 2$ mode. The major background at $m(p\pi) > 1250$ MeV/c^2 was found to be $K_s^0 \to \pi^+\pi^-$, where the π^+ was assigned the proton mass. To eliminate these events the authors made a cut requiring $m(\pi\pi) > 550$ MeV/c^2. This cut also would have eliminated many $\Xi^0 \to p\pi^-$, and thus decreased their sensitivity to the decay mode. The limit they obtained, based on a signal of zero events above a background of two events was $\Gamma(\Xi^0 \to p\pi^-)/\Gamma(\Xi^0 \to \Lambda\pi^0) < 3.6 \times 10^{-5}$ (90% confidence limit).

4 HIGH ENERGY HYPERON BEAMS

4.1 *Introduction*

As mentioned in Section 2, it was anticipated that an order of magnitude increase in laboratory hyperon energy would substantially improve the signal-to-noise ratio in hyperon beams. The Fermilab program, reviewed by Sanford (1976), has included plans for neutral and charged-hyperon beams since the first set of proposals was submitted to the laboratory. The neutral beam has been operating successfully for several years, and the charged beam is scheduled to come on in 1979. In the meantime a similar accelerator, the SPS, has been built at CERN, and a charged-hyperon beam has been brought into operation there. The currently available results from these beams are reviewed below.

4.2 *Fermilab Neutral Hyperon Beam*

4.2.1 EXPERIMENTAL APPARATUS The layout of the Fermilab neutral hyperon beam is shown in Figure 8. Some of the characteristics of the beam are listed in Table 4. The hyperons were produced by a secondary proton beam, diffracted from the main beam of the accelerator by the Meson Laboratory beryllium target, 450-m upstream of the apparatus.

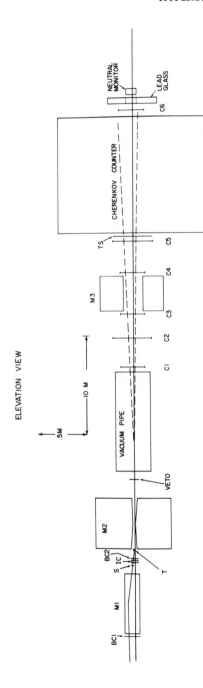

Figure 8 Layout of the Fermilab neutral hyperon beam. The magnet M1 restored the proton beam to the production target T. M2 was the neutral beam collimator and charged-particle sweeping magnet. Decays were observed after the veto by MWPCs C1–C6 and the analyzing magnet M3.

The proton energy equaled that of the accelerator—either 300 or 400 GeV. The intensity could be varied from 10^6–10^{10} protons per pulse. The production angle could be varied between 0 and ± 10 mrad by steering the proton beam in a vertical plane, keeping the axis of the collimator that defined the neutral beam fixed. The hyperons were produced in a 6-mm diameter metal target made of beryllium, copper, or lead. An ion chamber monitored proton intensity.

A brass collimator 5.3-m long in a vertical magnetic field defined the neutral beam and eliminated charged particles. The maximum $\int B\,dl$ was 13.6 Tm. The solid angle accepted by the collimator was 1.2 μsr. All neutrals produced in this solid angle were transmitted to the downstream detectors. A typical beam Λ had momentum of 150 GeV/c and a decay length of 10.4 m, thus suffering small decay loss in the channel.

A veto counter defined the beginning of the decay volume, an 11-m vacuum pipe. Decays $\Lambda \to p\pi^-$, $K_s^0 \to \pi^+\pi^-$, $\bar{\Lambda} \to \bar{p}\pi^+$ as well as $\gamma \to e^+e^-$ conversions and neutron interactions in the small amount of material in the beam were detected by a set of multiwire proportional chambers and an analyzing magnet. The device was designed to have high acceptance for $\Lambda \to p\pi^-$. A threshold helium-gas-filled Cerenkov counter could distinguish protons and pions for those protons with momenta less than 170 GeV/c. No other mass measurements were made on secondary particles. A lead glass wall at the downstream end of the apparatus was used to detect γ rays from the decay sequence $\Xi^0 \to \Lambda\pi^0$, $\pi^0 \to \gamma\gamma$, $\Lambda \to p\pi^-$. From the momentum vectors $\mathbf{p}_+$ and $\mathbf{p}_-$ measured by the spectrometer the invariant mass of the parent $M_{+-} = [m_+^2 + m_-^2 + 2E_+E_- - 2\mathbf{p}_+ \cdot \mathbf{p}_-]^{1/2}$ was calculated using hypotheses $(m_+, m_-) = (m_p, m_\pi), (m_\pi, m_\pi)$, and (m_π, m_p) to search for Λ, K, and $\bar{\Lambda}$ respectively.

4.2.2 INCLUSIVE CROSS SECTIONS Extensive measurements were made of the spectra of neutral particles produced by 300-GeV protons at various laboratory angles on beryllium, copper, and lead targets. The resulting invariant cross sections per nucleus for $p + Be \to \Lambda + X$ and $\bar{\Lambda} + X$ are shown in Figure 9 as a function of laboratory momentum for six angles between 0.25 and 8.8 mrad. The inclusive Λ spectra include daughter Λ's from Σ^0 decay, which could not be separated from "direct" Λ's. Note that at small angles the Λ spectra are rather flat as p_{lab} increases, which indicates a leading-particle effect, while the $\bar{\Lambda}$ spectra fall off very steeply with increasing p_{lab}. At the lowest measured momentum, 60 GeV/c, the ratio $\bar{\Lambda}/\Lambda = 0.1$. These data were fitted to smooth empirical functions of the scaling variables $(x, p_\perp)$, and the lines shown on the figures were calculated from these fits. The fitted parameters are given in Table VI of Skubic et al (1978). Production data subsequently taken at 400 GeV show that $E \, \mathrm{d}^3\sigma(x, p_\perp)/\mathrm{d}p^3$ is independent of energy in this region. The $p_\perp$ dependences of the Λ and $\overline{\Lambda}$ spectra are similar—essentially $\exp(-2.3p_\perp^2)$, while the x dependences are very different. At $p_\perp = 0$, an expression of the form $E \, \mathrm{d}^3\sigma/\mathrm{d}p^3 \sim (1-x)^n$ fits the data roughly, where $n_\Lambda = 0.6$, and $n_{\overline{\Lambda}} = 6.0$. These same data were compared to the triple Regge model by Devlin et al (1977).

Inclusive cross sections for K_s, Λ, $\overline{\Lambda}$, and n produced by 200-GeV π^-, K^-, $\bar{p}$, and p on beryllium were measured by Edwards et al (1978), who used the same beam line and detection apparatus. These results show remarkable similarities between the cross sections $\pi^- \to \Lambda$, $\pi^- \to \overline{\Lambda}$, and $p \to K_s$, between $\pi^- \to n$ and $K^- \to \Lambda$, and, to a lesser extent, between $\pi^- \to K_s$ and $p \to \Lambda$, and illustrate the insensitivity of the inclusive cross section to the particles involved.

4.2.3 A DEPENDENCE   Since invariant cross sections were measured for Λ production on copper and lead as well as beryllium targets, the dependence of the shape of the cross section on the atomic weight of the target nucleus (A dependence) could be studied. Interest in this subject has grown in recent years, because it may offer the possibility of studying the time development of a state of hadronic matter after a collision. A large number of particles are produced by a high energy interaction between two nucleons. Many of these particles are associated

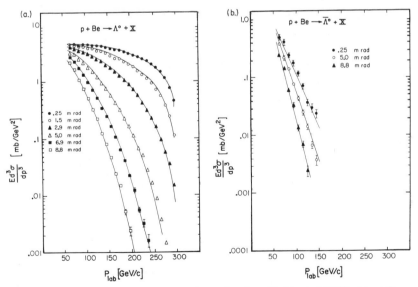

Figure 9 Invariant cross sections at 300 GeV taken from Skubic et al (1978): (*a*) $\mathrm{p} + \mathrm{Be} \to \Lambda/\Sigma^0 + \mathrm{X}$; (*b*) $\mathrm{p} + \mathrm{Be} \to \overline{\Lambda} + \mathrm{X}$.

with the incident projectile and are moving rapidly in the laboratory. In a heavy nucleus the time between collisions is so short that the definite number of particles observed in the asymptotic final state may not have time inside the nucleus to materialize. This picture was discussed by Gottfried (1974) and supported by the early work of Busza et al (1975), who showed that the forward angle multiplicity had very weak A dependence. Thus if a proton incident on a lead nucleus is excited by interaction with a target nucleon, this excited proton may travel through the rest of the nucleus and diffractively scatter, suffering a small loss in longitudinal momentum, and subsequently decay into the observed final-state particles outside the nuclear volume.

The forward inclusive production of Λ's by protons was analyzed by Heller et al (1977) from this point of view. Following Busza et al (1975), these authors define the average number of absorption mean free paths encountered by an incident hadron h on a nucleus A as $\bar{v} = A\sigma_{hN}/\sigma_{hA}$, where σ_{hN} is the absorption cross section on a single-nucleon target, and σ_{hA} is the corresponding cross section on nucleus A. Then taking experimental numbers $\sigma_{pN} = 33$ mb, $\sigma_{pBe} = 216$ mb, $\sigma_{pCu} = 812$ mb, and $\sigma_{pPb} = 1930$ mb they calculate the values $\bar{v}_{Be} = 1.4$, $\bar{v}_{Cu} = 2.6$, and $\bar{v}_{Pb} = 3.6$. Thus copper is approximately one absorption length thicker than beryllium, and lead is one absorption length thicker than copper.

Figure 10 shows the differential multiplicity—the invariant cross

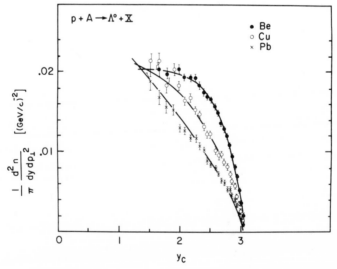

Figure 10 Differential multiplicities for Λ production at 0 mrad from Be, Cu, and Pb taken from the paper by Heller et al (1977).

section divided by σ_{hA}—for the three nuclear targets plotted as a function of rapidity y_c in the nucleon-nucleon center of mass. Note that as A increases the distribution at large y_c is depleted, as if the excited hadron lost energy by collision in leaving the nucleus. Although data were not taken between $y_c = 0$ and $y_c = 1.5$, the curves suggest an enhancement at small y_c as A increases, thus leaving the multiplicities, which are related to the areas under the curves, independent of A. If on the average the excited hadronic state makes one more collision in a copper nucleus than it does in a beryllium nucleus, then the difference between the shapes in Figure 10 should be explained by a function that predicts the probability of a final rapidity y after collision when the initial rapidity is y_0. This same $f(y-y_0)$ should also explain the difference between the copper shape and the lead shape. Heller et al (1977) found a suitable $f(y-y_0)$, quite similar in shape to the rapidity distribution of protons in $p+p \rightarrow p+X$, which satisfied these criteria, thus supporting the definition of $\bar{\nu}$ and the view that an excited proton-like object scatters through the nucleus before materializing into the observed final state.

4.2.4 POLARIZATION   Figure 8 shows that the Λ production angle could be varied in a vertical plane. A parity-conserving strong interaction could produce a Λ polarization in the direction $\hat{\mathbf{n}} = (\mathbf{p}_p \times \mathbf{p}_\Lambda)/|\mathbf{p}_p \times \mathbf{p}_\Lambda|$. A spin vector in this direction would be perpendicular to the magnetic field of the shield magnet M2 and would precess as the Λ moved through the collimator at a rate proportional to the magnetic moment.

The unexpected occurrence of substantial Λ polarization was first reported by Bunce et al (1976). This early work used the precession to eliminate systematic errors in searching for a polarization effect. Since parity is not conserved in the decay $\Lambda \rightarrow p\pi^-$, the proton distribution in the Λ rest frame is asymmetric if the parent Λ's are polarized, as discussed in Section 1. The relevant asymmetry parameter is given in Table 2. Reversing the direction of the magnetic field of M2 would reverse the longitudinal component of the polarization, without changing anything else, affording a technique to cancel apparatus bias. The original 300-GeV measurements of the polarization have since been augmented by work at 400 GeV, reported by Heller et al (1978) (Figure 11). The Λ polarization is plotted as a function of Λ transverse momentum for Λ's produced by protons at the fixed laboratory angle of 7.2 mrad. The polarization increases monotonically with increasing $p_\perp$, and vanishes at $p_\perp = 0$. The direction of the polarization is along $-\hat{n}$, defined above; P_Λ reaches -0.25 at $p_\perp = 2$ GeV/c. It is not caused by polarization of the incident proton beam, and it is consistent with a parity-conserving strong interaction in the reaction $p + Be \rightarrow \Lambda + X$. Preliminary data on $p+p \rightarrow$

$\Lambda + X$ from hydrogen, reported by Grobel et al (1979), show that the polarization is not a complex nuclear effect. The fixed-angle data of Figure 11 cannot be used to separate the kinematic dependent of P_Λ on x and $p_\perp$, since the two variables are uniquely related. Bunce et al (1976) observed no noticeable x dependence in the 300-GeV data, which were taken at several angles between 0 and 9 mrad. Subsequent careful measurements at several production angles have shown that P_Λ does increase slightly with increasing x, as if it were associated in some way with the leading particle.

Figure 11 also shows the companion data on the polarization of $\bar{\Lambda}$'s. The maximum transverse momentum is smaller, but it is clear that $P_{\bar{\Lambda}} = 0$ for $p_\perp \lesssim 1$ GeV/c, and that $P_{\bar{\Lambda}}$ does not resemble P_Λ. This strengthens the notion that the polarization is a leading-particle effect, since $\bar{\Lambda}$'s are not strongly associated with the incident proton. One more clue to the origin of this unexplained phenomenon has been supplied by Bunce et al (1979), who reported a polarization of Ξ^0 hyperons at 7.2-mrad production of $P_{\Xi^0} = -0.086 \pm 0.019$, in agreement in sign and magnitude with the average P_Λ.

4.2.5 MAGNETIC MOMENTS The precession of the polarized Λ's in the magnetic field of M2 could obviously be exploited for a precise measurement of the Λ magnetic moment. This was done by Schachinger et al (1978). For hyperons with velocity $\approx c$ (a 75-GeV Λ has $\beta = 0.99989$) the precession angle is given in terms of the field integral by the formula ϕ (degrees) $= 18.3 \, \mu_\Lambda \int B \, dl$ (Tm), where μ_Λ is measured in nuclear

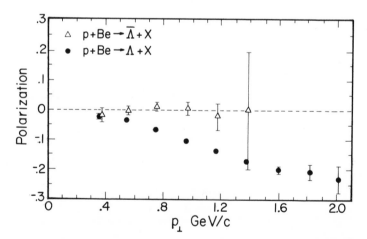

Figure 11 Polarization of Λ and $\bar{\Lambda}$ produced by 400-GeV protons on beryllium taken from the work of Heller et al (1978).

magnetons μ_N (units of $eh/2\,m_p c$). This angle was measured from the ratio of the transverse component to the longitudinal component of P_Λ after precession for various values of $\int B\,dl$. Equal amounts of data were taken with the proton beam steered onto the beryllium target from above and from below the Λ line of flight, thus reversing P_Λ in space. Asymmetries for the two signs of production angle were then subtracted to eliminate instrumental bias. The resulting precession angles ϕ versus $\int B\,dl$ are plotted in Figure 12. The slope of the straight line gives the magnetic moment $\mu_\Lambda = -0.6138 \pm 0.0047\ \mu_N$.

Λ hyperons from the decay $\Xi^0 \to \Lambda\pi^0$ were separated from Λ's produced by protons in the Be target by requiring the Λ momentum vector to intersect the plane of the collimator at a distance greater than 9 mm from its center. The actual collimator radius was 2 mm. Beam K_s^0 were used to estimate that the background from wide angle beam Λ's in this sample was $\sim 10\%$. The Λ's selected by this technique showed a growth curve in the decay vertex distribution characteristic of a parent-daughter decay. The 40,000 daughter Λ's selected in this manner were analyzed to measure spin direction in the same way as the beam Λ's. The daughter spin direction in turn gives the spin direction of the parent,

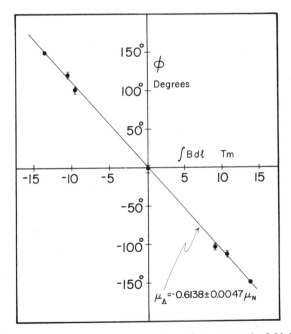

Figure 12 The Λ precession angle as a function of the magnetic field integral from Schachinger et al (1978).

and hence the precession angle of the Ξ^0's in the magnetic field. Analysis of ϕ versus $\int B \, dl$ gave the result $\mu_\Xi = -1.20 \pm 0.06 \ \mu_N$ (Bunce et al 1979).

The quark model discussed in Section 1 can be applied to the baryon magnetic moments. The baryons shown in Figure 1 are all composed of the three quarks (u, d, s). SU(6) gives a rule for combining the spins of the quarks to give the baryon spins: like quarks form spin 1. Thus the proton with spin up in this model would have the structure $p\uparrow = \sqrt{2/3}$ (u↑u↑d↓)$-\sqrt{1/3}$ [d↑(u↑u↓+u↓u↑)/$\sqrt{2}$]. The Σ^+, composed of (uus) has the same form. The Σ^0 (uds) is obtained by applying an I-spin-lowering operator to Σ^+, which results in (ud) in a triplet state. The Λ is also (uds), but is orthogonal to Σ^0, and hence has (ud) in a singlet state: $\Lambda\uparrow =$ s↑(u↑d↓−u↓d↑)/$\sqrt{2}$. In this model the spin and magnetic moment of the Λ arise entirely from the strange quark, and a measurement of μ_Λ is a measurement of μ_s. The magnetic moment of the u quark can be obtained from the proton moment by assuming that $\mu_u/\mu_d = -2$, the same as the charge ratio. This procedure gives a neutron moment about 3% smaller in magnitude than the experimental value. Taking $\mu_p =$ 2.7928 and $\mu_\Lambda = -0.6138$ gives $\mu_u = 1.862$ and $\mu_s = -0.6138$ in nuclear magnetons. The masses of the constituent quarks can be calculated if $\mu_q = e_q h/2m_q c$ is assumed, similar to Dirac "point" particles. The results $m_u = 336$ MeV/c^2 and $m_s = 510$ MeV/c^2 are in good agreement with quark mass differences and ratios calculated from observed splittings between particle masses in the various multiplets [see De Rujula, Georgi & Glashow (1975) and Lipkin (1978)]. Consistent numbers for constituent quark masses within the baryons do not resolve the question of quark masses in general, discussed by Lipkin (1978) and by Weinberg (1977). Free quarks may be infinitely massive (and hence unobservable), constituent quarks may have masses as quoted above, and current quarks may weigh only a few MeV/c^2, and yet all be manifestations of the same phenomena.

4.2.6 ELASTIC SCATTERING   A liquid-hydrogen target 91-cm long was inserted in the neutral beam just upstream of the veto in Figure 8 to study Λp elastic scattering. The neutral beam flux was monitored by counting neutron conversions in a thin CH_2 slab placed upstream of the target. Unscattered beam Λ's and Λp scattering events were then detected for a fixed number of neutral beam counts to determine the cross section. Recoil protons emitted from the hydrogen target at large angles to the neutral beam were detected in two pairs of multiwire proportional chambers, one next to the target on the left side of the beam, and one on the right.

Given the scattered Λ momentum vector determined by $\Lambda \to p\pi^-$

and the recoil-proton direction, the incident beam Λ momentum vector could be calculated, assuming the event to be elastic (a zero-constraint fit). Inelastic background was eliminated by requiring the reconstructed beam Λ to point back to the beryllium production target in the same way as true beam Λ's. The details of the event selection, reconstruction, and normalization are discussed by Martin (1977). Proton-proton scattering was measured with the same apparatus by turning off M2 and bringing the proton beam through the neutral collimator to check the geometry and efficiency of the detector.

Figure 13 shows the elastic differential cross sections $d\sigma/dt$ for Λp → Λp (average energy 216 GeV) and $\overline{\Lambda}$p → $\overline{\Lambda}$p (average energy 120 GeV). The $\overline{\Lambda}$p cross section has been divided by 10 to separate the points. The two cross sections are the same size, with essentially the same slope. The Λp cross section was fit by the expression $d\sigma/d|t| = A\exp(bt+ct^2)$, with $A = 50.2 \pm 0.75$ mb $(\text{GeV}/c)^{-2}$, $b = 10.06 \pm 0.11$ $(\text{GeV}/c)^{-2}$, and $c = 1.59 \pm 0.16$ $(\text{GeV}/c)^{-4}$. The $\overline{\Lambda}$p cross section was fit by $A = 54.3 \pm 7.5$ mb $(\text{GeV}/c)^{-2}$, $b = 10.3 \pm 0.7$ $(\text{GeV}/c)^{-2}$, and $c = 0$. If the lower energy value of $\sigma_T(\Lambda p) = 34.6 \pm 0.4$ mb reported by Gjesdal et al (1972) is used as a guide, $A_{\Lambda p} = 61 \pm 0.7$ mb $(\text{GeV}/c)^{-2}$ is predicted by the optical theorem, 11 ± 1 mb above the measured result. As discussed in Section 3.3.3 an increase in the slope at small $|t|$ would decrease the discrepancy.

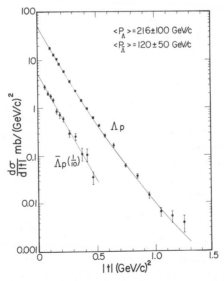

Figure 13   Λp and $\overline{\Lambda}$p elastic scattering cross sections from the Fermilab neutral hyperon beam. The $\overline{\Lambda}$p cross sections have been multiplied by 1/10 to facilitate plotting.

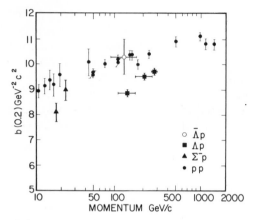

Figure 14 A compilation of elastic scattering *b* parameters evaluated at $t = 0.2$ $(GeV/c)^{-2}$ for hyperon proton scattering. Selected pp data is shown for comparison (Giacomelli 1976).

The Λp cross-section data were also divided into three separate momentum bins, centered at 134 GeV/c, 208 GeV/c, and 307 GeV/c, and the slopes were fit to each bin. These results for the effective slope at $|t| = 0.2$ GeV/c^2 are shown in Figure 14 together with the pp elastic cross-section data and the lower energy $\Sigma^- p$ data from the charged-hyperon beams.

4.3   *CERN SPS Charged-Hyperon Beam*

4.3.1   EXPERIMENTAL APPARATUS   The anticipated improvement in charged-hyperon beam fluxes at higher energy was recently demonstrated by the CERN SPS charged-hyperon beam (Bourquin et al 1979). At least three important new features were demonstrated by this beam: For the first time the feasibility of a positive hyperon beam was demonstrated; second, substantial fluxes of charged antihyperons were identified in the beam; and third, a flux of Ω^- sufficiently copious for detailed measurements of its properties was made available. Each of these features opens up a new experimental program.

This beam, which incorporated many design features of the earlier CERN beam, is shown in Figure 15*a*. Two superconducting quadrupole magnets and three bending magnets were used in the hyperon beam transport. Figure 15*b* shows the downstream detection apparatus, incorporating two DISC Cerenkov counters to count the hyperons directly and a magnetic spectrometer for the decay products. Under normal operating conditions only the SPS DISC counter was used for particle identification. At the downstream end of the SPS DISC, 12.6 m from the hyperon production target, the beam had a momentum bite

$\Delta P/P = 10\%$, FWHM, and a size of 1.5×2.0 cm ($h \times v$). A useful secondary beam intensity of about 10^6 particles could be achieved with 4×10^{10} protons incident on the hyperon production target. The background due to particles leaking through the shielding (mostly muons) was much less a problem than in the earlier charged-hyperon beams; the secondary beam flux was limited only by the rate capabilities of the experimental apparatus.

The magnetic spectrometer played an essential role in identifying the rarer hyperons through their decays. It consisted of an analyzing magnet with a field integral of 2.2 Tm with drift chambers on either side. The upstream chambers (DC1–DC4 of Figure 15b) had a drift space of only 5 mm to insure good resolution of close tracks as would occur from the decay $\Xi^- \to \pi^- \Lambda$ and the subsequent decay $\Lambda \to \pi^- p$. The chambers downstream of the magnet (DC5–DC8) had a 1-cm drift distance since the tracks were then further apart.

Another very important feature of the spectrometer was its excellent photon and lepton detection capabilities. Wide angle photons were detected by the lead scintillator hodoscope before the magnet and more

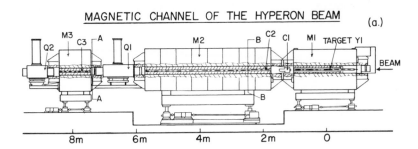

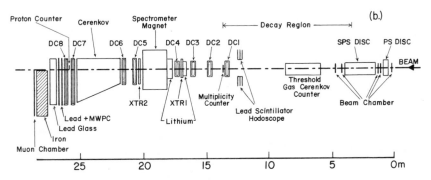

Figure 15   The CERN SPS charged-hyperon beam. The magnetic beam transport elements are shown in (*a*) and the layout of the magnetic spectrometer is shown in (*b*).

forward photons by lead glass and a lead and proportional chamber array behind the magnet. This allowed the reconstruction of the decay $\Sigma^+ \to \bar{p}\pi^0$ since both photons from the π^0 could be detected. The identification of Σ^+ in a much larger sample of Σ^- was thus possible.

4.3.2 HYPERON FLUXES   Figure 16 is a DISC pressure curve taken at $+100$ GeV/c and shows a clean peak at the Σ^+ mass. The small $\Sigma^+ - \bar{\Sigma}^-$ mass difference (8 MeV/c^2) is not resolved by the DISC. The flux of Σ^+ ($\bar{\Sigma}^-$) can be determined by reconstructing the decays Σ^+ ($\bar{\Sigma}^-$) $\to$ n ($\bar{\text{n}}$) π^+. Since the Σ^+ and $\bar{\Sigma}^-$ lifetimes differ by a factor of two (assuming $\tau_{\bar{\Sigma}} = \tau_{\Sigma}$), they will have different distributions of their decay vertices. Fitting to this vertex distribution and making cuts involving the $\Sigma^+ - \bar{\Sigma}^-$ mass difference allows a determination of the $\bar{\Sigma}^-/\Sigma^+$ production ratio. The flux of Ξ^- was too small to be seen as a peak in the DISC curve since the DISC background levels were 10^{-7} to 10^{-6} of the total beam flux through it. However, by triggering with the DISC set to the Ξ mass and then fitting to the decay $\Xi^- \to \Lambda\pi^+$ and $\bar{\Lambda} \to \bar{p}\pi^+$, a clean identification could be made. The corresponding reconstructed Ξ^- and Λ masses have widths of 6 and 4 MeV/c^2 (FWHM) respectively. The dashed curve of Figure 16 shows the expected Ξ^- signal based on events reconstructed in this way.

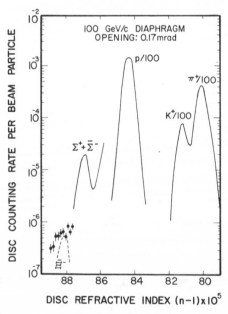

Figure 16   A DISC pressure curve taken for positive particles with the SPS hyperon beam.

The Ω^- flux, although also too small to be seen as a peak in the DISC pressure curve, could be readily measured by reconstructing the decay $\Omega \rightarrow K^- \Lambda$, $\Lambda \rightarrow \pi^- p$. The major background came from the topologically similar Ξ^- decay, but simple mass cuts allowed good separation. For 10^6 beam particles identified by the DISC, typical running conditions at

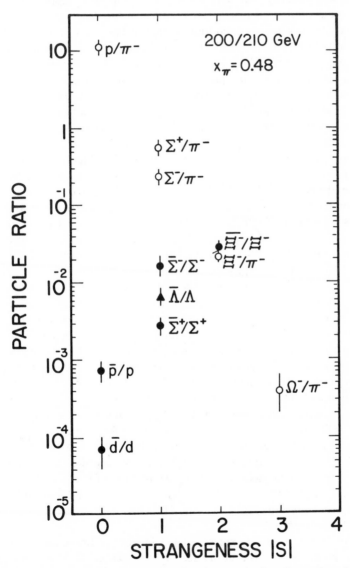

Figure 17   Particle production ratios as a function of strangeness from Bourquin et al (1979).

-100 GeV/c provided 4000 Σ^-, 2 $\overline{\Sigma^+}$, 400 Ξ^-, and 0.1 Ω^-. For $+100$ GeV/c, 46 Σ^+, 4 $\overline{\Sigma^-}$, and 2 $\overline{\Xi^-}$ were typical fluxes.

Bourquin et al compared particle production ratios as a function of strangeness at $x_\pi = 0.48$. The variable x_π is the Feynman x of the beam pions. The ratios are plotted in Figure 17 and display a number of interesting patterns. The decreasing production cross section of strange baryons as compared to pions is revealed in the decrease by a factor of 50–100 of this ratio for each unit change of baryon strangeness. As has been noted for Λ and $\overline{\Lambda}$ yields, two production mechanisms are probably at work. At large values of x the baryons are produced diffractively giving rise to the leading-particle effect, while at smaller x values the production of baryon-antibaryon pairs is significant. The figure shows that the diffractive production of strange baryons becomes progressively less dominant as the strangeness of the baryon increases.

4.3.3 Ω^- PROPERTIES   Measurements of the Ω^- lifetime by the early bubble chamber experiments were severely limited by statistics, but clustered about a value of 1.3×10^{-10} seconds. Three bubble chamber experiments with larger event samples have recently been reported. Deutschmann et al (1978), using 101 events produced by 10- and 16-GeV/c K$^-$, measured a lifetime of $\tau = 1.41^{+0.15}_{-0.24} \times 10^{-10}$ s in agreement with the earlier bubble chamber results. The measurement of Hemingway et al (1978) with 40 Ω^- produced by 4.2-GeV/c K$^-$ gives $\tau = 0.75^{+0.14}_{-0.11} \times 10^{-10}$ s and that of Baubillier et al (1978) using 41 events produced by 8.25-GeV/c K$^-$ yields $\tau = 0.80^{+0.16}_{-0.12} \times 10^{-10}$ s. Hemingway et al claim that their measurement is particularly free of biases because their data are just above the Ω^- production threshold and they used only events where the production kinematics were over constrained (no missing neutral particles) to help separate the Ω^- from the much more copious, and topologically similar, Ξ^- decays. This was not possible for the other measurements because of their higher energy.

The hyperon beam measurement of Bourquin et al (1978) used a sample of 1410 Ω^- that decayed by the mode $\Omega^- \to \Lambda K^-$ and $\Lambda \to \pi^- p$. They estimate their background as 23 events, about half of which arose from another Ω^- decay mode and the rest from misidentified Ξ^- decays. The best fit to the data yields a preliminary lifetime of $0.82 \pm 0.06 \times 10^{-10}$ s in good agreement with Hemingway et al (1978) and Baubillier et al (1978), strongly indicating that the Ω^- lifetime is considerably shorter than 1.3×10^{-10} s.

Bourquin et al (1978) measured the Ω^- decay branching ratios and the α parameter for the decay $\Omega^- \to \Lambda K^-$ (Table 2). The ratio of the decay $\Omega \to \Xi^0 \pi^-$ to $\Omega^- \to \Xi^- \pi^0$ is predicted to be two for a pure $\Delta I = 1/2$ amplitude; their measurements yield a value of 2.93 ± 0.50. This same group also find three possible candidates for the leptonic decay of the Ω^-. The analysis is still preliminary but these events would correspond to a $\sim 1\%$ leptonic branching ratio.

4.4 Fermilab Charged-Hyperon Beam

The charged-hyperon beam under construction at Fermilab is designed to target the primary 400-GeV proton beam at fluxes of up to 10^{12} protons per pulse. The main component of the hyperon channel is a large dipole magnet, 7 m in length, and capable of 3.5 T. The channel itself is designed to operate over a momentum range of 100–350 GeV/c. Figure 18 shows the expected positive and negative particle fluxes at a distance of 10 m from the production target. Flux estimates for this beam were made by Doroba (1978); the above figures incorporate the recent flux measurements of Bourquin et al (1979). The figures assume that the incident proton beam is adjusted to give 10^6 secondary particles emerging from the channel. These hyperon fluxes are very substantial and, in the high momentum range of the negative beam, hyperons become

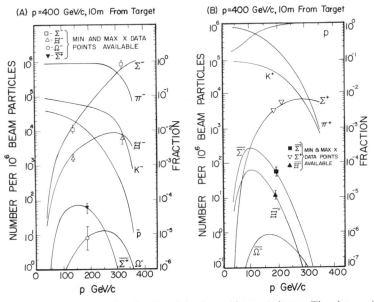

Figure 18 Flux estimates for the Fermilab charged-hyperon beam. The data points represent the range in x of measurements made with other charged-hyperon beams. See Doroba (1978) for details.

the dominant particles. The approved program for this beam will include a new particle search, a hyperon flux and polarization measurement, and a measurement of hyperon elastic scattering.

5 FUTURE PROSPECTS

At this time two hyperon beams are operational, a neutral beam at Fermilab and a charged beam at CERN. Within a year a new charged-hyperon beam will be completed at Fermilab and be capable of exploiting a considerably higher energy and intensity range than the existing beams. The utilization of these beams in the next few years should provide significant advances in a number of areas.

Systematic measurements of Λ and $\bar{\Lambda}$ production cross sections over a wide range of kinematic variables have been published, and a high statistics study of Ξ^0 production, including A dependence and polarization, has been completed. Similar measurements for the charged hyperons are either planned or already in progress. The behavior of the Λ polarization as a function of $p_\perp$ out to 6 GeV/c is planned for the Fermilab neutral beam.

Precision measurements of the static properties of the hyperons have been demonstrated by the recent hyperon beam measurements of the Ω^- lifetime and the Λ and Ξ^0 magnetic moments and have obvious extensions to the other hyperons. Only the high energy charged beams have fluxes of Ω^- sufficient to make significant advances in the study of the Ω^- branching ratios and other static properties. Almost all previous knowledge of antihyperon properties came from bubble chamber measurements and had limited statistical impact. The significant fluxes of antihyperons in the SPS and Fermilab high energy hyperon beams should allow measurements of their static properties and tests of CP and CPT with much greater precision.

The semileptonic decays of hyperons present a unique handle on the fundamental parameters of weak interaction theory. These decays are experimentally very challenging but measurements made with the early CERN and BNL beams demonstrate that they can be done. The much higher fluxes available with the higher energy beams coupled with modern detector technology should provide for better control of systematic errors and allow the extension of these measurements to a wider group of decays. A high statistics study of polarized $\Lambda^0 \rightarrow pe^- \bar{\nu}_e$ will be done in the Fermilab neutral beam. The existing discrepancy in the decay $\Sigma^- \rightarrow ne^- \bar{\nu}$ noted in Section 3.4.2 should be resolved and a much more accurate (and, it is hoped, consistent) set of decay parameters will be attainable. The possibility of making measurements of the Ω^- leptonic decays is particularly intriguing since no measurements of the weak decays of any member of the baryon decouplet now exist.

Hyperon beams provide the only means for measuring the strong interaction properties of the hyperons. Again the early hyperon beams showed the feasibility of measuring total and differential cross sections, and the diffractive excitation of hyperons. Because hyperons are convenient analyzers of the spin configuration of their final state, they have unique advantages in helping to understand hadronic excited states. This field is totally unexplored at high energies.

We are pleased to acknowledge helpful and frank discussions with many of our colleagues engaged in hyperon research.

Literature Cited

Allaby, J. V. et al. 1970. *CERN Rep. 70-12*
Arik, E. 1976. *Inclusive lambda production in sigma-proton collisions at 23 GeV/c.* PhD thesis. Univ. Pittsburgh, Penn. 58 pp.
Arik, E. et al. 1977. *Phys. Rev. Lett.* 38:1000
Armenteros, R., Barker, K. H., Butler, C. C., Cachor, A. 1951. *Philos. Mag.* 42:1113
Badier, J. et al. 1972a. *Phys. Lett. B* 39:414
Badier, J. et al. 1972b. *Phys. Lett. B* 41:387
Barbiellini, G. et al. 1972. *Phys. Lett. B* 39:663
Baubillier, M. et al. 1978. *Phys. Lett. B* 78:342
Blaising, J.-J. 1977. *Interactions hadroniques des hyperons sigma minus a 17 GeV.* PhD thesis. Universite Louis Pasteur, Strasbourg, France (In French)
Blaising, J.-J. et al. 1975. *Phys. Lett. B* 58:121
Bøggild, H., Ferbel, T. 1974. *Ann. Rev. Nucl. Sci.* 24:451
Bourquin, M. et al. 1978. Presented at *19th Int. Conf. on High Energy Physics, Tokyo* (*Session B7E*)
Bourquin, M. et al. 1979. *Nucl. Phys. B* 153:13
Bunce, G. et al. 1976. *Phys. Rev. Lett.* 36:1113
Bunce, G. et al. 1978. *Phys. Rev. D* 18:633
Bunce, G. et al. 1979. *Bull. Am. Phys. Soc.* 24:46
Burq, J. P. et al. 1978. *CERN EP Intern. Rep. 78-7*
Busza, W. et al. 1975. *Phys. Rev. Lett.* 34:836
Cabibbo, N. 1963. *Phys. Rev. Lett.* 10:531
Cabibbo, N., Gatto, R. 1960. *Nuovo Cimento* 15:159
Chinowsky, W. 1977. *Ann. Rev. Nucl. Sci.* 27:393
Coleman, S., Glashow, S. L. 1961. *Phys. Rev. Lett.* 6:423
Dalitz, R. H. 1957. *Rep. Prog. Phys.* 20:163
Decamp, D. et al. 1977. *Phys. Lett. B* 66:295

De Rujula, A., Georgi, H., Glashow, S. L. 1975. *Phys. Rev. D* 12:147
Deutschmann, M. et al. 1978. *Phys. Lett. B* 73:96
Devlin, T. et al. 1977. *Nucl. Phys. B* 123:1
Doroba, K. 1978. *Fermilab TM-818*
Dreitlein, J., Primakoff, H. 1962. *Phys. Rev.* 125:1671
Dydak, F. et al. 1977. *Nucl. Phys. B* 118:1
Edwards, R. T. et al. 1978. *Phys. Rev. D* 18:76
Eichten, T. et al. 1972. *Nucl. Phys. B* 44:333
Eisele, F. et al. 1976. *Phys. Lett. B* 60:297
Gell-Mann, M. 1953. *Phys. Rev.* 92:833
Gell-Mann, M. 1961. *Cal. Tech. Rep. CTSL-20*
Gell-Mann, M. 1964. *Phys. Lett.* 8:21
Gell-Mann, M., Rosenfeld, A. H. 1957. *Ann. Rev. Nucl. Sci.* 7:407
Geweniger, C. et al. 1974. *Phys. Lett. B* 48:483
Geweniger, C. et al. 1975. *Phys. Lett. B* 57:193
Giacomelli, G. 1976. *Phys. Rep. C* 23:123
Gjesdal, S. et al. 1972. *Phys. Lett. B* 40:152
Glauber, R. J. 1960. *International Conference on Nuclear Forces and the Few Nucleon Problem*, ed. T. C. Griffith, E. A. Power, Vol. 2. New York: Pergamon
Gottfried, K. 1974. *Phys. Rev. Lett.* 32:957
Grobel, R. et al. 1979. *Bull. Am. Phys. Soc.* 24:46
Harrington, D. R. 1964. *Phys. Rev. B* 135:358
Heller, K. et al. 1977. *Phys. Rev. D* 16:2737
Heller, K. et al. 1978. *Phys. Rev. Lett.* 41:607
Hemingway, R. et al. 1978. *Nucl. Phys. B* 142:205
Herbert, M. L. et al. 1978. *Phys. Rev. Lett.* 40:1230
Hungerbuhler, V. M. 1973. *Production of negative hyperon resonances and omega minus by sigma minus and chi minus at*

24.6 GeV/c. PhD thesis. Yale Univ., New Haven, Conn. 106 pp.

Hungerbuhler, V. et al. 1974a. Nucl. Instrum. Methods 115:221

Hungerbuhler, V. et al. 1974b. Phys. Rev. D 10:2051

Hungerbuhler, V. et al. 1975. Phys. Rev. D 12:1203

Jensen, D. A. et al. 1969. Phys. Rev. Lett. 23:615

Lach, J. 1977. Proc. 12th Rencontre De Moriond, Flaine 2:13–36

Lee, T. D., Yang, C. N. 1957. Phys. Rev. 108:1645

Lipkin, H. 1978. Phys. Rev. Lett. 41:1629

Lipkin, H. J., Scheck, F. 1966. Phys. Rev. Lett. 16:71

Litt, J., Meunier, R. 1973. Ann. Rev. Nucl. Sci. 23:1

Majka, R. D. 1974. Sigma minus-proton elastic scattering at 23 GeV/c. PhD thesis. Yale Univ., New Haven, Conn. 125 pp.

Majka, R. et al. 1976. Phys. Rev. Lett. 37:413

Marshak, R. E., Riazuddin, Ryan, C. P. 1969. Theory of Weak Interactions in Particle Physics. New York: Wiley-Interscience. 761 pp.

Martin, P. S. 1977. Lambda-proton elastic scattering at 23 GeV/c. PhD thesis. Univ. Wisconsin-Madison, Wisc. 113 pp.

Ne'eman, Y. 1961. Nucl. Phys. 26:222

Overseth, O. E., Pakvasa, S. 1969. Phys. Rev. 184:1663

Particle Data Group. 1978. Phys. Lett. B 75:1

Pritchett, P. L., Deshpande, N. G. 1972. Phys. Lett. B 41:311

Quigg, C., Rosner, J. L. 1976. Phys. Rev. D 14:160

Rochester, G. D., Butler, C. C. 1947. Nature 160:855

Samios, N. P., Goldberg, M., Meadows, B. T. 1974. Rev. Mod. Phys. 46:49

Sandweiss, J. 1971. Proc. Int. Conf. on Instrumentation for High Energy Physics, Dubna, p. 717

Sanford, J. R. 1976. Ann. Rev. Nucl. Sci. 26:151

Schachinger, L. et al. 1978. Phys. Rev. Lett. 41:1348

Skubic, P. et al. 1978. Phys. Rev. D 18:3115

Tanenbaum, W. et al. 1975. Phys. Rev. D 12:1871

Wang, C. L. 1970. Phys. Rev. Lett. 25:1068

Weinberg, S. 1958. Phys. Rev. 112:1375

Weinberg, S. 1977. Trans. NY Acad. Sci., p. 185

Willis, W. J. et al. 1971. Nucl. Instrum. Methods 91:33

Zweig, G. 1964. CERN Rep. 8409/TH412

Ann. Rev. Nucl. Part. Sci. 1979. 29:243–82

THE MUON $(g-2)$ EXPERIMENTS

✳5607

F. J. M. Farley

Royal Military College of Science, Shrivenham, England

E. Picasso

CERN, Geneva, Switzerland

CONTENTS

INTRODUCTION

It is now 21 years since a group of experimental physicists at CERN under Leon Lederman started to study the problem of the muon g-factor. The magnetic moment is $g(e/2mc)(h/2)$, where g is a dimensionless number. Since then, a number of measurements have been performed with higher and higher accuracy. At the same time a great deal of theoretical effort has been deployed to determine the theoretical value of g. If the muon obeys the simple Dirac equation for a particle of its mass (206 times heavier than an electron), then $g = 2$ exactly; but this is modified by the quantum fluctuations in the electromagnetic field around the muon, as specified by the rules of quantum electrodynamics (QED), making g

243

0163-8998/79/1201-0243$01.00

larger by about 1 part in 800. It requires further correction for the very rare fluctuations, which include virtual pion states and strongly interacting vector mesons. At present, theory and experiment agree at the level of 1 part in 10^8, and the muon g-factor, together with that of the electron, R_∞, c, and the frequency of the hydrogen maser, are the most accurately known constants of nature.

A number of reviews both theoretical (Lautrup et al 1972, Calmet et al 1977, Kinoshita 1978) and experimental (Farley 1964, 1968, 1975, Picasso 1967, Bailey & Picasso 1970, Combley & Picasso 1974, Field et al 1979) have already been published. In this article we give an overview of the programme as a whole, trying to set each measurement in its historical perspective, in order to show how one developed from another, and to relate each to the contemporary thinking about the muon and about quantum electrodynamics.

THEORY

The gyromagnetic ratio is increased from its primitive value of 2, arising from the Dirac equation, to $g = 2(1 + a_\mu)$, where $a_\mu \equiv (g-2)/2$ is defined as the anomalous magnetic moment or anomaly. In the QED theory the

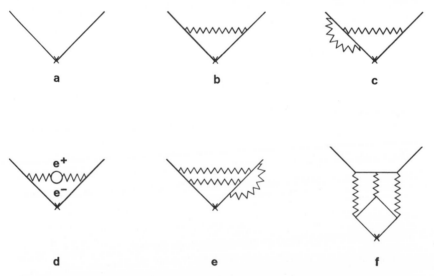

Figure 1 Feynman diagrams used in calculating a. The solid line represents the muon, which interacts with the laboratory magnetic field at X. The zigzag line represents a virtual photon, which is emitted and later reabsorbed. In (d) and (f) an $e^+ e^-$ pair, created and then annihilated, gives rise to the closed loop (*solid line*).

contributions to the anomalous moment are expressed as a power series in α/π

$$a_\mu^{\text{th}} = A(\alpha/\pi) + B(\alpha/\pi)^2 + C(\alpha/\pi)^3 + \dots . \qquad 1.$$

Typical Feynman diagrams, which contribute to the calculation of the theoretical value of a for the electron and muon, are shown in Figure 1, while a complete set of diagrams up to sixth order [terms in $(\alpha/\pi)^3$] is given by Lautrup et al (1972). These have all been calculated analytically, except for diagrams such as Figure 1f, including an electron loop with four electromagnetic vertices, which have been found by numerical integration. (Such diagrams are also involved in the scattering of light by light.) The coefficients in the expansion are listed in Table 1, together with rough estimates for the eighth- and tenth-order terms.

So far only the change in the gyromagnetic ratio due to the interaction of a particle with its own electromagnetic field has been mentioned. Any other field coupled to the particle should produce a similar effect, and the calculations have been made for scalar, pseudoscalar, vector, and axial-vector fields, using a coupling constant f (assumed small) to a boson of mass M (Berestetskii et al 1956, Cowland 1958). For example, for a vector field

$$\Delta a_\mu^{\text{V}} = \left(\frac{1}{3\pi}\right)\left(\frac{f^2}{M^2}\right) m_\mu^2. \qquad 2.$$

A precise measurement of a_μ could therefore reveal the presence of a new field, but first all known fields, including the weak and the strong inter-

Table 1 Summary of theoretical contributions[a] to a_μ

QED terms	Muon	Numerical values ($\times 10^9$)	
2nd order: A	0.5	Total QED:	1 165 852 (1.9)
4th order: B	0.765 782 23	Strong interactions:	66.7 (8.1)
6th order: C	24.452 (26)	Weak interactions:	2.1 (0.2)
8th order: D	135 (63)	Total theory:	1 165 921 (8.3)
10th order: E	420 (30)		

[a] Whereas the first two terms are obtained exactly from analytical results, in the sixth order (α^3) there remain some diagrams that must be calculated by numerical integration implying a small residual error in the coefficient (Cvitanovic & Kinoshita 1974, Barbieri & Remiddi 1975, Calmet & Petermann 1975a, Levine et al 1976, Levine & Roskies 1976, Samuel & Chlouber 1976, Lautrup & Samuel 1977). The α^4 and α^5 values are estimated by inserting electron loops in the sixth-order diagrams (Lautrup 1972, Lautrup & de Rafael 1974, Samuel 1974, Calmet & Petermann 1975b, Chlouber & Samuel 1977). The figures in parentheses following any figure indicate the estimated error in the final digits.

actions, must be taken into account. Strongly interacting particles do not couple directly to the muon, but if they are charged, they couple to the photon. Thus they can appear in the inner loops such as Figure 1d, with, for example, a pion pair replacing the e^+e^- pair. Because of the high mass of the pion, one would initially expect such amplitudes to be small, but there are strong resonances in the $\pi^+\pi^-$ system that enhance the effect. Only a vector resonance can contribute, because it alone can transform directly into the virtual photon that must have $J^{PC} = 1^{--}$ (one unit of angular momentum, negative parity, and negative charge conjugation).

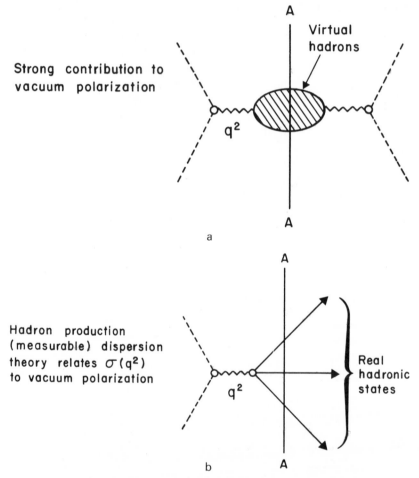

Figure 2   The photon propagator is modified by the creation of virtual hadrons (a). This is related by dispersion theory to real hadron production in e^+e^- collision (b).

To calculate this contribution it is necessary to specify the overall probability amplitude for a photon of a given q^2 to connect the two muon vertices shown in Figure 1b, with the effect of virtual hadron loops fully included. That is, one requires the propagator function of Figure 2a.

This cannot be calculated from theory, because not enough is known about hadrons. But fortunately the propagator Figure 2a can, in principle, be cut in half to obtain that of Figure 2b, which shows an e^+e^- pair annihilating to give real hadronic states. By using dispersion theory, the cross section for Figure 2b as a function of (total energy)^2, that is $\sigma(e^+e^- \to \text{hadrons})(t)$, can be related (Bouchiat & Michel 1961) to the propagator shown in Figure 2a and so to the anomalous moment arising from Figure 1d with hadron loops

$$\Delta a_\mu(\text{hadrons}) = (m_\mu^2/4\pi^3) \int_0^\infty \sigma(e^+e^- \to \text{hadrons})(t)\, g(t)\, \mathrm{d}t, \qquad 3.$$

where $t = q^2$, and

$$g(t) = \int_0^1 \frac{x^2(1-x)\,\mathrm{d}x}{m_\mu^2 x^2 + t(1-x)} \to \frac{1}{3t} \quad \text{at large } t. \qquad 4.$$

The process $e^+e^- \to$ hadrons has been extensively studied in electron-positron colliding beams and the cross section is rather well defined from near threshold up to about 3-GeV center-of-mass energy. Therefore the integration in Equation 3 can be carried out with fair confidence, the result being $\Delta a_\mu(\text{hadrons}) = (66.7 \pm 8.1) \times 10^{-9}$ (Barger et al 1975, Calmet et al 1976). This term is about 8 times the present experimental accuracy and its presence and order of magnitude have been confirmed.

The contribution of 4-fermion weak interaction is illustrated in Figure 3a. This is second order in the weak interaction and turns out to be negligibly small ($\sim 10^{-12}$). However, if the weak interaction is mediated by a charged intermediate boson $W^\pm$, the mechanism shown in Figure

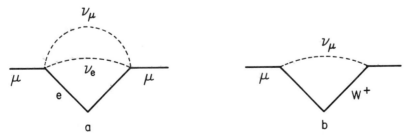

Figure 3 Contribution to a_μ by the weak interaction: (a) with 4-fermion interaction; (b) by the virtual production of an intermediate boson W^+.

$3b$ will contribute. The new renormalizable theory of weak interactions then leads to $\Delta a_\mu(\text{weak}) \sim 2 \times 10^{-9}$ (Bardeen et al 1972, Bars & Yoshimura 1972, Fujikawa et al 1972, Jackiw & Weinberg 1972, Primack & Quinn 1972). This is a small effect, at present masked by the uncertainty in the strong interaction contribution, and indeed in the sixth-order QED term; so it is unlikely to be detected.

Finally we must consider the effect on a_μ^{th} of a modification of QED. If the muon is not completely point-like in its behavior, but has a form factor $F(q)^2 = \Lambda_\mu^2/(q^2 + \Lambda_\mu^2)$, it can be shown that

$$\frac{\Delta a_\mu}{a_\mu} = \frac{-4m_\mu^2}{3\Lambda_\mu^2},$$

5.

implying, for example, a reduction in a_μ of 24 parts per million (ppm) if $\Lambda_\mu = 25$ GeV/c. Similarly a modification[1] of the photon propagator by the factor $\Lambda_\gamma^2/(q^2 + \Lambda_\gamma^2)$ implies

$$\frac{\Delta a_\mu}{a_\mu} = \frac{-2m_\mu^2}{3\Lambda_\gamma^2}.$$

6.

This result was first obtained by Berestetskii et al (1956), who emphasized the value of experiments on the muon; the high mass m_μ implies a significant correction to a_μ even when Λ_γ is large.

The theoretical predictions are summarized in Table 1, based on $\alpha^{-1} = 137.035\ 987\ (29)$ (Cohen & Taylor 1973, Hansch et al 1974). The figures in brackets following any value indicate the estimated error on the last two digits. We must emphasize that the figures relate not to g but to the small anomaly a_μ, a quantity that would be zero in the absence of quantum fluctuations.

We now turn to the experimental problem of measuring the anomaly for the muon.

$(g-2)$ Precession

Let us first consider a particle, longitudinally polarized, moving at slow speed in a uniform static magnetic field. The momentum vector turns at the cyclotron frequency f_c with

$$2\pi f_c = \frac{eB}{mc},$$

7.

while the spin precession frequency is the same as for a particle at rest,

---

[1] The violation of unitarity can be circumvented by introducing a negative metric (Lee & Wick 1969).

$$2\pi f_s = \frac{2\mu B}{h} = g\left(\frac{eB}{2mc}\right) = (1 + a_\mu)\left(\frac{eB}{mc}\right). \qquad 8.$$

If $g = 2$, then $f_s = f_c$ and the particle will always remain longitudinally polarized. But if $g > 2$ as predicted, the spin turns faster than the momentum vector. The laboratory rotation frequency f_a of the spin relative to the momentum vector is given by

$$2\pi f_a = 2\pi(f_s - f_c) = a\left(\frac{e}{mc}\right)B. \qquad 9.$$

This is the basic equation for the $(g-2)$ experiments: if the particle is kept turning in a known magnetic field B, and the angle between the spin and the direction of motion is measured as a function of time t, then a can be determined. The value of (e/mc) is obtained from the precession frequency of muons at rest (Equation 8). In fact, for the same magnetic field B, one has $f_a/f_s = a_\mu/(1 + a_\mu)$. In practice the fields are not the same in the two experiments, but are measured by proton magnetic resonance, and a proportional correction is applied.

Note that this gives a measure of the anomaly $a_\mu \equiv (g-2)/2$ instead of g itself; so the correction due to quantum fluctuations is measured directly. As $a_\mu \sim 1/800$, it follows from Equations 7 and 9 that the particle must take 800 turns in the field for the spin to make 801 turns, that is for the polarization to change gradually through $360°$. Clearly, if this is to be measured with any accuracy, the particle should make thousands of turns in the field, so that several cycles of the anomalous precession can be studied: the more cycles it is possible to record, the more accurate will be the measurement of frequency.

The fundamental formula (Equation 9) has been derived only in the limit of low velocities but it proves to be exactly true at any speed. This was demonstrated by Mendlowitz & Case (1955) and Carrassi (1958) using the Dirac equation, and by Bargmann et al (1959) using a covariant classical formulation of spin motion. Other treatments have been given by Farley (1968) and Fierz & Telegdi (1970), and reviewed by Farago (1965). Note that the $(g-2)$ precession is not slowed down by time dilation even for high velocity muons.

PART I: 1958–1962

By 1958, QED was an established theory of some 10 years standing, corroborated by accurate measurements of the Lamb shift. The g-factor of the electron was known through electron spin resonance (Franken & Liebes 1956) to one part per million (ppm); Karplus & Kroll (1950) had shown how to calculate the higher order corrections to g and a numerical error in their results had recently been corrected by Petermann (1957a,b), Sommerfield (1957), and Suura & Wichmann (1957), bringing theory into line with the experiment at the level of $(\alpha/\pi)^2$.

For the free electron a direct determination of the anomalous magnetic moment $a_e \equiv (g-2)/2$ was in progress at the University of Michigan (Nelson et al 1959, Schupp et al 1961) using the recently discovered principle of $(g-2)$ spin motion explained above. Equation 9 had been proved to hold for relativistic velocities.

Turning to the muon, the bremsstrahlung cross section at high energies had been measured with cosmic rays and shown to agree with a spin assignment of $\frac{1}{2}$ rather than $\frac{3}{2}$ (Mathews 1956, Mitra 1957, Hirokawa & Komori 1958). A similar conclusion followed from data on neutron production by cosmic-ray muons (de Pagter & Sard 1960). The angular distribution of electrons from the decay of polarized muons agreed with spin $\frac{1}{2}$ (Bouchiat & Michel 1957) and was inconsistent with spin $\frac{3}{2}$ (Brown & Telegdi 1958). Experiments with cosmic-ray and accelerator-generated muons were in progress to compare the electro-magnetic scattering of muons and electrons by nuclei.

Thus evidence was accumulating that the muon behaves as a heavy electron of spin $\frac{1}{2}$. Berestetskii et al (1956) had emphasized that QED theory implied an anomalous magnetic moment a_μ for the muon of the same order as for the electron, but as the typical invariant momentum transfer involved was $q^2 \sim m^2$ an experiment for the muon would test the theory at much shorter distances. Feynman (1962) felt that the divergences in QED could be limited by a real energy-momentum cutoff Λ, and it seemed reasonable to expect Λ to be of the order of the nucleon mass. This would imply a 0.5% effect in a_μ. On the other hand, it was thought (Schwinger 1957) that the muon should have an extra interaction that would distinguish it from the electron and give it its higher mass. This could be a coupling to a new massive field, or some specially mediated coupling to the nucleon. Whatever the source, the new field should have its own quantum fluctuations, and therefore give rise to an extra contribution to the anomalous moment a_μ. The $(g-2)$ experiment was recognized as a very sensitive test of the existence of such fields, and potentially a crucial signpost to the μ-e problem.

At this stage there was no prospect of such an experiment, but in 1957 parity violation was discovered (Lee & Yang 1956, Wu et al 1957), muon beams were found to be highly polarized, and better still it was found that the angular distribution of the decay electrons could indicate the spin direction of the muon as a function of time (Garwin et al 1957, Friedman & Telegdi 1957). A wide variety of muon precession and spin-resonance experiments were carried out in the next few years (for reviews see Feinberg & Lederman 1963, Farley 1964). The $(g-2)$ principle was invoked in the first paper on muon precession by Garwin et al (1957), who pointed out that g must be within 10% of 2.00, because although the muon trajectory had been deflected through 100° by the cyclotron magnetic field the muon polarization was still longitudinal.

The possibility of a $(g-2)$ experiment for muons was envisaged, and groups at Berkeley, Chicago, Columbia, and Dubna started to study the problem (Panofsky 1958). If the muon had a structure that gave a form factor less than one for photon interactions, the value of a_μ should be less than predicted. Compared to the measurement for the electron, the muon $(g-2)$ experiment was much more difficult because of the low intensity, diffuse nature, and high momentum of available muon sources. This implied large volumes of magnetic field; the lower value of (e/mc) made all precession frequencies 200 times smaller, but the time available for an experiment was limited by the decay lifetime, 2.2 μs. Hence large magnetic fields would be needed to give a reasonable number of precession cycles.

One solution was to scale up the method used at Ann Arbor for the electrons, using a large solenoid and injecting the muons spirally at one end (Schupp et al 1961). This was pursued at Berkeley and finally led to a 10% measurement (Henry et al 1969); see Table 2.

At CERN the work centered on the belief that it should be possible to store muons in a conventional bending magnet with a more or less uniform vertical field between roughly rectangular pole pieces. In a typical field of 1.5 T the muon orbit would make 440 turns during the lifetime of 2.2 μs. As $a_\mu \sim \alpha/2\pi \sim 1/800$, the angle between the spin and the momentum vector would develop 800 times more slowly, giving a change in beam polarization of about 180° to be studied.

The polarized muon beam from the CERN cyclotron could fairly easily be trapped inside a magnet. The particles were aimed at an absorber in the field; they lost energy and therefore turned more sharply and

Table 2 Experimental results[a] for a_μ

Charpak et al 1961a	μ^+	0.001 145 (22)
Charpak et al 1962, 1965	μ^+	0.001 162 (5)
Farley et al 1966	μ^-	0.001 165 (3)
Henry et al 1969	μ^+	0.001 060 (67)
Bailey et al 1968, 1972	$\mu^\pm$	0.001 166 16 (31)
Bailey et al 1975	μ^+	0.001 165 895 (27)
Bailey et al 1977a, 1979	$\mu^\pm$	0.001 165 924 (8.5)

[a] The error on the last digits is shown in parentheses.

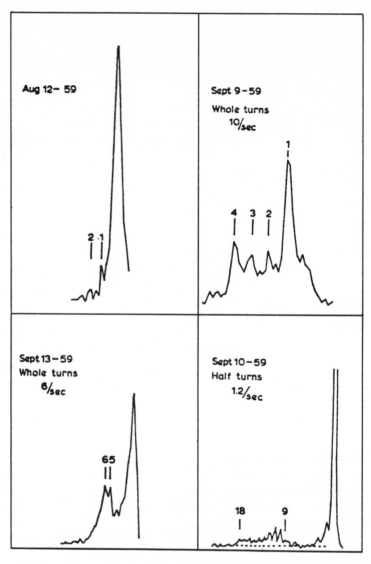

Figure 4 First evidence of muons making several turns in an experimental magnet. The time of arrival of the particles at a scintillator fixed inside the magnet is plotted horizontally (time increases to the left). The first (right-hand) peak coincides with the moment of injection. The equally spaced later peaks correspond to successive turns. Owing to the spread in orbit diameters and injection angles, some muons hit the counter after nine turns (lower right), while others take 18 turns to reach the same point (Charpak et al, unpublished).

remained inside the magnet. To prevent them reentering the absorber after one turn, a small transverse (y direction) gradient of the magnetic field was introduced, causing the orbits to drift sideways perpendicular to the gradient (x direction). Vertical focusing was added by means of a parabolic term in the field.

If the field is of the form

$$B_z = B_0(1 + ay + by^2),$$

10.

where a and b are small, an orbit of radius ρ moves over in the x direction a distance $s = a\pi\rho^2$ per turn (called the step size). On average, the wavelength of the vertical oscillations is $2\pi/b^{1/2}$. Figure 4 is of historical interest. It shows the first evidence of particles turning several times inside a small experimental magnet. These results gave the laboratory sufficient confidence to order a very long magnet for the experiment.

An overall view of the final storage system (Charpak et al 1961a,b, 1962, 1965) is shown in Figure 5. The magnet pole was 6-m long, 52-cm wide, and the gap was 14 cm. Muons entered on the left through a magnetically shielded iron channel and hit a beryllium absorber in the injection part of the field. Here the step size s was 1.2 cm. Then there was a transition to the long "storage region" where $s = 0.4$ cm with field gradient $a = (1/B)(dB/dy) = 3.9 \times 10^{-4}$ cm^{-1}. Finally, a smooth transition was made to the ejection gradient, where $s = 11$ cm per turn. After ejection the muons fell onto the "polarization analyzer" (Figure 6), where they were stopped and decayed to e$^+$. The time t a muon spent in the field was determined by recording the coincidences in counters 123 at the input, and counters 4566'7 at the output. The interval was measured with respect to a 10-MHz crystal.

The shimming of this large magnet to produce the correct gradients was a *tour de force*. This was assisted by the theorem that in weak gradients the flux through a wandering orbit is an invariant of the motion. Therefore, if the field along the center line of the magnet was constant, unwanted sideways excursions would be avoided, and this could be checked more exactly by moving a flux coil of the same diameter as the orbit all along the magnet.

However, the constant flux theorem implied that once the particle was trapped inside the magnet it would never emerge. This was seen as a major difficulty, because the final spin direction could only be measured in a weak or zero magnetic field: otherwise one would lose track of the spin direction, while waiting for the muon to decay. For weak gradients and slowly walking orbits, calculations of the orbit confirmed these doubts and some participants lost faith in the project. Fortunately it was found

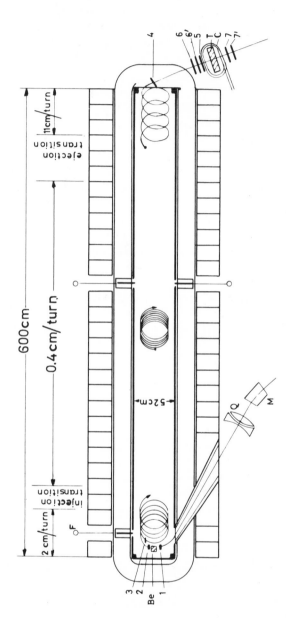

Figure 5   Storage of muons for up to 2000 turns in a 6-m bending magnet. The field gradient makes the orbit walk to the right. At the end a very large gradient is used to eject the muons so that they are stopped in the polarization analyzer. Coincidences 123 and 466'57', respectively, signal an injected and ejected muon. The coordinates used in the text are x (the long axis of the magnet), y (the transverse axis in the plane of the paper), and z (the axis perpendicular to the paper).

that in large gradients of order ±12% over the orbit diameter the particles were ejected successfully.

The muons were trapped in the magnet for times ranging from 2 to 8 μs depending on the location of the orbit center on the varying gradient given by Equation 10. About one muon per second was stopped finally

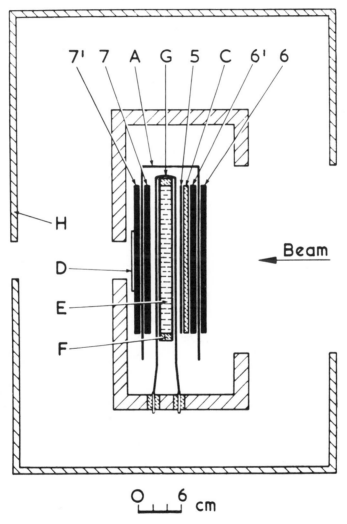

Figure 6 Polarization analyzer. When a muon stops in the liquid methylene iodide (*E*) a pulse of current in coil *G* is used to flip the spin through ±90°. Backward or forward decay electrons are detected in counter telescopes 66′ and 77′. The static magnetic field is kept small by the double iron shield and the mumetal shield *A*.

in the polarization analyzer, and the decay electron counting rate was 0.25 per second.

The spin direction can, in principle, be obtained from the ratio of two counting rates measured in different directions. But if two counter telescopes are used (say one forward and one backward relative to the direction of the arriving muons), it is not easy to ensure that they have equal efficiencies and solid angles. Therefore it is more reliable to use only one set of counters, but to move the muon spin direction after it has stopped. This can be done with a small constant magnetic field (cf muon precession at rest), but it is more efficient to turn the spin rapidly to a new position by applying a short sharp magnetic pulse, created by applying a pulse of current to a solenoid wound round the absorber in which the muon is stopped. This flipping was accomplished within 1 μs, before the gate that selected the decay electrons was opened.

In the apparatus shown in Figure 6, the electron counts c_+ and c_- in the forward telescope 77′ were recorded in separate runs with the spin flipped through $+90°$ and $-90°$, respectively. The asymmetry A of these counts, defined as $(c_+ - c_-)/(c_+ + c_-)$, was then related to the initial direction θ_s of the muon spin (before flipping) relative to the mean electron direction subtended by telescope 77′:

$$A \equiv \frac{(c_+ - c_-)}{(c_+ + c_-)} = A_0 \sin \theta_s. \qquad 11.$$

By flipping instead through $180°$ and $0°$, another ratio proportional to $A_0 \cos \theta_s$ was measured; so θ_s could be determined completely. Similar, but independent, calculations were made for the telescope 66′, which detected decay electrons emitted backwards.

This polarization analyzer was first used to study the muon beam available for injection. For muons that had been through the magnet the analyzer recorded the asymmetry A as a function of the time t the particle had spent in the field. This showed a sinusoidal variation due to the $(g-2)$ precession in the magnet. Using Equations 9 and 11 it follows that

$$A = A_0 \sin \theta_s = A_0 \sin \left\{ a_\mu \left(\frac{e}{mc} \right) Bt + \phi \right\}, \qquad 12.$$

where ϕ is an initial phase determined by measuring the initial polarization direction and the orientation of the analyzer relative to the muon beam.

The experimental data are given in Figure 7, together with the fitted line obtained by varying A_0 and a_μ in Equation 12. Full discussion of the precautions necessary to determine the mean field B seen by the muons,

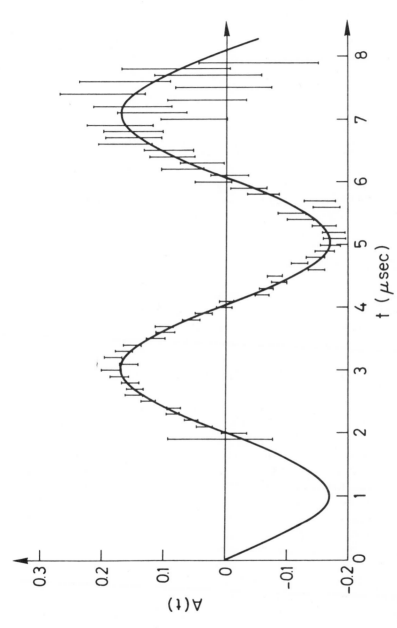

Figure 7   Asymmetry *A* of observed decay electron counts as a function of storage time *t*. The sinusoidal variation results from the $(g-2)$ precession; the frequency is measured to $\pm 0.4\%$.

and to avoid systematic errors in the initial phase ϕ, are given in Charpak et al (1962, 1965).

The results of this experiment are given in Table 2. Preliminary runs gave $\pm 2\%$ accuracy in a_μ, and this was later improved to $\pm 0.4\%$. The figures agreed with theory within experimental errors. The corresponding 95% confidence limit for the photon propagator cutoff (see Equation 6) was $\Lambda_\gamma > 1.0$ GeV, and for the muon vertex function (Equation 5) $\Lambda_\mu > 1.3$ GeV.

This was the first real evidence that the muon behaved so precisely like a heavy electron. The result was a surprise to many, because it was confidently expected that g would be perturbed by an extra interaction associated with the muon to account for its larger mass (Schwinger 1957, Kobzarev & Okun 1961). When nothing was observed at the 0.4% level, the muon became accepted as a structureless point-like QED particle, and the possibility of finding a clue to the μ-e mass difference seemed suddenly much more remote.

PART II: 1962–1968   MUON STORAGE RING I

By now the CERN Proton Synchrotron (PS) and Brookhaven Alternating Gradient Synchrotron (AGS) were operating, and the distinct properties of the two neutrinos ν_e and ν_μ had been established, further emphasizing the parallel but dual behavior of muon and electron (Danby et al 1962).

Muon-pair production by 1-GeV gamma rays on carbon was measured by Alberigi-Quaranta et al (1962) in agreement with theory. With this and the $(g-2)$ data, the evidence for point-like behavior was now much better for the muon than for the electron. The scattering of muons by lead and carbon (Masek et al 1961, 1963, Kim et al 1961, Citron et al 1962) agreed with the form factors deduced for electron scattering. Logically this was the best evidence for the point-like behavior of the electron, but was generally seen as another contribution to our knowledge of the muon. Knock-on electrons from 8-GeV muons confirmed the picture (Backenstoss 1963). Muonium formation in high pressure argon had been observed by Hughes et al (1960) and the hyperfine splitting of the ground state confirmed the theoretical picture to one part in 2000 (Ziock et al 1962). For this and subsequent muonium experiments the $(g-2)$ result was an essential input, not only for the g-factor, but also to deduce the muon mass from the precession frequency at rest, now determined to 16 ppm by Hutchinson et al (1963), see Equation 9.

The $(g-2)$ experiment was now the best test of QED at short distances. For this reason, and to search again for a new interaction, it was desirable to press the accuracy of the experiment to new levels. It would be essential to increase the number of $(g-2)$ cycles observed, either by increasing the field B or by lengthening the storage time. With the CERN PS available it was attractive to see what could be done by using high energy muons with relativistically dilated lifetimes. As there is no factor $\gamma \equiv (1 - v^2/c^2)^{-1/2}$ in Equation 9, the $(g-2)$ precession frequency would not be reduced and more cycles would be available before the muons decayed. But to store muons of GeV energy in a magnetic field and measure their polarization required totally new techniques. Farley (1962) proposed to measure the anomalous moment using a muon storage ring. As in the cyclotron, if the primary target was placed inside the magnet, muons would be produced by π-μ decay in flight and some of them would remain trapped in the field. With a pulsed accelerator, such as the PS, there should be no continuous background following the injection. Transfer of a pulse of protons from the PS to the muon ring could be achieved with the fast ejected beam already developed for the neutrino experiment. Estimates of the stored muon intensity and polarization looked favorable.

To determine the muon spin direction, decays in flight would be observed. The decay electrons would emerge on the inside of the ring and the detectors would respond only to the high energy particles emitted more or less forward in the muon rest frame. Thus as the spin rotated the electron counting rate would be modulated at the $(g-2)$ frequency.

It was later realized that at injection the muons would be localized in azimuth (injection time 10 ns, rotation time about 50 ns), so the counting rate would also be modulated at the mean rotation frequency. This would enable the mean radius of the stored muons to be calculated, leading to a precise knowledge of the corresponding magnetic field.

On the evening of October 21, 1963, a significant chance coincidence of time and place influenced the development of the project. The present authors, having first met that morning, found themselves filling a vacant evening in the same bar of the Hawthorne Hotel, Bristol, drifted into discussing physics, and thus initiated a 16-year collaboration.

The first Muon Storage Ring, Figure 8, (Bailey et al 1972) was a weak-focusing ring with $n = 0.13$, orbit diameter 5 m, a useful aperture of 4 cm $\times$ 8 cm (height $\times$ width), beam momentum 1.28 GeV/c corresponding to $\gamma = 12$ and a dilated muon lifetime of 27 μs. The mean field at the central orbit was $B = 72.852\ 7\ (36)$ proton MHz (1.711 T).

The injection of polarized muons was accomplished by the forward decay of pions produced when a target in the magnetic field was struck by 10.5-GeV/c protons from the CERN PS. The proton beam consisted of either two or three radiofrequency bunches (fast ejection), each $\sim$10-ns wide and spaced at $\sim$105 ns. As the rotation time in the ring was 52.5 ns, these bunches overlapped exactly inside the ring. Approximately 70% of the protons interacted, creating among other things pions of 1.3 GeV/c that started to turn around the ring. The pions made on an average four turns before hitting the target again, and in one turn about 20% of the pions decayed. The muons created in the exactly forward decay, together with undecayed pions and stable particles from the target, eventually hit the target and were lost. However, the decay of pions at small forward angles gave rise to muons of slightly lower momentum, and some of these fell into orbits that missed the target and remained permanently stored in the ring. Thus the perturbation, essential for inflection into any circular machine, was here achieved by the shrinking of the orbit arising from the change of momentum in π-μ decay and to some extent by the change in angle at the decay point, which could leave the muon with a smaller oscillation amplitude than its parent pion. The muons injected in this way were forward polarized, because they came from the forward decay of pions in flight. About 200 muons were stored per cycle of the PS. The muon injection was accomplished in a time much shorter than

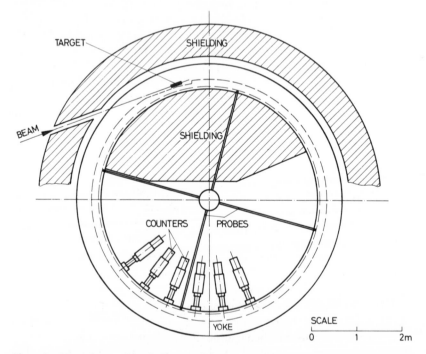

Figure 8 Muon Storage Ring I, diameter 5 m, muon momentum 1.3 GeV/c, time dilation factor 12. The injected pulse of 10-GeV protons produces pions at the target, which decay in flight to give muons.

both the dilated muon lifetime (27 μs), and the precession period of the anomalous moment (3.7 μs).

Unfortunately this simple injection system created a blast of particles inside the ring. Some of them were the desired pions, trapped for a few turns, but there were many more pions of higher momentum. Each had only a small probability of launching a muon into the storage aperture, but the overall contribution was significant. These muons were emitted at large angles in the pion rest frame so the average longitudinal polarization was around 26% compared to the 95% expected.

The method of injection used had the advantage of being very simple, but had the following disadvantages:

1. low muon polarization due to muons from a wide range of pion momenta;
2. high general background;
3. contamination by electrons at early times;
4. low average trapping efficiency.

For some time a magnetic horn was used around the target to concentrate pions of the correct energy in the forward direction. This gave a good muon polarization, but because of increased background was not finally adopted.

The muon precession was recorded by observing the decay in flight of muons in the ring magnet. The detectors responded to decay electrons of energy greater than a minimum value E_{min} (750 MeV). To obtain this high an energy in the laboratory, the electron had to be (a) near the top end of the β-spectrum in the muon rest frame [high asymmetry parameter (Bouchiat & Michel 1957)], and (b) emitted more or less forward in the muon rest frame. A counter with high energy threshold in the laboratory was equivalent, in the muon rest frame, to a telescope observing a small angular interval around the direction of motion. Therefore, as the muon spin rotated relative to its moment vector according to Equation 9, the observed counting rate (Figure 9) was modulated according to

$$N(t) = N_0 e^{-t/\tau} \{1 - A \sin (2\pi f_a t + \phi)\}$$ 13.

and the frequency f_a could be read from the data.

To calculate a_μ from the data using Equation 9 the value of $(e/mc)B$ is required. This was obtained from the magnetic field measurement in terms of the proton resonance frequency f_p and the known ratio $\lambda = f_s/f_p$ for muon and proton spin precession in the same field (Hutchinson et al 1963). From Equations 8 and 9,

$$\frac{a_\mu}{(1+a_\mu)} = \frac{f_a}{f_s} = \frac{f_a}{\lambda f_p(1+\varepsilon)},$$ 14.

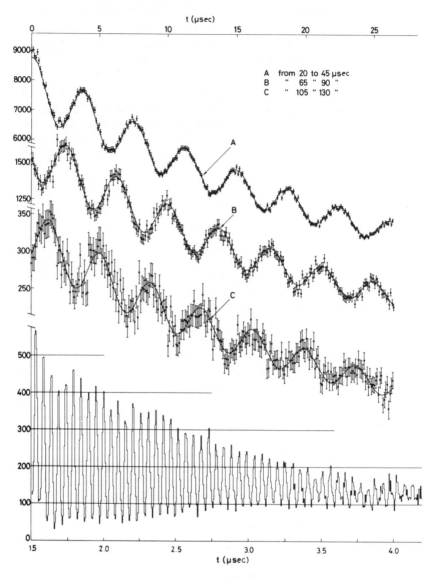

Figure 9 Muon Storage Ring 1: decay electron counts as a function of time after the injected pulse. The lower curve from 2 to 4.75 μs (*lower time scale*) shows 19-MHz modulation due to the rotation of the bunch of muons around the ring. As it spreads out the modulation dies away. This is used to determine the radial distribution of muon orbits. Curves *A*, *B*, and *C* are defined by the legend (*upper time scale*); they show various sections of the experimental decay (lifetime 27 μs!) modulated by the (g − 2) precession. The frequency is determined to 215 ppm, $\bar{B}$ to 160 ppm leading to 270 ppm in a_μ.

where ε is the small diamagnetic shielding correction (~ 26 ppm) to correct the measured field in water to the field in vacuum seen by the muons.

The magnetic field was surveyed in terms of the corresponding proton spin-resonance frequency f_p; measurements were taken at 288 azimuthal settings at each of 10 radii.

The radial magnetic gradient necessary for vertical focusing implied a field variation of $\pm 0.2\%$ over the full radial aperture (8 cm). Hence a major problem was to determine the mean radius of the ensemble of muons that contributed to the signal. This was obtained from measurements of the rotation frequency f_r of the muons. The injection pulse was only 5–10 ns long, and the rotation period of the muons, $T = 2\pi r/\beta c$, was about 52.5 ns, so the counting rate was initially modulated at the rotation frequency. The bunches of muons spread out uniformly around the ring after about 5 μs owing to the spread in radii so this modulation gradually diminished in amplitude and disappeared (see Figure 9). The analysis of the modulated record yielded the mean radius $\bar{r} = 2494.3\ (2.7)$ mm. Figure 10 shows the reconstructed number of muons as a function of radius compared with a theoretical prediction.

Unfortunately the determination of the mean radius was subject to systematic troubles. In the time interval during which the mean radius

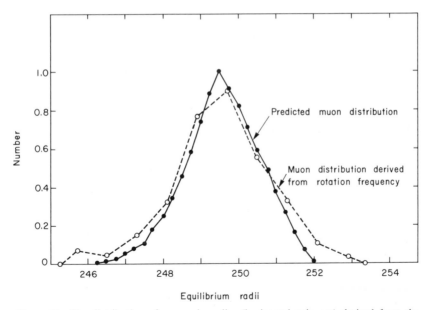

Figure 10 The distribution of muons in radius (horizontal axis, em) derived from the analysis of the decay electron events at early time. The muon rotation frequency has been analyzed from 1.8 μs to 5.5 μs.

could be determined, there was an excess of counts, caused partly by the fact that some muons were lost later, and partly by a nonrotating background produced by neutrons and other background created by the injection system. So numerous checks were needed to establish the validity of the radius measurement. A measurement of the rotation frequency was made, with reduced intensity, to minimize some of the systematic errors mentioned above. This experiment gave a value of the mean radius at very early times of 0.6–1.6 μs, $\bar{r} = 2492.5$ (2.1) mm, in good agreement with the value given above. Checks were also made to show that the mean radius did not change with time by more than ± 1.1 mm between 3 μs and 50 μs. A conservative overall error in the mean radius of ± 3 mm was assigned, implying an error of 160 ppm in the value of a_μ.

The statistical error in a_μ arising from the fit of Equation 13 to the counting data was $\pm 23 \times 10^{-8}$, and the fluctuations of the results of eight different runs about the mean gave $\chi^2 = 7.84$, compared to 6.35 expected. The error in the magnetic field corresponding to ± 3-mm uncertainty in radius contributed $\pm 19 \times 10^{-8}$ to a_μ. The two errors, combined in quadrature, gave the overall error in a_μ of $\pm 31 \times 10^{-8}$. The experimental result was

$$a_\mu = 1\ 166\ 16\ (31) \times 10^{-8}\ (270\ \text{ppm}).\qquad\qquad 15.$$

Initially this was nearly two standard deviations higher than the theoretical value, a sign perhaps that there was more to discover about the muon. In fact the discrepancy resulted from a defect in the theory. Theorists had originally speculated that the contribution of the photon-photon scattering diagrams (Figure 1f) to the $(\alpha/\pi)^3$ term in a_μ^{th} might be small or perhaps cancel exactly. The experimental result stimulated Aldins et al (1969, 1970) to examine this more carefully, obtaining a coefficient of 18.4! The situation then was

$$a_\mu^{\text{exp}} - a_\mu^{\text{th}} = 28\ (31) \times 10^{-8} = 240 \pm 270\ \text{ppm}.\qquad 16.$$

For the photon propagator cutoff this implied $\Lambda_\gamma > 5$ GeV, and for the muon vertex $\Lambda_\mu > 7$ GeV. The Einstein time dilation was confirmed to 1%.

PART III: 1969–1976   MUON STORAGE RING II

By 1969 an electron-electron colliding-beam experiment had demonstrated the point-like nature of the electron ($\Lambda_\gamma > 4$ GeV, $\Lambda_e > 6$ GeV) (Barber et al 1966), and e^+e^- storage rings were giving useful data on vector meson production (Augustin et al 1969, 1975, Auslander et al 1968). A comparison of ep and μp scattering, and experiments on e^+e^- and $\mu^+\mu^-$ pair production, wide-angle bremsstrahlung, and muon tridents were all in accord with theory. (For reviews see Lederman & Tannenbaum 1968, Brodsky 1969, Farley 1970, Picasso 1970, Brodsky & Drell 1970).

The pure quantum effects were less satisfactory. The Lamb shift data (Robiscoe 1968a,b, Robiscoe & Shyn 1970) were consistently higher than theory, but this was resolved by a recalculation of a small theoretical term by Appelquist & Brodsky (1970). The electron $(g-2)$ data of Wilkinson & Crane (1963) had been rediscussed by Farley (1968), Henry & Silver (1969), and Rich (1968), who concluded $a_e^{exp} - a_e^{th} = -(79 \pm 26)$ ppm.[2] This discrepancy was to be resolved in a new measurement by Wesley & Rich (1970, 1971). Thus QED was doing well, but in early 1969, a_μ, a_e, and the Lamb shift all showed uncomfortably large departures from theory. It could have been the beginning of something new.

The major motivations for carrying out a third measurement were therefore as follows:

1. to look for departures from standard QED;
2. to detect the contribution of strong interactions to a_μ through hadron loops in the vacuum polarization (see Table 1);
3. to search for new interactions of the muon.

The Third $(g-2)$ Experiment

A major difficulty in the previous experiment was the radial magnetic gradient necessary to provide the vertical focusing; this implied a magnetic field variation of $\pm 0.2\%$ over the aperture in which the muons were stored, and a corresponding radial dependence of f_a. Even if the mean radius was determined precisely after injection, uncertainties in radius would arise from uncontrolled muon losses. The central question for a new experiment was: Can the dependence of f_a on r be removed without destroying the vertical focusing? The answer is yes. The forces that hold the muon in its orbit and give focusing for small deviations from equilibrium arise from what appears in the muon rest frame as an electric field, while the spin precession arises from what appears there as a magnetic field. These two fields may be varied independently by applying suitable magnetic and electric fields in the laboratory frame.

The advantages of this method may be appreciated by writing the classical relativistic equations of motion of a charged particle with an anomalous magnetic moment in laboratory fields **B** and **E** (using cgs

---

[2] For later work on the pitch correction see Granger & Ford (1972, 1976), Farley (1972a), and Field & Fiorentini (1974).

units) (Bailey et al 1969, Bailey & Picasso 1970, Combley & Picasso 1974, Farley 1975, Jackson 1975):

$$d\boldsymbol{\beta}/dt = \boldsymbol{\omega}_c \times \boldsymbol{\beta}, \quad d\boldsymbol{\sigma}/dt = \boldsymbol{\omega}_s \times \boldsymbol{\sigma}$$

with

$$\boldsymbol{\omega}_c = \left(\frac{e}{mc}\right)\left[\frac{\mathbf{B}}{\gamma} - \left(\frac{\gamma}{\gamma^2-1}\right)\boldsymbol{\beta} \times \mathbf{E}\right]$$

and

$$\boldsymbol{\omega}_s = \left(\frac{e}{mc}\right)\left[\frac{\mathbf{B}}{\gamma} - \left(\frac{1}{\gamma+1}\right)\boldsymbol{\beta} \times \mathbf{E} + a_\mu(\mathbf{B} - \boldsymbol{\beta} \times \mathbf{E})\right]$$

Here we have assumed that $\boldsymbol{\beta}\cdot\mathbf{B} = \boldsymbol{\beta}\cdot\mathbf{E} = 0$ (muon charge is $-e$). The precession of the spin relative to the velocity vector is

$$\boldsymbol{\omega}_a \equiv \boldsymbol{\omega}_s - \boldsymbol{\omega}_c = \left(\frac{e}{mc}\right)\left[a_\mu\mathbf{B} + \left(\frac{1}{\gamma^2-1} - a_\mu\right)\boldsymbol{\beta} \times \mathbf{E}\right] \qquad 17.$$

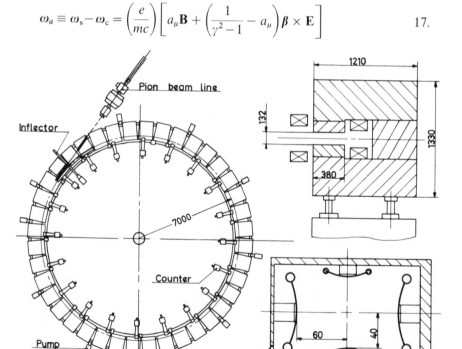

Figure 11   Muon Storage Ring II: diameter 14 m, muon momentum 3.094 GeV/c (magic energy), time dilation factor 29.3. The magnetic field is uniform and vertical focusing is by electric quadrupoles (shown on right). A pulse of 3.1-GeV/c pions is inflected to make nearly one turn. Their decay in flight leaves polarized muons stored in the field.

Thus the effect of the electric field on the precession of the polarization can be made zero if the particle energy satisfies $\gamma = [1+(1/a_\mu)]^{1/2} = 29.304$. For muons this is equivalent to a momentum of 3.094 GeV/c and this was the value chosen for the new storage ring (Figure 11) (Bailey et al 1969). The main idea, therefore, was to provide the vertical focusing by an electrostatic quadrupole field, to work at the "magic" value of $\gamma_0 = 29.304$, corresponding to a muon momentum $p_\mu = 3.094$ GeV/c, and to use a uniform magnetic field. As the magnetic field was to be independent of radius, and the electric field had no effect on spin, it would no longer be necessary to know the radius of the muon orbit.

This cancellation of the effect of the electric field on the spin motion would occur only for the central momentum of the muon sample; for the other equilibrium orbits a small correction (~ 1.5 ppm) would be necessary. The value of the electric field was chosen to give appropriate focusing, but was not needed for calculating a_μ.

The system for injecting muons into the ring was designed to give maximum muon polarization, minimum background, and as large an intensity as possible. A high value of the longitudinal polarization can be achieved by starting with a momentum-selected pion beam and only accepting those decay muons whose momenta lie in a narrow band close to that of the pion beam.

It was therefore decided to locate the primary target outside the Muon Storage Ring, and prepare a momentum-selected pion beam to be guided into the ring by a pulsed inflector. Because of the size of the inflector structure, the pions would only make one turn in the ring, and the useful aperture of the inflector would be very small. Loss of intensity due to these factors could however, be compensated by using special beam optics, which collected pions over a large solid angle and matched them to the acceptance of the storage ring.

The injector was in the form of a coaxial line in which a 10-μs current pulse of peak value 300 kA produced the required field of about 1.5 T between the inner and outer conductors. The great technical difficulty of this method of injection was outweighed by the increased pion flux and the high longitudinal polarization of the stored muons (95%). This was borne out by the large observed modulation of the decay electron spectra. The polarization direction was independent of the muon equilibrium radius and consequently any possible asymmetric muon losses could cause no significant shift in the measured spin precession frequency f_a. Finally the background was reduced considerably with respect to the previous experiment, in which the copper target was located in the ring, and therefore the electron detectors could be located all around the ring.

Other improvements can only be mentioned briefly in this review. The magnetic field was very stable, very reproducible, and uniform. Figure 12 shows a contour plot of the magnetic field strength in the muon aperture. This map was obtained by averaging a three-dimensional map in azimuth. The interval between the contours of equal field strength is 2 ppm or 3 μT. The stability and reproducibility of the magnetic field were achieved by very careful mechanical design of the yoke and magnet coils (Drumm et al 1979). Furthermore each of the 40 magnet blocks was separately stabilized (Borer 1977) with a nuclear magnetic resonance probe and a pick-up coil as sensors. The signals were used to determine automatically the current through additional compensating coils, which were wound around the yoke of the individual magnets close to each pole tip.

It is worth mentioning here the extreme insensitivity of the average value of the magnetic field ($\bar{B}$) computed for different assumed radial distributions of muons. Even in extreme cases the average magnetic field was the same within less than 2 ppm, compared with the 160 ppm uncertainty in $\bar{B}$ in the previous experiment.

To achieve the accuracy reached in this last experiment many technical problems had to be solved. The ability to find these solutions constituted part of the beauty of the experiment. A particular solution that deserves mention is the construction and understanding of the electric quadrupole

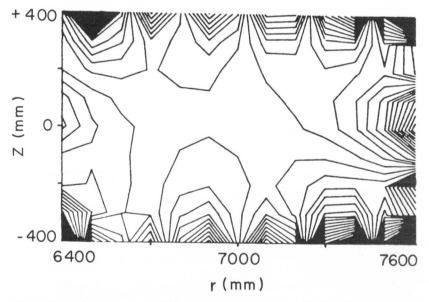

Figure 12 A contour line plot of the magnetic field strength in the muon storage aperture. This map is obtained by averaging a three-dimensional map in azimuth. The interval between the contours of equal field strength is 2 ppm or 3 μT.

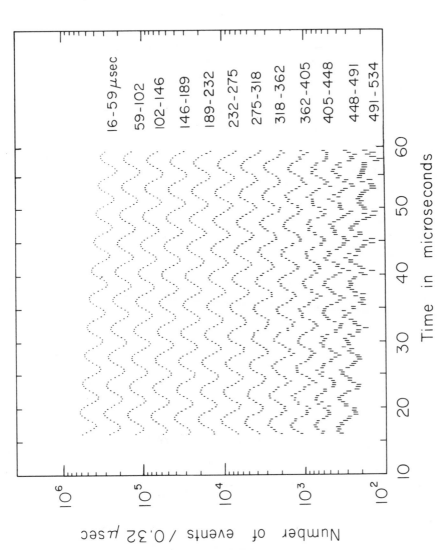

Figure 13 Muon Storage Ring II: decay electron counts versus time (in microseconds) after injection. Range of time for each line is shown on the right (in microseconds).

field in the presence of the high magnetic field. We invite the reader to consult the final report on the CERN Muon Storage Ring for these details (Bailey et al 1979; see also Flegel & Krienen 1973). Figure 11 shows the ring and focusing system, while Figure 13 gives a summary of the counting data.

EXPERIMENTAL RESULTS   Nine separate runs were made over a period of two years to measure the $(g-2)$ precession frequency f_a, the field being determined in terms of the proton resonance frequency f_p (Bailey et al 1975, 1977a, 1979). The ratio $R = f_a/f_p$ showed good consistency ($\chi^2 = 7.3$ for 8 degrees of freedom). The overall mean value is the principal result of the experiment:

$$R = f_a/f_p = 3.707\ 213\ (27) \times 10^{-3}\ (7\ \text{ppm}). \qquad 18.$$

Equation 14 allows us to calculate $a_\mu = R/(\lambda - R)$ if $\lambda = f_s/f_p$ is known.

The magnitude of λ has now been determined to about 1 ppm, directly from measurements of muon precession at rest (Crowe et al 1972, Camani et al 1978) and indirectly from the hyperfine splitting in muonium (Casperson et al 1977). The weighted average value of these measurements is

$$\lambda = 3.183\ 343\ 7\ (23).$$

This leads to the following results for the anomalous moment:

$$a_{\mu^+} = 1\ 165\ 911\ (11) \times 10^{-9}\ (10\ \text{ppm}), \qquad 19.$$

$$a_{\mu^-} = 1\ 165\ 937\ (12) \times 10^{-9}\ (10\ \text{ppm}), \qquad 20.$$

and for μ^+ and μ^- combined

$$a_\mu = 1\ 165\ 924\ (8.5) \times 10^{-9}\ (7\ \text{ppm}). \qquad 21.$$

COMPARISONS BETWEEN THEORY AND EXPERIMENT   The main six conclusions that can be drawn from this last measurement of the anomalous magnetic moment of the muon are the following:

1. The QED calculations of the muon anomaly are verified up to the sixth order, the experimental uncertainty being equivalent to 1.2×10^{-5} in A, 3.5×10^{-3} in B, or 4.7% in C (see Equation 1).

2. The hadronic contribution to the anomaly is confirmed to an accuracy of 20%. The existence of hadronic vacuum polarization has thus been established at the level of five standard deviations.

3. There is no evidence for a special coupling of the muon. The experimental range of possible values of an extra contribution to the moment is

$$-20 \times 10^{-9} < \Delta a_\mu < 26 \times 10^{-9} \qquad 22.$$

to 95% confidence. The limits implied for unknown boson fields then depend on the nature of the coupling and are given in Figure 14 (Bailey et al 1979).

4. With the advent of renormalizable gauge theories unifying the weak and the electromagnetic interactions, the calculation of the weak interaction contribution to the muon anomaly has become reliable. In general the weak contribution depends upon the parameters of the theory, such as the masses of the Higgs and intermediate vector bosons. To the extent that we do not yet know the correct form of the weak interaction Hamiltonian, the above results for a_μ (and also the result for a_e) can be used to restrict the range of possible models. Only in the simplest of such theories, that of Weinberg (1967) and Salam (1968), are the parameters sufficiently well determined experimentally to give a firm prediction of the expected value of the weak anomaly. Kinoshita (1978) has recently reviewed this subject. If the arguments of Weinberg on the Higgs boson are accepted, and the current limits on $\sin^2 \theta_W$ are taken into account, we obtain:

$$1.9 \times 10^{-9} \lesssim a_\mu \,(\text{weak}) \lesssim 2.3 \times 10^{-9}.$$

Clearly the precision of even the latest experiment is inadequate for testing this prediction.

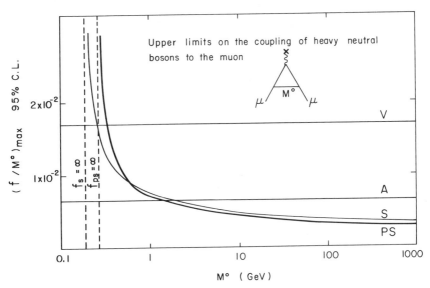

Figure 14 The upper limits on the coupling constant f of the muon to a heavy neutral boson of mass M^0, for vector (V), axial vector (A), scalar (S), and pseudoscalar (PS) coupling.

5. The range defined by the inequalities (Expression 22) may be used to set limits on single contributions to the muon anomaly from other sources including various models for breaking QED as discussed in the theoretical section. The following limits apply to 95% confidence:

(a) The muon may not behave like a point charge, but instead have a finite size, in analogy with the proton. This would show up as a form factor. The limit imposed on Λ_μ would be $\Lambda_\mu > 36$ GeV.

(b) The modification in the photon propagator leads to $\Lambda_\gamma > 20.7$ GeV.

(c) From the latest experiment it is possible to set a limit to the modification of the muon propagator (Kroll 1966) by a factor $(1 - q^4/\Lambda_{\text{prop}}^4)$. The value is $\Lambda_{\text{prop}} > 1.5$ GeV.

(d) A possible new, undiscovered lepton of mass M_L would contribute to the vacuum polarization through a mechanism such as diagrammed in Figure 1d. The value of the anomaly would depend on the ratio M_L/m_μ. The 95% confidence limit sets the limit $M_L \gtrsim 210$ MeV/c^2, which is not very interesting. In passing, the recently discovered heavy lepton τ (Perl et al 1975), with mass 1.8 GeV/c^2, gives a contribution,

$$\Delta a_\mu(\tau) \simeq 0.4 \times 10^{-9},$$

well below the present sensitivity.

6. Recently Kadyshevsky (1978) has given a new gauge formulation of the electromagnetic interaction theory, containing a "fundamental length" l as a universal scale constant as important as h and c. This new hypothetical constant l, together with h and c, is expected to regulate all microscopic phenomena. The quantity $M = h/lc$ plays the role of a fundamental mass. In the new approach the electromagnetic potential becomes a 5-vector associated with the de Sitter group O(4,1). Among the various predictions given are the value of the anomalous moment for a lepton of mass m_l

$$a_{\text{lepton}} \simeq \frac{m_l^2}{2M^2}$$

and the electric dipole moment (EDM)

$$\mathbf{d}_{\text{lepton}} \simeq \frac{el}{2}.$$

From the present experimental result one then obtains an upper bound for the fundamental length: $l < 2.6 \times 10^{-17}$ cm.

Electric Dipole Moment

An upper limit for the electric dipole moment (EDM) of the muon has been measured directly in the CERN Muon Storage Ring (Bailey et al

1978). For a particle with both magnetic and electric dipole moments the electromagnetic interaction Hamiltonian contains a term ($\boldsymbol{\mu} \cdot \mathbf{B} - \mathbf{d} \cdot \mathbf{E}$), where $\mathbf{B}$ and $\mathbf{E}$ are the magnetic and electric field strengths and $\boldsymbol{\mu}$ and $\mathbf{d}$ are the magnetic and electric dipole moment operators. Treating the electric dipole moment analogously to the magnetic dipole moment we can write

$$\boldsymbol{\mu} = g\left(\frac{e}{2mc}\right)\left(\frac{\hbar\boldsymbol{\sigma}}{2}\right) = g\mu_0\left(\frac{\boldsymbol{\sigma}}{2}\right)$$

$$\mathbf{d} = f\left(\frac{e}{2mc}\right)\left(\frac{\hbar\boldsymbol{\sigma}}{2}\right) = f\mu_0\left(\frac{\boldsymbol{\sigma}}{2}\right),$$

where μ_0 is the muon Bohr magneton $e\hbar/2mc$.

It is well known that the expectation value of the electric dipole moment $\mathbf{d}$ must be zero for a particle described by a state of well-defined parity. However, Purcell & Ramsey (1950) stressed that the existence of an EDM for particles should be treated as a purely experimental question, and they suggested possible physical mechanisms that could lead to a nonvanishing EDM. After the discovery of parity violation in the weak interactions, it was pointed out by Landau (1957) that even if P is violated, the existence of an EDM is still forbidden by T invariance, i.e. the existence of a nonvanishing EDM for a particle implies that both P and T are violated. See Field et al (1979) and Jackson (1977) for comprehensive reviews of the subject.

The technique used to measure the muon electric dipole moment follows from a suggestion originally made by Garwin & Lederman (1959). They pointed out that in the ($g-2$) precession experiments using magnetic mirror traps, the electron (or the muon) will experience in its rest frame an electric field proportional to the particle velocity, as a result of the Lorentz transformation of the laboratory magnetic field. This electric field is perpendicular to the magnetic field. If the EDM is not zero, the spin precession frequency relative to the momentum will pick up a component $\mathbf{f}_{\mathrm{EDM}}$ along the electric field direction in addition to the normal ($g-2$) frequency $\mathbf{f}_a$ along the magnetic field direction (Figure 15). The observed ($g-2$) frequency is then $f'_a = f_a[1 + \beta^2 f^2/4a^2]$. As a further consequence of this new precession component, the decay electrons from the muon will show a time-varying up-down asymmetry perpendicular to the plane of the orbit. Such an effect was explored in the first CERN muon ($g-2$) experiment (Charpak et al 1961b) in which the muons were brought to rest in a polarimeter. The value measured was

$$d_\mu = |\mathbf{d}| = (0.6 \pm 1.1) \times 10^{-17} \, e \cdot \mathrm{cm}.$$

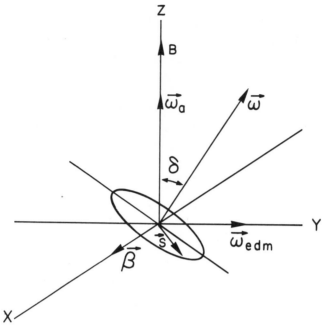

Figure 15   Precession of the spin relative to the momentum resulting from the combination of an anomalous magnetic moment and an electric dipole moment. The plane of precession is tilted through the angle $\delta = \beta f/2a$, see text.

A similar technique was used in the most recent Muon Storage Ring experiment by detecting separately the electrons emitted upwards and downwards from muon decay in flight. Separate measurements on μ^+ and μ^- (Bailey et al 1978) gave:

$$d_{\mu^+} = (8.6 \pm 4.5) \times 10^{-19} \ e \cdot cm$$
$$d_{\mu^-} = (0.8 \pm 4.3) \times 10^{-19} \ e \cdot cm.$$

Assuming opposite EDMs for the particle and antiparticle, the combined result was

$$d_{\mu} = (3.7 \pm 3.4) \times 10^{-19} \ e \cdot cm.$$

For comparison the current upper limits for the electron, proton, and neutron in units $e \cdot cm$ (Pais & Primack 1973) are electron $\lesssim 3 \times 10^{-24}$, proton $\lesssim 2 \times 10^{-20}$, and neutron $\lesssim 1 \times 10^{-23}$. That these limits are much lower than the limit of the muon largely reflects the fact that, unlike the muons, they are studied in neutral systems. The fundamental length l of Kadyshevsky can therefore not be greater than 2×10^{-18} cm (muon evidence) or 10^{-23} cm (electron evidence).

Muon Lifetime in Flight

Accurate measurements of the muon lifetime in a circular orbit provide a stringent test of Einstein's theory of special relativity. As a bonus it sheds light on the so-called twin paradox, gives an upper limit to the granularity of space time, and tests the CPT invariance of weak interaction.

The muon is an unstable particle, and can therefore be regarded as a clock and used to measure the time dilation predicted by special relativity. The existence of cosmic-ray muons at ground level supports the idea of time dilation, for, if the muon lifetime was not lengthened in flight, they would all decay in the upper atmosphere (Rossi & Hall 1941). Experiments verifying the time dilation in a straight path have also been made with high energy accelerators (see Bailey et al 1977b).

Recently Hafele & Keating (1972) loaded cesium atomic clocks onto a commercial aircraft on an around-the-world trip and verified the time dilation at low velocity with an accuracy of about 10%.

In the CERN Muon Storage Ring, the muon performs a round trip and so when compared with a muon at rest the experiment mimics closely the twin paradox already discussed in Einstein's first paper (Einstein 1905). The circulating muons, although they return again and again to the same place, should remain younger than their stay-at-home brothers. It is indeed observed that the moving muons live longer, in agreement to one part in a thousand with the predictions of special relativity. The stationary twin's time scale is given by the muon decay rate at rest determined in a separate experiment.

An accurate measurement of the muon lifetime in a circular orbit at $\gamma \simeq 29.3$ requires high orbit stability in a short time interval (a few hundred microseconds), for any loss of muons will set a limit to the accuracy of the measurement. The reported stability was achieved by using a scraping system that shifted the muon orbits at early times in order to "scrape off" those muons most likely to be lost.

The experiment consisted of measuring the decay electron counting rate $N(t)$ (see Equation 13) and the fitting procedure gave the value of $\tau = \gamma\tau_0$. The rotation frequency f_r of the muons obtained from the counting record at early times (see Figure 16) gave $\gamma = \lambda\bar{f}_p/(1+a)f_r = 29.327\,(4)$. The best value for the lifetime at rest is $2.197\,11\,(8)\,\mu s$ (Balandin et al 1974), which then gives $\tau_{th} = 64.435\,(9)\,\mu s$, compared with the experimental result $\tau_{exp} = 64.378\,(26)$. Thus the transformation of time is validated to an accuracy of $-(0.9\pm0.4) \times 10^{-3}$ (Bailey et al 1977b).

In the actual experiment, corrections were made for a residual small loss of stored muons, for variations of the photomultiplier gain accompanying the recovery from the initial flash, and for background counts due to stored protons (in the case of μ^+). In order to measure the muon lifetime with an accuracy of 0.1% it was necessary to study carefully these three effects, which could systematically distort the recorded time spectrum.

Another check on relativity theory can be obtained by comparing $(g-2)$ measurements carried out at different values of γ. For the electron this has been argued by Newman et al (1978), and discussed by Combley et al (1979) for both e and μ. Inevitably the conclusions are model dependent, but one can make a plausible case that these results confirm the relativistic transformation laws for magnetic field and mass, as well as for time.

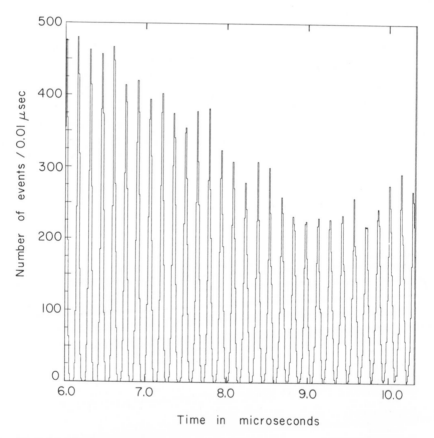

Figure 16 The fast rotation pattern. This is the count rate at early time which clearly shows the muon bunch rotating around the ring with a period of 147 ns.

Verification of the CPT Theorem

From CPT it follows that $g_{\mu^+} = g_{\mu^-}$. The measurements in the CERN Muon Storage Ring gave to 95% confidence

$$7 \times 10^{-9} > \frac{g_{\mu^+} - g_{\mu^-}}{g_\mu} > -58 \times 10^{-9}.$$

From CPT it follows also that $\tau_{\mu^+} = \tau_{\mu^-}$. The experimental data on the μ^+ and μ^- lifetime in flight give the best test of this equality (as τ_{μ^-} cannot be measured at rest because of muon capture). In this connection it should be noted that the Lorentz γ-factor is the same for μ^+ and μ^- to a much higher precision than the quoted lifetime errors. The limits are

$$3.0 \times 10^{-3} > \frac{\tau_{\mu^+} - \tau_{\mu^-}}{\tau_\mu} > -1.4 \times 10^{-3}.$$

Thus the theorem is validated for muons to very high accuracy for the electromagnetic interaction, and rather less accurately for the weak interaction.

CONCLUSION: THE SITUATION TODAY

The anomalous magnetic moment of the electron has now been measured to 0.2 ppm (Van Dyck et al 1977) in agreement with calculations to the order of $(\alpha/\pi)^3$ again confirming the QED series expansion. At present the theory is only good to 0.1 ppm. When this is improved the experiment will provide the most accurate and clearest measurement of the fine structure constant α (Kinoshita 1978). Any modification to the photon propagator or new coupling common to e and μ would imply a perturbation to a_μ a factor $(m_\mu/m_e)^2$ greater than for a_e. Therefore, barring possible couplings peculiar to the electron, the muon result ensures that a_e is a "pure QED quantity" to the order of three parts in 10^{10}. Another good route to α is via the hyperfine splitting in muonium. Here again the results for a_μ and a_e ensure that muonium is a "pure QED system" (Farley 1972b).

A major stride forward in QED at high energy has been made with e^+e^- colliding beams. All experiments agree with theory (except in the neighborhood of the new resonances Υ, ψ, ψ') and the corresponding cut-off limits are $\Lambda_e > 21$ GeV, $\Lambda_\mu > 27$ GeV (Hofstadter 1975) and $\Lambda_\gamma > 38$ GeV (Barber et al 1979). (See also Schwitters & Strauch 1976, Cords 1978, Hughes & Kinoshita 1977.)

Small parity-violating effects due to the weak interaction have been detected in the scattering of polarized electrons at high energies (Prescott et al 1978) in agreement with the predictions of the new unified theory.

The discovery of a new lepton of mass 1.8 GeV (Perl et al 1975) has changed the μ-e problem without providing any answers. Mass splittings in the lepton family are much larger than other mass splittings between similar particles. Although the $(g-2)$ result for the muon has shed no light on the problem, it nevertheless provides a serious constraint on the fantasies of theorists.

In first-order QED the self-mass $m = (m_0/137) \ln (\lambda_c/L)$ becomes significant when the cutoff distance L is of order 10^{-69} cm or less. The only other physical length of this order is the Schwarzchild radius of the electron ($\sim 10^{-54}$ cm). In principle any photons originating closer to the particle than this will not be able to reach the outside world, so in this sense gravitation provides a natural cutoff for QED. We already have a unified theory of the weak and electromagnetic interactions (Weinberg 1967, Salam 1968). If the attempt to include gravitation is successful it may in the end offer an explanation for the lepton masses (Isham et al 1971).

ACKNOWLEDGMENTS

It is a pleasure to acknowledge the contribution of Leon Lederman, who with G. Charpak, H. Sens, and A. Zichichi set up the first $(g-2)$ group at CERN.

All the experiments have been a team effort, with each member listed in the references making his unique personal contribution. We offer this summary as a souvenir of our shared endeavor. In particular John Bailey has been a constant and stimulating colleague through all the storage ring experiments.

We thank particularly Ms. M.-N. Pagès, Ms. M. Rabbinowitz, Ms. S. Vascotto, and Mrs. D. R. Baylis for their painstaking work on the manuscript.

Financial support of one of us (F.J.M.F.) by the Science Research Council is gratefully acknowledged.

Literature Cited

Alberigi-Quaranta, A., De Pretis, M., Marini, G., Odian, A. C., Stoppini, G., Tau, L. 1962. *Phys. Rev. Lett.* 9:226–29
Aldins, J., Kinoshita, T., Brodsky, S. J., Dufner, A. J. 1969. *Phys. Rev. Lett.* 23:441–43
Aldins, J., Brodsky, S. J., Dufner, A. J., Kinoshita, T. 1970. *Phys. Rev. D* 1:2378–95

Appelquist, T., Brodsky, S. J. 1970. *Phys. Rev. Lett.* 24:562–65
Augustin, J. E., Bizot, J. C., Buon, J., Haissinski, J., Lalanne, D., Marin, P., Nguyen Ngoc, H., Perez y Jorba, J., Rumpf, F., Silva, E., Tavernier, S. 1969. *Phys. Lett. B* 28:508–12
Augustin, J. E., Boyarski, A. M., Breidenbach,

M., Bulos, F., Dakin, J. T., Feldman, G. J., Fischer, G. E., Fryberger, D., Hanson, G., Jean-Marie, B., Larsen, R. R., Lüth, V., Lynch, H. L., Lyon, D., Morehouse, C. C., Paterson, J. M., Perl, M. L., Richter, B., Schwitters, R. F., Vannucci, F., Abrams, G.S., Briggs, D., Chinowsky, W., Friedberg, C. E., Goldhaber, G., Hollebeck, R. J., Kadyk, J. A., Trilling, G. H., Whitaker, J. S., Zipse, J. E. 1975. *Phys. Rev. Lett.* 34:233–36

Auslander, V. L., Budker, G. I., Pakhtusova, E. V., Pestov, Yu. N., Sidorov, V. A., Skrinskij, A. N., Khabakhasher, A. G. 1968. *Akad. Nauk SSSR, Sibirskee Otd. Preprint 243.* Novosibirsk

Backenstoss, G., Hyams, B. D., Knop, G., Marin, P. C., Stierlin, U. 1963. *Phys. Rev.* 129:2759–65

Bailey, J., Bartl, W., von Bochmann, G., Brown, R. C. A., Farley, F. J. M., Jöstlein, H., Picasso, E., Williams, R. W. 1968. *Phys. Lett. B* 28:287–90

Bailey, J., Farley, F. J. M., Jöstlein, H., Petrucci, G., Picasso, E., Wickens, F. 1969. *CERN proposal PH I/COM-69/20*

Bailey, J., Bartl, W., von Bochmann, G., Brown, R. C. A., Farley, F. J. M., Giesch, M., Jöstlein, H., van der Meer, S., Picasso, E., Williams, R. W. 1972. *Nuovo Cimento A* 9:369–432

Bailey, J., Borer, K., Combley, F., Drumm, H., Eck, C., Farley, F. J. M., Field, J. H., Flegel, W., Hattersley, P. M., Krienen, F., Lange, F., Petrucci, G., Picasso, E., Pizer, H. I., Rúnolfsson, O., Williams, R. W., Wojcicki, S. 1975. *Phys. Lett. B* 55:420–24

Bailey, J., Borer, K., Combley, F., Drumm, H., Farley, F. J. M., Field, J. H., Flegel, W., Hattersley, P. M., Krienen, F., Lange, F., Picasso, E., von Rüden, W. 1977a. *Phys. Lett. B* 68:191–96

Bailey, J., Borer, K., Combley, F., Drumm, H., Krienen, F., Lange, F., Picasso, E., von Rüden, W., Farley, F. J. M., Field, J. H., Flegel, W., Hattersley, P. M. 1977b. *Nature* 268:301–5

Bailey, J., Borer, K., Combley, F., Drumm, H., Farley, F. J. M., Field, J. H., Flegel, W., Hattersley, P. M., Krienen, F., Lange, F., Picasso, E., von Rüden, W. 1978. *J. Phys. G* 4:345–52

Bailey, J., Borer, K., Combley, F., Drumm, H., Eck, C., Farley, F. J. M., Field, J. H., Flegel, W., Hattersley, P. M., Krienen, F., Lange, F., Lebée, G., McMillan, E., Petrucci, G., Picasso, E., Rúnolfsson, O., von Rüden, W., Williams, R. W., Wojcicki, S. 1979. *Nucl. Phys. B* 150:1

Bailey, J., Picasso, E. 1970. *Progr. Nucl. Phys.* 12:43–75

Balandin, M. P., Grebenyuk, V. M., Zinov,

V. G., Konin, A. D., Ponomarev, A. N. 1974. *Sov. Phys. JETP* 40:811–14

Barber, D., Becker, U., Benda, H., Boehm, A., Branson, J. G., Bron, J., Buikman, D., Burger, J., Chang, C. C., Chen, M., Cheng, C. P., Chu, Y. S., Clare, R., Duinker, P., Fesefeldt, H., Fong, D., Fukishama, M., Ho, M. C., Hsu, T. T., Kadel, R., Luckey, D., Ma, C. M., Massaro, G., Matsuda, T., Newman, H., Paradiso, J., Revol, J. P., Rohde, M., Rykaczewski, H., Sinram, K., Tang, H. W., Ting, S. C. C., Tung, K. L., Vannucci, F., White, M., Wu, T. W., Yang, P. C., Yu, C. C. 1979. *Phys. Rev. Lett.* 42:1110–13

Barber, W. C., Gittelman, B., O'Neill, G. K., Richter, B. 1966. *Phys. Rev. Lett.* 16:1127–30

Barbieri, R., Remiddi, E. 1975. *Nucl. Phys. B* 90:233–66

Bardeen, W. A., Gastmans, R., Lautrup, B. E. 1972. *Nucl. Phys. B* 46:319–31

Barger, V., Long, W. F., Olsson, M. G. 1975. *Phys. Lett. B* 60:89–92

Bargmann, V., Michel, L., Telegdi, V. L. 1959. *Phys. Rev. Lett.* 2:453–36

Bars, I., Yoshimura, M. 1972. *Phys. Rev. D* 6:374–76

Berestetskii, V. B., Krokhin, O. N., Klebnikov, A. K. 1956. *Zh. Eksp. Teor. Fiz.* 30:788–89 (Transl. 1956. *Sov. Phys. JETP* 3:761–62)

Borer, K. 1977. *Nucl. Instrum. Methods* 143:203–18

Bouchiat, C., Michel, L. 1957. *Phys. Rev.* 106:170–72

Bouchiat, C., Michel, L. 1961. *J. Phys. Radium* 22:121

Brodsky, S. J. 1969. Quantum electrodynamics and the theory of the hydrogenic atom. *Proc. Brandeis Univ. Summer Inst. in Theoretical Physics,* ed. M. Chrétien, E. Lipworth, 1:91–169. New York: Gordon & Breach

Brodsky, S. J., Drell, S. D. 1970. *Ann. Rev. Nucl. Sci.* 20:147–94

Brown, L. M., Telegdi, V. L. 1958. *Nuovo Cimento* 7:698–705

Calmet, J., Narison, S., Perrottet, M., de Rafael, E. 1976. *Phys. Lett. B* 61:283–86

Calmet, J., Narison, S., Perrottet, M., de Rafael, E. 1977. *Rev. Mod. Phys.* 49:21–29

Calmet, J., Petermann, A. 1975a. *Phys. Lett. B* 56:383–84

Calmet, J., Petermann, A. 1975b. *Phys. Lett. B* 58:449–50

Camani, M., Gygax, F. N., Klempt, E., Ruegg, W., Schenck, A., Schilling, H., Schulze, R., Wolf, H. 1978. *Phys. Lett. B* 77:326–30

Carrassi, M. 1958. *Nuovo Cimento* 7:524–35

Casperson, D. E., Crane, T. W., Denison, A. B., Egan, P. O., Hughes, V. W., Mariam,

F. G., Orth, H., Reist, H. W., Souder, P. A., Stambaugh, R. D., Thompson, P. A., zu Putlitz, G. 1977. *Phys. Rev. Lett.* 38 : 956–59

Charpak, G., Farley, F. J. M., Garwin, R. L., Muller, T., Sens, J. C., Telegdi, V. L., Zichichi, A. 1961a. *Phys. Rev. Lett.* 6 : 128–32

Charpak, G., Farley, F. J. M., Garwin, R. L., Muller, T., Sens, J. C., Zichichi, A. 1961b. *Nuovo Cimento* 22 : 1043–50

Charpak, G., Farley, F. J. M., Garwin, R. L., Muller, T., Sens, J. C., Zichichi, A. 1962. *Phys. Lett.* 1 : 16–20

Charpak, G., Farley, F. J. M., Garwin, R. L., Muller, T., Sens, J. C., Zichichi, A. 1965. *Nuovo Cimento* 37 : 1241–363

Chlouber, C., Samuel, M. A. 1977. *Phys. Rev. D* 16 : 3596–601

Citron, A., Delorme, C., Fries, D., Goldzahl, L., Heintze, J., Michaelis, E. G., Richard, C., Øverå, H. 1962. *Phys. Lett.* 1 : 175–78

Cohen, E. R., Taylor, B. N. 1973. *J. Phys. Chem. Ref. Data* 2 : 663–734

Combley, F., Farley, F. J. M., Field, J. H., Picasso, E. 1979. *Phys. Rev. Lett.* 42 : 1383–85

Combley, F., Picasso, E. 1974. *Phys. Rep. C* 14 : 1–58

Cords, D. 1978. Results from e^+e^- interactions above 3 GeV. *DESY Rep.* 78/32

Cowland, W. S. 1958. *Nucl. Phys.* 8 : 397–401

Crowe, K. M., Hague, J. F., Rothberg, J. E., Schenck, A., Williams, D. L., Williams, R. W., Young, K. K. 1972. *Phys. Rev. D* 5 : 2145–61

Cvitanovic, P., Kinoshita, T. 1974. *Phys. Rev. D* 10 : 4007–31

Danby, G., Gaillard, J. M., Goulianos, K., Lederman, L. M., Mistry, N., Schwartz, M., Steinberger, J. 1962. *Phys. Rev. Lett.* 9 : 36–44

de Pagter, J., Sard, R. D. 1960. *Phys. Rev.* 118 : 1353–63

Drumm, H., Eck, C., Petrucci, G., Rúnolfsson, O. 1979. The storage ring magnet of the third muon $(g-2)$ experiment at CERN. *Nucl. Instrum. Methods* 158 : 347–62

Einstein, A. 1905. *Ann. Phys.* 17 : 891–921

Farago, P. S. 1965. *Adv. Electron. Electron Phys.* 21 : 1–66

Farley, F. J. M. 1962. Proposed high precision $(g-2)$ experiment. *CERN Intern. Rep. NP/4733*

Farley, F. J. M. 1964. *Progr. Nucl. Phys.* 9 : 259–93

Farley, F. J. M. 1968. *Cargèse Lect. Physics* 2 : 55–117

Farley, F. J. M. 1969. *Proc. First Int. Conf. Eur. Phys. Soc., Florence, 1969;* 1969. *Riv. Nuovo Cimento* 1 : 59–86

Farley, F. J. M. 1972a. *Phys. Lett. B* 42 : 66–68

Farley, F. J. M. 1972b. *Atomic Masses and*

Fundamental Constants, ed. J. H. Sanders, A. H. Wapstra, 4 : 504–8. New York : Plenum

Farley, F. J. M. 1975. *Contemp. Phys.* 16 : 413–41

Farley, F. J. M., Bailey, J., Brown, R. C. A., Giesch, M., Jöstlein, H., van der Meer, S., Picasso, E., Tannenbaum, M. 1966. *Nuovo Cimento* 45 : 281–86

Feinberg, G., Lederman, L. M. 1963. *Ann. Rev. Nucl. Sci.* 13 : 431–504

Feynman, R. P. 1962. La théorie quantique des champs. *Proc. Solvay. Conf. 12th, 1961*, p. 61. New York : Interscience & Brussels ; Stoop

Field, J. H., Fiorentini, G. 1974. *Nuovo Cimento A* 21 : 297–328

Field, J. H., Picasso, E., Combley, F. 1979. *Sov. Phys. Usp.* 127 : 553–92 (In Russian)

Fierz, M., Telegdi, V. L. 1970. In *Quanta*, ed. P. G. O. Freund, C. J. Goebel, Y. Nambu, pp. 196–208. Chicago Univ. Press, Ill.

Flegel, W., Krienen, F. 1973. *Nucl. Instrum. Methods* 113 : 549–560

Franken, P. A., Liebes, S. Jr. 1956. *Phys. Rev.* 104 : 1197–98

Friedman, J. I., Telegdi, V. L. 1957. *Phys. Rev.* 105 : 1681–82

Fujikawa, K., Lee, B. W., Sanda, A. S. 1972. *Phys. Rev. D* 6 : 2923–43

Garwin, R. L., Lederman, L. 1959. *Nuovo Cimento* 11 : 776–80

Garwin, R. L., Lederman, L., Weinrich, M. 1957. *Phys. Rev.* 105 : 1415–17

Granger, S., Ford, G. W. 1972. *Phys. Rev. Lett.* 28 : 1479–82

Granger, S., Ford, G. W. 1976. *Phys. Rev. D* 13 : 1897–1913

Hafele, J. C., Keating, R. E. 1972. *Science* 177 : 166–70

Hansch, T. W., Nayfeh, M. H., Lee, S. A., Curry, S. M., Shahin, I. S. 1974. *Phys. Rev. Lett.* 32 : 1336–40

Henry, G. R., Schrank, G., Swanson, R. A. 1969. *Nuovo Cimento A* 63 : 995–1000

Henry, G. R., Silver, J. E. 1969. *Phys. Rev.* 180 : 1262–63

Hirokawa, S., Komori, H. 1958. *Nuovo Cimento* 7 : 114–15

Hofstadter, R. 1975. Quantum electrodynamics in electron-positron systems. In *Proc. 7th Int. Symp. on Lepton and Photon Interactions at High Energies, Stanford Univ.*, ed. W. T. Kirk, pp. 869–913. Stanford, Calif. : SLAC

Hughes, V. W., Kinoshita, T. 1977. Electromagnetic properties and interactions of muons. In *Muon Physics*, ed. V. W. Hughes, C. S. Wu, 1 : 11–199. New York : Academic

Hughes, V. W., McColm, P. W., Ziock, K., Prepost, R. 1960. *Phys. Rev. Lett.* 5 : 63–65

Hutchinson, D. P., Menes, J., Patlach, A. M., Shapiro, G. 1963. *Phys. Rev.* 131:1351–67

Isham, C. J., Salam, A., Strathdee, J. 1971. *Phys. Rev. D* 3:1805–17

Jackiw, R., Weinberg, S. 1972. *Phys. Rev. D* 5:2396–98

Jackson, J. D. 1975. *Classical Electrodynamics*, p. 559. New York, London: Wiley

Jackson, J. D. 1977. *CERN Rep. 77-17*

Kadyshevsky, V. G. 1978. *Fermi Lab. Publ. 78/70 THY*

Karplus, R., Kroll, N. 1950. *Phys. Rev.* 77:536–49

Kim, C. Y., Kaneko, S., Kim, Y. B., Masek, G. E., Williams, R. W. 1961. *Phys. Rev.* 122:1641–45

Kinoshita, T. 1978. Recent developments of quantum electrodynamics. *Rep. CLNS-410*, presented at the 19th Int. Conf. High Energy Physics, Tokyo, Japan, Aug. 1978

Kobzarev, I. Yu. Okun, L. B. 1961. *Zh. Eksp. Teor. Fiz.* 41:1205–14 (Transl. 1962. *Sov. Phys. JETP* 14:859–65)

Kroll, N. 1966. *Nuovo Cimento* 45:65–92

Landau, L. 1957. *Nucl. Phys.* 3:127–31

Lautrup, B. E. 1972. *Phys. Lett. B* 38:408–10

Lautrup, B. E., Petermann, A., de Rafael, E. 1972. *Phys. Rep. C* 3:193–259

Lautrup, B. E., de Rafael, E. 1974. *Nucl. Phys. B* 70:317–50

Lautrup, B. E., Samuel, M. A. 1977. *Phys. Lett. B* 72:114–16

Lederman, L. M., Tannenbaum, M. J. 1968. *Adv. Part. Phys.* 1:1–67

Lee, T. D., Yang, C. N. 1956. *Phys. Rev.* 104:254–58

Lee, T. D., Wick, G. C. 1969. *Nucl. Phys.* B9:209–43

Levine, M. J., Perisho, R. C., Roskies, R. 1976. *Phys. Rev. D* 13:997–1002

Levine, M. J., Roskies, R. 1976. *Phys. Rev. D* 14:2191–92

Masek, G. E., Ewart, T. E., Toutonghi, J. P., Williams, R. W. 1963. *Phys. Rev. Lett.* 10:35–39

Masek, G. E., Heggie, L. D., Kim, Y. B., Williams, R. W. 1961. *Phys. Rev.* 122:937–48

Mathews, J. 1956. *Phys. Rev.* 102:270–74

Mendlowitz, H., Case, K. M. 1955. *Phys. Rev.* 97:33–38

Mitra, A. N. 1957. *Nucl. Phys.* 3:262–72

Nelson, D. F., Schupp, A. A., Pidd, R. W., Crane, H. R. 1959. *Phys. Rev. Lett.* 2:492–95

Newman, D., Ford, G. W., Rich, A., Sweetman, E. 1978. *Phys. Rev. Lett.* 40:1355–58

Pais, A., Primack, J. R. 1973. *Phys. Rev. D* 8:3063–74

Panofsky, W. K. H. 1958. *Proc. 8th Int.*

Conf. High Energy Physics, CERN, Geneva, ed. B. Ferretti, pp. 3–19. Geneva: CERN

Perl, M. L., Abrams, G. S., Boyarski, A. M., Breidenbach, M., Briggs, D. D., Bulos, F., Chinowsky, W., Dakin, J. T., Feldman, G. J., Friedberg, C. E., Fryberger, D., Goldhaber, G., Hanson, G., Heile, F. B., Jean-Marie, B., Kadyk, J. A., Larsen, R. R., Litke, A. M., Luke, D., Lulu, B. A., Lüth, V., Lyon, D., Morehouse, C. C., Paterson, J. M., Pierre, F. M., Pun, T. P., Rapidis, P. A., Richter, B., Sadoulet, B., Schwitters, R. F., Tanenbaum, W., Trilling, G. H., Vannucci, F., Whitaker, J. S., Winkelmann, F. C., Wiss, J. E. 1975. *Phys. Rev. Lett.* 35:1489–92

Petermann, A. 1957a. *Helv. Phys. Acta* 30:407–8

Petermann, A. 1957b. *Phys. Rev.* 105:1931

Picasso, E. 1967. *Methods Subnucl. Phys.* 3:499–539

Picasso, E. 1970. In *High Energy Physics and Nuclear Structure*, ed. S. Devons, pp. 615–35. New York: Plenum

Prescott, C. Y., Atwood, W. B., Cottrell, R. L. A., DeStaebler, H., Garwin, E. L., Gonidec, A., Miller, R. H., Rochester, L. S., Sato, T., Scherden, D. J., Sinclair, C. K., Stein, S., Taylor, R. E., Clendenin, J. E., Hughes, V. W., Sasao, N., Schüler, K. P., Borghini, M. G., Lübelsmeyer, K., Jentschke, W. 1978. *Phys. Lett. B* 77:347–52

Primack, J., Quinn, H. R. 1972. *Phys. Rev. D* 6:3171–78

Purcell, E. M., Ramsey, N. F. 1950. *Phys. Rev.* 78:807

Rich, A. 1968. *Phys. Rev. Lett.* 20:967–71

Robiscoe, R. T. 1968a. *Phys. Rev.* 168:4–11

Robiscoe, R. T. 1968b. *Cargèse Lect. Physics* 2:3–53

Robiscoe, R. T., Shyn, T. W. 1970. *Phys. Rev. Lett.* 24:559–62

Rossi, B., Hall, D. B. 1941. *Phys. Rev.* 59:223–28

Salam, A. 1968. In *Elementary Particles*, ed. N. Svartholm, pp. 367–77. Stockholm: Almqvist & Wiksells

Samuel, M. A. 1974. *Phys. Rev. D* 9:2913–19

Samuel, M. A., Chlouber, C. 1976. *Phys. Rev. Lett.* 36:442–46

Schupp, A. A., Pidd, R. W., Crane, H. R. 1961. *Phys. Rev.* 121:1–17

Schwinger, J. S. 1957. *Ann. Phys. NY* 2:407–34

Schwitters, R. F., Strauch, K. 1976. *Ann. Rev. Nucl. Sci.* 26:89–149

Sommerfield, C. M. 1957. *Phys. Rev.* 107:328–29

Suura, H., Wichmann, K. 1957. *Phys. Rev.* 105:1930

Van Dyck, R. S. Jr., Schwinger, P. B.,

Dehmelt, H. G. 1977. *Phys. Rev. Lett.* 38:310–14

Weinberg, S. 1967. *Phys. Rev. Lett.* 19: 1264–66

Wesley, J. C., Rich, A. 1970. *Phys. Rev. Lett.* 24:1320–25

Wesley, J. C., Rich, A. 1971. *Phys. Rev. A* 4:1341–63

Wilkinson, D. T., Crane, H. R. 1963. *Phys. Rev.* 130:852–63

Wu, C. S., Ambler, E., Hayward, R. W., Hoppes, D. D., Hudson, R. P. 1957. *Phys. Rev.* 105:1413–15

Ziock, K., Hughes, V. W., Prepost, R., Bailey, J. M., Cleland, W. E. 1962. *Phys. Rev. Lett.* 8:103–5

Ann. Rev. Nucl. Part. Sci. 1979. 29 : 283–312

DIAGNOSTIC TECHNIQUES   ✕5608
IN NUCLEAR MEDICINE

Ronald D. Neumann, M.D. and Alexander Gottschalk, M.D.

Section of Nuclear Medicine, Department of Diagnostic Radiology,
Yale University School of Medicine, New Haven, Connecticut 06510

CONTENTS

INTRODUCTION

Nuclear medicine is the medical specialty in which the physician employs radiopharmaceuticals in the diagnosis and treatment of disease. In the brief period of thirty years nuclear medicine has expanded from a few isolated radionuclide tracer studies in human beings to the present widespread clinical application of radiopharmaceuticals and imaging systems in virtually all human illnesses. The phenomenal growth of this medical specialty has resulted from the development of clinically useful radiopharmaceuticals and an evolution of imaging systems from the advanced technology of modern physics and engineering. The diversity of medical interest in this evolving specialty is exemplified by the establishment of the American Board of Nuclear Medicine, created through the efforts of physicians with prior certification in radiology, pathology, or internal medicine. The purpose of this conjoint specialty board is to insure proper training of physicians in the use of radiopharmaceuticals, to promote the

0163-8998/79/1201-0283$01.00

development of clinical procedures that will facilitate diagnosis and treatment of disease, and to stimulate research in the use of radionuclides in understanding basic pathophysiological mechanisms in the human being. Nuclear medicine specialists use the advances of modern technology to enhance the physician's interaction with the patient.

A radiopharmaceutical is any sterile, pyrogen-free preparation (organic or inorganic) containing a radionuclide that is used for diagnosis or therapy in the practice of nuclear medicine. Very few of the more than 1100 radioisotopes are in clinical use. A typical radiopharmaceutical should not exhibit physiologic or pharmacologic effects on the patient; it should act as a "tracer." The radioisotopes best suited for diagnostic nuclear medicine have gamma-ray energies between 50 and 500 keV and produce a usable photon flux without causing excessive tissue irradiation. Desirable properties also include suitable half-lives, easy availability, and relatively low cost. Physical decay characteristics of the radionuclide primarily determine the clinical detectability and patient irradiation. The chemical form of the tracer determines the physiologic distribution within the patient. Carrier-free radiopharmaceuticals contain no stable species of the radionuclide and are preferred because only nanogram amounts of the element are then introduced into patients. Preparations with high specific activity (activity per unit volume) are desired to keep administered volumes as small as possible while maintaining high photon fluxes for imaging. Radiopharmaceuticals are characterized by the physical half-life of decay of the radionuclide, biologic half-life of clearance from the body of the pharmaceutical form, and the effective half-life or time necessary for the radiopharmaceutical with its associated radioactivity to be reduced by one half in vivo. The general equation is

$$\frac{1}{T_{1/2} \text{ effective}} = \frac{1}{T_{1/2} \text{ biological}} + \frac{1}{T_{1/2} \text{ physical}}.$$

The basic scheme for detection systems in nuclear medicine consists of a shielded detector equipped with a collimator, a high voltage power supply, an amplifier, a pulse-height analyzer, and a display system. Current equipment is often interfaced to a dedicated computer. The detector shielding and the collimator are designed to eliminate most extraneous radiation, allowing detection of photons from a predetermined geometry. The detector requires a high voltage power supply and an amplifier to increase the output voltage from the detector. The amplifier output is sent through a pulse-height analyzer to assure that only photopeak photons are imaged. Many different display systems are in current use, usually with a recording device to produce a permanent record of the information.

INSTRUMENTATION FOR CURRENT DIAGNOSTIC TECHNIQUES

The Anger scintillation camera is the principal imaging device now employed in clinical nuclear medicine. The basic design of the camera was developed by H. O. Anger in the 1960s. Current commercial versions are not significantly different except for improvements in electronics and modification of the crystal size to enlarge the field of view. The camera's detector may be used with a variety of collimators depending upon the energy of the imaging photon and the relative resolution desired in the image.

The gamma ray coming from the patient is collimated and then allowed to interact with the thallium-activated sodium iodide crystal in the camera head (see Figure 1). This interaction produces a scintillation of light that is detected by an array of photomultiplier tubes optically coupled to the back of the crystal. The 3-eV photons from the scintillation are seen by multiple phototubes; those closest to the site of scintillation receive the most light while the tubes farthest away "see" the least amount of light. The individual photomultiplier tube produces a voltage pulse that is proportional to the amount of light detected by the tube. The pattern of these voltage pulses from the total array of photomultiplier tubes is designed to give the position coordinates for the location of the scintillation event in the detector crystal.

An analogue computer then analyzes the pattern of voltage pulses and produces three output signals. Two of these signals are termed the X and the Y positioning signals; their function is to deflect the beam of electrons in the cathode ray tube (CRT) of the readout device to reproduce the position of the original gamma-ray scintillation in the detector crystal. The third, or Z, signal is a sum of the individual pulses from all the photomultiplier tubes, and is proportional to the initial energy of the gamma-ray from the patient. The Z signal is passed through a pulse-

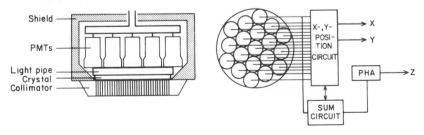

Figure 1   Schematic diagram of a gamma camera detector.

height analyzer, which is preset to pass only the photopeak energies of the isotope used in the radiopharmaceutical. For a scintillation event with the proper Z signal, the electron beam in the display cathode ray tube is activated and positioned according to the X and Y signals. A dot of light is seen on the face of the CRT in the exact position of the original scintillation in the NaI detector. The detector crystal is round as is the face of the CRT so that the light pattern produced by multiple events is a miniature display of the photons coming from the patient. The CRT display is then recorded, usually by any of a variety of film techniques. This final film "scintigram" reflects the distribution of the radiopharmaceutical in the patient as projected onto the detector crystal after collimation (Gottschalk & Hoffer 1976).

The collimator of the Anger scintillation camera is functionally analogous to the lens in a traditional light camera. A photographic lens changes the entering light rays by refraction but the collimator is constructed to allow radiation to enter the detector from a selected geometric field. The collimator merely allows gamma rays to pass through its single or multiple holes while absorbing most other radiation in the lead from which it is made. Although other high density (high atomic number) materials might make better collimators, lead is usually the best compromise between cost and absorption ability.

Four basic collimator designs are in current clinical use with the Anger camera. Single-hole (pinhole) collimators and converging (or multiple pinhole) collimators are used to magnify the organ image. These also produce the best resolution since the image is magnified with individual gamma rays separated over a larger area in the sodium iodide crystal. But these collimators are the least sensitive and, therefore, require longer imaging times. Parallel array, multihole collimators generally provide the best combination of resolution and sensitivity. Finally, multihole diverging collimators are designed to image large organs or areas in a single projection with the smaller crystal size cameras (ten to twelve inches in diameter). Diverging collimators have decreased sensitivity and resolution but the increased field of view is useful for some medical applications.

Collimators for the Anger camera are generally designed to have maximum resolution and sensitivity for a particular gamma-ray energy. Since 99m-technetium is the prevalent radionuclide in clinical use, most collimators are designed for the 140-keV gamma ray of technetium. But for patient studies using 131-iodine, 67-gallium, or 111-indium, collimators with thicker septa (more lead absorber) are necessary to prevent the higher energy gamma rays of these isotopes from entering the detector crystal uncollimated. Any septal penetration reduces image contrast. The collimator is currently the weakest link in the chain of components of the Anger camera; the collimator limits detector sensitivity, spatial resolution, and depth of field.

Commercial gamma cameras use thallium-activated sodium iodide as the scintillation material for the crystal. Pure sodium iodide is not an adequate scintillator so 0.1 to 0.5% thallium iodide is introduced into the crystalline structure to increase the light output. Single large crystals are grown from a melt and then fabricated into detector crystals. Since this material is hygroscopic the finished detector crystal is encased in a thin metallic container with special entrance and exit windows. The metallic shell excludes extrinsic light, prevents hygroscopic discoloration of the crystal, and protects the crystal from physical damage. The detector crystals used in commercial Anger cameras are generally between eleven and fifteen inches in diameter and one fourth to one half inch in thickness. These thin crystals, while not efficient for high energy gamma rays, have almost a 90% photopeak fraction for 99m-Tc; and like the collimator represent a compromise between structure and resolution (Lange 1973).

Approximately twenty to thirty 3-eV light photons are produced for each keV of energy transferred to the crystal (Hine 1967). These light photons pass freely through the NaI (Tl) crystal since it is relatively transparent to light photons of this energy. However, the number of visible light photons produced in the scintillation crystal represents only a portion of the original gamma photon energy transferred to the crystal.

The surface of the detector crystal abutting the photomultiplier tubes is sealed by an optically transparent material and is optically coupled to the photomultiplier tube (PMT) array usually by a disk of lucite. The lucite "light pipe" provides a directed path for the scintillation light from the crystal to the photocathodes of the photomultiplier tubes. Even with a highly polished crystal back, use of transparent cements of appropriate refraction indices, and reflective coatings on all other crystal surfaces, only about 30% of the light photons reach the photomultiplier tubes (Ramm 1966).

The photomultiplier tubes convert the 3-eV light photons from the scintillation into an electron stream that undergoes a series of multiplications to emerge as the output voltage pulse. Current photomultiplier tubes are vacuum tubes containing a photocathode such as cesium-antimony and a series of dynodes to multiply the number of electrons. At each dynode a number of secondary electrons are released for each incident electron, producing an overall gain of about 2×10^6 electrons per photoelectron. The high voltage supply to the photomultiplier tubes is rigidly controlled since the gain is dependent upon the applied voltage.

At the PMT anode the electrons are passed through a load resistor to form the output voltage pulse signal. Commercial Anger cameras now commonly employ arrays of 19 to 91 photomultiplier tubes arranged in a hexagonal array.

The output signals from the PMT array enter an electronic processing unit. It amplifies the shape of the voltage pulse to determine by pulse-height analysis whether or not an individual pulse represents an acceptable photon, and it records all significant detected photons. Preamplifiers may be present between the PMT outputs and the electronic processing unit. Pulse-shaping circuits are also present in the system to avoid "pulse pileup" at high count rates. On the basis of photon energy, the pulse-height analyzer discriminates between primary photons, those that have undergone predetection scatter, and background photons. Throughout the system electronic proportionality is maintained between energy transferred to the crystal by a photopeak gamma photon and the height of the voltage pulse reaching the pulse-height analyzer. The pulse-height analyzer has a lower level discriminator and an upper level discriminator bracketing the limits of acceptable voltage, i.e. creating a "window" about the photopeak. Various instrument manufacturers use either independent upper and lower discriminators or linked discriminators with a fixed window of acceptance. In most current cameras the window width is approximately plus or minus ten percent.

A variety of image processing systems have been developed and interfaced with detection systems used in nuclear medicine. An example is the dedicated minicomputer used to make point-to-point correction for camera non-uniformity over the entire cystal face. In this system a reference matrix is collected from a uniform activity source, stored, and then used to correct each matrix point in subsequent clinical images.

In general, analogue recording systems have been used to produce the scintigram interpreted by the nuclear medicine physician. The intensity gradients of analogue displays are usually proportional to the output, so that, for example, variation in the blackness of an organ image on radiographic film is analogous to the variation in the amount of gamma rays emanating from the organ. More recent trends toward computer manipulation and analysis of nuclear medicine procedures have led to digital signal display systems.

Several refinements and additions to the basic Anger gamma camera have extended the versatility of the instrument. A need for imaging larger areas of the body led to the production of wide-field-of-view (WFOV) scintillation cameras. These instruments are designed around a detector crystal approximately 39 cm in diameter, in contrast to the 25-cm diameter crystals used in the standard size cameras. The WFOV cameras have nearly twice the crystal area for photon detection, but most, like the smaller cameras, use 37 photomultiplier tubes with larger diameters. As a result, larger crystal volumes are monitored by each PMT. This results in slightly less intrinsic resolution in most WFOV systems as compared to the standard cameras. The minimal loss in resolution is acceptable since the WFOV cameras provide a significantly larger viewing area per image, which decreases imaging time for whole-body studies. These cameras also allow certain studies such as lung scans to include the entire organ in each image (Murphy & Burdine 1977).

Physiologic motion of organs during breathing can cause significant degradation of image quality. This is analogous to the image blurring in conventional photography when slow camera shutter speeds are used. The "shutter speed" of the gamma camera may be several minutes for a particular image in a clinical study. In 1971 Oppenheim described a digital method of correcting hepatic scintigrams for breathing motion (Oppenheim 1971). Hoffer et al (1972) designed an analogue version in which motion correction was achieved on-line by means of a relatively simple and inexpensive device. This motion-correction device is now commercially available. When used with the scintillation camera the device appears to be an effective method for improving hepatic scintigraphy, which is most degraded by respiratory movement (Turner et al 1978).

This device is attached to the gamma camera cathode ray tube. The Y-axis positioning signals are integrated over a short time interval. After count rate correction is made, a dc signal is produced with voltage proportional to the displacement of the centroid of liver radioactivity along the Y axis of the detector. A bias voltage proportional in magnitude, but opposite in polarity, is applied to the Y-axis deflection of incoming counts, which in effect displaces the counts in the direction opposite to that of the displacement of the hepatic activity centroid from the center of the detector. The net effect of this electronic motion correction is to hold the liver image stationary on the CRT display in spite of Y-axis motion of the liver caused by respiratory motion.

Mobile gamma cameras have been commercially available for several years. Most are modified versions on wheels of the standard-field-of-view camera. The performance characteristics are similar to the stationary units. But rapid clinical advances in such areas as cardiovascular nuclear medicine created the need for mobile cameras with on-board computer systems. Such devices can be taken to the bedside of a critically ill patient, and sophisticated physiologic studies of heart function can be undertaken. Within the limitations of electrical safety standards, maneuverability, and the delicate nature of computer disk read-write heads, commercial units are now available that provide comprehensive data acquisition, storage, and analysis at the patient's bedside. The detectors on portable scintillation cameras are supported by a mobile gantry, which permits proper positioning of the detector even when the patient cannot be moved. The data analysis systems are minicomputers with modest memories that are sufficient to store a small number of acquisition and analysis programs in addition to the patient data. Most units are available with magnetic tape or floppy disks for data storage. These mobile cameras can also be manufactured with only data collection and storage modules that record the data on tape or disk for transferral to larger central stationary computers where the data analysis is done.

Both the WFOV cameras and the mobile gamma camera–data analysis systems represent design modifications of the basic Anger scintillation camera to fill particular needs in clinical practice. But a fundamental problem of dimensionality remains: Anger cameras cannot quantify the actual amount of a radionuclide in three dimensions. Photon attenuation and superposition of activity in the target organ with activity in tissues above and below are intrinsic problems with two-dimensional images. Tomographic techniques for minimizing or removing this superposition of information emerged from the work of Kuhl & Edwards (1963) and Anger (1968). The recent introduction of practical computerized transmission tomography in conventional radiology has stimulated renewed interest in emission tomography for clinical nuclear medicine (Budinger 1977).

In the 1960s Anger designed a device capable of simultaneously producing images in multiple longitudinal tomographic planes. This design combined the focused collimator with Anger scintillation cameras that could move and thereby scan the patient in a fashion similar to the rectilinear scanner. The commercial version of this instrument (Searle Pho/Con) is now in clinical use and has found particular applicability in total-body gallium scanning (Hauser & Gottschalk 1977).

The Pho/Con has two detectors: Each is composed of an 8.5-inch diameter NaI (Tl) scintillation crystal that is one-inch thick rather than the half-inch crystal used in most gamma cameras. Each crystal has a coupled seven-PMT array. These units function like small scintillation cameras and have the electronics required to provide position coordinates and an energy value for each event detected. The intrinsic resolution of these detectors is not as good as a conventional camera, however, since the inch-thick crystals produce greater light divergence for each scintillation within the crystal, thereby degrading resolution. Each of the two detectors uses a focused collimator. The focal lengths of these collimators are equal. The two detectors are mounted in coupled opposition on arms that provide simultaneous movement in a rectilinear fashion. The patient lies on a table equidistance between the upper and lower detectors (see Figure 2). The detectors are mounted on worm gears equipped with shaft encoders that provide position signals to the electronic processing unit.

The Pho/Con is able to discriminate tomographic levels because of the following characteristics. The geometry of any focused collimator is such that the detector's field of view will decrease as a function of the collimator-to-source distance in the zone between the face of the collimator and its focal plane. Beyond the collimator's focal plane, the field of view for the detector will increase. Any source located in a plane either above or below the focal plane will come into the field of view of

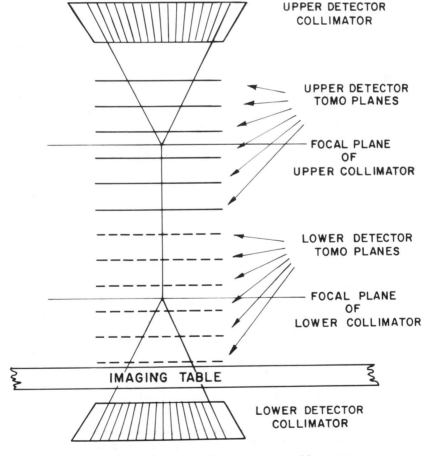

UPPER DETECTOR
COLLIMATOR

UPPER DETECTOR
TOMO PLANES

FOCAL PLANE
OF
UPPER COLLIMATOR

LOWER DETECTOR
TOMO PLANES

FOCAL PLANE
OF
LOWER COLLIMATOR

IMAGING TABLE

LOWER DETECTOR
COLLIMATOR

Figure 2   Schematic diagram of an Anger tomographic scanner.

different areas of the scintillation crystal when the detector is moving. For example, a point source that is located between the detector and the geometric focal plane of the collimator is first imaged by the leading edge of the crystal as the detector approaches the source. As this point source is traversed by the detector, its image moves across the crystal and disappears off the trailing edge. A point source located beyond the focal plane is seen first by the trailing edge of the crystal and "moves" across the crystal in a direction opposite to the detector motion. The speed with which this point source appears to move across the field of view of the crystal also depends upon the location of the source relative to the focal plane. The speed is greatest for a source located at the focal plane and becomes progressively slower as the point source is positioned either above or below the focal plane. Thus the speed of a source's image and the direction of its motion across the scintillation crystal are dependent upon the plane in which it is located. These principles permit image production in which a "slice" of desired thickness at a particular depth will produce a "sharp" or "in-focus" image, while activity sources in planes outside this slice will be deliberately "blurred."

The tomoscanner must also correctly locate the point source in the X-Y plane as well as in depth. This is achieved by a calculation that converts the coordinates of each detected event in the crystal into the true coordinates of photon origin from the patient. This calculation takes into account: (a) the position of the detector at the time the scintillation occurred within the crystal (determined from the shaft encoder signals), (b) the coordinates of the scintillation within the crystal (determined by PMT array signal), and (c) factors defining the tomographic plane within which the coordinates are to be located (determined by speed/depth relationships as well as direction of travel across the crystal).

The Pho/Con then uses this information in a multi-image format device with electronics allowing information display in twelve planar longitudinal tomographic images (six for each detector) simultaneously on dedicated separate sections of the CRT display (Rollo & Hoffer 1977). Because the upper and lower detectors can be adjusted in the vertical plane before a scan is started, the user can preselect the levels from which the twelve focal planes will be produced. Finally, pulse-height analysis is used in each detector to assure that photopeak photons are principally used for image formation.

One other ingenious detector system—the multicrystal camera developed by Bender & Blau (1963)—is in widespread clinical use, particularly in quantitative studies. This camera's detector assembly consists of a mosaic of 294 NaI (Tl) crystals; each crystal is $1\frac{1}{2}$-inches thick and $\frac{5}{16}$-inch square. The 294 crystals are arranged in a 14-by-21 rectangle with a complete size of 6-by-9 inches. Scintillation events occur within the crystals and the light is transferred by 35 separate light pipes. These lucite rods and spatulas are arranged in 14 rows and 21 columns such that each crystal in the mosaic is "viewed" uniquely by two of the 35 light pipes (see Figure 3). Since each light pipe terminates in an individual photo-multiplier tube, a scintillation event produces pulses simultaneously in a specific pair of phototubes. This direct light pipe design provides detailed

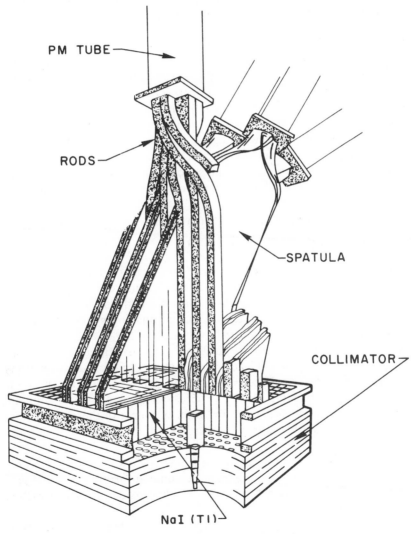

Figure 3 Detector assembly of the multicrystal camera.

positioning information. Compton scatter events are eliminated because counts arising from signals detected simultaneously in more than one crystal are rejected. Parallel-hole collimation is used routinely with the multicrystal camera.

The electronic signals from the PMTs are processed and stored in a dedicated minicomputer, which is part of this instrument. The computer controls data recording, dead-time correction, field uniformity in the detector, and data manipulation in playback. To produce static images, the multicrystal camera has a computer-controlled digital bed that is programmed to move 16 times under the crystal matrix area—2.78 mm per move. This allows an imaging area of 10-by-13 inches with 4704 individual data points. Data are displayed from computer memory on a cathode ray tube that reproduces the crystal matrix. The light intensity in each area on the CRT is proportional to the accumulated events that occurred in the corresponding crystal. The real advantage of the multicrystal camera is its ability to handle high count rates—up to 200,000 counts per second. The newer Anger-type scintillation cameras typically achieve 100,000 counts per second with a 20% window setting. The high count rate capacity is critical in selected dynamic function studies, particularly first-pass techniques in radionuclide angiocardiography.

DEVELOPMENTS IN RADIOPHARMACEUTICALS OF CLINICAL IMPORTANCE

Technetium (99m-Tc) is the most widely used clinical radionuclide. It is suitable because of its physical properties, which include a relative absence of particulate emissions, a six-hour half-life, and its 140-keV gamma photon, which is easily detected by current imaging systems. It decays by isomeric transition to 99-Tc, which has a half-life of 2.12×10^5 years. About 90% of the radiation from 99m-Tc consists of gamma photons with a mean energy of 140.5-keV.[1] Metastable 99m-Tc is also relatively inexpensive since it can be produced by simple separation from its molybdenum parent (99-Mo, $T_{1/2} = 66$ hours), an abundant uranium fission product in nuclear reactors. Another source of the molybdenum parent is production by neutron bombardment of enriched 98-molybdenum oxide.

Two methods of separating 99m-Tc from 99-Mo are currently feasible: solvent extraction and anion-exchange column absorption. The first procedure involves dissolving 99-Mo in sodium hydroxide and then extracting

[1] From the Evaluated Nuclear Structure Data File, Atomic Industrial Forum Steering Committee on Standards, May, 1977.

the 99m-Tc in methyl-ethyl-ketone. Removal of any trace 99-Mo contaminant is achieved by passing this solution through an alumina column. The methyl-ethyl-ketone is then evaporated and the 99m-Tc redissolved in 0.9% saline (physiological saline). The saline solution is sterilized by millipore filtration or autoclaving. The advantages of this solvent extraction separation include both high yields and high specific activity of 99m-Tc. The column absorption method separates 99-Mo from 99m-Tc by absorbing the 99-Mo on an alumina column and eluting the 99m-Tc as its pertechnetate ion with sodium chloride. This is the basis for the 99-Mo/99m-Tc isotopic generator.

Most clinics obtain commercially prepared 99m-technetium generators. These generators consist of a specially designed lead-shielded alumina column containing fission product–produced 99-Mo absorbed on the alumina, which releases 99m-Tc upon elution with sterile 0.9% saline. The lead shield has access ports at the top of the column, allowing aseptic elution and shielded storage (see Figure 4). Elution of the generator as required provides the short-lived 99m-Tc ($T_{1/2} = 6.03$ hr) from the 99-Mo ($T_{1/2} = 66.0$ hr). The activity of the 99m-Tc obtained in the initial elution depends on the original 99-Mo activity of the generator, while the activity obtained in each subsequent elution depends on the time interval between elutions. The United States Nuclear Regulatory Commission sets purity requirements for the fission product 99-Mo used in the generator and sets allowable contaminant levels in the 99m-Tc eluate.

In 1963 P. V. Harper first proposed 99m-technetium for diagnostic use as a thyroid-scanning agent. It has since become the primary radionuclide in diagnostic nuclear medicine. The biologic distribution of the pertechnetate ion of 99m-Tc is similar to that of iodide ion. 99m-TcO$_4^-$ is trapped by the thyroid gland, salivary glands, choroid plexus of the brain, and

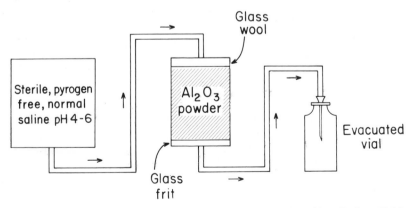

Figure 4   Schematic diagram of the alumina column for separation of 99m-Tc from 99-Mo.

gastric mucosa. This distribution can be altered by pretreatment of the patient with perchlorate either as $NaClO_4$ or $KClO_4$ (Welch et al 1969). Since perchlorate is a negative monovalent ion of approximately the same ionic size as $99m\text{-}TcO_4^-$, the perchlorate blocks tissue uptake of pertechnetate by competitive inhibition. The 99m-Tc pertechnetate is usually administered intravenously but will also be rapidly absorbed from the gastrointestinal tract if given orally. Once in the blood, $99m\text{-}TcO_4^-$ is loosely bound to plasma proteins and subsequently equilibrates within the extravascular space (Lathrop & Harper 1972).

The technetium (as TcO_4^-) is excreted by the kidneys and gastrointestinal system; approximately 30% of an intravenous dose appears in the urine within 24 hours, and slower fecal excretion removes another 40% by four days. Exact percentages of urinary and fecal excretion vary widely between individuals (Lathrop & Harper 1972). The large intestine receives the largest irradiation when pertechnetate is used clinically; approximately 1–2 rads per 10 mCi of 99m-Tc pertechnetate administered.

As a radiopharmaceutical, 99m-technetium is employed in a number of forms, which are discussed as they are currently used to study particular organs in the human. Nearly all current, clinically used radiopharmaceuticals are formulated from nonradioactive kits that are lyophilized, sterile, apyrogenic, and sealed under nitrogen. Appropriate amounts of $99m\text{-}TcO_4^-$ are added to the compounds in the kit to produce a specific radiopharmaceutical labeled with the specified activity of the added pertechnetate. Several compounds of 99m-Tc in use localize in tissues according to the pharmacodynamics of the chemical moiety to which the 99m-Tc is attached. Others are simply particulate matter to which 99m-Tc is attached. These localize by virtue of the particle size and/or phagocytosis by the reticuloendothelial system in the body.

Thyroid and Salivary Glands

Imaging of the thyroid and salivary glands is accomplished by simple trapping of the 99m-Tc pertechnetate ion in the thyroid cells or salivary gland cells. In the thyroid, pertechnetate ion acts as an analogue of iodide; it is taken into the thyroid cell but does not undergo hormonogenesis into tyrosine as would iodide. Twenty minutes after injection, enough $99m\text{-}TcO_4^-$ remains in the thyroid to permit imaging of the gland with a distribution pattern that depicts a variety of pathologic conditions. A similar process occurs in the salivary glands, where the 99m-Tc pertechnetate is secreted in the saliva. Salivary tissue can be imaged to assess structural abnormalities or ectopic location in the body. When physiologic stimulation is added (for example, sucking a lemon drop), the clinician can determine patency of the salivary gland ducts leading to the mouth. Concentration of 99m-TcO$_4^-$ in gastric mucosa allows detection of Meckel's diverticula. These are embryologic remnants attached to the small intestine; they may contain heterotopic gastric mucosa, which may cause ulceration with intestinal bleeding or symptoms resembling acute appendicitis.

In addition to the structural thyroid images made with 99m-Tc pertechnetate, tests of thyroid function are routinely done by the nuclear medicine physician. The radioactive iodine uptake test (RAIU) was one of the earliest medical uses for radionuclides and has continued as a common test of thyroid function. The RAIU test is based upon the obligatory uptake of iodine by the thyroid gland for hormone production. Thyroid tissue has a high affinity for iodine and a concentrating mechanism capable of generating follicular-cell-to-plasma-iodide ratios in the hundreds. The trapped iodide ion is oxidized and incorporated into tyrosine during the hormonogenesis. Enzymatic reactions couple molecules of monoiodotyrosine and diiodotyrosine to form thyroxine. The thyroid follicular cells cannot discriminate between nonradioactive 127-iodine in the diet and the radioactive isotopes administered for diagnostic or therapeutic purposes.

For many years 131-iodine was commonly employed for thyroid function testing and scanning. But its major gamma photon of 364 keV is somewhat high for gamma cameras with standard collimators and crystals. More importantly, 131-I, as it decays by beta-particle emission, gives a high radiation dose to the thyroid. Scanning doses of 50 μCi of 131-I give the thyroid gland between 50 and 100 rads. More expensive cyclotron-produced 125-iodine and 123-iodine are replacing 131-iodine for diagnostic tests. When 125-iodine decays by electron capture, it produces relatively low energy photons that are easily collimated and detected. But its long physical half-life of 60 days results in nearly 50 rads to the thyroid from a typical imaging dose. 123-Iodine has emerged as the best radionuclide for thyroid function tests. The physical half-life of 123-I is 13 hr and its 159-keV photon is suitable for detectors and gamma cameras. Current cyclotron production methods for 123-I have nearly eliminated other long-lived radioiodine contaminants. With scanning doses of 300 μCi 123-I, total radiation to the thyroid is less than 5 rads. Many centers use 123-I in 100-μCi capsules for the radioactive iodine uptake test only and use 99m-TcO$_4^-$ to produce the thyroid scan. This technique further decreases thyroid irradiation and uses the superior imaging properties of 99m-technetium for optimum resolution in the thyroid scan.

Brain

Radionuclide brain scans are valuable for detecting focal neurologic disease by a combination of static and dynamic imaging. We can use 99m-Tc pertechnetate as the agent since its initial vascular phase can be imaged after a bolus of the radionuclide is injected intravenously. Rapid sequence images (about three seconds each) are obtained as the bolus traverses the arteries, capillaries, and veins of the cranium. Suspected vascular lesions may be detected as alterations from the normal flow pattern. Most brain lesions cause a defect in the "blood brain barrier," which allows the radionuclide to enter the extravascular space of the lesion in greater amounts than in the normal brain parenchyma. This relative difference between the lesion and its surroundings permits detection of the lesion as a "hot spot" on the static image.

Since the normal accumulation of pertechnetate by the parotid salivary glands and choroid plexus in the brain ventricles would confuse interpretation of the cranial images, perchlorate is given prior to the 99m-Tc pertechnetate to block these tissues. Static images are usually made several hours after injection of the radionuclide to allow time for maximal lesion-to-brain contrast ratios of radioactivity. A wide variety of radiopharmaceuticals other than 99m-Tc pertechnetate have been used for brain imaging, but the diagnostic accuracy of the procedure has remained unchanged. Approximately 85% of verified brain neoplasms are detected by radionuclide imaging. The recent introduction of computerized x-ray transmission techniques into clinical medicine has had a mixed impact on the use of the radionuclide brain scan. It must be remembered that while radionuclide brain scans are relatively sensitive for lesion detection they are nonspecific. Neoplasms, vascular lesions, malformations, and inflammatory processes all present areas of increased activity on the scan. The computerized reconstruction techniques of transmission x-ray scanners (CAT scan, ACTA scan, etc) provide better anatomic images with greater specificity but may prove to have comparable sensitivity. That is, the improved theoretical anatomic resolution has not always proven to give better results (in terms of percentage) of lesion detection in actual patient care situations.

Bone

Bone-scanning radiopharmaceuticals depend upon the chemical moiety to which 99m-Tc is attached and upon relative blood flow to detect lesions. Early studies used "bone seekers" such as 85-strontium, 87m-strontium, and 18-fluorine that relied on an ion exchange at the surface of the hydroxyapatite crystal in bone matrix. Increased localization of these radionuclides was noted in almost all osseous lesions including neoplasms, metabolic disorders, osteomyelitis, and healing fractures. A major development occurred in 1972 when Subramanian was able to form a 99m-Tc sodium tripolyphosphate complex by using stannous chloride as the reducing agent (Subramanian et al 1972). The use of technetium-labeled compounds allowed better use of Anger camera imaging with its increased resolution and faster images. The previously used radionuclides with their gamma photons of high energy required thick collimator septa and thick detector crystals (usually only available on rectilinear scanners); the Anger cameras have a marked sensitivity loss with the high energy photons. Subsequently, other inorganic (pyrophosphate) and organic (diphosphonate) compounds have been labeled with 99m-Tc and used for clinical bone imaging.

The mechanism of bone uptake of the technetium-labeled agents is known to be related to osteogenic activity and relative blood perfusion but little more is proven. Refinement of these newer bone agents has increased blood and soft tissue clearance via urinary excretion so as to provide the highest bone-to-background ratios in the shortest possible time after intravenous injection.

Heart

The polyphosphate and phosphonate bone-scanning agents were found to concentrate in abnormal heart muscle after acute experimentally induced myocardial infarction (Bonte et al 1974). The exact mechanism of localization was not clear but the observations made in the experimental animals suggested that these phosphate bone-scanning agents might be used for positive infarct imaging in humans. The 99m-Tc pyrophosphate concentrates in damaged myocardium more than ten times as avidly as in normal tissue. Infarcts can be imaged 4–12 hours after onset of infarction and scans will remain positive for 7–10 days afterward. Widespread clinical trials in patients soon proved the technique sensitive for detecting myocardial infarcts, but somewhat nonspecific since positive scans were reported in patients with unstable angina, cardiac neoplasms, carditis, and calcified cardiac aneurysms or valves (Donskey et al 1976, Mason et al 1977, Ahmad et al 1976).

Dual radionuclide studies of patients with acute myocardial infarction help to overcome the problem of nonspecificity. This technique involves administration of 201-thallium chloride to delineate normally perfused myocardium, since 201-Tl depends upon myocardial blood flow and cell viability for localization (Strauss et al 1975, Zaret et al 1976). A true acute myocardial infarct should produce a negative zone ("cold spot") in the 201-Tl image corresponding to a positive zone ("hot spot") in the patient's 99m-Tc pyrophosphate study. Such combined radionuclide techniques approach 100% sensitivity for acute myocardial infarct detection (Berger et al 1978a).

The thallium and pyrophosphate studies rely on radionuclides that concentrate in the myocardium. A second group of cardiovascular imaging techniques deal with the evaluation of dynamic cardiac performance. These cardiovascular nuclear medicine procedures use tracers designed to remain within the intravascular space during the time required to obtain data. Two distinct methodologies have evolved. In first-pass quantitative radionuclide angiocardiography, the radionuclide bolus is recorded as it initially passes through the heart and great vessels. The second approach, gated cardiac blood pool imaging, uses the physiologic signal provided by the patient's electrocardiogram to control the temporal sequence of images for data acquisition. Both of these techniques depend upon interface of stationary detector systems (for example, the gamma camera) with dedicated computer systems for both data acquisition and analysis.

The earliest first-pass studies of cardiovascular function used sequential scintillation camera images for visual assessment of cardiac chamber size, anatomic relationships of cardiac chambers and great vessels, and the presence of intracardiac shunts (Kriss et al 1971). Computer techniques were soon added to allow the data manipulation necessary to quantify blood flow anomalies caused by intracardiac shunts (Alazraki et al 1972) and to evaluate left ventricular pumping functions (Van Dyke et al 1972). For these types of calculations, regions of interest are selected and time-activity curves generated by the computer. Because of statistical limitations high activity bolus injections must be recorded to produce valid measurements. Conventional current-generation Anger scintillation cameras are limited by inability to record such high count rates with statistical reliability, but the multicrystal scintillation camera has no such limitations. Marshall et al (1977) used the multicrystal camera and a technique of summing several sequential cardiac cycles to compute a composite representative cardiac systole and diastole. One of the most important measurements of cardiac performance is the left ventricular ejection fraction—the amount of blood pumped into the peripheral circulation by each heart beat. After appropriate correction for background activity, the left ventricular ejection fraction can be calculated from the end-diastolic (maximum volume) and end-systolic (minimum volume) counts as determined from the representative cycle computed in a first-pass study. Left ventricular ejection fractions measured by this radionuclide method have been standardized against the invasive cardiac catheterization measurements that have been the standard in clinical practice (Steele et al 1974, Jengo et al 1978).

In addition to calculation of the left ventricular ejection fraction, the first-pass technique allows determination of mean circumferential fiber shortening (Steele et al 1976), mean left ventricular ejection rate (Marshall et al 1977), and regional cardiac wall motion during left ventricular contraction (Jengo et al 1978). Each first-pass study requires a separate injection of 99m-technetium tracer; this limits the time between sequential studies because of the six-hour half-life of the technetium. By using 99m-Tc sulfur colloid for the initial study, hepatic phagocytosis will clear most of the activity from the circulation so that a second study may be done with 99m-Tc pertechnetate. Radionuclides with an extremely short half-life, such as 181m-iridium (Treves et al 1976) or 178-tantalum (Holman et al 1978), may permit sequential studies within a short period of time.

First-pass radionuclide angiocardiography is suitable for sequential evaluation of cardiac patients because the tests can be performed with low intrinsic variability (Marshall et al 1978). The technique may also be used to assess the effect of drug therapy on cardiac performance (Alexander et al 1979), stress-induced abnormalities due to coronary artery disease (Berger et al 1979), cardiac effects of lung disease (Berger et al 1978b), and residual cardiac function after acute myocardial infarction (Reduto et al 1978).

Gated cardiac blood pool imaging permits measurement of cardiac performance parameters after complete equilibration of a radionuclide tracer within the intravascular blood pool. For this purpose, 99m-technetium is attached to the patient's own red blood cells or to human serum albumin. After the radiotracer is in equilibrium the imaging procedure begins, using the patient's electrocardiogram to sort temporally nuclear image data according to the cardiac function cycle. The data are collected over several hundred cardiac cycles and composite images formed representing specific segments of the cardiac cycle. The diastolic-to-diastolic interval (R-wave-to-R-wave interval on the electrocardiogram) is divided into 24 to 64 segments and counts collected from each interval recorded in the appropriate "bin" during each heart beat. When data acquisition is completed, computer techniques are used to calculate a summed relative ventricular volume curve. After correction for background activity, calculated ejection fractions correspond closely with results from first-pass radionuclide angiography and contrast x-ray angiography (Borer et al 1977, Burrow et al 1977, Green et al 1978).

Cardiac images obtained from generation of the representative gated cardiac cycle can be displayed sequentially in a cinematic format. This allows assessment of regional wall motion abnormalities (Rigo et al 1974b). Gated cardiac scintiphotography may also be used to examine effects of myocardial infarction (Rigo et al 1974a) and congestive heart failure (Nichols et al 1978a,b, Borer et al 1978a).

The gated radionuclide angiographic techniques have been used in conjunction with exercise stress testing (Kent et al 1978), to assess cardiac valvular disease (Borer et al 1978b), and to demonstrate functional improvement from coronary artery bypass surgery (Kent et al 1978).

Another modification of cardiovascular nuclear medicine techniques involves the use of a counting probe controlled by a microprocessor. This is a nonimaging device gated by the patient's electrocardiogram to generate a time-activity curve from the left ventricle. From these data the ventricular ejection fraction and ejection rate can be calculated but this instrument produces no images to use for evaluating wall motion and regional function (Wagner et al 1976, Bacharach et al 1977).

The rapidly developing area of cardiovascular nuclear medicine has provided safer specific techniques for visual diagnosis and quantitative measurement of cardiac performance by using modern nuclear detectors and dedicated minicomputers.

Kidneys

Evaluation of kidney function and renal scanning is now commonly done with 99m-Tc-labeled compounds that are chelating agents. The most common test for the measurement of renal function is the glomerular filtration rate because this parameter remains stable over wide ranges of kidney perfusion pressure and renal blood flow. The renal glomerulus is akin to a semipermeable membrane that allows only materials of small molecular size to enter the urine. But the urine is further adjusted by reabsorption and tubular secretion in the part of the nephron distal to the glomerulus. Accurate measurement of renal function thus becomes a complex and sophisticated procedure. The rate at which the kidney handles 99m-Tc-diethylenetriamine pentaacetic acid (DTPA) chelate approximates the glomerular filtration rate and allows images to be made with Anger cameras. This radiopharmaceutical is administered intravenously as a bolus. Rapid sequence images are made during the initial arterial perfusion of the kideys to give some assessment of the blood flow through the renal arteries. From these data relative blood flow to the right and left kidney, or a transplanted kidney, can be ascertained. Appearance of the 99m-Tc-DTPA in the urine gives a measure of glomerular and renal tubular function.

The metal chelate 2,3-dimercaptosuccinic acid (DMSA) tagged with

99m-Tc has a higher renal parenchymal concentration than the other 99m-Tc complexes and is useful for structural renal imaging. Simple 99m-Tc pertechnetate can be used to give information on renal perfusion but is not concentrated sufficiently in the renal tubules to give useful images of the kidney parenchyma. Widely used "compromise" agents include the 99m-Tc-labeled saccharides, particularly glucoheptonate, whose characteristics allow adequate perfusion and parenchymal studies. Suffice it to say that the optimum radiopharmaceutical for renal studies with Anger camera systems has yet to be developed.

Liver and Spleen

The largest organ in the body—the liver—and the largest lymph node—the spleen— both present a continuing challenge to the nuclear medicine physician, who needs to assess the structure and function of each organ. Structural images are vital in modern clinical management of cancer since the liver in particular is an early and favored site for metastases. The nuclear medicine physician is asked to detect metastatic disease and monitor any effect of chemotherapy or immunotherapy on the metastatic lesions. Spleen imaging is required for patients with lymphomas, leukemias, melanomas, hematopoietic diseases, and abdominal trauma. Functional assessment of the spleen is limited to monitoring the spleen's role in red blood cell destruction, which may be a manifestation of a number of diseases. Liver function studies are even more limited with current focus on hepatobiliary imaging.

Structural images of both the liver and spleen are commonly made by recording the distribution of radionuclide-labeled colloids that are phagocytized by the reticuloendothelial cells located in these organs. This technique is perhaps the most common noninvasive procedure for evaluation of liver and spleen morphology. The diagnostic accuracy of the test has been reported to be between 70 and 90% depending upon the type of disease one is asked to detect (Covington 1970, Lunia et al 1975, Ruiter et al 1977). The commonly used agent is 99m-technetium sulfur colloid; the probable form is 99m-Tc heptasulfide coprecipitated with colloidal sulfur and stabilized in gelatin. This preparation is injected intravenously and undergoes rapid blood clearance with a circulatory half-time of approximately three minutes. Reticuloendothelial cells phago-cytize the colloid, and in a normal person about 90% of the injected material localizes in the liver. The remaining 10% of colloid is distributed equally between the spleen and bone marrow (Smith 1965).

The relative distribution between liver, spleen, and bone marrow may change in disease or with variable particle size in the colloid. Since the pattern of relative distribution is used as a diagnostic parameter every effort is made to insure uniform particle size during preparation of the radiopharmaceutical. Images of the liver and spleen are made in various anatomic projections. Abnormalities may be demonstrated as focal "cold" zones, i.e. voids in normal colloid distribution, or as diffuse alterations of normal morphology and reticuloendothelial cell geography. For patients with abdominal trauma, radiocolloid liver and spleen scans are reliable for the diagnosis of laceration and/or hematoma in these organs. Sequential scans are used as a means to monitor resolution of such injuries (Gilday & Alderson 1974, Harris et al 1977).

Selective images of splenic tissue alone can be made by using 99m-Tc-tagged red blood cells. This technique uses one of the spleen's natural functions—sequestration of damaged red blood cells—to locate ectopic splenic tissue or to provide an image of the spleen without overlapping liver tissue. The 99m-technetium may be attached to erythrocytes by injecting stannous ion from the FDA-approved pyrophosphate pharmaceutical kit. When 99m-Tc pertechnetate is added, the $99m\text{-}TcO_4^-$ is reduced by the stannous chloride and becomes associated with erythrocyte hemoglobin. If this is done by injecting the 99m-Tc pertechnetate intravenously, a large number of circulating erythrocytes are "tagged," and blood pool images are obtained. But if a small sample of blood is withdrawn after the stannous ion exposure, tagged with $99m\text{-}TcO_4^-$ in vitro, heat damaged, and then reinjected, these damaged, labeled cells will be sequestered by any functioning splenic tissue (R. R. Armas, in preparation). This tissue can then be imaged by Anger camera techniques. Selective splenic images are useful in patients with splenic trauma, ectopic splenic tissue, splenosis, or when the normal liver anatomy obscures the spleen with the conventional colloid scan (Atkins et al 1972, Pearson et al 1978).

Eighty-five percent of the cells in the liver are hepatocytes, which perform the metabolic functions of the liver. Current functional nuclear medicine techniques are directed toward evaluation of bile production and excretion by these cells. The standard test is the 131-iodine rose bengal study in which rose bengal (a phenolphthalein derivative) is selectively extracted from the blood by hepatocytes and excreted into the bile being produced. In practice, rose bengal studies were most often used to assess patency of the biliary duct system and gallbladder function. But because of the high energy photons of 131-iodine, gamma cameras cannot produce optimal images; also, patients receive excessive irradiation from the beta decay of 131-iodine.

Exciting advances in hepatobiliary imaging have recently been made possible by the introduction of 99m-Tc-labeled chemical derivatives of iminodiacetic acid and pyridocylidine glutamate. These new radiopharmaceuticals used in multimillicurie doses permit visualization of the biliary tract in detail unattainable with 200–300-μCi doses of 131-I rose bengal (Weismann et al 1979).

The technetium-labeled hepatobiliary agents are extracted from the blood by hepatocytes, excreted with the bile, accumulate in the gall-bladder, and enter the intestine minutes after intravenous administration. Firnau (1976) described the essential chemical and structural characteristics for hepatobiliary uptake and excretion: a molecular weight of 300–1000, strong polar groups, two or more ring systems in different geometric planes, and a strong affinity for albumin. A number of compounds meet these criteria and are currently undergoing clinical trials to determine which may be best for hepatobiliary imaging (Stadalnik et al 1976, Rosenthall et al 1978, Pauwels et al 1978).

Lung

Ventilation and perfusion lung studies with radionuclides are used to investigate regional lung function and to evaluate patients with suspected pulmonary embolism. By far the largest number of lung scans are done to detect pulmonary emboli. Over 600,000 persons have clinically significant pulmonary emboli each year and nearly 100,000 die from this disease. Perfusion lung scans were introduced in the 1960s and became the primary study in evaluating patients with suspected pulmonary emboli (Wagner et al 1964, Taplin et al 1964).

Perfusion lung scanning with 99m-Tc is accomplished by binding the technetium to carriers with a particle size sufficient to allow particle trapping in the lung precapillaries (35 μm in diameter) or capillaries (8 μm in diameter). Radiopharmaceuticals made of albumin macroaggregates or albumin microspheres have the necessary particle size range and provide uniform particle distribution when injected intravenously. These particles eventually break down to smaller sizes, leave the lungs, and are phagocytized by the cells of the reticuloendothelial system primarily in the liver. This perfusion imaging procedure is safe since only one in several hundred-thousand pulmonary capillaries and one in about 1500 lung arterioles are transiently occluded in the procedure. Regions of the lung without blood flow due to pulmonary emboli will appear as "cold" spots on the perfusion lung images. Multiple views in different projections are done to improve the detection of emboli in the various lung segments (Caride et al 1976).

A normal perfusion lung scan virtually excludes the possibility of a clinically significant pulmonary embolism. But perfectly normal perfusion scans are rarely seen in clinical practice since many lung diseases other than embolism affect pulmonary perfusion. Therefore, most patients require a ventilation study in conjunction with the perfusion scan. Ventilation studies are also used to assess airway disease in patients with emphysema, obstructive pulmonary airway lesions, and central respiratory control abnormalities. Inert radioactive gases are used for ventilation scans. Two isotopes of xenon—133-Xe and 127-Xe—are commonly used. Because 133-xenon has a photon energy (80 keV) less than 99m-Tc (140 keV) the ventilation scan must be done first. The isotope 133-xenon in air is commercially available for use in closed respirometer systems. The patient breathes the mixture through a face mask and serial gamma camera images are taken to show the distribution of the 133-Xe from a single breath, during a rebreathing equilibrium phase, and, after the patient begins to rebreathe room air, early and delayed washout phases. Areas of poor ventilation are identified by absence of activity in the first or single-breath image and by delayed clearance of the gas during the washout phase, when retained activity stands out from the normally ventilated lung (which rapidly clears away the radioxenon).

Recently, inert tracers with photon energies greater than that of 99m-Tc have become available to allow ventilation imaging after the perfusion study. This is a better clinical procedure since perfusion abnormalities may be identified first and then the nuclear physician can select the best image position to study ventilation in the abnormal zone. With energies of 172 keV, 203 keV, 375 keV and a 36.3 day half-life (long shelf life), 127-xenon is advantageous for hospitals that do few ventilation studies but require ready availability for emergency perfusion-ventilation studies. Both 133-Xe ($T_{1/2} = 5.3$ days) and 127-Xe must either be trapped in shielded containers after use, or discharged directly to the outside air via a duct system allowing no recirculation. Because 81m-krypton has a 13-second half-life and a 190-keV photon energy, it is eminently suitable for ventilation studies. But like 127-Xe, 81m-Kr is expensive and because of its short half-life is available only to hospitals close to cyclotron production facilities.

Several clinical studies comparing 127-Xe and 81m-Kr to 133-Xe have shown the potential advantages of these newer gases (Atkins et al 1977, Goris et al 1977, Weber et al 1977).

Total-Body Imaging

The role of serendipity in the progress of nuclear medicine is well demonstrated by the story of 67-gallium as an imaging agent (Hoffer et al 1978). The initial studies of medical uses of this isotope of gallium were done to evaluate its use for bone scanning. Though 67-gallium was found to be a mediocre bone seeker, the carrier-free form showed significant potential as a tumor-imaging radiopharmaceutical (Edwards & Hayes 1969).

Currently, 67-Ga citrate is effectively used as a tumor-localizing agent for Hodgkin's disease, certain non-Hodgkin's lymphomas, lung carcinoma, hepatocellular carcinomas, melanoma, testicular seminomas, and Burkitt's lymphoma. In addition, 67-gallium citrate has been useful, but inconsistent in detecting several other tumors. In general, 67-Ga is a useful adjunct for clinical staging of many tumors; and it is occasionally indispensible for detecting recurrent cancer.

As 67-gallium began to be used widely for tumor imaging, several investigators noted that gallium not only localized in malignant lesions but in inflamed tissues as well (Lavender et al 1971, Littenberg et al 1973). The use of 67-gallium to detect occult inflammatory lesions in the abdomen, lung, bone, kidney, etc, has now been described, and is standard clinical practice (Hoffer et al 1978).

The proton bombardment of 68-zinc-enriched zinc oxide produces 67-gallium. It has a half-life of about 78 hours and decays to stable 67-zinc by electron capture; its principle gamma photons have approximate energies of 93 keV, 195 keV, 296 keV, and 389 keV. After intravenous administration, carrier-free 67-gallium is bound primarily to the plasma protein transferrin. It undergoes equilibration with other plasma proteins and is distributed throughout the body. During the first 24 hours about 10–15% is excreted in the urine and 10% in the feces. Over the next 5–7 days an additional 35% of the total dose administered is excreted primarily via the colon. Normal gallium accumulation occurs in the lacrimal glands, nasopharynx, liver, spleen, kidney, bowel, bone, and bone marrow. Gallium will also accumulate in the breast tissue during hormone stimulation, lactation, and in the postpartum period.

The mechanism for concentration of gallium in tumors and inflammatory lesions is not known. Currently popular hypotheses propose that 67-Ga acts in part as an iron analogue since it binds to proteins such as transferrin, the normal iron transport protein. Transferrin is also known to act as a carrier protein for 67-Ga in the circulation, transporting it to sites of cellular localization. Ito and co-workers (1971) and Winchell (1976) postulated that 67-Ga concentrates in tumors as the result of binding with intracellular tumor proteins, perhaps intracellular ferritin (Clausen et al 1974), or a glycoprotein (Hayes & Carlton 1973), such as lactoferrin (Hoffer et al 1977).

Gallium has proven to be a valuable radiopharmaceutical but is plagued by nonspecificity. Normal uptake occurs in a number of organs and abnormal concentration occurs in both neoplastic and inflammatory lesions. Nonspecificity continues to be the bane of current nuclear medicine techniques.

McAfee & Thakur (1976a,b) made a dramatic breakthrough in a search for an optimum tracer to label leukocytes. Autologous cell imaging gives the nuclear medicine physician a powerful tool. The technique uses the body's own defensive cellular reaction for diagnosis. Oxine proved to be the most attractive agent for leukocyte labeling (Thakur 1977). The commercially available cyclotron-produced radionuclide 111-indium, with a 67-hour half-life and two photons (173 keV and 247 keV), is suitable for gamma camera imaging. Autologous leukocytes labeled with 111-indium oxine have been used successfully in the diagnosis of inflammatory lesions in clinical practice (Doherty & Goodwin 1978). While leukocyte imaging appears to be both specific and sensitive, it does have the disadvantage of requiring preparation of labeled cells that must remain viable and function normally. Leukocytes labeled with 111-indium are currently being studied as a potential acute myocardial infarct–imaging technique (Thakur et al 1979). The same technique has also been employed to tag platelets as well as leukocytes in order to detect the lesion of infective endocarditis in rabbits (Riba et al 1978), and acute coronary artery thrombosis in dogs (A. L. Riba, in preparation). These methods may soon be applicable in humans.

CONCLUSION

Current diagnostic techniques in nuclear medicine are centered around 99m-technetium and the Anger scintillation camera. This instrument in its many forms has replaced the rectilinear scanner as the "workhorse" in most nuclear medicine clinics. The multiplicity of radionuclides used in the early decades of nuclear medicine have now been replaced by a multiplicity of radiopharmaceuticals that contain 99m-technetium.

Two major developments in the practice of nuclear medicine seem likely in the near future. First, the instrumentation and procedures developed to measure cardiovascular function will find broad application in practice at all levels. The incidence of cardiovascular disease in modern society is great and nuclear diagnosis will be used for more patients each year. Cardiovascular diagnostic procedures outlined in this chapter are applicable in both the initial assessment of the type and extent of disease in an individual, and can be used sequentially to monitor the effects of medical or surgical therapy. The instrumentation, radiopharmaceuticals, techniques, and trained personnel are no longer confined to university research centers.

The second major growth will occur with the commercial production of emission-computed tomographic instruments and hospital cyclotrons for production of short-lived positron-emitting compounds, i.e. positron decay with subsequent annihilation photons of 511 keV after interaction with tissue chemical electron. Now the cost of these instruments restricts such investigation to a limited number of facilities. But their potential value for the study of in vivo metabolism in patients is enormous. The historical development of clinical techniques in nuclear medicine has always been the study of form and function. The future holds the promise of functional diagnoses from the metabolic alterations that foretell the pathophysiology of disease.

ACKNOWLEDGMENTS

The authors wish to express their gratitude to Paul B. Hoffer, Vincent Caride, Robert C. Lange, and Harvey J. Berger for their assistance with materials and comments. We wish to thank Patty DeStefano, Rose Mason, and Gael Maher for preparation of the manuscript.

Literature Cited

Ahmad, M., Dubiel, J. P., Verdon, T. A., Martin, R. H. 1976. Technetium-99m stannous pyrophosphate myocardial imaging in patients with and without left ventricular aneurysm. *Circulation* 53:833

Alazraki, N. P., Ashburn, W. L., Hagan, A., Friedman, W. F. 1972. Detection of left-to-right cardiac shunts with the scintillation camera pulmonary dilution curve. *J. Nucl. Med.* 13:142

Alexander, J., Dainiak, N., Berger, H. J. et al. 1979. Serial assessment of doxorubicin cardiotoxicity with quantitative radionuclide angiocardiography. *N. Engl. J. Med.* 300(6):278

Anger, H. O. 1968. *Fundamental Problems in Scanning*, ed. A. Gottschalk, R. N. Beck, pp. 192–211, Springfield, Ill: Thomas

Atkins, H. L., Eckelman, W. C., Hauser, W., Klopper, J. F., Richards, P. 1972. Splenic sequestration of 99m-Tc-labeled red blood cells. *J. Nucl. Med.* 13:811

Atkins, H. L., Susskind, H., Klopper, J. F., Ansari, A. N., Richards, P., Fairchild, R. G. 1977. A clinical comparison of 127-Xe and 133-Xe for ventilation studies. *J. Nucl. Med.* 18:626

Bacharach, S. L., Green, M. V., Borer, J. S., Ostrow, H. G., Redwood, D. R., Johnston, G. S. 1977. ECG-gated scintillation probe measurement of left ventricular function. *J. Nucl. Med.* 18:1176

Bender, M. A., Blau, M. 1963. The autofluoroscope. *Nucleonics* 21:52

Berger, H. J., Gottschalk, A., Zaret, B. L. 1978a. Dual radionuclide study of acute myocardial infarction: comparison of thallium-201 and technetium-99m stannous pyrophosphate imaging in man. *Ann. Intern. Med.* 88:145

Berger, H. J., Matthay, R. A., Loke, J. et al. 1978b. Assessment of cardiac performance with quantitative radionuclide angiocardiography: right ventricular ejection fraction with reference to findings in chronic obstructive pulmonary disease. *Am. J. Cardiol.* 41:897

Berger, H. J., Reduto, L. A., Johnstone, D. E. et al. 1979. Global and regional left ventricular response to bicycle exercise in coronary artery disease: assessment by quantitative radionuclide angiocardiography. *Am. J. Med.* 66:13

Bonte, F. J., Parkey, R. W., Graham, K. D. et al. 1974. A new method for radionuclide imaging of myocardial infarcts. *Radiology* 110:476

Borer, J. S., Bacharach, S. L., Green, M. V. et al. 1977. Real-time radionuclide cineangiography in the noninvasive evaluation of global and regional left ventricular function at rest and during exercise in patients with coronary artery disease. *N. Engl. J. Med.* 296:839

Borer, J. S., Bacharach, S. L., Green, M. V. et al. 1978a. Effect of nitroglycerin on exercise-induced abnormalities of left ventricular regional function and ejection fraction in coronary artery disease: assessment by radionuclide cineangiography in symptomatic and asymptomatic patients. *Circulation* 57:314

Borer, J. S., Bacharach, S. L., Green, M. V. et al. 1978b. Exercise-induced left ventricular dysfunction in symptomatic and asymptomatic patients with aortic regurgitation: assessment with radionuclide cineangiography. *Am. J. Cardiol.* 42:351

Budinger, T. F. 1977. Instrumentation trends in nuclear medicine. *Semin. Nucl. Med.* 7:285

Burrow, R. D., Strauss, H. W., Singleton, R. et al. 1977. Analysis of left ventricular function from multiple gated acquisition cardiac blood pool imaging. *Circulation* 56:1024

Caride, V. J., Puri, S., Slavin, J. D. et al. 1976. The usefulness of posterior oblique views in perfusion lung imaging. *Radiology* 121:669

Clausen, J., Edeling, C. J., Fogh, J. 1974. 67-Ga binding to human serum proteins and tumor components. *Cancer Res.* 34:1931

Covington, E. E. 1970. The accuracy of liver photoscans. *Am. J. Roentgenol.* 109:742

Doherty, P., Goodwin, D. A. 1978. Indium-111 oxine labeled autologous leukocytes in the diagnosis of inflammatory disease. *J. Nucl. Med.* 19(6):742 (Abstr.)

Donsky, M. S., Curry, C. G., Parkey, R. W. et al. 1976. Unstable angina pectoris: clinical angiographic, and myocardial scintigraphic observations. *Br. Heart J.* 38:257

Edwards, C. L., Hayes, R. L. 1969. Tumor scanning with gallium citrate. *J. Nucl. Med.* 10:103

Firnau, G. 1976. Why do 99m-Tc chelates work for cholescintigraphy? *Eur. J. Nucl. Med.* 1:137

Gilday, D. L., Alderson, P. O. 1974. Scintigraphic evaluation of liver and spleen injury. *Semin. Nucl. Med.* 4:357

Goris, M. L., Daspit, S. G., Walter, J. P. et al. 1977. Applications of ventilation lung imaging with 81m-Krypton. *Radiology* 122:339

Gottschalk, A., Hoffer, P. B. 1976. *Diagnostic Nuclear Medicine*, ed. A. Gottschalk, E. J. Potchen, Chap. 11. Baltimore: Williams & Wilkins

Green, M. V., Brody, W. R., Douglas, M. A. et al. 1978. Ejection fraction by count rate from gated images. *J. Nucl. Med.* 19:880

Harris, B. H., Morse, T. S., Weidenmire, C. H. et al. 1977. Radioisotope diagnosis of splenic trauma. *J. Pediatr. Surg.* 12:385

Hauser, M. F., Gottschalk, A. 1977. Comparison of the Anger tomographic scanner and the 15-inch scintillation camera in gallium imaging. *J. Nucl. Med.* 18:603 (Abstr.)

Hayes, R. L., Carlton, J. E. 1973. A study of macromolecular binding of 67-Ga in normal and malignant animal tissue. *Cancer Res.* 33:3265

Hine, G. J. 1967. *Instrumentation in Nuclear Medicine*, Vol. 1, Chap. 6, New York: Academic

Hoffer, P. B., Oppenheim, B. E., Yasillo, N. J., Sterling, M. L. 1972. Motion correction in liver scanning: description of new device and preliminary clinical results. *J. Nucl. Med.* 13:437 (Abstr.)

Hoffer, P. B., Huberty, J. P., Dhayam-Bashi, H. 1977. The association of 67-Ga and lactoferrin. *J. Nucl. Med.* 18:713

Hoffer, P. B., Beckerman, C., Henkin, R. E. 1978. *Gallium-67 Imaging*, Part 1, pp. 4–6. New York: Wiley

Holman, B. L., Harris, G. I., Neirinckx, R. D. et al. 1978. Tantalum-178: a short-lived nuclide for nuclear medicine: Production of the parent W-178. *J. Nucl. Med.* 19:510

Ito, Y., Okuyama, S., Sato, K. et al. 1971. 67-Ga tumor scanning and its mechanisms studied in rabbits. *Radiology* 100:357

Jengo, J. A., Mena, I., Blaufuss, A., Criley, J. M. 1978. Evaluation of left ventricular function (ejection fraction and segmental wall motion) by single pass radioisotope angiography. *Circulation* 57:326

Kent, K. M., Borer, J. S., Green, M. V. et al. 1978. Effects of coronary artery bypass on global and regional left ventricular function during exercise. *N. Engl. J. Med.* 298:1434

Kriss, J. P., Enright, L. P., Hayden, W. G. et al. 1971. Radioisotopic angiocardiography: wide scope of applicability in diagnosis and evaluation of therapy in diseases of the heart and great vessels. *Circulation* 43:792

Kuhl, D. E., Edwards, R. Q. 1963. Image separation radioisotope scanning. *Radiology* 80:653

Lange, R. C. 1973. *Nuclear Medicine for Technicians*, Chap. 7. Chicago: Year Book Medical

Lathrop, K. A., Harper, P. V. 1972. *Progress in Nuclear Medicine, Neuronuclear Medicine*, Vol. 1, p. 145. Baltimore: University Park

Lavender, J. P., Barker, J. R., Chaudhri, M. A. 1971. 67-Gallium citrate scanning in neoplastic and inflammatory lesions. *Br. J. Radiol.* 44:361

Littenberg, R. L., Taketa, R. M., Alazraki, N. P. et al. 1973. 67-Gallium for the localization of septic lesions. *Ann. Intern. Med.* 79:403

Lunia, S., Parthasarathy, K. L., Bakshi, S., Bender, M. A. 1975. An evaluation of 99m-Tc sulfur colloid liver scintiscans and their usefulness in metastatic workup. *J. Nucl. Med.* 16:62

Marshall, R. C., Berger, H. J., Costin, J. C. et al. 1977. Assessment of cardiac performance with quantitative radionuclide angiocardiography: sequential left ventricular ejection fraction, normalized left ventricular ejection rate, and regional wall motion. *Circulation* 56:820

Marshall, R. C., Berger, H. J., Reduto, L. A. et al. 1978. Variability in sequential measures of left ventricular performance assessed with radionuclide angiocardiography. *Am. J. Cardiol.* 41:531

Mason, J. W., Myers, R. W., Alderman, E. L. et al. 1977. Technetium-99m pyrophosphate myocardial uptake in patients with stable angina pectoris. *Am. J. Cardiol.* 40:1

McAfee, J. G., Thakur, M. L. 1976a. Survey of radioactive agents for in vivo labeling of phagocytic leukocytes. I. Soluble agents. *J. Nucl. Med.* 17:480

McAfee, J. G., Thakur, M. L. 1976b. Survey of radioactive agents for in vivo labeling of phagocytic leukocytes. II. Particles. *J. Nucl. Med.* 17:488

Murphy, P. H., Burdine, J. A. 1977. Large-field-of-view(LFOV)scintillation cameras. *Semin. Nucl. Med.* 7(4):305

Nichols, A. B., McKusick, K. A., Strauss, H. W. et al. 1978a. Clinical utility of gated blood pool imaging in congestive left heart failure. *Am. J. Med.* 65:785

Nichols, A. B., Pohost, G. M., Gold, H. K. et al. 1978b. Left ventricular function during intra-aortic balloon pumping assessed by multigated cardiac blood pool imaging. *Circulation* 58:176

Oppenheim, B. E. 1971. A method using a digital computer for reducing respiratory artifact on liver scans made with a camera. *J. Nucl. Med.* 12:625

Pauwels, S., Steels, M., Piret, L., Beckess, C. 1978. Clinical evaluation of Tc-99m-diethyl-IDA in hepatobiliary disorders. *J. Nucl. Med.* 19:783

Pearson, H. A., Johnston, D., Smith, K. A., Touloukian, R. J. 1978. The born-again spleen. Return of splenic function after splenectomy for trauma. *New Engl. J. Med.* 298:1889

Ramm, W. I. 1966. *Radiation Dosimetry*, ed. F. H. Attix, W. C. Roesch, Vol. 2, Chap. 11, New York: Academic. 2nd ed.

Reduto, L. A., Berger, H. J., Cohen, L. S. et al. 1978. Sequential radionuclide assessment of left and right ventricular performance after acute myocardial infarc-tion. *Ann. Intern. Med.* 89:441

Riba, A. L., Thakur, M. L., Gottschalk, A. et al. 1978. Cellular imaging of experimental endocarditis. *Circulation* 58:131

Rigo, P., Murray, M., Strauss, H. W. et al. 1974a. Left ventricular function in acute myocardial infarction evaluated by gated scintiphotography. *Circulation* 50:678

Rigo, P., Murray, M., Strauss, H. W., Pitt, B. 1974. Scintiphotographic evaluation of patients with suspected left ventricular aneurysm. *Circulation* 50:985

Rollo, F. D., Hoffer, P. B. 1977. Comparison of whole-body imaging methods. *Semin. Nucl. Med.* 7:315

Rosenthall, L., Shaffer, E. A., Lisbona, R., Pare, P. 1978. Diagnosis of hepatobiliary disease by 99m-Tc-HIDA cholescintigraphy. *Radiology* 126:467

Ruiter, D. J., Byck, W., Pauwels, E. K. J. et al. 1977. Correlation of scintigraphy with short interval autopsy in malignant focal liver disease: a study of 59 cases. *Cancer* 39:172

Smith, E. M. 1965. Internal dose calculation for 99m-Tc. *J. Nucl. Med.* 6:231

Stadalnik, R. C., Matolo, N. M., Jansholt, A. L. et al. 1976. Technetium-99m pyridoxylidene glutamate cholescintigraphy. *Radiology* 121:657

Steele, P., Kirch, D., Matthews, M., Davies, H. 1974. Measurement of left heart ejection fraction and end-diastolic volume by a computerized, scintigraphic technique using a wedged pulmonary artery catheter. *Am. J. Cardiol.* 34:179

Steele, P., LeFree, M., Kirch, D. 1976. Measurement of left ventricular mean circumferential fiber shortening velocity and systolic ejection rate by computerized radionuclide angiocardiography. *Am. J. Cardiol.* 37:388

Strauss, H. W., Harrison, K., Langan, J. K. et al. 1975. Thallium-201 in regional myocardial imaging. Relation of thallium-201 to regional myocardial perfusion. *Circulation* 51:641

Subramanian, G., McAfee, J. G., Bell, E. G. et al. 1972. 99m-Tc-labeled polyphosphate as a skeletal imaging agent. *Radiology* 102:701

Taplin, G. V., Johnson, D. E., Dore, E. K. et al. 1964. Lung photoscans with macroaggregates of human serum radioalbumin. *Health Phys.* 10:1219

Thakur, M. L. 1977. Indium-111: a new radioactive tracer for leukocytes. *Exp. Hematol. Copenhagen* 5:145

Thakur, M. L., Gottschalk, A., Zaret, B. L. 1979. Imaging experimental myocardial infarction with indium-111 labeled autologous leukocytes: effects of infarct age and residual regional myocardial blood flow. *Circulation.* Aug.

Treves, S., Kulprathipanja, S., Hnatowich, D. J. 1976. Angiocardiography with iridium-191m: an ultrashort-lived radio-nuclide ($T_{1/2} = 4.9$ seconds). *Circulation* 54:275

Turner, D. A., Fordham, E. W., Amjad, A. et al. 1978. Motion corrected hepatic scintigraphy: an objective clinical evaluation. *J. Nucl. Med.* 19:142

Van Dyke, D., Anger, H. O., Sullivan, R. W. et al. 1972. Cardiac evaluation from radioisotope dynamics. *J. Nucl. Med.* 13:585

Wagner, H. N. Jr., Sabiston, D. C. Jr., McAfee, J. G. et al. 1964. Diagnosis of massive pulmonary embolism in man by radioisotope scanning. *N. Engl. J. Med.* 271:377

Wagner, H. N., Wake, R., Nickoloff, E., Natarajan, T. K. 1976. The nuclear stethoscope: a simple device for generation of left ventricular volume curves. *Am. J. Cardiol.* 38:747

Weber, P. M., dos Remedios, L. V., Schor, R. et al. 1977. 81m-Kr versus 133-Xe for ventilation imaging in suspected pulmonary embolism. *J. Nucl. Med.* 18:625

Weismann, H. S., Frank, M., Rosenblatt, R. et al. 1979. Cholescintigraphy, ultrasonography, and computerized tomography in the evaluation of biliary tract disorders. *Semin. Nucl. Med.* 9:22

Welch, M. J., Adatepe, M., Potchen, E. J. 1969. An analysis of technetium kinetics: the effect of perchlorate and iodine pretreatment. *Int.J. Appl. Radiat. Isot.* 20:437

Winchell, H. S. 1976. Mechanisms for localization of radiopharmaceuticals in neoplasms. *Semin. Nucl. Med.* 6:371

Zaret, B. L., DiCola, V. C., Donabedian, R. K. et al. 1976. Dual radionuclide study of myocardial infarction: relationships between myocardial uptake of potassium-43, technetium-99m stannous pyrophosphate, regional myocardial blood flow, and creatinine phosphokinase depletion. *Circulation* 53:422

Ann. Rev. Nucl. Part. Sci. 1979. 29: 313–37

COSMOLOGY CONFRONTS   ×5609
PARTICLE PHYSICS

Gary Steigman

Bartol Research Foundation of the Franklin Institute, University of Delaware, Newark, Delaware 19711[1]

CONTENTS

1   INTRODUCTION

These are exciting times. The progress that has led to the unification of the weak and electromagnetic forces provides encouragement that full unification, including the strong force and, possibly, gravity, may soon be achieved. These developments in particle physics are followed with great interest by astrophysicists who must use the best available basic physics in their work. At present, there is a growing symbiotic relationship between particle physics and astrophysics. Astronomers are consumers of new ideas in particle physics and, in return, astrophysical considera-

[1] Visiting Scholar, Institute for Plasma Research and Stanford Linear Accelerator Center, Stanford University, Stanford, California 94305.

0163-8998/79/1201-0313$01.00

tions can provide probes of, and constraints on, these new ideas. Indeed, the Universe is a supplement to the accelerator as a laboratory in which new theories may be tested. This approach is quite economical, and often there is no other way to probe specific aspects of particle physics (for example, enormously high energies $\gtrsim 10^{15}$ GeV). To paraphrase Zeldovich[2] (1970), the early Universe is a laboratory for the particle physicist.

It is possible to approach particle physics via cosmology because the temperature and density were very high and collisions were exceedingly frequent during the early evolution of the Universe. A variety of particles were produced sufficiently early and, through rapid interactions, quickly came into equilibrium with each other. Included here are particles that may have avoided discovery thus far because their masses are too high and/or they interact too weakly to have been produced by current accelerators. The early Universe never faced the budgetary constraints that confront accelerator physics!

As the Universe expanded and cooled, the reaction rates began to lag behind the expansion rate; various particle species dropped out of equilibrium, leaving behind "relics" of those early epochs. How many survive and how they affect the subsequent evolution depend on masses, lifetimes and interaction strengths. If, indeed, some "footprints" do remain, the cosmology may be used to probe particle physics. Fortunately, some clues do exist.

Stable or long-lived ($\tau \gtrsim 10^{10}$ yr) particles surviving from the early Universe will be present today. Direct evidence for such particles may be provided by searches for anomalous nuclei, free quarks, and stable charged leptons. This approach is of most value in probing the physics of hadrons, quarks, and charged leptons. Searches for these particles have the greatest chance for success; negative results will lead to the most severe constraints.

All particles interact gravitationally. Therefore, stable or long-lived relic particles will contribute to the present cosmological mass density ρ_0. Observational limits to ρ_0 ($\lesssim 2 \times 10^{-29}$ g cm^{-3}) lead to significant constraints on the masses of neutral leptons ("light" and "heavy" neutrinos) and charged leptons.

Primordial nucleosynthesis provides another, very powerful, probe. The primordial abundance of ^{4}He depends sensitively on the competition between the weak interaction rate and the expansion rate. The expansion rate depends on the number of types ("flavors") of light particles ($M \lesssim T$)

---

[2] Zeldovich's article was entitled "The Universe as a Hot Laboratory for the Nuclear and Particle Physicist."

present when $T \gtrsim 1$ MeV. This approach leads to strong limits to the number of flavors of weakly or superweakly interacting light particles ($M \lesssim 1$ MeV) with lifetimes $\tau \gtrsim 1$ s.

This review explores the approach to particle physics through cosmology; the intent is pedagogic rather than encyclopedic. The goal is to describe the simple basic physics that underlies this approach so that recent and future developments will be comprehensible. The general approach is illustrated with specific examples. However, this is not intended to be a complete survey of the current literature. Rather, a balance is struck, which if successful will provide sufficient breadth to stimulate further inquiry by the reader and sufficient depth to ensure that that effort will be rewarded. It is, for example, beyond the scope of this review to consider in any detail those constraints that may be derived from stellar astrophysics or cosmic ray physics. I apologize in advance to those of my colleagues whose excellent work is not cited here as a result of my effort to provide a thorough introduction rather than an exhaustive survey.

Section 2 is the heart of this review. A quick introduction to cosmology is provided in order that the early evolution of the Universe may be calculated. The conditions under which equilibrium is attained are described and the reader is reminded of some simple results from equilibrium statistical mechanics. Having assembled the necessary machinery, we proceed to calculate particle production and survival (Section 2.3). Virtually everything in this review and, indeed, in this subject follows from the results of Section 2.3. In Section 3 the various cosmological probes are outlined and in Section 4 a brief review is given of the constraints that have emerged from this approach. The salient points of the approach and the interesting results attained thus far are summarized in Section 5.

This review relies to a great extent on the recent article by Schramm & Wagoner (1977) to which the reader is referred for details (and for an excellent review of primordial nucleosynthesis). For a further discussion of cosmology, the reader is referred to the excellent, popular book by Weinberg (1977) and to the author's earlier lecture notes (Steigman 1973). A detailed introduction to cosmology is to be found in Weinberg (1972).

2   PHYSICS OF THE EARLY UNIVERSE

2.1   *Cosmology*

In the cosmological context, a unified treatment of massive and massless, charged and neutral, leptons and hadrons is possible because thermal equilibrium was established in the very earliest epochs. During the subsequent evolution, crucial differences developed among the various particle species as, one by one, they departed from equilibrium. A brief review of the physics of the early Universe will provide a basis for the subsequent discussion.

On sufficiently large scales, the Universe is observed to be remarkably homogeneous and isotropic (see, for example, Schramm & Wagoner 1977). The isotropy of the black body radiation provides evidence that this was true, too, in the distant past. These observations, along with the Copernican principle, have led to the *Cosmological Principle*: The Universe is, and always was, homogeneous and isotropic. An equivalent rephrasing is that, at any fixed epoch, the Universe (on sufficiently large scales) looks the same to all observers. On sufficiently small scales, of course, the Universe deviates from isotropy and homogeneity. It is assumed that these smaller scale, more local departures may be treated as perturbations on the homogeneous isotropic cosmological model in the early Universe.

A Universe satisfying the *Cosmological Principle* is described by a unique metric (see Weinberg 1972), the Robertson-Walker (R-W) metric.

$$ds^2 = c^2\,dt^2 - a^2(t)\left[\frac{dr^2}{(1-kr^2)} + r^2(d\theta^2 + \sin^2\theta\,d\phi^2)\right]. \qquad 1.$$

The expansion is described by the time-dependent scale factor $a(t)$; r, θ, and ϕ are co-moving coordinates; the 3-space curvature is related to the constant k. The Friedman models (or the Robertson-Walker-Friedman models) are described by two independent differential equations that emerge when the R-W metric is substituted in the Einstein equations:

$$\frac{1}{2}\left(\frac{da}{dt}\right)^2 = \frac{GM}{a} + kc^2 + \frac{\Lambda a^2}{3}; \qquad M = \frac{4\pi}{3}a^3\rho, \qquad 2a.$$

$$d(\rho c^2 V) + p\,dV = 0; \qquad V \propto a^3. \qquad 2b.$$

In Equation 2, ρ is the mass-energy density, p is the isotropic pressure, Λ is the cosmological constant, and V is an arbitrary (but definite) co-moving volume. There are three functions of time (a, ρ, p); to close the set, one further equation is required. For our cosmological considerations, two simple equations of state, $p = p(\rho)$, are of most interest.

For an ideal gas of nonrelativistic (NR) particles (i.e. $kT \ll mc^2$),

$$p = nkT, \qquad \rho \simeq mn \Rightarrow p \ll \rho c^2. \qquad 3.$$

Using this equation of state in Equation 2b leads to $\rho_{NR} \propto a^{-3}$.

For an ideal gas of extremely relativistic (ER) particles (i.e. $mc^2 \ll kT$),

$$3p = \rho c^2, \qquad \rho c^2 \simeq 3nkT \propto T^4. \qquad 4.$$

With this equation of state, it follows from Equation 2b that $\rho_{ER} \propto a^{-4}$.

This simple analysis yields a very important result: the early Universe is dominated by ER particles. That is, for $a \to 0$, $\rho_{NR}/\rho_{ER} \propto a \to 0$. It is conventional to call such epochs radiation dominated (RD).

For early epochs, Equation 2a simplifies considerably; the first term on the right-hand side dominates and neither the curvature nor the cosmological constant plays a role in the early evolution of the Universe. With $\rho \simeq \rho_{ER} \propto a^{-4}$, Equation 2a is easily integrated:

$$\left(\frac{32\pi}{3}\right) G\rho t^2 = 1. \qquad 5.$$

From conservation of entropy[3] it follows that $T(t) \propto [a(t)]^{-1}$. Since the scale factor is unobservable, it is preferable to describe the expansion by the age versus temperature relation: $t = t(T)$. Once the $\rho(T)$ dependence is known, the t versus T relation may be obtained from Equation 5. During the RD era, relativistic particles ($mc^2 < kT$) dominate the total density; using the result reviewed in Section 2.2, we may write

$$\rho = \left(\frac{g}{2}\right)\rho_\gamma; \qquad g \equiv g_B + \frac{7}{8}g_F; \qquad \rho_\gamma = \left(\frac{\pi^2}{15}\right)\frac{(kT)^4}{(\hbar^3 c^5)}. \qquad 6.$$

Here, g is the "effective" number of degrees of freedom; g_B (g_F) is the total number of boson (fermion) degrees of freedom; and ρ_γ is the photon density. Notice that the g's are temperature dependent; the higher the temperature, the more particles contribute (i.e. the more particles for which $mc^2 < kT$). During the RD era then, the age t (in seconds) as a function of the temperature is,

$$t = 2.42 \times 10^{-6} g^{-1/2} T^{-2}. \qquad 7.$$

[Unless otherwise noted, energies, masses, and temperatures are in GeV.]

A simple example will serve to illustrate these results. Consider the time when the temperature was 1 MeV. Photons were in equilibrium with electron-positron pairs and with three kinds of left-handed neutrinos: $g_B = g_\gamma = 2$; $g_F = g_e + g_{ve} + g_{v\mu} + g_{v\tau} = 10$. Then g (1 MeV) = 43/4 and t (1 MeV) = 0.74 s.

By comparing the expansion rate (t^{-1}) with reaction rates, we can determine those conditions under which statistical equilibrium is established. Consider a general reaction with cross section $\sigma(E)$ and write $\beta(T) = \langle \sigma v \rangle$ for the thermally averaged product of the cross section and the relative velocity. With a target number density

---

[3] For a relativistic gas, the entropy density varies as the cube of the temperature; recall that the size of the co-moving volume varies as the cube of the scale factor.

$n \propto a^{-3} \propto T^3$, the reaction rate is $\Gamma(t) = \beta(T)n(T) \propto T^3\beta(T)$. A measure of the number of collisions is $\Gamma(t)t(T) \propto T\beta(T)$, and equilibrium is established and maintained when $\Gamma t \gg 1$.

For low energy exothermic reactions, for example, pair annihilation, $\sigma \propto v^{-1}$ so that β is independent of temperature. At energies below the mass of the intermediate vector boson M_W, weak interaction cross sections vary as $\sigma(E) \propto E^2$ so that $\beta(T) \propto T^2$. Above M_W the behavior changes to $\sigma(E) \propto E^{-2}$ leading to $\beta(T) \propto T^{-2}$.

The following general picture emerges. Very early, at high temperature and density, equilibrium was established ($\Gamma t \gg 1$). As the Universe expanded and cooled, reactions became less frequent ($\Gamma t \approx 1$). Still later, collisions occurred too infrequently to maintain equilibrium ($\Gamma t \ll 1$), or to change the abundances.

Equilibrium permits an enormous simplification. The physical quantities of interest (number density, energy density, pressure, etc) depend on the temperature (and on the chemical potentials associated with conserved quantities such as electric charge). It is not necessary to investigate the detailed reactions that lead to equilibrium nor to know the prior evolution of the Universe. Keep in mind, though, the possibility that the actual values of the chemical potentials may be determined by specific particle reactions during those earlier epochs.

2.2   Thermal Equilibrium

Consider an arbitrary volume V in thermal equilibrium with a heat bath at temperature T. As they are in the cosmological context, V and T may be time dependent. The number of particles of type i and their energy (see, for example, Steigman 1973) are

$$N_i = V\left(\frac{g_i}{2\pi^2}\right)\left(\frac{kT}{\hbar c}\right)^3 \int_0^\infty \left[\exp\left(\frac{\varepsilon}{kT}\right) \pm 1\right]^{-1} z^2 \, dz, \qquad 8.$$

$$E_i = V\left(\frac{g_i}{2\pi^2}\right)\left(\frac{kT}{\hbar c}\right)^3 kT \int_0^\infty \left[\exp\left(\frac{\varepsilon}{kT}\right) \pm 1\right]^{-1}\left(\frac{\varepsilon}{kT}\right) z^2 \, dz, \qquad 9.$$

where g_i is the number of spin states; $z = pc/kT$; and $\varepsilon = [(pc)^2 + (m_i c^2)^2]^{1/2}$. Note that for fermions (bosons) the $+$ ($-$) sign applies. The above integrals simplify in the high temperature (or low mass) limit ($kT \gg m_i c^2$) and in the low temperature (or high mass) limit ($kT \ll m_i c^2$).

First, consider photons (bosons with $g_i = 2$, $m_i = 0$). Their number and energy densities are

$$n_\gamma = \frac{2\zeta(3)}{\pi^2}\left(\frac{kT}{\hbar c}\right)^3 = \frac{2.404}{\pi^2}\left(\frac{kT}{\hbar c}\right)^3 \simeq 31.8 \, T_{\text{Gev}}^3 \, (\text{fm}^{-3}), \qquad 10.$$

$$\rho_\gamma = \frac{6\zeta(4)}{\pi^2}\left(\frac{kT}{\hbar c}\right)^3\left(\frac{kT}{c^2}\right) = \left(\frac{\pi^2}{15}\right)\frac{(kT)^4}{(\hbar^3 c^5)} \simeq 2.7\left(\frac{kT}{c^2}\right)n_\gamma. \qquad 11.$$

We may compare all other densities with those of photons. In the high temperature limit we have for bosons,

$$n_B = \left(\frac{g_B}{2}\right)n_\gamma, \qquad \rho_B = \left(\frac{g_B}{2}\right)\rho_\gamma; \qquad 12.$$

and, for fermions,

$$n_F = \frac{3}{8}g_F n_\gamma \qquad \rho_F = \frac{7}{16}g_F\rho_\gamma. \qquad 13.$$

In the low temperature limit the distinction between fermions and bosons disappears if the chemical potential can be neglected.

$$n_i = \left[\frac{g_i}{(2\pi)^{3/2}}\right]\left(\frac{kT}{\hbar c}\right)^3\left[\left(\frac{m_i c^2}{kT}\right)^{3/2}\exp\left(\frac{-m_i c^2}{kT}\right)\right]; \qquad \rho_i = m_i n_i. \qquad 14.$$

In cosmology, relativistic particles dominate during the RD era so that,

$$\rho = \rho_B + \rho_F = \left(\frac{g}{2}\right)\rho_\gamma; \qquad g(T) = g_B(T) + \frac{7}{8}g_F(T). \qquad 15.$$

The "effective" number of degrees of freedom, $g(T)$, is temperature dependent through two separate effects. At higher temperatures, more particles are relativistic, which leads to a slow increase of g with T. In addition, relativistic particles may be present, which, because of their weak interactions, have dropped out of equilibrium (for example, neutrinos, gravitons, etc). Such particles still interact gravitationally, but their contribution to the density is reduced by a factor $(T'/T)^4$, where T' is the temperature of the decoupled particles (Steigman, Olive & Schramm 1979).

Many potentially complicated calculations are greatly simplified when entropy conservation is accounted for. By restricting our attention to those relativistic particles in a specific but arbitrary co-moving[4] volume that are still interacting, the entropy (divided by the Boltzmann constant) is (see, for example, Weinberg 1972)

$$\frac{S_I}{k} = \frac{4}{3}\left(\frac{\rho_I c^2}{kT}\right)V = \frac{2}{3}g_I(T)\left(\frac{\rho_\gamma c^2}{kT}\right)V \sim g_I(T)N_\gamma(T). \qquad 16.$$

[4] During the epochs in which baryon number is conserved, a co-moving volume may be defined by the net baryon number in it.

The subscript I is to remind us of the restriction to *interacting* particles; g_I differs from the effective number of degrees of freedom g. As the temperature and volume vary (as the Universe expands), the product $g_I(T)N_\gamma(T)$ is preserved. This has application to cosmology in the following situation.

As the temperature drops from $T > m_i$ to $T < m_i$, the energy released in i-$\bar{i}$ annihilation heats the remaining *interacting* particles in V. The heating leads to the production of more photons and other interacting particles. Remember that in the absence of this effect, the number of relativistic particles in a co-moving volume is preserved ($N_{ER} = n_{ER} V \propto T^3 \times T^{-3} = $ constant). An application of entropy conservation makes the calculation of the extra photons trivial:

$$\frac{N_\gamma(T < m_i)}{N_\gamma(T > m_i)} = \frac{g_I(T > m_i)}{g_I(T < m_i)}. \qquad 17.$$

For example, consider $T \simeq 1$ MeV when photons are in equilibrium with $e^\pm$ pairs and with e-, μ-, and τ-neutrinos. Then, $g_I(1 \text{ MeV}) = g_\gamma + (7/8)(g_e + 3g_\nu) = 43/4$ and, for $T > 1$ MeV,

$$\frac{N_\gamma(1 \text{ MeV})}{N_\gamma(T)} = \frac{4}{43} g_I(T). \qquad 18.$$

If we neglect the contribution to ρ from gravitons and, if there are no yet-to-be-discovered superweakly interacting particles, then for $T \gtrsim 1$ MeV, $g(T) = g_I(T)$. However, below ~ 1 MeV, the e-, μ-, and τ-neutrinos decouple so that when $e^\pm$ pairs annihilate, the neutrinos do not share in the energy released:

$$g_I(\gtrsim m_e) = g_\gamma + \frac{7}{8}(g_e) = \frac{11}{2}; \qquad g_I(\lesssim m_e) = g_\gamma = 2, \qquad 19a.$$

$$\frac{N_{\gamma 0}}{N_\gamma(m_e)} = \frac{g_I(\gtrsim m_e)}{g_I(\lesssim m_e)} = \frac{11}{4}. \qquad 19b.$$

Here, $N_{\gamma 0}$ is the number of photons in our co-moving volume for $T \ll m_e$. In particular, the present number of photons is approximately $N_{\gamma 0}$. Since no new photons were created between decoupling at $T \simeq 1$ MeV and annihilation at $T \simeq m_e$, it follows (for $T > m_e$) that

$$\frac{N_{\gamma 0}}{N_\gamma(T)} = \frac{11}{43} g_I(T). \qquad 20.$$

Furthermore, since $N_\nu = n_\nu V \sim T_\nu^3 V$ and $N_\gamma = n_\gamma V \sim T^3 V$, the ratio of temperatures (below m_e) is $T_\nu/T_\gamma = (4/11)^{1/3}$. The total effective number

Table 1 Degrees of freedom, photon numbers, and the expansion rate

T_d^a	$4g_1(T_d)$	$N_{\gamma 0}/N_{\gamma d}{}^b$	$t(s)T^2(MeV)^c$
m_e-m_μ	43	2.75	0.74
$m_\mu-m_\pi$	57	3.65	0.64
$m_\pi-T_c$	69	4.41	0.58
T_c-m_s	205	13.1	0.34
m_s-m_c	247	15.8	0.31
m_c-m_τ	289	18.5	0.28
$m_\tau-m_b$	303	19.4	0.28
m_b-m_t	345	22.1	0.26
m_t-m_W	387	24.8	0.25
$>m_W$	423	27.1	0.24

[a] The decoupling temperature, T_d, may be anywhere in the range indicated. For quark masses it is assumed that $m_u \approx m_d \lesssim T_c \lesssim m_s$. T_c is the temperature at which the quark-hadron transition occurs ($T_c \approx 0.2$–0.4 GeV, Wagoner & Steigman 1979).
[b] The ratio of photons now in a co-moving volume to those present at decoupling (see Equations 17–21).
[c] The time-temperature relation (see Equation 7).

of degrees of freedom are

$$g(\gtrsim m_e) = g_\gamma + \frac{7}{8}(g_e + 3g_\nu) = \frac{43}{4},$$ 21a.

$$g(\lesssim m_e) = g_\gamma + \frac{7}{8} \times 3g_\nu \left(\frac{T_\nu}{T_\gamma}\right)^4 = 3.36.$$ 21b.

For $T > m_e$, $g_1 \simeq g$ is a slowly varying function of temperature. With increasing temperature, more particles are relativistic ($T > m_i$) and g increases from 43/4 at $T \simeq 1$ MeV to 423/4 at $T \simeq 100$ GeV (Steigman, Olive & Schramm 1979).

These results are collected in Table 1 where, as a function of the decoupling temperature (T_d), are shown the number of degrees of freedom (g_1), the ratio of photons now in our co-moving volume to those present at decoupling ($N_{\gamma 0}/N_{\gamma d}$) and the time-versus-temperature relation (Equation 7).

2.3 Particle Production and Survival

The creation, annihilation, and survival of particles in the cosmological context was first analyzed by Zeldovich (1965) and Chiu (1966), who considered nucleons in the early Universe. All subsequent work in this area has been an extension and generalization of their discussions.

There is no necessity to restrict our exposition to cosmology, the problem is more general. Think again of a specific but arbitrary volume in a heat bath at some temperature and consider the effects of arbitrary changes in V and/or T. Our application, of course, is to cosmology where, the changes in V and T are related ($V \propto T^{-3}$) and the time dependence is fixed (in the RD era, $t \propto T^{-2}$).

The goal is to calculate the number of particles (and antiparticles) of type i (arbitrary) in V as a function of time. Assume, for convenience, that there are equal numbers of particles and antiparticles ($N_i = N_{\bar{i}}$); this constraint will be relaxed later. The number of particles ($N = \bar{N} = nV$) is determined by the competition between production and annihilation,

$$\frac{dN}{dt} = \psi_{\text{prod}} V - \beta_{\text{ann}} n^2 V. \qquad 22.$$

To simplify the notation, the subscript "i" has been omitted; subsequently we also drop the subscripts "prod" and "ann."

Recently, Lee & Weinberg (1977) and, independently, Wolfram (1979) have numerically integrated Equation 22 for various choices of M_i and β. The purpose of our discussion here is to obtain simple, general, analytic expressions that are in reasonable agreement with the numerical results. In the process, it is to be hoped, the physics of the survival of relic particles will be clarified.

In the low temperature limit ($T < m$) the thermally averaged annihilation rate coefficient, β, is independent of temperature. Detailed balancing relates the production rate per unit volume, ψ, to β. That is, "stop" the expansion at temperature T and wait for equilibrium to be achieved. In equilibrium, $dN/dt = 0$ so that

$$\psi(T) = \beta(T) n_{\text{eq}}^2(T), \qquad 23.$$

where $n_{\text{eq}}(T)$ is the density (of particle i) in equilibrium at temperature T (see Section 2.2). Knowing ψ, we may recast Equation 22 in the convenient form

$$\frac{dN}{dt} = \beta(n_{\text{eq}} + n)(N_{\text{eq}} - N). \qquad 24.$$

In place of the temperature, introduce the dimensionless ratio $x(t) = mc^2/kT = m(\text{GeV})/T(\text{GeV})$ and recall from Section 2.2 that $N_{\text{eq}} = $ constant for $x \lesssim 1$ and $N_{\text{eq}} \propto x^{3/2} e^{-x}$ for $x \gtrsim 1$.

At high temperatures ($x \lesssim 1$), $dN_{\text{eq}}/dt = 0$, so that $N = N_{\text{eq}}$ is a solution of Equation 24. Thus, at sufficiently high temperatures, our particles will be in equilibrium and there will be a constant number of them in our co-moving volume. Notice that for massless particles ($x = 0$), $N = N_{eq}$ is always a solution. Independent of interactions, relativistic particles are conserved (we have neglected the heating effects due to the annihilation of the other particles present).

For $x \gtrsim 1$, N_{eq} is no longer a solution and deviations from equilibrium occur. Let us measure this deviation by

$$\Delta \equiv \frac{N - N_{eq}}{N_{eq}} = -(\beta n_{eq} t)^{-1} \left(1 + \frac{n}{n_{eq}}\right)^{-1} \left[\frac{t}{N_{eq}} \frac{dN}{dt}\right]. \qquad 25.$$

Recall that $t \propto T^{-2} \propto x^2$ so that $dt/t = 2dx/x$. Now, even for $x \gtrsim 1$, reactions tend to maintain equilibrium so that $\Delta \ll 1$ and we may rewrite Equation 25 as

$$\Delta(2 + \Delta) \approx -(\Gamma t)^{-1} \frac{d(\ln N_{eq})}{d(\ln t)} \approx \frac{x}{2\Gamma t}, \qquad 26.$$

where $\Gamma \equiv \beta n_{eq}$. As x increases (T decreases), Δ begins to grow exponentially and the deviations become large. Choose a specific value of $\Delta = \Delta_*$ (shortly, we will choose $\Delta_* = 1/2$); this fixes $x = x_*$. For $x \lesssim x_*$, $N \approx N_{eq}$ whereas for $x \gtrsim x_*$, annihilation dominates:

$$x \gtrsim x_*, \qquad \frac{dN}{dt} \approx -\beta n^2 V = -\frac{\beta N^2}{V}. \qquad 27.$$

Equation 27 may be integrated subject to the boundary condition that, at $x = x_*$, $N_* = (1 + \Delta_*) N_{eq*}$. The volume V may be specified by giving the number of photons $N_{\gamma*}$ in V at x_*: $V = V_*(t/t_*)^{3/2}$, $V_* = N_{\gamma*}/n_{\gamma*} = N_*/n_*$. Equation 27 is easily integrated to give for $N_0 = N(t \to \infty)$

$$N_0 = \frac{N_{\gamma*}}{2\beta n_{\gamma*} t_*} = \frac{N_*}{2\beta n_* t_*} = \left(\frac{1}{1 + \Delta_*}\right)\left(\frac{N_*}{2\Gamma_* t_*}\right) = \left[\frac{\Delta_*(2 + \Delta_*)}{(1 + \Delta_*)}\right]\frac{N_*}{x_*}. \qquad 28.$$

Recall that, $\Delta_*(2 + \Delta_*) = x_*(2\Gamma_* t_*)^{-1} \propto x_*^{1/2} e^{x_*}$ so that x_* depends logarithmically on the choice of Δ_*. With $\Delta_* = 1/2$ we find,

$$x_*^{1/2} e^{x_*} = 5.0 \times 10^{19} \, g_i Z; \qquad Z = \frac{m_i \tilde{\beta}}{g_*^{1/2}}, \qquad 29.$$

where $\tilde{\beta} = 10^{15} \beta$ (cm^{3}s^{-1}) and m_i is in GeV. Comparing N_0 with the present number of photons in our volume we have for the ratio of relic particles to photons,

$$\left(\frac{N}{N_\gamma}\right)_0 = \left(\frac{\bar{N}}{N_\gamma}\right)_0 = 6.5 \times 10^{-21} \left(\frac{x_*}{Z}\right)\left(\frac{N_{\gamma*}}{N_{\gamma 0}}\right). \qquad 30.$$

From Equation 20 we may relate $N_{\gamma*}/N_{\gamma 0}$ to g_*: $N_{\gamma 0}/N_{\gamma*} = 11g_*/43$ so that,

$$\left(\frac{N+\bar{N}}{N_\gamma}\right)_0 \approx \frac{5.1 \times 10^{-20} \, x_*}{m_i \, \tilde{\beta} \, g_*^{1/2}}. \tag{31.}$$

With $n_{\gamma 0} \approx 400 \, \text{cm}^{-3}$ ($T_0 \approx 2.7$ K), the present number and mass densities of our relic particles are

$$(n+\bar{n})_0 \approx 2.0 \times 10^{-17} x_* (m_i \tilde{\beta} g_*^{1/2})^{-1} \, (\text{cm}^{-3}), \tag{32a.}$$

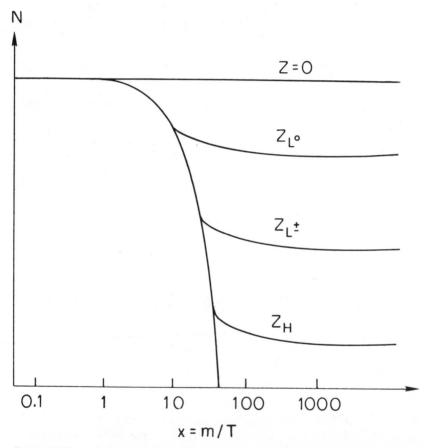

Figure 1 The evolution of particles in the early Universe. The number of particles of various types (leptons, hadrons) is shown as a function of the ratio of the mass to the temperature. The various particles are distinguished by the values of $Z \propto m \langle \sigma v \rangle$.

$$\rho_0 = m_i(n+\bar{n})_0 \approx 3.4 \times 10^{-41} x_* (\tilde{\beta} g_*^{1/2})^{-1} \; (\text{g cm}^{-3}), \qquad\qquad 32\text{b.}$$

where

$$x_* + 1/2 \ln x_* = 45.4 + \ln(g_i Z). \qquad\qquad 32\text{c.}$$

Notice that departures from equilibrium become significant for $T \lesssim T_* = m_i x_*^{-1}$ when $\Gamma_* t_* = [2\Delta_*(2+\Delta_*)]^{-1} x_* \approx 0.4 x_* \gg 1$.

Before proceeding to a more detailed discussion, a summary will help set the scene. For $T > m_i$, the particles i and $\bar{i}$ are relativistic and roughly as abundant as photons. Those massless or very weakly interacting particles that depart from equilibrium while they are still relativistic are always as abundant (roughly) as photons; their number in a co-moving volume is conserved. For massive particles, collisions may maintain equilibrium for $T < m_i$ and their number in a co-moving volume decreases exponentially with decreasing temperature [$N_i \propto \exp(-m_i/T)$]. Eventually ($T \lesssim T_*$), equilibrium can no longer be maintained; no new particles are created and those present cease to annihilate. Thereafter ($T \ll T_*$), the number of particles in a co-moving volume is conserved. These results are illustrated in Figure 1 and specific order-of-magnitude estimates are provided in Table 2.

The general trend is clear. The stronger the interaction, the longer the particles remain in equilibrium (larger x_*) and the fewer survive ($\sim e^{-x_*}$). Although in the symmetric case ($N_i = N_{\bar{i}}$) few relic hadrons should be present today, their strong interaction renders them directly observable in sufficiently sensitive experiments. More charged leptons will have survived and, if they are sufficiently heavy ($\gtrsim 1$ TeV), they might contribute significantly to the present mass density. Many more neutral leptons will emerge from the big bang and the constraint that they not dominate the present mass density places severe restrictions on the range of allowed masses.

Table 2 Order-of-magnitude estimates for stable relic particles

Particle type	T_*^a	$n_0(\text{cm}^{-3})^b$	$\rho_0(\text{g cm}^{-3})^c$
Hadrons	$\sim m/45$	$\sim 10^{-16} m_{\text{GeV}}^{-1}$	$\sim 10^{-40}$
Charged leptons	$\sim m/30$	$\sim 10^{-10} m_{\text{GeV}}$	$\sim 10^{-34} m_{\text{GeV}}^2$
Heavy neutral leptons ($m \gtrsim 1$ MeV)	$\sim m/20$	$\sim 10^{-4} m_{\text{GeV}}^{-3}$	$\sim 10^{-28} m_{\text{GeV}}^{-2}$
Light neutral leptons ($m \lesssim 1$ MeV)	~ 1 MeV	$\sim 10^2$	$\sim 10^{-31} m_{\text{eV}}$

a The "freeze-out" temperature; see Equation 32c ($T_* = m x_*^{-1}$).
b The present number density of stable relic particles.
c The present mass density of stable relic particles.

3   COSMOLOGICAL CONSTRAINTS

3.1   *Cosmological Mass Density*

Relic particles that are stable ($\tau \gtrsim 10^{10}$ yr) will contribute to the present cosmological mass density. Constraints on their masses follow from the requirement that their contributions not exceed observational limits. Hypothetical particles exceeding these limits either must not exist or must be unstable.

The present cosmological density is very uncertain (see, for example, Schramm & Wagoner 1977, Gott et al 1974, Kazanas, Schramm & Hainebach 1978). Current estimates suggest a low density Universe with $\rho_0 \approx 2 \times 10^{-31}$ g cm^{-3}. Although it cannot be excluded that ρ_0 may be much higher,[5] it is likely that the mass associated with galaxies and clusters of galaxies contributes $\gtrsim 2 \times 10^{-30}$ g cm^{-3}. For our comparisons we may take as a rather firm upper limit, $\rho_0 \gtrsim 2 \times 10^{-29}$ g cm^{-3}.

3.2   *Abundances of Anomalous Nuclei*

Although few strongly interacting particles should survive as relics of the big bang, their strong interaction leaves open the possibility that very small abundances of such particles may be observable. Free relic quarks should be easy to find if they have nonintegral charge. Prior to the introduction of color and color confinement into hadron physics, the absence of relic quarks was a puzzle (Zeldovich, Okun & Pikel'ner 1965, Zeldovich 1970, Frautschi, Steigman & Bahcall 1972).

Jones (1977) has provided an excellent review of quark searches. At present, there is conflicting evidence concerning the existence of free quarks with nonintegral charge. The levitometer experiments of LaRue et al (1977, 1979a,b) imply an abundance of $\gtrsim 10^{-20}$ quarks per nucleon in niobium. Much smaller upper limits have been claimed in other searches but their interpretation is not straightforward (Jones 1977).

Exotic heavy hadrons should also be easy to find (provided that they are stable). For masses less than 16 GeV, bounds on heavy isotopes of hydrogen are exceedingly low, $\sim 10^{-18}$, relative to ordinary hydrogen (Muller et al 1977, Alvager & Naumann 1967). Heavy hadrons may, however, reside preferentially in $Z > 1$ nuclei (Dover, Gaisser & Steigman 1979) where bounds on anomalous isotopes are much less severe (Schiffer 1974). Nonetheless, specifically designed experiments can have sensitivities as great as one part in 10^{16} or greater (Bennett et al 1978, Muller 1979). Middleton et al (1979) and colleagues have begun a search for stable heavy

---

[5] The present critical density separating open ($\rho_0 < \rho_c$) from closed ($\rho_0 > \rho_c$) models is $\rho_c \lesssim 2 \times 10^{-29}$ g cm^{-3}.

hadrons in oxygen. No anomalous oxygen with $M < 60$ GeV has been found at an abundance greater than 10^{-17}. Stable heavy hadrons or massive charged leptons from the big bang should be found in such searches. Their absence will set limits to their masses and/or lifetimes.

3.3   Primordial Nucleosynthesis

The physics of primordial nucleosynthesis and the primordial abundances of the elements were discussed in the excellent review article by Schramm & Wagoner (1977). Here, we briefly review the salient features of nucleosynthesis that permit us to probe particle physics. For further details and references the reader is urged to consult Schramm & Wagoner (1977) and Steigman, Schramm & Gunn (1977).

The primordial abundance of ^{4}He is most sensitive to the competition between the expansion rate (t^{-1}) and the rates (Γ_{wk}) of the standard, charged current, weak interactions. The presence of "light" particles $(\lesssim 1$ MeV) affects the expansion rate during the RD era: $t^{-1} \propto [g(T)]^{1/2}$. Thus ^{4}He is a probe of the early expansion rate, and specifically therefore for $g_{nuc} \approx g\,(T \approx 1$ MeV). Any "new" light particles (with $\tau \gtrsim 1$ s) will contribute to g_{nuc} and, therefore, will hasten the expansion. Let us see how this affects the ^{4}He abundance.

For $T \gtrsim 1$ MeV, the weak interactions $p + e^- \rightleftarrows n + \nu_e$, $p + \bar{\nu}_e \rightleftarrows n + e^+$, and $n \rightleftarrows p + e^- + \bar{\nu}_e$ are rapid enough to keep the neutron-to-proton ratio in equilibrium: $N_n/N_p = \exp(-\Delta mc^2/kT)$. For $T \lesssim 1$ MeV, the weak interactions are too slow and the fraction of neutrons decreases very slowly from $X_n = N_n/N_n + N_p \approx 0.2$. All the while deuterons are produced and photodissociated: $n + p \rightleftarrows D + \gamma$. For $T \gtrsim 0.1$ MeV, photodissociation is so rapid that the deuterium abundance is negligibly small and this provides a bottleneck to further nucleosynthesis. Only for $T \lesssim 0.1$ MeV is there sufficient deuterium present to permit the building up of heavier nuclei. There is, however, a "gap" (no stable nucleus) at mass-5 and this prevents the synthesis of significant amounts of the elements heavier than ^{4}He. Virtually all neutrons that were present when nucleosynthesis began in earnest $(T \simeq 0.1$ MeV, $X_n \simeq 1/7)$ are incorporated into ^{4}He; the abundance by mass of ^{4}He is $Y \approx 2X_n \approx 0.3$. Notice that the neutron abundance has decreased since "freeze-out."

The primordial abundance of ^{4}He increases slowly with increasing nucleon density, for $2 \times 10^{-31} \lesssim \rho_0 \lesssim 2 \times 10^{-29}$ (g cm^{-3}) and with $g_{nuc} = 43/4$ $(\gamma,\ e^{\pm},\ \nu_e,\ \nu_\mu,\ \nu_\tau)$; $0.25 \lesssim Y \lesssim 0.29$ (Yang et al 1979). As discussed in Schramm & Wagoner (1977) and, in more detail in Yang et al (1979), observations suggest that the primordial abundance of ^{4}He was $\lesssim 0.25$. This leaves very little room for "new" particles; we return to this constraint in Sections 4.6 and 4.7.

Deuterium is the other light element produced in significant quantities in the big bang. In contrast with ^{4}He, the deuterium abundance decreases rapidly with increasing nucleon density. Deuterium, then, is an excellent probe for ρ_0 but a poor probe of the early expansion rate; the uncertainties in ρ_0 swamp the small effects in the deuterium abundance due to "new" particles.

4   COSMOLOGY CONFRONTS PARTICLE PHYSICS

4.1   *Nucleons*

The present density of nucleons is uncertain. With ρ_0 in the range from $\gtrsim 2 \times 10^{-31}$ g cm^{-3} to $\lesssim 2 \times 10^{-29}$ g cm^{-3}, we find $(n_N)_0 \simeq 10^{-6 \pm 1}$ cm^{-3}. This estimate does not distinguish antinucleons from nucleons and therefore is an estimate of their sum. The photon density is better known. For a 2.7-K black body, $(n_\gamma)_0 \simeq 400$ cm^{-3}. The observed nucleon-to-photon ratio is

$$\left(\frac{N_N + N_{\bar{N}}}{N_\gamma} \right)_0 \simeq 3 \times 10^{-9 \pm 1}. \qquad\qquad 33.$$

Now let us compare this result with the predicted ratio of relic nucleons-to-photons (refer to Section 2.3). With $m_N \simeq 0.94$ GeV and $\tilde{\beta}_N \simeq 1$, $x_* \simeq 44$ and the "freeze-out" temperature is $T_* \simeq 22$ MeV. The expected nucleon-to-photon ratio is predicted to be

$$\left(\frac{N_N + N_{\bar{N}}}{N_\gamma} \right)_0 \simeq 7 \times 10^{-19}. \qquad\qquad 34.$$

Our first effort to confront particle physics fails by some ten orders of magnitude!

It is, of course, not really a failure. We learn that one of the assumptions underlying the calculations of Section 2.3 must be invalid for nucleons. It is obvious that the constraint of particle-antiparticle symmetry is too restrictive for nucleons. Indeed, there is no evidence whatever for any large amounts of antimatter anywhere in the Universe (for recent detailed reviews of the antimatter question, see Steigman 1976, 1979).

Notice that for $T \gtrsim 37$ MeV, $N_N/N_\gamma \gtrsim 10^{-9}$ so that nucleon-antinucleon asymmetry must have been established earlier than $t \simeq 10^{-3}$ s. It is, unfortunately, beyond the scope of this review to consider the recent, exciting suggestions that baryon asymmetry may have been established exceedingly early $(t \lesssim 10^{-37}$ s$)$ at enormously high temperatures $(T \gtrsim 10^{15}$ GeV$)$ by baryon nonconserving, CP-violating processes that never were in equilibrium (Dimopoulos & Susskind 1978a,b, Toussaint et al 1979, Ellis et al 1979, Weinberg 1979).

In subsequent sections we often prefer to use nucleons rather than photons for comparison. For that purpose we adopt, as a best guess, a present nucleon-to-photon ratio $(N_N/N_\gamma)_0 \simeq 10^{-9}$. Keep in mind, though, the uncertainty in this estimate.

4.2   Heavy Hadrons

Recently, Dover et al (1979) and Wolfram (1979) have considered the possibility that "heavy hadrons" may have survived as relics of the big bang. The idea is that there might be new, massive, stable quarks (for example, color-sextet quarks) that were produced in the early Universe. During the quark-hadron confining transition, new heavy-quark-bearing hadrons would have been formed. In estimating the present abundance of heavy hadrons there are two possibilities.

1. In the previous section we saw dramatic evidence that nucleon-antinucleon symmetry is badly broken. If the same processes responsible for nucleon asymmetry operate for the new heavy quarks, we might expect (within a few orders of magnitude) there to be as many heavy hadrons (H) as nucleons (N). This is clearly not true.

2. In contrast, the smallest number of relic heavy hadrons survive if baryon-antibaryon symmetry is assumed. For this situation, Dover et al (1979) estimate $(N_H/N_N)_0 \gtrsim 2 \times 10^{-11}$. Such a high abundance should be easily observable. Indeed, for masses less than 16 GeV, there are no anomalous isotopes of hydrogen at an abundance relative to protons of $\sim 10^{-18}$ (Muller et al 1977); for masses less than 44 GeV there are no anomalous isotopes of oxygen at an abundance greater than 10^{-17} (Middleton et al 1979). Such negative results are, however, somewhat inconclusive. Perhaps m_H is very large, or more likely, perhaps such heavy hadrons are to be found preferentially in heavier nuclei such as Fe. On the other hand, negative results in thorough searches may well provide strong constraints on new models of unified theories (i.e. perhaps there are no *stable* heavy quarks).

4.3   Relic Quarks

Zeldovich et al (1965, 1966) noted that, if free quarks exist, they would have survived the big bang in appreciable numbers. Zeldovich et al chose $\tilde{\beta} \simeq 1$, which leads to a freeze-out temperature T_* (MeV) $\simeq 20 \, M_{GeV}$ and a predicted abundance (relative to nucleons) $(N_q/N_N)_0 \simeq 6 \times 10^{-10} M_{GeV}^{-1}$. The evidence referred to in Section 3.2 suggests that $N_q/N_N \lesssim 10^{-20}$. It is clear that for any reasonable mass, the naive estimate fails by many orders of magnitude.

As Frautschi, Steigman & Bahcall (1972) noted, the abundance of free relic quarks depends sensitively on the highest temperature reached during the cosmological evolution. For example, in the Hagedorn (1970) cosmology with a limiting temperature T_{max}, the predicted abundance of relic quarks is $(N_q/N_N)_0 \approx 10^{27} \exp(-2\ M_q/T_{max})$ (Frautschi, Steigman & Bahcall 1972). For $T_{max} \approx 0.16$ GeV, the expected abundance is consistent with the data for $M_q > 9$ GeV.

Even in the standard cosmology, the discussion of Zeldovich et al (1965, 1966) must be modified for $T_* \gtrsim 200$ MeV. At such high temperatures the density is very large and hadrons overlap and "dissolve" into a "hot quark soup" of free quarks and gluons. In this regime ($T_* \gtrsim 0.2$ GeV), the calculation of Zeldovich et al (1965, 1966) no longer applies; their results are valid for $T_* \lesssim 0.2$ GeV, which corresponds to a quark mass $M_q \lesssim 10$ GeV and a predicted abundance $(N_q/N_N)_0 \gtrsim 10^{-10}$.

Color and color confinement introduce a crucial new ingredient. If confinement is absolute then, of course, there will be no free quarks. If, however, confinement is incomplete then the possibility exists that during the quark-hadron transition in the early Universe $[0.2 \lesssim T_c$ (GeV) $\lesssim 0.4]$, some quarks escaped confinement (Wagoner & Steigman 1979). Wagoner & Steigman (1979) find (for $M_q \gtrsim 10$ GeV)

$$\left(\frac{N_q + N_{\bar{q}}}{N_N}\right)_0 \simeq 10^9 \left(\frac{M_q}{T_c}\right)^2 \exp\left(\frac{-M_q}{T_c}\right). \qquad 35.$$

Notice that for $M_q/T_c \gtrsim 75$, the quark-to-nucleon ratio is below $\sim 10^{-20}$, which suggests that $M_q \gtrsim 15$–30 GeV.

4.4 Charged Leptons

It is likely that any heavy charged leptons will decay via the weak interaction. Still, it may be of interest to consider the constraints on a stable, heavy, charged lepton. By numerical integration of the evolution equations (see Section 2.3), Wolfram (1979) has calculated the present density of charged leptons.

$$n_L \approx 10^{-11}\ \text{cm}^{-3},\ m_L \gtrsim 10\ \text{GeV}, \qquad 36a.$$

$$n_L \approx 10^{-12}\ m_L\ \text{cm}^{-3},\ m_L \gtrsim 10\ \text{GeV}. \qquad 36b.$$

Wolfram's estimates are some one to two orders of magnitude lower than the very rough value in Table 2 due mostly to a larger estimate of the L^+-L^- annihilation cross section. Thus, any $L^\pm$ produced in the early Universe should be relatively abundant today. Notice that if $m_L \gtrsim 10$ GeV the present mass density contributed by charged leptons might be appreciable: $\rho_{L0} \approx 2 \times 10^{-35}\ m_L^2\ \text{g cm}^{-3}$.

4.5   *Heavy Neutral Leptons*

Neutral leptons interact very weakly. They therefore drop out of equilibrium very early in the expansion of the Universe when their relative abundance is high. Lee & Weinberg (1977) and Dicus, Kolb & Teplitz (1977) have followed the evolution of heavy ($\gtrsim 1$ MeV) neutral leptons and have calculated their present density (provided they are stable). In rough agreement with their more accurate numerical results, we may obtain a simple estimate by applying the results of Section 2.3.

With $\bar{\beta} \simeq 3.4 \times 10^{-12} \, m_{\text{GeV}}^2$ (see Lee & Weinberg 1977) and $g_*^{1/2} \simeq 4$ (i.e. $T_* \lesssim 0.2$ GeV), we find that $x_* \simeq 17 \, [T_* \, (\text{MeV}) \simeq 59 \, m_{\text{GeV}}]$ and

$$\left(\frac{N_\nu + N_{\bar{\nu}}}{N_\gamma}\right)_0 \simeq \frac{6 \times 10^{-8}}{m_{\text{GeV}}^3} \qquad\qquad 37a.$$

$$\rho_{\nu 0} \simeq \frac{4.5 \times 10^{-29}}{m_{\text{GeV}}^2} \, \text{g cm}^{-3}. \qquad\qquad 37b.$$

As Lee & Weinberg (1977) noted, a stable heavy neutrino with $m_{\text{GeV}} \lesssim 2$ is in conflict with the observational limits to ρ_0 (see Section 3.1).

Gunn et al (1978) and Steigman et al (1978) have investigated in more detail the astrophysical consequences of a stable heavy neutrino. The weak interaction of stable heavy neutrinos prevents them from participating fully in astrophysical clustering but they would form "halos" around galaxies and be associated with clusters of galaxies. Unless $m_\nu \gtrsim 10$ GeV, they would contribute too much mass to galaxies and clusters of galaxies. It may be of interest to note that, were our galaxy surrounded by a heavy neutrino-antineutrino halo, the gamma rays produced in the occasional annihilations might be detectable (Gunn et al 1978, Stecker 1978).

Dicus et al (1977, 1978a,b) considered the consequences of an unstable heavy neutrino, and Gunn et al (1978) pursued this question further. Although the constraints on unstable neutrinos are less severe than those found above, astrophysical considerations can serve to restrict large areas in the "mass vs lifetime" plane.

Basically, neutrinos in the mass range ~ 50 eV to ~ 10 GeV must be unstable (the lower end of this range is discussed in the next section). If the neutrino decays result in photons, either directly or through charged leptons, limits from the observed radiation backgrounds (γ rays, x rays etc) can severely restrict the lifetime: $\tau \gtrsim 10^6$ s (and $\tau \gtrsim t_0$, the present age of the Universe) is excluded (Dicus et al 1978a,b, Gunn et al 1978). Considerations of the effect of the entropy produced by decaying neutrinos on the primordial synthesis of deuterium further reduce the allowed lifetime (Dicus et al 1978b), $\tau \lesssim 10^3$ s.

Finally, it should be noted that similar constraints may be found for other weakly interacting particles, such as axions (Dicus et al 1978c, Schramm & Falk 1978). The best limits on the axion mass come not from cosmology but from stellar astrophysics. From the evolution of helium-burning stars, Dicus et al (1978c) conclude that $m_A \gtrsim 0.2$ MeV; from the energetics of supernovae, Schramm & Falk (1978) raised this limit to $m_A \gtrsim 10$ MeV.

4.6 Light Neutral Leptons

Some of the most interesting and restricting constraints on particle physics follow from the cosmology of long-lived ($\tau \gtrsim 1$ s), light ($\lesssim 1$ MeV), neutral leptons.

First, following the lead of Cowsik & McClelland (1972), consider

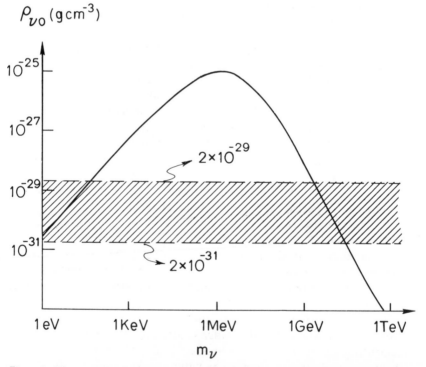

Figure 2 The present mass density in relic neutrinos. For stable massive neutral leptons the present mass density is shown as the function of the neutrino mass. The shaded area indicates the range for the estimates of the present total cosmological mass density.

the constraints from the present cosmological mass density (in this case we must have $\tau > t_0$; see Goldman & Stephenson 1977). Light neutrinos decoupled early ($T \simeq 1$ MeV) while they were still relativistic and therefore as abundant as photons:

$$\left(\frac{N_v}{N_\gamma}\right)_{1\ \text{MeV}} = \frac{3}{8} g_v. \qquad\qquad 38.$$

When $e^\pm$ pairs annihilate, extra photons are produced so that the present neutrino-to-photon ratio is

$$\left(\frac{N_v}{N_\gamma}\right)_0 = \frac{3}{22} g_v, \qquad\qquad 39.$$

and the mass density contributed by such neutrinos is

$$\rho_{v0} = 10^{-31} g_v m_v\ (\text{eV})\ \text{g cm}^{-3}. \qquad\qquad 40.$$

The restriction that $\rho_{v0} < 2 \times 10^{-29}$ leads (with $g_v = 4$ for each massive neutrino-antineutrino type) to the constraint

$$m_{ve} + m_{v\mu} + m_{v\tau} + \cdots \lesssim 50\,\text{eV}. \qquad\qquad 41.$$

For stable massive neutrinos, their present mass density as a function of their mass is shown in Figure 2.

The primordial abundance of ^{4}He provides another important constraint. In Section 3.3 (see also, Schramm & Wagoner 1977) we saw that the abundance by mass Y, of ^{4}He depends sensitively on the number of relativistic particles present for $T \gtrsim 1$ MeV. With three two-component neutrinos (v_e, v_μ, v_τ) and with $\rho_0 \gtrsim 2 \times 10^{-31}$ g cm^{-3}, Yang et al (1979) find $Y \gtrsim 0.25$. For each new two-component neutrino, the predicted helium abundance increases: $\Delta Y \simeq 0.01\ (\Delta g_v/2)$. The observational evidence reviewed by Yang et al (1979) suggests that the primordial abundance of ^{4}He was $\lesssim 0.25$, which suggests that there are no new neutrinos to be found. For example, if there were right-handed neutrinos associated with the known left-handed neutrinos, we would find $\Delta g_v = 6$ and $\Delta Y \simeq 0.03$; this is in apparent conflict with the data. It should be noted that these predictions based on cosmological considerations may be tested in accelerator experiments of the type proposed by Ma & Okada (1978).

4.7 Superweak Particles

In the preceding section it was seen that the primordial abundance of ^{4}He severely restricts the number of weakly interacting, light particles.

Suppose, however, there were superweakly interacting particles (for example, gravitons, gravitinos, right-handed neutrinos) that decoupled very early in the evolution ($T_d \gg 1$ MeV). Such particles would not have been heated by the annihilation of various pairs as the Universe cooled. Thus, at nucleosynthesis, their contribution to the mass-energy density is less than what it would have been had they remained in equilibrium. That is, below the decoupling temperature, T_d, the temperature of these decoupled particles, T_F (T_B) for fermions (bosons), is less than the photon temperature. From our entropy considerations in Section 2.2 it follows that

$$\left(\frac{T_{F,B}}{T_\gamma}\right)_T = \left[\frac{N_\gamma(T_d)}{N_\gamma(T)}\right]^{1/3} = \left[\frac{g_1(T)}{g_1(T_d)}\right]^{1/3}. \qquad 42.$$

At nucleosynthesis, $T \simeq 1$ MeV, $g_1(T) \simeq 43/4$ and $\rho_F = (7/16)g_F(T_F/T_\gamma)^4$, $\rho_B = (1/2)g_B(T_B/T_\gamma)^4$. The "new" total density is

$$\rho' = \rho + \Delta\rho_F + \Delta\rho_B = \left[\frac{43}{8} + \frac{7}{16}g_F\left(\frac{T_F}{T_\gamma}\right)^4 + \frac{1}{2}g_F\left(\frac{T_B}{T_\gamma}\right)^4\right]\rho_\gamma. \qquad 43.$$

The expansion is speeded-up (see Equation 5), the ratio of expansion rates being

$$\xi^2 \equiv \left[\frac{t(T)}{t'(T)}\right]^2 = \frac{\rho'(T)}{\rho(T)} = 1 + \frac{4}{43}\Delta g, \qquad 44a.$$

with

$$\Delta g = g_B\left(\frac{T_B}{T_\gamma}\right)^4 + \frac{7}{8}g_F\left(\frac{T_F}{T_\gamma}\right)^4. \qquad 44b.$$

The effect on the expansion rate due to these decoupled particles is diluted by the factors $(T_{F,B}/T_\gamma)^4 < 1$. Therefore, the restrictions on $g_{F,B}$ will be less severe. Steigman et al (1979) and Olive et al (1979) have explored this question in some detail; for $Y \lesssim 0.25$ and $\rho_0 \gtrsim 2 \times 10^{-31}$ g cm^{-3} they find that $\xi \lesssim 1.074$, which implies $\Delta g \lesssim 1.65$.

For neutrinos coupled to a heavier W', T_d (MeV) $\simeq (M_{W'}/M_W)^{4/3}$. If $T_d \lesssim 100$ MeV ($M_{W'} \lesssim 32 M_W$), $g_1(T_d) = 43/4$, and $\Delta g_\nu \lesssim 2$; at most one new neutrino is permitted. For $T_d \simeq 100$ GeV ($M_{W'} \simeq 5.6 \times 10^3 M_W$), $g_1(T_d) = 423/4$ and $\Delta g_\nu \lesssim 40$; at most twenty new two-component neutrinos are allowed. A somewhat smaller (by a factor of 7/8) limit applies to gravitons. Such constraints may be of some value in limiting the viable unified theories.

5 SUMMARY

The early Universe was very hot and very dense, providing an environment in which particles were produced copiously and interacted frequently. As the Universe expanded and cooled, the densities became so low that equilibrium could no longer be maintained. Some "relic" particles will have survived these early stages to affect the subsequent evolution of the Universe; stable relic particles will be present today.

The *Cosmological Principle* (isotropy and homogeneity) plus General Relativity ensures that, provided the basic particle physics is known, the physics of the early Universe is quite simple. As a result, if "footprints" remain from the early epochs, cosmology may be used to probe particle physics. The theory and results of this approach have been the subject of this review.

A quick introduction to cosmology was presented in Section 2.1 and it was seen that the early Universe evolves according to $t \sim \rho^{-1/2} \sim T^{-2}$. Comparing the expansion rate (t^{-1}) with various reaction rates (Γ) shows that virtually all particles will have come to equilibrium quite early. In Section 2.2 we were reminded of some simple but valuable results from equilibrium statistical mechanics. As the Universe evolved, collisions became less frequent and equilibrium was harder to maintain. The departures from equilibrium, the heart of this entire approach to particle physics via cosmology, were studied in Section 2.3. Those particles that interact most weakly (gravitons, neutral leptons) decouple earliest, before their abundances decreased significantly. Such particles remain abundant throughout the subsequent evolution of the Universe. For example, they would have been present during nucleosynthesis. The abundance of ^{4}He constrains the number of types of long-lived $(\tau \gtrsim 1 \text{ s})$, light $(\lesssim 1 \text{ MeV})$ neutral leptons to $\lesssim 3$–4. If they are stable $(\tau \gtrsim t_0)$, neutral leptons will be present today and will contribute to the cosmological mass density. Limits to $\rho_0 (\lesssim 2 \times 10^{-29} \text{ g cm}^{-3})$ constrain the mass of such neutrinos to $\gtrsim 2 \text{ GeV}$ or $\lesssim 50 \text{ eV}$. Neutrinos with masses in the excluded range ($\sim 50 \text{ eV}$ to $\sim 2 \text{ GeV}$) must be unstable. A variety of other cosmological constraints suggest that, if unstable, $\tau_\nu \lesssim 10^3$ s.

More strongly interacting particles, such as charged leptons, remain in equilibrium longer, reducing the surviving abundance. Nonetheless, stable massive charged leptons should be easy to find in direct searches and, if $m_{L^\pm} \gtrsim 100 \text{ GeV}$, they would contribute significantly to the present mass density.

Finally, the most strongly interacting particles are maintained in equilibrium longest; this reduces considerably the abundance of relic hadrons. Indeed, when we applied the analysis of Section 2.3 to relic nucleons, the predicted abundance found was some ten orders of magnitude below the observed abundance. This, however, was not a failure. We learned an important lesson: The early Universe had a net baryon number. Predicting the baryon number (for example, the nucleon-to-photon ratio) on the basis of baryon-violating, CP-nonconserving, non-equilibrium processes at very high temperatures ($\gtrsim 10^{15}$ GeV) in the very early Universe ($\lesssim 10^{-37}$ s) is a challenge confronting super-unified theories.

As with massive stable charged leptons, relic heavy hadrons or free quarks should be easily detected, even at very low concentrations. Their absence constrains the possible models of hadron physics.

We have seen, I am certain, only the tip of the iceberg. The approach to particle physics via cosmology is and will continue to be a powerful and often indispensable supplement to accelerator physics.

NOTE ADDED IN PROOF

In Section 3.3 it was shown that the abundance of ^{4}He produced in primordial nucleosynthesis can lead to a significant constraint on the number of types of light neutral leptons. The results of Yang et al (1979) described in Section 4.6 suggest that there are no more than 3–4 two-component neutrinos. Recently, Linde (1979) and Dimopoulos & Feinberg (1979) have challenged this conclusion. Their point is that if the electron neutrinos (or antineutrinos) are degenerate, then the neutron-to-proton ratio is changed and, ultimately, so is the resulting abundance of ^{4}He. Appropriate degeneracy may then compensate for the effect of extra neutrino types and still leave $Y \approx 0.25$. However, the "appropriate" degeneracy is extreme and unnatural (Dimopoulos & Feinberg 1979, Nanopoulos et al 1979, Schramm & Steigman 1979). Let us explore this in a little detail.

For temperatures $\gtrsim 1$ MeV, the neutron-to-proton ratio is maintained in chemical equilibrium by the reactions $p + e^- \rightleftarrows n + \nu_e$, $n + e^+ \rightleftarrows p + \bar{\nu}_e$, $p \rightarrow n + e^+ + \nu_e$. Allowing for degeneracy, the chemical potentials of the electrons and the electron neutrinos are nonzero and the neutron-to-proton ratio in equilibrium is

$$\frac{n}{p} = \exp\left(-\frac{\Delta mc^2}{kT} - \frac{\mu_{ve}}{kT} - \frac{\mu_e}{kT}\right).$$

The excess of electrons over positrons is equal to the excess of protons over antiprotons, which is roughly equal to the baryon number: $\Delta N_e \approx \Delta N_B \ll N_\gamma$. Since N_B is some nine orders of magnitude smaller than N_p, $\mu_e \ll kT$ and so

$$\frac{n}{p} \approx \exp\left(\frac{-\Delta mc^2}{kT} - \frac{\mu_{ve}}{kT}\right).$$

Because there exist no significant constraints on the v (or $\bar{v}$) excess, $|\mu_{ve}|$ may, in principle, be large and n/p may be very different from our previous, nondegenerate estimate. The reactions listed above reveal that a neutrino excess tends to decrease the neutron abundance and hence the ^{4}He abundance; an antineutrino excess has the opposite effect. This was first noted in the paper of Wagoner et al (1967) (see also Schramm & Wagoner 1977), and most recently by Yahil & Beaudet (1976). A secondary effect of degeneracy of any neutrino type is that the extra relativistic particles accelerate the expansion rate and, hence, the weak interactions drop out of equilibrium at a higher temperature. In order to retain the agreement between observation and the predicted abundances of D, He, and Li, different degeneracies must be assumed for the different neutrino types (Yahil & Beaudet 1976, Schramm & Steigman 1979).

To modify significantly the nondegenerate neutron abundance requires $|\mu_{ve}| \gtrsim kT$. But, the ratio of excess neutrinos to photons, $\Delta N_{ve}/N_\gamma$, is of order $(\mu_{ve}/kT)^3$ so that $\mu_{ve} \gtrsim kT$ implies $\Delta N_{ve} \gtrsim N_\gamma$. We easily see then that the required neutrino excess is some nine orders of magnitude larger than the baryon or electron excess: $\Delta N_{ve} \gtrsim N_\gamma \approx 10^9 \Delta N_B$. Such an imbalance is very unnatural in any baryon and lepton number violating grand unified theory and would require very unlikely accidental cancellations (Dimopoulos & Feinberg 1979).

In summary, an enormous neutrino degeneracy could modify the primordial abundance of ^{4}He and invalidate the constraint on the number of neutrino types. Although such a large degeneracy cannot be excluded, there is no natural explanation of its origin. Indeed, the excellent agreement between the observed abundances and the predictions of the "standard" (nondegenerate) hot big bang model is evidence that this model correctly describes the early evolution of the universe.

ACKNOWLEDGMENTS

It is not possible to thank individually all my friends in particle physics and cosmology from whom I have learned so much. I have benefited particularly from my close collaboration with T. K. Gaisser and D. N. Schramm. This review was written while I was a Visiting Scholar at Stanford University. I wish to thank Bob Wagoner (Physics) and Peter Sturrock, Vahé Petrosian, and the other members of the Institute for Plasma Research for their hospitality. Special thanks go to Louise Meyers and Evelyn Mitchell for expert typing and unlimited patience. Support at Stanford was, in part, through NASA Grant NGR 05-020-668. Finally, I am deeply indebted to Tom Gaisser, Vahé Petrosian, Dave Schramm and Bob Wagoner for their critical reading of this manuscript.

Literature Cited

Alvager, T., Naumann, R. 1967. *Phys. Rev. Lett. B* 24:647
Bennett, C. L., Beukens, R. P., Clover, M. R., Elmore, D., Gove, H. E., Kilius, L., Litherland, A. E., Purser, K. H. 1978. *Science* 201:345
Chiu, H. Y. 1966. *Phys. Rev. Lett.* 17:712
Cowsik, R., McClelland, J. 1972. *Phys. Rev. Lett.* 29:669
Dicus, D. A., Kolb, E. N., Teplitz, V. 1977. *Phys. Rev. Lett.* 39:168
Dicus, D. A., Kolb, E. N., Teplitz, V. 1978a. *Astrophys. J.* 221:327
Dicus, D. A., Kolb, E. N., Teplitz, V., Wagoner, R. V. 1978b. *Phys. Rev. D* 17:1529
Dicus, D. A., Kolb, E. N., Teplitz, V., Wagoner, R. V. 1978c. *Phys. Rev. D* 18:1829
Dimopoulos, S., Feinberg, G. 1979. Preprint CU-TP-159
Dimopoulos, S., Susskind, L. 1978a. *Phys. Rev. D* 18:4500
Dimopoulos, S., Susskind, L. 1978b. *Stanford Univ. Rep. No. ITP-616*
Dover, C. B., Gaisser, T. K., Steigman, G. 1979. *Phys. Rev. Lett.* 42:1117
Ellis, J., Gaillard, M. K., Nanopoulos, D. V. 1979. *Phys. Lett. B* 80:30
Frautschi, S., Steigman, G., Bahcall, J. 1972. *Astrophys. J.* 175:307
Goldman, T., Stephenson, G. J. Jr. 1977. *Phys. Rev. D* 16:2256
Gott, J. R., Gunn, J. E., Schramm, D. N., Tinsley, B. M. 1974. *Astrophys. J.* 194:543
Gunn, J. E., Lee, B. W., Lerche, I., Schramm, D. N., Steigman, G. 1978. *Astrophys. J.* 223:1015
Hagedorn, R. 1970. *Astron. Astrophys.* 5:184
Jones, L. W. 1977. *Rev. Mod. Phys.* 49:717
Kazanas, D., Schramm, D. N., Hainebach, K. 1978. *Enrico Fermi Inst. Preprint No. 78-18*
La Rue, G. S., Fairbank, W. M., Hebard, A. F. 1977. *Phys. Rev. Lett.* 38:1011
La Rue, G. S., Fairbank, W. M., Phillips, J. D. 1979a. *Phys. Rev. Lett.* 42:142
La Rue, G. S., Fairbank, W. M., Phillips, J. D. 1979b. *Phys. Rev. Lett.* 42:1019
Lee, B. W., Weinberg, S. 1977. *Phys. Rev. Lett.* 39:165
Linde, A. D. 1979. *Phys. Lett. B* 83:311
Ma, E., Okada, J. 1978. *Phys. Rev. Lett.* 41:287
Middleton, R. et al. 1979. *Phys. Rev. Lett.* 43:429
Muller, R. A. 1979. *Phys. Today* (Feb.) p. 23

Muller, R. A., Alvarez, L. W., Holley, W. R., Stephenson, E. J. 1977. *Science* 196:521
Nanopoulos, D. V., Sutherland, D., Yildiz, A. 1979. *Phys. Lett.* In press
Olive, K. A., Schramm, D. N., Steigman, G. 1979. In preparation
Schiffer, J. 1974. Workshop on BeV/Nucleon Collisions of Heavy Ions, Bear Mountain, NY. *Brookhaven Natl. Lab. Publ. 50445*, p. 61
Schramm, D. N., Falk, S. 1978. *Phys. Lett. B* 79:511
Schramm, D. N., Steigman, G. 1979. *Phys. Lett.* In press
Schramm, D. N., Wagoner, R. V. 1977. *Ann. Rev. Nucl. Sci.* 27:37
Stecker, F. W. 1978. *Astrophys. J.* 223:1032
Steigman, G. 1973. *Cargèse Lect. Phys.* 6:505
Steigman, G. 1976. *Ann. Rev. Astron. Astrophys.* 14:339
Steigman, G. 1979. *Stanford Univ. Inst. Plasma Res. Rep. No. 782*
Steigman, G., Olive, K. A., Schramm, D. N. 1979. *Phys. Rev. Lett.* 43:239
Steigman, G., Schramm, D. N., Gunn, J. E. 1977. *Phys. Lett. B* 66:202
Steigman, G., Sarazin, C. L., Quintana, H., Faulkner, J. 1978. *Astron. J.* 83:1050
Toussaint, B., Treiman, S. B., Wilczek, F., Zee, A. 1979. *Phys. Rev. D* 19:1036
Wagoner, R. V., Fowler, W. A., Hoyle, F. 1967. *Science* 155:1369
Wagoner, R. V., Steigman, G. 1979. *Stanford Univ. Rep. No. ITP-633*; *Phys. Rev. D.* In press
Weinberg, S. 1972. *Gravitation and Cosmology: Principles and Applications of the General Theory of Relativity*. New York: Wiley. 657 pp.
Weinberg, S. 1977. *The First Three Minutes*. New York: Basic. 188 pp.
Weinberg, S. 1979. *Phys. Rev. Lett.* 42:850
Wolfram, S. 1979. *Phys. Lett. B* 82:65
Yahil, A., Beaudet, G. 1976. *Astrophys. J.* 206:26
Yang, J., Schramm, D. N., Steigman, G., Rood, R. T. 1979. *Astrophys. J.* 227:697
Zel'dovich, Ya. B. 1965. *Adv. Astron. Astrophys.* 3:241
Zel'dovich, Ya. B. 1970. *Comments Astrophys. Space Phys.* 2:12
Zel'dovich, Ya. B., Okun, Pikel'ner, S. B. 1965. *Phys. Lett.* 17:164
Zel'dovich, Ya. B., Okun, L. B., Pikel'ner, S. B. 1966. *Sov. Phys. Uspekhi* 8:702

Ann. Rev. Nucl. Part. Sci. 1979. 29: 339–93

LIGHT HADRONIC SPECTROSCOPY: Experimental and Quark Model Interpretations[1]

*5610

S. D. Protopopescu and N. P. Samios

Brookhaven National Laboratory, Upton, New York 11973

CONTENTS

1 INTRODUCTION

Since the recent discovery of charmed particles, the quark model has become one of the fundamental ingredients of elementary particle

[1] The submitted manuscript has been authored under contract EY-76-C-02-0016 with the US Department of Energy. Accordingly, the US Government retains a nonexclusive, royalty-free license to publish or reproduce the published form of this contribution, or allow others to do so, for US Government purposes.

339

physics. A measure of the fruitfulness of any dynamical scheme of quark interactions is obviously how well it can describe the observed hadrons. First attempts have been made to actually calculate hadronic masses, given the quark configurations, with some degree of success (see, for example, DeRujula et al 1975). For any exercises of this kind it is imperative to be provided with an adequate data base, i.e. well-determined masses and quantum numbers. Progress in spectroscopy without charm has been slow but steady in the past few years, and it now seems to be an opportune time to give a general overview of the field, particularly how the known particles fulfill quark model expectations, what still remains to be discovered, and what are the promising avenues to pursue. In a review of this field, which has grown enormously since the early 1960s, it is, of course, impossible to cover all topics of interest in great depth. Rather than dwell on experimental techniques and formalism, we have chosen to concentrate on final results and particularly to follow in some detail the results from partial wave analyses, which essentially give the most information one can hope to extract from experimental data. The time honored technique of looking for bumps in mass spectra is still very much alive and is not likely to go out of style, but it has its limitations; the presence or absence of bumps need not be correlated with the presence or absence of resonances, particularly when dealing with very broad resonances.

First we discuss what has been learned about meson states from two-body decays, and then, more complex decay modes. A special section is devoted to vector mesons, particularly results from photoproduction and e^+e^- annihilations concerning states with masses less than 2.0 GeV. In the second part of this paper we review the baryons emphasizing results from partial wave analysis of formation (s-channel) experiments excepting, of course, states with strangeness (S) less than -1 that are not accessible via formation processes. Before concluding, we compare the presently known states with quark model expectations. The theoretical tools for making detailed quantitative comparisons with the data are not yet fully developed, but in a naive qualitative comparison the model holds up quite well. Most of the difficulties so far seem to be with the data rather than the model. In spite of the great proliferation of states, there are still many gaps to be filled; this is particularly true for un-natural spin parity $(J^P = 0^-, 1^+, 2^-, \ldots)$ mesons and for baryon decuplets. In the latter case the problems are understood to be mainly experimental, Ξ^* and Ω^* resonances $(S < -1)$ can only be studied via baryon exchange processes, which have small cross sections and their decays generally involve many particles compounding the difficult task of finding them. For unnatural spin parity mesons the difficulties are deeper. Excepting the ground states, 0^-, all of the other unnatural spin parity nonets are incomplete and, even given the fact that some members are missing, there seems to be no obvious pattern indicating which states should be grouped together. There is no clear-cut reason why they should not be produced just as plentifully in meson exchange processes as the natural spin parity mesons; in fact, well established 1^+ states such as the B and D have comparable production cross sections. These gaps are, therefore, a source of some embarrassment for spectroscopists. It may be that higher statistics data and more refined methods of analysis will finally reveal them or it may be that the quark model itself needs some subtle refinements. It is unlikely, given its success in many other areas, that these difficulties will lead to a major revision of the quark model.

Another topic that we cover is that of exotics, i.e. hadrons that cannot be explained as $q\bar{q}$ or qqq states. Evidence is mounting for the existence of six-quark states; these are not unexpected, one can think of them as relatives of the deuteron. Another interesting possibility is the existence of $qq\bar{q}\bar{q}$ or $qqqq\bar{q}$ states, nothing in the quark model forbids them. The question is whether or not such configurations can stay together long enough to form recognizable resonant states. Invoking $qq\bar{q}\bar{q}$ states seem to help some of the problems in classifying known meson states, particularly the 0^+ states. They are also required to preserve the notion of duality in nucleon-antinucleon ($N\bar{N}$) interactions (i.e. that interactions can be equally described in terms of formation in the s-channel or exchanges in the t-channel). It is therefore important to find them. However, this may prove to be a difficult experimental task; they may be very broad and massive and thus not easy to distinguish from background processes. So far there are many prospective candidates but no conclusive proof that these types of exotics actually exist.

One major aspect of particle spectroscopy that we have neglected is branching ratios and, in particular, the important topic of electromagnetic decays. Our only excuse is that we had to satisfy some boundary conditions; these are topics worthy of a separate article.

2   MESON STATES

2.1 0^-0^- Systems

Practically all we know about meson resonances (excepting the special case of vector mesons) has been gathered from studies of two-body or quasi two-body decays. In particular the studies of the 0^-0^- decays have played a very important role. Despite their being plagued by ambiguities and analysis uncertainties, they are at present the best-understood systems. Their main shortcoming is that they can yield information only about resonances belonging to the natural spin parity series $(0^+, 1^-, 2^+, \ldots)$. This shortcoming is of course a blessing from the point of view of analysis.

2.1.1 $\pi\pi$ CHANNEL   Because of its simplicity (both theoretically and experimentally), the $\pi\pi$ system has been one of the most studied and also one of the most fruitful systems. The $\rho(1^-)$, $f^0(2^+)$, $g(3^-)$, and most recently the $h(4^+)$ were first observed and their spin parity determined in this system. It was noted very early by Chew & Low (1959) that reactions of the type $\pi N \to \pi\pi N$ and $\pi N \to \pi\pi\Delta$ would be dominated by π-exchange at small $|t|$ and that by extrapolating the data in t to the π-pole one could obtain the amplitudes for the reaction $\pi\pi \to \pi\pi$. The above resonances lie on leading Regge trajectories and are the most noticeable feature of $\pi\pi$ scattering. As such they are relatively easy to distinguish in any $\pi\pi$ phase shift analysis, provided there are sufficient statistics and energy for their copious production. Hidden underneath these leading resonances other states (daughter trajectories) were expected to exist, but only a detailed phase shift analysis would be able to reveal them. Earlier theoretical conjectures such as the Veneziano model (Veneziano 1968) predicted the existence of a 0^+ resonance under the ρ (referred to as ε), a 0^+ (ε'), and a 1^- $[\rho'(1250)]$ state under the $f^0(1270)$, etc. These daughter resonances are also expected in the quark models as $q\bar{q}$ radial excitations, although in this case there is no a priori reason for low spin resonances to be degenerate in mass with higher spin ones.

The first studies of $\pi^+\pi^-$ scattering showed that, below 1.0 GeV, it was mainly elastic and dominated by the P-wave (ρ-resonance) with significant S-wave (0^+) background (Gutay et al 1969, Oh et al 1970, Colton et al 1971). These investigations were, however, hampered by low statistics. With higher statistics and information from other channels (such as $\pi^+\pi^+$, $\pi^-\pi^0$, etc) it became clear that most of the S-wave background was $I = 0$, saturating the unitarity limit near the ρ-peak, and could be described by two solutions: one indicated the possible existence of a resonance with roughly the same width as the ρ (up solution); the other showed no rapid phase variation and could be consistent with a very broad resonance $\gtrsim 300$ MeV or no resonance at all (down solution). Given the uncertainties in the extrapolation to the π-pole (absorption, exchanges other than π, etc), this ambiguity is in practice impossible to resolve near the ρ-peak (between 600 and 900 MeV) (Baton et al 1970, Baillon et al 1972).

The up-down ambiguity was resolved by the observation that when one crosses the $K\bar{K}$ threshold (1 GeV) the $\pi\pi$ S-wave falls rapidly and by an amount indicating that it must have been near the maximum allowed by unitarity (Flatté et al 1972), a behavior incompatible with the up solution. This phenomenon seems to be well explained by the existence of a 0^+ resonance, S*, very near the $K\overline{K}$ threshold and coupling more strongly to $K\overline{K}$ than to $\pi\pi$ (Protopopescu et al 1973, Grayer et al 1974). The sudden decrease of the $I = 0$, $\pi\pi$ S-wave is matched by a rapid rise of the $K\overline{K}$ S-wave at threshold. The S* is a good example of the need for careful investigation of the amplitudes (and not just searching for bumps in cross sections) in order to detect resonances. By now the S* has been observed in reactions not involving π-exchange, such as $\pi N \rightarrow \pi\pi N$ near threshold (Binnie et al 1973), $K^- p \rightarrow \pi\pi\Lambda(\Sigma)$ (Brandenburg et al 1976a), $e^+ e^- \rightarrow \psi \rightarrow \phi\pi\pi$ (Vanucci et al 1977), and can be considered well established.

More information on the S-wave can be obtained studying the $\pi^0\pi^0$ channel (in reactions such as $\pi^- p \rightarrow \pi^0\pi^0 n$ or $\pi^+ p \rightarrow \pi^0\pi^0\Delta^{++}$), which has the advantage of being a much simpler channel in theory (only even waves can contribute) but is much more difficult experimentally than the $\pi^+\pi^-$ channel. Although there is disagreement in detail between the various sets of data available at present, most high statistics experiments seem to be compatible with the down solution, rejecting the possibility of a narrow ε (Skuja et al 1973, Apel et al 1972, Braun et al 1973, Riester et al 1975, Apel et al 1978).

Beyond 1.0 GeV many detailed phase shifts analyses have been performed using the very high statistics data from the CERN-Munich experiment (Hyams et al 1973, Estabrooks & Martin 1974, Hyams et al 1975). Although there are some discrepancies in the fine details of the phase shifts obtained by very different methods, there is no disagreement on the overall features. We concentrate here mostly on the results of Hyams et al (1975), as they seem to be the most complete. There are two solutions between 1.0 at 1.2 GeV and four solutions at higher masses. All solutions clearly show the f^0-meson dominating the D-wave and g-meson dominating the F-wave. Fits to the phase shifts give for the f^0 a mass ~ 1275 MeV and a width $\Gamma \sim 185$ MeV, while the respective values for the g-meson are $m \sim 1720$ MeV and $\Gamma \sim 240$ MeV. All four solutions give values fairly close to the ones given above; what is worth noting is that the masses tend to be higher and widths larger than typical values obtained by simply fitting the mass spectrum.

When it comes to the nonleading partial waves, no definitive statements can be made. The present status of knowledge of the $I = 0,1$ S and P $\pi\pi$ partial waves above 1.0 GeV is summarized in Figure 1. There are some general features common to all solutions worth pointing out. In all cases the S-wave is close to the unitarity limit under the f^0, i.e. $\delta_s^0 \sim 90^0$ and $\eta_s^0 \gtrsim 0.75$, very similar to the S-wave under the ρ. All solutions are compatible with a fairly broad ε', in some cases between 200-

and 300-MeV wide, in others 400-MeV wide or more. The interpretation of the S-wave under the g-meson is not straightforward, but it seems again that every solution is compatible with a broad S-wave resonance in that region. We return to the S-waves when discussing $K\bar{K}$ partial waves. None of the solutions shows the presence of a P-wave resonance near 1250 MeV, although two of them are compatible with very inelastic and broad P-wave resonances near 1380 MeV and 1600 MeV. Finally, there is no hint in any of the solutions of a D-wave resonance under the g-meson. Thus, the question of daughter resonances (or radial excitations) cannot be answered with present $\pi^+\pi^-$ data and no increase in statistics is likely to remove the ambiguities. Nevertheless, some conclusions about resonances on nonleading Regge trajectories can be reached. The possible 0^+ resonances (excluding the S*) must be broad and mostly elastic, i.e. couple more strongly to $\pi\pi$ than any other channel, while the 1^- and/or

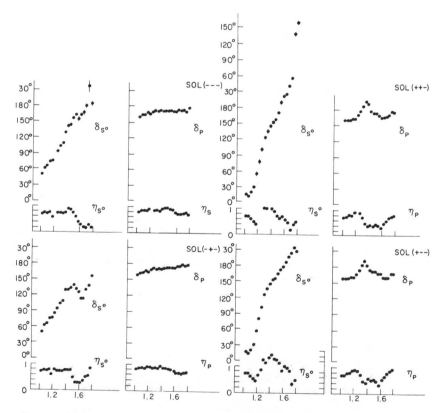

Figure 1 Phases and inelasticities of S and P $\pi\pi$ scattering waves as a function of mass for the four solutions above 1.0 GeV found by Hyams et al (1975).

2^+ resonances (if they exist at all) must have relatively small couplings to the $\pi\pi$ system.

Any further progress along these lines requires the study of other channels such as $\pi^0\pi^0$, $K\overline{K}$, $\pi\omega$, etc. The $\pi^0\pi^0$ channel is particularly important as the even partial waves observed in $\pi^+\pi^-$ can be trivially related to $\pi^0\pi^0$ amplitudes and odd waves cannot contribute. A high statistics phase shift analysis, of reactions of the type $\pi^-p \to \pi^0\pi^0N$ or $\pi^+p \to \pi^0\pi^0\Delta^{++}$ for $m_{\pi\pi} > 1.06$ GeV, should help reduce the ambiguities but may not necessarily lead to a unique solution. Although data exist, such an analysis has not yet been performed. One result from high energy $\pi^0\pi^0$ data is the existence of a 4^+ resonance (h-meson) at $m \sim 2.02$ GeV and $\Gamma \sim 180$ MeV (Apel et al 1975).

2.1.2 $K\overline{K}$ CHANNEL The dominant states in the $K\overline{K}$ channel are the $\phi(1020)$ and $f'(1514)$ in K-induced reactions. Their spin and parity are well determined as 1^- and 2^+, respectively. These states are not as prominent in π-induced reactions for which important new information is now available from reactions of the type $\pi^-p \to K\overline{K}N$ (Wetzel et al 1976, Cason et al 1976, Pawlicki et al 1977, Baldi et al 1978). Particularly interesting would be the study of final states K_SK_S and K_SK_L. The first can only have contributions from even spin states, while the second only from odd spins. Unfortunately, no high statistics sample of data exists for K_SK_L. Just as much can be learned, however, using K_SK_S and K^+K^- data (as both spins contribute to the latter).

The prominent features of the K_SK_S mass spectrum (below 1.8 GeV) are the S* at threshold and a broad peak near 1.4 GeV. The broad peak also dominates the $\langle Y_4^0 \rangle$ moment and is well explained by f^0-f' interference (with additional A_2 contribution at high t). Another noticeable feature of this system is a very large and rapidly varying $\langle Y_2^0 \rangle$ moment between 1.2 and 1.4 GeV, which can only be explained by interferences between S- and D-waves. The experimental moments of the two highest statistics experiments at present (Cason et al 1976 at Argonne, Wetzel et al 1976 at CERN) are in excellent agreement (Figure 2). However, each group reaches different conclusions concerning the S-wave amplitude. Both find the S* at threshold, but Wetzel et al favor a solution with a broad, structureless S-wave under the f^0-f', while Cason et al obtain two solutions, one similar to Wetzel et al, while the other shows a Breit-Wigner-like phase variation indicating an S-wave resonance of mass ~ 1275 MeV and width $\Gamma \sim 80$ MeV. Although there is no compelling reason to do so, Cason et al favored the second solution. From a more detailed analysis as a function of t, the ANL group concluded this S-wave resonance is not produced by π-exchange and is most likely an $I = 1$ state, which they named δ' (recurrence of the δ).

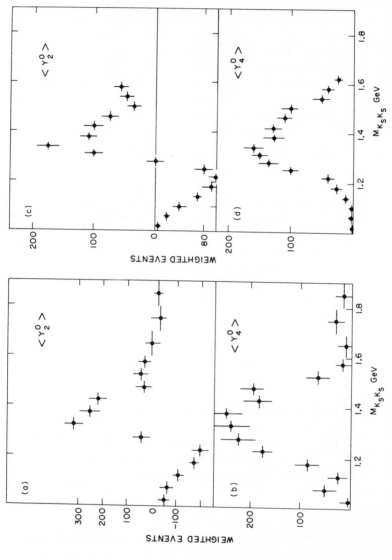

Figure 2 The $K_S K_S \langle Y_2^0 \rangle$ and $\langle Y_4^0 \rangle$ unnormalized moments from the reaction $K^- p \rightarrow K_S K_{SN}$: (*a*) and (*b*) found by Cason et al (1976), (*c*) and (*d*) found by Wetzel et al 1976).

The controversy seems to have been partially resolved by a high statistics experiment with the Argonne Effective Mass Spectrometer (EMS) that studied reactions (Pawlicki et al 1977):

$$\pi^- p \rightarrow K^- K^+ n \qquad\qquad 1.$$

$$\pi^+ p \rightarrow K^- K^+ p \qquad\qquad 2.$$

The above reactions allow one to clearly separate $I = 0$ and $I = 1$ amplitudes as the $I = 1$ amplitude changes sign between reactions 1 and 2. The EMS analysis shows the expected D-wave from f^0-f' resonances and the S* effect in the S-wave below 1200 MeV. Above 1200 MeV Pawlicki et al obtain two solutions for the size of the S- and P-waves. One shows the S-wave peaking near 1300 MeV with a small P-wave, the other solution is exactly reversed with a large P-wave and small S-wave. The analysis was done at small momentum transfer $|t|$ ($|t| < 0.08$ GeV^2), so these partial waves must be produced mostly by π-exchange. The solution with small S-wave is, in fact, incompatible with the $K_S K_S$ data as it predicts a $\langle Y_0^2 \rangle$ moment that is much too small. Thus, the large S-wave solution is selected.

When one then proceeds to solve for the amplitudes and attempts to separate $I = 0$ and $I = 1$ components, both solutions observed by Cason et al (1976) agree with the $K^- K^+$ data. However, the solution having a large $I = 0$ S-wave with slowly varying phase requires an $I = 1$ P-wave that is in excellent agreement with expectations from the tail of the $\rho \rightarrow K^+ K^-$, with a $\rho K \bar{K}$ coupling as predicted from SU(3). None of the other solutions has this last property, so the EMS group tends to favor it. Analysis at large $|t|$ again shows large S-wave contribution. The analysis is not yet complete since no separation of $I = 0$ and $I = 1$ waves has been done, but indications are that both are large. At higher masses there is evidence from the reaction $\pi^- p \rightarrow K^- K^+ n$ at 18.4 GeV/c for a $J^P = 4^+$ resonance with mass ~ 2050 MeV and width 225 MeV (Blum et al 1975). From its t distribution it was deduced to be $I = 0$ and thus an alternate decay mode of the h(2020).

In the charged mode $K^- K^0$ there are now high statistics data available from a counter experiment at CERN studying the reaction $\pi^- p \rightarrow K^- K^0 p$ at 10 GeV/c (Baldi et al 1978). A study of the moments and mass spectrum between 1.4 GeV and 2.1 GeV shows strong evidence for a $J^P = 4^+$ state at 1950 MeV with a width of about 200 MeV. An amplitude analysis of the same data from threshold up to 2.0 GeV (Martin et al 1978) shows in addition to the above state, the A$_2$, the g, and a significant S-wave under the A$_2$. The phase variation does not unambiguously show that this enhancement is indeed a resonance.

The main conclusions from present $K\overline{K}$ data are: D-waves are dominated by f^0 and f' resonances (A_2 is also present at large $|t|$), a small P-wave component exists, the S-wave shows effects of the S* resonance at threshold, and a broad $I = 0$ S-wave with a slowly varying phase that peaks near 1300 MeV exists. There are also indications of an $I = 1$ S-wave peaking near the same mass. What is missing as yet from the analysis of $K\overline{K}$ data are attempts to extrapolate to the π-pole and thereby extract precisely what is the contribution from $\pi\pi \to K\overline{K}$ for each partial wave. This should help choose among the various $\pi\pi$ solutions, at least below ~ 1400 MeV. For example, the solutions showing resonant-like behavior in the P-wave amplitude near 1380 and 1600 MeV require an almost purely elastic S-wave saturating unitarity around 1300 MeV. This is clearly incompatible with a large S-wave contribution to the $K\overline{K}$ channel.

Cason et al (1978) extrapolated $\pi^- p \to K_S K_S n$ data to the π-pole to obtain the $\pi\pi \to K\overline{K}$ cross section. They found that it could be well explained with the S* and f^0 alone, which would favor a purely elastic S-wave solution near 1300 MeV. However, the calculated cross section for the S* is significantly smaller than that expected from $\pi^+\pi^-$ phase shift analyses and K^+K^- data. This question needs further study to sort out contradictory claims.

2.1.3 $K\pi$ CHANNEL   Information on the $K\pi$ channel has been accumulated somewhat more slowly than in the $\pi\pi$ case, in part because it is harder to obtain high statistics in this channel. However, substantial progress has been made in the last few years. The leading resonances K* (890) (1^-) and K*(1420) (2^+) were observed very early (Alston-Garnjost et al 1961, Badier et al 1965, Chung et al 1965), but it is only recently that the 3^- K*(1780) has been firmly established (Baldi et al 1976, Chung et al 1978) and there is still some question concerning its width.

Early results on $K\pi$ phase shifts were plagued by the same up-down ambiguity in the mass region under the K*(890) as exists in $\pi\pi$ under the $\rho(750)$. Higher statistics and more detailed studies (Matison et al 1974, Chung et al 1972) resolved the ambiguity in favor of the down solution, i.e. no narrow 0^+ resonance hidden under the K*(890). At present the most detailed study of $K\pi$ phase shifts is the one of Estabrooks et al (1978) and as such we concentrate mainly on their results. Concerning resonances on the leading Regge trajectory, the results of Estabrooks et al (1978) on the K*(890) are in agreement with earlier experiments. On the other hand, they report a higher mass than previously estimated for the D-wave K*(1420) ~ 1435 MeV. This is mainly owing to the large S-wave contribution that peaks below 1.4 GeV and results in a fit to the phase

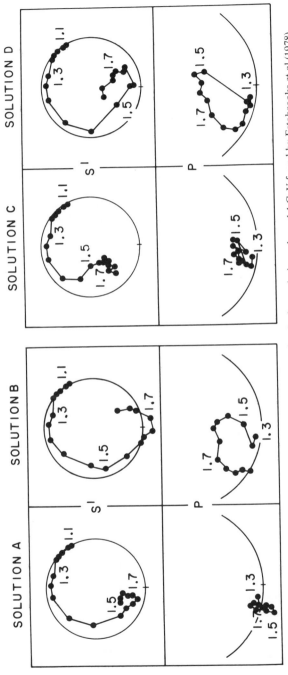

Figure 3   Argand diagram of the S- and P-Kπ scattering waves for the four solutions above 1.1 GeV found by Estabrooks et al (1978).

shifts that tends to give a higher mass than a simple fit to the mass spectrum. The F-wave shows the K*(1780) and indicates that it is highly inelastic, thus one should expect to see large contributions of this resonance in other channels.

Below 1.4 GeV the only additional wave contributing significantly is the S-wave, which is mostly elastic with a very slowly varying phase, reaching 90^0 slightly above 1300 MeV. It is debatable whether this S-wave could be considered a resonance (usually referred as κ). If so, the slow phase shift variation implies a width of ~ 900 MeV. Above 1.4 GeV there are four possible solutions (see Figure 3 for S- and P-waves). All solutions show the presence of an S-wave resonance (κ') at 1.45 GeV with a width of 200 to 300 GeV (relatively narrow compared to the κ). In two of the solutions there is a P-wave resonance at 1.65 GeV. These four solutions predict quite different moments in the $K^0\pi^0$ channel, so that a high statistics experiment studying this channel at small t could lead to a unique solution up to 1.8 GeV.

It is remarkable how similar the $K\pi$ and $\pi\pi$ systems are. In both cases the leading resonances are P, D, F, ...; there is a large featureless S-wave background and probably an S-wave resonance under the D-wave resonance; practically no background from $L > 0$ partial waves; and two of the four possible solutions at high masses show a P-wave resonance close to the F-wave resonance. Apart from special effects coming from the strong coupling of pieces of the $\pi\pi$ S-wave to $K\bar{K}$ (i.e. the S* resonance), the detailed dynamics of $\pi\pi$ and $K\pi$ channels seem to be almost identical. This similarity extends also to the exotic channels $I = 2$ $\pi\pi$ and $I = 3/2$ $K\pi$ states, in which an elastic S-wave with negative phase shift dominates and slowly decreases as function of mass (Hoogland et al 1974, Wagner 1974). Present information on $\pi\pi$, $K\pi$, and $K\bar{K}$ channels is probably of sufficiently high quality to put stringent constraints on dynamical models of mutual π and K scattering.

2.1.4 OTHER 0^-0^- CHANNELS   Knowledge of 0^-0^- channels other than $K\pi$, $\pi\pi$, and $K\bar{K}$ is still quite sketchy. At present there is no evidence for any mesons decaying to $\eta\eta$ or $K\eta$ final states, nor any two-body decays with η' as one of the final products. It was known for some time that the A_2 had $\sim 10\%$ branching ratio into $\eta\pi$, and there was also low statistics evidence for a relatively narrow $\delta(970) \to \eta\pi$ (Chung et al 1968, Defoix et al 1968). There is now quite conclusive evidence for this state in $K^-p \to \Sigma^+ (1385)\eta\pi^-$ (Gay et al 1976a) and in the reaction $\pi^-p \to \pi p\eta\pi\pi$ (Grässler et al 1977). The mass and width obtained for the δ are 974 ± 9 MeV and 55 ± 15 MeV. Study of the $\eta\pi$ angular distribution favors a 0^+ interpretation. Given its mass and the fact that it is significantly above

$\eta\pi$ threshold, this resonance seems curiously narrow. In both of the above observations of the $\delta(970)$ it was also noted that there is a significant enhancement in the $K^{\pm}K^0$ channel near its threshold. If this enhancement is indeed associated with the $\delta(970)$ it leads to a natural explanation of this observed narrow width since it implies a $\delta^{\pm}(970)$ with a strong coupling to $K\overline{K}$, in the same manner as the strong coupling of the S* to $K\overline{K}$ makes this latter state appear narrow in the $\pi\pi$ channel (Flatté 1976). This explanation leads to interesting complications when one attempts to interpret the 0^+ nonet with the quark model (see Sections 2.4 and 4).

2.2   Quasi Two-Body Decays

The two-body decays discussed in the previous section clearly show the existence of resonances belonging to the natural spin parity series 0^+, $1^-, 2^+, \ldots$; most of which fit nicely with quark model expectations. The quark model also predicts the existence of resonances belonging to the unnatural spin parity series $1^+, 2^-, \ldots$, which requires the study of more complicated, i.e. three-body systems. One major ingredient in an analysis of systems such as that consisting of 3π's is the isobar model, i.e. all partial waves proceed via quasi two-body decays. Practically all partial wave analyses with three or more particles in the final state make this assumption. How to introduce a pure three-body decay is still a matter of controversy (particularly how to make certain that the resulting amplitudes satisfy unitarity).

Quasi two-body decays such as $A_2 \to \rho\pi$ ($\rho \to \pi\pi$) are much more complex to analyze than the 0^-0^- decays, since there are parameters not only for production and decay, but also for the decay of the secondary resonance. In spite of its complexity, the possibility of interference between the decay products introduces many constraints that make three-body analysis somewhat less dependent than the two-body decays on assumptions about production mechanism.

2.2.1 3π SYSTEM   The first thorough partial wave analysis of the $(3\pi)^{\pm}$ system, which set the standard for all subsequent analyses of this kind, was that of Ascoli and collaborators at the University of Illinois (Ascoli et al 1970), who used reactions of the type $\pi^{\pm}p \to (3\pi)^{\pm}p$. The analysis showed that the leading partial waves are $1^+, 2^+$, and 2^- (A_1, A_2, A_3) up to a mass of 1.6 GeV. The 1^+ wave is mainly $\rho\pi$ S-wave and peaks near 1.15 GeV, the 2^+ mostly $\rho\pi$ D-wave peaking near 1.32 GeV, and the 2^- wave is dominated by S-wave $f^0\pi$. These results have been subsequently confirmed by analyses of higher statistics data by the Illinois group and others (Ascoli et al 1973, Antipov et al 1973, Thompson et al 1974, Otter et al 1974).

The interpretation of the data, however, is not trivial. The phase variation of the 2^+ partial waves with respect to background partial waves beautifully exhibits the rapid change expected for a resonance, but the 1^+ and 2^- states do not; in fact, they show practically no phase variation as a function of mass. A substantial fraction of the 1^+ $\rho\pi$ S-wave was expected to consist of a $\rho\pi$ threshold enhancement. A Reggeized Deck (Ascoli et al 1974) model indeed shows very little phase variation as a function of the 3π mass, in qualitative agreement with observation (but with some quantitative discrepancies). This model also correctly predicts the observed phase between the 1^+ and 2^+ waves. The surprise was that there was no hint of any Breit-Wigner-like phase motion superimposed on a large and slowly varying background. The similar behavior of the 2^- wave indicates that it is dominated by an $f^0\pi$ threshold enhancement. This view is supported by the fact that no enhancement is observed in the 2^- $\rho\pi$ P-wave. Two questions arise naturally: are the solutions unique and are slow phase variations incompatible with the existence of resonances? Much effort has been spent on the second question.

Bowler et al (1975) showed that it is possible to have a slow phase variation and an A_1 resonance with mass ~ 1.3 GeV (instead of 1.1 GeV where the 1^+ $\rho\pi$ S-wave peaks) and at least 250 MeV wide, if one introduces an ad hoc $40°$ phase between Deck and resonant terms. The Reggeized Deck model used to explain the 1^+ wave has been criticized because it does not satisfy unitarity constraints. An analysis by Longacre & Aaron (1977) imposing unitarity led them to conclude that it is not possible to explain the very slow phase variation of the 1^+ partial wave with only a Deck mechanism; it requires in addition an A_1 resonance of mass 1450 MeV with a 300 MeV width.

A more detailed analysis by Basdevant & Berger (1977), allowing for possible couplings of the A_1 to $K^*\bar{K}$, showed that the inelasticity of the A_1 into $\rho\pi$ could be directly responsible for modest phase variation. They found that the data could accommodate four possible solutions for an A_1 resonance, ranging in mass from 1190 to 1490 MeV and with quite large widths, between 400 and 500 MeV. Their solutions require a large 1^+ S-wave in the $K^*\bar{K}$ channel near threshold, which is indeed observed (Otter et al 1975), although higher statistics in this channel are needed for a detailed quantitative comparison with the predictions of Basdevant & Berger (1977). More recent data from $\pi^+p \to K^+K^-\pi^+p$ at 13 GeV/c from the SLAC Hybrid Bubble Chamber (HBC) facility (Leedom et al 1978) seem to indicate that the K^*K channel does not rise as rapidly as expected by the above analysis.

The above studies utilized the Ascoli-type analysis (Ascoli et al 1970, 1973, Antipov et al 1973, Thompson et al 1974, Otter et al 1974) as input, assuming implicitly that the partial wave solutions are unique. A re-analysis of $\pi^{\pm}p \to (3\pi)^{\pm}p$ data by Shult & Wyld (1977) imposing unitarity constraints found a solution that differed very little from the previous one, but Shult & Wyld observed that below 1.2 GeV there is in fact an unresolvable ambiguity that leads to quite different 0^- S-$(\varepsilon\pi)$ and 0^- P-$(\rho\pi)$ waves. By jumping from one solution to the other, one could have a rapidly varying phase for the 1^+ S$(\rho\pi)$ and 1^+ P$(\varepsilon\pi)$ wave between 1.0 and 1.2 GeV. Therefore, the data are not necessarily inconsistent with a 1^+ resonance below 1.2 GeV. The only remaining stumbling block arguing against such a resonance is that both 1^+ S $(\rho\pi)$ and 1^+ P $(\varepsilon\pi)$ must be resonant, since the relative phase between these waves remains constant as a function of mass, but unlike the 1^+ S $(\rho\pi)$ wave, the 1^+ P $(\varepsilon\pi)$ wave intensity shows no maximum.

A partial wave analysis of the 3π system produced coherently off nuclei $(\pi^- A \to \pi^+\pi^-\pi^- A)$ [which has the advantage of requiring fewer partial waves than $\pi^{\pm}p \to (3\pi)^{\pm}p$] also found two possible solutions (Pernegr et al 1978). Below 1.4 GeV only four waves are required: 1^+S, 1^+P, 0^-S, and 0^-P. Unlike incoherent production, however, both the 1^+S and the 1^+P wave peak near the same mass. Furthermore, in one solution, the 1^+ waves resonate versus the 0^-P$(\rho\pi)$ wave and in the other versus the 0^-S$(\varepsilon\pi)$ wave. Thus, the authors conclude that regardless of which solution is chosen, both the 1^+S and 1^+P wave have resonances (they need not be the same resonance). They estimate the mass of the 1^+ resonance (S) to be somewhat below 1.1 GeV and its width to be of the order of 300 MeV. In addition, in the A_3 region, their analysis shows the presence of a 2^- resonance of mass 1.65 GeV and width 400 MeV with two decay modes: $f^0\pi$ S-wave and $\rho\pi$ P-wave. There is also an indication of a 1^+D$(\rho\pi)$ resonance degenerate with the above mass. On balance, then, one can state that the (3π) data certainly allow the possibility of a 1^+ resonance with mass below 1.2 GeV, but the evidence is not conclusive.

Obviously, much of the difficulty in obtaining clear-cut results on resonances belonging to the unnatural spin parity series with the $(3\pi)^{\pm}$ final state is the considerable background coming from diffraction processes. Such background is absent in charge-exchange reactions. The first partial wave analysis of the $\pi^+\pi^-\pi^0$ final state obtained from reaction $\pi^+p \to \pi^+\pi^-\pi^0\Delta^{++}$ showed, in addition to the $\omega(780)$ and the A_2^0, a 3^-, $I = 0$ resonance, the $\omega(1675)$ with $\Gamma \sim 170$ MeV. However, there was no sign of either an A_1 or A_3 resonance (Wagner et al 1975). These results were confirmed by Baltay et al (1978) and Corden et al (1978a), studying reactions $\pi^-p \to \pi^+\pi^-\pi^0n$. The latter, in addition, find evidence for a wide 4^+ resonance $(0.5 \pm 0.2$ GeV) at 2.03 GeV, which they call $A_{\frac{3}{2}}^*(2030)$. The failure to observe unnatural spin parity states in charge-exchange reactions is disturbing, but fairly reasonable arguments can be invoked for their being strongly suppressed (see, for example, Basdevant & Berger 1978). Thus, it cannot be considered strong evidence against their existence.

A more promising avenue seems to be backward meson resonant production. Enhancements near 1.1 GeV are observed in reactions $\pi^- p \to p\pi^+ \pi^- \pi^-$ (Ferrer et al 1978) and $K^- p \to \Sigma^- \pi^+ \pi^+ \pi^-$ (Gavillet et al 1977) with cross sections of the same order of magnitude as A_2 production. Gavillet et al (1977) were also able to show that the enhancement is in the 1^+ $S(\rho\pi)$ wave. For both reactions Breit-Wigner fits to the mass spectrum give masses near 1.04 GeV and widths of the order of 200 MeV, somewhat lower in mass and narrower than that obtained from partial wave analysis of forward production. However, in the backward production data, the statistics are not, at present, sufficient to study phase variations. A thorough partial wave analysis of backward $(3\pi)^{\pm}$ production will probably be required before one can conclusively prove that the observed enhancements are resonant.

Another quite interesting result is the observation of the decay $\tau^{\pm} \to \nu(3\pi)^{\pm}$ and particularly that the resulting 3π mass spectrum peaks near 1.1 GeV (Jaros et al 1978, Alexander et al 1978). Although this seems strong evidence for the existence of the A_1, the present statistics do not lead to a very clear determination of the A_1 mass and width (Basdevant & Berger 1978).

In spite of all the effort that has been expended on the search for the A_1, one must sadly conclude that the situation at present is still somewhat unsettled. There are now many pieces of evidence pointing toward the existence of an A_1 near 1.1 GeV but no conclusive proof.

2.2.2 $K\pi\pi$ SYSTEM   Preliminary partial wave analysis of the $K\pi\pi$ system was again done using the Ascoli et al (1970) programs on the reaction $K^- p \to K^- \pi^+ \pi^- p$. There is a large broad enhancement between threshold and 1.4 GeV, which was shown to be mostly 1^+ $K\rho$ and $K^*\pi$ S-wave, with some significant $2^+ K^*\pi$ [$K^*(1420)$]. The statistics of these first analyses were too small and precluded the study of fine details; the overall behavior could be explained by Deck models (as in the 3π system) (Deutschmann et al 1974, Antipov et al 1975, Otter et al 1975, Tovey et al 1975, Hansen et al 1974).

Much higher statistics work at SLAC (Brandenburg et al 1976b) showed rich structure in the partial wave content of this enhancement. The observed amplitudes and phases could be explained by introducing, in addition to the $2^+ K^*(1420)$, a 0^- $\{K'\}$ state with $m \sim 1400$ MeV and $\Gamma \sim 250$ MeV decaying mostly to εK and some coupling to $K^*\pi$, and two 1^+ resonances; one decaying to $K\rho$ with $m \sim 1300$ MeV and $\Gamma \sim 200$ MeV $\{Q_1\}$, another decaying mostly to $K^*\pi$ with $m \sim 1380$ MeV and $\Gamma \sim 160$ MeV $\{Q_2\}$. The two 1^+ resonances could be a mixture of the expected companions to the A- and B-mesons (Q_A and Q_B), while the last one would be the first example of a radial excitation of the K-meson (radial excitation of the other members of the 0^- nonet are still to be found). There is, of course, in addition to these resonances a very large 1^+ $K^*\pi$ Deck background that complicates the interpretation of the partial waves. It is very difficult to prove the uniqueness of the partial wave solutions, or their interpretation. Additional confirmation is crucial. An analysis of the same reaction by Otter et al (1976) gives evidence for a Q_1 resonance, while the evidence for the Q_2 is weaker.

A good way to remove the diffractive background is obviously to study the charge-exchange reaction $K^-p \to (K\pi\pi)^0 n$. A partial wave analysis of this reaction was performed by Vergeest et al (1976) using data from a high statistics (80 events/μb) exposure of the CERN 2-m HBC to a K^- beam of 4.2 GeV/c. The results confirm the presence of a 1^+ resonance at $m \sim 1400$ MeV and a very clean K*(1420) decaying into $K^*\pi$ and $K\rho$ is observed. Although the 1^+ $K\rho$ wave is also present, it shows no peaking near 1300 MeV nor any phase variation relative to the background characteristic of a resonance. Also, significant contributions were seen from a 0^- S-wave $\kappa\pi$ and 1^- P-wave $K^*\pi$, but again without the rapid phase variation expected for a resonance. Analysis of the same reaction at 10 GeV/c from data taken at the Ω spectrometer is in good agreement with the above solutions and, in addition, finds good evidence for a 3^- resonance near 1800 MeV, consistent with it being a $K\pi\pi$ decay mode of the K*(1780) (Beusch et al 1978). Thus, one would conclude from the above that the 1^+ $K^*\pi$ resonance $\{Q_2\}$ is well established but not the 1^+ $K\rho$ $\{Q_1\}$.

There is, in fact, very impressive evidence for the existence of the Q_1 from backward meson production, this time involving a study of the reaction $K^-p \to \Xi^- (K\pi\pi)^+$ by Gavillet et al (1978) (see Figure 4). A fit to the mass spectrum gives $m = 1275 \pm 10$ MeV and $\Gamma = 75 \pm 15$ MeV, and a partial wave analysis shows the enhancement to be mostly $K\rho$ 1^+ S-wave with some $K^*\pi$ S-wave. The statistics are not adequate to draw conclusions on any $K^*\pi$ resonance near 1400 MeV. Although the width disagrees with that obtained by Brandenburg et al (1976b) in their partial wave analysis, it is striking confirmation of a $1^+ K\rho$ resonance near 1280 MeV. Additional evidence for the Q_1 comes from reactions $\pi^-p \to (K\pi\pi)^0 \Lambda$ at 3.9 GeV/c (BCCMS collaboration 1977) and $\bar{p}p \to K^+K^-\pi^+\pi^-$ at 2.32 GeV (Chen et al 1977), again with narrower width than that observed by the partial wave analysis of diffractive data. Concerning higher mass states, one analysis of the $K^-\pi^+\pi^-$ system between 1.5 and 2.0 GeV (diffractively produced) shows some indication of a 2^- resonance with mass ~ 1.58 GeV and width ~ 0.11 GeV that decays into $K^*(890)\pi$ (Otter et al 1979), while the well-known L-region enhancement (~ 1.75) is shown to

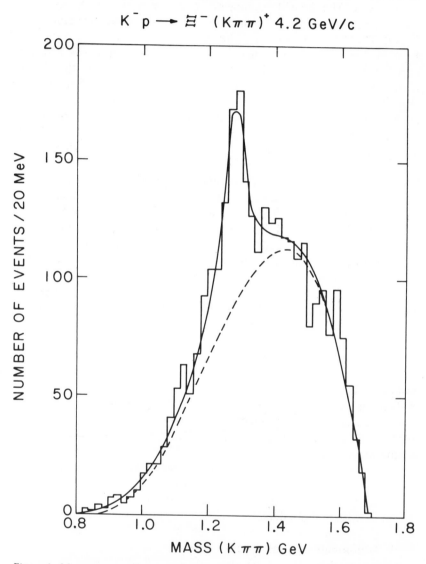

Figure 4 Mass spectrum from reaction $K^-p \to \Xi^- (K\pi\pi)^+$ at 4.2 GeV/c. The solid line represents the result of a fit to a Breit-Wigner resonance plus polynomial background (*dotted line*) (Gavillet et al 1978).

be a superposition of many partial waves with a large K*(1420)π S-wave component, having the expected characteristics of a threshold enhancement.

In conclusion, from the detailed partial wave analysis of K$\pi\pi$ systems comes quite solid evidence for two 1^+ resonances Q_1 and Q_2, one coupling mostly to Kρ and the other to K*π. These states are presumably a mixture of the Q_A or Q_B [the SU(3) members of the A_1 and B nonets, respectively]. Claims for additional new states require confirmation.

2.2.3 K$\bar{\text{K}}\pi$ SYSTEM   Analysis of the K$\bar{\text{K}}\pi$ system has not yet been carried out to the level of detail and precision that exists for the 3π and K$\pi\pi$ systems. From the bubble chamber data of Otter et al (1975) it is known that the reaction $\pi^\pm\text{p} \to \text{K}^+\text{K}^-\pi^\pm\text{p}$ is dominated by 1^+S K*K states with mass between 1.4 and 1.6 GeV, which is consistent with being diffractively produced (see Section 2.2.1 on the 3π system for possible implications on partial wave analysis of that system).

From the study of K$\bar{\text{K}}\pi$ systems produced nondiffractively, such as $\pi^-\text{p} \to \text{K}\bar{\text{K}}\pi\text{n}$ and $\bar{\text{p}}\text{p} \to \text{K}\bar{\text{K}}\pi\pi$, K$\bar{\text{K}}\pi\pi\pi$, K$\bar{\text{K}}\pi\pi\pi\pi$, a fairly narrow state [the D(1285) with mass 1282 ± 2 MeV and width 28 ± 7 MeV] shows up quite clearly (D'Andlau et al 1965, 1968, Miller et al 1965, Dahl et al 1967, Nacash et al 1978). The dominant decay mode of D into K$\bar{\text{K}}\pi$ is found to be $\delta(970)\pi$, with $\delta(970) \to$ K$\bar{\text{K}}$ (threshold enhancement), consistent with the observation of D $\to \delta(970)\pi$, $\delta(970) \to \eta\pi$ discussed below. The favored spin parity value for this state is 1^+ and isospin is well established as $I = 0$. One additional state decaying to K$\bar{\text{K}}\pi$ had been observed until recently only in $\bar{\text{p}}$p annihilations, the E(1420) (Baillon et al 1967, Nacash et al 1978). The mass and width for this enhancement are given as ~ 1420 MeV and 50 ± 20 MeV, respectively. Analysis favors $I^G J^P = 0^- 1^+$ (Vuillemin et al 1975); however, the background was large and the resonant interpretation of the enhancement was open to question as it has been observed only at low momentum $\bar{\text{p}}$p annihilation at the peak of phase space. Recently the E has also been seen in the reaction $\pi^-\text{p} \to \text{K}^\pm\text{K}^0\pi^\pm\text{n}$ by two experiments, one at the Omega Spectrometer (Gordon et al 1978) and a bubble chamber experiment at 3.9 GeV/c (BCCFMS collaboration 1978), and can be considered well established.

2.2.4 OTHER QUASI TWO-BODY FINAL STATES   Study of final states other than the ones given above has generally been meager. Some high statistics information exists on $\pi\omega$ final states, in particular on the B-meson, its mass and width are now well determined to be 1230 ± 10 MeV and 140 ± 20 MeV, respectively, and its quantum numbers $I^G J^P = 1^+ 1^+$ (Chung et al

358 PROTOPOPESCU & SAMIOS

1973, 1975, Chaloupka et al 1974). The observation of an enhancement in $\gamma p \rightarrow \pi^0 \omega p$ leads to the possibility that there might be a $J^P = 1^-$ resonance in the same mass region as the B, the $\rho'(1250)$ also decaying to $\pi \omega$ (Ballam et al 1974). Partial wave analyses of the $\pi^{\pm} \omega$ systems have

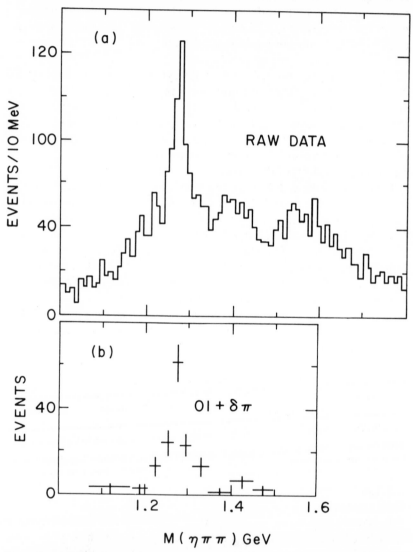

Figure 5 (a) The $\eta \pi \pi$ mass spectrum observed in reaction $\pi^- p \rightarrow \eta \pi \pi n$ at 8.5 GeV/c; (b) Contribution to the $\eta \pi \pi$ mass spectrum of the 1^+ M = 0 $\delta \pi$ partial wave (Stanton et al 1979).

failed to reveal any such state and, if it exists, its production cross section in π-induced interactions must be substantially smaller than the one for the B. The $\pi\omega$ final state is ideal to look for $I = 0$, $J^{PC} = 2^{-+}$ and $I = 1$, $J^{PC} = 2^{--}$ states, both expected in the quark model. High statistics data in this channel for partial wave analysis are sorely needed.

Recently some detailed analysis has been done on high statistics $\eta\pi\pi$ data obtained in the reaction $\pi^- p \to \eta\pi^+\pi^- n$ (Corden et al 1978b, Stanton et al 1979). The most noticeable feature in this final state is the D(1280) decaying into $\delta\pi$, with $\delta \to \pi\eta$. The analysis by Stanton et al (1979) shows quite convincingly that the D is indeed 1^+ (see Figure 5). In addition, two more states are claimed (which need confirmation): a 100-MeV wide object at 1.26 GeV with quantum numbers $I^G J^P = 0^+ 0^-$, and a second, broader pseudoscalar at 1.4 GeV decaying into $\varepsilon\eta$. It is not clear at present whether or not the solution showing the existence of these new states is unique. The possible presence of 0^- resonance degenerate with the D-meson means that the D has a narrower width (~ 10 MeV) than previously estimated.

2.3   Vector Mesons

Because vector mesons made of a quark and its corresponding antiquark can couple directly to the photon and thus predominate in hadronic production via electromagnetic processes such as photoproduction, $e^+ e^-$ annihilation, etc, they can be studied in more detail than other states and deserve a section of their own. Since the discovery of the J/ψ and the family of higher mass relatives, most of which are explained as radial excitations of the J/ψ, it has been expected that with the ρ, ω, and ϕ should also come a wide spectrum of states. The discovery of such states and a study of their properties should bring much needed understanding of the dynamics of low mass quarks. As we see in the following sections, there is still a great deal of uncertainty concerning these possible states.

2.3.1   PHOTOPRODUCTION   Relatively high statistics experiments on the 82-inch HBC at SLAC with polarized photons between 4.0 and 9.0 GeV/c showed early on that vector meson production is substantial and that channels such as $\gamma p \to \pi^+ \pi^- p$, $\pi^+ \pi^- \pi^0 p$, and $K\bar{K}p$ are dominated by the ρ, ω, and ϕ, respectively (Ballam et al 1973). An interesting observation from these reactions is that there is no evidence for higher mass vector mesons decaying into $\pi^+ \pi^-$, $\pi^+ \pi^- \pi^0$, or $K\bar{K}$ to levels ~ 50 times smaller than the ρ cross section. Thus, the couplings for the ρ, ω, and ϕ radial excitations to those channels must be significantly suppressed. The first solid evidence for a higher mass vector meson came from photoproduction in the reaction $\gamma p \to \pi^+ \pi^- \pi^+ \pi^- p$ (Bingham et al 1972). Analysis of the above reaction showed that there is a wide ($\sim 0.6 \pm 0.1$ GeV) 1^- enhancement near 1.6 GeV decaying to $\rho^0 \pi^+ \pi^-$ in which the $\pi^+ \pi^-$ are most likely in an $I = 0$ S-wave state. The favored decay for this $\rho'(1600)$ thus seems to be $\rho\varepsilon$, although such a decay mode cannot be distinguished clearly from an $A_1\pi$ decay (particularly given the uncertainties on the A_1). Study of the reaction $\gamma p \to \pi^+ \pi^- MMp$ (where MM stands for missing mass) selecting $0.3 < M(\pi^+ \pi^-) < 0.6$ GeV shows a very clear bump (≤ 150-MeV wide) at 1250 MeV. By studying the angular correlations of those events it can be proven quite conclusively that it is a state decaying to $\pi^0 \omega$ (Ballam et al 1974). Without any π^0 detection the data lack the analyzing power to distinguish a 1^+ from a 1^- state. It is therefore not possible to tell whether the observed bump is the well-established B-meson or a $\rho'(1250)$. Analysis of $\pi\omega$ data produced in π-induced interactions (see Section 2.2.4) shows no indication of a 1^- state with the same width as the B-meson.

A recent DESY-Frascati collaboration has searched for vector mesons in the reaction $\gamma p \to e^+ e^- p$ (virtual Compton scattering) with a two-arm spectrometer using the $e^+ e^-$ decay as a unique signature for vector mesons (Bartolucci et al 1977, Berger 1977). A problem in this reaction is the very large background coming from Bethe-Heitler pair production. The invariant mass spectrum clearly shows a peak in the ρ-ω region and at the ϕ mass; anything else is not clearly distinguishable from background. However, when the calculated QED background is subtracted, there is a clear excess of events above 1 GeV that cannot be explained by the ρ, ω, and ϕ alone. A more sensitive way to look for structures is to plot the asymmetry between e^+ and e^- as a function of mass, this essentially shows the interference between Bethe-Heitler and the virtual Compton scattering amplitude. The results of this type of study are quite interesting. In addition to clear bumps at the ρ-ω and ϕ masses, one sees a relatively narrow structure (~ 50 MeV) at 1100 MeV and a very broad enhancement extending from 1200 MeV to beyond 1800 MeV, reaching a maximum near 1400 MeV. This broad structure cannot be explained in any simple way by the states observed in $e^+ e^-$ annihilations (discussed below). Keep in mind that in $e^+ e^-$ annihilations one is observing an amplitude squared while in this case one sees directly an amplitude as it interferes with the Bethe-Heitler amplitude.

The DESY-Frascati collaboration interpret their data as evidence for a new, relatively narrow, vector meson at 1100 MeV and the broad enhancement above 1.2 GeV as a superposition of at least four resonances among them the $\rho'(1250)$ and $\rho'(1600)$. Interpreting the data is obviously not straightforward and is quite model dependent. It is possible, for instance, to interpret the structure at 1100 MeV as a manifestation of a

$\rho'(1250)$ interfering with Bethe-Heitler background and effects due to the opening of the $\pi\omega$ channel, rather than due to a narrow 1100-MeV state (Budner 1977). The existence of this latter state would probably be quite difficult to explain in the quark model.

2.3.2 e^+e^- ANNIHILATIONS Results coming from storage rings at energies above 3.0 GeV have been among the most spectacular in recent years, ever since the discovery of the J/ψ, and supplied new impetus for the search of radial excitations of its low mass relatives (for a review of results above 3.0 GeV see Chinowsky 1977). In contrast the region between 1.0 and 2.0 GeV remains still relatively unexplored and high statistics experiments in that energy interval are likely to have a major impact on light quark spectroscopy. (Early results at low energies were reviewed by Schwitters & Strauch 1976.)

Below 1.0 GeV fairly detailed studies have been done by experiments on the low energy e^+e^- storage ring ACO at Orsay. These experiments have obtained many new results on radiative decays of vector mesons such as $\omega \to \pi^0\gamma$, $\phi \to \pi^0\gamma$, $\phi \to \eta\gamma$, etc (Benaksas et al 1972). In addition, from ACO comes new information on ρ-ω interference in $e^+e^- \to \pi^+\pi^-$, evidence on ω-ϕ interference in $e^+e^- \to \pi^+\pi^-\pi^0$, and also good evidence that the $\phi \to \pi^+\pi^-\pi^0$ decay mode proceeds via a $\rho\pi$ intermediate state (Jullian 1976).

Beyond 1.0 GeV there is now good data available up to 1.34 GeV coming from the e^+e^- storage ring VEPP-2M at Novosibirsk. Of interest are the rising cross sections of $e^+e^- \to \pi^+\pi^-\pi^+\pi^-$ and $e^+e^- \to \pi^+\pi^-\pi^0\pi^0$, in particular the very rapid rise of the latter compared to the all-charged state beyond the $\omega\pi$ threshold (Budner 1977). It has been shown that the $\pi^+\pi^-\pi^0\pi^0$ final state is completely dominated by an $\omega\pi^0$ intermediate state. This rapid rise of the $\omega\pi$ cross section is accompanied by a very noticeable deviation of the $e^+e^- \to \pi^+\pi^-$ cross section (see Figure 6) from what would be expected from the ρ tail (Bukin et al 1978). The interpretation of the data is not straightforward. The behavior of the $\pi\pi$ and $\pi\omega$ cross sections between 1.0 and 1.2 GeV most likely have the same underlying cause. One plausible explanation is that the ρ has a strong coupling to $\omega\pi$ that produces the rapid rise of the $\omega\pi$ cross section: What we see in the $\pi\pi$ cross section is then simply an effect due to unitarity (Sidorov 1976, Costa de Beauregard et al 1977). This interpretation can be accommodated within the range of present $\pi\pi$ P-wave phase shifts. Another interpretation favored by many authors is that both phenomena indicate the existence of a $\rho'(1250)$ resonance, this interpretation is also favored by the analysis of some low statistics data from the storage ring ADONE at Frascati (Conversi et al 1974). This alternative explanation suffers from two (not insurmountable) problems. One, it is difficult to accommodate a $\rho'(1250)$ resonance with present $\pi\pi$ P-wave phase shifts. Two, if the $\rho'(1250)$ exists, its total contribution to the cross section for $e^+e^- \rightarrow$ hadrons must be at least a factor of two less than the corresponding cross section of the higher mass state $\rho'(1600)$.

Above 1.3 GeV early data from ADONE at Frascati already indicated that the hadronic cross section was fairly large, peaking in the neighborhood of 1600 MeV (Conversi et al 1974, Barbarino et al 1972, Grilli et al 1973, Ceradini et al 1973). Preliminary results from the new storage ring DCI at Orsay, when combined with data from VEPP-2M, show quite clearly the existence of a broad $\rho'(1600)$ decaying into $\pi^+\pi^-\pi^+\pi^-$ and $\pi^+\pi^-\pi^0\pi^0$, but it also shows that one cannot explain those channels by fitting them with a single Breit-Wigner form (Figure 7) (Laplanche et al 1977, Cosme et al 1979). It is possible to describe the data adequately invoking two $I = 1$ resonances, one near 1550 MeV and the other near 1700 MeV, both about 200 MeV wide. The 1700-MeV resonance could

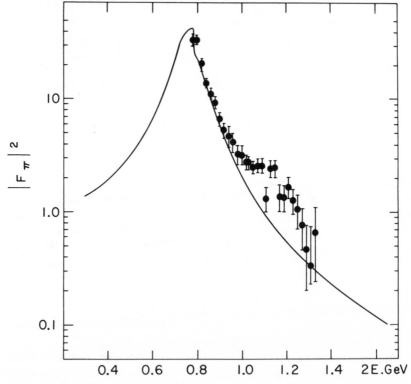

Figure 6   The pion form factor obtained from reaction $e^+e^- \rightarrow \pi^+\pi^-$ (Bukin et al 1978).

couple more strongly to channels that end up as $\pi^+\pi^-\pi^0\pi^0$. Given the present statistics and limitations of the DCI experiment, such an interpretation is by no means unique. Note that there are large gaps in the data between 1350 and 1550 MeV (Figure 7). Filling those gaps could produce surprises and completely destroy the fits. The marked deviations from a single Breit-Wigner form may be a reflection of new channels opening up, channels to which the $\rho'(1600)$ is strongly coupled (for example $\rho^+\rho^-$, $\pi\omega'$, which contribute to $\pi^+\pi^-\pi^0\pi^0$ but not to $\pi^+\pi^-\pi^+\pi^-$). Similarly, the effects between 1.0 and 1.2 GeV in $\pi^+\pi^-$ and $\pi\omega$ may be a reflection of strong $\rho\omega\pi$ coupling.

In addition to the possibility of as many as three relatively broad

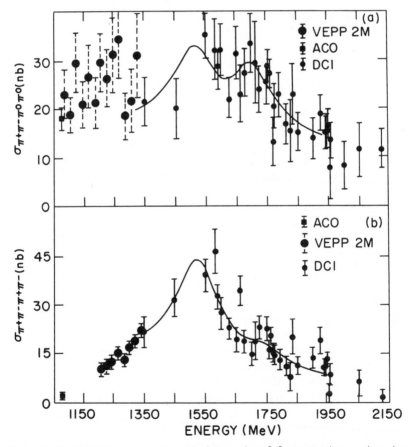

Figure 7   Cross sections for reactions (*a*) $e^+e^- \to \pi^+\pi^-\pi^0\pi^0$ and (*b*) $e^+e^- \to \pi^+\pi^-\pi^+\pi^-$ obtained at ACO, VEPP 2M, and DCI (Cosme et al 1979). The solid line is a fit to two interfering Breit-Wigner resonances plus linear background.

$I = 1$ resonances between 1.0 and 2.0 GeV, there have been, in the last two years, claims for the existence of five more states in that mass region, all of them less than 50-MeV wide. Two were observed at DCI and clearly identified as $I = 0$ states, one near 1770 MeV decaying mostly into $\pi^+\pi^-\pi^+\pi^-\pi^0$ (Laplanche et al 1977, Cosme et al 1976, 1979), and the other near 1650 MeV decaying both into $\pi^+\pi^-\pi^0$ (predominantly) and into $\pi^+\pi^-\pi^+\pi^-\pi^0$ (Laplanche et al 1977, Cosme et al 1979). The most likely interpretation is that they are radial excitations of the ω; if so, they are surprisingly narrow, particularly the one at 1650 MeV, which has such large branching ratio into $\pi^+\pi^-\pi^0$. Another disturbing feature of the Orsay data is that in the 1650-MeV region there is a 2.5-σ upward fluctuation in the $\pi^+\pi^-\pi^+\pi^-$ channel, much too large to be explained by a ρ'-ω' interference analogous to ρ-ω. If this effect turns out not to be simply a statistical fluctuation, then the interpretation of this state as an ω radial excitation will have to be reexamined.

Three more states are claimed by three experimental groups at the ADONE storage ring (Bemporad 1977, Esposito et al 1977, Ambrosio et al 1977, Bacci et al 1977). One of the states was observed by only one of the groups as an excess of K*(890) events near 2130 MeV (Esposito et al 1978). The statistics are too small to determine the width, but it seems to be less than 30-MeV wide. The fact that this state seems to decay via a K* intermediate state is a clear signature for a ϕ-like resonance. The other two states were observed by all three groups (with generally low statistics, signals are typically less than 4 σ), one at 1830 MeV with a width $\sim$ 25 MeV and the other at 1498 MeV with a width $\lesssim$ 4 MeV. They were detected essentially as an increase in the number of hadronic events observed, with very little information on the final states into which they decay. So, apart from their mass and width, very little is known about them, not even their I-spin.

Their narrow width can be understood if they are ϕ-like resonances coupling mostly to K*$\bar{\text{K}}$ or K*$\bar{\text{K}}$*; in which case they would be close to threshold. At present there is no evidence for substantial K production at these resonances and, in fact, in the 1830-MeV state there is some evidence against it. One of the groups observes the enhancement only in final states with γ's (presumably from π^0's), not at all what one would expect from K*$\bar{\text{K}}$ or K*$\bar{\text{K}}$* decays, in which final states without π^0's predominate. The situation in this region has been muddied even further by the observation of a narrow enhancement at 1.8 GeV decaying into $\pi^+\pi^-\pi^+\pi^-$ and possibly K$^+$K$^-\pi^+\pi^-$ in electroproduction (e$^-$p $\rightarrow$ e$^-\pi^+\pi^-\pi^+\pi^-$p) (Killian et al 1979). This electroproduction experiment also observed an enhancement near 2.1 GeV in K$^+$K$^-\pi^+\pi^-$. Although there is no proof that these enhancements are 1$^-$ states, one would expect virtual photon production to dominate.

To summarize the e^+e^- annihilation data, there are many candidates for the radial excitations of the ρ, ω, and ϕ but little evidence on any of them that could be called conclusive. The $I = 1$ candidates are fairly broad (> 150 MeV): $\rho'(1250)$, $\rho''(1550)$, and $\rho'''(1700)$. Of these only one, the $\rho''(1550)$, can be considered established and even in this case the width is quite uncertain as it depends critically on whether the other two states exist. There are two ω' candidates, at 1650 MeV and 1770 MeV, both no more than 50-MeV wide; and one ϕ' candidate at 2130 MeV. There is not sufficient information on the states at 1498 MeV and 1830 MeV to determine where they fit in quark spectroscopy. It must be emphasized that for all the narrow states between 1.0 and 2.0 GeV the data are of relatively low statistical significance and require confirmation before they can be considered established. Clearly, there is great need for high statistics and good quality e^+e^- annihilation data in this energy region.

2.4   *Exotic Mesons*

From the quark model point of view the term exotic state is well defined; it is any state that cannot be described as made of a quark-antiquark pair [q$\bar{\text{q}}$] or three quarks [qqq] or antiquarks [$\bar{\text{q}}\bar{\text{q}}\bar{\text{q}}$]. The simplest exotic state would be made by adding an additional quark-antiquark pair, thus the simplest exotic mesons would have two quarks and two antiquarks [qq$\bar{\text{q}}\bar{\text{q}}$] and the simplest exotic baryon four quarks and an antiquark [qqqq$\bar{\text{q}}$]. The interest in finding qq$\bar{\text{q}}\bar{\text{q}}$ states is not only that the quark model allows their existence but that they are, in fact, required if one is to preserve the notion of duality between Regge exchange in the t-channel and resonance formation in the s-channel in baryon-antibaryon (B$\bar{\text{B}}$) interactions.

The strongest evidence for the existence of exotic meson states would be the observation of a doubly charged or doubly strange meson resonance. So far all searches for such states have been fruitless. Best limits on the production cross section of relatively narrow doubly charged mesons come at present from studying backward production in the reactions $\pi^-\text{n} \to \text{p}_f$ p$\bar{\text{p}}\pi^-\pi^-$ [$\sigma \lesssim 10$ nb for $M(\text{p}\bar{\text{p}}\pi^-\pi^-) < 3.0$ GeV] and $\pi^-\text{n} \to$ p$\pi^+\pi^-\pi^-\pi^-$ [$\sigma \lesssim 50$ nb for $M(\pi^+\pi^-\pi^-\pi^-) < 3.0$ GeV] (Boucrot et al 1977). These limits are an order of magnitude lower than some theoretical predictions for exotic meson production via baryon exchange (Jacob & Weyers 1970, Lipkin 1973), if such states are narrow. The significance of the baryon-exchange process for the search of exotic mesons will become evident below.

A q$\bar{\text{q}}$ state decays into two mesons basically by creating an additional quark-antiquark pair from the vacuum. A similar process clearly leads from qq$\bar{\text{q}}\bar{\text{q}}$ to qqq$\bar{\text{q}}\bar{\text{q}}\bar{\text{q}}$, i.e. a baryon-antibaryon final state. Of course, a much simpler process is to simply fission into two $q\bar{q}$ pairs. According to one model (the MIT bag model), low mass $qq\bar{q}\bar{q}$ states would do just that, so they are extremely short-lived and appear as very broad resonances. Possible examples, proposed by R. Jaffe (1977a), are the 0^+ resonances, ε, κ, δ, and S*. The ε and κ are very broad, so broad, in fact, that there is some question as to whether they are resonant states, and what keeps the δ and S* relatively narrow is that they lie below $K\bar{K}$ threshold and contain an $s\bar{s}$ pair. From the point of view of quantum numbers alone it is not possible to tell whether these states belong to a regular $q\bar{q}$ nonet or are indeed $qq\bar{q}\bar{q}$ states. It was shown by Morgan (1974) that the 0^+ states could be accommodated as a conventional $q\bar{q}$ nonet (with somewhat unconventional mixing angles) if the ε and ε' are, in fact, one super broad state. This scheme is very much in dispute.

In the Jaffe model the ε, κ, δ, and S* would be $qq\bar{q}\bar{q}$ 0^+ states. One member is still missing, which should have mass ~ 1400 MeV, be relatively broad, and decay mostly to $\eta\eta$. The $q\bar{q}$ 0^+ nonet should actually consist of the ε', the $\kappa'(1450)$ seen by Estabrooks et al (1978), the possible $\delta'(1300)$, and some as yet to be found $I = 0$ state coupling mostly to $K\bar{K}$ and having mass around 1600 MeV. This scheme makes the 0^+ nonet nearly degenerate in mass with the 2^+ nonet, bringing it in closer agreement with bag model expectations. The most compelling argument for a $qq\bar{q}\bar{q}$ interpretation of the ε, κ, S*, and δ is that one cannot explain a strong coupling to $K\bar{K}$ for the δ state if it is a conventional $q\bar{q}$ state, since its I-spin is 1. Thus, these 0^+ states would be examples of low spin $qq\bar{q}\bar{q}$ states. Higher spin $qq\bar{q}\bar{q}$ resonances can be prevented from fissioning by keeping the qq pair spatially separated from the $\bar{q}\bar{q}$ pair, their major mode of decay would then be as for ordinary mesons by the creation of a $q\bar{q}$ pair leading to a baryon-antibaryon final state (Jaffe 1978). Other models, such as the string model (Rossi & Veneziano 1977, Chew & Rosenzweig 1976), lead to very similar conclusions.

Therefore, a good hunting ground for these exotic states is $N\bar{N}$ interactions. The first $\bar{p}p$ state observed was the S(1930), found by Carroll et al (1974), and it seems reasonably well established (Kalogeropoulos & Tzanakos 1975, Chaloupka et al 1976, Brückner et al 1977, Yakamoto et al 1978). This state is particularly interesting in that most of the observed signal in the total cross section seems to come from the elastic $\bar{p}p$ channel, which indicates small branching ratios to purely mesonic states. Examination of the charge-exchange reaction $\bar{p}p \rightarrow \bar{n}n$, shows no sign of the S(1930) enhancement (Alston-Garnjost et al 1975). This can be interpreted as a result of a cancellation of two superimposed $I = 0$, $I = 1$ states or as a negative comment on the existence of the S(1930). A recent high statistics measurement of the total $\bar{p}p$ cross section in the region of the S(1930) does not observe this state in direct contradiction with the above experiments (Nicholson et al 1979). Therefore there is now some doubt concerning its existence, and further experiments are required to clarify the situation.

The existence of higher mass states has been claimed based on detailed amplitude analysis of two-body formation experiments such as $\bar{p}p \rightarrow \pi^+ \pi^-$ (Carter et al 1977), $\bar{p}p \rightarrow K^+ K^-$ (Carter 1978), and $\bar{p}p \rightarrow \pi^0 \pi^0$ (Dulude et al 1978). The $\bar{p}p \rightarrow \pi^+ \pi^-$ data are consistent with the presence of three relatively broad resonances (between 200- and 300-MeV wide) at 2150, 2310, and 2480 MeV with $J^P = 3^-$, 4^+, and 5^-, respectively. Because of the symmetry constraints of the $\pi\pi$ system the 2150- and 2480-MeV states must be $I = 1$, while the 2310-MeV state is $I = 0$ or $I = 2$. Such constraints are relaxed in the $K^+ K^-$ system. Data from the reaction $\bar{p}p \rightarrow K^+ K^-$ seem to be consistent with the above results and, in addition, give some indication that the $J^P = 3^-$ and 5^- may, in fact, be two nearby mass degenerate states with $I = 0$ and $I = 1$ (Carter 1978). However, an analysis of $\bar{p}p \rightarrow \pi^0 \pi^0$ fails to support the above claims (Dulude et al 1978). One could accommodate a $J^P = 4^+$ state at 2310 MeV, although it is not needed by the data, but more importantly in order to explain the $\bar{p}p \rightarrow \pi^0 \pi^0$ data one needs to invoke a $J^P = 2^+$ state at 2150 MeV. Clearly, before one can reach any definitive conclusions on the possible states observed in these reactions, an analysis combining all channels is required in order to obtain one unique consistent solution. If any of the above states are confirmed, they will be an important new ingredient in meson spectroscopy as they do not lie on leading Regge trajectories. Whether or not they are $qq\bar{q}\bar{q}$ states is a completely open question.

There may also be states that couple to $\bar{N}N$ but lie below threshold. Such states are expected not only from $qq\bar{q}\bar{q}$ models but also from nuclear potential calculations starting from NN interactions, which then are related to $N\bar{N}$ interactions via a G-parity transformation. This approach predicts a rich spectrum of $N\bar{N}$ states (Dalkarov et al 1970, Dover 1975, Shapiro 1978). One such state may have been found in $\bar{p}n$ at 1794 ± 1.4 MeV. Such a state is reached by looking at relatively high momentum spectator protons (>250 MeV) in the reactions $\bar{p}d \rightarrow p +$ many π's (Gray et al 1971). This state seems to be of odd G parity as it is seen to decay mostly into five or seven π's. Three additional bound states are claimed based upon results of a search for discrete γ rays in $\bar{p}p$ annihilations, at 1684, 1646, and 1395 MeV, respectively (Pavlopoulos et al 1978). None of the above claims has been confirmed as yet by other experiments.

Another way to search for $N\bar{N}$ states is in baryon-exchange processes. Formation of these states then occurs with the antibaryon off the mass shell. A study of the reaction $\pi^- p \to \pi^- p\, \bar{p}p$ in which a Δ^0 or N* is produced forward shows evidence for narrow states (< 30 MeV) decaying to $\bar{p}p$ at 2.02 and 2.2 GeV (Benkheiri et al 1977). These states are unusual in two respects: they are exceptionally narrow given that they are not close to $\bar{p}p$ threshold, and they represent a substantial fraction of the $\bar{p}p$ backward production cross section, which does not seem to be the case in $\bar{p}p$ formation. Detailed studies indicate that these states are produced mostly by nucleon ($I = 1/2$), rather than Δ ($I = 3/2$), exchange.

The same experiment searched for charged decay modes of these states studying reactions $\pi^- p \to p_f \bar{p}n$ and $\pi^- p \to p_f \bar{p}p\pi$, where p_f is the forwardly produced proton (Benkheiri et al 1979). No evidence for such decay modes was found and a natural conclusion is that they are $I = 0$ states. Although one cannot prove that they are $qq\bar{q}\bar{q}$ states, their narrow width and the failure to observe decays into purely mesonic final states argues against interpreting them as $q\bar{q}$ states. Their characteristics are quite similar to that of the S(1930) and it is a natural temptation to group them together. If so, it is surprising that the 2.02- and 2.2-GeV states have not yet been observed in $\bar{p}p$ elastic scattering. A possibility is that they are very narrow and past experiments may have missed them or did not have the required sensitivity. Elastic $\bar{p}p$ experiments in the near future may answer this puzzle. It should be noted that $qq\bar{q}\bar{q}$ states were postulated originally for salvaging duality in $\overline{B}B$ interactions. The above narrow states can only make a negligible contribution to that problem as their couplings are simply too small (Pennington 1978). The $\bar{p}p$ states at 2.02 and 2.2 GeV may have also been observed (with very low statistics) in reaction $e^- p \to e^- \bar{p}pp$ (Gibbard et al 1979). This again is a rather unusual observation as they are not expected to be 1^- mesons.

3 BARYONS

The number of well-established baryon states is considerably larger than for mesons, mainly because they can be studied via formation experiments, in addition to production experiments, which makes it much easier to accumulate good quality, high statistics data for phase shift analysis. The exceptions are Ξ and Ω states, which are inaccessible by the former method. In fact a large number of Ξ^* and Ω^* states have yet to be found (no Ω^* state has been observed so far). Unlike strangeness 0 and -1 states they can only be produced via baryon-exchange processes, which have small cross sections, and the higher mass states decay into high multiplicity final states that are difficult to detect with counter experiments.

In the following sections we concentrate only on results from formation experiments, excepting of course, observation of Ξ^* states. We refer to resonances in formation experiments by their orbital angular momentum in the elastic channel with two subscripts; the first one gives the isospin (multiplied by two in the case of N*'s and Δ's) and the second twice the spin. Thus, for example, a $J^P = 7/2^+$ Δ state is also called an F_{37} resonance.

3.1 πN Phase Shifts (Elastic and Charge Exchange)

The available data on πN elastic and charge-exchange scattering is considerable and increasing year by year. Although new polarization data indicate that there is still work to be done on the subject, there is an impressive consensus on the existence of many N* and Δ resonances. In Tables 1 and 2 we summarize the resonances found from the three largest phase shift analyses performed so far (Almehed & Lovelace 1972, Ayed 1976, and Pietarinen 1977). This last analysis is an energy-independent phase shift analysis simultaneously fitting data at fixed t and fixed center-of-mass angle, constrained by dispersion relation predictions to obtain a single solution (see Pietarinen 1974). The agreement between these three different analyses is quite good, considering the complexity and large number of parameters involved. In some cases masses can differ by as much as a 100 MeV and widths vary by almost a factor of two, but these discrepancies occur on states with small elastic branching ratios or very large widths. There are two N* candidates near 1500 MeV (P_{11} and P_{13}) that appear only in the analysis of Ayed (1976), one Δ candidate at 1680 (P_{33}) observed only by Almehed & Lovelace (1972), and a Δ candidate near 2170 (G_{19}) observed only by Ayed (1976). These states probably do not exist. Among the Δ resonances there is a D_{35} state near 1890 MeV that seems to be present only in Ayed's analysis. The region near 1900 MeV was very carefully reanalyzed by Cutkosky et al (1976a), using a parametrization based on hyperbolic dispersion relations (see Chao et al 1975). They confirm the F_{35}, F_{37}, P_{31}, and D_{35} resonances with the values for the masses and widths given by Ayed, except for the P_{31} state where they obtain a mass ~ 1940 MeV, which is in better agreement with the analyses of Pietarinen.

There is some recent polarization data, with high statistics, from the reaction $\pi^- p \rightarrow \pi^0 n$ (Brown et al 1978) that is in excellent agreement with some previous data obtained by Shannon et al (1974). Brown et al compare their observations with predictions from πN partial wave analysis. There is poor agreement with the old analysis of Almehed & Lovelace, mainly because they limited the number of fitted partial waves.

There is general qualitative agreement with the Saclay results (Ayed 1976); at some energies the agreement is strikingly good, but at most points the predictions for the backward direction do not match the data. It is therefore important to redo the partial wave analysis imposing the polarization data as an additional constraint. This is likely to give better parameters for the main resonances and confirm or discard the more doubtful states.

Table 1   N* states from πN phase shifts

States[a]		Almehed & Lovelace (1972)			Ayed (1976)			Pietarinen (1977)		
L_{2I2J}		M (MeV)	Γ (MeV)	Γ_{el}/Γ_{tot}	M (MeV)	Γ (MeV)	Γ_{el}/Γ_{tot}	M (MeV)	Γ (MeV)	Γ_{el}/Γ_{tot}
	****	1500	50	0.25	1520	80	0.36	1517	108	0.38
S_{11}	****	1670	120	0.5	1672	180	0.59	1690	160	0.54
	*	2100	200	0.5	2283	310	0.14	2055	205	0.16
	****	1470	220	0.65	1413	190	0.55	1447	190	0.55
P_{11}		—	—	—	1532	90	0.12	—	—	—
	***	1720	160	0.20	1729	220	0.19	1745	155	0.19
P_{13}		—	—	—	1530	80	0.08	—	—	—
	***	1850	300	0.25	1697	120	0.14	1788	224	0.17
	****	1520	120	0.58	1525	122	0.56	1517	128	0.56
D_{13}	***	—	—	—	1710	100	0.09	1800	218	0.08
	**	2075	150	0.30	2030	120	0.10	2078	263	0.11
D_{15}	****	1683	150	0.45	1660	150	0.41	1660	172	0.38
	*	2100	150	0.20	2100	220	0.08	2171	280	0.09
F_{15}	****	1688	140	0.65	1680	130	0.59	1684	146	0.66
	*	2175	150	0.25	1990	180	0.08	2067	310	0.12
F_{17}	**	2050	120	0.06	2000	200	0.15	1962	260	0.05
G_{17}	***	2225	150	0.35	2140	240	0.16	2123	250	0.10
G_{19}	***	—	—	—	2133	193	0.10	2150	400	0.08
H_{19}	***	—	—	—	2250	350	0.20	2116	380	0.19

[a] The number of * indicates the status attributed to these states by the Particle Data Group compilation (1978):

    **** Clear and unmistakable
    *** Good evidence, but not absolutely certain
    ** Needs confirmation
    * Weak evidence.

3.2 πN Phase Shifts (Inelastic Channels)

Most of the information on inelastic channels in πN scattering comes from detailed studies of $\pi N \to \pi\pi N$. For this three-body final state the basic assumption is made that all decays proceed via quasi two-body reactions (as in Section 2.2). The first detailed analysis of quasi two-body final states was that of Herndon et al (1975). They assumed that only $\Delta\pi$, ρN, and εN (where ε is a broad $\pi\pi$ S-wave) were needed. This assumption is probably inadequate at higher masses where $N^*\pi$ decays are possible. This analysis basically confirms the existence of states that have small couplings to πN configuration, particularly the P_{11} at 1730 MeV, the P_{13} at 1700 MeV, and the D_{13} at 1700 MeV for N^*'s; and P_{31}, P_{33}, and F_{35} near 1900 MeV, and D_{33} near 1700 MeV for Δ's. The analysis was criticized by Novoseller (1978) in a more detailed study, but the overall major features remain unchanged. A K-matrix fit by Longacre & Dolbeau (1977) to a partial wave analysis of $\pi N \to \pi\pi N$ data by Dolbeau et al (1976) in the mass region 1360 to 1760 MeV, plus the πN

Table 2 Δ States from πN phase shifts

States[a]		Almehed & Lovelace (1972)			Ayed (1976)			Pietarinen (1977)		
L_{2I2J}		M (MeV)	Γ (MeV)	Γ_{el}/Γ_{tot}	M (MeV)	Γ (MeV)	Γ_{el}/Γ_{tot}	M (MeV)	Γ (MeV)	Γ_{el}/Γ_{tot}
S_{31}	****	1620	140	0.35	1623	160	0.31	1616	160	0.31
	*	—	—	—	2000	310	0.08	1920	228	0.17
P_{31}	***	1900	200	0.33	1790	222	0.16	1932	285	0.22
P_{33}	****	1235	129	1.0	1234	82	1.0	1232	—	1.0
	*	1680	220	0.1	—	—	—	—	—	—
		2150	200	0.3	1900	204	0.19	1955	200	0.10
D_{33}	***	1700	260	0.16	1723	190	0.17	1660	236	0.21
D_{35}	**	$\sim$2200	$\sim$600	0.25	1890	120	0.08	—	—	—
F_{35}	***	1875	250	0.18	1870	255	0.14	1860	305	0.13
F_{37}	****	1925	200	0.4	1928	237	0.41	1910	259	0.41
G_{39}		—	—	—	2174	210	0.05	—	—	—
$H_{3,11}$	***	—	—	—	2392	290	0.15	2420	—	0.10

[a] Asterisks are used as in Table 1.

phase shifts of (Ayed 1976) confirms the above results below 1800 MeV. However, all the obtained masses tend to be shifted systematically towards lower masses, sometimes substantially so. For example, the well-known 1234 is at 1211 MeV, the 1450 P_{11} is at 1360 MeV, the 1620 S_{31} is at 1575 MeV, and the 1700 MeV D_{33} is at 1600 MeV. This seems to be a general feature of parameters obtained from T-matrix poles as opposed to Breit-Wigner-plus-background fits to the amplitudes. Very similar results were obtained in an earlier K-matrix fit using the data of Herndon et al (Longacre et al 1975).

The reaction $\pi^- p \to K^0 \Lambda^0$ is an interesting inelastic channel to study, as only $I = 1/2$ states can be formed with the Λ polarizations imposing additional constraints. The most recent analysis, using a world data compilation from threshold to ~ 2300 MeV, is that of Baker et al (1977). They find the S_{11} state near 1700 MeV with about the same mass and width as the elastic phase shifts, while the $P_{11}(1700)$, $P_{13}(1700–1800)$, and $D_{13}(1700–1800)$ are about 100 MeV lower in mass. There is also evidence for a D_{15} effect near 2000 MeV. It shows up as a resonance in the energy-dependent analysis but not in the energy-independent one. There is no evidence in this channel for any additional states beyond the ones described above.

3.3 $\bar{K}N$ Phase Shifts

Except for the $\Sigma(1385)$ and $\Lambda(1405)$ resonances, which are below threshold, all the other Λ and Σ resonances can be observed in $\bar{K}N$ formation experiments. There have been many $\bar{K}N$ partial wave analyses in the past; these have included not only the elastic and charge-exchange channels but also two-body inelastic channels such as $\Lambda\pi$ and $\Sigma\pi$. In recent years, new high statistics data have become available from $K^- p$ and $K^- d$ interactions at low energies, both from bubble chamber and electronic counter experiments. These new data have led to a number of partial wave analyses that are a great improvement over older ones. These analyses usually contain all the available world data plus new contributions from their own experimental group. Among the notable recent analyses are

(a) the Rutherford Laboratory–Imperial College collaboration (RL-IC), which used $\bar{K}N \to \bar{K}N$, $\bar{K}N \to \Lambda\pi$, and $\bar{K}N \to \Sigma\pi$ channels (Gopal et al 1977) and recently expanded their analysis to include quasi two-body final states from reactions $K^- p \to \Lambda(1520\pi$, $\Sigma(1385)\pi$, and K^*N (Cameron et al 1978);

(b) a K-matrix multichannel analysis by Martin et al (1977);

(c) an analysis of elastic and charge-exchange channels by Hansen et al

Table 3 Λ States below 2.2 GeV from K̄N partial wave analysis

Final states L_{I2J}	Gopal et al (1977) M, Γ^c K̄N, $\Sigma\pi$	Martin et al (1978)[a] M, Γ K̄N, $\Sigma\pi$	Hansen et al (1976) M, Γ K̄N	Alston-Garnjost et al (1978a)[b] M, Γ K̄N
S_{01} ****d	1670, 45 0.2, −0.31	1664, 12 0.15, −0.13	1660–1700	1671, 29 0.17
S_{01} **	1825, 230 0.37, −0.08	1767, 435 1842, 473 1.21, −0.74 0.7, −0.43	1840–1910	1725, 185 0.28
P_{01} *	1573, 147 0.24, −0.16	1572, 247 1617, 271 0.30, −0.39 0.29, −0.39	out of range	
P_{01} **	1853, 166 0.21, −0.24	1861, 535 1953, 585 0.52, 0.25 0.49, 0.23		1703, 593 0.14
P_{03} ***	1900, 72 0.18, 0.09	1856, 191 1868, 193 0.36, 0.15 0.34, 0.14	1850–1890, 177 0.42	1908, 119 0.34
D_{03} ****	1519, 15 0.47, 0.46	out of range	out of range	1520, 15 0.45
D_{03} ****	1690, 60 0.24, −0.25	1688, 62 0.27, −0.29 1962, 182 1971, 183 0.33, 0.10 0.29, 0.08	1680–1690, 74 0.18 ~1815, −60 0.08	1692, 64 0.22
D_{05} ****	1825, 94	1817, 56 0.04, −0.17	~1740, 120 0.05	[1825, 90] 0.02
F_{05} ****	1822, 81 0.57, −0.28	1818, 76 0.59, −0.25	1815–1821, 105 0.53	1777, 116 0.37
F_{05} **	2100, 200 0.07, 0.10	out of range	~2180	out of range
G_{07} ****	2110, 250	out of range	2085–2110, 257 0.19	out of range
G_{09}	1808, 27 0.04	—	—	—

[a] Two solutions were found in this analysis.
[b] Quantities in brackets were held constant for the fits.
[c] Masses and width given in MeV.
[d] Asterisks are used as in Table 1.

Table 4 Σ States below 2.2 GeV from $\bar{K}N$ partial wave analysis

Final states[a] L_{I2J}	Gopal et al (1977) M, Γ $\bar{K}N$, Σπ, Λπ	Martin et al (1977) M, Γ $\bar{K}N$, Σπ, Λπ	Hansen et al (1976) M, Γ $\bar{K}N$	Alston-Garnjost et al (1978a) M, Γ $\bar{K}N$
S_{11}				
***	1770, 60	1800, 117	—	1770, 161
	0.15, −0.09, 0.04	0.06, 0.06, −0.1		0.33
		1813, 119		
		0.05, 0.06, −0.09		
		1755, 413		
		0.62, 0.26, 0.19		
		1834, 450		
		0.57, 0.24, −0.18		
*	1955, 170	out of range	2010–2050, 59	out of range
	0.44, 0.20, 0.08		0.17	
P_{11}				
**	—	1565, 202	—	—
		0.27, −0.34, −0.10		
		1597, 217		
		0.29, −0.37, −0.11		
**	1676, 120	—	—	1678, 33
	0, −0.16, 0	—		0.14
*	1738, 72	—	1710–1750, 192	—
	0.14, 0, 0	—	0.14	
**	1847, 216	1863, 220	1860–1910, 213	—
	0.27, 0.30, −0.24	0.27, 0.29, −0.24	0.35	

P_{13}	— —	1800, 93 0.0, −0.04, 0.03 —	1890–1910, 213 0.35 —	— —
D_{13}	**** 1670, 50 0.08, 0.21, 0.10 *** 1920, 300 0, −0.08, −0.06	1668, 46 0.07, 0.18, 0.08 1886, 157 0.14, 0.16, −0.15 1893, 159 0.13, 0.16, −0.14	1660–1690, 70 0.05 1930–1950, 100 0.13	1679, 56 0.11 —
D_{15}	**** 1774, 130 0.41, 0.13, −0.28	1792, 102 0.37, 0.08, −0.29 1777, 103 0.36, 0.08, −0.28	1765–1775, 116 0.38	1777, 116 0.37
F_{15}	**** 1920, 130 0.05, −0.19, −0.09	1925, 171 0.08, −0.05, −0.09 1933, 173 0.08, −0.05, −0.09	~1970, 80 0.06	1934, 132 0.13
F_{17}	**** 2040, 190 0.24, −0.15, 0.16	out of range —	2045–2075, 155 0.22	out of range —
G_{17}	*	out of range	~2160	out of range

[a] Asterisks used as in Table 1.

(1976) that used the fixed t analysis method of Pietarinen (1974) to remove ambiguous solutions; and

(d) an analysis by Alston-Garnjost et al (1978a) that also used only the elastic and charge-exchange channels. This analysis utilizes some very high precision data from a counter experiment studying the reaction $K^-p \rightarrow \bar{K}^0 n$ below 1.1 GeV/c (Alston-Garnjost et al 1977).

Tables 3 and 4 summarize the results from these four separate efforts. When comparing the number in the tables, one should keep in mind that the very high statistics $K^-p \rightarrow \bar{K}^0 n$ data of Alston-Garnjost et al (1977)

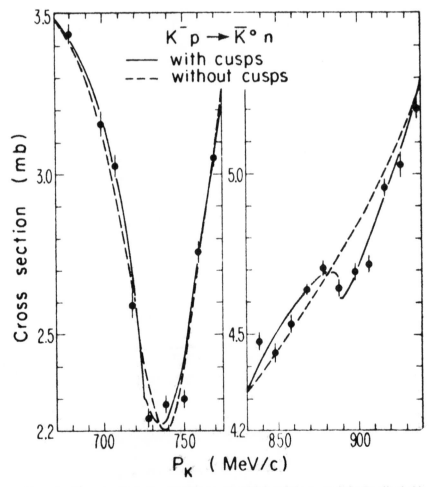

Figure 8 Charge-exchange cross sections in the vicinity of the $\Lambda\eta$ and the $\Sigma\eta$ thresholds showing the best overall fit obtained with and without cusps (Alston-Garnjost et al 1978a).

were not available to the other groups. One notable feature from these data is the clear and unmistakable need for cusps near the $\Lambda\eta$ and $\Sigma\eta$ thresholds (Figure 8). These cusps are noticeable because of nearby S-wave resonances, which strongly couple to these channels: the $S_{01}(1670)$ to $\Lambda\eta$ and the $S_{11}(1770)$ to $\Sigma\eta$. Both resonances are only a few MeV above threshold.

The agreement between the various phase shift analyses is not as good as in the πN case. States considered well established by the Particle Data Group (1978) compilation are clearly seen by all four analyses and the derived parameters are reasonably close to one another, except for the $F_{05}(1800)$ for which the mass ranges from 1777 to 1822. Hansen et al (1976) obtain a rather low value for the mass of the $D_{05}(1820)$, ~ 1740 MeV. However this result should not be taken too seriously as they only study the $\bar{K}N$ channel to which this resonance has a small coupling. The evidence for Λ states $P_{01}(1570)$, $D_{03}(1800-1960)$, and $G_{09}(1808)$ is very weak; it is quite likely that they do not exist. On the other hand the $P_{03}(1900)$ can now be considered well established and so is the existence of an S_{01} state somewhere between 1700 and 1850 MeV, even though there is considerable variation in the mass and width obtained by the various groups.

There is very little consensus on the S- and P-wave Σ states (see Figure 9), except for the $S_{11}(1770)$, which is curiously missed by Hansen et al (1976). Given the statistical significance of the cusp near $\Sigma\eta$ threshold in the data of Alston-Garnjost et al (1977) (Figure 8) and also the observation of this resonance in $K^-p \rightarrow \Sigma^0\eta$ by Jones (1974), this state should be considered well established. All other S- and P-wave Σ candidates require more evidence. Using new high statistics bubble chamber data from K^-d interactions between 680 and 840 MeV/c, Hepp et al (1976) studied the reaction $K^-n \rightarrow (\Sigma\pi)^-$ at center-of-mass energies between 1600 and 1730 MeV. They observe the $S_{01}(1670)$, $D_{03}(1690)$, and the $D_{13}(1670)$ (all well-established states) and no other, in particular no evidence for a $P_{11}(1680)$. The RL-IC expanded analysis using quasi two-body final states (Cameron et al 1978) reveals, in addition to the well-established states, some evidence for the $P_{11}(1850)$ coupling to K^*N and for a new state $S_{01}(2030)$ also coupling to K^*N.

Beyond 2.0 GeV, the $\bar{K}N$ data are not very accurate and recently only one group has attempted to study the channels $\bar{K}N$, $\Sigma\pi$, and $\Lambda\pi$ between 2.0 and 2.4 GeV (DeBellefon et al 1975, 1976). They observe the $F_{05}(2100)$ and $G_{07}(2100)$, also seen by Gopal et al and Hansen et al, but do not find the $F_{17}(2040)$. In addition they claim four new states: D_{03} at $m = 2340$ MeV with $\Gamma = 177$ MeV, H_{09} at $m = 2370$ MeV with $\Gamma = 100-200$ MeV, D_{15} at $m = 2270$ MeV with $\Gamma = 70$ MeV, and G_{19} at

$m = 2210$ MeV with $\Gamma = 80$ MeV. Needless to say, these states need confirmation.

Interesting new results were obtained searching for bumps (rather than performing phase shifts analysis) in the total cross sections by Carrol et al (1976). They accumulated very high statistics on the total cross section for K^-p and K^-d interactions between 410 and 1070 MeV/c, corresponding to a center-of-mass energy range of 1526 to 1826 MeV. Given K^-p and K^-d total cross sections, one can deduce the K^-n

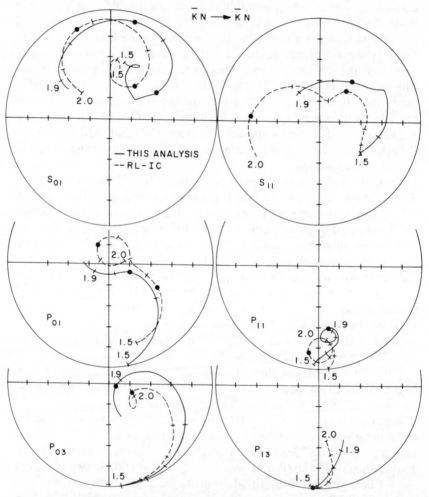

Figure 9 Argand diagram showing the solutions for S- and P-waves for two $\overline{K}N$ phase shifts. Dotted lines from Gopal et al (1977), solid lines from Alston-Garnjost et al (1978a).

total cross section that is pure $I = 1$ and in turn obtain the $I = 0$ cross section by subtracting $\sigma(K^-n)$ from $2\sigma(K^-p)$. The $I = 0$ cross section shows, very clearly, the $\Lambda(1520)$, $\Lambda(1820)$, and $\Lambda(1690)$. There is no way of determining their spin, but from their contribution to the cross section and branching ratios observed in partial wave analysis it is clear that the bumps come mostly from the well-established D_{03} and F_{05} states. In addition they observe statistically significant peaks at 1645 and 1735 MeV. These are relatively narrow (20 and 28 MeV) and couple rather weakly to $\bar{K}N$, their contributions to the $I = 0$ cross section are respectively one tenth times and one half that of the $\Lambda(1690)$, which already has much smaller coupling than the $\Lambda(1520)$ and the $\Lambda(1820)$.

Similarly, in the $I = 1$ cross section they observe, in addition to the $\Sigma(1670)$ (D_{13}) and the $\Sigma(1770)$ (D_{15}), four possible narrow structures (all with $\Gamma \leq 15$ MeV) at 1583, 1608, 1633, and 1715 MeV. These have cross sections between one half and one quarter that of the $\Sigma(1670)$ (which is already fairly small, $\sigma \sim 6.5$ mb) except for the 1715-MeV candidate, which has about the same cross section as the $\Sigma(1670)$. These narrow bumps are not seen in the high statistics $K^-p \rightarrow \bar{K}^0n$ data of Alston-Garnjost et al (1977) nor in some recent high statistics data of backward K^-p elastic scattering (Alston-Garnjost et al 1978b). It should be pointed out, however, that these structures are not observed in K^-p or K^-d total cross sections either; they appear only after separating $I = 0$ and $I = 1$ cross sections. From the partial wave analysis of Alston-Garnjost et al (1978a) one can only conclude that if these narrow structures exist they cannot be S- or P-wave resonances.

One of these narrow structures has been confirmed in a reanalysis of data on the reaction $K^-p \rightarrow \Lambda\pi^0$ (Litchfield 1974). This analysis provides evidence for the existence of the $\Sigma(1580)$ and shows it is a D_{13} state $(J^P = 3/2^-)$. All the other narrow states found by Carrol et al (1976) require confirmation.

3.4 $S < -1$ Baryon Resonances

The difficulties of finding stranger baryon resonances, Ξ's and Ω's, are numerous. The major difficulty has to do with the fact that such states are accessible only by production experiments (formation or s-channel modes not being viable mechanisms since there are no $S = -2$ mesons, where S is strangeness). In addition, the cross section for producing Ξ's is of the order of 100 μbarns, and for a given state, it decreases rapidly with the beam momentum of the initiating particle, proportional to p^{-4}. The cross section for Ω's is at least one order of magnitude smaller, ≈ 2 μb. Nevertheless quite a few Ξ's have been reported, some being on more solid experimental footing than others. Among the firmly established

Ξ states are the $\Xi(1320)$ (Fowler et al 1961), which decays only weakly, and the $\Xi^*(1530)$, which was also found quite early on (Bertanza et al 1962 and Pjerrou et al 1962). The determination of its spin parity followed soon afterwards (Schlein et al 1963, Button-Shafer et al 1966).

There is reasonable evidence, $\geq 3\sigma$, for two more resonances: the $\Xi(1820)$, which mainly decays into ΛK or ΣK (Smith et al 1965, Badier et al 1965, Alitti et al 1969), and the $\Xi(1940)$, which mainly couples to $\Xi\pi$ (Badier et al 1965, Alitti et al 1968). In fact there is some evidence that there may be more than one resonance in these mass regions. Of lesser experimental significance were the claims for three other states: (a) The $\Xi(1630)$ (Apsel et al 1969, Ross et al 1972, Briefel et al 1977). One additional difficulty with this state is its low mass and the absence of any obvious SU(3) companions. (b) The $\Xi(2030)$ (Alitti et al 1969, Bartsch et al 1969). (c) The $\Xi(2430)$ (Alitti et al 1969, Bartsch et al 1969).

Progress occurred in the last few years, thanks to a very high statistics (for a bubble chamber) experiment, 128 events/μb, with a 4.2-GeV/c K^- beam incident on the 2-m HBC at CERN. The analysis of these data has been performed by the Amsterdam-CERN-Nijmegen-Oxford collaboration (ACNO 1978). First, this group clearly established the existence of a $\Xi^*(1820)$ decaying equally into $\Lambda\bar{K}$ and $\Xi^*(1530)\pi$, with some branching into $\Sigma\bar{K}$ and no visible $\Xi\pi$ decay mode. Its mass was determined to be 1823 ± 2 MeV and its width 21 ± 5 MeV (Gay et al 1976b). Its spin was later found to be 3/2 (Teodoro et al 1978) but the data were still not adequate to establish the parity. The existence of the Ξ^* at $m = 2024\pm2$ MeV and with $\Gamma = 16\pm5$ MeV was confirmed quite conclusively. This state decays predominantly into $\Sigma\bar{K}$ and some $\Lambda\bar{K}$, but no evidence for this state is seen in $\Xi\pi$ or $\Xi\pi\pi$ channels (Hemingway et al 1977). Study of the moments suggests that this state has $J \geq 5/2$. There was also some evidence for an enhancement at 2120 MeV. However, its u distribution seems anomalous, so the resonance interpretation is in doubt. The ACNO collaboration also observes a statistically significant enhancement at the $\Sigma\bar{K}$ threshold (at least six standard deviations in the Σ^+K^- channel, about three standard deviations in the Σ^0K^- channel). The $\Lambda\bar{K}$ channels are consistent with there being a resonance near the $\Sigma\bar{K}$ threshold that couples mostly to the $\Sigma\bar{K}$ channel (ACNO 1978). A coupled channel fit suggests a resonance with $m = 1690\pm10$ MeV and $\Gamma = 23\pm7$ MeV. The fact that the number of $\Sigma\bar{K}$ events reaches a maximum within less than 20 MeV of the threshold indicates that this state must decay via an S-wave, thus it must be $1/2^-$. The ACNO collaboration so far has found no evidence for a state near 1630 MeV nor near 1940 MeV.

To summarize, there are now in addition to the $\Xi^*(1530)$ two more well-established Ξ resonances, at 1820 MeV and 2024 MeV. The first has $J = 3/2$ and (or) its mass makes it a natural candidate for the missing member of the $3/2^-$ octet. The evidence on the second points to $J \geq 5/2$; its mass is very close to that expected for the $5/2^-$ octet member. There is also some evidence for a $1/2^-$ resonance near the $\Sigma\bar{K}$ threshold, at 1690 MeV. It is worth noting that, except for the $\Xi^*(1530)$ (which happens to be below $\Lambda\bar{K}$ threshold) and probably the $\Xi(1940)$, all the resonances observed so far have small branching ratios to $\Xi\pi$, their two-body decays tend to be mostly $\Lambda\bar{K}$ or $\Sigma\bar{K}$.

The Ω^- hyperon, discovered in 1964 (Barnes et al 1964) was the key finding in establishing SU(3) as the symmetry for hadrons (Gell-Mann 1961, Ne'eman 1961). Its mass is well determined to be 1672 MeV to an accuracy of $\simeq 1$ MeV and several decay modes have been observed, namely ΛK^-, $\Xi^0\pi^-$, and $\Xi^-\pi^0$. There is reasonable evidence that the Ω^- spin is $\geq 3/2$ (Deutschmann et al 1977) and until recently there was a discrepancy concerning its lifetime (1.4 versus 0.8×10^{-10} s) (Hemingway et al 1978). This has been clarified by a beautiful experiment (Bourquin et al 1978) that accumulated approximately 2,000 Ω^- decays; previously only about 100 decays were available. These events yielded precise values for the Ω^- lifetime ($0.82 \pm 0.06 \times 10^{-10}$ s), decay rates for the ΛK^-, $\Xi^0\pi^-$, and $\Xi^-\pi^0$ (ΛK^- accounts for about 2/3 of the total decays), and the decay asymmetry parameter α for $\Omega^- \to \Lambda K^-$ ($\alpha = 0.06 \pm 0.14$, essentially consistent with parity conservation). There has been no evidence presented for any excited Ω^* state. The anticipated mass region for such resonances is 2–2.5 GeV with expected strong decay modes such as $\Omega\pi\pi$ and Ξ^*K. It would be comforting to find at least one such excited state to maintain our confidence in our understanding of hadron spectroscopy.

3.5 *Baryon Exotics*

We denote here as baryon exotics all the states with baryon number (N_B) greater than zero that cannot be described as being made of only three quarks. Under this definition all of nuclear physics becomes exotic. However, for the sake of brevity we concern ourselves only with baryons having $N_B = 1$ and 2 (dibaryons).

The simplest signature for an exotic baryon is a state with strangeness $S = +1$; this immediately implies it must contain an $\bar{s}$-quark. The first evidence for the possible existence of such states came from an experiment of Cool et al (1966), which showed bumps in the K^+p and K^+d total cross section both in $I = 1$ (Z_1^*) and $I = 0$ (Z_0^*) states. In particular, there was a prominent bump in the $I = 1$ total cross section at 1910 MeV and in the $I = 0$ total cross section at 1750 MeV. These bumps are now understood to come from the superposition of different channels; the $I = 1$ bump comes mostly from the $K\Delta$ channel and the $I = 0$ bump from the K^*N channels. To show whether these bumps are in any way related to resonances requires a multichannel partial wave analysis.

The simplest channel to study for Z_1^* resonances is K^+p, which is pure $I = 1$. Phase shift analyses of K^+p data find two types of solutions. The first type favored by the majority has an S_{11} wave that becomes inelastic above 1750 MeV (Arndt et al 1974, Martin 1975, Cutkosky et al 1976b); in the second type the S-wave becomes small and stays elastic (Fich et al 1974). The particular wave of interest is P_{13}, which shows strong threshold effects associated with the onset of Δ and K^* production in all solutions. Martin (1975) and Cutkosky et al (1976b) concluded that although a Z_1^* resonance could be accommodated in their phase shift solutions, it was not necessary to invoke one and thus the evidence was not compelling. On the other hand, Arndt et al (1974), did a two-channel K-matrix fit, in which the inelastic channel was $K\Delta$. They found a pole in the complex energy plane at 1787–i100 MeV (the real part gives the mass and the imaginary part half the width of a resonance).They redid the analysis (Arndt et al 1978) with more data, including polarization data at four momenta and found two solutions, both of which exhibit poles near 1797–i110 MeV. Thus, although there is still need for confirmation, there seems to be rather good evidence for a 200-MeV wide $P_{13}(Z_1^*)$ resonance near 1797 MeV.

Evidence for a possible Z_0^* resonance comes from a partial wave analysis performed by the Bologna-Glasgow-Roma-Trieste (BGRTO) collaboration (Giacomelli et al 1974), which used all world data available at the time on the K^+n elastic and charge-exchange reactions. They found three solutions (A, C, and D). Solution A shows no resonances, solutions C and D exhibit resonance behavior in the P_{01} wave ~ 1780 MeV. Solution A cannot fit the only polarization data presently available from the reaction $K^+n \rightarrow K^0p$ at 600 MeV/c (Ray et al 1968). However this is only one measurement with relatively low statistics, and as such it is not sufficient for conclusively ruling out this nonresonant solution A. Martin (1975) performed another phase shift analysis using additional data, the same $K^+n \rightarrow K^0p$ polarization data and also $K^+p \rightarrow K^+p$ data. A single solution was obtained that differed in detail from the BGRT solutions but had a P_{01} partial wave quite similar to that of solutions C and D. Analyses comparing the BGRT solution to more recent data in $K^+n \rightarrow K^0p$ reactions between 700 and 900 MeV/c (Sakitt et al 1977) and below 600 MeV/c (Glasser et al 1977) again tend to favor the C and D solutions over A, but the differences are not very significant.

It was suggested by London (1974) that it would be profitable to study the channel $K_L p \rightarrow K_S p$ since it permits one to look at the interferences between the $I = 0,1$, $S = +1$ amplitudes and the $I = 1$, $S = -1$ amplitudes. Results from studies of this channel (Alexander et al 1975, Bigi et al 1976, 1978, Corden et al 1978b) are somewhat inconclusive, in part because of uncertainties in the $I = 1$, $S = -1$ amplitudes. They tend to favor the resonant solutions C and D, but no one solution fits the data perfectly. Overall, the evidence seems to point toward the existence of Z_0^* resonance, $P_{01}(1780)$. The resonance solutions predict large and negative (positive) polarization in the reactions $K^+ n \rightarrow K^+ n$ ($K^0 p$), as opposed to large positive (zero) polarization for the nonresonant solution. Thus, a good polarization measurement could conclusively rule out or establish such a resonance.

For dibaryons the first observation of another possible state besides the deuteron was made quite early (Dahl et al 1961), although not interpreted as such, in the reaction $K^- d \rightarrow \Lambda p \pi^-$. Dahl et al (1961) observed an enhancement in Λp very near the ΣN threshold, at a mass of 2130 MeV. This enhancement has since been seen by many others. The best data and most detailed analysis of this reaction currently are those of Braun et al (1977). Since the enhancement occurs so close to the ΣN threshold there was some question as to its being purely a cusp effect. Detailed analysis shows that it is a $^3S_1$ state (like the deuteron), peripherally produced, and that the effect is too strong and narrow to be explained without a nearby pole. Depending on the strength of its coupling to ΣN it could be quite broad (100 MeV or more) even though its observed width is only 6 MeV. The existence of such a state would be expected from six-quark bag model calculations (Jaffe 1977b). These calculations also predict the existence of a $\Lambda\Lambda$ state (H) that may be below threshold and thus stable against strong decays. A search for such a state produced in the reaction $pp \rightarrow K^+ K^+ X$ was performed, where X scanned the missing-mass range 2.0 to 2.5 GeV (Carroll et al 1978). Such an object was not found; however, an upper limit on the production cross section for such a resonance was determined to be of the order of 100 nb.

Recently, a series of measurements with polarized beams impingent on polarized targets at Argonne have shown a striking energy dependence in the difference between pp total cross sections for parallel and antiparallel spin states. The first measurements were done with spins transverse to the beam direction and $P_{\text{Lab}} \geq 2.0$ GeV/c. The most noticeable effect was a rapid rise near 2.0 GeV/c (de Boer et al 1975). Another experiment looking at polarization parallel to the beam direction observed rapid variations near 1.5 GeV/c and 1.2 GeV/c (Auer et al 1977a,b,

1978a). The rapid variation near 1.5 GeV/c has been interpreted as evidence for a $J^P = 3^-$ ($^3F_3$) resonance around 2260 MeV with width of 200 MeV (Hidaka et al 1977). It was confirmed by studying pp cross sections with polarizations transverse to the beam direction by Biegest et al (1978); they also confirmed the effect near 2.0 GeV/c seen by de Boer et al (1975). The results of Auer et al (1977a,b, 1978a) were further refined by studying the spin-spin correlation parameter C_{LL}, which can be obtained using the differential cross sections for parallel and anti-parallel spin states. For center-of-mass angles near 90°, this parameter exhibits dramatic changes around 1.2 GeV/c, 1.5 GeV/c, and 2.0 GeV/c, corresponding to pp masses of 2150 MeV, 2250 MeV, and 2430 MeV (Auer et al 1978b). This is additional evidence for the $^3F_3$ state and indication of possible $^1D_2$ and $^1G_4$ resonances.

4 COMPARISON WITH QUARK MODEL

The quark model so far has been quite successful in aiding the classification of meson and baryon resonances. Considered merely a mnemonic until not long ago, it has acquired reality since the discovery of charmed particles. It provides for us the best framework for summarizing what is presently known in particle spectroscopy. Neglecting the charm quark (c) the appropriate symmetry is SU(3) based on the three quark flavors u, d, and s.

The simplest systems are the mesons that can be described as $q\bar{q}$ states, the three quarks imply that these states will come in octets and singlet families. As described below the octet and singlet are intimately mixed so it is more appropriate to refer to them as nonet families. A nonet can be quickly described by grouping its members by I-spin. The $I = 1/2$ members are u$\bar{s}$, d$\bar{s}$, s$\bar{u}$, and s$\bar{d}$; the $I = 1$ members are u$\bar{d}$, d$\bar{u}$, and $1/\sqrt{2}$ (u$\bar{u}$-d$\bar{d}$); the $I = 0$ members are the octet $1/\sqrt{6}$ (u$\bar{u}$ + d$\bar{d}$ − 2s$\bar{s}$) and the singlet $1/\sqrt{3}$ (u$\bar{u}$ + d$\bar{d}$ + s$\bar{s}$).

For illustration we shall use as a guide the 1^- ground states (which can at present be said to be the best-understood systems). The $q\bar{q}$ pairs are in an $L = 0$, $S = 1$ configuration so the charge conjugation is $C = -1$.

If we denote the octet and singlet states as $|8\rangle$ and $|1\rangle$, the actual observed states, $|\omega\rangle$ and $|\phi\rangle$, are related to the above by a mixing angle θ

$$|\omega\rangle = \cos\theta\,|8\rangle + \sin\theta\,|1\rangle$$

$$|\phi\rangle = -\sin\theta\,|8\rangle + \cos\theta\,|1\rangle.$$

At the root of SU(3) symmetry breaking are the quark masses, in particular the difference in mass between the s, u, and d quarks. Note that if $\tan\theta = 1/\sqrt{2}$ the ϕ becomes pure s$\bar{s}$ and the ω pure u$\bar{u}$ + d$\bar{d}$, which would explain why the ϕ couples mostly to $K\bar{K}$. Why aren't the observed states pure $u\bar{u}$ and $d\bar{d}$? This would imply physical states of $\rho_0 + \omega$ and $\rho_0 - \omega$ thereby violating I-spin conservation in strong interactions. That I-spin is conserved is intimately connected to the fact that u and d quarks are close in mass. From this it also follows that the members of the nonet with no s quarks should be very close in mass, as is indeed the case for the ρ and ω. The Gell-Mann Okubo formula for the masses of a meson octet can be written as $m_0 = (4m_{1/2} - m_1)/3$ where the subscript stands for the I-spin of the octet member. Using this formula and assuming $\tan\theta = 1/\sqrt{2}$ and that the masses of the $I = 1$ and $I = 0$ states with no strange quarks are equal, one obtains a very simple mass relation for the $s\bar{s}$ member: $m(s\bar{s}) = 2m_{1/2} - m_1$. Applying this simple formula to the 1^- states and using the latest mass values gives $m_\phi = 1022$ MeV. Considering that the Gell-Mann Okubo formula is a first-order approximation valid only to the extent that the mass differences in a nonet are smaller than the average mass, the result is better than one has a right to expect. It is amusing to use the above formula on the SU(3) nonet made up of u, d, and c (instead of s) quarks. The $I = 1/2$ state's mass is then 2010 MeV and one predicts for the J/ψ a mass of 3250 MeV; given the enormous mass differences, this is a remarkable result.

We can now proceed to classify the known mesons in terms of the quark model according to their spin and orbital quantum numbers. This is summarized in Table 5 including expected values for the $m(s\bar{s})$ based on the very naive assumptions described above. The difference in mass between the $I = 1$ and the $I = 0$ ($u\bar{u} + d\bar{d}$) member gives an indication of how good we can expect the calculation of $m(s\bar{s})$ to be. The $J^{PC} = 0^{-+}$ nonet is exceptional in that both η and η' must mix strange and non-strange pairs. The 1^{--} and 2^{++} are in great shape and are prime examples of nonets satisfying naive expectations. For the 3^{--} nonet only the $s\bar{s}$ member is missing; given the closeness in mass between the $\omega(1670)$ and the g(1690) it would be surprising if $m(s\bar{s})$ does not turn out to be very near the expected value of 1870 MeV.

For the 0^{++} we follow Jaffe's (1977a) suggestion and do not include the lower mass states ε, κ, S*, and δ in the table. The S* and δ badly violate the naive mass formula and the strong coupling of δ to $K\bar{K}$ cannot be explained if it is considered to be a $q\bar{q}$ state. Thus, one expects a 0^{++} state near 1600 MeV decaying mostly to $K\bar{K}$. One should look for it in K-induced reactions; a partial wave analysis with high statistics of $K^-p \rightarrow K_sK_s\Lambda$ would probably be the best way to find it. The strange members of the 4^{++} nonet are still missing and it is not clear at present which of $I = 1, 4^{++}$ claimed states is the correct one. Using an exchange

Table 5 Mesons classified as qq̄ states

LS J^{PC}	$I = 1$	$I = 1/2$	Masses in MeV $I = 0$ $(u\bar{u} + d\bar{d})$	$I = 0$ $s\bar{s}$
00 0^{-+}	$\pi(139)^a$	K(495)	Mix[$\eta(540)$, $\eta'(958)$]	Mix[$\eta(540)$, $\eta'(958)$]
01 1^{--}	$\rho(770)$	K*(895)	$\omega(780)$	predicted 1022 $\phi(1019)$
10 1^{+-}	B(1230)	Mix[$Q_1(1290)$, $Q_2(1400)$]		
11 2^{++}	$A_2(1310)$	K*(1430)	$f^0(1270)$	predicted 1550 f'(1515)
1^{++}	$A_1(1100)?^b$	Mix[$Q_1(1290)$, $Q_2(1400)$]?	D(1285)?	
0^{++}	$\delta'(1300)$	$\kappa'(1450)$	$\varepsilon'(1300)$	predicted 1600
20 2^{-+}	$A_3(1650)$	Q(1750)?		
21 3^{--}	g(1690)	K*(1780)	$\omega(1670)$	predicted 1870
2^{--}	$F_1(1540)?$	Q(1580)?		
1^{--}				
31 4^{++}	$A_4(1950)?$ $A_4(2030)?$	predicted 2090	h(2020)	predicted 2230 predicted 2140

^{a} Well-established states are underlined.
^{b} A question mark indicates states that may be improperly classified.

degenerate linear Regge trajectory one can predict a K*(2090) and in turn an s$\bar{s}$ state near 2200 MeV. The natural spin parity series seems to be living up to expectations, but the unnatural spin parity series is in sad shape. We have one 1^{+-}, $I = 1$ well-established state, the B(1230), and one 1^{++}, $I = 0$ well-established state, the D(1285). In addition, the evidence is good for two 1^+ strange mesons, $Q_1(1290)$ and $Q_2(1400)$; these are presumably a mixture of the expected 1^{++} and 1^{+-} states. The mass of the $I = 1$, 1^{++} candidate (A_1) is still in dispute, but the best evidence seems to point towards 1100 MeV.

Out of these ingredients it is difficult to come up with nonets that have the same systematics as the natural spin parity ones. The decay of the D mostly to $\delta\pi$ precludes it being considered an s$\bar{s}$ state, yet its large mass difference from an A_1 at 1100 MeV runs against expectations from a pure u$\bar{u}$ + d$\bar{d}$ state. There is also no simple explanation for its narrow width.

Furthermore, if we take seriously the suggestion that the δ is a $qq\bar{q}\bar{q}$ state, then the D becomes a strange animal indeed.

Another problem is that in the natural parity series typical splittings between $I = 1/2$ and $I = 1$ states are of the order of a π-mass or less; the masses of the Q_1 and Q_2 are too high then to get a reasonable mix maintaining this split for both the A_1 and the B, unless the A_1 mass is closer to 1300 MeV than to 1100 MeV. Maybe the basic configuration model for the unnatural spin parity states should not be the simple 1^{--} and 2^{++} nonets but rather the 0^{-+} nonet. It is clearly an important task for meson spectroscopy to find the missing members of the 1^+ nonets. The simple fact that it is very difficult to find and produce many of these states in hadronic interactions already indicates that there is something intrinsically different in the underlying dynamics.

Note that it is possible to accommodate the 0^{++} nonet with the other natural spin parity families by invoking the existence of $qq\bar{q}\bar{q}$ states. They may be responsible also for the present confusion in the 1^+ nonets, particularly if mixing between $q\bar{q}$ and $qq\bar{q}\bar{q}$ states occurs. Demonstrating that $qq\bar{q}\bar{q}$ states truly exist is another important challenge to particle physics spectroscopists. In general, they are expected to be broad states fissioning immediately into ordinary mesons except maybe for high spin states, which may have large branching ratios into baryon-antibaryon final states. There are now quite a few fairly convincing claims for $\bar{p}p$ states that need further confirmation and are quite interesting in themselves, but are probably not sufficient to settle the question of the existence of $qq\bar{q}\bar{q}$ states. For this it is important to find doubly charged or doubly strange meson states or states with $J^{PC} = 0^{--}$, 0^{+-}, or 1^{-+} which cannot be accommodated in the simple $q\bar{q}$ model. They may be very broad and massive resonances (Jaffe 1977a) and thus may be difficult to see. Partial wave analysis of exotic channels may be the only way to find them and such analyses should not be neglected.

We have not included radial excitations in Table 5 because they are still in a state of great confusion. There are many candidates from photoproduction and e^+e^- annihilations: $\rho'(1250)$, $\rho''(1550)$, $\rho'''(1700)$, $\omega'\,(1650)$, $\omega''(1770)$, $\phi'''\,(2030)$. With the exception of the $\rho''(1550)$ none of these can be considered well established. There presently exists also some evidence (in need of confirmation) for the existence of radial excitations of the 0^{-+} nonet, a K$'(1400)$, and two $I = 0$ states: at 1.26 GeV decaying to $\delta^0\pi^0$ and at 1.4 GeV decaying to $\varepsilon\eta$.

When it comes to the baryons a very naive approach similar to the one used above for mesons is not very fruitful. One needs to introduce an additional ingredient, color (which must come in three varieties), to satisfy Fermi statistics. Imposing the condition that color is not directly observable, i.e. all physical states are color singlets, forces all observable states to be either qqq or q$\bar{q}$ or multiples thereof. Having the baryons made up of only qqq limits the SU(3) multiplets to singlets, octets, and decuplets. The octets and singlet can mix, but there does not seem to be any simple mixing pattern as for the natural spin-parity mesons. Until recently only the ground-states $1/2^+$ octet and $3/2^+$ decuplets were fully filled. For some of the other octets there was enough information from N*, Λ, and Σ states (masses and branching ratios) to predict with some confidence the mass of the corresponding Ξ^* and its main decay. There are now two more octets that only lack complete information on the spin-parity of the Ξ^* candidate to be declared complete:

(a) The $3/2^-$ with N*(1520), Λ_8(1690), Λ_1(1518), Σ(1670), and Ξ^*(1820) (only the parity of the Ξ^* has not yet been determined) with the Ξ^*(1820) decaying mostly to $\Lambda\bar{K}$ basically as expected (Samios et al 1974). The mixing angle between octet and singlet is fairly small, $\sim 20°$.

(b) The $5/2^+$ octet with the N*(1685), Λ_8(1815), Λ_1(2100), Σ(1930) and the Ξ^*(2024) (for which one only knows that $J \geq 5/2$). The masses are such that this octet satisfies the Gell-Mann Okubo mass formula with practically no mixing ($\theta \lesssim 10°$). The Ξ^*(2024) decays mostly to $\Sigma\bar{K}$, again as expected (Samios et al 1974). It is, of course, not known for sure that the Λ(2100) is a singlet, or whether all other members for an octet are missing.

There is also good evidence for a $1/2^-$ Ξ^*(1690). The octet that emerges trying to associate this state with the other known $1/2^-$ states is rather peculiar: N*(1535), Λ(1405), Λ(1670), and Σ(1770). The Σ state has higher mass than the Ξ state and the mixing angle comes out quite high ($\sim 41°$). There is also a $1/2^-$ Δ state near 1620, the mass difference between it and the Σ(1770) is of the order expected for a decuplet. It is probably more likely that the Σ(1770) belongs to a $1/2^-$ decuplet and that the corresponding Σ for the $1/2^-$ octet remains to be found.

For the decuplets there has been essentially no progress in many years, mainly because the corresponding Ξ^* and Ω^* states have yet to be seen. Of the many Δ and Σ states observed only two seem to have a mass difference similar to the one in the $3/2^+$ decuplet and thus may belong together: the $1/2^-$ Δ(1620) and Σ(1770) and the $7/2^+$ Δ(1930) and Σ(2030). If they indeed are part of the same multiplets, then the following states should be found: for Ξ a $1/2^-$ at 1920 MeV and a $7/2^+$ at 2130 MeV; for Ω^- a $1/2^-$ at 2070 MeV and a $7/2^+$ at 2280 MeV. The $3/2^-$ Δ(1670) and the Σ(1920) may also be part of the same decuplet, although in this case the mass difference is rather large, the corresponding

Ξ and Ω would then be at 2170 and 2420 MeV. It is clearly important to find these Ξ and Ω states. At present we cannot even answer the question of whether the $3/2^+$ decuplet mass splittings are typical or exceptional.

5 CONCLUSION

In recent years there has been steady progress in light hadronic spectroscopy. There are now three reasonably well-established baryon octet families ($J^P = 1/2^+$, $3/2^-$, and $5/2^+$); they seem to mix little, if at all, with singlet states. For decuplets there is still only one complete example ($3/2^+$). It would certainly increase our confidence in our understanding of hadron spectroscopy to have at least one more example. The natural spin parity mesons seem to be falling into a simple pattern of nonets fulfilling naive quark model expectations. For completeness it would be satisfying to find more ϕ-like states, i.e. states that have as dominant decay modes $K\bar{K}$ or $K\bar{K}^*$ final states. In particular a 0^+ state with mass near 1600 MeV and a 3^- state with mass near 1870 MeV are expected. One of the major current problems is that of the unnatural spin parity mesons. The only complete nonet of this type is the ground state 0^- that has a different mixing pattern between octet and singlet than the other completed nonets (η, η' are mixtures of $u\bar{u}$, $d\bar{d}$, and $s\bar{s}$ states, unlike ω and ϕ which are pure $u\bar{u} + d\bar{d}$ and $s\bar{s}$, respectively). It is not at all obvious what pattern the 1^+ mesons will follow and it is clearly of interest to find out whether or not unnatural spin parity ϕ-like mesons exist. A promising avenue to look for ϕ-like states is the study of reactions of the type $K^-p \rightarrow K\bar{K}\Lambda$ or $K\bar{K}^*\Lambda$. Until recently, no analysis with high statistics of these channels had been done.

There is still a great deal of uncertainty concerning possible radially excited states. Since the discovery of excited states of the J/ψ in e^+e^- annihilation, it is generally believed that at lower center-of-mass energies one should observe radially excited relatives of the ρ, ω, and ϕ. Experimental results to date suffer from lack of statistics and, although there are many claimed states, none except for the $\rho'(1550)$ can be considered well established.

A topic that has become increasingly popular in recent years is that of exotic states, i.e. hadrons that cannot be explained as being made of a $q\bar{q}$ pair or three quarks. Multiples of these configurations such as $qq\bar{q}\bar{q}$, $qqq q\bar{q}$, 6-quarks, etc, are allowed in the quark model. Six-quark states seem to have been observed, they can be thought of as relatives of the deuteron and lie at the boundary between nuclear and elementary particle physics. The evidence for the existence of Z^* (strangeness $= +1$, $qqqq\bar{q}$) states is good but not conclusive. It is plausible that in the near future such states will be firmly established. The status of $qq\bar{q}\bar{q}$ states is somewhat less firm. There are dynamical arguments favoring the interpretation of some of the observed meson states as $qq\bar{q}\bar{q}$ states, the most notable examples are some of the 0^+ states and some states that seem to couple more strongly to $\bar{p}p$ than to mesons. In terms of quantum numbers alone, however, they cannot be distinguished from $q\bar{q}$ states. No example of a well-established meson state currently exists that is impossible to interpret as a $q\bar{q}$ state. A clear-cut example would be a doubly charged or doubly strange meson. There are good reasons for believing that baryon exchange processes should favor production of such states. Therefore, what may be needed are partial wave analyses of doubly charged or doubly strange final states produced via baryon exchange, a nontrivial task.

Overall, the quark model provides a very good framework for classifying hadronic states. There still remain, as described above, important and interesting questions but it is unlikely that the answers will shake the quark model's foundations. Rather, they should illuminate the dynamics of light quarks, which is currently poorly understood.

Acknowledgments

One of us (SDP) wishes to thank Dr. Frank Paige for stimulating discussions. We also want to thank Ms. Rae Greenberg, Ms. Donna Earley, and Ms. Robin Maragioglio for carefully typing the manuscript. This research supported by the US Department of Energy under contract EY-76-C-02-0016.

Literature Cited

ACNO Collaboration (Amsterdam-CERN-Nijwegen-Oxford). 1978. Paper submitted to the *XIX Int. Conf. on High Energy Physics, Tokyo*

Alexander, G. et al. 1978. *Phys. Lett. B* 73:99

Alexander, G. et al. 1975. *Phys. Lett.* 58B: 484

Alitti, J. et al. 1968. *Phys. Rev. Lett.* 21:1119

Alitti, J. et al. 1969. *Phys. Rev. Lett.* 22:79

Almehed, S., Lovelace, C. 1972. *Nuc. Phys. B* 40:157

Alston-Garnjost, M. et al. 1961. *Phys. Rev. Lett.* 6:300

Alston-Garnjost, M. et al. 1975. *Phys. Rev. Lett.* 35:1685

Alston-Garnjost, M. et al. 1977. *Phys. Rev. Lett.* 38:1007

Alston-Garnjost, M. et al. 1978a. *Phys. Rev. D* 18:182

Alston-Garnjost, M. et al. 1978b. Paper submitted to the *XIX Int. Conf. on High Energy Physics, Tokyo*

Ambrosio, M. et al. 1977. *Phys. Lett.* 68B: 397

Antipov, Yu M. et al. 1973. *Nucl. Phys. B* 63:141, 153

Antipov, Yu M. et al. 1975. *Nucl. Phys. B* 86:381

Apel, W. D. et al. 1975. *Phys. Lett.* 57B:398

Apel, W. D. et al. 1972. *Phys. Lett.* 41B:542

Apel, W. D. et al. 1978. Paper submitted to the *XIX Int. Conf. on High Energy Physics, Tokyo*

Apsel, S. et al. 1969. *Phys. Rev. Lett.* 23:884

Arndt, R. et al. 1974. *Phys. Rev. Lett.* 33:987

Arndt, R. et al. 1978. Paper submitted to the *XIX Int. Conf. on High Energy Physics, Tokyo*

Ascoli, G. et al. 1970. *Phys. Rev. Lett.* 25:962

Ascoli, G. et al. 1973. *Phys. Rev.* D7:669
Ascoli, G. et al. 1974. *Phys. Rev.* D9:1963
Auer, I. P. et al. 1977a. *Phys. Lett.* 67B:113
Auer, I. P. et al. 1977b. *Phys. Lett.* 70B:475
Auer, I. P. et al. 1978a. *Phys. Rev. Lett.* 41: 354
Auer, I. P. et al. 1978b. *Phys. Rev. Lett.* 41: 1436
Ayed, R. 1976. PhD thesis, Saclay, France
Bacci, C. et al. 1977. *Phys. Lett.* 68B:393
Badier, J. et al. 1965. *Phys. Lett.* 19:612
Badier, J. et al. 1965. *Phys. Lett.* 16:171
Baillon, P. et al. 1972. *Phys. Lett.* 38B:555
Baillon, P. et al. 1967. *Nuovo Cimento* 50A: 393
Baker, R. D. et al. 1977. *Nucl. Phys.* B126: 365
Baldi, R. et al. 1978. *Phys. Lett.* 74B:413
Baldi, R. et al. 1976. *Phys. Lett.* 63B:344
Ballam, J. et al. 1973. *Phys. Rev.* D7:3150
Ballam, J. et al. 1974. *Nucl. Phys.* B76:376
Baltay, C. et al. 1978. *Phys. Rev. Lett.* 40:87
Barbarino, G. et al. 1972. *Lett. Nuovo Cimento* 3:689
Barnes, V. E. et al. 1964. *Phys. Rev. Lett.* 12: 204
Bartolucci, S. et al. 1977. *Lett. Nuovo Cimento* 39A:374
Bartsch, J. et al. 1969. *Phys. Lett.* 28B:439
Basdevant, J. L., Berger, E. L. 1977. *Phys. Rev.* D16:657
Basdevant, J. L., Berger, E. L. 1978. *Phys. Rev. Lett.* 40:994
Baton, J. P. et al. 1970. *Phys. Lett.* 33B:528
BCCFMS (Bergen-CERN-College de France-Madrid-Stockholm Collaboration). 1978. Paper contributed to the *XIX Int. Conf. on High Energy Physics, Tokyo*
BCCMS Collaboration (Bergen-CERN-California-Madrid-Stockholm). 1977. *Preprint CERN/DP/PHYS 77-31*
Bemporad, C. 1977. *Int. Conf. on Lepton and Photon Interactions, Hamburg*
Benaksas, D. et al. 1972. *Phys. Lett.* 42B:511
Benkheiri, P. et al. 1977. *Phys. Lett.* 68B:483
Benkheiri, P. et al. 1979. *Phys. Lett.* 81B:380
Berger, Ch. 1977. *Eur. Conf. on Particle Physics, Budapest*, p. 793
Bertanza, L. et al. 1962. *Phys. Rev. Lett.* 9: 180
Beusch, W. et al. 1978. *Phys. Lett.* 74B:282
Biegest, E. K. et al. 1978. *Phys. Lett.* 73B:235
Bigi, A. et al. (BEGPR: Bologna-Edinburgh, Glasgow, Pisa, Rutherford). 1976. *Nucl. Phys.* B110:25
Bigi, A. et al. (BEGPR: Bologna-Edinburgh, Glasgow, Pisa, Rutherford). 1978. *Nucl. Phys.* B132:189
Bingham, H. et al. 1972. *Phys. Lett.* 41B:635
Binnie, D. M. et al. 1973. *Phys. Rev. Lett.* 31:1534

Blum, W. et al. 1975. *Phys. Lett.* 57B:403
Boucrot, J. et al. 1977. *Nucl. Phys.* B121:251
Bourquin, M. et al. 1978. *CERN preprint.* Geneva
Bowler, M. et al. 1975. *Nucl. Phys.* B97:227
Brandenburg, G. et al. 1976a. *Nucl. Phys.* B104:413
Brandenburg, G. et al. 1976b. *Phys. Rev. Lett.* 36:703, 1239
Braun, O. et al. 1973. *Phys. Rev.* D8:3794
Braun, O. et al. 1977. *Nucl. Phys.* B124:45
Briefel, E. et al. 1977. *Phys. Rev.* D16:2706
Brown, R. M. et al. 1978. *Nucl. Phys.* B144: 287
Brückner, W. et al. 1977. *Phys. Lett.* 67B:222
Budner, N. M. 1977. *Eur. Conf. on Particle Physics, Budapest*, p. 771
Bukin, A. D. et al. 1978. *Phys. Lett.* 73B:226
Button-Shafer, J. et al. 1966. *Phys. Rev.* 142: 883
Cameron, W. et al. 1978. Paper submitted to *Int. Conf. on High Energy Physics, Tokyo*
Carroll, A. S. et al. 1974. *Phys. Rev. Lett.* 32:247
Carroll, A. S. et al. 1976. *Phys. Rev. Lett.* 37:806
Carroll, A. S. et al. 1978. *Phys. Rev. Lett.* 41:777
Carter, A. A. et al. 1977. *Phys. Lett.* 67B:117
Carter, A. A. 1978. *Nuc. Phys.* B141:467
Cason, N. M. et al. 1976. *Phys. Rev. Lett.* 36:1485
Cason, N. M. et al. 1978. *Phys. Rev. Lett.* 41:271
Ceradini, F. et al. 1973. *Phys. Lett.* 43B:341
Chaloupka, V. et al. 1974. *Phys. Lett.* 51B: 407
Chaloupka, V. et al. 1976. *Phys. Lett.* 61B: 487
Chao, Y. A. et al. 1975. *Phys. Lett.* 57B:150
Chen, C. K. et al. 1977. *Preprint ANL-HEP-PR*, p. 22
Chew, G. F., Low, F. 1959. *Phys. Rev. Lett.* 113:1640.
Chew, G. F., Rosenzweig, C. 1976. *Nucl. Phys.* B104:290
Chinowsky, W. 1977. *Ann. Rev. Nucl. Sci.* 27:393
Chung, S. U. et al. 1965. *Phys. Rev. Lett.* 15:325
Chung, S. U. et al. 1978. *Phys. Rev. Lett.* 40:355
Chung, S. U. et al. 1972. *Phys. Rev. Lett.* 29:1570
Chung, S. U. et al. 1968. *Phys. Rev.* 165:1491
Chung, S. U. et al. 1973. *Phys. Lett.* 47B:526
Chung, S. U. et al. 1975. *Phys. Rev.* D11:2426
Colton, E. et al. 1971. *Phys. Rev.* D3:2033
Conversi, M. et al 1974. *Phys. Lett.* 52B:493
Cool, R. L. et al. 1966. *Phys. Rev. Lett.* 17: 102

Corden, M. J. et al. 1978a. *Nucl. Phys.* B136:77
Corden, M. J. et al. 1978b. *Nucl. Phys.* B144:253
Cosme, G. et al. 1976. *Phys. Lett.* 63B:350
Cosme, G. et al. 1979. *Nucl. Phys.* In press
Costa de Beauregard, B. et al. 1977. *Phys. Lett.* 67B:213
Cutkosky, R. E. et al. 1976a. *Phys. Rev. Lett.* 37:645
Cutkosky, R. E. et al. 1976b. *Nucl. Phys.* B102:139
Dahl, O. I. et al. 1967. *Phys. Rev.* 163:1377
Dahl, O. I. et al. 1961. *Phys. Rev. Lett.* 6:142
Dalkarov, O. D. et al. 1970. *Nucl. Phys.* B21:88
D'Andlau, Ch. et al. 1965. *Phys. Lett.* 17:367
D'Andlau, Ch. et al. 1968. *Nucl. Phys.* B5:693
DeBellefon, A. et al. 1975. *Nucl. Phys.* B90:1
DeBellefon, A. et al. 1976. *Proc. Topical Conf. on Baryon Resonances, Oxford*, p. 295
de Boer, W. et al. 1975. *Phys. Rev. Lett.* 34:558
Defoix, C. et al. 1968. *Phys. Lett.* 28B:353
DeRujula, A. et al. 1975. *Phys. Rev.* D12:147
Deutschmann, M. et al. 1974. *Phys. Lett.* 49B:388
Deutschmann, M. et al. 1977. Paper presented at the *Eur. Conf. on Particle Physics, Budapest*
Dolbeau, J. et al. 1976. *Nucl. Phys.* B108:365
Dover, C. B. 1975. *Proc. 4th Int. Conf. on NN Interactions*, Syracuse University, New York, Vol. 2, p. 37
Dulude, R. S. et al. 1978. *Phys. Lett.* 79B:335
Esposito, B. et al. 1977. *Phys. Lett.* 68B:389
Esposito, B. et al. 1978. *Lett. Nuovo Cimento* 22:305
Estabrooks, P. Martin, A. D. 1974. *Phys. Lett.* 53B:253
Estabrooks, P. et al. 1978. *Nucl. Phys.* B133:490
Ferrer, A. et al. 1978. *Phys. Lett.* 74B:287
Fich, O. et al. 1974. *Proc. London Conf. on Elementary Particle Physics*
Flatté, S. et al. 1972. *Phys. Lett.* 38B:232
Flatté, S. 1976. *Phys. Lett.* 63B:224
Fowler, W. B. et al. 1961. *Phys. Rev. Lett.* 6:134
Gavillet, Ph. et al. 1977. *Phys. Lett.* 69B:119
Gavillet, Ph. et al. 1978. *Phys. Lett.* 76B:517
Gay, J. B. et al. 1976a. *Phys. Lett.* 63B:220
Gay, J. B. et al. 1976b. *Phys. Lett.* 62B:477
Gell-Mann, M. 1961. *Calif. Inst. Technol. Intern. Rep. No. CTSL-20*; 1962. *Phys. Rev.* 125:1067
Giacomelli, G. et al. 1974. *Nucl. Phys.* B71:138
Gibbard, B. G. et al. 1979. *Phys. Rev. Lett.* 42:1593
Glasser, R. G. et al. 1977. *Phys. Rev.* D15:1200

Gopal, G. P. et al. 1977. *Nucl. Phys.* B119:362
Gordon, M. J. et al. 1978. Contributed paper to the *XIX Int. Conf. on High Energy Physics, Tokyo*
Gray, L. et al. 1971. *Phys. Rev. Lett.* 26:1491
Grayer, G. et al. 1974. *Nucl. Phys.* B75:189
Grässler, H. et al. 1977. *Nucl. Phys.* B121:189
Grilli, M. et al. 1973. *Nuovo Cimento* 13A:593
Gutay, L. et al. 1969. *Nucl. Phys.* B12:31
Hansen, J. et al. 1974. *Nucl. Phys.* B81:403
Hansen, P. N. et al. 1976. *Proc. Topical Conf. on Baryon Resonances, Oxford*, p. 275
Hemingway, R. J. et al. 1977. *Phys. Lett.* 68B:197
Hemingway, R. J. et al. 1978. *Nucl. Phys.* B142:205
Hepp, V. et al. 1976. *Phys. Lett.* 65B:487
Herndon, D. J. et al. 1975. *Phys. Rev.* D11:11
Hidaka, K. et al. 1977. *Phys. Lett.* 70B:479
Hoogland, W. et al. 1974. *Nucl. Phys.* B69:266 and references herein.
Hyams, B. et al. 1973. *Nucl. Phys.* B64:134
Hyams, B. et al. 1975. *Nucl. Phys.* B100:205
Jacob, M., Weyers, H. 1970. *Nuovo Cimento* 69A:521; 70A:285
Jaffe, R. L. 1977a. *Phys. Rev.* D15:267
Jaffe, R. L. 1977b. *Phys. Rev. Lett.* 38:195, 617
Jaffe, R. L. 1978. *Phys. Rev.* D17:1444
Jaros, J. A. et al. 1978. *Phys. Rev. Lett.* 40:1120
Jones, M. 1974. *Nucl. Phys.* B73:141
Jullian, S. 1976. *Int. Conf. on High Energy Physics, Tblisi, USSR*, p. B13
Kalogeropoulos, T. E., Tzanakos, G. S. 1975. *Phys. Rev. Lett.* 34:1047
Killian, T. J. et al. 1979. *Phys. Rev.* In press
Laplanche, F. et al. 1977. *Hamburg Int. Conf. on Lepton and Photon Interactions*
Leedom, I. D. et al. 1978. Contributed paper to the *XIX Int. Conf. on High Energy Physics, Tokyo*
Lipkin, H. J. 1973. *Phys. Rev.* D7:237, 2262
Litchfield, P. J. 1974. *Phys. Lett.* 54B:509
London, G. W. 1974. *Phys. Rev.* D9:1569
Longacre, R. S., Aaron, R. 1977. *Phys. Rev. Lett.* 38:1509
Longacre, R. S., Dolbeau, J. 1977. *Nucl. Phys.* B122:493
Longacre, R. S. et al. 1975. *Phys. Lett.* 55B:415
Martin, A. D. et al. 1978. *Phys. Lett.* 74B:417
Martin, B. R. et al. 1977. Paper submitted to the *Eur. Conf. on Particle Physics, Budapest*
Martin, B. R. 1975. *Nucl. Phys.* B102:413
Matison, M. et al. 1974. *Phys. Rev. Lett.* D9:1872
Miller, D. H. et al. 1965. *Phys. Rev. Lett.*

14:1074
Morgan, D. 1974. *Phys. Lett.* 51B:71
Nacasch, R. et al. 1978. *Nucl. Phys.* B135:203
Ne'eman, Y. 1961. *Nucl. Phys.* 26:222
Nicholson, H. et al. 1979. Presented at the Spring Meet. Am. Phys. Soc., Washington, DC
Novoseller, D. E. 1978. *Nucl. Phys.* B137:445
Oh, B. Y. et al. 1970. *Phys. Rev. Lett.* D1:2494
Otter, G. et al. 1974. *Nucl. Phys.* B80:1
Otter, G. et al. 1975. *Nucl. Phys.* B96:365
Otter, G. et al. 1976. *Nucl. Phys.* B106:77
Otter, G. et al. 1979. *Nucl. Phys.* B147:1
Particle Data Group. 1978. *Phys. Lett.* 75B:1
Pavlopoulos, P. et al. 1978. *Phys. Lett.* 72B:415
Pawlicki, A. J. et al. 1977. *Phys. Rev.* D15:3196
Pennington, M. R. 1978. *Nucl. Phys.* B137:77
Pernegr, J. et al. 1978. *Nucl. Phys.* B134:436
Pietarinen, E. 1977. Paper submitted to the *Eur. Conf. on Particle Physics, Budapest*
Pietarinen, E. 1974. *Nucl. Phys.* B76:231
Pjerrou, G. M. et al. 1962. *Phys. Rev. Lett.* 19:114
Protopopescu, S. D. et al. 1973. *Phys. Rev. Lett.* D7:1279
Ray, A. K. et al. 1968. *Phys. Rev.* 183:1183
Riester, J. L. et al. 1975. *Nucl. Phys.* B96:407
Ross, R. T. et al. 1972. *Phys. Lett.* 38B:177
Rossi, G. C., Veneziano, G. 1977. *Nucl. Phys.* B123:507

Sakitt, M. et al. 1977. *Phys. Rev.* D15:1846
Samios, N. P. et al. 1974. *Rev. Mod. Phys.* 46:49
Schlein, P. et al. 1963. *Phys. Rev. Lett.* 11:167
Schwitters, R. F., Strauch, K. 1976. *Ann. Rev. Nucl. Sci.* 26:89
Shannon, S. R. et al. 1974. *Phys. Rev. Lett.* 33:237
Shapiro, I. S. 1978. *Phys. Rev.* C35:129
Shult, R. L., Wyld, H. W. 1977. *Phys. Rev.* D16:62
Sidorov, V. A. 1976. *Int. Conf. on High Energy Physics, Tblisi, USSR*, p. B13
Skuja, A. et al. 1973. *Phys. Rev. Lett.* 31:653
Smith, G. A. et al. 1965. *Phys. Rev. Lett.* 14:25
Stanton, N. R. et al. 1979. *Phys. Rev. Lett.* 42:346
Teodoro, D. et al. 1978. *Phys. Lett.* 77B:451
Thompson, G. et al. 1974. *Phys. Rev.* D9:560
Tovey, S. et al. 1975. *Nucl. Phys.* B95:109
Vanucci, F. et al. 1977. *Phys. Rev.* D15:1814
Veneziano, G. 1968. *Nuovo Cimento* 57A:190
Vergeest, J. et al. 1976. *Phys. Lett.* 62B:471
Vuillemin, V. et al. 1975. *Lett. Nuovo Cimento* 14:165
Wagner, F. 1974. *Int. Conf. on High Energy Physics, London* II:27
Wagner, F. et al. 1975. *Phys. Lett.* 58B:201
Wetzel, W. et al. 1976. *Nucl. Phys.* B115:208
Yakamoto, S. et al. 1978. Contributed paper to the *XIX Int. Conf. on High Energy Physics, Tokyo*

Ann. Rev. Nucl. Part. Sci. 1979. 29:395–410

EXPERIMENTAL NEUTRINO ASTROPHYSICS

✖5611

Kenneth Lande

Department of Physics, University of Pennsylvania, Philadelphia, Pennsylvania 19104

CONTENTS

INTRODUCTION

Neutrinos provide a very powerful probe of the various stages of stellar evolution. Because they only interact weakly with matter, neutrinos are able to travel through large amounts of matter without alteration. They thus permit us to view inner regions of stars and other dense objects that are not visible by electromagnetic radiation.

The primary present experimental interests in neutrinos of astrophysical origin are

1. Probe of nuclear fusion reactions in the interior of the Sun;
2. Study of the dynamics of neutronization of massive stars during their final gravitational collapse into neutron stars or black holes;
3. Search for localized extraterrestrial sources of high energy neutrinos; and
4. Search for oscillation phenomena in cosmic ray neutrinos as they traverse the earth.

395

0163-8998/79/1201-0395$01.00

In this paper I will briefly review the properties of neutrinos, the early ideas for neutrino astrophysics, and the various experiments that have been carried out and are being planned. Since there are many excellent articles on the various speculations about extraterrestrial neutrino sources, I will not dwell on these topics in this article.

PROPERTIES OF THE NEUTRINO

The neutrino as originally hypothesized by Pauli (1933) was a massless neutral stable particle with spin 1/2; it was emitted in association with an electron in nuclear beta decay. The first direct detection of neutrinos was accomplished by Reines & Cowan in 1953. They observed the reaction $\bar{\nu}_e + p \rightarrow n + e^-$ using $\bar{\nu}_e$ emitted by the Savannah River reactor. Shortly thereafter Davis (1955) demonstrated that ν_e are distinctive from $\bar{\nu}_e$ by showing that reactor neutrinos will not drive a $^{37}Cl \rightarrow {}^{37}Ar$ transition ($\bar{\nu}_e + {}^{37}Cl \rightarrow {}^{37}Ar + e^-$).

In 1962 in a beautiful experiment, Danby et al demonstrated that a similar but different particle was emitted in association with a muon in the decay of charged pions. This pairing of leptons with neutrinos, electron with ν_e and muon with ν_μ, was recently expanded by the discovery of a third lepton, the τ (Perl et al 1975). Although the ν_τ has not yet been observed, there is no reason to doubt that the associative neutrino-lepton pattern observed for the electron and muon will not continue. It is also possible that more massive leptons and their associated neutrinos remain to be discovered. The existence of several species of neutrinos and the distinction between neutrinos and anti-neutrinos has led to the assignment of lepton quantum numbers to these neutrinos and their associated charged leptons. This procedure involves the creation of a separate lepton number for each species, L_e, L_μ, L_τ, etc, with $+1$ assigned to particles and -1 to antiparticles. Although it has not yet been demonstrated experimentally, it is assumed that the sum of each of these lepton numbers is separately conserved in any reaction. The additive nature of this conservation law and the possibility of small violations of such are now under experimental investigation.

NEUTRINO PROBE OF SOLAR FUSION REACTIONS

The source of energy that drives the stars has long been a quest of astrophysics. In 1939 Bethe considered the possibility that the fusion of four protons into a helium nucleus with the release of the requisite binding energy was the prime source of the energy radiated by the sun. In such a process two of the protons must be converted into neutrons accompanied by positrons and neutrinos. The observation of these neutrinos would provide a confirmation of this hypothesis, and a measurement of the energy spectrum and flux of such neutrinos would specify the reaction chain involved in this fusion process. The present standard model of the sun (Bahcall 1978) predicts the reactions and associated fluxes given in Table 1.

An ideal solar neutrino program would first establish the existence of nuclear fusion in the sun by observing the v_e emitted by the p–p reaction; it would then study the details of the fusion reaction chains by exploring the v_e energy and flux spectrum.

There are two approaches to the detection of low energy v_e. One involves the reaction $v_e + e^- \rightarrow v_e + e^-$. This reaction has no threshold energy and so is well suited to the problem at hand. The cross section for this process, however, is extremely small, $1.7 \times 10^{-44} E_v$ cm^2 (E_v in MeV), so that massive targets would be required. The reaction is closely mimicked by Compton scattering, $\gamma + e^- \rightarrow \gamma + e^-$.

The nineteen orders of magnitude between Compton scattering and neutrino-electron scattering make very stringent demands on the rejection of γ rays in the neutrino detector itself and in its surroundings. To date no electron-scattering solar neutrino detectors have successfully operated, although several are now under consideration (Chen 1978, Lande 1978).

The second approach to v_e detection involves the reaction $v_e + n \rightarrow p + e^-$ with neutrons in appropriate nuclei. There is a dual requirement in the reaction $v_e + $ nucleus (1) $\rightarrow$ nucleus (2) $+ e^-$; namely, that the mass difference of nucleus (2) $-$ nucleus (1) be small enough to permit neutrinos with energies appropriate to the solar process to drive the reaction, and that the reaction have a large enough cross section (small ft value) to permit a reasonable size detector. There is an additional

Table 1 Neutrino-producing reactions in the solar energy–generating process and energy range of emitted neutrinos

Reaction	$E(v_e)$(MeV)	Flux (v_e)[a](cm^{-2} s^{-1})
^{1}H $+ ^1$H $\rightarrow ^2$H $+ e^+ + v_e$	0–0.42	6.0×10^{10}
^{1}H $+ ^1$H $+ e^- \rightarrow ^2$H $+ v_e$	1.44	1.5×10^8
^{7}Be $+ e^- \rightarrow ^7$Li $+ v_e$	0.86 (90%)	2.7×10^9
	0.34 (10%)	3.0×10^8
^{8}B $\rightarrow ^8$Be $+ e^+ + v_e$	0–14	3.0×10^6
^{13}N $\rightarrow$ C$^{13} + e^+ + v_e$	0–1.20	3.0×10^8
^{15}O $\rightarrow$ N$^{15} + e^+ + v_e$	0–1.74	2.0×10^8

[a] From the present standard solar model (Bahcall 1978).

constraint on target material cost and availability since even for super-allowed transitions multi-ton detectors are required to give an adequate counting rate.

The first appropriate detector, ^{37}Cl, was proposed by Pontecorvo in 1946 and independently by Alvarez in 1949. The reaction involved, $v_e + {}^{37}$Cl $\rightarrow {}^{37}$Ar $+ e^-$, has a threshold of 0.81 MeV. However, as Bahcall (1946) pointed out, it has a particularly large cross section above 6 MeV. The ready availability of liquid chlorine compounds, as well as the relative ease of extracting a noble gas from a liquid have made this detection approach particularly attractive. Davis and co-workers (1968) utilizing a detector of C_2Cl_4 in a beautifully planned and executed program, have measured a total flux of v_e from the Sun of 1.75 ± 0.4 SNU (Davis et al 1978) (1 SNU $= 10^{-36}$ interactions per target nucleus per second). (See Figure 1.) This measurement, which is the first observation of neutrinos from an extraterrestrial source, is considerably less than the 4.7 SNU signal predicted for ^{37}Cl by the fluxes of the standard solar model. The disagreement between expectation and observation has given rise to considerable speculation about either unexpected behavior of nuclear fusion within the Sun or transformation of v_e into undetected species during the transit from the solar core to the neutrino detector (Pontecorvo 1967).

One complication with the ^{37}Cl detector is that it is primarily sensitive to the high energy end of the solar neutrino spectrum, that due to ^{8}B decay. The ^{8}B decay neutrinos constitute an extremely small fraction, $\sim 10^{-4}$, of the solar emission. Rather than providing a good test of the nuclear fusion hypothesis, they provide a critical test of the details of the solar fusion model.

In order to test the basic fusion hypothesis without undue sensitivity to the specifics of the fusion reaction chain, it is necessary to utilize a detector with a threshold less than 0.4 MeV and observe the p–p reaction neutrinos. Two such detectors are now being studied; one involves ^{71}Ga, $v_e + {}^{71}$Ga $\rightarrow {}^{71}$Ge $+ e^-$, E (threshold) $= 0.236$ MeV (Bahcall et al 1978), and the other utilizes ^{115}In, $v_e + {}^{115}$In $\rightarrow {}^{115}$Sn$^* + e^-$, E (threshold) $= 0.128$ MeV (Raghavan 1976).

The gallium will be in the form of a solution of $GaCl_3$ in hydrochloric acid. At a temperature of 60°C, $GeCl_4$, the neutrino reaction product, can be distilled out of the gallium chloride solution. This characteristic of the detector, the separation of a gas from a liquid, resembles the chlorine experiment. The technical aspects of the detector have been successfully demonstrated with a 20-kg prototype system. A 1.5-ton scale model detector is now under construction and should be operational late in 1979. Present plans call for a complete 50-ton gallium detector by 1984.

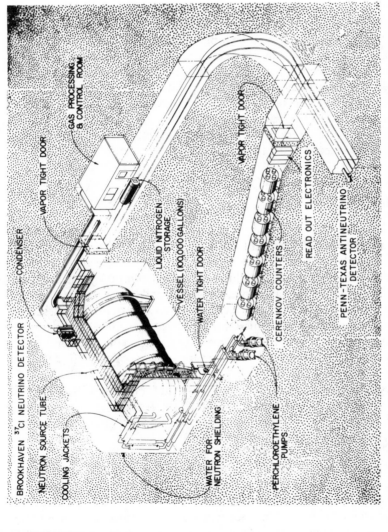

Figure 1 View of the Homestake Neutrino Astrophysics observatory showing the ^{37}Cl Solar Neutrino Detector and Cerenkov counters involved in an early neutrino burst detector.

Whereas the gallium detector is readily cast into the image of its predecessor, the chlorine detector, the indium detector presents a whole host of new and challenging problems. The foremost problem is that ^{115}In is unstable with a lifetime of 5×10^{14} years. The same mass of indium that will capture one solar v_e per day will also give rise to 10^{11} indium decays per day. Fortunately, the v_e capture converts ^{115}In into an excited state of ^{115}Sn (614 keV). This excited state then decays to the ground state of ^{115}Sn with a $T_{1/2}$ of 3.26 μs via the emission of two γ's 116 keV and 498 keV. The triple coincidence of e$^-$ followed by two γ's provides a distinctive signature of neutrino capture. It only remains to divide the indium detector into n separate units such that three-fold accidental coincidences from ^{115}In beta decay in each cell occur at a rate less than solar neutrino capture. Detailed considerations of this problem lead to $n = 10^6$. Work is now proceeding on a prototype of such a detector (Pfeiffer et al 1978).

FINAL GRAVITATION COLLAPSE OF MASSIVE STARS

Theory

For massive stars the central temperature in the late stages of evolution is sufficiently high ($\sim 10^{10}$ K) to permit e$^-$ + p $\rightarrow$ n + v_e (Zeldovich & Guseinov 1965, Arnett 1967). This neutronization process, which spreads rapidly through the stellar core, triggers a collapse of the star into a neutron star or possibly a black hole. One striking external manifestation of this occurrence is the emission of a burst of v_e. In addition to this prompt v_e burst there is also expected to be radiation of various neutrinos (v_e, $\bar{v}_e$, v_μ, $\bar{v}_\mu$, etc) associated with the subsequent cooling of the collapsed star (Bludman 1975).

Recent calculations by Wilson (1974) and Freedman, Schramm & Tubbs (1977) indicate that the v_e burst has a rise time of $\lesssim 2$ ms and carries away 2–3×10^{53} ergs. This v_e burst provides a clear indication of a gravitational stellar collapse.

Subsequent to the collapse, the hot compressed stellar core rapidly cools by the emission of pairs of various species of neutrinos and anti-neutrinos (v_e, v_μ, v_τ, etc). The thermal neutrino emission begins a few milliseconds after the neutronization burst and attenuates in a few seconds. The maximum of the thermal neutrino emission occurs about 30 ms after the neutronization neutrino emission peak and reaches a luminosity of $\approx 2 \times 10^{53}$ ergs s^{-1}.

The emitted neutrinos are predicted to have a mean energy in the range of 10–20 MeV.

Observation of such a neutrino burst and correlation of this event with subsequent optical and radio detection of a supernova and/or a pulsar would give a complete picture of the collapse scenario. It is also possible that there are "quiet" collapses, ones that involve neutronization of the stellar core and neutrino emission without a subsequent optical supernova display.

Detector Characteristics

Given the expected neutrino burst intensity and time structure, it is possible to specify the parameters of a suitable detector. For a source at the Galactic center, 3×10^{22} cm away, we expect a burst of 10^{12} v_e per cm^2. These v_e can be detected either by

$$v_e + e^- \to v_e + e^- ; \sigma = 1.7 \times 10^{-44} E_v \, \text{cm}^2$$

or by

$$v_e + n \to p + e^- ; \sigma = 7.5 \times 10^{-44} E_v^2 \, \text{cm}^2$$

where E_v is given in MeV.

For detection via neutrino–electron scattering, a v_e burst will give one electron–neutrino scattering per 4×10^{30} electrons in the detector. In order to ensure a recognizeable signal (10 scattered electrons) we require a detector with at least 4×10^{31} electrons or about 100 tons.

The neutron target, for the second of the above reactions, requires a choice of suitable nuclei in which the transition from A $(N, Z) \to A$ $(N-1, Z+1)$ involves a highly favored (super-allowed) transition. The target mass for 10 secondary electrons per neutrino burst will depend on the target nucleus, but will be at least several hundred tons. This class of detector is represented by the presently operating and planned solar neutrino detectors. Unfortunately, these radiochemical detectors integrate signals over periods of many days and are not instrumented for burst detection.

The thermal neutrino emission is most readily detected by looking for $\bar{v}_e$ via the reaction

$$\bar{v}_e + p \to n + e^+ ; \sigma = 7.5 \times 10^{-44} E_v^2 \, \text{cm}^2.$$

The expected flux of these $\bar{v}_e$ is expected to be about one tenth of the collapse v_e or 10^{11} per cm^2 at the Earth for a source at the Galactic center. The signal for this reaction is then one secondary e^+ per 6×10^{29} free protons in the detector.

The specifications for a suitable detector require that it be composed of material with a ratio of electrons to free protons of about 5, that it be readily available and of reasonable cost in quantities of hundreds of tons, and that it be capable of recognizing positrons and electrons with $E \sim 10$ MeV. These conditions are admirably met by either a water Cerenkov counter or by a liquid scintillation counter. Three such detectors are now in operation, a water Cerenkov detector in the Homestake Gold Mine, South Dakota, USA, and two liquid scintillation detectors in the Soviet Union, one in a salt mine near Artyomovsk, Ukraine, and another in a tunnel in the Baksan Valley, Caucasus.

The Homestake Detector, Figure 2, is a 500-m^3 counter hodoscope in the form of a closed hollow box, 20 meters long, 10 meters wide, and 6 meters high, that surrounds the Brookhaven ^{37}Cl Solar Neutrino Detector (Deakyne et al 1978a,b). The counter elements that form the sides and bottom of the hodoscope are water Cerenkov counters, $2\,m \times 2\,m \times 1.2\,m$ and those that form the top of the hodoscope are liquid scintillation counters, $1.5\,m \times 1.5\,m$. Each water Cerenkov module is viewed by four specially designed hemispherical photomultipliers,

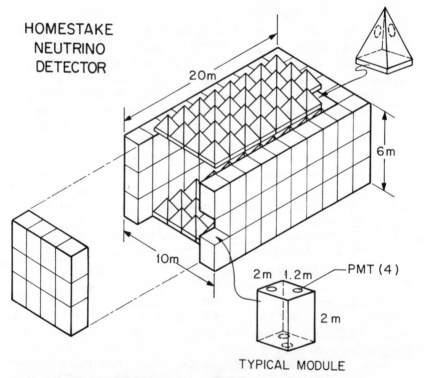

Figure 2   The Homestake Neutrino Detector counter hodoscope is constructed around the ^{37}Cl vessel shown in Figure 1. The sides of the detector are water Cerenkov counters while the top counters are filled with liquid scintillator.

EMI D312, that are 13 cm in diameter. The readout electronic logic is designed to record bursts with instantaneous rates as high as two counts per μs.

In order to increase the useful Cerenkov emission from electron and positron secondaries a wavelength shifter is added to the water. This shifter absorbs Cerenkov radiation in the ultraviolet range (2000 Å $\leq \lambda \leq$ 3500 Å) and emits light in the wavelength region where the photomultiplier photocathode is most sensitive.

The Artyomovsk detector is a single large cylindrical counter filled with a 100-m^3 liquid scintillator located at a depth of 600 meters water equivalent in a salt mine in the Ukraine, USSR (Beresnev et al 1978). The detector is a cylinder with a 5.6-m diameter, a 5.6-m height, and 160 photomultipliers viewing the liquid scintillator through portholes in the cylinder. The shallow depth of the detector results in a cosmic ray flux of several muons per second passing through the detector. Considerable electronic efforts are required to properly label these cosmic rays and their progeny.

A major neutrino astrophysics laboratory is under construction in the Baksan Valley of the Northern Caucasus, USSR. This deep underground laboratory is being constructed along and at the end of a 4-km tunnel that is being bored into the side of Mt. Andyrchi, a 4000-m high peak near Mt. Elbrus. At the center of the mountain the tunnel will have an overburden of 1300 m of rock or about 4000 meters water equivalent.

This laboratory is expected to house a number of neutrino detectors directed at the Sun and at other extraterrestrial neutrino sources. Some of the apparatus planned for the central mountain laboratory are two solar neutrino detectors, one containing 3000 tons of C_2Cl_4 and the other one about 50 tons of Ga, and a 20,000-ton scintillation counter hodoscope for other extraterrestrial neutrinos. These detectors are expected to be in operation by 1985.

One major detector, a 330-ton liquid scintillation counter hodoscope (Figure 3) located in the front part of the tunnel at a depth of 850 meters water equivalent has already been put into operation (Chudakov & Ryazhskaya 1978). This hodoscope consists of 3200 individual counters arrayed in 8 layers of 400 counters each. Six of these layers, each 14 m × 14 m, form the sides of a box. Two additional horizontal layers are situated inside the box. The multiple layers of counters provide several signals for a traversing particle. A vertical cosmic ray muon will generate pulses in each of the four horizontal counter layers. Because of this multiple label characteristic, the various layers serve as anti-coincidence shields for each other. This aspect may be of importance in distinguishing bursts of low energy neutrinos from various background processes.

Backgrounds

An essential characteristic of neutrino burst detectors is that the background counting rate be low enough to permit the unambiguous recognition of the neutrino burst when it occurs. Since the expected neutrino burst rate is one per several years it is desirable to limit the simulation of neutrino bursts by background processes to less than 10^{-2} per year. For a 500-ton detector, such as that at Homestake, a neutrino burst should give about 50 v_e-induced counts in a few milliseconds followed by a similar number of $\bar{v}_e$-produced counts in the following several seconds.

We can thus specify the single count rate, R, in the detector by $R^n \tau^{n-1} < 3 \times 10^{-10}$ per second $= 10^{-2}$ per year, where τ is the burst width and n is the expected number of pulses in the burst. For the Homestake detector, $n \approx 50$, and so $R < 300$ counts per second for $\tau = 2 \times 10^{-3}$ s (neutronization burst) and $R < 0.64$ counts per second for $\tau \approx 1$ s (thermalization burst).

These constraints on R require that the detector be deep underground

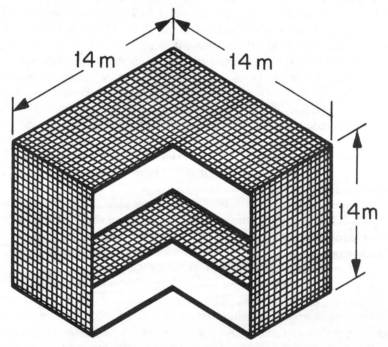

Figure 3 The Baksan Scintillation Counter Hodoscope is a 3200 counter hodoscope formed into a closed box 14 m on each side, with two extra horizontal layers of counters. The total volume of scintillater is 330 m³.

to reduce the cosmic ray muon flux and that the detector either be shielded from the effects of local radioactivity or be insensitive to this background. The earth shield above the Homestake Detector, 4400 meters water equivalent, reduces the cosmic ray muon flux to $4 \times 10^{-9} \, \mu \, \mathrm{cm}^{-2} \, \mathrm{s}^{-1} \, \mathrm{sr}^{-1}$ and thus gives R (cosmic rays) $= 2 \times 10^{-3}$ per second. For a Cerenkov counter the major local radioactivity background arises from Compton scattering of gamma rays, which give rise to electron secondaries. Gamma rays from natural radioactivity are mainly below 3 MeV and thus do not constitute a serious background to the detection of electrons and positrons with $E \geq 10$ MeV.

An additional constraint that can be imposed on the detection of neutrino bursts is that the signal be detected by two or more widely separated detectors. For a pair of detectors on opposite sides of the Earth, τ must be twice the neutrino burst transit time of the Earth, 80 ms. For this constraint, an accidental rate of one per 10^2 years can be achieved by having less than one neutrino burst candidate per day in each detector. In addition to permitting a higher single detector accidental rate, the requirement of coincident observation in widely separated detectors avoids the possibility that an apparent neutrino burst signal at a single detector site is caused by a very rare complex coherent cosmic ray or local background process.

The present ongoing search for neutrino bursts is a coordinated program involving the above detectors. The clocks associated with each of these detectors are correlated with each other via the internationally available UTC (Universal Coordinated Time). At present the clock synchronization is better than 100 μs. In addition to providing a means for establishing coincident observations between these detectors, the measurement of the relative arrival time of the neutrino burst wave front at the various observatories provides the means for determining the neutrino burst source location. The angular resolution of this technique is $\approx \tau$ (pulse rise time)/Δt (neutrino transit time between detectors). For the present detectors we anticipate $\tau = 0.5$ ms and $\Delta t = 30$ ms, so that the angular resolution could be as good as 0.02 radian. With three detectors there is a two-fold ambiguity about source location. A fourth detector would provide a unique solution.

HIGH ENERGY NEUTRINOS

Cosmic Ray Neutrino Secondaries

In the mid-1960s two major efforts to investigate the natural high energy neutrino environment were undertaken. One of these was carried out in the Kolar Gold Field Mines, Heathcote Shaft, of southern India and the other in the East Rand Proprietary Gold Mine near Johannesburg, South Africa. Although these two experiments differed somewhat in approach, they were both directed toward the same physics goals:

1. measuring the neutrino-induced muon flux within the Earth;
2. seeking evidence for the existence of the intermediate vector boson; and
3. searching for localized extraterrestrial sources of high energy neutrinos.

Both of these experiments were placed at great depths, the South African one at 8800 meters water equivalent (Reines et al 1971) and the Indian one at 7500 meters water equivalent (Krishnaswamy et al 1971). At these depths the cosmic ray muon flux is reduced to 1.8×10^{-7} $m^{-2} s^{-1} sr^{-1}$ and $1.1 \times 10^{-6} m^{-2} s^{-1} sr^{-1}$, respectively, compared to a muon flux at the Earth's surface of about $10^2 m^{-2} s^{-1} sr^{-1}$. This suppression of the cosmic ray muon flux is still not sufficient to permit the ν-induced muon flux to stand out. The rapid variation of cosmic ray muon flux with zenith angle at these depths, $\sim \cos^9 \theta$, permits the further necessary reduction in cosmic ray background by requiring that the incident muon have a zenith angle greater than $50°$. With this in mind, both of these detectors were designed to look for horizontal muons.

The South African detector consisted of two vertical walls of liquid scintillation detectors 1.9 m high × 54 m long, separated by 1.8 m. Each of the walls consisted of counter elements that were 0.6 m high × 5.5 m long × 0.13 m thick. Neutrino-produced secondaries were required to produce coincidences between the two walls of counters. A later version of the South African detector added flash tubes to the detector. This permitted a more detailed view of multiple secondary events and improved the zenith angle measurement. No distinction was made in the South African experiment between particles moving upward and those moving downward. All the data were folded about a zenith angle of $90°$ (horizontal).

The Indian experiment consisted of a complex of seven scintillation counter telescopes, two of which had solid iron magnets in their centers. In addition, each of these telescopes also had an array of flash tubes and some absorber within its fiducial volume. The Indian detector was smaller in aperture than the South African one, but was able to measure additional parameters of the neutrino secondaries such as multiplicity and charge; and in some cases it could determine whether the neutrino was moving upward or downward.

These two experiments established similar conclusions:

1. The horizontal flux of neutrino-induced muons passing through a plane deep within the Earth is $4 \times 10^{-13} \mu \, cm^{-2} s^{-1} sr^{-1}$.

2. This flux is consistent with the standard assumption about neutrino origins from the decay of pion and kaon cosmic ray secondaries and a neutrino–nucleon cross section that rises linearly with energy for $E(v) > 1$ GeV.
3. The intermediate vector boson, if it exists, has a mass greater than 3 GeV/c^2.
4. There is no evidence for localized extraterrestrial sources of high energy neutrinos with intensities greater than 10% of the cosmic ray–induced intensity.

A third underground neutrino experiment was begun in the late 1960s by the Utah group. This experiment reported five events, all upward. The analysis of this experiment permitted a limit on the boson mass ≥ 5 GeV/c^2 (Bergeson et al 1973).

One interesting observation by the Indian group (Krishnaswamy et al 1976) was that five of their twenty neutrino-induced events had a topology suggestive of a massive charged short-lived particle that decayed into three particles, usually muons, in their detector. This particle has a mass $\gtrsim 2$ GeV/c^2 and $\tau \sim 10^{-9}$ s. It was suggested that this particle might be leptonic in character and so could be produced by a neutrino with the same lepton number. In four of the five cases cited the neutrino was moving upward (zenith angle $> 90°$), while the 15 normal neutrino events involved predominantly downward moving neutrinos. There are two interesting aspects to these results. One is that the five unusual events, except for the relatively long lifetime, resemble τ meson decays. The second is that the angular distribution pattern is suggestive of what might be expected from neutrino oscillations, with $v_\mu \to v_\tau$ for flight paths $\gtrsim 10^2$–10^3 km. Unfortunately, the amount of data is much too limited to reach any meaningful conclusions and we will have to wait for experiments with increased statistics before the possibilities of neutrino oscillations can be properly explored.

Interest in extraterrestrial high energy neutrinos and the possibility of using them to search for sources of high energy cosmic rays have been renewed by some recent astronomical discoveries and by new speculations about unusual stellar and galactic phenomena. The most specific of these suggested neutrino sources is pulsars, the neutron star remnants of gravitational collapse. If indeed the pulsar environment is a region of acceleration of cosmic ray primaries, then it is also likely to be a region in which these high energy particles can interact and produce secondaries such as pions and kaons that will decay into neutrinos. The charged cosmic ray primaries may have considerable difficulty in escaping from the accelerator environment because of trapping magnetic fields and possibly appreciable column densities of surrounding shell material. Those that do escape will follow a twisting path, defined by the galactic magnetic field, from their source to the Earth. Since neutrinos are unaffected by magnetic fields and find most aggregates of matter transparent, they provide the perfect vehicle with which to locate and identify the natural particle accelerators.

The interaction characteristics of high energy neutrinos have now been well established at particle accelerators such as Fermilab (Barish et al 1977). The cross section for neutrino interactions are

$$\sigma(v_\mu + n \rightarrow \mu^- + x) = 0.7 \times 10^{-38} E_v (\text{GeV}) \text{cm}^2$$

with $\langle E_\mu \rangle = 0.5 \langle E_v \rangle$

$$\sigma(\bar{v}_\mu + n \rightarrow \mu^+ + x) = 0.3 \times 10^{-38} E_{\bar{v}} (\text{GeV}) \text{cm}^2$$

with $\langle E_\mu \rangle = 0.7 \langle E_{\bar{v}} \rangle$.

For muon energies below 1 TeV the muon energy loss in matter is approximately 2 MeV per g cm^{-3}, or Range $(\mu) = 3 \times 10^{23} E_\mu (\text{MeV})$ nucleons cm^{-2}. Thus the probability that a neutrino interacts in this column is

$$\sigma(v)R(\mu) = \{0.7 \times 10^{-38} E_v (\text{GeV})\} \times \{3 \times 10^{26} E_\mu (\text{GeV})\}$$

$$\approx 1.0 \times 10^{-12} E_v^2 (\text{GeV}).$$

Similarly, for $\bar{v}_\mu$

$$\sigma(\bar{v})R(\mu) \approx 0.6 \times 10^{-12} E_{\bar{v}}^2 (\text{GeV}).$$

This effect, where the neutrino cross section grows with increasing energy and where the sensitive detection volume grows with increasing muon energy (which in turn is proportional to the incident neutrino energy), is termed rock amplification. As an example, a neutrino of 1-TeV energy will produce a muon secondary with a range of about 1 km in rock. Rock amplification produces an enormous increase in neutrino sensitivity for deep underground neutrino detectors.

For the Homestake Neutrino Detector with a surface area of 2×10^6 cm^2 per face, a stellar source that produces $2 \times 10^{-7} v$ cm^{-2} s^{-1} of 1 TeV at the Earth will result in 10 neutrino-induced muons per year from one small region of sky. The cosmic ray neutrino background from that same region of sky will give less than one muon secondary per year. For a source near the center of the Galaxy this corresponds to a radiated energy in neutrinos of 2×10^{39} ergs s^{-1}, a reasonable value for a young pulsar. Since a neutrino detector constantly views the entire sky, it provides an excellent scanning tool for locating high energy cosmic ray sources.

Possibilities for Detector Expansion

Because of the rock amplifying process the power of a neutrino telescope is determined by the surface area of the detector. The two large operating detectors, Homestake and Baksan, both have faces with areas of $2 \times 10^6 \, cm^2$. One could construct underground detectors with surface areas of 10^8–$10^9 \, cm^2$ without undue demands on standard mining techniques. Indeed, the future plans for the Baksan observatory central laboratory include provisions for a detector of this size.

Another proposed approach to the establishment of much larger neutrino detectors is to segregate a large volume of water in a deep ocean locale and to look for neutrino-induced muon signals in the defined water region. The DUMAND proposal calls for a detector of a cubic kilometer at a depth of 4.5 to 5 km in the Pacific Ocean near Hawaii. Such a detector would provide a surface area per face of $10^{10} \, cm^2$ (Roberts 1978).

Unlike underground detectors where conventional particle detection techniques are employed, the DUMAND environment calls for radically new and different approaches to particle detection, data collection, and apparatus maintenance. In addition, there is the general problem of a potentially hostile biological environment. The two suggested approaches to particle detection involve the use of hydrophones to detect the acoustic pulses produced by the passage of charged particles, and optical detectors to collect the Cerenkov radiation emitted by the particles.

In addition to the problem of choosing a particle detection approach for which the particle-produced signal is clearly above the instrumentational and biological background, there are the problems of deployment of a cubic kilometer detector in the ocean, the distribution of power to the various detector elements, the collection of signals from these elements, and the transmission of the data to the shore or ship control center. There are also the problems of designing and constructing detector components that can function in a pressure environment of 500 atmospheres and the procedure for maintaining the apparatus and replacing malfunctioning elements of the detector array. Impressive technical progress has been made in many of these areas in the last several years.

Acknowledgments

I am deeply grateful for many stimulating and provocative discussions with my colleagues and collaborators S. Bludman, R. Davis, E. Fenyves, W. Frati, C. K. Lee, and R. Steinberg. This research has been supported in part by NSF grant 78-08669 and DOE contract EY-76-C-02-3071.

Literature Cited

Alvarez, L. W. 1949. *UCRL Rep.* 328
Arnett, W. D. 1967. *Can. J. Phys.* 45:1621
Bahcall, J. N. 1946. *Phys. Rev. B* 135:137
Bahcall, J. N. 1978. *Rev. Mod Phys.* 50:881
Bahcall, J. N. et al. 1978. *Phys. Rev. Lett.* 40:1351
Barish, B. C. et al. 1977. *Phys. Rev. Lett.* 39:1595
Beresnev, V. I. et al. 1968. See Davis, Evans & Cleveland 1978, p. 881
Bergeson, H. E., Cassiday, G. L., Hendricks, M. B. 1973. *Phys. Rev. Lett.* 31:66
Bethe, H. 1939. *Phys. Rev.* 55:434
Bludman, S. A. 1975. *Ann. NY Acad. Sci.* 262:181
Chen, H. H. 1978. Proc. Informal Conf. on Status and Future of Solar Neutrino Research, p. 55. Brookhaven Upton, NY: Natl. Lab.
Chudakov, A. E., Ryazhskaya, O. B. 1978. See Deakyne et al 1978a, p. 155
Danby, G., Gaillard, J. M., Goulianos, K., Lederman, L. M., Mistry, N., Schwartz, M., Steinberger, J. 1962. *Phys. Rev. Lett.* 9:36
Davis, R. Jr. 1955. *Phys. Rev.* 97:766
Davis, R., Harmer, D. S., Hoffman, K. C. 1968. *Phys. Rev. Lett.* 20:1205
Davis, R., Evans, J. C., Cleveland, B. 1978. *Neutrino-1978 Conference Proceedings*, p. 53. Lafayette, Ind: Purdue University
Deakyne, M., Frati, W., Lande, K., Lee, C. K., Steinberg, R. I., Fenyves, E. 1978a, Moscow: Publ Off "Navka," p. 170
Deakyne, M., Frati, W., Lande, K., Lee, C. K., Steinberg, R. I., Fenyves, E., 1978b.

See Davis, Evans & Cleveland 1978, p. 887
Freedman. D. Z., Schramm, D. N., Tubbs, D. L. 1977. *Ann. Rev. Nucl. Sci.* 27:167–207
Krishnaswamy, M. R., Menon, M. G. K., Narasimhan, V. S., Hinotani, K., Ito, N., Miyake, S., Osborne, J. L., Parsons, A. J., Wolfendale, A. W. 1971. *Proc. R. Soc. London* 323:489
Krishnaswamy, M. R. et al. 1976. *Proc. Int. Neutrino Conf. Aachen, 1976*, p. 197
Lande, K. 1978. See Chen 1978, p. 79
Pauli, W. 1933. *Rapports du Septeme Conseil de Physique Solvay, Brussels* (Gauthier-Villars, Paris, 1934).
Perl, M. et al. 1975. *Phys. Rev. Lett.* 35:1489
Pfeiffer, L., Mills, A. P., Raghavan, R. S., Chandross, E. A. 1978. *Phys. Rev. Lett.* 41:63
Pontecorvo, B. 1946. *Natl. Res. Counc. Can. Rep. P.D.* 205
Pontecorvo, B. 1967. *Sov Phys. JETP* 53:1771
Raghavan, R. S. 1976. *Phys. Rev. Lett* 37:259
Reines, F., Cowan, C. L. 1953. *Phys. Rev.* 92:830
Reines, F., Kropp, W. R., Sobel, H. W., Gurr, H. S., Lathrop, J., Crouch, M. F., Sellschrop, J. P. F., Meyer, B. S. 1971. *Phys. Rev. D* 4:80
Roberts, A., ed. 1978. *Proc. 1978 DUMAND Summer Workshop*, DUMAND Scripps Inst Oceanogr
Wilson, J. R. 1974. *Phys. Rev. Lett.* 32:849
Zeldovich, Ya., Guseinov, O. 1965. *Sov. Phys. JETP Lett.* 1:109

Ann. Rev. Nucl. Part. Sci. 1979. 29 : 411–54

APPLICATIONS OF LASERS ✕5612
TO NUCLEAR PHYSICS

Daniel E. Murnick
Bell Laboratories, Murray Hill, New Jersey 07974

Michael S. Feld
Spectroscopy Laboratory and Physics Department,
Massachusetts Institute of Technology, Cambridge, Massachusetts 02139

CONTENTS

1 INTRODUCTION

1.1 *History*

Atomic and molecular spectroscopy has had a major impact on our understanding of nuclear structure. Optical hyperfine structure was

411

0163-8998/79/1201-0411$01.00

observed in the earliest interference spectroscopy investigations (Michelson 1891, Fabry & Perot 1897), and by 1914 extensive hyperfine structure tabulations had been published (Wali-Mohammed). W. Pauli (1924) first suggested that hyperfine structure was caused by a small magnetic moment associated with the atomic nucleus, and subsequent high resolution studies led to the determination of the spins and magnetic moments of many nuclei (Kopfermann 1958). Systematic data on nuclear spins and moments have provided fundamental information for development of the nuclear shell model.

In 1931 the observed intensity alternation of successive rotational-vibrational lines of the spectrum of the nitrogen molecule ($^{14}N_2$) led to the conclusion that nuclei with even spin obey Bose-Einstein statistics, and abandonment of the hypothesis that nuclei consist of protons and electrons (Ehrenfest & Oppenheimer 1931). In 1935 Schüler & Schmidt showed that anomalies in hyperfine structure splitting were caused by the existence of an intrinsic nuclear quadrupole moment, and Casimir (1936) developed a formalism from which quadrupole moment values could be determined.

The isotope shift, which refers to the energy change in a transition in an atom of fixed atomic number (Z) when the atomic mass (A) is varied, was shown in 1932 (Rosenthal & Breit) to be caused by nuclear mass and volume effects. Large isotope shifts in the rare earth region have led to the predictions of large intrinsic nuclear deformations (Brix & Kopfermann 1949), important in the development of nuclear collective models.

Other atomic and molecular physics techniques have also been important in improving our knowledge of nuclear parameters. For example, ground-state hyperfine splittings can also be determined by nuclear magnetic resonance techniques, often with much higher precision than from optical spectra (Abragam 1961).

1.2   Effects of Nuclear Structure in Atomic and Molecular Spectra

The quantum numbers that determine atomic energy levels are n (Bohr's principal quantum number) and L, S, and J, the orbital, spin, and total angular momentum, respectively. In addition, for atoms with nuclei having non-zero spin, I, energy levels with different values of total angular momentum $F = I + J$ are split by small amounts, giving rise to hyperfine structure in the optical spectral lines. These splittings, primarily due to interaction of the nuclear magnetic and quadrupole moments with the atomic electrons, are usually in the range ~ 10–$1000\,MHz$. Thus, conventional optical spectroscopic techniques, which are limited in resolution by Doppler broadening ($\sim$0.5–5 GHz), are often incapable of resolving them.

Figure 1*a* shows the hyperfine splittings and allowed transitions of the low-lying energy levels of ^{23}Na, which has nuclear spin $I = 3/2$. Figure 1*b* illustrates progressive improvements in spectral resolution of the ^{23}Na D-lines, enabling first the fine structure and then the hyperfine structure to be resolved with increasingly greater precision. Note the dramatic improvement in the laser spectra, especially the complete elimination of the Doppler broadening, which permits direct resolution of the hyperfine splittings. The unprecedented resolution obtainable with lasers, coupled with the sub-Doppler line narrowing techniques described

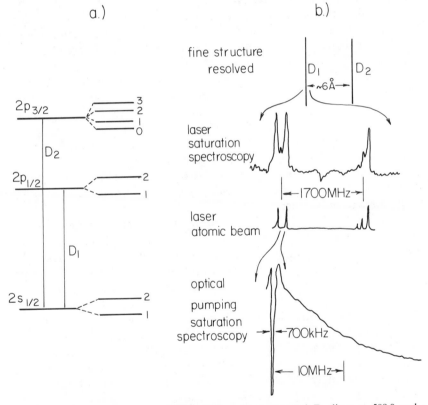

Figure 1   (*a*) Atomic structure of ^{23}Na giving rise to D_1 and D_2 lines at 589.0 and 589.6 nm. (*b*) Representative examples of experimental spectra of atomic sodium demonstrating advances in resolution, which from top to bottom are 1. Fine structure resolved (Back & Landé 1925). 2. D_1 hyperfine structure observed with laser saturation spectroscopy (Hänsch et al 1971). 3. Laser-atomic beam spectroscopy (Huber et al 1978b). 4. Saturation spectroscopy with optical pumping (Murnick et al 1979c).

below, makes possible precision hyperfine structure studies in a wide range of systems.

Once the hyperfine structure is resolved the nuclear spin can be established from the number of hyperfine components, and the magnetic dipole, electric quadrupole, and higher nuclear moments can be determined from the frequency splittings between components. The reduction of atomic hyperfine spectral data to obtain nuclear spins and moments is discussed in detail by Kopfermann (1958).

For a series of atomic isotopes, even with zero nuclear spin, precise comparison of spectral line frequencies yields important nuclear information. This splitting, known as the isotope shift, is of the same order as the hyperfine structure separation. Isotope shifts are produced by two physical mechanisms, a nuclear mass effect and a volume effect, sometimes called the field effect, caused by the change in mean squared nuclear charge radius as the neutron number changes. Experimentally, of course, the two contributions cannot be separated, and since the mass effect cannot be reliably calculated, empirical estimation must be used to determine nuclear size changes (Heilig & Steudel 1974). The splitting due to the mass effect is proportional to $(A_2 - A_1)/A_1 A_2$ for atomic masses A_1 and A_2; hence it is small for large A ($A \gtrsim 100$). For lighter elements isotope shift deviations from this proportionality can be taken as evidence for nuclear volume effects. Alternatively, if a series of spectral lines can be measured for a given set of isotopes the volume effect, which is independent of electronic configuration, may also be separated in certain cases.

The isomer shift, the counterpart of the isotope shift for different nuclear levels of a single isotope, is purely a nuclear volume effect. With the improved resolution and sensitivity brought about by laser techniques, isotope and isomer shift measurements are becoming increasingly useful spectroscopic tools for obtaining nuclear structure information. The important example of the mercury isotopes is discussed in the chapter in this volume by P. G. Hansen. Other cases are described in Section 3 below.

The "hyperfine anomaly" is another interesting nuclear structure effect that can be resolved with high resolution spectroscopic techniques. The hyperfine anomaly, Δ, obtained by studying the frequency shift of a spectral line between two isotopes, is defined as

$$\Delta = \frac{\text{ratio of magnetic hyperfine interaction constants}}{\text{nuclear moment ratio}} - 1$$

(Bohr & Weisskopf 1950). The hyperfine anomaly arises from the non-uniform distribution of nuclear magnetism over the nuclear volume. It is nonzero only for atomic resonance transitions involving s electrons, since the probability distributions of other electrons vanish at the nucleus. Thus far, only limited nuclear structure information has been obtained from hyperfine anomaly data (Stroke et al 1961, Moskowitz & Lombardi 1973).

Analogous isotope shift and hyperfine structure splittings occur in electronic (visible–UV) and vibrational (infrared) transitions in molecules. Molecular isotope shifts, although large, are of little interest as far as nuclear structure is concerned, since they are dominated by changes in reduced molecular mass and insensitive to nuclear volume effects (Herzberg 1950). As an example, in a diatomic molecule the fractional shift in vibrational frequency is $\Delta\omega/\omega = (\mu_1/\mu_2)^{\frac{1}{2}} - 1$; here μ_1 and μ_2 are the reduced molecular masses of the two isotopic species. Thus, the $H^{35}Cl$-$H^{37}Cl$ vibrational isotope shift is over 100 times larger than the infrared Doppler width. The large frequency shift produced by a relatively small change in nuclear mass also opens the possibility of observing *isomer* shifts in molecular spectral lines (Letokhov 1973a).

Molecular hyperfine splittings are smaller than those in atoms. In most molecules the nuclear quadrupole structure dominates (~ 0.1–10 MHz), the magnetic moment contributions being much smaller (~ 1–100 kHz). Hence, in most cases the hyperfine splitting is smaller than the Doppler width, typically ~ 10–100 MHz for infrared vibrational transitions. Appropriate high resolution laser techniques can resolve sub-Doppler structure at this level and, in fact, several observations of molecular hyperfine splittings have been reported. Examples are $CH_3{}^{35}Cl$ (Meyer et al 1973), CH_4 (Bordé & Hall 1973), and $^{189}OsO_4$ (Kompanetz et al 1976).

Unfortunately, the reduction of molecular hyperfine splitting data to obtain nuclear moments is more difficult than in the atomic case because of complexities introduced by new internal degrees of freedom and, in most cases, the accompanying lack of accurate molecular wave functions. Hence, in most analyses of nuclear hyperfine splittings the data reduction process is reversed and, for example, known nuclear quadrupole moments are used to obtain information about electric field gradients at the nucleus, rather than the other way around. The elements of molecular hyperfine structure are treated by Townes & Schawlow (1955).

Advances in instrumentation techniques and theoretical methods have consistently improved the quality and quantity of optical spectroscopic information of value to nuclear physics. For example, experiments can now be performed on rare and separated isotopes, as well as on short-lived species produced in reactors or at particle accelerators. A recent review by Jacquinot & Klapisch (1979) discusses hyperfine spectroscopy of radioactive atoms, and a paper in this volume by Hansen reviews recent on-line studies, including optical spectroscopy of radioactive isotopes.

1.3 Laser Properties

Lasers have made a major impact on nuclear structure studies. The monochromaticity of laser radiation (typically from $1:10^7$ to $1:10^{10}$) makes possible unsurpassed spectral resolution. The high power available [continuous wave (CW) powers of milliwatts to watts are common] leads to greatly enhanced sensitivity and opens the possibility of detecting low atomic densities and even single atoms. In nuclear physics terminology 10 mW of laser power at 600 nm is equivalent to 5 "milliamperes" of photons (3×10^{15} photon/s). In addition, interaction cross sections are on the order of 100 Mbarns. Lasers are widely tunable—at present the entire visible and near-UV spectrum can be continuously scanned with appropriate sources, and the useful tuning range is continually being extended. The phase coherence of laser radiation leads to beams with narrow divergence, which can be focused to extremely small spot sizes ($\sim \lambda^2$); this makes possible extremely large energy densities. The inherent complete polarization of laser radiation is another property that can be usefully exploited. Applications to ultra-high resolution polarization spectroscopy, laser optical pumping, and nuclear orientation are discussed below.

Lasers can be profitably applied to almost all of conventional hyperfine structure spectroscopy, with greatly enhanced sensitivity and, in some cases, improved resolution. In addition, lasers have led to the development of several important new techniques to study nuclear hyperfine structure, including measurements in beams and cells in which the resolution limitation of Doppler broadening is overcome. Other unprecedented new applications, perhaps among the most important, are made possible by the fact that laser radiation can significantly alter the state of an atomic (or molecular) system with which it interacts. This represents a substantial departure from the past, where optical radiation has been used primarily to probe the state of a system.

Section 2 of this paper reviews the physical basis of laser applications to nuclear science. Methods to achieve natural linewidth resolution are emphasized. These techniques include modified atomic beams and laser saturation spectroscopy. Techniques are discussed in which the high power and monochromaticity of laser radiation permit spectroscopic studies of extremely dilute samples, and provide practical, efficient isotope separation schemes. Other applications are presented in which laser coherence and polarization properties are used to align and orient atomic nuclei and to alter gamma ray angular and spectral distributions.

Selected examples and applications are given in Section 3, with an emphasis on the most recent and novel cases. Experiments to probe fundamental interactions impossible without the use of lasers are described in Section 3.1 and several laser spectroscopic studies yielding nuclear structure information are given in Sections 3.2 and 3.3. Sections 3.4 and 3.5 on the proton monochromator and gamma ray lasers are examples of laser applications to nuclear instrumentation and technology. In the conclusion, Section 4, we comment on where the field now stands and present our thoughts on important future directions and potential areas for exploration.

In addition to specific references in the text, recent specialized reviews by Huber (1977), Letokhov (1977), and Kluge (1979) may be of interest.

2 TECHNIQUES

2.1 *Conventional Optical Spectroscopy*

Optical detection of nuclear effects in atomic and molecular spectra have historically been resolution limited. Most fruitful studies have been carried out in the vapor phase, where atoms and molecules respond individually and the small splittings due to nuclear structure are unaffected by broadening caused by collective interactions. Lifetimes of excited electronic states, determined by radiative decay, are typically $\sim 3\text{--}30$ ns. Hence, in the visible spectrum, natural linewidths lie in the $\sim 5\text{--}50$ MHz range. Linewidths of forbidden transitions can be several orders of magnitude narrower. Radiative lifetimes of molecular vibrational transitions, which fall in the infrared, generally lie in the kilohertz range, and so at conveniently low pressures ($\sim 10\text{--}100$ mtorr) "homogeneous" linewidths are usually determined by collisional de-excitation and range from ~ 100 kHz to 3 MHz.

Because of the Doppler effect, however, observed spectral linewidths are much broader, ranging from 50 MHz in the infrared to several gigahertz in the near-UV. A Doppler-broadened spectral profile can be thought of as a Maxwellian distribution of narrow naturally broadened components, each Doppler shifted in frequency by a different amount $\mathbf{k} \cdot \mathbf{v}$, where $\mathbf{v}$ is the molecular velocity and $\mathbf{k}$ is the radiation wave vector. The net Doppler-broadened linewidth is thus

$$\Delta\omega_{\mathrm{D}} = ku = \frac{u\omega}{c}, \qquad\qquad 1.$$

with $\omega = ck$ the center frequency of the transition,

$$u = \left(\frac{2\kappa T}{M}\right)^{1/2} \qquad\qquad 2.$$

the rms velocity, and M the molecular mass. This limits spectral resolution (at room temperature) to $u/c \simeq 1:10^6$.

Considerable effort is required to attain Doppler-limited resolution by means of conventional spectroscopic techniques. The best optical and infrared spectrometers can achieve $1:10^6$ resolution under optimal conditions. Etalons and similar interference devices can easily resolve frequency differences with the required resolution. These approaches have been used successfully to resolve structure in systems exhibiting large hyperfine splittings (Longhurst 1967). Various techniques employing optical pumping have also been used to resolve atomic hyperfine structure at the Doppler limit (Bernheim 1965).

The incorporation of laser sources in conventional spectrometers is an important straightforward extension of a series of developments in the field of spectroscopy that have occurred over the past eighty years. Readily available tunable lasers are now being used in routine spectroscopic studies of samples that are smaller and more dilute than was heretofore possible. A few examples of numerous recent studies are given in Section 3.2.

2.2 Conventional Sub-Doppler Techniques

As discussed earlier, spectral features associated with isotope and isomer effects and hyperfine interactions usually lie within the Doppler profile of a spectral line, except for heavy elements such as mercury and lead. Thus, extremely monochromatic laser radiation cannot be effectively utilized unless the Doppler width can be suppressed.

Atomic interference effects such as level crossing (Franken 1961) and the Hanle effect (Mitchell & Zemansky 1971) can be used to circumvent the limitation of Doppler broadening. These methods are based on the fact that the angular distribution of the radiation scattered from an atomic system composed of two coupled, closely spaced transitions changes dramatically when the transitions are Zeeman or Stark tuned to approach each other within a natural width. Related transient techniques, such as the quantum beat effect (Dodd et al 1964, Alexandrov 1964), have also been used to attain high resolution.

Double resonance techniques of various types can also be used to circumvent Doppler broadening (Kastler & Brossel 1949). For example, small changes in the populations of closely spaced levels induced by radiofrequency or microwave radiation can be studied by monitoring the transmission or scattering of light at a coupled optical transition. In these techniques the perturbing radiation is low frequency, hence free of the effects of Doppler broadening.

The above techniques were all developed for use with conventional sources. Their effectiveness is considerably extended when laser sources are used (Walther 1976). Lasers also make possible stimulated emission counterparts, such as stimulated level crossing (Feld et al 1973).

Doppler broadening can be reduced by altering the velocity distribution in straightforward ways. The Doppler width can be narrowed somewhat by cooling the vapor (since $\Delta\omega_D \propto T^{1/2}$), but reduced vapor pressures and condensation limit the effectiveness of this approach.

A more general technique, in use long before the advent of lasers, employs a collimated atomic (or molecular) beam. If light is observed in a direction perpendicular to the motion of the beam [as in beam foil spectroscopy (Bashkin 1975)], or the beam is excited by light incident at right angles, the first-order Doppler shift vanishes, hence the Doppler broadening is eliminated. Many hyperfine interaction studies have utilized crossed atomic and laser beams (see Sections 3.2 and 3.3). Two draw-backs of this technique are the loss of sensitivity engendered by the limited interaction volume for crossed beams, and the unavoidable collimation losses inherent in producing thermal atomic beams. Hence, laser-atomic beam techniques are best suited to studies of abundant stable isotopes and optical transitions with large oscillator strengths.

An important variation, which is particularly useful when sensitivity is limited, has been developed by Kaufman and his collaborators (1976, 1978, and Anton et al 1978a,b) and independently by Wing et al (1976). In this approach an ion beam is accelerated to a high velocity, v_i,

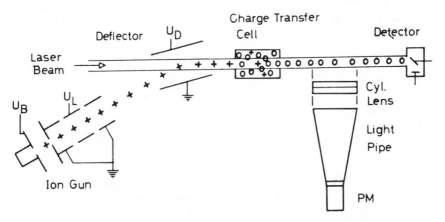

Figure 2 Schematic diagram of apparatus for collinear laser-atomic beam spectroscopy on neutralized accelerated beams (Anton et al 1978b).

neutralized, and then probed by means of a *collinear* laser beam (Figure 2). Since the energy spread of the accelerated beam remains thermal, the velocity spread is reduced by a factor $u/v_i \simeq 10^{-3}$, leading to a corresponding reduction in Doppler broadening. Note the enhanced sensitivity made possible by the increased interaction length. Another advantage is the large Doppler shift, $\omega v_i/c$, which permits frequency tuning by varying beam velocity as well as laser frequency. If suitable transitions exist, the ion resonance line may be studied directly, and losses caused by small neutralization cross sections are eliminated.

2.3 *Techniques of Laser Saturation Spectroscopy*

The monochromaticity and high intensity of laser radiation make possible a set of spectroscopic techniques in atomic and molecular gases, the resolution of which is determined by the natural or homogeneous width rather than the Doppler width. Such "laser saturation" techniques can be used to resolve hyperfine and other sub-Doppler structure with the same precision as that obtainable using collimated beams. Hence they provide a precise means of extracting nuclear structure information in stable nuclei. When combined with the techniques of laser-induced nuclear orientation (Section 2.5), measurements can be made in samples of extremely low density and on relatively short-lived nuclei. Extensive discussions of laser saturation spectroscopy are given by Letokhov & Chebotayev (1977) and by Feld (1973).

Laser saturation techniques are based on the principle that an intense monochromatic beam of laser radiation can selectively saturate an atomic (or molecular) Doppler-broadened resonance with which it interacts. The intense field induces rapid transitions between the atomic levels, and hence tends to equalize or "saturate" the level populations. (The condition on the laser intensity I for saturation to occur is

$$\sigma I/h\omega \gtrsim \gamma, \qquad\qquad 3.$$

with σ the homogeneously broadened atomic absorption cross section and γ the natural width.) However, these changes occur only over a narrow section of the velocity distribution (Figure 3a), since only those atoms that are Doppler shifted into resonance with the applied field can resonantly interact with it. Thus, measured along the propagation direction of the laser field, the thermal velocity distribution of upper and lower levels is altered over a narrow range. If the laser frequency, Ω, is detuned from the transition center frequency by an amount $\Delta = \Omega - \omega$, these changes will be centered at a velocity

$$v_0 = \Delta/k \qquad\qquad 4.$$

$(k = \omega/c)$, and extend over a velocity band

$$\Delta v = \gamma/k. \hspace{4cm} 5.$$

These selective population changes can lead to correspondingly narrow spectral resonances. Consider the transmission of a small portion I' of the intense beam reflected back upon itself through the sample cell (Figure 3b). This weak reflected beam acts as a probe of the saturated transition. When the applied field is well detuned from the atomic center frequency $(|\Delta| \gtrsim \gamma)$, the transmitted signal will trace out the broad Doppler profile of the transition, since the oppositely propagating fields resonate with two distinct velocity groups symmetrically located on opposite sides of the velocity profile. However, when the applied field approaches the atomic center frequency, the probe field begins to interact with the same atoms saturated by the intense field. Thus a narrow resonance of width γ occurs at the center of the Doppler profile. The Doppler broadening is thereby eliminated, making possible a considerable improvement in the location of the atomic center frequency. This central tuning feature is often referred to as the Lamb dip, since it was first predicted by Lamb (1964) in a slightly different context.

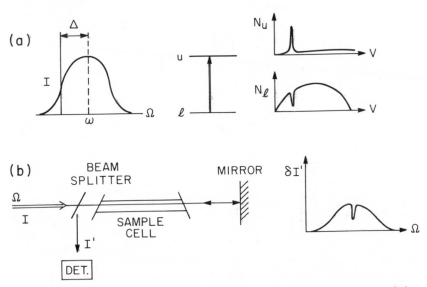

Figure 3 (a) Schematic representation of laser saturation of a narrow portion of the velocity distribution. A laser detuned Δ from the center of the u–l transition will transfer a narrow band of atoms from the velocity distribution as illustrated. (b) Observation of narrow resonance in transmission of back reflected beam.

Useful variations of this technique that enhance the signal-to-noise and contrast ratios have been developed. One scheme employs incident and reflected waves of comparable intensities, and monitors the side fluorescence from one of the levels of the saturated transition to a third, lower level (Freed & Javan 1970). As the laser frequency is tuned through the Doppler profile a narrow resonant change in fluorescence intensity is observed at line center. Another related technique, called Doppler-free polarization spectroscopy, uses laser-induced polarization rotation to increase sensitivity (Wieman & Hänsch 1976).

Narrow saturation resonances can also be induced in three-level systems. Consider the fluorescence arising from a transition formed by either of the levels of the saturated transition and a third lower level (Figure 4a). Viewed along the axis of the laser field, the narrow change in the velocity distribution produces a narrow resonance superposed on the broad Doppler background (Figure 4b). A similar narrow resonance will occur in transmission (i.e. stimulated emission or absorption) by probing the coupled transition by a weak tunable monochromatic field collinear with the saturating field. Laser saturation spectroscopy in three-level systems is known as laser-induced line narrowing (Feld 1973 and references therein).

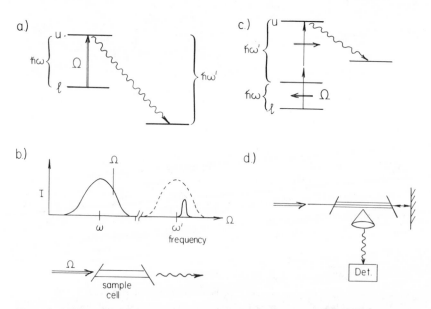

Figure 4   (*a*) Three-level system with u–l laser resonance. (*b*) Change signal in fluorescence from level u. (*c*) Three-level system with nonresonant intermediate state. (*d*) Detection of side fluorescence from state u.

Saturation resonances of closely related origin can also be observed in three-level systems in which the intermediate level is *nonresonant*. Consider the cascade system of Figure 4c, composed of a ground state and an excited state radiatively coupled to a common level lying between them (not necessarily halfway). The system is subjected to a standing wave laser field of frequency Ω, which is about half that of the two-photon transition. Detection is achieved by monitoring the side fluorescence formed by the excited state and a convenient lower-lying state (Figure 4d). Fluorescence is emitted as the laser field is tuned to a frequency exactly half that of the two-photon transition, over a tuning range equal to the width of the ground and excited states. There is no Doppler background. The resonance results from a two-photon absorption of the oppositely directed waves. The resonance condition is given by

$$\omega + \omega' = (\Omega - kv) + (\Omega + kv) = 2\Omega, \qquad\qquad 6.$$

where ω and ω' are the atomic center frequencies of the coupled transitions. Since the opposite Doppler shifts exactly cancel, atoms of all velocities can simultaneously interact with the applied fields. This enhances the effect and, at the same time, makes it independent of Doppler broadening. This effect, known as Doppler-free two-photon spectroscopy, allows us to study hyperfine structure of high-lying atomic states (Vasilenko et al 1970, Bloembergen & Levenson 1976).

Essentially the same effect can be observed in a folded system consisting of two closely spaced low-lying levels (such as hyperfine levels) coupled to a common upper level forming two transitions, ω and ω'. If a pair of co-propagating applied fields of frequency Ω and Ω' interacts nonresonantly with the atomic system ($|\Omega - \omega| \gg ku$), a narrow resonant feature will occur when

$$\omega - \omega' = (\Omega - kv) - (\Omega' - k'v) \simeq \Omega - \Omega', \qquad\qquad 7.$$

where the frequency detuning is small enough to render $(k - k')v$ negligible. This effect is actually an example of stimulated Raman scattering (Weber et al 1967, Schuler 1968). Note that in contrast to the cascade case co-propagating fields are required to eliminate the Doppler effect.

2.4   Few-Atom Spectroscopy

The high selectivity and sensitivity of laser radiation permit measurements to be made on extremely dilute samples (Letokhov 1978). This feature is particularly relevant for nuclear physics studies since rare species, often with short lifetimes, are of major current interest and are likely to become even more important with the new generation of heavy ion accelerators.

Nuclear physicists routinely study evanescent systems and systems that produce weak signals. Nuclear counting techniques are being integrated into laser spectrometers to enhance detection of trace amounts of atoms and molecules. The most straightforward scheme, laser-induced resonance fluorescence detected by single-photon counting, is limited by background light due to the huge number of photons of the same wavelength present in the exciting laser beam. The signal-to-noise ratio can be improved when branching to another transition permits resonance detection of photons of different wavelength or polarization. Sensitivity is reduced in this situation, however, since atoms decaying via the non-resonant branch cannot, in general, interact with the pump laser again.

Further improvements are possible through the use of coincidence and time correlation techniques. If an atom interacts with a saturating laser field for a time T it will re-emit about $T/2\tau$ photons, where τ is the excited-state spontaneous emission lifetime. The Minnesota group (Kaufman et al 1978) has been developing this scheme for studying weak beams of short-lived radioisotopes; they eliminate more abundant isotopic species of the same element by means of an isotope separator.

Another technique, developed by Hurst et al (1975) and Bekov et al (1978), employs single-*atom* detection via ionization spectroscopy (Figure 5). Here, an atom is resonantly excited by one laser field and photo-ionized by a second, broadband laser field. (A related scheme for separating isotopes is discussed in Section 2.6.) Single-ion counting in a dc electric field with channeltron multipliers or proportional counters provides high detection sensitivity. If the energy of the intermediate

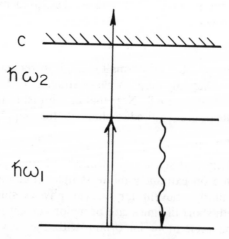

Figure 5   Schematic representation of laser resonance ionization spectroscopy.

state is greater than half the ionization energy, a single laser field can be used for both steps. Experiments of this type require high saturation and ionization efficiencies, and so generally utilize high power pulsed lasers. Coincidence counting in this case further improves sensitivity.

Other possibilities for few-atom spectroscopy include schemes based on detection of nuclear radiation (Section 2.5) and techniques that employ ion and atom traps (Neuhauser et al 1978). The latter approach, though extremely useful in certain applications, is unlikely to be of major importance in nuclear spectroscopy because the lifetimes of relevant nuclear species are short compared to the relatively long times required to trap and cool atoms or ions. However, for (possibly) long-lived rare atoms such as the proposed bound quark system, a laser spectrometer combined with a suitably designed trap could be used to provide a definitive existence test. Another scheme for detecting sodium-quark atoms has been proposed by Fairbank et al (1975).

2.5   *Laser-Induced Nuclear Orientation*

The techniques discussed in previous sections utilize the mono-chromaticity and intensity of laser radiation. If the radiation is polarized, additional applications are possible (Burns et al 1977, 1978). Thus, the angular momentum transferred to a resonantly absorbing atomic (or molecular) transition by polarized laser radiation can orient the atomic nuclei along a given axis. If the nuclei are unstable the angular distribution of the subsequent nuclear radiation will be anisotropic. This spatial anisotropy can be a sensitive probe for resolving hyperfine splittings and other nuclear structure effects, especially when coupled with laser saturation or atomic-molecular beam techniques. The new physics to be studied includes precision measurements of isomer shifts and excited state nuclear moments. The technique can also be used for nuclear isomer separation (Section 2.6), production of narrow gamma ray resonances (Section 2.7), and weak interaction studies.

Consider a vapor containing isomeric atoms, that is, atoms whose nuclei are in an excited state (Figure 6a). Since the concentration of such atoms is always very low, the optical spectrum either in absorption or spontaneous emission is exceedingly weak and difficult to resolve. An enormous enhancement in sensitivity can be obtained by studying the nuclear radiation rather than optical photons. Although this radiation is ordinarily isotropic, its angular distribution can be altered by a circularly polarized laser field that is tuned into coincidence with one of the hyperfine transitions of the isomeric atoms.[1] The circularly polarized

---

[1] Nuclear orientation may also be induced by a linearly polarized laser beam.

426   MURNICK & FELD

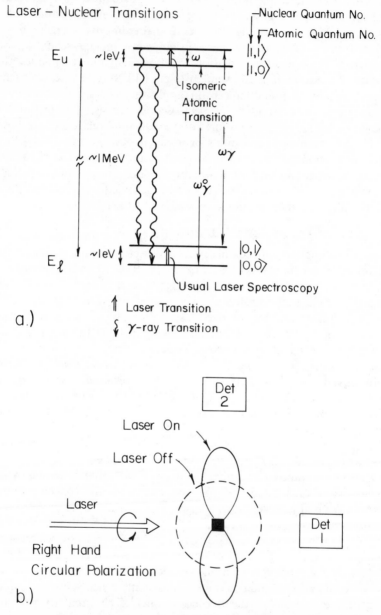

Figure 6 (*a*) Schematic representation of coupled nuclear and atomic transitions. (*b*) Schematic representation of nuclear radiation anisotropy induced by laser optical pumping.

laser photons exert a torque on the atomic electrons, which tends to align their angular momenta along the propagation axis of the laser beam. Since the electrons and nuclei are coupled by the hyperfine interaction, the atomic nuclei can also become aligned, and hence radiate anisotropically (Figure 6b). In this way, as the laser is tuned the hyperfine structure of the isomeric atoms will manifest itself as a series of resonant changes in the angular distribution of the nuclear radiation.

The degree of nuclear orientation can be greatly enhanced by taking advantage of optical pumping techniques. The optical pumping effect (Bernheim 1965) gives rise to a net transfer of atoms to higher or lower orientation quantum numbers, depending on the laser polarization. If the laser is right circularly polarized, for example, the selection rule for absorption is $\Delta M_F = +1$, whereas for spontaneous re-emission $\Delta M_F = 0, \pm 1$ is generally allowed. Hence the pumping cycle tends to shift the population to higher M_F levels, leading to a nuclear polarization in ground and excited atomic states. The degree of polarization depends on the laser intensity and the oscillator strength of the atomic transition, as well as on the extent of relaxation effects, such as collisional de-orientation. It is worth noting that the presence of optical pumping in the laser saturation process leads to greatly enhanced nuclear orientation efficiency at reduced laser intensity levels (Murnick et al 1979c).

Under suitable conditions (relatively low pressure, so that the transition is Doppler broadened), laser optical pumping of the isomeric transition can lead to a high degree of nuclear orientation over a narrow range of velocities. Accordingly, in combination with the laser saturation techniques described in Section 2.4, resolution at the natural width level can be achieved. Laser-induced nuclear orientation can also be combined with rf double-resonance techniques of the type used by Capeller & Mazurkewitz (1973).

Laser-induced nuclear orientation should be applicable to unstable nuclei with lifetimes as short as ~ 1 ns, the lower limit being determined by the optical pumping cycle and by uncertainty broadening introduced into the atomic resonance line. In optimal cases the single-particle sensitivity of nuclear counting techniques allows experiments to be performed on equilibrium concentrations as low as ~ 1 atom cm^{-3}.

Laser-induced orientation and polarization of nuclei, both stable and unstable, may have other interesting applications. Possibilities include production of polarized beams and targets and, in certain unstable systems, polarized beta and gamma radiation. If the polarized nuclei are beta unstable, beta–gamma coincidence experiments are also feasible. As in low temperature nuclear orientation experiments, the hyperfine interaction in the beta-unstable ground state, as well as in gamma-emitting intermediate states, can be studied. The application of this technique to produce narrow-band gamma radiation is discussed in Section 2.7.

2.6   *Isotope and Isomer Separation*

The field of isotope separation is undergoing a revolution as a result of the impact of lasers, whose monochromaticity and high spectral intensity have led to efficient new separation schemes based on laser photophysical and photochemical processes. Benefits to nuclear physics include making available large quantities of purified rare isotopes and providing a means to extract trace amounts of nuclides of interest from samples with large impurity backgrounds. Industrial applications include separation of radioactive components in nuclear waste, production of enriched isotopes of deuterium, uranium, and other elements for use in fission and fusion reactors, and nuclear medicine applications. This section describes two of the most important laser isotope separation schemes. An extensive review of the basic principles and recent applications has been given by Letokhov & Moore (1976).

Laser isotope separation techniques are based on the small but significant isotope shifts that occur in many spectral lines of almost all atomic and molecular vapors. Atomic schemes utilize ground-state spectral lines in which the isotope shifts exceed the Doppler width. In the most common approach an isotopically mixed sample is simultaneously subjected to pulses from two tunable dye lasers; a monochromatic laser and a broadband laser. The monochromatic laser is tuned to a ground-state transition of one isotope, A_1, which does not overlap with transitions of neighboring isotopes and hence can selectively promote atoms of species A_1 to the excited state, A_1^* (Figure 5). The broadband laser is tuned to photoionize A_1^* atoms. The ionized atoms may then be removed and collected either physically (i.e. electrostatically) or chemically, by forming compounds with scavenger molecules placed in the sample cell.

The laser requirements for this process are fairly straightforward. The linewidth of the "monochromatic" laser must be narrow enough to discriminate between ground-state transitions of neighboring isotopes, but broad enough to cover the entire Doppler linewidth. The broadband laser need not be isotope selective, merely narrow enough so as not to produce appreciable ground-state photoionization of unwanted atoms. Typically, microsecond pulses in the 1–10-mJ range are required, and dye lasers of this type are readily available.

The most successful molecular scheme at present is based on multiphoton dissociation of polyatomic molecules. In this process an infrared laser pulse resonantly excites a particular vibrational mode, inducing multiple quantum transitions to high vibrational levels at which the molecule can dissociate. The dissociation fragments can then recombine with a scavenger gas (such as H_2) placed in the sample cell to form compounds that can be physically or chemically removed.

Isotopic selectivity in this case requires that the isotope shift for the mode being pumped exceed the width of the vibrational band. Pressures must be kept low (a fraction of a torr) and laser pulses short (~ 100 ns) to prevent collisional excitation transfer between isotopic species. Laser pulses of the order of a joule are required for efficient dissociation. Until now almost all such experiments used pulses from a CO_2 laser, which operates in the 9–10-μm spectral range and coincides with modes of many polyatomic molecules. Spectral coverage of the band is achieved by the power broadening inherent at these power levels (~ 10 MW).

A variety of other atomic and molecular laser isotope separation schemes are reviewed by Letokhov & Moore. It should be noted that all such schemes may be used to remove trace amounts of a substance of interest from a sample with high concentrations of unwanted species.

Variations of these techniques can also be applied to separation of nuclear isomers. Very long-lived isomers can, of course, be treated in the same manner as isotopes. For short-lived and very rare isomeric species whose spectral lines may be impossible to detect optically (because of nearby lines of the more abundant stable species) the laser-induced nuclear orientation scheme (Section 2.5) can be used to locate the ground-state transition of the isomeric atom by monitoring the angular distribution of the emitted gamma rays. Two-step photoionization may then be used to remove selectively the isomeric atoms. Samples of separated nuclear isomers may be important in possible biomedical applications and in gamma ray laser development.

2.7   Use of Lasers to Influence Nuclear Radiation

2.7.1   TUNABLE NARROW-BAND GAMMA RAYS   Gamma rays emitted from a gaseous sample of isomeric atoms are Doppler broadened, and thus distributed over a spectral range $\sim \omega_\gamma u/c$ (Equation 1), with $\hbar\omega_\gamma$ the energy separation between initial and final nuclear states (Figure 6a). Since at room temperature $u/c \sim 10^{-6}$, the Doppler widths of gamma radiation in the 100-keV to 1-MeV range are ~ 0.1–1 eV. The gamma ray absorption profile of the gaseous sample is similarly broadened. Furthermore, because of the momentum transfer between the gamma rays and the recoiling nuclei, absorption and emission frequencies do not coincide. The center frequency of the emission line is downshifted to

$$\omega_{em}^0 = \omega_\gamma(1 - R),$$
8.

and the absorption center frequency is upshifted to

$$\omega_{abs}^{0} = \omega_{\gamma}(1+R), \qquad\qquad 9.$$

with

$$R = \frac{\hbar\omega_{\gamma}}{2Mc^{2}}, \qquad\qquad 10.$$

and M the nuclear mass. The fractional shift $2R$ between emission and absorption resonances is typically $\sim 10^{-5}$–10^{-6} ($2R\omega_{\gamma} \sim 0.1$–$10\,\text{eV}$).

It should be possible to use laser saturation techniques to enhance or suppress a narrow portion of the gamma ray Doppler profile and thus produce sharp gamma emission and absorption resonances that can be tuned over the entire Doppler profile, a tuning range which exceeds that of the Mössbauer effect by one or two orders of magnitude. As discussed in Section 2.3, an intense monochromatic laser field propagating in the $\hat{x}$ direction selectively saturates a Doppler-broadened atomic or molecular resonance over a narrow section Δv (Equation 5) of the $\hat{x}$ component of the velocity profile centered at v_{0} (Equation 4). Consider now the gamma radiation emitted from such a system. The relevant energy levels and transitions are shown in Figure 6a, in which $|0,0\rangle$ and $|0,1\rangle$ denote ground and excited electronic states of the atomic-nuclear system with the nucleus in its ground state, and $|1,0\rangle$ and $|1,1\rangle$ the corresponding electronic states with the nucleus excited. Since the coupling between nucleus and electrons is very weak, in a transition between states the atomic and nuclear quantum numbers cannot both change. Therefore there will be two atomic transitions and two gamma transitions, as indicated in the figure. If the laser field selectively interacts with the isomeric atomic transition, center frequency ω, a build-up of population in the $|1,1\rangle$ state will occur over a narrow range of velocities with a corresponding depletion in the $|1,0\rangle$ velocity profile. Thus, viewed along the $\hat{x}$ axis the gamma rays emitted from the $|1,1\rangle$ state will be centered at frequency

$$\omega_{em} = \omega_{\gamma}(1-R+v_{0}/c) = \omega_{em}^{0} + \frac{\omega_{\gamma}}{\omega}\Delta, \qquad\qquad 11.$$

where $\Delta = \Omega - \omega$ is the laser frequency offset, with a spectral linewidth

$$\Delta\omega \simeq \omega_{\gamma}\frac{\Delta v}{c} \simeq \frac{\omega_{\gamma}}{\omega}\gamma. \qquad\qquad 12.$$

Narrow gamma absorption resonances can be similarly obtained if the laser interacts with the $|0,0\rangle \rightarrow |0,1\rangle$ transition. For typical allowed atomic transitions $\gamma \sim 25\,\text{MHz} \simeq 10^{-7}\,\text{eV}$, giving $\Delta\omega \simeq 10^{-2}$–$10^{-3}\,\text{eV}$

$(80-800 \text{ cm}^{-1})$. Weaker transitions can be narrower by a factor of 1000, giving correspondingly smaller values of $\Delta\omega$. Similarly, a narrow depletion will occur in the gamma emission arising from the $|1,0\rangle$ state.

Unfortunately, these narrow features are not ordinarily observable, since splitting between ground and isomeric atomic transitions is small $(\sim 10^{-5}-10^{-6} \text{ eV})$. In the net gamma emission from the isomeric atom, the narrow gamma peak from the $|1,1\rangle$ level will be compensated for by the depletion in the $|1,0\rangle$ emission profile. To observe the narrow features requires changing the velocity distribution of one state, for example, by adding a buffer gas that preferentially induces velocity-changing collisions in one state. Another approach would be to selectively photoionize and electrostatically remove the $|1,1\rangle$ excited state atoms by the techniques presented in Section 2.6. However, the methods discussed below are probably of greater feasibility and interest.

2.7.2 RECOIL-INDUCED GAMMA RAY SIDEBANDS Letokhov (1975) has pointed out that the gamma ray spectrum considered above can be modified if some of the recoil energy associated with the gamma ray transition is imparted to internal degrees of freedom of the molecule. Consider again Figure 6a, in which we now let levels $|0,1\rangle$ and $|1,1\rangle$ represent electronic, vibrational, or rotational levels of a molecule, as well as atomic electronic levels. Consider a transition in which the isomeric atomic or molecular system, initially in a state of energy E_u (either $|1,1\rangle$ or $|1,0\rangle$), emits a gamma photon and is de-excited to a state of energy E_l (either $|0,1\rangle$ or $|0,0\rangle$). The frequency of the emitted gamma ray is determined by the momentum and energy conservation relations:

$$M\mathbf{v}_0 = M\mathbf{v} + \hbar\mathbf{k}_{em}, \qquad\qquad 13.$$

$$E_u + \frac{Mv_0^2}{2} = E_l + \frac{Mv^2}{2} + \frac{\hbar\omega'_{em}}{c}, \qquad\qquad 14.$$

with $\mathbf{k}_{em} = (\omega'_{em}/c)\hat{\mathbf{k}}_{em}$ the wave vector of the emitted photon, and $\mathbf{v}_0$ and $\mathbf{v}$ the velocities of the molecular center of mass before and after emission. Neglecting small terms, one obtains

$$\omega'_{em} = \omega_\gamma\left(1 - R + \frac{\hat{\mathbf{k}}_{em}\cdot\mathbf{v}_0}{c}\right) + \delta\omega, \qquad\qquad 15.$$

where

$$E_u - E_l = \hbar(\omega_\gamma^0 + \delta\omega). \qquad\qquad 16.$$

Here, $\hbar\omega_\gamma^0$ is the gamma ray transition energy in the absence of any atomic-molecular internal energy, and $\hbar\delta\omega$ is the net internal energy change in the transition. Equation 15 shows that the "forbidden" transitions $|1,1\rangle \to |0,0\rangle$ and $|1,0\rangle \to |0,1\rangle$ are energetically possible. To see that they can also have allowed matrix elements when recoil is included one need only consider a diatomic molecule, initially at rest, that emits a gamma ray from one nucleus in a direction perpendicular to the internuclear axis and starts to spin. (This would be an event of the type $|1,0\rangle \to |0,1\rangle$.) It is clear that a substantial amount of rotational energy can be developed in this process, resulting in a final state in which the molecule is highly excited. Thus, in addition to the unshifted emission frequency (Equation 11), a set of sidebands (Equation 15) can occur. In this case the sidebands are rotational and the frequency spacing is determined by the net energy change of the system.

Narrow laser-induced gamma resonances of the type discussed in Section 2.7.1 can be induced in these gamma satellites. These sidebands may be resolved if they are sufficiently detuned from the central peak, thus overcoming the problem of overlap of the narrow peak and narrow depletion mentioned in the last section.

Letokhov and his collaborators (Letokhov 1977 and references therein) have performed quantum mechanical calculations to estimate the relative transition probabilities for recoil sidebands of various types. Rotational sidebands are found to have a large probability (over 50% in some cases); the probability of vibrational sidebands is usually less than 10%; and electronic sidebands in both atoms and molecules have very small probabilities, $\sim 10^{-4}$ or less. Unfortunately, pure rotational sidebands appear to be below the limits of resolution achievable even in very narrow laser-induced gamma resonances, and so are probably unobservable. Vibrational (vibrational-rotational) satellites offer the most promise, since their splittings can be larger than the gamma ray Doppler widths. However, the smallness of their amplitudes, combined with the difficulty of finding appropriate gamma sources to probe and detect them, make their observation difficult. A further difficulty in observing laser-narrowed vibrational satellites in molecules is that the laser interacts with only a single vibrational-rotational transition, hence only a small fraction of the total molecular density can be excited.

2.7.3 NARROW GAMMA RAY RESONANCES USING LASER-INDUCED NU-CLEAR ORIENTATION   Laser-induced nuclear orientation (see Section 2.6) provides another possible method for obtaining narrow gamma ray resonances. Consider again Figure 6a in which the $|1,0\rangle$-$|1,1\rangle$ transition represents an atomic transition of the isomeric atom. As discussed in Section 2.5, under suitable conditions laser optical pumping can lead to a high degree of orientation over a narrow range of velocities. Depending on the polarization state of the laser field (circular or linear), and the F values of the transition being optically pumped ($\Delta F = +1$, 0, or -1), population will be transferred either to $M_F = \pm F$ or $M_F \simeq 0$ ground-state levels. Depending on the nuclear spin change in the subsequent gamma emission, the gamma ray angular distribution will be sharply peaked, either along the laser propagation direction or at right angles to it.

The combination of population changes in the M_F states induced only over a narrow velocity range with sharply peaked angular distribution patterns overcomes the overlap problem mentioned in Section 2.7.2 and makes possible narrow tunable gamma resonances. For example, if the atomic transition is a $\Delta F = -1$ transition pumped by linearly polarized laser light (Figure 7), then within the narrow velocity band selected population will be transferred to the $M_F = \pm 2$ states from all the other M_F states. The gamma radiation emitted will be sharply peaked along the laser propagation direction. The center frequency and linewidth of this radiation are given by Equations 11 and 12, respectively. The gamma radiation emitted from velocities outside this frequency band will be

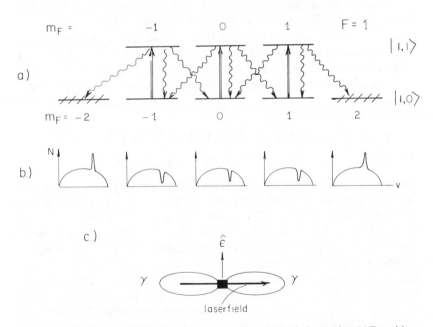

Figure 7 Anisotropic tunable gamma radiation induced by optical pumping. (*a*) Transitions using linearly polarized light to pump a 2–1 transition. (*b*) Relative population for M_F levels as a function of velocity. (*c*) Schematic angular distribution of gamma radiation from oriented atoms.

isotropic since the M_F levels are equally populated at all other velocities. Hence the net gamma radiation spectrum in the forward direction will consist of a large narrow-frequency peak superimposed on a broad Doppler background.

With other laser polarizations, ΔF transitions, etc, one can similarly obtain a broad emission band with a narrow *depletion* occurring at a frequency corresponding to the resonant velocity group. Similarly, by optically pumping the ground state $(|0,0\rangle\text{-}|0,1\rangle)$ atomic transition, narrow gamma absorption resonances can also be induced.

2.7.4 OBSERVATION OF NARROW GAMMA RESONANCES   Observation of narrow gamma resonances using conventional gamma ray techniques is likely to be difficult because of the absence of detectors of sufficient frequency resolution. A Mössbauer absorber of the appropriate frequency would be a suitable detector, but finding a frequency coincidence between a source of interest and a Mössbauer absorber may prove to be difficult. An absorbing sample of the same type as the source can be used if the recoil shift is smaller than the Doppler width, so that source and absorber can be tuned to the same frequency. It should be noted, however, that direct detection of absorption is not feasible, because nuclear absorption cross sections are small and the gas density is low. Observation of conversion electrons or reradiated absorbed gammas emitted from the sides of the absorber may be a useful means of detection, though the Compton effect and Rayleigh scattering will produce a large unavoidable background.

3   APPLICATIONS AND CASE STUDIES

3.1   *Laser Tests of Fundamental Interactions*

The unified gauge theory of weak and electromagnetic interactions of Weinberg (1967) and Salam (1968) implies the existence of a small degree of parity nonconservation (PNC) in the electron nucleon interaction. The PNC interaction is mediated by weak neutral currents, that is, weak processes induced by the exchange of a neutral heavy boson. Bouchiat & Bouchiat (1974) first pointed out that weak neutral currents would lead to small subtle effects in atomic structure, such as a tiny admixture between close-lying states of opposite parity. The Bouchiats also recognized that laser properties of monochromaticity, tunability, intensity, and polarization could be used to search for predicted effects too small to be observed by other experimental techniques. The method suggested, which has been used in one form or another by several research groups, entails searching for circular dichroism, i.e. rotation of the plane of polarization, as a linearly polarized laser field is resonantly tuned through an M1 transition in a heavy atom. Because of the PNC admixture of an opposite parity state, the transition will have a small E1 component, leading to a weak circular polarization of the order $1:10^4$ in favorable cases.

Experiments have been carried out in bismuth and thallium. Unfortunately, conflicting results were obtained in the Bi experiments; negative results were reported by groups in Oxford and Seattle (Lewis et al 1977, Baird et al 1977) and positive results were reported by a Novosibirsk group (Bekov & Zolotorev 1978). A group in California (Conti et al 1979) has obtained results in Tl consistent with the Weinberg-Salam model. All of these experiments have been complicated by the theoretical problems entailed in calculating the multi-electron atomic wave functions necessary for proper evaluation of the results.

In an independent high energy physics experiment confirming the Weinberg-Salam model, Prescott et al (1978) measured PNC in inelastic electron scattering from deuterium at 20 GeV. Here, too, a laser was crucial in providing an intense beam of polarized electrons via the photoeffect from polarized laser radiation in a GaAs crystal.

As indicated above, fundamental studies using heavy atoms are hampered by our less than perfect knowledge of multi-electron wave functions. For the hydrogen atom and hydrogen-like ions, however, the relativistic wave equation can be solved exactly, and precise tests of certain fundamental interactions are possible. In addition to PNC studies (Lewis & Williams 1975), several groups are applying laser techniques to measure the Lamb shift as a probe of the range of validity of quantum electrodynamics (QED).

The Lamb shift, or splitting between the $2^2s_{1/2}$ and $2^2p_{1/2}$ energy levels, is calculated in quantum electrodynamics as a power series expansion in α, the fine structure constant, times Z, the atomic number. Terms of order $(\alpha Z)^6$ have been computed to date. Measurements of the Lamb shift for high Z hydrogenic ions test the limits of validity of the QED series expansion, as well as recent attempts to separate the Z dependence from a perturbation expansion in α (Kugel & Murnick 1977). A current experiment on the Cl^{16+} ion is illustrated in Figure 8. Cl^{16+} ions in the $2s_{1/2}$ level are formed via electron pickup by 200-MeV Cl^{17+} ions provided by a tandem van de Graaff accelerator. A high power (2.5 kW) CO_2 laser excites the $2s_{1/2}$-$2p_{1/2}$ transition, from which excess Lyman α photons at 2.97 keV are detected. The resonance is traced out by discretely tuning the laser through various rotational-vibrational transitions over the range 900–1000 cm^{-1} and by Doppler-shift tuning that is obtained through varying the angle of intersection of the laser beam with the relativistic particle beam (Murnick et al 1979a). Such an

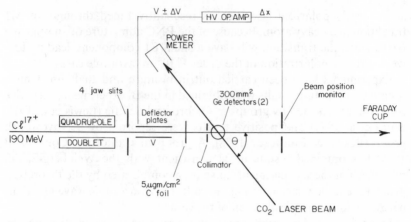

Figure 8 Diagram of experimental system for laser Lamb shift resonance measurement on 190 MeV Cl^{16+}

experiment would be impossible without the high power, coherence, and monochromaticity of the CO_2 laser.

Lamb shift measurements on the hydrogen atom ($Z = 1$) have become so precise that the experimental uncertainty is of the same order as the calculational uncertainty due to the proton size. In principle, then, a more precise Lamb shift measurement might be used to obtain an improved value for the proton radius, assuming QED to be valid. An elegant laser spectroscopy experiment has exploited this approach in the $\mu^- He^{++}$ ion (Bertin et al 1975). In that system the ~0.5% Lamb shift calculational uncertainty arises almost entirely from uncertainty in nuclear size, as the muon is much more tightly bound that an electron would be. Assuming QED to be valid, we find an experimental value of 1.644 ± 0.005 fm for the ^4He rms charge radius, which is an order of magnitude more precise than the value obtained from electron scattering data. By using this result, a precise He^+ Lamb shift experiment may now be used to test QED to greater accuracy than heretofore possible.

Another laser Lamb shift measurement in hydrogen used Doppler-free two-photon spectroscopy (see Section 2.3) to obtain an accurate value for the 1s-2s energy splitting (Hänsch, Schawlow & Series 1979), from which the $2s_{1/2}$-$2p_{1/2}$ Lamb shift as well as the $1s_{1/2}$ "Lamb shift" is obtained. This value, however, is not as precise as that obtained from microwave techniques.

3.2   Stable Isotope Studies

Laser spectroscopic techniques are now being widely used to study nuclear structure effects in a large variety of systems—visible-UV atomic transitions, and vibrational (infrared) and electronic (visible-UV) molecular transitions. While most visible studies have employed tunable dye lasers, infrared studies have utilized a variety of sources, both fixed and tunable. Reviews of laser sources can be found in Arecchi & Schulz-Dubois (1972) and Hinkley et al (1976).

Most high resolution laser studies have employed atomic beam (Section 2.2) or laser saturation spectroscopy (Section 2.3) techniques. A large number of isotope shifts and hyperfine structure splittings have been observed in both atoms and molecules. Lists of systems studied are given by Walther (1976). Unfortunately, relatively few of these studies have yielded information about nuclear moments. This situation is understandable in molecules, where extraction of nuclear parameters requires accurate molecular wave functions, but it is surprising that more nuclear moment information has not been obtained in atoms thus far. The following discussions review a few cases where nuclear parameters have been obtained.

Beam techniques have been employed most often to study atomic transitions. The advent of broadband tunable dye lasers enables a wide range of atomic species to be studied. Table 1 from Fairbank et al (1975) lists the most promising resonance lines for laser spectroscopy for most elements of the periodic table. The present wavelength range of dye lasers is limited by the availability of suitable dyes. As laser dyes are improved, the more difficult species, those with resonance lines in the blue to UV range, will undoubtedly be subject to studies of the type now carried out routinely in the 600-nm spectral region. Research in this field is very active, with new results published at a rapid pace, so that no review of the subject can be complete. Instead, a few representative examples will be described.

Isotopes in the rare earth and actinide series are especially interesting because in these regions of changing nuclear properties a large number of isotopes are available, many of which exhibit relatively strong atomic resonance transitions in the visible wavelength range. The Minnesota group (Broadhurst et al 1974, Clark et al 1978) has used atomic beam techniques to study the 597-nm atomic transition of several dysprosium isotopes. A thermal atomic beam, well collimated to reduce the linewidth to close to the natural broadened limit, was illuminated at right angles by dye laser radiation. The spectral features were monitored by collecting the resonance fluorescence induced as the laser was tuned through the spectral structure. The data of Figure 9 are typical of the large number of hyperfine transitions resolvable.

The hyperfine structure information obtained in this study and a similar one in ytterbium at 555.6 nm by the same group (Clark et al 1979) yields

Table 1 Ground-state transitions (in angstroms) of the 87 elements whose atomic energy levels are known. Only the most promising transition for use in the resonance fluorescence method is shown for each atom (Fairbank, Hänsch & Schawlow 1975).

Future (800–2300)	Doubled, mixed dye (1780–2650)	Commercial doubled dye (2650–3400)	Pulsed dye (3400–4300)	CW dye (4200–5400)	CW dye (easy) (5400–6500)		CW dye (6500–7800)
Ar 1067	As 1937	Ag 3383	Al 3961	Co 4234	Ac 6360	Sc 6211	Ca 6573
Br 1541	B 2497	Au 2676	Ga 4033	Cr 4290	Am 6055	Sm 5626	K 7665
Cl 1396	Be 2349	Bi 3068	Ho 4104	Cs 4593	Ba 5535	Ta 5403	Li 6708
F 956	Hg 2537	C 2967	In 4105	Fe 5110	Dy 5988	Tc 5925	Rb 7800
H 1216	I 1783	Cd 3261	Ir 3800	Nb 5252	Er 5827	Th 5761	
He 591	Po 2558	Cu 3274	Mo 3903	Os 4420	Eu 6018	Tm 5971	
Kr 1250	Rn 1786	Ge 2652	Ni 3625	Ra 4826	Gd 5709	U 5915	
N 1201	S 1915	Mg 2852	Rh 3692	Re 5276	Hf 6185	Y 6223	
Ne 744	Sb 2311	Pb 2833	Ru 3926	Sr 4607	La 5501	Yb 5556	
O 1359	Si 2514	Pd 2763	Tl 3776	Tb 5376	Lu 5737	Zr 6135	
P 1775		Pt 3064		Ti 5174	Mn 5433		
Se 2075		Sn 2863		V 4851	Na 5890		
Te 2259		Zn 3076		W 4983	Nd 6149		
Xe 1491					Pu 5865		

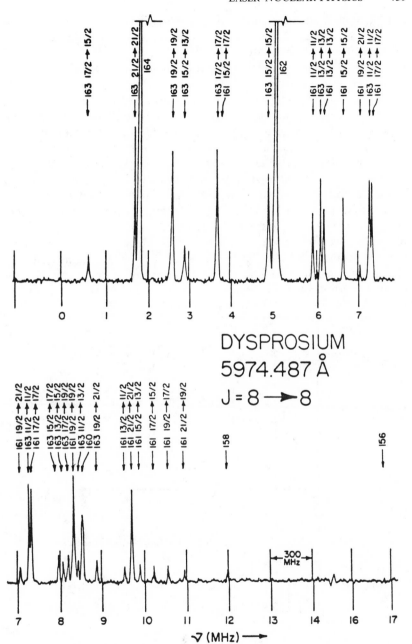

Figure 9   Laser scan of natural Dy beam near 577.4 nm. The marker spacing is 300.35 MHz (Clark et al 1978).

isotope shifts and, most interestingly, measurements of the hyperfine anomaly parameter Δ (Section 2.1). Comparison of the latter data with calculations based on Nilsson wave functions for ^{161,163}Dy gives poor agreement. As more such data becomes available this heretofore "exotic" parameter should receive more scrutiny by nuclear theorists.

A novel atomic beam technique was developed by Childs et al (1979a,b,c) to study hyperfine structure of rare earth and other isotopes. The approach, which in some ways resembles the classical atomic beam magnetic resonance technique, is schematically illustrated in Figure 10. The laser field intersects the atomic beam at an initial point and saturates a particular hyperfine transition, depopulating its lower level. An rf field in the intermediate region of the beam can induce transitions between hyperfine levels when the rf tuning condition is satisfied, repopulating the depleted level. The laser field is reflected back and intersects a third region of the beam, inducing resonance fluorescence. Since the fluorescence intensity depends on the lower level population, the resonant changes in this signal as a function of rf tuning provide a precise indication of the hyperfine structure. Precision is determined by the rf linewidths

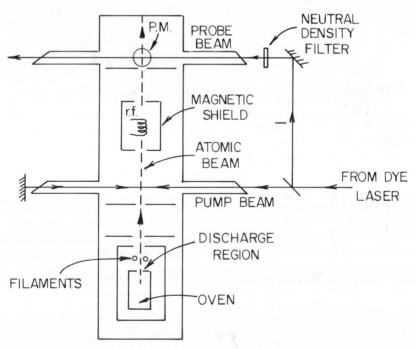

Figure 10 Experimental setup for laser-rf double-resonance spectroscopy (Childs, Poulsen & Goodman 1979b).

and is in the 10-kHz region. The results have led to improved data on hyperfine splittings, isotope shifts, and quadrupole moments in Sm and U. Similar studies have been carried out on lighter isotopes by other groups (Penselin 1979).

Laser saturation techniques (Section 2.3) have been used to observe nuclear structure effects in many atomic and molecular transitions. Indeed,

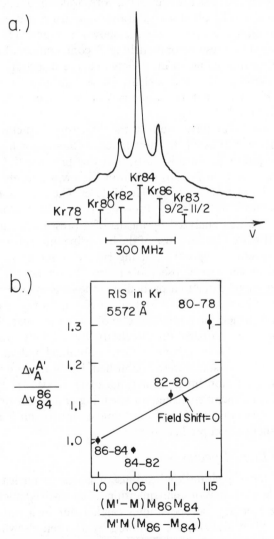

Figure 11 (*a*) Saturation spectrum of even Kr isotopes. (*b*) Relative isotope shifts plotted against the mass shift ratio (Gerhardt et al 1977).

the first laser saturation spectroscopy experiment (Szöke & Javan 1963), a Lamb dip study in a helium-neon laser, measured the ^{20}Ne-^{22}Ne isotope shift at 1.15 μm. Described below are two cases where nuclear parameters have been extracted from saturation experiments.

Gerhardt, Wenz & Matthias (1977) studied Lamb dips in the 557-nm transitions ($1s_5$-$2p_3$) of even isotopes of naturally abundant krypton. The metastable $1s_5$ level, in a sample at 0.2 torr, was populated by means of an rf discharge. The observed saturation spectrum and experimental values of the relative isotope shifts are shown in Figure 11. Deviations from the straight line result from a field shift contribution to the isotope shift, and reflect the change in the average charge distribution $\delta \langle r^2 \rangle$ as neutron pairs are added to the nucleus. These data imply a much larger change in $\delta \langle r^2 \rangle$ as the first half of the $g_{9/2}$ neutron shell is filled than as the second half is filled.

Ducas et al (1972) have studied the hyperfine structure of ^{21}Ne (nuclear spin $I = 3/2$) using the laser-induced line-narrowing technique described in Section 2.3. In this study, in a low pressure (0.1 torr) neon discharge sample cell, the $2s_2$-$2p_4$ neon transition was saturated by monochromatic radiation from a 1.15-μm He-Ne laser. The spectrum of cascade fluorescence emitted at the $2p_4$-$1s_4$ transition ($\lambda = 610$ nm) along the laser propagation axis was analyzed by means of a pressure scanned Fabry-Perot etalon (Figure 12a). Because of the hyperfine interaction, each level is split into several components, giving rise to a set of 18 three-level systems, eight of which give observable narrow resonances. The position of a narrow resonance due to a particular three-level system depends on the hyperfine splitting at both 1.15-μm and 610-nm transitions, as well as the direction in which the fluorescence is observed (i.e. parallel or antiparallel to the laser propagation direction). By carefully analyzing the data (Figure 12b and c), a theoretical fit was obtained and values for the hyperfine A and B constants were established. Using the known value of the ^{21}Ne magnetic moment gives a value of $Q = +0.1029 \pm 0.0075$ for the quadrupole moment, corrected for shielding and antishielding effects. This value is very close to that of the "almost" mirror nucleus ^{23}Na, in accord with shell-model predictions.

3.3 Short-Lived Species

Some of the most exciting applications of lasers to nuclear structure physics have been studies of small quantities of short-lived nuclei far from stability. The mercury isotope studies carried out in a cell on-line at ISOLDE (Kühl et al 1977) are reviewed by Hansen elsewhere in this volume. For lighter isotopes, atomic beam or saturation spectroscopy techniques must be used to achieve sub-Doppler resolution.

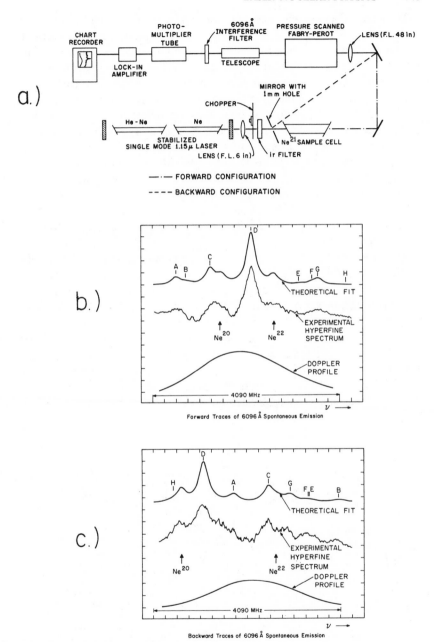

Figure 12 (a) Experimental arrangement of [21]Ne experiment. (b) and (c) Experimental data and theoretical fits to data (Ducas et al 1972).

Huber et al (1978a,b) reported impressive results on a chain of 11 sodium isotopes and 11 cesium isotopes by using an atomic beam on-line at a 20-GeV proton synchrotron (for Na) and at ISOLDE on-line at the CERN synchrocyclotron (for Cs). Figure 13 illustrates the arrangement for the Na experiment, which is a laser analog of the classical atomic beam magnetic resonance technique. Thermalized neutralized sodium nuclei produced by spallation reactions on uranium form an atomic beam that is irradiated by a tunable CW dye laser in a region of weak magnetic field. As the laser is tuned into resonance with one of the four hyperfine components of the D_1 line, the M_F population distribution changes. The beam atoms then enter a region of strong magnetic field, where I and J are decoupled. A sextupole magnet focuses only $M_J = 1/2$ atoms on the detector. Optical resonances are observed, as indicated, by monitoring the number of focused atoms of a particular mass as a function of laser frequency.

Data obtained in this experiment are tabulated in Table 2. The experimenters were able to show clearly the presence of a volume isotope shift in their data, making sodium the lightest element for which this was so. Their results indicate prolate deformation for $^{21-24}$Na and $^{28-31}$Na, with $^{25-27}$Na being almost spherical. Further studies of the D_2-line hyperfine structure should permit determination of Na quadrupole moments. Huber et al have also obtained similar extensive data for Cs, Rb, and Fr isotopes (Liberman et al 1979).

Schinzler et al (1978) have studied the cesium isotopes 133 and 137–139 on-line at the Mainz reactor by using the collinear laser-fast atomic beam

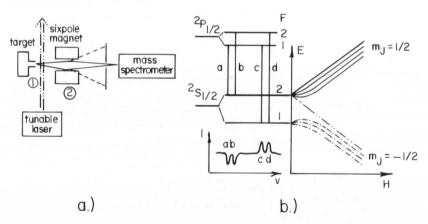

Figure 13 (*a*) Schematic of laser-Na atomic beam experiment. (*b*) Signals with D_1 line for $I = 3/2$ (Huber et al 1978b).

Table 2 Spins, hyperfine structure, magnetic moments, and isotope shifts of $^{21-31}$Na. The quoted errors are one standard deviation (Huber et al 1978a,b).

^{x}Na	$T_{1/2}{}^a$	I	$A(^2S_{1/2})$(MHz)	$\mu_I(\mu_N)$ corrected	$A(^2P_{1/2})$(MHz)	$IS^{23,x}$(MHz)
^{21}Na	22.5 s	$\frac{3}{2}{}^b$	953.7(2.0)		102.6(1.8)	−1596.7(2.3)
			953.233(11)^c	2.38612(10)^c		
^{22}Na	2.60 y	3^b	349.3(1.0)		37.5(1.0)	−756.9(1.9)
			348.75(1)^d	1.746(3)^d	37.9(1)^e	−758.5(7)^e
^{23}Na	stable	$\frac{3}{2}{}^b$	885.8130644(5)^f	2.2175203(22)^f	94.25(15)^e	
^{24}Na	15.02 h	4^b	253.2(2.7)		28.2(2.7)	706.4(6.2)
			253.185018(23)^g	1.6902(5)^g		
^{25}Na	60.0 s	$\frac{5}{2}{}^b$	882.7(9)		94.5(5)	1347.2(1.3)
			882.8(1.0)^h	3.683(4)^h		
^{26}Na	1.07 s	3	569.4(3)	2.851(2)	61.0(3)	1397.5(9)
^{27}Na	290 ms	$\frac{5}{2}$	933.6(1.1)	3.895(5)	100.2(1.1)	2481.3(2.0)
^{28}Na	30.5 ms	1	1453.4(2.9)	2.426(3)	156.0(2.7)	2985.8(2.7)
^{29}Na	43 ms	$\frac{3}{2}$	978.3(3.0)	2.449(8)	104.4(3.0)	3446.2(3.8)
^{30}Na	53 ms	2	624.0(3.0)	2.083(10)	66.2(2.8)	3883.5(6.0)
^{31}Na	17 ms	$\frac{3}{2}$	912(15)	2.283(38)		4286(16)

^{a} F. W. Walker, G. J. Kirouac, F. M. Rourke. *Chart of the Nuclides*, 12th edition, revised to April 1977, distributed by Educational Relations, General Electric Company, Schenectady, NY 12345.

^{b} V. S. Shirley, C. M. Lederer 1975. In *Hyperfine Interactions Studied in Nuclear Reactions and Decay*, edited by E. Karlsson and R. Wäppling (Almquist & Wiksell, Stockholm).

^{c} O. Ames, E. A. Phillips, S. S. Glickstein 1965. *Phys. Rev.* 137: B1157.

^{d} L. Davis Jr., D. E. Nagle, J. R. Zacharias 1949. *Phys. Rev.* 76: 1068.

^{e} Pescht, Gerhardt & Matthias (1977).

^{f} A. Beckmann, K. D. Böklen, D. Elke 1974. *Z. Phys.* 270: 173.

^{g} W. Chan, V. W. Cohen, M. Lipsicas 1966. *Phys. Rev.* 150: 933; V. W. Cohen 1973. *Bull. Am. Phys. Soc.* 18: 727.

^{h} M. Deimling, R. Neugart, H. Schweikert 1975. *Z. Phys. A* 273: 15.

technique described in Section 2.2. Nuclear magnetic and quadrupole moments and values for $\delta\langle r^2 \rangle$ have been obtained.

A straightforward laser resonance fluorescence study on an atomic beam of radioactive atoms was performed by Nowicki et al (1977, 1978) on ^{128}Ba and ^{131}Ba produced by (d,xn) reactions on enriched ^{130}Ba and ^{134}Ba and subsequent β^+ decay. Combined with previous results (Figure 14), isotope shift data have now been obtained for 11 barium isotopes. This experiment and the cesium study indicate a very important application of laser spectroscopy: studying trends in nuclear structure data through systematic study of long chains of isotopes.

In a few cases "conventional" laser saturation spectroscopy techniques have been applied to unstable species. A good example is the Lamb dip studies of Matthias and his collaborators on atomic transitions of long-lived radionuclides. This group has used single-mode dye laser radiation to study optical transitions in ^{22}Na (2.6 y) (Pescht et al 1978), ^{85}Kr (10.7 y) (E. Matthias, private communication), and ^{135}Cs (3 × 10^6 y) and ^{137}Cs (30 y) (Gerhardt et al 1978). The alkali samples require special cell preparation techniques because even relatively large amounts of activity (several mCi are used in the experiments) correspond to relatively low atomic densities, and sticking of the atoms to the cell walls can drastically reduce the number of atoms in the vapor phase. To reduce this effect the cell walls were passivated with alkali atoms prior to introducing the activity. The resulting vapor densities were rather low, typically ∼3 × 10^8 cm^{-3}, but still sufficient to observe the saturation effect in the side fluorescence with a good signal-to-noise ratio.

No such wall problem arises in the case of krypton, a noble gas, but the densities were correspondingly low because in the discharge only a

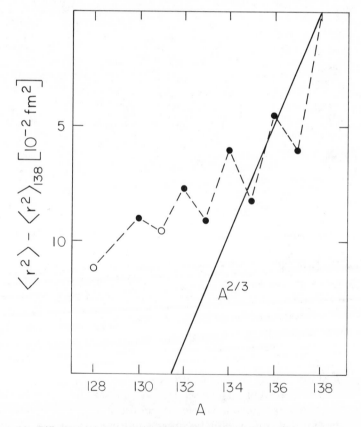

Figure 14 Differences of rms radii of light barium isotopes from optical isotope shifts. The straight line represents the dependence expected for the standard homogeneous sphere (Nowicki et al 1977).

small fraction of the population resides in the krypton $1s_5$ metastable level, the lower level of the transition under study.

Resolution of the cesium hyperfine splittings was not high enough to improve the accurate values obtained from earlier level-crossing and double-resonance studies, but the isotope shift results give new information about contributions due to the field effect. Nuclear structure information was likewise extracted from the ^{22}Na and ^{85}Kr spectra.

Experiments using laser-induced nuclear orientation (Section 2.5) to obtain nuclear parameters of short-lived states are being pursued actively at several laboratories. Both on-line and sample cell approaches are being explored. ^{137m}Ba is under study at the MIT Spectroscopy Laboratory (Burns et al 1977). In these experiments several μCi of ^{137}Cs (30 y) and a small amount of argon buffer are prepared in a sample cell. The cesium undergoes β decay to form ^{137m}Ba, which has a lifetime of 2.6 min and decays via 661-keV M4 gamma radiation. A single-mode dye laser optically pumps and orients the barium $^1P_1$-$^1S_0$ resonance transition ($\lambda = 553.5$ nm), which is split into several components by the hyperfine interaction with the ^{137m}Ba nucleus, the $I = 11/2$ spin of which is believed to be a result of an $h_{11/2}$ neutron hole. The hyperfine structure splitting is estimated to be a few hundred megahertz. The isomer shift with respect to the $I = 3/2$ barium ground state may be greater than 1 GHz.

In initial studies a small laser-induced gamma anisotropy has been observed. Related studies of stable barium atoms (Pappas et al 1979) have shown that essentially complete nuclear orientation can be obtained at surprisingly low intensity levels.

An on-line study of ^{24m}Na in progress at Bell Laboratories (Murnick et al 1979b) uses a 4-MV van de Graaff accelerator. The nuclear isomer is populated by the reaction ^{23}Na (d,p) ^{24m}Na at deuteron energies of about 2 MeV, and the 20-ms nuclear lifetime obviates the long confinement times necessary for ^{137m}Ba. Optical pumping is achieved by means of dye laser radiation at 589 nm. In preliminary measurements a small gamma ray anisotropy has been observed with a small isomer shift relative to the ^{24}Na ground state (Feld & Murnick 1979). Calculations of the structure of the low-lying levels of ^{24}Na based on the Lawson-Nilsson model may be tested by this experiment.

A novel related experiment is underway at Oak Ridge National Laboratory (C. E. Bemis, private communication) in which laser-induced nuclear orientation is detected by fission fragment anisotropy from ^{240m}Am. The fission isomers produced in a ^{7}Li (^{238}U, 5n) ^{240m}Am reaction are stopped in Ar buffer gas and irradiated with polarized laser light at 641 nm. At resonance, a fission fragment anisotropy should be observed. Nuclear theory predicts large deformations in this case that could yield an isomer shift as large as 0.5 nm!

3.4   *Proton Monochromator*

The preceding sections have primarily emphasized applications of lasers to nuclear spectroscopy. Lasers have great potential in other areas of nuclear science. For example, the high degree of nuclear orientation (or polarization) that can be achieved by laser optical pumping can be used to provide sources for weak interaction studies, and polarized beams and targets for nuclear reaction experiments.

Letokhov & Minogin (1978, Letokhov 1979) have proposed a novel way to produce beams of completely polarized, highly monoenergetic relativistic protons. Their scheme combines the laser saturation effect (Section 2.3), laser optical pumping, and multistep ionization by tunable laser radiation (Section 2.6). The method of achieving mono-chromatization is schematically indicated in Figure 15. A neutralized beam of relativistic protons (or H^- ions) is subjected to an overlapping beam of monochromatic laser radiation propagating in the reverse direction. The laser beam is resonant with the Lyman α transition ($\lambda = 121.5$ nm), but because of the Doppler shift the laser beam need not be tuned so far to the UV region. For a 600-MeV proton beam, 352-nm laser photons can be used, a far more convenient wavelength for present lasers. As described in Section 2.3, the laser field selectively saturates atoms over a narrow range of velocities. Laser photoionization of the excited hydrogen atoms permits essentially complete selection of the resonant velocity group.

Letokhov et al estimate that the energy spread of a 600-MeV proton beam can be narrowed by a factor of 100 by using a laser beam with an intensity of 500 W cm^{-2} and an angular divergence of 2.1 mrad. The intensity of the monoenergetic beam is reduced to a few percent of the

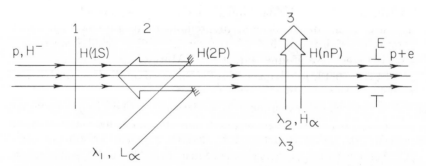

Figure 15   Schematic representation of laser proton monochromator (Letokhov 1979). Neutralization (1) is followed by velocity selective ionization (2) and stepwise photo-ionization (3).

initial proton beam intensity. Measurement of the laser wavelength should provide an absolute energy measurement of the monochromatized protons accurate to at least 10 ppm.

A variation of this scheme in which the monoenergetic protons are also polarized is based on two-step selective hyperfine excitation with circularly polarized light followed by photoionization. Complete (100%) polarization should be achievable by exciting atoms in the $2s_{1/2}$ $(F = 1, M_F = 1)$ state to the $10p_{3/2}$ $(F = 2, M_F = 2)$ Rydberg state by means of 379.8-nm circularly polarized laser radiation. Subsequent photoionization can be achieved by using a CO_2 laser.

3.5 Prospects for a Gamma Ray Laser

Coherent gamma ray laser radiation would have diverse applications in medical tomography, in providing precise information about the structure of biomolecules, greatly improving the accuracy of fundamental nuclear constants, and developing the field of nonlinear γ ray optics. According to Baldwin & Khoklov (1975), some general assertions can already be made about the form that gamma ray lasers will take: They will operate without mirrors, they will be pumped by neutron capture, they will use Mössbauer transitions, and their photon energy will be in the 5–200-keV range. Extensive references on this subject are given in the Baldwin & Khoklov article and by Letokhov (1973b).

The basic requirement for a gamma ray laser is no different from that of lasers of conventional wavelength: the gain must exceed the loss. If efficient reflectors can be developed at gamma ray wavelengths, resonator-type lasers can be considered. [An example is the novel single crystal device envisioned by Denne (1978).] Since developing such reflectors may be extremely difficult, it is probably more realistic to base preliminary considerations on lasers of the single-pass variety (MacGillivray & Feld 1977). In that case, the build-up of oscillation requires that the net single-pass gain exceed about 10:

$$n(\sigma - \kappa)L > 10, \qquad\qquad 17.$$

where n is the inversion density, σ the stimulated emission cross section, κ the nonresonant absorption cross section, and L the sample length.

Because of its narrow linewidth the most favorable type of transition for gamma ray–stimulated emission is a Mössbauer transition, the peak gain cross section of which is given by

$$\sigma_0 = \frac{\lambda^2}{2\pi} \frac{f}{1+\alpha}, \qquad\qquad 18.$$

with f the Debye-Waller factor and α the internal conversion coefficient.

As an example, consider a transition wavelength of $\lambda = 0.1$ Å (120 keV). For a favorable case $f \simeq 0.1$ and $\alpha \simeq 1$, giving $\sigma_0 = 10^{-20}$ cm^2. This is much larger than the photoelectric absorption cross section, $\kappa \sim 10^{-22}$ cm^2, which is the major source of nonresonant loss for elements of medium atomic number and energies below 200 keV.

For an active length $L = 1$ cm the gain condition (17) then requires $n = 10^{21}$ cm^{-3}, which is a small fraction of typical crystal densities, and hence a reasonable value to hope to achieve.

The estimate of the stimulated emission cross section of Equation 18 assumes the Mössbauer line to be radiatively broadened. In general

$$\sigma = \frac{\sigma_0}{\Delta\omega\tau}, \qquad\qquad 19.$$

with $\Delta\omega$ the observed linewidth and τ the effective lifetime of the transition $\{\tau^{-1} = \gamma_u + \gamma_l$, where $\gamma_{u(l)}$ is the radiative decay rate of the upper (lower) level$\}$. Due to a variety of perturbations, the narrowest Mössbauer linewidths presently achievable are $\Delta\omega \sim 10^{-5}$ s^{-1}. Thus, for lifetimes shorter than $\sim 10^{-6}$ s the product $\Delta\omega\tau \simeq 1$, whereas for long-lived transitions, $\Delta\omega\tau \gg 1$.

Proposed gamma laser approaches thus fall into two categories.

1. Schemes using long-lived isomers. This approach has the advantage that for isomers with sufficiently long lifetimes it is relatively easy to produce the required inversion density. However, in this case $\Delta\omega\tau \gg 1$, leading to a reduced value of σ. Thus, this approach requires development of new methods to reduce $\Delta\omega$. Considerable theoretical attention is being given to this problem.

2. Schemes using short-lived isomers. In this case stimulated emission cross sections are large, since $\Delta\omega\tau \simeq 1$ and $\sigma \simeq \sigma_0$. However, achieving a sufficient inversion density is then difficult, since the required density must be produced within the mean lifetime of the transition.

The minimum number of excited nuclei required is determined by the condition that diffraction loss be negligible. For a cylindrical geometry (filament) of radius a, negligible loss requires a large Fresnel number:

$$\frac{a^2}{\lambda L} \gtrsim 3. \qquad\qquad 20.$$

For the parameters used above this gives a minimum excited-state number

$$N = \pi a^2 L n \simeq 10^{11}, \qquad\qquad 21.$$

which implies an excited-state production rate $N/\tau \simeq 10^{17}$ s^{-1} for

$\tau = 1\,\mu s$. Production rates of this magnitude would require extremely high neutron fluxes to excite the Mössbauer filament.

In conclusion, the development of gamma ray lasers using schemes envisioned thus far requires advances in several technical areas including (a) efficient gamma ray reflectors, (b) rapid production of isomeric nuclei, (c) on-line methods for growing crystals of excited Mössbauer nuclei, (d) methods of reducing Mössbauer linewidths, and (e) intense neutron sources.

4   CONCLUSIONS

4.1   New Directions and Possible Applications

Laser technology is still in its infancy and it is reasonable to expect advances in several areas, notably wavelength range, tunability, efficiency, and average power. As lasers improve, new applications to nuclear (and other branches of) science will undoubtedly follow. Spectroscopic studies of the type described in Section 3 will be greatly expanded, especially to atomic and molecular systems having resonance lines in the near and far-UV region. Studies of rare nuclei and short-lived excited nuclear states will become more common, leading to systematic studies of isotope shifts of single particle and highly excited collective states (for example, high spin yrast levels), hyperfine anomalies, and precision values for magnetic and quadrupole moments.

Single-atom detection for nuclear reactions of extremely small cross section should become a viable nuclear technique. For example, Hurst et al (1979) propose to use laser resonance ionization spectroscopy to count ^{7}Li atoms produced by electron capture in ^{7}Be that is formed by solar neutrino capture. The same group has used a similar technique to count Cs products of ^{252}Cf spontaneous fission (Kramer et al 1978).

The increased resolution of hyperfine splittings obtainable with laser techniques may lead to measurement of higher order nuclear multipole moments and finer details of nuclear structure and interactions. Larson 1978), for example, used a laser-rf double-resonance technique to study the hyperfine structure of ^{85}Rb in magnetic fields of 75 kG. He has determined a diamagnetic shift of $1.65 \pm 0.16 \times 10^{-8}\,\mathrm{Hz\,G^{-2}}$, which includes a contribution due to nuclear susceptibility, a parameter related to the distribution and motion of charge in the nucleus. Data of this type and hyperfine anomaly information provide additional constraints on, and possibly increased knowledge of, nuclear wave functions.

Other areas of laser-nuclear interactions likely to develop are laser optical pumping to prepare polarized (aligned) targets and beams, tunable narrow-band sources of polarized gamma radiation, and fundamental interaction studies.

Laser applications to nuclear power include isotope and isomer separation for fuel enrichment and waste disposal (Section 2.6), laser-induced fusion and laser plasma diagnostics (Chen 1979, Wharton 1979). Each of these areas is the subject of intense research and development, with laser diagnostics already an invaluable tool.

Another active area is nuclear pumped lasers (Marcum 1978). The ultimate goal of this research is to achieve high power by pumping a laser with a nuclear reaction process (fission or fusion) and to convert nuclear power to laser power efficiently.

4.2 *Summary*

This article has reviewed the areas of nuclear science in which laser techniques have had a substantial impact. As a narrow-band tunable light source, the laser has had immediate application to hyperfine spectroscopy, particularly of rare species. Because of high available powers and the use of saturation techniques, novel experiments of greatly improved resolution and sensitivity have been carried out both on individual species and on chains of isotopes such as sodium, barium, cesium, and mercury.

Lasers have been used to study parity nonconservation in the electromagnetic interaction and the Lamb shift in heavy ions. Research has been proposed and is underway on laser-induced nuclear orientation, narrow-band gamma ray sources, and isotope and isomer separation. All of these areas can be expected to expand, and new areas to open as laser technology is improved and more of the nuclear science community turns to this field of research.

ACKNOWLEDGMENTS

This work was supported in part by the US Department of Energy.

Literature Cited

Abragam, A. 1961. *The Principles of Nuclear Magnetism*. Oxford: Clarendon
Alexandrov, E. G. 1964. *Opt. Spectrosc.* 17: 957
Anton, K. R. et al. 1978a. *Phys. Rev. Lett.* 40: 642
Anton, K. R., Kaufman, S. L., Klempt, W., Neugart, R., Otten, E. W. Schinzler, B. 1978b. Narrowed optical lines observed in laser method for use with accelerated beams. In *Hyperfine Interactions IV*, ed.
R. S. Raghavan, D. E. Murnick, pp. 87–90. Amsterdam: North-Holland
Arecchi, F. T., Schulz-Dubois, E. O. 1972. *Laser Handbook*, Vol. 1. Amsterdam: North Holland
Back, E., Landé, A. 1925. *Zeeman-Effekt und Multiplettstruktur der Spektrallinien*. Berlin: Springer
Baird, P. E., Brimicombe, M. W. S. N., Hunt, R. G., Roberts, G. V., Sandars, P. G. M., Dacy, D. V. 1977. *Phys. Rev.*

Lett. 39:798–801
Baldwin, G. C., Khoklov, R. V. 1975. *Phys. Today (Feb.)*, pp. 33–39
Bashkin, S., ed. 1975. *Topics in Modern Physics, Vol. 1: Beam Foil Spectroscopy.* Berlin: Springer
Bekov, G. I., Letokhov, V. S., Mishin, V. I. 1978. *JETP Lett.* 27:47–51
Bekov, L. M., Zolotorev, M. S. 1978. *JETP Lett.* 26:379
Bertin, A. et al. 1975. *Phys. Lett. B.* 55:411–14
Bernheim, R. 1965. *Optical Pumping.* New York: Benjamin
Bloembergen, N., Levenson, M. D. 1976. In *High Resolution Laser Spectroscopy*, ed. K. Shimoda, p. 315. Berlin: Springer
Bohr, A., Weisskopf, V. F. 1950. *Phys. Rev.* 77:94
Bordé, C., Hall, J. L. 1973. *Phys. Rev. Lett.* 30:1101
Bouchiat, M. A., Bouchiat, C. 1974. *J. Phys.* 35:899–926
Brix, P., Kopfermann, H. 1949. *Z. Phys.* 126:347
Broadhurst, J. H. et al. 1974. *J. Phys. B* 7:L5131
Burns, M., Pappas, P., Feld, M. S., Murnick, D. E. 1978. See Anton et al 1978b, p. 50
Burns, M., Pappas, P., Feld, M. S., Murnick, D. E. 1977. *Nucl. Instrum. Methods* 141:429–32
Capeller, V., Mazurkewitz, W. 1973. *J. Magn. Reson.* 10:15–21
Casimir, H. 1936. *Teylors Tweede Genootshap* 11:255
Chen, F. F. 1979. *Phys. Today* 32(5):36
Childs, W. J., Poulsen, O., Goodman, L. S. 1979a,b,c. *Opt. Lett.* 4:35; *Phys. Rev. A* 19:160–67; *Opt. Lett.* 4:63–65
Clark, D. L., Cage, M. E., Greenlees, G. W. 1978. *Hyperfine Interactions* 4:83–86
Clark, D. L., Cage, M. E., Lewis, D. A., Greenlees, G. W. 1979. *Phys. Rev. A.* In press
Conti, R., Bucksbaum, P., Chu, S., Commins, E., Hunter, L. 1979. *Phys. Rev. Lett.* 42:343–45
Denne, W. A. 1978. *Acta Crystallogr. A* 34:1028–29
Dodd, J. N., Kane, R. D., Warrington, D. M. 1964. *Proc. Phys. Soc.* 84:176
Ducas, T. W., Feld, M. S., Ryan, L. W. Jr., Skribanowitz, N., Javan, A. 1972. *Phys. Rev. A* 5:1036–43
Ehrenfest, P., Oppenheimer, J. R. 1931. *Phys. Rev.* 37:333
Fabry, C., Perot, A. 1897. *Ann. Chem. Phys.* (7)12:459
Fairbank, W. M. Jr., Hänsch, T. W., Schawlow, A. L. 1975. *J. Opt. Soc. Am.* 65:199
Feld, M. S. 1973. In *Fundamental and*

Applied Laser Physics, ed. M. S. Feld, A. Javan, N. A. Kurnit, p. 369. New York: Wiley-Interscience
Feld, M. S., Sanchez, A., Javan, A., Feldman, B. J. 1973. *Proc. Aussois Conf. High-Resolution Mol. Spectrosc.* Colloq. Int. CNRS, p. 87; and erratum in Ref. 25 of B. J. Feldman, M. S. Feld. *Phys. Rev. A* 12:1013
Feld, M. S., Murnick, D. E. 1979. See Liberman et al 1979
Franken, P. 1961. *Phys. Rev.* 121:508
Freed, C., Javan, A. 1970. *Appl. Phys. Lett.* 17:53
Gerhardt, H., Wenz, R., Matthias, E. 1977. *Phys. Lett. A* 61:377–79
Gerhardt, H., Matthias, E., Schneider, F., Timmermann, A. 1978. *Z. Phys. A* 288:327–33
Hänsch, T. W., Shahin, I. S., Schawlow, A. 1971. *Phys. Rev. Lett.* 27:707
Hänsch, T. W., Schawlow, A. L., Series, G. W. 1979. *Sci. Am.* 240(3):94
Heilig, K., Steudel, A. 1974. *At. Data Nucl. Data Tables* 14:639–54
Herzberg, G. 1950. *Molecular Spectra and Molecular Structure*, Vols. 1 and 2. New York: Van Nostrand
Hinkley, E. D., Nill, K. W., Blum, F. A. 1976. In *Laser Spectroscopy of Atoms and Molecules*, ed. H. Walther, p. 125. Berlin: Springer
Huber, G. 1977. In *Laser Spectroscopy Applied to Hyperfine Structure Problems in Atomic Physics in Nuclear Experiments*, ed. B. Rosner, R. Kalish, pp. 335–72. Bristol and New York: Hilger and AIP
Huber, G. et al. 1978a. *Phys. Rev. C* 18:2342–54
Huber, G. et al. 1978b. *Phys. Rev. Lett.* 41:459–60
Hurst, G. S., Payne, M. G., Nayfeh, M. H., Judish, J. P., Wagner, E. B. 1975. *Phys. Rev. Lett.* 35:82
Hurst, G. S., Kramer, S. D., Payne, M. G., Young, J. P. 1979. *IEEE Nucl. Trans.* In press
Jacquinot, P., Klapisch, R. 1979. *Rep. Prog. Phys.* In press
Kastler, A., Brossel, J. 1949. *Compt. Rend.* 229:1213
Kaufman, S. L. 1976. *Opt. Commun.* 17:309
Kaufman, S. L., Greenlees, G. W., Lewis, D. A., Tonn, J. F., Broadhurst, J. H., Clark, D. L. 1978. *Hyperfine Interactions* 4:921–25
Kluge, H. J. 1979. In *Progress in Atomic Spectroscopy*, ed. W. Hanle, H. Kleinpopen. New York: Plenum. In press
Kompanetz, O. N., Kukudzhanov, A. R., Letokhov, V. S., Minogin, V. G., Hikhailov, E. L. 1976. *Sov. Phys. JETP* 42:15

Kopfermann, H. 1958. *Nuclear Moments.* New York: Academic
Kramer, S. D., Bemis, C. E. Jr., Young, J. P., Hurst, G. S. 1978. *Opt. Lett.* 3: 16–18
Kugel, H. W., Murnick, D. E. 1977. *Rep. Prog. Phys.* 40: 297–343
Kühl, T., Dabkiewicz, P., Duke, C., Fischer, H., Kluge, H.-J., Krammling, H., Otten, E. W. 1977. *Phys. Rev. Lett.* 39: 180–83
Lamb, W. E. Jr. 1964. *Phys. Rev. A* 134: 1429
Larson, D. J. 1978. See Anton et al 1978b, pp. 73–78
Letokhov, V. S. 1973a. *Opt. Commun.* 7: 59
Letokhov, V. S. 1973b. *Sov. Phys. JETP* 37: 787–93
Letokhov, V. S. 1975. *Phys. Rev. A* 12: 1954–1965
Letokhov, V. S., Moore, B. 1976. *Sov. J. Quantum Electron.* pp. 129–50 and pp. 259–79
Letokhov, V. S. 1977. In *Frontiers in Laser Spectroscopy*, ed. R. Balian et al, pp. 717–70. Amsterdam: North Holland
Letokhov, V. S., Chebotayev, V. P. 1977. *Nonlinear Laser Spectroscopy.* Berlin: Springer
Letokhov, V. S. 1978. *Opt. Laser Technol.* 10: 175–83
Letokhov, V. S., Minogin, V. G. 1978. *Phys. Rev. Lett.* 41: 775–77
Letokhov, V. S. 1979. *Comments At. Mol. Phys.* 8: 103–12
Lewis, R. R., Williams, W. L. 1975. *Phys. Lett. B* 59: 70–72
Lewis, L. L., Hollister, J. H., Soreide, D. C., Lindahl, E. G., Fortson, E. N. 1977. *Phys. Rev. Lett.* 39: 795–98
Liberman, S. et al. 1979. In *Laser Spectroscopy IV*, ed. H. Walther, K. W. Rothe. Berlin: Springer. In press
Longhurst, R. S. 1967. *Geometrical and Physical Optics.* London: Longman
MacGillivray, J. C., Feld, M. S. 1977. *Appl. Phys. Lett.* 31: 74–76
Marcum, S. D. 1978. *Laser Focus* 14(10): 12–24
Meyer, T. W., Brilando, J. F., Rhodes, C. K. 1973. *Chem. Phys. Lett.* 18: 382
Michelson, A. A. 1891. *Philos. Mag.* 31: 338
Mitchell, A. C. G., Zemansky, M. W. 1971. *Resonance Radiation and Excited Atoms* London: Cambridge Univ. Press
Moskowitz, P. A., Lombardi, M. 1973. *Phys. Lett. B* 46: 334–36
Murnick, D. E., Patel, C. K. N., Leventhal, M., Wood, O. R. II, Kugel, H. W. 1979a. *J. Phys.* 40: 34–37
Murnick, D. E., Gibbs, H. M., Wood, O. R. II, Burns, M., Pappas, P., Kühl,

T., Feld, M. S. 1979b. *Bull. Am. Phys. Soc.* 24: 43
Murnick, D. E., Feld, M. S., Burns, M. M., Kühl, T. U., Pappas, P. G. 1979c. See Liberman et al 1979
Neuhauser, W., Hohenstatt, M., Toschek, P., Dehmelt, M. 1978. *Phys. Rev. Lett.* 41: 233
Nowicki, G., Bekk, K., Gering, S., Hanser, M., Rebel, H., Schatz, G. 1977. *Phys. Rev. Lett.* 39: 332–34; 1978. *Phys. Rev. C* 18: 2369–79
Pappas, P. G., Burns, M. M., Hinshelwood. D. O., Feld, M. S., Murnick, D. E. 1979. *Phys. Rev. A.* In press
Pauli, W. 1924. *Naturwissenschaften* 12: 741
Penselin, S. 1979. See Liberman et al 1979
Pescht, R., Gerhardt, H., Matthias, E. 1978. *Z. Phys. A* 281: 199–204
Prescott, C. Y. et al. 1978. *Phys. Lett. B* 77: 347–52
Rosenthal, J. E., Breit, G. 1932. *Phys. Rev.* 41: 459
Salam, A. 1968. In *Elementary Particle Theory*, ed. N. Svarthaom. Stockholm: Tomquist & Ferlag
Schinzler, B., Klempt, W., Kaufman, S. L., Lochmann, H., Moruzzi, G., Neugart, R., Otten, E.-W., Bonn, J., von Reisky, L., Spath, K. P. C., Steinacher, J., Weskott, D. 1978. *Phys. Lett. B*, pp. 209–12
Schüler, M., Schmidt, T. 1935. *Z. Phys.* 94: 457
Schuler, C. J. 1968. *Progress in Nuclear Energy, Ser. 9, Analytical Chemistry*, Vol. 8, Part 2. Oxford: Pergamon Press
Stroke, H. H., Blin-Stoyle, R. J., Jaccarino, V. 1961. *Phys. Rev.* 123: 1326–48
Szöke, A., Javan, A. 1963. *Phys. Rev. Lett.* 10: 521
Townes, C. H., Schawlow, A. L. 1955. *Microwave Spectroscopy*, New York: McGraw-Hill
Vasilenko, L. S., Chebotayev, V. P., Shishaev, A. V. 1970. *JETP Lett.* 12: 113
Wali-Mohammed, Ch. 1914. *Astrophys. J.* 39: 185
Walther, H., ed. 1976. *Laser Spectroscopy of Atoms and Molecules*, Ch. 1. Berlin: Springer
Weber, A., Porto, S. P. S., Cheesman, L. E., Barrett, J. J. 1967. *J. Opt. Soc. Am.* 57: 19
Weinberg, S. 1967. *Phys. Rev. Lett.* 19: 1264
Wharton, C. B. 1979. *Phys. Today* 32(5): 52
Wieman, C., Hänsch, T. W. 1976. *Phys. Rev. Lett.* 36: 1170
Wing, W. H., Ruff, G. A., Lamb, W. E. Jr., Spezeski, J. J. 1976. *Phys. Rev. Lett.* 41: 1488–91

AUTHOR INDEX

CUMULATIVE INDEXES

CONTRIBUTING AUTHORS VOLUMES 20–29

CHAPTER TITLES, VOLUMES 20–29

472

PARTICLE INTERACTIONS AT HIGH ENERGIES

PARTICLE SPECTROSCOPY

THE GROWTH OF
WHITE-COLLAR
UNIONISM

BY

GEORGE SAYERS BAIN

Research Fellow in Industrial Relations
Nuffield College, Oxford

OXFORD
AT THE CLARENDON PRESS
1970

Oxford University Press, Ely House, London W. 1

GLASGOW NEW YORK TORONTO MELBOURNE WELLINGTON
CAPE TOWN SALISBURY IBADAN NAIROBI DAR ES SALAAM LUSAKA ADDIS ABABA
BOMBAY CALCUTTA MADRAS KARACHI LAHORE DACCA
KUALA LUMPUR SINGAPORE HONG KONG TOKYO

PRINTED IN GREAT BRITAIN

FOR MY
MOTHER AND FATHER

PREFACE

THE purpose of this book is to isolate the major factors which determine the growth of white-collar unionism. It attempts to do this primarily by accounting for the variations in the industrial and occupational pattern of white-collar unionism in Great Britain. This membership pattern was constructed during 1964, the year in which the research for this project began, and generally indicates the position as of the beginning of that year. The unions' records were not sufficiently detailed to allow a systematic pattern to be constructed for previous years. Since the object is to explain the 1964 pattern, the situation is generally described as it existed in that and earlier years. Major changes which have occurred since then are generally indicated in footnotes. But no attempt has been made to provide a completely up-to-date account. Nor has any attempt been made to give a detailed history of the growth of white-collar unionism in Britain. In general, description has been provided not for its own sake, but only in so far as it contributes to the analysis.

Anyone who undertakes a study of this nature accumulates a large number of debts. These become particularly numerous if, as in this case, the institutions which the investigator is examining are not those of his own country. I am particularly indebted to the officials of all the trade unions, trade union federations, companies, employer associations, and government departments who granted interviews, completed questionnaires, answered letters, gave access to documents in their possession, and provided unpublished data. Some of these officials were also kind enough to read sections of the manuscript and comment on them. Without their assistance this study would not have been possible. In view of the numbers involved, I hope they will excuse me if I thank them here without mentioning them by name. I also hope that those who disagree with my conclusions will not feel that in arriving at them I have failed to take their views into account.

By electing me first to a Research Studentship and later to a Research Fellowship, the Warden and Fellows of Nuffield College made it possible for me to devote my time to research, and for this I am most grateful. The Research Fellowship was partially financed out of a grant from the Leverhulme Trustees to the College for research into industrial relations, and I should like to record my thanks to them. My thanks are also due to two of the College's secretaries, Jenny Bond and Lyn Yates, who typed various drafts of the manuscript with great efficiency and unfailing good humour.

A considerable amount of this study's argument is supported by statistical data and analysis. I would like to thank Beryl Cuthbertson who helped to collect much of the basic data, Ann Black who patiently and efficiently steered it through the computer, and Emiel van Broekhoven who designed the regression programme which has proved so useful to myself and other social scientists in Oxford. My special thanks go to Bob Bacon who convinced me of the contribution which statistical techniques could make to this study, instructed me in their use, and guided me in interpreting the results.

Parts of Chapters II and III were originally published in the *British Journal of Industrial Relations*, iv (November 1966), pp. 304–35, and I am obliged to the editor of this journal for permission to reproduce this material. This same material as well as parts of Chapters VIII and IX appeared in *Trade Union Growth and Recognition*, a research paper I prepared for, and at the request of, the Royal Commission on Trade Unions and Employers' Associations; this material is reproduced here with the permission of the Controller of Her Majesty's Stationery Office. I am particularly grateful to the Royal Commission for giving me the opportunity to undertake the research upon which Chapter VIII is based, and to its secretariat and staff for all the help they gave me in carrying out this research.

I also benefited greatly from the comments of those who read all or part of the manuscript. They include David Coates, Robert Currie, Valerie Ellis, Alan Fox, Arthur Gillman, Jackie Johns, Archie Kleingartner, Donald Robertson, Inga Taylor (who also helped to correct the proofs), and Alex Wedderburn. They helped to correct numerous errors of fact and interpretation and to remove many infelicities of style. Those that remain are a result of my obstinacy rather than their lack of critical judgement.

My greatest debt is to my teachers. Clare Pentland, my first teacher of labour economics and industrial relations, inspired my interest in the theory of union growth. Hugh Clegg and Allan Flanders taught me most of what I know about industrial relations in Britain, provided much helpful guidance especially during the early stages of the study, suffered through several drafts of the manuscript, and made helpful suggestions on almost every page. To these three individuals, I owe far more than this brief acknowledgement can suggest.

Finally, a special note of gratitude is due to my wife not only for her sustained encouragement throughout my long preoccupation with this study but also for the many improvements she suggested.

The help of all these people was invaluable and greatly improved the finished product. For any shortcomings which may remain in spite of their efforts and for all the opinions expressed, I alone am responsible.

GEORGE BAIN

Nuffield College, Oxford
December 1968

CONTENTS

LIST OF FIGURES

LIST OF TABLES

KEY TO ABBREVIATIONS

ABT   Association of Building Technicians.

ACTT   Association of Cinematograph, Television and Allied Technicians.

AEU   Amalgamated Engineering Union. This union merged during 1967 with the Amalgamated Union of Foundry Workers to form the Amalgamated Union of Engineering and Foundry Workers (AEF). The designation AEU is used in this study as the merger did not take place until after the end of the period under review.

AScW   Association of Scientific Workers. This union merged during 1968 with the Association of Supervisory Staffs, Executives and Technicians to form Division A of the Association of Scientific, Technical and Managerial Staffs. The designation AScW is used in this study as the merger did not take place until after the end of the period under review.

ASSET   Association of Supervisory Staffs, Executives and Technicians. This union merged during 1968 with the Association of Scientific Workers to form Division I of the Association of Scientific, Technical and Managerial Staffs. The designation ASSET is used in this study as the merger did not take place until after the end of the period under review.

ASTMS   Association of Scientific, Technical and Managerial Staffs. See also AScW and ASSET.

BEC   British Employers' Confederation. This organization merged in 1965 with the Federation of British Industry and the National Association of British Manufacturers to form the Confederation of British Industry.

BISAKTA   British Iron, Steel and Kindred Trades Association. This union is also referred to as the Iron and Steel Trades Confederation.

BISMA   British Iron and Steel Management Association.

CAWU   Clerical and Administrative Workers' Union.

CBI   Confederation of British Industry.

CIR   Commission on Industrial Relations.

CSEU   Confederation of Shipbuilding and Engineering Unions.

DATA   Draughtsmen's and Allied Technicians' Association.

EEF   Engineering Employers' Federation.

FSMBS   Foremen and Staff Mutual Benefit Society.

IDT   Industrial Disputes Tribunal.

IOJ   Institute of Journalists.

ISTC   Iron and Steel Trades Confederation. This union is also referred to as the British Iron, Steel and Kindred Trades Association.

JIC   Joint Industrial Council.

NACSS   National Association of Clerical and Supervisory Staffs. This is the
            white-collar section of the Transport and General Workers' Union.
NATSOPA   National Society of Operative Printers and Assistants. This union
            merged during 1966 with the National Union of Printing, Book-
            binding and Paper Workers to form Division I of the Society of
            Graphical and Allied Trades.
NEJTM   National Engineering Joint Trades Movement.
NFPW   National Federation of Professional Workers.
NPA   Newspaper Proprietors Association.
NUBE   National Union of Bank Employees.
NUGMW   National Union of General and Municipal Workers. In practice, this
            union is sometimes referred to as the General and Municipal Workers'
            Union (GMWU).
NUJ   National Union of Journalists.
NUPBPW   National Union of Printing Bookbinding and Paper Workers.
            This union merged during 1966 with the National Society of Opera-
            tive Printers and Assistants to form Division A of the Society of
            Graphical and Allied Trades.
PKTF   Printing and Kindred Trades Federation.
SOGAT   Society of Graphical and Allied Trades. See also NATSOPA and
            NUPBPW.
TGWU   Transport and General Workers' Union.
TUC   Trades Union Congress.
USDAW   Union of Shop, Distributive and Allied Workers.

Great Britain—refers to England, Wales, and Scotland.

United Kingdom—refers to England, Wales, Scotland, and Northern Ireland.

All references to *Written Evidence* or *Minutes of Evidence* is to that given to the
Royal Commission on Trade Unions and Employers' Associations under the
chairmanship of Lord Donovan.

I

INTRODUCTION

THIS study attempts to discover the major factors which promote or hinder the growth of trade unionism among white-collar workers, particularly those employed in manufacturing industries, in Great Britain. There are two reasons for wishing to do this.

The first reason is practical. While the number of manual workers in Britain is declining, the number of white-collar workers is increasing so rapidly that they will soon be in the majority. If the trade union movement is to maintain its relative position in the power structure of this country and to continue to play an effective role in the industrial relations system, it will have to recruit these white-collar workers. If it does not or cannot, the best it will achieve is numerical stability within an increasingly narrow band of the occupational distribution, and its ability to advance even the interests of its manual membership will be seriously impaired. From the point of view of those who believe that an effective trade union movement is essential to the successful functioning of a democratic society, it is obviously important to isolate the factors which promote or hinder its growth among white-collar workers.

The second reason is theoretical. There is a fairly extensive literature on the theory of union growth.[1] Unfortunately, very little of it is useful in giving insights into the process of union growth among white-collar workers. Many of the theories are concerned solely with manual unionism. More important, many of the theories are not oriented to empirical research and are couched in such a way as to preclude any chance of verification. They are more in the tradition of social philosophy than social science and have as their objective the 'interpretation' of the labour movement. They are, as one writer has aptly noted, a 'heritage of curiously fascinating and conflicting admixture of restrained or explosive polemic, implicit or patent advocacy, muddied metaphor, mild expressions of faith, fiery depositions of dogma, and occasional

---

[1] It is not possible or necessary to review here the large literature, most of which is American in origin, on the theory of union growth. This has been done by John T. Dunlop, 'The Development of Labor Organization: A Theoretical Framework', *Insights Into Labor Issues*, Richard A. Lester and Joseph Shister, editors (New York: Macmillan, 1949), pp. 163–93, and Mark Perlman, *Labor Union Theories in America* (Evanston: Row, Peterson & Co., 1958). See also H. B. Davis, 'The Theory of Union Growth', *Quarterly Journal of Economics*, lv (August 1941), pp. 611–37; J. Shister, 'The Logic of Union Growth', *Journal of Political Economy*, lxi (October 1953), pp. 413–33; and Irving Bernstein, 'The Growth of American Unions', *American Economic Review*, xliv (June 1954), pp. 301–18. This is by no means a complete list.

flashes of brilliant insight, which make all the rest so much easier to take'.[1]
By attempting to provide generalizations about the growth of white-collar
unionism which are basically operational in nature, this study departs from
the traditions of much of the literature in this field. But by so doing, it hopes
to be able to make some contribution to both the substance and the method of
the theory of union growth.

The practical and the theoretical also explain why manufacturing has been
singled out for special attention in this study. On the one hand, manufacturing industries are the major 'commanding height' of the economy and offer
the trade union movement its largest untapped potential of white-collar
members. On the other hand, statistical data are more plentiful for this sector
of the economy than for others, and this makes it easier to test generalizations
about union growth.

SOME METHODOLOGICAL CONSIDERATIONS

Those who have written on the subject of union growth have often defined it
very broadly. Besides trying to explain the growth of union membership, they
have also tried to explain the development of union structure as well as what
has variously been referred to as union personality or character.[2] The present
study takes a much more restricted view of union growth and confines itself
to trying to explain membership growth only.[3] This is not because these other
aspects of trade unions are uninteresting or irrelevant to an understanding of
their behaviour. Rather, it is because there is no good reason for assuming
that the factors which explain the growth of union membership necessarily
also explain the development of union structure, personality, or character.
Granted this, these aspects of trade unionism are too complex to be explained
adequately within the confines of a study that focuses on the growth of union
membership, and too important to be explained inadequately.[4]

[1] Abraham J. Siegel, 'Method and Substance in Theorizing About Worker Protest',
Aspects of Labor Economics (A Report of the National Bureau of Economic Research, New
York, and published by Princeton University Press, Princeton, 1962), p. 27.

[2] See, for example, Irving Bernstein, 'Union Growth and Structural Cycles', *Proceedings
of the Industrial Relations Research Association*, vii (December 1954), pp. 202–46; G. W.
Brooks, 'Reflections on the Changing Character of American Labor Unions', *Proceedings
of the Industrial Relations Research Association*, ix (December 1956), pp. 33–43; Richard A.
Lester, *As Unions Mature: An Analysis of the Evolution of American Unionism* (Princeton:
Princeton University Press, 1958); Arthur B. Shostak, *America's Forgotten Labor Organiza-
tion* (Princeton: Industrial Relations Section, Princeton University, 1962); David Lock-
wood, *The Blackcoated Worker* (London: Allen & Unwin, 1958), chap. 5; H. A. Turner,
Trade Union Growth, Structure and Policy (London: Allen & Unwin, 1962); and R. M.
Blackburn, *Union Character and Social Class* (London: Batsford, 1967).

[3] Although this study does not try to explain the growth and development of union
structure, personality, or character, it does, of course, consider the impact which these
factors may have upon union growth.

[4] Blackburn (op. cit., p. 10) has argued that it is not useful or valid to consider the
membership of a union separately from its character. Whether or not it is useful, the reader
can judge for himself after reading this study. Whether or not it is valid, there is not room

As used in this study, union growth is a quantitative concept; it is something which can be measured. There are primarily two ways in which this can be done: in terms either of *actual* union membership or of the *density* of union membership. The density of union membership is given by the following formula:

$$\frac{\text{Actual Union Membership}}{\text{Potential Union Membership}} \times 100$$

This study uses this latter measure of union growth on the grounds of methodological expediency.[1] By using the density of union membership rather than actual union membership, one of the most obvious causes of union growth, changes in potential union membership, can be excluded from the subsequent inquiry into the factors which influence the growth of white-collar unionism.

There are basically two types of data on the density of aggregate union membership which might be studied in making inferences about union growth: time series data and cross-section data. Time series data consist of observations on the density of union membership at different points in time while cross-section data consist of observations on the density of union membership at a single point in time but over different occupations, firms, industries, or geographical regions. Ideally, both types of data should be examined. This study does this to some extent, but it relies mainly upon cross-section data—the occupational and industrial pattern of white-collar unionism in Britain.[2] This is simply because of the difficulty, if not the impossibility, of constructing a time series containing a significant number of observations on actual white-collar union membership, not to speak of the density of white-collar union membership, in manufacturing industries or even the economy as a whole. It is a monumental task, as Chapter III will reveal, to obtain such figures even for selected post-war years.

The occupational and industrial pattern of white-collar unionism in Britain is the dependent variable of this study. In attempting to isolate the explanatory variables which account for this pattern of membership, it is not possible to carry out controlled experiments such as those used in the physical sciences whereby each factor is varied in turn while all the others are kept constant.

to debate here. But it is the author's contention that Blackburn's view is mistaken, and this contention will be supported by argument in a study entitled *Social Stratification and White-Collar Unionism* which the author, together with David Coates and Valerie Ellis, is preparing for publication in 1970.

[1] The density of union membership has also been referred to as the percentage organized, real membership, degree of unionization, and completeness. These terms are used synonymously throughout this study, but the density of unionization is the term which is generally used as it is the accepted British terminology. Blackburn (op. cit., p. 16 n. 1) notwithstanding, density is not 'the incorrect term'.

[2] Figures are also provided on the geographical pattern of white-collar unionism, but not much attention is devoted to it in the following analysis partly because of the unsatisfactory nature of the data and partly because there is not sufficient quantitative or even qualitative information available on a geographical basis to help account for this pattern.

In one or two places in the following analysis, it is possible by means of quantitative techniques to study simultaneously the effects of a few important variables. But generally, the examination of variables is step-wise. That is, the effect of each factor on the dependent variable is considered separately with only the most general allowances being made for the fact that 'all other things' may not be equal or constant. There is consequently a danger that other variables which affect the growth of white-collar unionism may obscure or distort the influence of the factor under consideration. But in order to make any progress at all, this study, like most social science investigations, has had to adopt this procedure.

SOME CONCEPTUAL CONSIDERATIONS

Before a pattern of white-collar union membership can be constructed, two concepts—a 'white-collar employee'[1] and a 'trade union'—have to be defined.

However the term 'white-collar employee' is defined, it is unlikely to be entirely satisfactory as there are almost bound to be some borderline cases.[2] Nevertheless, this study suggests that it is useful to define 'white-collar employee' to cover the members of the following occupational groups: foremen, overlookers, and supervisors; scientists, technologists, and technicians; clerical and administrative workers; security personnel; professionals; salesmen, commercial travellers, and shop assistants; government administrators and executive officials; and specially 'creative' occupations such as artists, musicians, and entertainers.

In a study on industrial relations, the justification for treating these various occupational groups as a collectivity is that this reflects general industrial practice. In industry, as Burns has noted, 'the line between the management and the worker, the widest social barrier, is so drawn that it includes in management all but the rank-and-file workmen'.[3] The members of the occupational groups listed above generally see themselves as belonging more with management than with manual workers, and are generally regarded by manual workers as one of 'them' rather than one of 'us' and by employer-managers as part of the 'staff' rather than part of the 'works'. The definition of 'white-collar employee' used in this study is significant and useful precisely because it is thought to be so by industry itself.

There are a whole complex of factors which explain why the line between manual and non-manual workers should be drawn in this way. But one of

---

[1] In this study the terms 'white-collar employee', 'non-manual worker', 'blackcoated worker', 'salaried employee', and 'staff worker' are used interchangeably.

[2] For a history and analysis of the major attempts to provide a logically consistent definition of a 'white-collar employee' see Fritz Croner, *Soziologie der Angestellten* (Cologne: Kiepenheuer & Witsch, 1962), chap. 2, and by the same author 'Salaried Employees in Modern Society', *International Labour Review*, lxix (February 1954), pp. 97–110.

[3] Tom Burns, 'The Cold Class War', *New Statesman* (7 April 1956), p. 331.

the most important, especially in private industry, is the relations of authority within the enterprise. Regardless of the white-collar workers' position in society or in the authority structure of the enterprise, they are linked with employer-managers 'by being associated with that part of the productive process where authority is exercised and decisions are taken'.[1] Even in non-industrial sectors of the economy such as the distributive trades and national and local government, the white-collar worker is the person who takes your money in shops and gives you orders in offices; in Lockwood's words, the person 'on the other side of the desk who is somehow associated with authority'.[2]

This study has followed the Ministry of Labour[3] and defined a 'trade union' to mean 'all organisations of employees—including those of salaried and professional workers, as well as those of manual wage earners—which are known to include in their objects that of negotiating with employers with a view to regulating the wages and working conditions of their members'.[4] The Ministry lists all the organizations which it feels are covered by this definition in its *Directory of Employers' Associations, Trade Unions, Joint Organizations, etc.*, and it is these bodies which have generally been taken as trade unions for the purpose of this study. This might be objected to on the grounds that it excludes certain organizations commonly referred to as 'professional associations' and 'staff associations' which have similar functions to trade unions. While this is true, it does not create serious difficulties given the scope and purpose of this study.

Professional associations which engage in collective bargaining are covered by the Ministry's definition, and, in general, they are listed in the *Directory*. There are certain exceptions, the most prominent of which is the British Medical Association. It is not listed even though it bargains with the Ministry of Health over the pay and conditions of general practitioners employed under the National Health Service.[5] Moreover, collective bargaining is not the only method of job regulation. This can also be accomplished by means of unilateral regulation, and some professional associations such as the Law Society use this method rather than collective bargaining.[6] But most professionals who rely upon collective bargaining are employed in the public sector

[1] W. G. Runciman, *Relative Deprivation and Social Justice* (London: Routledge & Kegan Paul, 1966), p. 47.

[2] Op. cit., p. 132.

[3] The Ministry of Labour is now known as the Department of Employment and Productivity, but its former designation is used here because the change did not occur until after the end of the period under review.

[4] 'Membership of Trade Unions in 1964', *Ministry of Labour Gazette*, lxxiii (November 1965), p. 480.

[5] See Harry Eckstein, *Pressure Group Politics: The Case of the British Medical Association* (London: Allen & Unwin, 1960).

[6] See Allan Flanders, *Industrial Relations: What Is Wrong With the System?* (London: Faber, 1965), chaps. 2 and 3 for a discussion of the various forms of job regulation.

of the economy, while most of those who depend upon unilateral regulation are concentrated among independent fee-paid practitioners in the service sector of the economy. Few are to be found in manufacturing, the sector of major concern to this study. In private industry, as Prandy has pointed out, the major functions of professional associations are education and certification as opposed to professional protection and collective bargaining.[1] Even granted the omission of some professional associations which should have been taken into account in computing density figures, they would not be of much quantitative significance. In manufacturing, professional associations are largely restricted to scientists and technologists, and this group accounts for only 4 per cent of this sector's total white-collar labour force.[2]

Few internal staff associations or staff committees are included in the *Directory*, regardless of whether or not they engage in collective bargaining, but this is of small consequence for this study. What evidence there is suggests that the vast majority of these organizations are sponsored, influenced, or dominated by employers.[3] They are therefore more indicative of the behaviour of employers than of employees, and it is with the latter that this study is primarily concerned. Moreover, many of these organizations do not have to recruit members; in companies in which they exist all employees often become members automatically. Hence the concept of membership density cannot be fruitfully applied to them. While there may be some staff associations which are both independent of employers and required to recruit members, these would seem to be few and far between,[4] and the omission of any which might exist would not significantly affect membership density figures. Although organizations of employees which are sponsored, influenced, or dominated by employers are not considered to be trade unions, they are a device which employers use to discourage the growth of trade unions among their white-collar employees. As such they are discussed in Chapter VIII.

The criterion of correctness cannot be applied to a definition; specific purposes require specific definitions. It is not suggested that the definitions of 'white-collar employee' and 'trade union' which are used in this study are universally valid or useful. Nor is it suggested that they cover every borderline case. All that is claimed on their behalf is that they are useful for analysing the growth of trade unionism among white-collar workers in manufacturing industries.

---

[1] Ken Prandy, 'Professional Organization in Great Britain', *Industrial Relations*, v (October 1965), pp. 67–79.

[2] See Table 2.7.

[3] See *infra*, pp. 132–3.

[4] The only major exception to this is in banking and insurance. Here the membership of staff associations has generally ceased to be automatic and they have become more or less independent of employers, if still favoured by them. They are listed in the *Directory* and were taken into account in computing density figures.

TECHNIQUES

Most of the usual techniques of social science are used in this study including the examination of documents and direct observation by means of
questionnaires and interviews. The details of the various documents and
surveys are given in the text, footnotes, and appendices, and there is no need
to elaborate further upon them here. But this study also uses quantitative
methods to analyse a relatively large amount of data. Given that union growth
is by its very nature a quantitative concept and that many of the factors which
might explain it are also quantitative in nature, this should not be surprising.
The use of quantitative data and methods is now common in most areas of
social science, but they are not all that frequently employed in industrial
relations and labour history. Some comment on their use in this study is
therefore called for.

There are two widely divergent views of statistics which are currently
popular. 'One view is that published statistics are themselves vested with some
quality of meaning not unlike the qualities ascribed to numbers by the
Pythagoreans, and that they enjoy such a degree of infallibility that they may
be accepted without question.' The other and yet more popular view is that
'statistics can be made to prove anything and therefore, by implication, that
in fact they can prove nothing'.[1] Neither of these views is subscribed to here.
The statistics used in this study should enable the reader to gain a better
understanding of the process of union growth. This does not mean that these
statistics are completely accurate. They are not. And if the generalizations
which are based upon them are to be properly evaluated, their shortcomings
must be fully understood. For this reason, considerable detail on the source,
method of calculation, and reliability of all the statistics used in this study is
provided in Appendix A, and, to a lesser extent, in the text.

This is not the place to become involved in a lengthy discussion of the
advantages and disadvantages of using quantitative methods in social science.
This question has been very fully discussed elsewhere,[2] and little that is new
could be added here. But it must be emphasized that the use of these methods
in this study is not an unnecessary complication. On the contrary, they have
given deeper insights into the effect of certain factors on union growth and

---

[1] W. J. Reichmann, *Use and Abuse of Statistics* (Harmondsworth, Middlesex: Penguin
Books, 1964), p. 11.

[2] Almost every textbook on the methodology of the social sciences discusses this question.
See, for example, Abraham Kaplan, *The Conduct of Inquiry* (San Francisco: Chandler
Publishing Co., 1964). Economic history is a field in which the application of quantitative
methods raises similar problems to those in industrial relations or labour history, and the
reader might find the following references particularly useful: Alfred H. Conrad and John
R. Meyer, 'Economic Theory, Statistical Inference, and Economic History', *Studies in
Econometric History* (London: Chapman & Hall, 1965), chap. 1, and Robert William
Fogel, *Railroads and American Economic Growth: Essays in Econometric History* (Baltimore: Johns Hopkins Press, 1964), pp. 237–49.

have permitted generalizations to be made with greater confidence than would have been possible by visual assessment alone.

What has been said so far regarding quantitative methods is not intended to give the impression that these are the only techniques used in this study. They are only employed where the factors under consideration readily lend themselves to quantification, and many of the factors examined in this study do not. Where this is the case, more qualitative methods of analysis are used. In fact, besides adding, subtracting, multiplying, and dividing, the only statistical technique used in the following pages is regression analysis. The way in which this type of analysis is used in this study is explained in Appendix B, but it might be helpful if a few more general comments are made here.

Regression analysis is simply a statistical method for investigating the relationship or degree of correlation between variables.[1] When this method is used to investigate the relationship between a dependent variable and a single independent or explanatory variable, it is referred to as a simple regression analysis. When it is used to investigate the relationship between a dependent variable and more than one independent variable, it is referred to as a multiple regression analysis.

The strength of the association between the variables is given by correlation coefficients which are calculated mathematically from the available data in the form of numbers. They are expressed in values ranging between -1 and $+1$. The nearer the value is to either of these extremes the stronger is the correlation between the variables. If the value is positive, then the correlation is direct; high values of one variable are associated with high values of the other. If the value of the coefficient is negative, then the correlation is inverse; high values of one variable are associated with low values of the other. The closer the coefficient's value is to zero, the less is the correlation between the variables. The 'significance' of the correlation coefficients generated by this study is judged at the usual level of 5 per cent. If a correlation coefficient or any other result is said to be significant at this level, this means that there is only one possibility in twenty of the result arising by chance.

If variables are not correlated and the data is reliable, then it is fairly safe to conclude that there is no causal connection between them. But if variables are correlated, even very strongly, it does not necessarily follow that there is a causal relationship between them. Correlations are sometimes observed between factors that could not conceivably be causally related such as soil erosion in Alaska and the amount of alcohol consumed in South America. To establish a causal relationship, it is not sufficient to show that variables are highly correlated. Something more is required: strong reasons for expecting them to be correlated. Although correlation must not be confused

---

[1] For an introduction to regression analysis see Reichmann, op. cit., chap. 10, and J. J. Moroney, *Facts From Figures* (Harmondsworth, Middlesex: Penguin Books, 1956), chap. 16.

with causal relation, the former often helps to provide the pointers to the latter.

Where regression analysis has been used in this study, the relationship between the variables is generally first investigated by means of more qualitative methods. In addition, an attempt has been made to translate the significance of the results into English which is free of statistical jargon. But in one or two places this has not proved possible, and the author can only apologize in advance to the less statistically inclined reader.

THE ANALYTICAL FRAMEWORK

Every scientific investigation has some analytical framework to guide the selection of relevant data from the infinite mass of available material and to help organize this data in a meaningful and coherent form once it has been gathered. The analytical framework used in this study is reflected by the organization of the chapters.

Chapter II gives an account of the extent and nature of the white-collar labour force. Chapter III describes the aggregate pattern of white-collar union membership which exists in Britain.

All the remaining chapters analyse the factors which this study considers worthy of examination in searching for an explanation of this pattern. Chapter IV considers such socio-demographic characteristics of white-collar workers as their sex, social origins, age, and status in the community. Chapter V examines the white-collar workers' economic position, and Chapter VI analyses their work situation. The role which trade unions and employers play in union growth is explored in Chapters VII and VIII respectively while the influence of the government and the social climate is investigated in Chapter IX. Chapter X draws the various parts of the analysis together, and tries to produce a few generalizations regarding the growth of white-collar unionism.

The analytical framework used in this study does not provide a separate category for the attitudes of white-collar workers towards trade unionism, and no systematic attempt has been made to ascertain what these attitudes are by means of questionnaires or interviews. This is not an oversight; it is deliberate. If this study were concerned with discovering why one worker in a given environment joins a trade union while another worker in the same environment does not, then the explanation would have to take these attitudes into account as an independent variable. In such an analysis all the environmental influences surrounding individual workers would be held constant, and it would be safe to assume that any differences in their attitudes were mainly attributable to differences in their personality structures. But this study is concerned with explaining the behaviour of groups rather than of individuals. It is not interested in explaining why the propensity to unionize varies from one individual to another, but why it varies from one occupational group or industry to another. In such an analysis it is safe to assume that the attitudes of workers towards trade unionism are not an independent variable, but are primarily dependent upon one or more of the factors analysed in Chapters IV to IX.

II

THE PATTERN OF WHITE-COLLAR EMPLOYMENT

THE ECONOMY AS A WHOLE

THE growth of the white-collar labour force is one of the most outstanding characteristics of the economic and social development of the twentieth century. This growth is both absolute and relative; not only is the total number of white-collar workers increasing, but so also is the proportion of these workers in the labour force as a whole.

While the growth of the white-collar labour force has not been as large in Britain as in some other countries,[1] it has nevertheless been significant as Table 2.1 shows.[2] Between 1911 and 1961 the number of white-collar workers increased by 147 per cent, while the number of manual workers increased by only 2 per cent having actually decreased since 1931. The disparate growth of these two groups is reflected in the increasing relative importance of the white-collar occupations. The white-collar section of the labour force increased from 18·7 per cent to 35·9 per cent of the total between 1911 and 1961 while the manual share decreased from 74·6 per cent to 59·3 per cent. During this same period the remaining section of the labour force, the employers and proprietors, showed a slight tendency to decline, this decline being balanced to some extent by an increase in the number of managers and administrators.[3]

Occupational composition[4]

Although the white-collar labour force as a whole has increased enormously, there are significant differences in the growth of its constituent occupational

[1] See Guy Routh, *Occupation and Pay in Great Britain* (Cambridge: Cambridge University Press, 1965), p. 12, table 3 for international comparisons.

[2] Throughout this study comments on the source, method of calculation, and reliability of all statistical tables and graphs appear in Appendix A.

[3] This decline in the employer and proprietor group should be interpreted with caution. Although there is a legal distinction between an employer and a manager, in social science the dividing line is more imaginary than real, for an employer becomes a manager as soon as his business is incorporated. The trend towards the incorporation of business enterprises is at least part of the explanation for the decline in employers and proprietors and the increase in managers and administrators. On this point see Routh, op. cit., pp. 19–21.

[4] For a much more detailed occupational breakdown of the white-collar labour force from 1881 to 1931 see A. L. Bowley, 'Notes on the Increase in Middle Class Occupations', *Wages and Income in the United Kingdom Since 1860* (Cambridge: Cambridge University Press, 1937), Appendix E. For a Marxist analysis of white-collar labour force trends during 1851–1931 see F. D. Klingender, *The Condition of Clerical Labour in Britain* (London: Martin Lawrence, 1935), pp. xi–xxii.

TABLE 2.1

The Occupied Population of Great Britain by Major Occupational Groups, 1911–61

Occupational groups	Number of persons in major occupational groups, 1911–61 (000s)					Major occupational groups as a percentage of total occupied population, 1911–61					Growth indices of major occupational groups, 1911–61 (1911 = 100)				
	1911	1921	1931	1951	1961	1911	1921	1931	1951	1961	1911	1921	1931	1951	1961
1. Employers and proprietors	1,232	1,318	1,407	1,117	1,139	6·7	6·8	6·7	5·0	4·7	100	107	114	91	92
2. All white-collar workers	3,433	4,094	4,841	6,948	8,480	18·7	21·2	23·0	30·9	35·9	100	119	141	202	247
(a) Managers and administrators	631	704	770	1,245	1,268	3·4	3·6	3·7	5·5	5·4	100	112	122	197	201
(b) Higher professionals	184	196	240	435	718	1·0	1·0	1·1	1·9	3·0	100	107	130	236	390
(c) Lower professionals and technicians	560	679	728	1,059	1,418	3·1	3·5	3·5	4·7	6·0	100	121	130	189	253
(d) Foremen and inspectors	237	279	323	590	682	1·3	1·4	1·5	2·6	2·9	100	118	136	249	288
(e) Clerks	832	1,256	1,404	2,341	2,996	4·5	6·5	6·7	10·4	12·7	100	151	169	281	360
(f) Salesmen and shop assistants	989	980	1,376	1,278	1,398	5·4	5·1	6·5	5·7	5·9	100	99	139	129	141
3. All manual workers	13,685	13,920	14,776	14,450	14,020	74·6	72·0	70·3	64·2	59·3	100	102	108	106	102
4. Total occupied population	18,350	19,332	21,024	22,515	23,639	100·0	100·0	100·0	100·0	100·0	100	105	115	123	129

groups. Table 2.1 shows that the clerks have claimed most of the ground yielded by the manual workers. During the period under review, clerical occupations grew by 260 per cent and increased their share of the total labour force from 4·5 per cent to 12·7 per cent. The growth in the proportionate share of the other white-collar occupational groups has been more moderate: the share of shop assistants remained remarkably constant; that of foremen and inspectors increased from 1·3 per cent to 2·9 per cent; that of managers and administrators from 3·4 per cent to 5·4 per cent; that of lower professionals and technicians from 3·1 per cent to 6·0 per cent; and that of higher professionals from 1 per cent to 3 per cent.

The very broad occupational classification of Table 2.1 tends to obscure the extraordinary increase in the number of scientific and technical employees. Although the total number of such workers is relatively small, they are increasing more rapidly than any other component of the white-collar labour force (see Table 2.2). The Census of Population of 1921 was the first to consider draughtsmen and laboratory assistants sufficiently important groups to merit a separate classification. By 1961 the number of draughtsmen had increased by 376 per cent, professional scientists and engineers by 688 per cent, and laboratory technicians by 1,820 per cent.[1] If these high growth-rates continue, the occupational composition of the future white-collar labour force will be considerably changed.

TABLE 2.2

The Growth of Scientists and Engineers, Draughtsmen, and Laboratory Technicians in Great Britain, 1921–61

Year	Scientists and engineers		Draughtsmen		Laboratory technicians	
	Number	Growth indices	Number	Growth indices	Number	Growth indices
1921	48	100	38	100	5	100
1931	71	148	59	155	11	220
1951	187	390	130	342	69	1,380
1961	378	788	181	476	96	1,920

NOTE: All numbers are in thousands; for growth indices, 1921 = 100.

---

[1] For a more detailed discussion of the trends in scientific and technical manpower see the following publications of the British government: *Scientific Manpower*, Cmd. 6824, 1946; *Scientific and Engineering Manpower in Great Britain* (London: HMSO, 1956); *Scientific and Engineering Manpower in Great Britain 1959*, Cmnd. 902, 1959; *The Long-Term Demand for Scientific Manpower*, Cmnd. 1490, 1961; *Scientific and Technological Manpower in Great Britain 1963*, Cmnd. 2146, 1963; 'Survey of the Employment of Women Scientists and Engineers', *Ministry of Labour Gazette*, lxviii (September 1960), pp. 256–7; 'Survey of the Employment of Technicians in the Chemical and Engineering Industries', *Ministry of Labour Gazette*, lxviii (December 1960), pp. 464–7. See also Kenneth Prandy, *Professional Employees: A Study of Scientists and Engineers* (London: Faber, 1965), chap. 3.

Sex composition

The large and growing proportion of women in the white-collar labour force is one of its most noticeable characteristics as Table 2.3 demonstrates. Between 1911 and 1961 the proportion of women in white-collar jobs increased from 29·8 per cent to 44·5 per cent. Although there were relatively few women in the higher professions or the managerial and supervisory grades, by 1951 they formed a majority among the lower professionals,[1] shop assistants, and clerical workers. Table 2.3 indicates that the most significant substitution of women for men occurred among clerical grades during the First World War.

TABLE 2.3

The Percentage of Female Workers in Major Occupational Groups in Great Britain, 1911–61

Occupational group	1911	1921	1931	1951	1961
1. Employers and proprietors	18·8	20·5	19·8	20·0	20·4
2. All white-collar workers	29·8	37·6	35·8	42·3	44·5
(a) Managers and administrators	19·8	17·0	13·0	15·2	15·5
(b) Higher professionals	6·0	5·1	7·5	8·3	9·7
(c) Lower professionals and technicians	62·9	59·4	58·8	53·5	50·8
(d) Foremen and inspectors	4·2	6·5	8·7	13·4	10·3
(e) Clerks	21·4	44·6	46·0	60·2	65·1
(f) Salesmen and shop assistants	35·2	43·6	37·2	51·6	54·9
3. All manual workers	30·5	27·9	28·8	26·1	26·0
4. Total occupied population	29·6	29·5	29·8	30·8	32·4

Between 1911 and 1921 the number of male clerks increased a little more slowly than the occupied population while the number of female clerks increased more than three times. In general terms, the increased number of women in white-collar occupations is explained by the increased demand for white-collar skills in the face of relatively full employment, shorter hours, earlier marriage, mechanization of housekeeping, improved educational opportunity, and the particular attraction and suitability of many of these occupations for women.

---

[1] The high proportion of women among the lower professionals over the whole period is explained by the preponderance of the traditional female occupations—teaching and nursing—in this occupational group. Likewise, the decline in the proportion of women in this group over the years is largely explained by the influx of men into these 'female' occupations. Men accounted for almost two-thirds of the entire increase in the number of teachers between 1911 and 1961. Even in nursing, men accounted for almost 10 per cent of the total in 1961.

Industrial composition[1]

The changing occupational structure of any society can be explained by two related causes: technical changes within industries leading to changes in the type of skills required, and differences in the relative rates of growth of industries. So far, only the horizontal distribution of employment has been examined, but changes in the vertical or industrial distribution are also important.

Part of the increase in white-collar employment can be explained by the shift of total employment from the primary sector, and to a much lesser extent from the secondary sector, to the tertiary or service sector of the economy—that sector with the highest proportion of white-collar employment. As can be seen from Table 2.4, the primary or agricultural sector of the economy has steadily declined while the service sector has increased. During the period 1881 to 1951, gains in the service sector were made primarily at the expense of the agricultural labour force. Contrary to popular belief, neither the rise of white-collar employment nor the decline of manual employment can be explained by any serious contraction in the secondary or manufacturing sector, the traditional manual stronghold. Although manufacturing employment as a percentage of total employment has fluctuated, on balance it has held up remarkably well.

TABLE 2.4

The Percentage Distribution of the Occupied Population in Great Britain by Economic Sectors, 1881–1951

Sector	1881	1891	1901	1911	1921	1931	1951
Primary	13	11	9	9	7	6	5
Secondary	50	49	47	51	49	47	49
Tertiary	37	40	44	40	44	47	46
Total	100	100	100	100	100	100	100

While manufacturing employment as a whole has remained fairly constant, there are considerable differences between the growth-rates of one manufacturing industry and another. For example, the growth of the chemical industry which has a very high proportion of white-collar employees, and the decline of the clothing and footwear industry which has a very low proportion, have obviously worked in favour of increased white-collar employment.[2] One scholar who has analysed the effects of industrial change on

[1] Except for manufacturing industries (*infra*, pp. 16–20) no industrial breakdown of the white-collar labour force on the basis of the 1958 Standard Industrial Classification exists. The most detailed and reliable industrial analysis of the labour force on the basis of the 1948 Standard Industrial Classification appears in Routh, op. cit., appendix B.

[2] *Infra*, Table 2.6 and Table 3.2.

occupational distribution in much greater detail than is permitted by the scope of the present study concluded that

For both lower and higher professionals, the growth of industries has been more potent than their proportions within each industry; for clerical workers and foremen, the reverse has been true—it is their increased proportions within industries that have given the strongest impetus to their growth.[1]

MANUFACTURING INDUSTRIES

Manufacturing industries as a whole

The growing importance of white-collar workers within the manufacturing sector of the economy is demonstrated by Table 2.5. Between 1907 and 1963 the white-collar work force in manufacturing increased by 377 per cent, whereas the manual work force grew by only 32 per cent having actually decreased in numbers since 1954. The growth of white-collar occupations in manufacturing is also reflected in the fact that over this period their share of the labour force increased from 8 per cent to 23·8 per cent.

TABLE 2.5

Employment in the Manufacturing Industries of the United Kingdom, 1907–63

Year	Numbers (000s)			Percentage of total employment			Growth indices (1907 = 100)		
	White-collar	Manual	Total	White-collar	Manual	Total	White-collar	Manual	Total
1907	394	4,557	4,951	8·0	92·0	100·0	100	100	100
1924	512	4,345	4,857	10·5	89·5	100·0	130	95	98
1930	590	4,285	4,875	12·1	87·9	100·0	150	94	98
1935	696	4,679	5,375	12·9	87·1	100·0	177	103	109
1949	1,178	5,873	7,051	16·7	83·3	100·0	299	129	142
1958	1,673	6,108	7,781	21·5	78·5	100·0	425	134	157
1963	1,881	6,017	7,898	23·8	76·2	100·0	477	132	160

Individual manufacturing industries

The system of industrial classification has been changed so often since the turn of the century that it is not possible to obtain a picture of the changing composition of the work force in individual manufacturing industries over any length of time.[2] Table 2.6, however, shows the growing relative importance of the white-collar occupations in the various manufacturing industries between 1948 and 1963. This growth was most marked in the chemical industry and to a lesser extent in the engineering; vehicles; and paper, print-

[1] Routh, op. cit., pp. 41–2. Salesmen and shop assistants were not classified separately so it is not possible to determine from Routh's study the industrial effect on this group.

[2] For the percentage of white-collar workers in individual manufacturing industries in 1924, 1930, and 1935 on the basis of the pre-war industrial classification system, see *Manpower* (London: Political and Economic Planning, 1951), p. 38, table 6.

ing, and publishing industries. It was least marked in clothing and footwear; textiles; leather, leather goods, and fur.[1]

TABLE 2.6

White-Collar Employment as a Percentage of Total Employment by Manufacturing Industry in Great Britain, 1948–64

Industry	1948	1959	1964
Food, drink, and tobacco	17·7	18·8	20·2
Chemicals and allied	25·7	32·3	35·5
Metal manufacture	13·7	18·4	20·7
Engineering and electrical goods	..	28·2	29·8
Shipbuilding and marine engineering	..	14·8	18·3
Vehicles	18·1	24·1	26·6
Metal goods N.E.S.	..	17·4	19·0
Textiles	8·8	12·1	13·5
Leather, leather goods, and fur	13·0	13·8	14·3
Clothing and footwear	9·9	11·0	11·7
Bricks, pottery, glass, cement, etc.	10·9	16·0	17·7
Timber, furniture, etc.	12·8	15·7	18·2
Paper, printing, and publishing	18·5	23·0	24·0
Other manufacturing	..	21·1	23·3
ALL MANUFACTURING INDUSTRIES	16·0	21·1	23·1

Occupational composition[2]

The most outstanding characteristic of the occupational distribution shown in Table 2.7 is the overwhelming numerical importance of clerks in the white-collar labour force of every manufacturing industry. In manufacturing as a whole, clerks comprise approximately 50 per cent of total white-collar employment: their share ranges from a low of 31·6 per cent in shipbuilding to over 60 per cent in leather, leather goods, and fur; timber and furniture; and paper, printing, and publishing. The rest of the labour force is divided fairly equally between foremen (16·2 per cent); scientists, technologists, and technicians (17·8 per cent); and other white-collar workers (16·5 per cent). There are significant variations from these over-all trends in the different industries.

[1] For an analysis of the factors promoting the growth of the white-collar labour force see P. Galambos, 'On the Growth of the Employment of Non-Manual Workers in the British Manufacturing Industries, 1948–1962', *Bulletin of the Oxford University Institute of Economics and Statistics*, xxvi (November 1964), pp. 369–87; Seymour Melman, 'The Rise of Administrative Overhead in the Manufacturing Industries of the United States, 1899–1947', *Oxford Economic Papers*, iii (February 1951), pp. 62–112, and by the same author, *Dynamic Factors in Industrial Productivity* (Oxford: Blackwell, 1956), chaps. 10–14.

[2] Since the main purpose of the following discussion is to demonstrate the nature and extent of the potential for unionization in private industry, the managerial grades have been excluded from the analysis. In modern, large-scale private industry, it is the managers who generally control the operation of the business and direct the labour force. Functionally, therefore, they perform the role of employer and cannot be realistically considered part of the trade union potential. To date, only managerial grades in the public sector have shown any general desire to join trade unions.

TABLE 2.7

Employment in Various White-Collar Occupations as a Percentage of Total White-Collar Employment in Manufacturing Industries in Great Britain, 1964

Occupational group	Food, drink, tobacco	Chemical	Metal manuf.	Metal N.E.S.	Eng. elect.	Ship. M.E.	Vehicles	Textiles	Leather, fur	Clothing, footwear	Bricks, etc.	Timber, furn., etc.	Paper, print, pub.	Other manuf.	All manuf.
1. Foremen	18·1	13·2	20·4	18·1	11·5	31·0	14·0	31·6	23·4	22·8	21·5	22·0	12·4	17·8	16·2
2. All scientists, technologists, technicians	4·8	25·3	16·4	10·2	26·6	23·8	24·6	10·8	4·0	4·3	11·9	4·9	3·5	10·9	17·8
(a) Scientists, technologists	1·7	10·1	4·1	1·6	5·6	1·8	3·2	2·8	1·6	0·3	3·2	0·1	1·2	3·0	4·1
(b) All technicians	3·0	15·3	12·3	8·6	20·9	22·0	21·3	8·1	2·4	4·0	8·7	4·7	2·3	7·9	13·8
(i) Draughtsmen	0·7	1·8	3·7	5·5	11·0	16·2	9·4	1·5	0·1	0·5	4·1	3·1	0·4	2·6	6·0
(ii) Other technicians	2·3	13·5	8·6	3·1	9·9	5·8	11·9	6·6	2·3	3·5	4·5	1·7	2·2	5·4	7·7
3. Clerks	53·1	43·5	49·3	56·5	46·6	31·6	48·3	47·2	60·0	55·5	51·3	60·2	60·3	49·1	49·5
4. Other white-collar workers	24·0	18·0	13·9	15·2	15·3	13·6	13·1	10·4	12·6	17·3	15·4	12·9	23·8	22·2	16·5
5. All white-collar workers	100·0	100·0	100·0	100·0	100·0	100·0	100·0	100·0	100·0	100·0	100·0	100·0	100·0	100·0	100·0

TABLE 2.8

The Female White-Collar Labour Force as a Percentage of the Total White-Collar Labour Force in Manufacturing Industries in Great Britain, 1964

Occupational group	Food, drink, tobacco	Chemical	Metal manuf.	Metal N.E.S.	Eng. elect.	Ship. M.E.	Vehicles	Textiles	Leather, fur	Clothing, footwear	Bricks, etc.	Timber, furn., etc.	Paper, print, pub.	Other manuf.	All manuf.
1. Foremen	19·1	9·0	1·4	8·4	6·4	0·2	1·3	12·8	16·5	55·0	7·0	3·9	16·5	13·4	10·6
2. All scientists, technologists, technicians	16·4	10·6	3·5	3·2	2·5	2·3	2·2	18·4	5·4	35·4	4·7	6·3	7·1	8·4	5·0
(a) Scientists, technologists	11·9	7·0	2·8	5·0	2·2	0·0	1·1	8·3	6·7	35·0	5·6	0·0	5·8	2·4	4·1
(b) All technicians	19·0	13·0	2·1	2·8	2·6	2·6	2·4	21·9	4·5	35·5	4·3	6·4	7·7	10·6	5·3
(i) Draughtsmen	0·0	0·3	0·9	1·3	2·0	2·5	1·4	0·6	0·0	35·5	1·3	2·1	4·1	1·1	1·8
(ii) Other technicians	25·0	14·7	4·9	1·7	3·3	2·8	3·2	26·6	4·8	35·5	7·1	14·5	8·4	15·2	7·9
3. Clerks	68·5	68·5	53·2	67·6	61·9	44·0	48·6	66·1	83·9	82·6	59·7	68·1	63·7	67·6	62·8
4. Other white-collar workers	13·4	14·6	16·9	24·5	20·2	26·4	19·1	25·8	3·4	46·4	16·1	12·8	15·0	21·5	19·1
5. All white-collar workers	43·8	36·3	29·4	43·8	33·3	18·1	26·7	39·9	54·9	68·0	35·2	43·8	44·3	41·3	36·8

Scientists, technologists, and technicians are of much greater importance in engineering, chemicals, vehicles, and shipbuilding than in the other industries; foremen are of greater importance in textiles and shipbuilding than elsewhere.

Sex composition

As in the economy as a whole, women form a significant proportion of the white-collar labour force in manufacturing, but here they are largely restricted to one occupational group—clerks. In 1964, 36·8 per cent of all white-collar employees in manufacturing were female (see Table 2.8) and over 84 per cent of these were engaged in clerical work.[1] Women are relatively unimportant in the scientific, technical, and supervisory occupations, except in clothing and footwear, and, to a lesser extent, in textiles, where there are a high proportion of females in all the white-collar occupations.

CONCLUSIONS

The number of white-collar workers in Britain is rapidly increasing. Already almost four out of ten workers are white-collar employees. There is every likelihood that this trend will continue and that the labour force will soon be dominated by white-collar workers. The American economy has already reached a point where the white-collar employees outnumber the manual, and, if present occupational trends continue in Britain, this point will be reached here during the 1980s.[2] If the trade union movement is to continue as a dynamic and effective force in British society, it must recruit these white-collar workers. The extent to which it has already done so is considered in the next chapter.

[1] The latter figure was calculated from Table 2A.4 in Appendix A.

[2] Automation is unlikely to reverse or even greatly slow this trend. At most, it will probably simply change the composition of the white-collar labour force: more technologists, fewer clerks. See *infra*, pp. 70–1 for a discussion of the effect of automation on the demand for white-collar workers.

III

THE PATTERN OF WHITE-COLLAR UNIONISM

THE GROWTH OF TOTAL UNIONISM, 1892–1964[1]

VIEWED over the long run the growth of British trade unionism is most impressive. Although total union membership has fluctuated widely with changes in the social and economic environment, the long-run trend has been steadily upwards (see Table 3.1). Between 1892 and 1964 trade union membership increased from 1·5 million to slightly over 10 million while the number of employees increased from 14·1 million to 23·6 million. Union membership therefore increased by 539 per cent while potential union membership increased by only 67 per cent. As a result, the over-all density of unionization increased from a little over 11 per cent to almost 43 per cent.

Viewed over the immediate short run the growth of British trade unionism is much less impressive. Lately many signs have appeared which suggest that an era of stabilization is following upon the last great upsurge of union growth which began around 1933. During the fifteen years between 1933 and 1948, actual union membership increased by 113 per cent while potential union membership increased by only 6·5 per cent. But during the sixteen years between 1948 and 1964 union membership increased by only 8 per cent while potential union membership increased by 14 per cent.

These disparate increases in actual and potential union membership are reflected by changes in the density of unionization. Although union density figures are not continuously available prior to 1948, Table 3.1 makes it fairly clear that union density increased steadily from 1933 to 1946–7 when it reached a peak of 45–7 per cent. Since that time there has been a gradual but certain decline in union density. For density to increase, actual union membership must grow faster than potential union membership. This condition only existed in five of the sixteen years from 1948 to 1964, and, for the period as a whole, potential union membership increased almost twice as

---

[1] Although some reference will be made throughout this chapter to the development of British union membership in general, the main emphasis will be placed on the growth of white-collar union membership. For a more general analysis of membership growth see: Keith Hindell, *Trade Union Membership* (London: Political and Economic Planning, 1962); B. C. Roberts, 'The Trends of Union Membership', *Trade Union Government and Administration in Great Britain* (London: Bell, 1956), Appendix 1; and Guy Routh, 'Future Trade Union Membership', *Industrial Relations: Contemporary Problems and Perspectives*, B. C. Roberts, editor (London: Methuen, 1962), pp. 62–82.

much as actual union membership. Consequently, the over-all density of unionization declined from 45·1 per cent in 1948 to 42·6 per cent in 1964.[1]

TABLE 3.1

Total Union Membership in the United Kingdom, 1892–1964

Year	Labour force (000s)	Annual % change in labour force	Total union membership (000s)	Annual % change in union membership	Density of union membership (%)
1892	14,126	..	1,576	..	11·2
1901	15,795	..	2,025	..	12·8
1911	17,555	..	3,139	..	17·9
1921	17,618	..	6,633	..	37·6
1931	19,328	..	4,624	..	23·9
1933	19,498	..	4,392	..	22·5
1938	20,258	..	6,053	..	29·9
1948	20,767	..	9,362	..	45·1
1949	20,818	+0·2	9,318	−0·5	44·8
1950	21,096	+1·3	9,289	−0·3	44·0
1951	21,222	+0·6	9,535	+2·6	44·9
1952	21,322	+0·5	9,588	+0·6	45·0
1953	21,401	+0·4	9,527	−0·6	44·5
1954	21,718	+1·5	9,566	+0·4	44·0
1955	21,990	+1·3	9,738	+1·8	44·3
1956	22,230	+1·1	9,776	+0·4	44·0
1957	22,382	+0·7	9,827	+0·5	43·9
1958	22,346	−0·2	9,636	−1·9	43·1
1959	22,404	+0·3	9,621	−0·2	42·9
1960	22,764	+1·6	9,832	+2·2	43·2
1961	23,037	+1·2	9,893	+0·6	42·9
1962	23,354	+1·4	9,883	−0·1	42·3
1963	23,470	+0·5	9,928	+0·5	42·3
1964	23,616	+0·6	10,065	+1·4	42·6

There are two major reasons for the decline in the growth-rate of unionism and in over-all union density. The first is the changing pattern of employment which was described in Chapter II and is further illustrated by Table 3.2. This table shows that there has been a shrinkage of employment in a number of basic industries which have a long tradition of union activity and the highest density of membership—railways, coal-mining, national government, cotton, and manual employment in general.[2] At the same time there has been a steady expansion of employment in those areas which have proved

---

[1] Not only has there been a decline in the density of total union membership, but also in the density of that section of the total which is affiliated to the TUC. In spite of a number of new affiliations and an increase of 834,000 in its membership between 1948 and 1964, the density of TUC membership declined from 38·2 per cent to 37·1 per cent.

[2] The fact that employment has declined in these industries does not necessarily mean that union density has also declined. In spite of a decline in employment in cotton, agriculture, coal-mining, and national government, the density of unionization in these industries has increased.

most difficult to organize and have a relatively low density of unionization—
professional and business services; insurance, banking, and finance; distribution; chemicals; food, drink, and tobacco; and white-collar occupations in
all industries. But not all the industrial redistribution of employment has
worked against the unions. Employment in agriculture, a low-density industry,

TABLE 3.2

*Changing Employment and Density of Union Membership by
Industry in the United Kingdom, 1948–64*

Industry	Employees (000s)			Density (%)	
	1948	1964	% change 1948–64	1960	% change 1948–60
1. Education	521	1,094	+110	50	−11
2. Professional and business services	806	1,268	+57	24	−5
3. Insurance, banking, and finance	432	637	+48	31	+10
4. Distribution	2,093	3,026	+45	15	−2
5. Paper, printing, and pub- lishing	472	632	+34	57	+2
6. Gas, electricity, and water	329	413	+26	51	−9
7. Building	1,375	1,708	+24	37	−6
8. Metals and engineering	3,739	4,537	+21	54	−1
9. Chemicals and allied	447	515	+15	20	+1
10. Food, drink, and tobacco	731	842	+15	11	−5
11. Other transport and com- munications	1,221	1,320	+8	75	+2
12. Local government	720	776	+8	84	+16
13. Theatres, cinemas, sport, etc.	238	251	+6	39	+4
14. Furniture, timber, etc.	294	296	+1	37	−4
15. Footwear	116	116	0	63	−1
16. Clothing	498	453	−9	30	−3
17. Textiles other than cotton	708	613	−13	21	−3
18. Cotton	293	228	−22	75	+2
19. National government	717	550	−23	83	+19
20. Coal-mining	803	596	−26	89	+10
21. Railways	576	396	−31	84	−5
22. Agriculture, forestry, and fishing	868	551	−37	27	+5

declined while employment in metals and engineering and in paper, printing,
and publishing expanded, although a large proportion of this increase was
composed of white-collar employees. Nevertheless, Table 3.2 does demonstrate that on balance the industrial tide has been running against the trade
union movement. In the ten areas where employment expanded most rapidly
between 1948 and 1964, density of unionization was in every case less than
60 per cent and in three of these areas it was 20 per cent or less. In the ten
industries where employment expanded the least or declined, density of

unionization was in every case over 20 per cent and in five of these areas it was over 60 per cent. Clearly, union density is highest in the declining industries and lowest in the expanding industries, and this is causing the over-all density of unionization to fall.

The second factor explaining the diminishing growth-rate of unionism and the decrease in over-all density is that the unions are not recruiting members quickly enough among the expanding areas of employment, in particular among the white-collar occupations, to offset the decline in the traditional industries and among manual workers generally. As can be seen from Table 3.2, density of unionization declined in eight of the fourteen expanding industries between 1948 and 1960. Moreover, in spite of the increase in trade unionism among white-collar employees during the period 1948–64 being over thirty times greater than the increase among manual workers,[1] this growth was not sufficient to increase or even maintain the over-all density of unionization. In order to maintain, let alone extend, its numerical strength, the trade union movement must increase even further its rate of growth among white-collar employees.

THE GROWTH OF WHITE-COLLAR UNIONISM, 1948–1964

The growth of white-collar unionism in the economy as a whole

To assess the growth and extent of white-collar unionism in Britain is a most difficult task. The membership figures of each of almost 600 unions must be classified into manual and white-collar categories. Moreover, more than 20 per cent of total white-collar union membership belongs to partially white-collar unions,[2] and they do not always compile separate figures for their white-collar membership. In a sense, almost every manual union in Britain is a partially white-collar union because most of them take foremen into membership. Unfortunately, very few of these unions keep separate membership figures for this occupational category. In spite of all these difficulties some conclusions can be drawn regarding the growth and extent of white-collar unionism.

Of the 591 unions operating in the United Kingdom in 1964, there were approximately 280 purely white-collar unions and at least 19 partially white-collar unions.[3] Forty-three of the purely white-collar unions and all the partially white-collar unions were affiliated to the TUC. Total white-collar union membership in 1964 was 2,623,000 and close to 1,711,000 of this total

---

[1] *Infra*, Table 3.3.

[2] A partially white-collar union caters for both manual and white-collar employees while a purely white-collar union caters solely for white-collar employees. See Appendix A, Table 3A.1, for a complete list of all the purely and partially white-collar unions affiliated to the TUC.

[3] Only those partially white-collar unions which could estimate their white-collar membership are included in this figure.

was affiliated to the TUC (see Table 3.3); this represented almost 20 per cent of total TUC membership. In short, approximately one in four trade unionists were white-collar employees and slightly more than 65 per cent of them were affiliated to the TUC.

TABLE 3.3

The Growth of Total White-Collar and Manual Unionism in the United Kingdom, 1948–64

Type of union membership	1948 (000s)	1964 (000s)	% change 1948–64
1. Total white-collar unionism	1,964	2,623	+33·6
(a) Adjusted membership of TUC white-collar unions	1,257	1,711	+36·2
(b) Adjusted membership of non-TUC white-collar unions	707	912	+29·0
2. Total manual unionism	7,398	7,442	+0·6

The growth of white-collar unionism has been an extremely important factor in the post-war development of the TUC. In fact, almost the entire expansion of the TUC since the war has been due to the increase in its affiliated white-collar membership. Between 1948 and 1964 the affiliated membership of the TUC expanded by 11 per cent. This average over-all expansion resulted from an increase of 79 per cent in the affiliated membership of purely white-collar unions, and an increase of only 4 per cent in the affiliated membership of manual and partially white-collar unions.[1] To look at it another way, if there had been no purely white-collar unions affiliated to the TUC between 1948 and 1964, then instead of expanding by 11 per cent the TUC would have expanded by only 4 per cent. If it were possible to segregate the white-collar membership of the partially white-collar unions prior to 1964, then the importance of white-collar membership to the TUC would be even more striking. If it is assumed that the white-collar

---

[1] White-collar membership figures prior to 1964 could not be obtained for all the partially white-collar unions. Consequently, the white-collar membership of these unions had to be grouped with the manual membership.

The increase in the membership of unions affiliated to the TUC in the post-war period is composed of three elements. First, the increase in size between 1948 and 1964 of those unions which were affiliated to the TUC prior to 1948. Second, the initial affiliating membership of those unions which affiliated to the TUC between 1948 and 1964. Third, the increase in the size of these unions between the time they affiliated and 1964. Since seven purely white-collar unions affiliated to the TUC between 1948 and 1964, the growth-rate of the TUC's purely white-collar unions is considerably inflated by the initial increase in membership occurring at the time of affiliation. If the membership of post-1948 affiliated unions is included even for the years they were not affiliated, and the membership of all unions which left or were expelled during the period are excluded (except for the Electrical Trades Union) for the whole period, then the growth-rate of the purely white-collar unions becomes 36 per cent rather than 79 per cent. Membership figures obtained in this way are referred to as 'adjusted' membership. For all calculations see Appendix A, Table 3A.1.

membership of the partially white-collar unions expanded at the same rate as the purely white-collar unions[1] and that none of this had been affiliated to the TUC between 1948 and 1964, then the TUC would have expanded by only 2·5 per cent, and during the latter half of this period, 1955 to 1964, it would actually have shrunk in size by 0·6 per cent.

Although the above TUC figures exaggerate the actual growth of white-collar unionism, it has nevertheless increased substantially in total amount since 1948. Table 3.3 demonstrates that between 1948 and 1964, the 'adjusted' white-collar membership affiliated to the TUC expanded by 36·2 per cent while that portion of the total which remained unaffiliated grew by 29 per cent. Total white-collar unionism increased by 33·6 per cent as opposed to an increase in total manual unionism of only 0·6 per cent.

Taken by themselves these white-collar growth figures are most impressive. To determine their real significance, however, changes in the white-collar labour force must also be taken into account. Although the government does not publish figures of the number of white-collar employees in the economy as a whole, some rough estimates can be obtained by performing a few arithmetical manipulations.[2] During the period 1948–64 total white-collar union membership increased by 33·6 per cent while the white-collar labour force increased by 32·4 per cent. During this same period manual union membership increased by 0·6 per cent while the manual labour force increased by 4·6 per cent. In other words, the over-all density of white-collar unionism increased only very slightly while the over-all density of manual unionism fell slightly.

In fact, as Table 3.4 shows, the density of total white-collar unionism only increased from 28·8 per cent to 29 per cent between 1948 and 1964, while the density of manual unionism declined from 53·1 per cent to 51 per cent. Table 3.4 also reveals a significant difference in the density of unionism between men and women. Density of unionization among both manual and white-collar female workers is considerably less than among male workers. Moreover, the density of unionization among female white-collar workers has remained more or less constant since 1948.

Because of lack of detail in the systems of classifying both labour force and union membership figures, only rough estimates of the real growth of

---

[1] The 'adjusted' growth-rate of 36 per cent was used. If anything, this figure understates the growth of the partially white-collar unions. Appendix A, Table 3A.1, indicates that the white-collar membership of most of these unions grew by amounts substantially in excess of 36 per cent.

[2] The total number of white-collar persons gainfully occupied in Great Britain in 1951 and 1961 as a percentage of the total gainfully occupied (excluding employers, but including self-employed and the armed forces) was 32·5 per cent and 37·7 per cent respectively (see Table 2.1). The total number of employees in the United Kingdom in 1948 and 1964 (see Table 3.1) were distributed on the same basis as the gainfully occupied population in 1951 and 1961 respectively. This gave 6,749,000 white-collar and 14,018,000 manual employees in 1948 and 8,903,000 white-collar and 14,713,000 manual employees in 1964.

white-collar unionism could be obtained. Nevertheless, the relative changes in the size of white-collar union membership and the white-collar labour force are so nearly equal that, even granted an error of a few percentage points, at the very most the density of white-collar unionism in the economy as a whole could have increased only very slightly.

TABLE 3.4

The Density of Total White-Collar and Manual Unionism in the United Kingdom, 1948–64

Type of union membership	1948			1964		
	Male	Female	Total	Male	Female	Total
1. Total white-collar unionism	33·6	22·8	28·8	34·9	22·6	29·0
(*a*) Adjusted membership of TUC white-collar unions	22·4	13·5	18·4	23·1	14·4	18·9
(*b*) Adjusted membership of non-TUC white-collar unions	11·3	9·3	10·4	11·9	8·2	10·1
2. Total manual unionism	63·9	25·3	53·1	60·3	28·0	51·0

The growth of union membership in the post-war period will hardly excite or reassure a realistic supporter of trade unionism. The trade union movement must do much better than simply keep up with changes in the labour force. Even if both white-collar and manual membership had increased sufficiently to maintain their respective densities, the density of total union membership would still have declined because the high-density manual sector of the labour force was contracting while the low-density white-collar sector was expanding. Thus the trade union movement is at present faced with the paradoxical situation that in order simply to mark time, it must advance.

The growth of white-collar unionism in various industrial sectors

It is extremely difficult to assess accurately the density of unionization in particular industries or occupations. The membership of most British unions conforms neither to an industrial nor to an occupational pattern but sprawls across a wide variety of industries and occupations. Many unions do not classify their membership figures, and, even if they do, it is generally by negotiating group rather than by the official systems of classifying occupations and industries.[1] Even if someone undertook the formidable task of rationalizing union membership figures with the official systems of classification, it would still be difficult to calculate white-collar membership densities throughout

[1] Central Statistical Office, *Standard Industrial Classification* (revised edition; London: HMSO, 1958); General Register Office, *Classification of Occupations, 1960* (London: HMSO, 1960).

every industrial sector because the government's manpower figures give the number of white-collar employees only in manufacturing industries. Nevertheless, enough published data exist to give some idea of the extent of white-collar unionism in various industries.

A very rough estimate of the extent of white-collar unionism in various sectors can be acquired by looking at the total union density figures of those industries in which white-collar workers predominate. Table 3.5 gives the total union density figures for such industries and also for manufacturing for which a special survey was undertaken.[1]

TABLE 3.5

The Density of White-Collar Unionism by Industrial Sector in the United Kingdom, 1960

Industry	Total union density 1960 (%)	Change in union density 1948–60 (%)
1. Local government	84	+16
2. National government	83	+19
3. Education	50	−11[a]
4. Theatres, cinemas, sport, etc.	39	+4
5. Insurance, banking, and finance	31[b]	+10
6. Distribution	15	−2
7. Manufacturing industries	12[c]	..

[a] The density of unionism among teaching staff has not decreased but the number of non-teaching ancillary staff such as secretaries and caterers, who are not generally unionized as yet, has greatly increased and this has caused an over-all decline in the density of unionization in the education sector.

[b] The density of unionization among bank clerks is much higher than this figure would suggest. As of 1 January 1962 the density of unionization among bank clerks was 76 per cent. Even excluding the staff associations, the density of the National Union of Bank Employees was 31 per cent. See R. M. Blackburn, *Union Character and Social Class* (London: Batsford, 1967), Appendix A.

[c] This figure is for 1963 and only covers Great Britain. Moreover, the labour-force figure used to calculate the density excludes the unemployed. Consequently, this figure is slightly inflated relative to the other figures in the table.

Although, with the exception of manufacturing, the figures are little more than estimates within a wide margin of error, at least two conclusions emerge clearly from this table. First, there is obviously a strong relationship between density of unionization and whether the industry is publicly or privately owned. National and local government not only have a very high density of unionization, but they have also increased their density substantially in the post-war period. This relationship between public ownership and high union density is also apparent in industries in which manual workers predominate.

[1] For the details of this survey see the notes to Table 3.8 in Appendix A.

In 1960 in the nationalized coal-mining industry, union density was 89 per cent and in the nationalized railways 84 per cent, while in paper, printing, and publishing and in the engineering and metal industries, density of unionization was respectively only 57 per cent and 54 per cent in spite of the power and maturity of the unions in these areas.[1] Second, although the approximate nature of the statistics does not permit a detailed ranking of the various industries in terms of the degree of white-collar unionization, the differences in density are so vast that it is safe to conclude that with the possible exception of the distributive trades, manufacturing industries have the lowest density of white-collar unionism.

THE GROWTH OF WHITE-COLLAR UNIONISM IN MANUFACTURING INDUSTRIES, 1948-1964

The union pattern

There were eight major unions catering for the various categories of white-collar employees in manufacturing industries between 1948 and 1964. Like most British unions their membership was generally not confined to one occupation or industry. The names of these unions and some of the pertinent details regarding the occupational and industrial distribution of their membership are given in Table 3.6.[2] The Clerical and Administrative Workers' Union (CAWU), the Draughtsmen's and Allied Technicians' Association (DATA), and the National Union of Journalists (NUJ) are purely white-collar unions, as were the Association of Scientific Workers (AScW) and the Association of Supervisory Staffs, Executives and Technicians (ASSET).[3] The British Iron, Steel and Kindred Trades Association (BISAKTA)[4] and the Transport and General Workers' Union (TGWU)[5] are partially white-collar unions, as was the National Society of Operative Printers and Assistants (NATSOPA).[6] Except for the two industrial unions, BISAKTA and NATSOPA, all the unions had membership in both the manufacturing and non-manufacturing sectors of the economy. In each of these unions, however,

[1] See Table 3.2.

[2] The exact occupational and industrial distribution of each of these unions' membership is known. For fairly obvious reasons, however, most of the unions requested that their exact strength within any one occupation or industry should not be published. Consequently, in Table 3.6 only the strength of each union in all manufacturing industries is given and the individual industries in which it has membership are merely listed in order of importance. The exact industrial and occupational distribution of total white-collar unionism in manufacturing is given in Table 3.8 but in such a way that it is not possible to determine the industrial or occupational strength of most of the individual unions.

[3] The AScW and ASSET merged in 1968 to form the Association of Scientific, Technical and Managerial Staffs (ASTMS).

[4] This union is also known as the Iron and Steel Trades Confederation (ISTC).

[5] The white-collar section of the TGWU is known as the National Association of Clerical and Supervisory Staffs (NACSS).

[6] NATSOPA merged during 1966 with the National Union of Printing, Bookbinding and Paper Workers to form the Society of Graphical and Allied Trades (SOGAT).

TABLE 3.6

Characteristics of the Major Unions Catering for White-Collar Workers in Manufacturing Industries in Great Britain, 1963–4

Unions	Total membership	Total W-C membership	W-C membership as percentage of total membership	Total W-C membership in manufacturing industries	W-C membership in manufacturing as a percentage of total W-C membership	Occupational composition of W-C membership	Major industries in which manufacturing W-C membership was concentrated (listed in order of importance)	Major industries in which non-manufacturing W-C membership was concentrated (listed in order of importance)
ASeW	19,098	19,098	100	9,127	48	Roughly 25% composed of qualified scientists, engineers, and technologists. The remainder was made up of laboratory technicians	1. Engineering and electrical goods 2. Chemicals	1. National Health Services 2. Universities and Technical Colleges
ASSET	29,939	29,939	100	24,251	81	Roughly 50% composed of foremen; the other half was made up of technicians other than draughtsmen and laboratory assistants	1. Engineering and electrical goods 2. Vehicles 3. Metal manufacture 4. Rubber and plastics	1. Civil air transport
CAWU	74,529	74,529	100	51,043	69	Entirely composed of clerical and administrative employees	1. Engineering and electrical goods 2. Vehicles 3. Metal manufacture 4. Food and drink	1. Coal-mining 2. Co-operatives 3. Trade union staffs 4. Electricity supply
DATA	61,446	61,446	100	58,715	96	Composed of draughtsmen, tracers, and other technicians allied to design	1. Engineering and electrical goods 2. Vehicles 3. Shipbuilding and marine engineering 4. Metal manufacture	1. Construction
BISAKTA	107,205	7,520	7	7,520	100	Roughly 63% were clerks, 19% were technicians, 16% were foremen, and 2% were in other grades	1. Entirely metal manufacture	
NATSOPA	45,832	11,308	25	11,308	100	Composed almost entirely of clerical and administrative grades	1. Entirely paper, printing, and publishing	

NUJ	17,826	17,826	100	90	Composed entirely of journalists, press photographers, and publicists	1. Printing and publishing	1. Freelance-journalists 2. Radio and T.V. 3. Public relations
TGWU	1,412,603	51,337	4	55	Composed almost entirely of clerical and supervisory grades. Almost all the supervisors were employed in non-manufacturing industries	1. Vehicles 2. Engineering and electrical goods 3. Metal manufacture 4. Food and drink 5. Rubber and plastics 6. Textiles 7. Chemicals	1. Road and water transport

their white-collar membership in manufacturing made up roughly half or more of their total white-collar membership in 1964, ranging from a low of 48 per cent for the AScW to a high of 96 per cent for DATA. Manufacturing therefore provided the membership base for all these unions. Occupationally, the unions are difficult to classify except that in very general terms DATA is predominantly a technicians' union, as were the AScW and ASSET;[1] the CAWU, BISAKTA, and the TGWU (Nacss) are predominantly clerks' unions, as was NATSOPA; and the NUJ is a journalists' union.

In addition to these eight major unions, there were a number of other unions with a small white-collar membership in manufacturing in 1964. Almost every manual workers' union in manufacturing has some white-collar membership by virtue of having foremen and supervisors among its members. Of these unions, however, only the Electrical Trades Union (2,200)[2] and the Amalgamated Engineering Union (700) kept separate membership records for foremen in 1964. In the textile industry there are a number of unions catering strictly for overlookers, foremen, and supervisors: the General Union of Associations of Loom Overlookers (4,700), the Yorkshire Association of Power Loom Overlookers (1,800), the Managers' and Overlookers' Society (1,800), the Textile Officials' Association (700), and the National Federation of Scribbling Overlookers (130).[3] The National Union of General and Municipal Workers had approximately 12,000 white-collar members, mostly among clerks in the gas- and electricity-supply industries, but a few of them were employed in engineering, rubber, chemicals, and food. The British Association of Chemists had 1,800 qualified chemists in the chemical industry, while the Institute of Journalists with 2,400 members competed with the NUJ in the printing and publishing industry. There were also approximately 2,000 circulation representatives in this industry in membership of the National Union of Printing, Bookbinding and Paperworkers.[4]

The growth of manufacturing white-collar unionism

The growth of white-collar union membership in manufacturing industries between 1948 and 1964 was more than twice as great as that in the economy

[1] Historically, ASSET was a foreman's union. But after 1942 it began to concentrate much more on recruiting technicians. In 1964 approximately half of its membership was composed of technicians.

[2] The figures in brackets give the size of the total white-collar membership of the various unions in 1964. Besides the unions given here there are several others with a little white-collar membership. They are too numerous to list here, but they are included in Table 3.8 and are listed in the notes to this table in Appendix A.

[3] Strictly speaking, only spinning overlookers are white-collar workers. Weaving overlookers are simply highly skilled craftsmen who are responsible for the maintenance of the looms, and unlike the spinning overlookers few, if any, are expected to perform a supervisory function. However, both spinning and weaving overlooker union memberships have been included in this survey in order to make the union membership figures comparable to the *Census of Population* 'foremen' figures.   [4] *Supra*, p. 29 n. 6.

as a whole. Table 3.7 shows that during this period the major white-collar unions operating in manufacturing industries increased their membership by 77 per cent[1] while the increase in white-collar unionism for the economy as a whole was only 34 per cent. Not only did total white-collar unionism in manufacturing industries increase, but in addition each of the individual unions catering for white-collar workers in manufacturing increased their membership by substantial amounts ranging from 34·1 per cent for the AScW to 107·1 per cent for ASSET.

White-collar union density in manufacturing has also increased, but not as impressively as white-collar union membership. During the period 1948–64 white-collar union membership in manufacturing increased by 77 per cent while the white-collar union potential increased by 58 per cent.[2] Thus the density of white-collar unionism in manufacturing probably did not rise by more than 2–3 per cent between 1948 and 1964.

The industrial and occupational pattern of manufacturing white-collar unionism

In 1964 approximately 225,000 out of the potential of 1,900,000 white-collar employees in manufacturing industries, or 12 per cent of the potential, were unionized. This average figure covers a number of significant industrial and occupational variations, however, as Table 3.8 demonstrates. This table presents the dependent variable of the study—the density of white-collar unionism in manufacturing industries by occupation and by industry. Since the remainder of the study will largely be an attempt to explain the variations in the density of manufacturing white-collar unionism, this central set of statistics should be carefully studied and its limitations fully understood.[3]

In very general terms, Table 3.8 can be quickly summarized. Industrially, the most striking aspect of the table is the low level of white-collar union density in all the manufacturing industries relative to other areas of the economy. Even the degree of white-collar unionism in the best-organized manufacturing industry—23 per cent in paper, printing, and publishing—is much lower than that prevailing in most other industrial sectors.[4] Relative to each other the various manufacturing industries can be classified into three categories in terms of their density of white-collar unionism. First, there are the relatively high-density industries which fall into two groups: the paper,

[1] As was shown in Table 3.6, some of the white-collar membership of these unions is outside manufacturing industries. However, on average, the great majority of their total white-collar membership—75 per cent in 1964—is in these industries. It is assumed here that the non-manufacturing white-collar unionism in these totals does not significantly affect the rate of growth of the totals.

[2] The actual figures for the white-collar labour force in manufacturing industries in Great Britain from which the white-collar labour force growth-rate was calculated are given in Appendix A, Table 2A.2.

[3] For a discussion of the methods used in compiling this table see the notes to Table 3.8 in Appendix A.

[4] Cf. Table 3.5.

D

TABLE 3.7

The Growth of the Major Unions Catering for White-Collar Workers in Manufacturing Industries in the United Kingdom, 1948–64

Union	1948	1950	1955	1960	1964	% increase 1948–64	Ranking by % increase 1948–64	% increase 1955–64	Ranking by % increase 1955–64
AScW	15,521	13,264	11,911	12,645	20,809	+34·1	8	+74·7	2
ASSET	15,709	12,630	16,010	21,776	32,540	+107·1	1	+103·2	1
CAWU	38,493	33,150	55,921	59,545	79,177	+105·7	2	+41·6	5
DATA	45,049	45,039	51,806	60,740	62,048	+37·7	7	+19·8	7
BISAKTA	4,774	4,797	6,306	7,167	9,039	+89·3	4	+43·3	4
NATSOPA	6,846	7,879	10,462	10,800	12,250	+78·9	5	+17·1	8
NUJ	10,684	11,684	13,364	15,780	18,526	+73·4	6	+38·6	6
TGWU (Nacss)	27,620	29,133	36,525	44,491	56,541	+104·7	3	+54·8	3
Total	164,696	157,576	202,305	232,944	290,930	+76·6		+43·8	
Growth index (1948 = 100)	100	96	123	141	177				
Growth index (1955 = 100)			100	115	144				

TABLE 3.8

The Density of White-Collar Unionism in Manufacturing Industries in Great Britain by Industry and by Occupation, 1964

Occupational group	Food, drink, tobacco	Chemical	Metal manuf.	Metal N.E.S.	Eng. elect.	Ship. M.E.	Vehicles	Textiles	Leather, fur	Clothing, footwear	Bricks, etc.	Timber, furn., etc.	Paper, print, pub.	Other manuf.	All manuf.
1. Foremen	—[a]	1·6	8·3	3·2	13·6	2·9	10·8	28·2	1·1	—	2·0	3·8	n/a[b]	4·7	8·8
2. All scientists, technologists, technicians	1·7	9·6	26·4	2·6	28·2	53·5	38·0	2·6	0·5	0·9	8·3	12·6	0·4	14·1	24·0
(a) Scientists, technologists	1·2	12·5	3·0	—	3·9	13·5	2·3	1·7	—	3·0	4·4	—	0·3	3·3	5·3
(b) All technicians	2·0	7·7	34·3	3·1	34·7	56·8	43·6	3·0	0·9	0·8	9·8	12·8	0·5	18·1	29·6
(i) Draughtsmen	—	5·8	50·3	—	50·2	67·1	80·0	—	—	—	—	—	—	12·1	48·7
(ii) Other technicians	2·7	7·9	27·4	8·5	17·4	27·9	14·7	3·6	1·0	0·9	18·8	36·8	0·6	21·0	14·7
3. Clerks	5·4	2·3	21·3	2·4	12·3	5·6	22·3	3·4	—	0·6	5·2	0·4	15·2	10·2	10·5
4. Other white-collar workers	—		1·5	—	1·8	3·7	2·9						59·5[c]	0·1	7·1
5. All white-collar workers	3·0	3·7	16·8	2·2	15·0	15·9	22·0	10·8	0·3	0·3	4·1	1·7	23·4	7·4	12·1

[a] The symbol (—) is used where density is less than 0·1 per cent.

[b] The symbol 'n/a' means 'not available'. Unfortunately, none of the unions in this industry keep their membership figures in such a way that it is possible to determine the number of foremen in membership. It is generally known, however, that a relatively large number of employees in this industry retain their membership in a union when promoted to supervisory positions.

[c] A large proportion of the labour force in this category is composed of journalists and almost all the union membership is among the journalists. A very small amount of the total membership is found among circulation representatives.

printing, and publishing group (23·4 per cent), and the metal group of industries—vehicles (22 per cent), metal manufacture (16·8 per cent), shipbuilding and marine engineering (15·9 per cent), and engineering and electrical goods (15 per cent). Then there are the medium-density industries: textiles (10 per cent) and other manufacturing (7·4 per cent), mainly rubber and plastics. All the remaining industries fall into the low-density category with densities ranging from a high of 4 per cent in bricks, pottery, glass, cement, etc., to a low of 0·3 per cent in leather and fur, and clothing and footwear.

Occupationally, the most striking aspect of Table 3.8 is the very high density of unionization among draughtsmen (48·7 per cent) and journalists (90 per cent),[1] and the very low levels of density among all the other occupational groups. Here also there are significant differences between industries. Draughtsmen are over 80 per cent unionized in vehicles, but only 5·8 per cent organized in chemicals; clerks are over 22 per cent organized in vehicles but have a negligible degree of unionization in leather and fur, timber and furniture, and clothing and footwear; foremen are over 28 per cent unionized in textiles but are relatively poorly organized in most of the other industries.

In order to obtain comparable manpower statistics, union membership figures had to be arranged according to the official systems of industrial and occupational classification.[2] These systems of classification may be useful for classifying over-all social and economic organization, but they are not always useful for classifying trade union membership. In some cases the real significance of white-collar unionization in a particular area is obscured because an area of high density is combined with an area of low density to give an average figure.

In paper, printing, and publishing, for example, over 90 per cent of the clerical membership of NATSOPA was concentrated among the clerks of the major national dailies in Fleet Street. The number of clerks employed in Fleet Street is not known exactly, but union officials and employers estimated that union density among these workers is well over 90 per cent as opposed to the average of 15·2 per cent for the industry as a whole. In the same industry journalists are grouped under the heading 'other white-collar workers' for which the average density figure is 59·5 per cent. The density of unionization among journalists alone, however, is over 90 per cent and in Fleet Street it is very close to 100 per cent.[2] In metal manufacture most of the white-collar membership is concentrated in the heavy-steel trade. The government does

---

[1] The total membership of the NUJ and the Institute of Journalists in 1964 was approximately 21,000. The NUJ estimated that the potential number of journalists, photographers, etc., eligible for recruitment in 1964 was 22,000. Even allowing for dual membership between the two organizations, this makes the density of unionization over 90 per cent. See George Viner, *Basic Statistics on Journalism in the British Isles* (London: The NUJ, 1965), p. 2. (Mimeographed.)

[2] *Supra*, p. 27 n. 1.

not publish separate manpower figures for this section of the industry, but the density figures for the heavy-steel trade would certainly be higher than those given for metal manufacture generally. A large proportion of clerical membership in the food, drink, and tobacco industry is concentrated in the chocolate trade, and density of unionization here would be substantially above the 5·4 per cent for clerks in the industry as a whole. Finally, the manufacture of rubber and plastic goods is grouped with a number of other industries under the heading 'other manufacturing' to give an over-all density of white-collar unionism of 7·4 per cent. If it were possible to segregate the manpower figures of these two industries from the rest, their densities would be much more significant. If some of the other industry groups could be broken down into smaller units, no doubt additional variations would emerge.

The geographical pattern

Although there are no statistics available on the geographical distribution of membership in Britain, it is commonly believed that there are considerable regional variations in the membership pattern. As can be seen from Table 3.9, the geographical distribution of white-collar union membership in manufacturing tends to support this belief. Unfortunately, a great deal of the regional variation in the density of manufacturing white-collar unionism is hidden by the over-all averages which result from the government's very broad geographical groupings, the lack of separate white-collar labour force figures for the major conurbations, and the lack of a reliable occupational breakdown.[1] About the only conclusion which can be drawn from Table 3.9 is that the density of white-collar unionism tends to be highest in Wales and to a lesser extent in the northern and north-western areas of the country and lowest around London and the eastern and southern areas.[2]

CONCLUSIONS

The degree of unionization among white-collar employees is considerably less than that found among manual workers. In Britain only three out of ten white-collar workers belong to a union whereas five out of ten manual workers are members.

Since 1948 the absolute amount of white-collar unionism has increased greatly. This has prompted many people to speak of a boom in white-collar unionism. Such people are suffering from a growth illusion which results from considering changes in union membership in isolation from changes in the labour force. In real terms this membership boom is non-existent. In spite of the phenomenal growth of some white-collar unions, white-collar unionism in general has done little more than keep abreast of the increasing white-collar

[1] For a comment on the system of classification see notes to Table 3.9 in Appendix A.

[2] Trade unionism among bank clerks also tends to be stronger in Wales and in the northern areas than in the south. See Blackburn, op. cit., pp. 116–20.

TABLE 3.9

The Density of White-Collar Unionism in Manufacturing Industries in Great Britain by Industry and by Region, 1964

Region[a]	Food, drink, tobacco	Chemical	Metal manuf.	Metal N.E.S.	Eng. elect.	Ship M.E.	Vehicles	Textiles	Leather, fur	Clothing, footwear	Bricks, etc.	Timber, furn., etc.	Paper, print, pub.	Other manuf.	All manuf.
Scotland	1·2	3·0	14·3	1·2	20·9	32·7	27·8	2·3	1·8	—	4·0	0·6	15·6	9·3	13·3
Northern	1·7	3·5	40·2	7·8	24·4	24·2	8·3	8·3	—	—	—	1·7	35·1	9·2	16·6
North-west	1·3	4·1	19·8	1·9	22·0	15·1	25·8	21·6	—	0·2	6·7	1·0	19·7	8·4	14·3
East and West Ridings of Yorkshire and east and west Midlands	6·7	3·7	10·6	1·1	16·8	26·8	21·9	8·8	0·6	0·4	2·3	0·6	13·4	10·1	11·4
Wales	1·5	2·5	38·6	10·6	24·2	n/a	51·4	11·6	—	4·1	4·4	22·0	24·4	11·1	22·9
Eastern and southern	2·1	1·3	6·4	9·2	11·1	15·2	19·9	0·1	—	—	3·3	0·8	10·7	5·8	9·5
London and south-east	2·2	4·2	3·4	0·6	8·6	4·0	13·8	0·7	—	0·3	2·5	1·3	29·3	2·2	9·1
South-western	4·6	6·8	0·8	—	12·9	2·4	26·2	8·3	—	—	17·7	0·5	10·7	19·8	13·3
ALL REGIONS[b]	3·1	3·6	16·8	2·0	14·7	20·7	21·7	10·7	0·2	0·3	3·7	1·5	21·8	7·5	11·8

a For definitions of the various regions see notes to this table in Appendix A.
b These figures are slightly different from the ones given in Table 3.8 as the manpower figures are from a different source and include managers but not foremen. The difference in percentage terms is not great for most industries. For a more complete discussion of the source of these statistics see Appendix A.

labour force, and the density of white-collar unionism has not increased significantly during the post-war period. Of even greater importance, the growth of trade unionism among white-collar employees has not been sufficient to offset a decline in the density of manual unionism or to prevent a decline in the density of total unionism. Thus despite all the recruiting activity of white-collar unions during the post-war period, the real membership strength of white-collar unions in general is roughly the same today as it was in 1948, while the real membership strength of manual unionism and the trade union movement as a whole has actually decreased.

But of greater interest than the over-all level and growth-rate of white-collar unionism, are the occupational and industrial variations in the degree of white-collar unionism. The density of white-collar unionism in the public sector of the economy is over 80 per cent while in the private sector it is just slightly over 10 per cent. Even within the private sector there are great variations in the degree of white-collar unionism: printing and the metal group of industries are relatively highly unionized, the rest are not; journalists and draughtsmen have joined unions in large numbers, other white-collar groups have not. Even within the same occupational group there is considerable variation in the degree of unionization from one industry to another: draughtsmen are 80 per cent unionized in vehicles, but only 50 per cent in engineering and less than 6 per cent in chemicals. Clearly, there are great variations in the density of white-collar unionism, and, as Lockwood has remarked, 'it cannot be assumed that these variations are purely random'.[1] What then is their explanation? This is the question which the remainder of this study attempts to answer.

[1] David Lockwood, *The Blackcoated Worker* (London: Allen & Unwin, 1958), p. 138.

IV

SOCIO-DEMOGRAPHIC CHARACTERISTICS

PERHAPS the most obvious place to begin looking for an explanation of the pattern of white-collar unionism is at the white-collar workers themselves—at their sex, social origins, age, and status in the community. Social scientists have demonstrated that these personal characteristics determine many aspects of the individuals' outlook and behaviour; they may also determine their response to trade unionism.

SEX

Women make up a large proportion of the white-collar labour force. Approximately 45 per cent of all white-collar employees are female, and, in such white-collar groups as lower professionals, shop assistants, and clerks, women outnumber men.[1] The major characteristics of female employment are well known: most women do not participate continuously in the labour market because of marriage and family responsibilities, and they generally are supplementary earners in the sense that their pay is not the family's main source of income but merely supplements the earnings of their husbands. It is often suggested that these characteristics tend to reduce women's commitment to work thereby increasing their indifference to trade unionism,[2] and that the large proportion of women among white-collar workers therefore helps to account for their generally low degree of unionization.

This view has been challenged by Lockwood. He found that the proportion of women in most of the major clerical unions is 'roughly equal to their representation in the field of employment which the unions seek to organise' and this led him to conclude that 'differences in the degree of unionization are therefore to be attributed to something other than differences in the sex ratio'.[3] Lockwood's findings are sufficient to indicate the inadequacy of generalizations about women having a dampening effect on the level of unionization, but they are not sufficient to discount this argument completely.

It is common knowledge that the density of unionization is much less among women than among men. Density of unionization among manual workers is approximately 60 per cent for males as opposed to 28 per cent for females,

---

[1] See Table 2.3.

[2] See, for example, R. M. Blackburn, *Union Character and Social Class* (London: Batsford, 1967), p. 56.

[3] *The Blackcoated Worker* (London: Allen & Unwin, 1958), p. 151.

while that among white-collar workers is 35 per cent for males as compared to 23 per cent for females.[1] Obviously, the unions Lockwood examined are not representative of trade unionism in general. Moreover, it is also noticeable from comparing Tables 2.8 and 3.8 that there is some relationship between the density of unionization and the proportion of women among white-collar workers in manufacturing industries. Industries which have a large proportion of women, such as textiles, leather and fur, clothing and footwear, and timber and furniture, also have a low degree of white-collar unionism while those which have a small proportion of women, such as metal manufacture, engineering and electrical goods, shipbuilding and marine engineering, and vehicles, also have a relatively high degree of white-collar unionism. Similarly, clerical occupations have a large proportion of women and are poorly organized while technical occupations have a small proportion of women and are quite highly unionized. A regression analysis of Tables 2.8 and 3.8 confirms that the density of unionization and the proportion of women among white-collar workers in manufacturing industries are correlated, although not quite significantly at the 5 per cent level.[2]

But although Lockwood does not provide sufficient evidence to substantiate his conclusion that differences in the degree of unionization are to be attributed to something other than differences in the sex ratio, it is nevertheless largely correct. The fact that a low degree of unionization is associated with a high proportion of women and that women generally are not as highly unionized as men can be accounted for by differences in the way males and females are distributed across firms. Chapter VI will demonstrate that density of unionization is higher in areas where the average size of establishments is large and employment concentrated than in areas where the average size of establishments is small and employment diffused.[3] That the distribution of female employment is skewed in the direction of small establishments is obvious from Table 4.1. The proportion of women is highest in the smaller establishments and lowest in the larger establishments. In fact, only 25 per cent of the women employed in manufacturing industries work in establishments with 1,000 or more employees compared with 39 per cent of males while 43 per cent of the women work in establishments with less than 250 employees as compared with 33 per cent of males.[4]

The close association between female employment and establishment size is also indicated by the data presented in Tables 2.8 and 6.1 on the proportion

----

[1] See Table 3.4.

[2] Regressing density of unionization on proportion of women and allowing for the level of the equation to shift with occupation produces a $\bar{R}^2$ of 0·109 and a t value for the proportion of women coefficient of $-1·79$.

[3] See *infra*, pp. 72–81 and also p. 42 n. 2.

[4] If establishments with less than eleven employees could be included in this table, it would be seen that the distribution of female employment is even more skewed in the direction of small establishments. See Viola Klein, *Britain's Married Women Workers* (London: Routledge & Kegan Paul, 1965), pp. 94–5.

of women and the degree of employment concentration among white-collar workers in manufacturing industries. Even a superficial glance at these tables demonstrates that industries which have a high degree of employment concentration, such as engineering and electrical goods, shipbuilding and marine engineering, metal manufacture, and vehicles, also have a low proportion of women while those which have a low degree of employment concentration, such as textiles, leather and fur, clothing and footwear, and timber and furniture, also have a high proportion of women. A regression analysis of Tables 2.8 and 6.1 confirms that the proportion of women and the degree of employment concentration among white-collar workers in manufacturing industries are very highly correlated.[1]

TABLE 4.1

Employment in Manufacturing Industries in Great Britain by Size of Establishment

Size of establishment (no. of employees)	Males as a percentage of total males	Females as a percentage of total females	Females as a percentage of total employment
11–24	2·6	2·9	34·6
25–49	5·9	7·2	36·5
50–99	10·0	12·9	38·1
100–249	14·7	19·7	38·7
250–499	13·8	17·0	36·9
500–999	13·8	14·9	33·7
1,000–1,999	13·8	11·7	28·7
2,000–4,999	14·6	8·3	21·4
5,000 or more	10·4	4·9	18·2
Totals	100·0	100·0	32·1

The very high correlation between proportion of women and employment concentration suggests that the former will exert little influence upon density of unionization independently of the latter. This is, in fact, the case. Chapter VI will demonstrate that density of unionization and the degree of employment concentration among white-collar workers in manufacturing industries are correlated very highly.[2] This chapter has already shown that the density

[1] Regressing employment concentration on proportion of women and allowing for the level of the equation to shift with occupation produces a $\bar{R}^2$ of 0·392. Regressing proportion of women on employment concentration and allowing for the level of the equation to shift with occupation produces a $\bar{R}^2$ of 0·835. Both $\bar{R}^2$s are significant at the 5 per cent level.

[2] Regressing density of unionization on employment concentration and allowing for the level of the equation to shift with occupation produces a $\bar{R}^2$ of 0·177 and a t value for the employment concentration coefficient of 2·96. Allowing for the slope of the equation to shift with occupation produces a $\bar{R}^2$ of 0·641 and a t value for the employment concentration coefficient of 2·14. Allowing for both the level and the slope of the equation to shift with occupation produces a $\bar{R}^2$ of 0·669 and a t value for the employment concentration coefficient of 2·28. These are all significant at the 5 per cent level. See *infra*, pp. 72–81.

of unionization and the proportion of women among white-collar workers in manufacturing industries are correlated, although not quite significantly at the 5 per cent level.[1] But if density of unionization is regressed simultaneously on both employment concentration and proportion of women, then employment concentration remains very significantly associated with density of unionization but proportion of women does not.[2] This suggests that density of unionization is influenced by proportion of women mainly because it is correlated with employment concentration. Once the effect of employment concentration upon proportion of women has been allowed for, proportion of women as such exerts little influence on density of unionization. To put it another way, density of unionization and proportion of women have no significant connection with each other except through their separate relationships to a third variable, the degree of employment concentration.

The conclusion to be drawn from this rather technical discussion is a simple one. Female employees appear to have no inherent characteristics which make them more difficult to unionize than men, or, at least if they have, unions have been able to overcome them. 'Where men are well organised in a particular plant', as the TUC has noted, 'generally women are too. The fact that the proportion of women in employment who belong to trade unions is only about half that of men is mainly to be accounted for by differences in their industrial and occupational distribution.'[3] In short, the proportion of women has not been in itself a significant determinant of the pattern of manual or white-collar unionism in Britain.

SOCIAL ORIGINS

Sociologists have argued that the social origins of workers determine many aspects of their behaviour. In particular, they have hypothesized that those who come from the homes of manual or unionized workers are more likely to join trade unions than those who come from the homes of white-collar or non-unionized workers. The suggested explanation for this is that white-collar or non-unionized parents are likely to possess anti-union attitudes while manual or unionized parents are likely to possess pro-union attitudes, and these attitudes are transmitted to their children.[4]

This is a difficult hypothesis to test because of lack of adequate data.

---

[1] See *supra*, p. 41 n. 2.

[2] Allowing for the level of the equation to shift with occupation produces a $\bar{R}^2$ of 0·174 and a t value for the employment concentration coefficient of 2·46 and for the proportion of women coefficient of −0·859.

[3] *Selected Written Evidence Submitted to the Royal Commission* (London: HMSO, 1968), p. 185.

[4] See, for example, Seymour M. Lipset and Joan Gordon, 'Mobility and Trade Union Membership', *Class, Status and Power*, Reinhard Bendix and Seymour M. Lipset, editors (London: Routledge & Kegan Paul, 1954), pp. 491–500.

Although a considerable amount of information has been collected on the social origins of various white-collar groups,[1] the various sets of data are not comparable with each other because of differences in sample size, reliability, and date, nor are they comparable with the data on union membership which has been collected for this study. But several social surveys have generated comparable data on union membership and social origins, and scholars have used it to test the above hypothesis.

Lipset and Gordon analysed data from a 1949 sample of 953 Californian manual workers and found that workers whose fathers had non-manual occupations were the least likely to belong to trade unions.[2] Kornhauser analysed the data generated by a national public opinion survey in 1952 and also by a survey of labour mobility in six major American cities in 1951 and found no relationship between the social origins of manual workers and union membership in either sample.[3] But Lipset also analysed the data provided by the six-city sample and found that clerical workers and shop assistants who came from manual homes were much more likely to be union members than those who did not.[4] Goldstein and Indik could find no significant correlation between social origins and union membership among the 705 professional engineers they surveyed, but did find that union members were significantly more likely to have had fathers who were also union members.[5] In his sample of 6,115 professional engineers and 596 technicians, Kleingartner found that union members were not significantly more likely than non-members to have had fathers who belonged to a trade union.[6]

In Britain Blackburn found that among his sample of thirty-five male bank clerks there was a tendency for those from higher-status homes to be in the staff associations or nothing, while the lower the status of their background, the more likely they were to be in NUBE.[7] But this relationship did not hold among his sample of sixty-six female bank clerks. Nor did it hold for either males or females among his second sample of fifty-eight bank clerks. Many

[1] See David Glass (ed.), *Social Mobility in Britain* (London: Routledge & Kegan Paul, 1954); for clerical workers, see Lockwood, op. cit., pp. 106–16 and J. R. Dale, *The Clerk in Industry* (Liverpool: Liverpool University Press, 1962), p. 33; for draughtsmen, see Guy Routh, 'The Social Co-ordinates of Design Technicians', *The Draughtsman* (September 1961), p. 7; for engineers, see J. E. Gerstl and S. P. Hutton, *Engineers: The Anatomy of a Profession* (London: Tavistock Publications, 1966), chap. 3; and for civil servants, see R. K. Kelsall, *Higher Civil Servants in Britain* (London: Routledge & Kegan Paul, 1955), chap. 7.

[2] Op. cit., p. 492, table 1.

[3] Ruth Kornhauser, 'Some Social Determinants and Consequences of Union Membership', *Labor History*, ii (Winter 1961), p. 43.

[4] Seymour M. Lipset, 'The Future of Non-Manual Unionism' (an unpublished paper, Institute of Industrial Relations, University of California, Berkeley, 1961), pp. 21–2.

[5] Bernard Goldstein and Bernard P. Indik, 'Unionism as a Social Choice: The Engineers' Case', *Monthly Labor Review*, lxxxvi (April 1963), pp. 366–7.

[6] A. Kleingartner, 'The Organization of White-Collar Workers', *British Journal of Industrial Relations*, vi (March 1968), pp. 85–6.

[7] Op. cit., pp. 197–8.

draughtsmen come from manual homes and many of them start their working lives as manual workers.[1] Phillipson analysed the data provided by his sample of 182 white-collar workers and found that draughtsmen who had manual fathers were more likely to be union members than those who did not. But he could find no relationship between fathers' union membership and union membership among draughtsmen, nor could he find any connection between fathers' occupation or union membership and union membership among clerical and administrative workers.[2]

It is difficult to draw any firm conclusions from these studies, not only because of their contradictory findings, but also because of their methodology. The American samples were large and random, but the British ones were small and something less than random.[3] More serious, none of the studies controlled for all the other factors which might influence workers to join unions. For example, none of the studies controlled for size of firm or for the way in which the firm was administered, and none but Blackburn's even controlled for industry. It may be that many of the non-unionists were working in small paternalistically-administered firms in which they had close interpersonal relations with their employers, and that this rather than their social origins explains their reluctance to join trade unions.[4]

About all that can be said until such time as there are better empirical studies is that there is not generally a connection between social origins and union membership. This conclusion is reinforced by the data on union membership which has been collected for this study. The variations in the density of white-collar unionism reflected by Table 3.8 would seem to be too large to be explained by variations in social origins. The density of unionism among draughtsmen is 80 per cent in vehicles, 50 per cent in engineering and electrical goods, and less than 6 per cent in chemicals. It is extremely doubtful if the social origins of draughtsmen vary sufficiently from industry to industry to account for these variations in union density. Nor is it very likely that the high degree of clerical unionism in the public sector and the low degree in the private sector is accounted for by differences in social origins. Most foremen come from manual homes and almost all of them would have once had manual jobs, but they are one of the most poorly organized groups in private industry. It is very probable that the social origins of journalists are more middle class than those of clerks, yet the former are much more highly unionized than the latter. It is unnecessary to pursue this line of argument in order to conclude

[1] Routh found that in his sample of 941 DATA members, 63 per cent had come from manual homes and 43 per cent had entered the profession after serving a craft apprenticeship (op. cit., pp. 7–8).

[2] C. M. Phillipson, 'A Study of the Attitudes Towards and Participation in Trade Union Activities of Selected Groups of Non-Manual Workers' (unpublished M.A. thesis, University of Nottingham, 1964), pp. 279–80.

[3] The researchers themselves admit this. See ibid., pp. 164–70, and Blackburn, op. cit., pp. 64–5.

[4] Lipset and Gordon frankly admit this (op. cit., p. 705 n. 14).

that there is no clear connection between the density of unionization and social origins.[1]

AGE

The age distribution of an occupation seldom is mentioned as a factor which might affect the growth of trade unionism. But at least one scholar has claimed that younger workers are likely to show a greater propensity to unionize than older workers.[2] He advances several reasons in support of this proposition. Firstly, younger workers usually have shorter tenure of service with a firm and therefore feel less 'loyalty' towards it than older workers. Secondly, if victimized because of their union activity, younger workers stand to lose much less than older workers in terms of any company benefits which may exist, and, in addition, it is easier for them to find a job elsewhere and adjust themselves to it. Thirdly, younger workers are likely to be better educated than older ones, and therefore more resentful of arbitrary treatment by management and better able to give leadership at the rank and file level. Finally, younger workers have been nurtured in an era when unionism has become an integral part of the institutional framework and they look upon it, therefore, not as something new and exceptional, but as the 'natural' way of handling worker problems.

The empirical evidence is rather fragmentary, but what there is does not support this *a priori* reasoning. The percentage age distribution for selected white-collar occupations is given in Table 4.2. It is clear that laboratory

TABLE 4.2

Percentage Age Distribution for Selected White-Collar Occupational Groups, England and Wales, 1961

Occupational group	Total	15–19	20–4	25–9	30–4	35–44	45–54	55–9	60–4	65+
All gainfully occupied	100	10·8	10·3	9·1	9·4	21·1	21·4	9·0	5·9	3·2
Foremen	100	0·2	1·7	4·6	9·0	28·5	32·7	13·6	8·0	1·8
Clerks	100	18·8	15·6	10·1	8·8	18·2	15·8	6·4	4·2	2·0
All professionals and technicians	100	6·8	13·9	13·0	12·1	21·2	19·2	7·3	3·9	2·6
Authors and journalists	100	4·8	10·1	13·1	12·8	20·8	20·8	7·3	5·2	5·1
Draughtsmen	100	12·7	22·0	18·8	14·5	18·1	8·3	3·0	1·9	0·8
Laboratory assistants	100	23·4	24·1	13·4	9·6	14·2	9·0	3·5	2·1	0·6

[1] The amount and type of education which white-collar workers receive is another personal characteristic which might influence their propensity to unionize. It has not been analysed in this chapter because of lack of adequate data. But it is well known that the amount and type of education received by occupational groups is correlated very highly with their social origins (see Lockwood, op. cit., pp. 119–20, and Jean Floud, 'The Educational Experience of the Adult Population of England and Wales at July 1949', *Social Mobility in Britain*, op. cit., pp. 120–39). Hence the influence of education upon the propensity to unionize is probably much the same as the influence of social origins, namely—little, if any. Goldstein and Indik (op. cit., p. 366) could find no significant relationship between education and union membership among professional engineers.

[2] Joseph Shister, 'The Logic of Union Growth', *Journal of Political Economy*, lxi (October 1953), pp. 421–2.

technicians, draughtsmen, and clerks are among the younger occupations: the proportion under thirty-five years of age is 70·5 per cent for laboratory technicians, 68 per cent for draughtsmen, 53·4 per cent for clerks, 45·9 per cent for professionals and technicians as a group, 40·7 per cent for authors and journalists, 39·6 per cent for all the gainfully occupied, and 15·5 per cent for foremen. Yet the members of these young occupations have had widely differing responses to trade unionism: draughtsmen are highly unionized, laboratory technicians and clerks are not. Moreover, journalists are a relatively old occupation, yet almost all of them belong to a union.

The only white-collar group for which the age distribution is available on a detailed industrial basis is foremen, and this is given in Table 4.3.[1] A regression analysis of Tables 3.8 and 4.3 revealed that there was no significant connection between the age distribution of foremen and the degree to which they are unionized.[2]

TABLE 4.3

Percentage Age Distribution of Foremen in Manufacturing Industries in England and Wales, 1961

Industry	Total	15–19	20–4	25–9	30–4	35–44	45–54	55–9	60–4	65+
Food, drink, and tobacco	100	0·2	2·4	5·6	8·5	28·1	33·3	13·4	7·1	1·3
Chemical and allied	100	0·2	1·5	4·3	8·6	32·1	33·7	12·1	6·7	0·9
Metal manufacture	100	0·1	0·9	3·7	7·8	29·7	32·8	14·5	8·5	2·0
Metal goods N.E.S.	100	0·1	2·0	5·2	10·0	33·0	29·2	11·8	6·4	2·2
Engineering and electrical goods	100	0·1	1·7	5·1	11·3	35·3	28·2	10·8	6·1	1·5
Shipbuilding and marine engineering	100	0·0	0·3	1·5	6·5	26·1	29·8	20·1	12·7	3·1
Vehicles	100	0·1	0·4	3·1	8·5	33·5	32·1	13·0	7·9	1·3
Textiles	100	0·3	2·8	6·1	8·3	22·5	33·0	15·3	8·8	2·9
Leather, leather goods, and fur	100	0·5	2·0	3·0	8·0	27·4	30·3	14·9	8·5	5·5
Clothing and footwear	100	0·9	5·5	6·8	10·2	24·8	30·2	11·6	6·5	3·5
Bricks, pottery, glass, etc.	100	0·5	1·5	5·0	10·6	28·3	34·0	12·0	6·3	1·9
Timber, furniture, etc.	100	0·1	1·6	6·5	10·9	27·6	36·4	10·2	5·0	1·6
Paper, printing, and publishing	100	0·1	2·2	5·3	7·7	26·7	34·9	13·3	7·3	2·4
Other manufacturing	100	0·3	2·8	5·9	13·5	32·1	30·8	9·3	4·2	1·0
All manufacturing	100	0·2	1·9	4·9	9·5	30·4	31·5	12·6	7·1	1·8

[1] These figures are derived from Industry Table 13 of the 1961 *Census of Population*. For each Minimum List Heading only the 'principal occupations' (those with 500 or more males or females in the sample figure) are given. 'Foremen' is the only occupational group of interest to this study which exceeds this 'threshold' in every industry (Minimum List Heading) and hence the only group for which an industrial age distribution can be calculated with any accuracy.

[2] The first row (foremen) of Table 3.8 was regressed on each column of Table 4.3. Then each column of Table 4.3 was cumulated one at a time from left to right and then right to left, and the first row of Table 3.8 was regressed on each of the resulting columns (that is, on 15–19, 15–24, 15–29, etc., and on 65+, 60–5+, 55–65+, etc. All the resulting correlation coefficients were very much below the figure required for significance at the 5 per cent level. The highest correlation coefficient was 0·296. To be significant at the 5 per cent level with 12 degrees of freedom, the correlation coefficient must be at least 0·532. Of course, other combinations of the age distribution (for example, 20–34 or 35–64) could have been used, but this did not seem worth while in view of the unpromising results obtained above.

STATUS

The most common explanation for white-collar resistance to trade unionism is what various writers have referred to as 'false class consciousness', the 'psychology of prestige striving', or plain old-fashioned 'snobbishness'.[1] They argue that white-collar workers have possessed considerable social aspirations but have had limited means for achieving them. 'Unlike the members of the upper class they could not claim prestige as their birthright; nor could they, like the captains of industry, base it on power and authority.' Hence 'they sought it in the only way left open to them—by concentrating on social differences'. They claimed higher prestige on the basis of such factors as their social origins, manner of dress and speech, tastes, education, working conditions, and association with the 'managerial cadre'.

But 'prestige involves at least two persons: one to claim it and another to honor the claim'. In the case of white-collar workers, the community in general and employers in particular were prepared to honour the claims. Indeed, they even fostered them by encouraging white-collar workers 'to identify their interests with those of the employers and to regard themselves as having a personal relationship with them'. To emphasize this, white-collar workers were called staff, not hands; they were paid salaries, not wages; they received better terms and conditions of employment; and they were provided with their own entrances and had different starting and finishing times.[2] 'The separation of staff from workmen was built into the structures of industrial organizations as if the two represented different castes.'

The supposed result of all this is that white-collar workers 'formed an image of themselves which bore little resemblance to economic realities. They saw themselves as individuals, superior to manual workers and able to progress through society unaided and without protection.' Most of them felt that unions were only for those who did not have sufficient determination and merit to succeed on their own. Even those who did not feel this way were afraid to join unions for fear of identifying themselves with manual workers and thereby losing their middle-class status.

But fortunately for the supporters of trade unionism this argument has a happy ending. 'Every basis on which the prestige claims of the bulk of the white-collar employees have historically rested has been declining in firmness and stability.' Such trends as the spread of educational opportunity, the narrowing of the manual/white-collar pay differential and the granting of staff

[1] The major American exposition of this thesis is found in C. Wright Mills, *White Collar* (New York: Oxford University Press, 1956), chaps. 11 and 14. The major British exposition is found in V. L. Allen, 'White-Collar Revolt?', *The Listener*, lxvi (30 November 1961), pp. 895–7. The statement of the argument given below is paraphrased largely from these two sources.

[2] Many employers retain these distinctions today because they believe that such status symbols discourage white-collar workers from joining trade unions. See 'Status Has Its Compensations', *Personnel Magazine* (November 1965), p. 13.

status to manual workers, the concentration of white-collar employees in big work places and their resulting separation from management, the recruitment of white-collar employees from lower social strata, and the vast increase in the total numbers of white-collar people are all 'tearing away the foundations of the white-collar rejection of unions on the basis of prestige'. In short, the 'status proletarianization' of white-collar workers will reduce their traditional hostility towards trade unionism.

There can be little doubt that most white-collar workers generally think of themselves as being socially superior to manual workers, as belonging more with management and the middle class than with manual workers and the working class.[1] Nor can there be much doubt that the status of almost all white-collar groups has been declining over the course of the twentieth century.[2] But this does not necessarily mean that the first fact prevents white-collar workers from joining unions or that the second will encourage them to do so.

Lockwood has demonstrated the fundamental weakness with 'snobbishness' as an explanation of white-collar union growth, and it is worth while to quote him at some length:

Has snobbishness varied through time? If so, assuming that it has lessened, say between 1921 and 1951, why has the proportion of commercial and industrial clerks in unions hardly increased at all? If not, how is the fact to be explained that twice as many blackcoated workers are unionized nowadays as thirty years ago? Is snobbishness a factor which operates 'all along the line', something displayed by all clerks equally? If so, why are there very significant differences in the degree of clerical unionization from one field to another? Alternatively, is snobbishness connected with relative social status among blackcoats? If so, why have certain groups of clerks with high social status been highly unionized, and others with a relatively low social status poorly unionized? Most perplexing of all, why have two groups of clerical workers, both with a relatively low standing in the blackcoated world—railway clerks and industrial clerks—joined their respective unions to such radically different degrees? It is unnecessary to pursue this line of argument in order to conclude that no clear connection can be established between a factor such as 'snobbishness' and the empirical variations in clerical trade unionism. Whatever influence snobbishness has on the mutual relations of clerks and manual workers

[1] Studies in this and other countries support such a contention, although they indicate that among the lower level white-collar groups an increasing number see themselves as members of the working class. In general see Richard F. Hamilton, 'The Marginal Middle Class: A Reconsideration', *American Sociological Review*, xxxi (April 1966), pp. 192–9. For Britain see John Bonham, *The Middle Class Vote* (London: Faber, 1954), p. 60; F. M. Martin, 'Some Subjective Aspects of Status', *Social Mobility in Britain*, D. V. Glass, op. cit., p. 56; Lockwood, op. cit., p. 127; Dale, op. cit., pp. 28–9; and W. G. Runciman, *Relative Deprivation and Social Justice* (London: Routledge & Kegan Paul, 1966), p. 158, table 1. For America see Richard Centers, *The Psychology of Social Classes* (Princeton: Princeton University Press, 1949), p. 86. For France see Natalie Rogoff, 'Social Stratification in France and the United States', *Class, Status and Power*, op. cit., p. 585.

[2] If documentation is thought necessary see Runciman, op. cit., chap 5; Lockwood, op. cit., chap. 4; and Dale, op. cit., chap. 3.

it does not seem to have prevented the former from organising themselves in trade unions.[1]

While Lockwood argues his case only in respect of clerical trade unionism, it obviously applies to other types of trade unionism as well. Technicians and journalists enjoy a higher social standing in the community than clerical workers, and yet the former are more highly unionized than the latter. Craftsmen are obviously very status conscious, but this has not prevented them from organizing some of the strongest and most militant unions in the country.[2] Clearly, there is no general correlation between social status and the extent of trade unionism.

Even granting for the moment that the white-collar workers' concern for their middle-class status has prevented them from joining unions in the past, this does not necessarily mean that they will become more receptive to unionism as their status declines. Mills himself has suggested that as white-collar workers are increasingly proletarianized there may be a 'status panic' —a frantic drive to protect the remaining bases for separate consideration. The white-collar worker may 'seize upon minute distinctions as bases for status' and this may 'operate against any status solidarity among the mass of employees . . . lead to status estrangement from work associates, and to increased status competition'.[3] If this occurs, then by this argument's own logic white-collar unionism is unlikely to be the result.

CONCLUSIONS

In writings on white-collar unionism, much has been made of the personal characteristics of white-collar workers. Such features as their sex, social origins, and generally higher status in the community are supposed to produce a 'psychology' or 'mentality' which is unsympathetic to trade unionism. A review of the available evidence lends little support to this view. It suggests that white-collar workers do not possess any intrinsic qualities which make them less receptive to trade unionism than manual workers. At least the personal characteristics of white-collar workers do not account for variations in the density of white-collar unionism. If the white-collar workers' decision to join or not to join trade unions is not explained by the nature of their psyches, then perhaps it can be explained by such 'external' factors as their economic position or work situation. The influence of such factors is considered in the following chapters.

---

[1] Op. cit., pp. 150–1.

[2] See E. J. Hobsbawm, 'The Labour Aristocracy in Nineteenth-Century Britain', *Labouring Men* (London: Weidenfeld & Nicolson, 1964), pp. 272–315.

[3] Op. cit., p. 254.

V

THE ECONOMIC POSITION

THAT workers will join trade unions if they are dissatisfied with their terms and conditions of employment and the insecurity of their jobs, or refrain from joining if they are not, would seem to be common sense. But what appears to be common sense sometimes is revealed, on closer examination, to be little more than nonsense. Hence this chapter will systematically analyse the most important aspects of the white-collar workers' economic position in order to ascertain what influence it has, if any, on their decision to join or not to join trade unions.

EARNINGS

There always has been, and probably always will be, some overlap between the levels of white-collar and manual earnings. But, historically, white-collar workers have generally earned more than manual workers. This is still true of many white-collar workers today. Evidence to support these statements is presented in Table 5.1. The average annual earnings of all manual workers in 1922–4 were £149 for males and £93 for females; female industrial clerks were the only white-collar occupational group to earn less than these amounts. By 1960 the average annual earnings of all manual workers had increased to £663 for males and £343 for females; with the exception of male laboratory technicians and male clerks in the civil service, all the white-collar groups earned more than these amounts.

Not only have most white-collar workers generally earned more than manual workers as a whole, but many also have earned more than the best paid manual workers—the skilled. The only white-collar groups not to earn more than skilled manual workers in 1922–4 were male and female industrial clerks and male bank clerks. By 1960 all the male clerical groups were earning less than the skilled as were male laboratory technicians, but the rest of the white-collar groups were still better off.

The white-collar workers' economic position relative to that of manual workers is actually more favourable than these earnings figures suggest. Even leaving aside the question of fringe benefits,[1] there are a number of factors which tend to increase the economic disparity between the two groups of workers. In remissions from the Exchequer, which include both

---

[1] This is discussed *infra*, pp. 63–7.

TABLE 5.1

The Average Annual Earnings of Various White-Collar and Manual Occupational Groups in Great Britain, 1922/4–1960
(£ per annum)

Occupational group	Sex	1922–4	1935–6	1955–6	1960
1. All higher professionals	M	582	634	1,541	2,034
(a) Engineers	M	468	..	1,497	1,973
(b) Chemists	M	556	512	1,373	1,717
2. All lower professionals	M	320	308	610	847
and technicians	F	214	211	438	606
(a) Draughtsmen	M	250	253	679	905
(b) Laboratory technicians	M	201	186	420	536
3. Foremen	M	268	273	784	1,015
	F	154	156	477	602
4. All clerks	M	182	192	523	682
	F	106	99	317	427
(a) Railway clerks	M	221	224	559	751
(b) Industrial clerks	M	153	..	506	669
	F	87	..	305	412
(c) Bank clerks	M	174	223	627	746
	F	162	150	366	440
(d) Civil-service clerks	M	284	260	503	661
	F	171	155	360	520
5. All manual workers	M	149	159	527	663
	F	93	94	273	343
(a) Skilled manual workers	M	180	195	622	796
	F	87	86	317	395

deductible expenses and allowances for retirement schemes, white-collar workers gain much more than manual workers.[1] Professor Titmuss and others have shown that considerably higher state 'welfare' benefits can accrue to persons with higher incomes.[2] White-collar workers also generally have greater net worth and personal assets than manual workers. The Oxford Savings Survey of 1953 found that although the mean gross income of clerical and sales workers was lower than that of skilled manual workers—£403 as against £466—the former were better off by nearly a third both in terms of mean net worth—£394 as against £299—and of mean personal assets—£479 as against £368.[3] The same survey also demonstrated that the position of

[1] W. G. Runciman, *Relative Deprivation and Social Justice* (London: Routledge & Kegan Paul, 1966), p. 87.

[2] Titmuss gives an example based on the rates per annum for 1955–6 for child awards: a man earning £2,000 a year with two children under fifteen receives £97 while a man earning £400 receives £28. Over the lives of the two families, the first will receive a total of £1,455 while the second will receive a total of £422. See R. M. Titmuss, *Essays on 'The Welfare State'* (2nd ed.; London: Unwin University Books, 1958), p. 47.

[3] K. H. Straw, 'Consumers' Net Worth: The 1953 Savings Survey', *Bulletin of the Oxford University Institute of Statistics*, xviii (February 1956), pp. 12 and 14, tables 7 and 8, cited by Runciman, op. cit., p. 88.

white-collar workers was more favourable in terms of the ownership of stocks and shares. The mean among skilled manual workers was £4, with less than 1 per cent owning stocks and shares worth more than £100; the mean among clerical and sales workers was £50, with 2·6 per cent owning more than £100.[1] In respect of all these items, unskilled manual workers were even worse off while technical and managerial workers were even better off.

But although many, if not most, white-collar workers continue to receive larger incomes than manual workers, the relative pay of most white-collar groups has been reduced considerably over the past few decades. Table 5.2 shows that in spite of a slight increase in their relative earnings during the

TABLE 5.2

The Change in the Average Annual Earnings of Various White-Collar Occupational Groups Relative to the Change in the Average Annual Earnings of all Manual Workers in Great Britain, 1922/4–1960

(1922–4 = 100)

Occupational group	Sex	1922–4	1935–6	1955–6	1960
1. All higher professionals	M	100	102	75	78
(a) Engineers	M	100	..	90	95
(b) Chemists	M	100	86	70	69
2. All lower professionals	M	100	90	54	60
and technicians	F	100	98	70	77
(a) Draughtsmen	M	100	94	77	81
(b) Laboratory technicians	M	100	87	59	60
3. Foremen	M	100	95	83	85
	F	100	100	105	106
4. All clerks	M	100	98	81	84
	F	100	92	102	109
(a) Railway clerks	M	100	94	71	76
(b) Industrial clerks	M	100	..	94	98
	F	100	..	119	128
(c) Bank clerks	M	100	125	102	96
	F	100	92	77	74
(d) Civil-service clerks	M	100	86	50	52
	F	100	90	72	82

last few years, the average annual earnings of all the male white-collar groups relative to those of all male manual workers declined between 1922–4 and 1960 by amounts ranging from 2 per cent for industrial clerks to 48 per cent for civil-service clerks. During the same period, the earnings of fore-women, female industrial clerks, and female clerks as a whole relative to the earnings of all female manual workers increased by 6 per cent, 28 per cent,

[1] Ibid.

and 9 per cent respectively. But the relative earnings of female clerks in the civil service, female lower professionals and technicians, and female bank clerks declined by 18 per cent, 23 per cent, and 26 per cent respectively.

The narrowing of the white-collar manual earnings differential has been one of the most striking changes in pay structure during the twentieth century. Perhaps not surprisingly, therefore, many scholars,[1] as well as almost all the union and management officials interviewed during the course of this study, have claimed that this has been a major factor encouraging white-collar workers to join trade unions.

But the empirical evidence does not support this argument. White-collar workers whose earnings relative to those of manual workers have declined the most are not necessarily those who have been most ready to join trade unions. It is true that groups such as male and female civil-service clerks whose relative earnings have been reduced very greatly between 1922–4 and 1960 (see Table 5.2) are unionized very highly, and that other groups such as male and female industrial clerks whose relative earnings have deteriorated very little or even improved are unionized very poorly. But the reverse is equally true: the relative earnings of male laboratory technicians and chemists have been eroded very seriously and yet they are organized very poorly while the relative earnings of male bank clerks have been reduced very little and yet they are organized quite highly.

Nor does there appear to be any relationship between the absolute level of white-collar workers' earnings and the degree to which these workers are unionized. As can be seen from Table 5.1, draughtsmen are better paid than male clerks, yet the former are unionized much more highly than the latter. Male and female bank clerks, male railway clerks, and female civil-service clerks earn more than their opposite numbers in industry, yet industrial clerks are the most poorly organized clerical group. The lowest paid white-collar workers, like the lowest paid manual workers, are not generally among the most highly unionized.

Data on the earnings of white-collar workers are also available for the various manufacturing industries. The Ministry of Labour has been computing the average earnings of male and female weekly paid and monthly paid white-collar workers in manufacturing industries since 1959, and roughly comparable data can be obtained from Board of Trade sources from 1954 onwards. In addition, both the Ministry of Labour and the Board of Trade collect similar series on the earnings of manual workers. The figures are not given in sufficient detail to allow any conclusions to be drawn

[1] See, for example, G. D. H. Cole, 'Non-Manual Trade Unionism', *North American Review*, ccxv (January 1922), pp. 38–9; F. D. Klingender, *The Condition of Clerical Labour in Britain* (London: Martin Lawrence, 1935), *passim*; B. C. Roberts, *Trade Unions in a Free Society* (2nd ed.; London: Institute of Economic Affairs, 1962), p. 109; and E. M. Kassalow, 'White-Collar Unionism in Western Europe', *Monthly Labor Review*, lxxxvi (July 1963), p. 768.

regarding the absolute level of white-collar earnings prevailing in these industries.[1] But by employing the same methods as were used in the construction of Table 5.2, indices can be derived which give a fairly good idea of the change in white-collar earnings relative to the change in manual earnings in manufacturing industries between 1959–63 and 1954–63. These indices are given in Table 5.3.

As can be seen by comparing Tables 3.8 and 5.3, there is no obvious relationship between the degree to which white-collar workers in manufacturing industries have been unionized and the extent to which the differential between their earnings and those of manual workers has been narrowed. For example, the average annual earnings of all white-collar workers relative to the average annual earnings of all manual workers in other manufacturing industries; leather, leather goods, and fur; and clothing and footwear have increased between 1959 and 1963 and these are industries in which white-collar workers are very poorly unionized. But the relative earnings of all white-collar workers in metal manufacture and paper, printing, and publishing also have increased between 1959 and 1963 and these are industries in which the density of white-collar unionism is relatively high. Similarly, white-collar workers in shipbuilding and marine engineering are organized fairly well and their earnings have decreased relative to those of manual workers. But so have the earnings of white-collar workers in food, drink, and tobacco; timber and furniture; and chemicals; yet these workers are very poorly organized.

It is difficult, however, to discover the exact nature of the relationship between the pattern of unionization among white-collar workers in manufacturing industries and their pattern of relative earnings merely by a visual comparison of Tables 3.8 and 5.3. This can most easily be accomplished by undertaking a simple regression analysis. Consequently, each column of Table 3.8 was regressed on each row of Table 5.3.[2] The results of this regression analysis support the conclusion suggested by the visual assessment. The density of unionism among white-collar workers in manufacturing industries did not correlate significantly with the degree to which the differential between their earnings and those of manual workers had been reduced.[3]

[1] Since the proportion of white-collar employees in the various occupational and age-groups varies considerably by industry, the over-all averages given by the Ministry of Labour and the Board of Trade can not be used to compare the absolute level of earnings in one industry with another.

[2] Foremen were omitted from this analysis because the data presented in Table 5.3 excludes the earnings of all foremen except works foremen.

[3] Too many regressions were undertaken to give the results here in any detail. But well over 90 per cent of the correlation coefficients were very much below the figure required for significance at the 5 per cent level. Of those few which were significant, several possessed the wrong sign. That is, they had a positive sign, indicating that instead of the density of white-collar unionism being high where the white-collar/manual earnings differential had been narrowed the most, density was high where the differential had been narrowed the least. Scatter diagrams revealed that the remaining significant correlation coefficients were

TABLE 5.3

The Change in the Average Annual Earnings of White-Collar Workers Relative to the Change in the Average Annual Earnings of Manual Workers in Manufacturing Industries in the United Kingdom, 1959-63 (1959 = 100) and 1954-63 (1954 = 100)

| Industry | 1959-63 | | | | | | | 1954-63 |
| | Males | | | Females | | | All | All |
	Weekly paid	Monthly paid	Weekly and monthly paid	Weekly paid	Monthly paid	Weekly and monthly paid	Weekly and monthly paid	Weekly and monthly paid
Food, drink, and tobacco	93·5	89·9	95·8	96·5	93·5	100·3	96·8	89·2
Chemicals and allied	92·2	92·6	96·4	101·0	96·4	101·8	97·9	91·7
Metal manufacture	99·1	93·2	100·7	101·8	97·9	103·3	101·8	92·0
Metal N.E.S.	97·1	94·7	99·7	102·1	99·5	103·1	100·1	87·5
Engineering and electrical goods	98·1	96·6	102·3	104·0	102·7	105·9	104·5	100·8
Shipbuilding and marine engineering	95·6	85·9	96·0	91·8	89·8	92·2	95·6	97·9
Vehicles	98·8	98·0	103·4	99·6	89·9	100·4	103·9	94·5
Textiles	93·9	93·8	98·0	99·9	99·5	101·4	97·3	81·9
Leather, leather goods, and fur	102·0	96·5	103·7	102·7	96·5	102·8	101·9	86·5
Clothing and footwear	99·5	95·3	103·1	99·2	98·6	101·4	105·1	87·5
Bricks, pottery, glass, cement, etc.	90·9	90·8	95·8	98·4	98·3	101·6	96·7	86·9
Timber and furniture	93·1	95·8	99·4	99·3	94·4	99·2	97·1	83·0
Paper, printing, and publishing	99·8	95·1	103·3	101·9	98·1	104·6	104·2	85·9
Other manufacturing	97·0	97·3	99·7	96·4	100·4	98·4	101·4	87·5

Of course, the absence of significant correlation coefficients between the change in relative earnings of white-collar workers in manufacturing industries and the degree to which these workers are unionized does not necessarily 'prove' that a relationship between these two variables does not, in fact, exist. The above analysis contains too many statistical and methodological shortcomings to allow such a definite conclusion to be drawn.

The white-collar earnings statistics are of limited value. They are not exactly comparable with the union density figures for they exclude the earnings of all foremen except works foremen but include those of all managers and directors except those paid by fee only. This obviously causes the white-collar earnings to be higher than they otherwise would, and, more important, it may also affect the rate of change of these figures. In addition, these figures are averages for industries as a whole and give no information about the dispersion of earnings, earnings in particular occupations within each industry, or variations in earnings according to skill and age. While the figures are nevertheless of some use for the purpose at hand—to give a general indication of the change in white-collar earnings relative to the change in manual earnings—they would be even more useful and allow firmer conclusions to be drawn if they were given in greater detail.

Even if this analysis were free of these statistical difficulties, it would still be characterized by methodological shortcomings. Although there does not seem to be any relationship between the *change* in relative white-collar earnings and the *absolute* level of white-collar unionism, there might be a significant relationship between the *change* in relative white-collar earnings and the *change* in the level of white-collar unionism. Unfortunately, detailed figures on the density of white-collar unionism exist only for the year 1964, and it is thus impossible to calculate the change in union density in particular industries over time. It seems most improbable, however, that the trend of white-collar union density over the course of the twentieth century has followed faithfully that of relative white-collar earnings.

There is also the possibility that earnings and density of unionization may be mutually dependent variables. That is, not only may earnings influence the density of unionization, but the density of unionization may also influence earnings. If this is the case, then the nature of the relationship between them could be ascertained only by the construction of a simultaneous model.[1] In view of the limitations of the earnings data, the returns from building such a model probably would not justify the effort involved, and,

---

affected by one or two 'outliers'. That is, a very large proportion of the variance was caused by one or two points, and hence any regression line passing near them inevitably produces a very high correlation.

[1] That is, by simultaneously solving two equations: one in which the density of white-collar unionism is a function of the change in relative white-collar earnings and other variables, and another in which the change in relative white-collar earnings is a function of the density of white-collar unionism and other variables.

consequently, it has not been attempted. In any case, if there is some inter-dependence between these two variables, it would not seem to be very strong. The increase in white-collar earnings between 1948 and 1964 has been considerable; yet, as Chapter III demonstrated, the density of white-collar unionism held more or less constant during this period.

Thus the empirical evidence and analysis only suggest that there is no relationship between white-collar unionism and relative earnings; it does not prove this. But there are other reasons for believing that the narrowing of the white-collar/manual earnings differential has not been a major factor encouraging white-collar workers to join trade unions.

To begin with, whether or not it can even be said that the earnings differential has generally been narrowed depends upon the time period chosen. If the baseline is taken as 1922–4 or even 1935–6, it can be demonstrated (see Table 5.2) that the relative earnings of most white-collar groups have been reduced.[1] But if 1955 is taken as the baseline, it is obvious from Table 5.2 that the relative earnings of most white-collar groups have increased. Data from the Central Statistical Office confirms this trend. It indicates that between 1948 and 1963 the average annual earnings of all manual workers increased by 137 per cent while those of white-collar workers increased by only 118 per cent. But during the period 1955–63 white-collar earnings increased by 55 per cent while manual earnings increased by only 48 per cent.[2]

The reduction in the relative earnings of white-collar workers has 'not occurred gradually, as a result of a general tendency, but suddenly, within short periods and owing to an extraordinary conjuncture of circumstances'.[3] It was primarily during and immediately after the Second World War that the relative pay of white-collar workers was reduced; during the latter part of the post-war period, relative white-collar pay has tended to increase. Hence it is only by the standards of pre-war Britain that most white-collar workers can be said to have lost ground to manual workers. But 1938 is a long way off, and it is unlikely that most present-day white-collar workers know how their relative earnings of today compare with their relative earnings before the war, or, more exactly in view of the large proportion of white-collar workers who are now under thirty years of age,[4] how their relative earnings of today compare with those of the individuals who were filling their posts before the war. They are much more likely to know how their present

---

[1] The relative earnings of most white-collar workers probably were higher between 1922 and 1936 than at any other time during the period 1906–60. See Guy Routh, *Occupation and Pay in Great Britain 1906–60* (Cambridge: Cambridge University Press, 1965), chap. 2, especially table 48.

[2] Computed from unpublished data supplied by the Central Statistical Office. Ministry of Labour data is available for the period 1955–63 and agrees with that from the CSO. It indicates that white-collar earnings increased by 55·8 per cent while average weekly manual earnings increased by 48·9 per cent. See *Statistics on Incomes, Prices, Employment and Production*, No. 20 (March 1967), p. 8, table B.I.

[3] Routh, op. cit., p. x.   [4] See *supra*, pp. 46–7.

economic position compares with that of 1955 or later. If this is so, then the very phenomenon which so many people have relied upon to explain the growth of white-collar unionism in the post-war period—the narrowing of the white-collar/manual earnings differential—has, in fact, not even occurred.

Of course, while the relative earnings of most white-collar groups have increased over the past decade, those of some white-collar groups have not. But this does not mean that the relative earnings of all or even most of the individual workers within these groups also have fallen. Manual workers generally reach their final grade early in life, quickly rise to a peak of earnings within this grade, and then earn less towards retiring age.[1] White-collar workers, on the other hand, take much longer to reach their final grade,[2] and, even after they have reached it, their earnings are likely to increase right up to retirement because of the practice of awarding them annual increments based on age or length of service. Thus even when the relative earnings of a white-collar group are falling, those of individual white-collar workers within this group may be increasing because of promotion and age or length of service increments. For example, between 1947 and 1960 the average annual earnings of Civil Service Executive Officers as a class, as measured by the average of their pay scale, increased by only 97 per cent. But by 1960 the average annual earnings of an individual Civil Service Executive Officer who started working in 1947 at the age of eighteen would have increased by 362 per cent, even if he had not been promoted to a higher grade.[3] This is a rate of increase which few manual workers could match. Most white-collar workers, unlike their manual colleagues, are fortunate in possessing a number of escape routes from the 'tyranny of the "rate for the job"'.[4]

Although the number of white-collar workers whose earnings have been reduced relative to those of manual workers has been grossly exaggerated, there are no doubt some white-collar workers who have lost ground. There are probably also other white-collar workers who, whatever the realities of the situation, believe that their relative earnings have been reduced. But even granted this, it does not necessarily follow that these workers will respond by joining trade unions.

[1] For unskilled labourers the peak seems to be around age 30, and the drop in earnings from then to age 55–64 averages 15–20 per cent. For skilled men the peak is often around age 40 and the drop much smaller, around 10–15 per cent. See M. P. Fogarty, 'The White-Collar Pay Structure in Britain', *Economic Journal*, lxix (March 1959), p. 57. There are exceptions to this, notably in the steel industry, where a manual worker's earnings will increase markedly as a result of promotion within the range of manual work (Runciman, op. cit., p. 47).

[2] Male clerks, for example, seem most often to reach it about age 35–44 (Fogarty, loc. cit.).

[3] Computed from data provided in Guy Routh, 'Future Trade Union Membership', *Industrial Relations: Contemporary Problems and Perspectives*, B. C. Roberts, editor (London: Methuen, 1962), p. 67, table 2.

[4] Ibid., p. 65.

Those who have argued this have assumed that the more white-collar workers are proletarianized in terms of income, the more likely they are to follow the proletarian's example and join trade unions in order to improve or at least maintain their economic position. But an equally plausible assumption is that as white-collar workers become increasingly identified with manual workers in economic terms, they may react by emphasizing even more strongly those ways in which they are still different and upon which a claim to special prestige may be based.[1] One way of doing this is by refusing to join those organizations which have been reserved largely for manual workers—trade unions.

A similar point has been made by Runciman.[2] He has argued that workers may react to a relative deprivation of earnings primarily in one of two ways —'egoistically' or 'fraternalistically'. Egoists have no sense of common cause with others in a like situation; they want to 'better themselves', even at the expense of the group. Fraternalists, on the other hand, want to better their own situation by improving that of the group to which they belong. Runciman's research led him to conclude that the relative deprivation felt by most manual workers is likely to be 'fraternalistic' while that felt by most white-collar workers is likely to be 'egoistic'. If he is correct, then few white-collar workers are going to join trade unions simply because they have suffered a relative deprivation of earnings.

The response of white-collar workers to unions will also be influenced by whether they see management or unions as the cause of their relative disadvantage. Is management at fault for being insensitive to the needs and aspirations of white-collar employees and unappreciative of the contribution which they feel they make to the company's success? Or, are the unions at fault for forcing management to give part of the white-collar employee's share of the firm's earnings to manual workers? Those white-collar workers who adopt the latter interpretation of events may refuse to embrace those organizations which they see as being responsible for their predicament.

It must be remembered also that it is not at all certain that white-collar workers take manual workers as their comparative reference group on questions of earnings. At least those who have so claimed have offered little or no evidence in support of their contention. Besides, there is some evidence to the contrary. Runciman's research suggests that the reference group comparisons of most workers on economic matters have been and still are very much restricted: most manual workers compare themselves to other manual workers while most white-collar workers compare themselves to other white-collar workers.[3] If this is generally true, then most white-collar workers will not feel relatively deprived because their earnings have been

[1] See *supra*, p. 50.
[2] Op. cit., especially pp. 32–5 and 50.
[3] Ibid., chap. 4, and also pp. 192–208; see especially p. 196, table 20.

reduced relative to those of manual workers, and they are unlikely to join trade unions as a result.

If most white-collar workers do judge their earnings as satisfactory or unsatisfactory by comparison with those of other white-collar workers, it does not seem that those whose earnings compare least favourably are those who are most likely to join a union. Table 5.4 shows the change in earnings of white-collar workers in each manufacturing industry relative to the change in earnings of white-collar workers in the economy as a whole between 1959 and 1963. It is obvious that white-collar workers in almost every manufacturing industry have fared worse in terms of earnings than white-collar workers in most other sectors of the economy. Yet, as Chapter III made clear, white-collar workers in manufacturing industries are the most poorly unionized of all white-collar groups.

Nor does there appear to be any relationship within the manufacturing sector itself between the pattern of unionization of white-collar workers and the extent to which their earnings have been reduced relative to those of white-collar workers as a whole. A simple regression analysis, similar to that undertaken on Tables 3.8 and 5.3, was carried out on Tables 3.8 and 5.4. The result was much the same. The degree of unionism among white-collar workers in manufacturing industries generally was not significantly correlated with the extent to which the differential between their earnings and those of white-collar workers as a whole had been narrowed.[1]

The empirical evidence and analysis presented in the preceding pages suggests a twofold conclusion: white-collar workers have less reason to be dissatisfied with their salaries than is generally imagined, and even those who are unhappy will not necessarily respond by joining a trade union. The findings of several attitude surveys among white-collar workers support this conclusion.[2] Dale found that only 26 per cent of the 208 male industrial

---

[1] Once again, the overwhelming majority of the correlation coefficients were very much below the figure required for significance at the 5 per cent level. In fact, the only correlation coefficients which were significant at this level were those produced by regressing: (a) the density of unionization among draughtsmen on the change in the relative earnings of monthly paid, weekly paid, and monthly and weekly paid females, and (b) the density of unionization among other technicians on the change in the relative earnings of weekly paid males.

It is most surprising that there should be any relationship between female relative earnings and the degree of unionization among draughtsmen since less than 2 per cent of this occupational category are female, and it can only be assumed that the high correlation produced by (a) was the result of chance factors. The significant correlation coefficient produced by (b) cannot be dispensed with by such a priori considerations. But even if this correlation coefficient does indicate a significant relationship—and, in view of the statistical and methodological difficulties associated with the analysis, it would be unwise to draw such a firm conclusion—this particular case certainly does not provide the basis for a general argument that those white-collar workers whose earnings have been most seriously eroded relative to those of white-collar workers as a whole are those workers who are most likely to join trade unions.

[2] The findings of these surveys should be treated with caution not only because the

TABLE 5.4

The Change in the Average Annual Earnings of White-Collar Workers in Manufacturing Industries Relative to the Change in the Average Annual Earnings of White-Collar Workers in the Economy as a Whole in the United Kingdom, 1959–63 (1959 = 100)

Industry	Males			Females			All	
	Weekly paid	Monthly paid	Weekly and monthly paid	Weekly paid	Monthly paid	Weekly and monthly paid	Weekly and monthly paid	Weekly and monthly paid
Food, drink, and tobacco	96·4	92·7	98·8	94·4	91·4	98·1	98·1
Chemicals and allied	94·6	95·0	98·9	97·4	93·0	98·2	99·3
Metal manufacture	94·3	88·7	95·9	93·6	90·1	95·1	96·4
Metal N.E.S.	94·6	92·2	97·1	95·9	93·5	96·9	97·7
Engineering and electrical goods	94·6	93·1	98·6	94·8	93·6	96·6	98·8
Shipbuilding and marine engineering	93·8	84·3	94·2	88·9	86·9	89·3	93·4
Vehicles	95·8	95·0	100·2	93·9	84·8	94·8	100·2
Textiles	93·8	93·6	97·8	95·2	94·8	96·6	96·4
Leather, leather goods, and fur	100·0	94·7	101·7	97·1	91·2	97·2	98·4
Clothing and footwear	97·4	93·3	100·9	94·5	93·9	96·5	99·0
Bricks, pottery, glass, cement, etc.	92·5	92·5	97·6	95·6	95·4	98·6	98·4
Timber and furniture	92·4	95·1	98·7	97·2	92·4	97·1	96·6
Paper, printing, and publishing	96·7	92·1	100·1	95·4	91·8	97·9	100·2
Other manufacturing	96·4	96·8	99·1	92·7	96·6	94·7	100·2

clerks he interviewed considered the salary they received was not a 'fair' one.[1] Prandy found that only 30 per cent of his sample of 44 engineers and 40 per cent of his sample of 49 AScW members expressed any dissatisfaction with their salaries.[2] He also found that only 17 per cent of his sample of 286 metallurgists were dissatisfied with their salaries by comparison with those of other groups, and that the figure fell to 11 per cent when they compared their salaries with those of other metallurgists.[3] When Phillipson asked the 112 union members in his sample of clerks and technicians their reasons for membership, only 13 per cent of the answers he received specifically concerned salaries.[4] Similarly, when Prandy asked his sample of 49 AScW members their reasons for joining the union, none specifically mentioned dissatisfaction with salaries or other terms and conditions of employment.[5] In short, neither macro- nor micro-analysis reveals any significant connection between white-collar unionization and salaries.

OTHER TERMS AND CONDITIONS
OF EMPLOYMENT

Not only have white-collar workers generally earned more than manual workers, but they have also enjoyed a privileged position with respect to other terms and conditions of employment. Writing in 1916, the editor of *The Clerk* noted that 'although by reason of their unorganized state, clerks suffer many economic disabilities, yet they have a great many economic advantages not enjoyed by manual workers'. Among them he cited 'permanency of employment, periodical increases of salary, payment of salary during sickness and holidays, comparatively reasonable hours of work, and in certain sections superannuation'.[6] Writing in 1965, a journalist was still able to report that most 'office staff get longer holidays, better pensions and better sickness pay, work shorter hours and have longer notice than their fellows in the factories'. He also observed that the 'two groups come to work through different gates, park their cars in separate places, eat in separate places and excrete in separate places'.[7]

samples are small and something less than random, but also because they fail to control for other variables which might influence white-collar workers' attitudes towards trade unions. See *supra*, p. 45.
   [1] J. R. Dale, *The Clerk in Industry* (Liverpool: Liverpool University Press, 1962), pp. 23–4, table 13.
   [2] K. Prandy, *Professional Employees: A Study of Scientists and Engineers* (London: Faber, 1965), p. 115, table 3, and p. 164.
   [3] Ibid., p. 97, table 5.
   [4] C. M. Phillipson, 'A Study of the Attitudes Towards and Participation in Trade Union Activities of Selected Groups of Non-Manual Workers' (unpublished M.A. thesis, University of Nottingham, 1964), pp. 223–4, table 18.
   [5] Op. cit., p. 160, table 2.
   [6] Cited by David Lockwood, *The Blackcoated Worker* (London: Allen & Unwin, 1958), p. 40.
   [7] Jeremy Bugler, 'Shopfloor Struggle for Status', *New Society* (25 November 1965), p. 19.

Manual workers may have been able to narrow the differential between their earnings and those of white-collar workers and, in some cases, even reverse it in their own favour, but they have been obliged to work a longer day and a longer year in order to do so. Although the hours worked by manual workers have declined, they were still averaging over 44 hours a week in 1966[1] compared with roughly 37½ for white-collar workers.[2] White-collar workers had grown accustomed to holidays with pay, mostly of a fortnight, long before the First World War,[3] but holidays with pay for manual workers, usually of a week's duration, were only generally introduced towards the end of the inter-war period.[4] Since then, holiday entitlement has increased for both groups of employees, but manual workers still tend to lag behind. A study undertaken in the early 1960s found that manual workers were the 'least favoured group' in 67 per cent of the firms studied and that the maximum holiday granted to any manual worker was still only two weeks in over 46 per cent of the firms. In contrast, a maximum of two weeks for white-collar workers was found in only 13 per cent of the firms.[5] It is also unusual for manual workers to be paid for any time off for domestic reasons. A survey in 1965 by the Industrial Society of 180 of its member-firms indicated that only 27 per cent of them granted paid leave of absence for reasons other than sickness to manual workers, but 65 per cent extended this facility to white-collar employees.[6]

Another way in which white-collar and manual workers traditionally have differed is in terms of method of payment. Manual workers have generally

[1] *Statistics on Incomes, Prices, Employment and Production*, No. 20 (March 1967), p. 66, table D.1.

[2] *Status and Benefits in Industry* (London: Industrial Society, 1966), p. 41, and Philip Marsh, 'Recent Developments in Office Staff Practices', *Office Management* (Spring 1966), p. 40.

[3] For example, an inquiry among typists in 1906 found that holidays were a grievance 'as only a fortnight is normally allowed and paid for'. See B. L. Hutchins, 'An Enquiry Into the Salaries and Hours of Work of Typists and Shorthand Writers', *Economic Journal*, xvi (September 1906), p. 449.

[4] G. L. Cameron, 'The Growth of Holidays with Pay in Britain', *Fringe Benefits, Labour Costs and Social Security*, G. L. Reid and D. J. Robertson, editors (London: Allen & Unwin, 1965), pp. 273–85.

[5] *Holidays: Current Practice and Trends* (London: Industrial Welfare Society, 1963), pp. 18, 4, and 6.

[6] *Status and Benefits in Industry*, op. cit., p. 44. This study is one of the few comprehensive and up-to-date sources of information on both manual and white-collar 'fringe benefits'. It is thus extremely useful. But its findings should be treated with some caution as the sample is almost certainly biased towards the practice in 'better' firms. For instance, 84 per cent of manual employees covered by the sample firms were included in a pension scheme, compared with 45 per cent of all manual workers as shown by the Government Actuary's 1965 inquiry (*infra*, p. 66). There was also a higher coverage of manual workers by sickpay schemes among the sample firms than shown in the Ministry of Labour's inquiry of 1961 (*infra*, pp. 66–7). A great deal of information on the pay and conditions of manual and white-collar workers in engineering is contained in the National Board for Prices and Incomes' Report No. 49 and its supplement. See Cmnd. 3495 and 3495–I.

been paid a fluctuating wage at weekly intervals which is based at least partly on the number of hours worked. White-collar workers, on the other hand, have generally received a fixed salary at monthly intervals. Related to this, manual workers have been required to record the number of hours worked by clocking in and out, have been paid for overtime, and have generally received one week's notice, while white-collar workers have not been required to clock in and out, have not been paid overtime, and have generally received one month's notice.

Recently, however, there has been a tendency to reduce the differences between the two groups of workers in some of these respects. The Industrial Society found that although most of the firms in its sample still pay their manual employees a fluctuating wage based on payment-by-results systems, a considerable proportion (30 per cent) now pay them a standard salary.[1] A weekly pay interval is now the usual practice for both groups of employees, with all the sample firms following this practice for manual workers and 81 per cent following it for white-collar workers.[2] An example of the levelling-up process working in reverse is given by practice in respect of payment for overtime: 72 per cent of the firms in the Industrial Society's sample now pay their white-collar employees for overtime (as compared with 99 per cent for manuals) while the remainder give themt ime off in lieu.[3] In spite of the passing of the Contracts of Employment Act in 1963, white-collar workers still appear to be better provided for than manual workers. As a result of the Act, both groups of workers can now claim one, two, or four weeks' notice depending solely upon the number of years of continuous service with their employer. But the Industrial Society found that while notice of termination for manual workers was limited to the provisions of the Act in 76 per cent of the sample firms, this was true for white-collar employees in only 54 per cent of the firms.[4] In regard to clocking, the traditional pattern has been maintained. Only 33 per cent of the Industrial Society's sample firms used this method for recording the attendances of white-collar employees, but 93 per cent still used it for manual employees. Moreover, in 96 per cent of the firms manual employees forfeit payment when late while in only 11 per cent of the firms do white-collar employees do likewise.[5]

[1] Ibid., pp. 41–2. Only 2 of the 180 companies paid their white-collar employees a fluctuating wage.

[2] Ibid., p. 42. A survey by the Institute of Office Management in 1965 of 485 firms employing 108,380 office workers also found that roughly 80 per cent of the establishments paid their clerical employees by the week (Marsh, op. cit., p. 38).

[3] Loc. cit.   [4] Op. cit., p. 43.

[5] Ibid., pp. 42–3. Of course, many of the firms used other methods for recording the attendances of white-collar employees. While only 33 per cent used clocking, 16 per cent used signature books and 11 per cent used recording by officials. But in 46 per cent of the firms there was no recording at all for staff employees. The Institute of Office Management survey (Marsh, op. cit., p. 44) supports this: only 50 per cent of the firms in its sample maintained time-keeping records for staff employees in 1965 (as compared with 76 per cent in 1952).

White-collar workers have been, and continue to be, better placed for retirement than manual workers, although the latter are catching up in this respect. The surveys by the Government Actuary, some of the results of which are presented in Table 5.5, show that although pension coverage for manual workers is increasing more quickly than for white-collar workers, the gap remains wide. Not only are manual workers at a quantitative disadvantage in terms of pensions, they are also at a qualitative disadvantage. The Industrial Society survey shows that in only 37 per cent of the firms in which pension schemes were provided for both manual and white-collar workers were the provisions the same for both groups, with a strong presumption that they were less favourable for manual workers.[1] In addition, Carr-Saunders and his colleagues have noted that whereas many pension schemes for white-collar workers are based on a capital payment and a pension tied to peak earnings, arrangements for manual workers are generally based on length of service only.[2]

TABLE 5.5

Proportion of Employees in Private Industry in the United Kingdom Covered by Occupational Pension Schemes, 1956 and 1963

	1956	1963
White-collar employees		
Male	71	80
Female	34	40
Manual employees		
Male	38	55
Female	23	15[a]

[a] The proportion of women manual workers covered by occupational pension schemes has fallen, presumably as a result of the introduction of the National Insurance Graduated Pension Scheme.

In line with general trends, an increasing number of sick-pay schemes are being made available to manual workers, not only by individual employers but also by the state.[3] But manual workers are still at a disadvantage in this respect compared with white-collar workers. A survey by the Ministry of Labour in 1961 revealed that 86 per cent of white-collar workers in private industry were covered by sick-pay schemes compared with 33 per cent of manual workers.[4] There was also data which showed that the amount of pay and the length of time for which it was made available for manual workers are less than for white-collar workers. This is supported by data from the

[1] Op. cit., p. 43.

[2] A. M. Carr-Saunders *et al.*, *A Survey of Social Conditions in England and Wales as Illustrated by Statistics* (Oxford: Oxford University Press, 1958), p. 198.

[3] A state earnings-related sick-pay scheme came into effect in October 1966 and applies equally to both manual and white-collar workers.

[4] *Sick Pay Schemes: A Report* (London: HMSO, 1964), p. 5, table A. See also G. L. Reid, 'Sick Pay', *Fringe Benefits, Labour Costs and Social Security*, op. cit., pp. 218–22.

Industrial Society's survey. Although 69 per cent of the sample firms provided sick-pay schemes for their manual employees, the benefits provided were the same as those for white-collar employees in only 21 per cent of these firms.[1]

Differences between white-collar and manual workers are also being reduced in respect of so-called 'company welfare benefits'—canteens, playing-fields, social clubs, and the like. Ninety-four per cent of the firms in the Industrial Society's sample provided canteen facilities for both manual and white-collar employees, and in the majority of these firms (60 per cent) facilities were the same for both groups. Similarly, 83 per cent of the firms provided recreational facilities, and in every case manual and white-collar workers shared the same facilities.[2]

Two conclusions emerge fairly clearly from the information which has been presented on white-collar fringe benefits. First, white-collar workers have, and continue to enjoy, a privileged position compared with manual workers. Second, the white-collar workers' relative advantage has dwindled somewhat during the post-war period. There is every likelihood that their relative advantage will be reduced even further, as the social trend in industry is against the maintenance of these differentials. What impact this has had and is likely to have in the future on white-collar unionism, it is difficult to say with any exactness. But for similar reasons to those advanced in respect of earnings,[3] it is very doubtful if either the absolute or relative level of white-collar fringe benefits is an important factor in explaining the pattern of white-collar unionism in this country. This conclusion is strengthened when it is remembered that fringe benefits, unlike earnings, tend to be very similar for most white-collar occupations within a given industry. They are, therefore, unlikely to explain a phenomenon such as white-collar unionism which shows considerable occupational variation even within a particular industry.

EMPLOYMENT SECURITY

Lockwood has argued that security of employment 'was perhaps the most significant difference between manual and non-manual work, for, although it fell short of the full independence which comes with property, job-security did constitute a partial alternative to ownership, conferring on the clerk a relative immunity from those hazards of the labour market which were the lot of the working classes'.[4] Most people would agree with him. Indeed, the security of white-collar employment was its major attraction in the eyes of manual workers. 'If,' commented the Pilgrim Trust investigators in 1938, 'working men and women seem to be unduly anxious to make their sons and

[1] Loc. cit.
[2] Ibid., p. 44. See also A. G. P. Elliott, 'Company Welfare Benefits', *Fringe Benefits, Labour Costs and Social Security*, op. cit., pp. 300–12.
[3] See *supra*, pp. 58–63.   [4] Op. cit., p. 204.

daughters into clerks, the anxiety behind it is not for more money but for greater security.'[1]

Prior to the First World War most white-collar workers could safely assume that they would be 'permanently employed, provided they were efficient and their character good'.[2] Even during the decades of the twenties and thirties when being out of work was the normal situation for a large proportion of the population, white-collar workers were less affected by unemployment than manual workers. Those on the railways as well as those in public service and banking[3] hardly suffered from unemployment at all. Those in private industry suffered somewhat, but even they were affected less seriously by unemployment than manual workers. Using data provided by the 1931 Census, Colin Clark has calculated that unemployment among male unskilled manual workers was 30·5 per cent and among skilled and semi-skilled manual workers 14·4 per cent, while that among shop assistants was 7·9 per cent, clerks and typists 5·5 per cent, professionals 5·5 per cent, and higher office workers 5·1 per cent.[4] Generally speaking, those higher in occupational status suffered less than those lower down.

While unemployment during the twenties and thirties was less among white-collar than among manual workers, it was nevertheless sufficiently high 'to destroy the traditional association of security and blackcoat employment'.[5] The General Secretary of the National Federation of Professional Workers estimated the total number of white-collar unemployed in 1934 at between 300 and 400 thousand.[6] Older and more senior white-collar workers were particularly hard hit. They often fell outside the National Insurance limit and were therefore ineligible for unemployment relief payments.[7] Older men also found it hardest to obtain another job and thus their period of unemployment was often of considerable duration. Even when they found another job, it was usually at work very much below their previous status and remuneration.[8] In short, although white-collar workers were not affected as badly by the depression as manual workers 'in bare quantitative terms', to quote Lockwood once again, 'they suffered as acutely as almost any other group due to the lack of communal provision for their plight and the conventional expectations of their position'.[9]

[1] Pilgrim Trust, *Men Without Work* (Cambridge: Cambridge University Press, 1938), p. 144.
[2] Memorandum submitted to the Royal Commission on Unemployment Insurance by Herbert Elvin, General Secretary of the CAWU, *The Clerk* (October 1931), p. 46, cited by Lockwood, op. cit., p. 56.
[3] Lockwood (ibid., p. 55) cites the *Bank Officer* (June 1932), p. 7, to the effect that in 1932 there were only twelve or fourteen bank clerks unemployed out of a total membership of 21,000.
[4] Colin Clark, *National Income and Outlay* (London: Macmillan, 1938), p. 46, table 19.
[5] Lockwood, op. cit., p. 57.
[6] Letter in *Manchester Guardian*, 25 April 1934, cited by Klingender, op. cit., p. 92.
[7] *Report of the Unemployment Insurance Statutory Committee on the Remuneration Limit for Insurance on Non-Manual Workers*, 1936.
[8] Klingender, op. cit., pp. 91–9, and Lockwood, op. cit., pp. 56–7.   [9] Ibid., p. 57.

What role the experiences of the twenties and thirties play in determining the behaviour of present-day white-collar workers, it is difficult to say. But it is probably small.[1] To have been affected by the unemployment of the thirties, never mind that of the twenties, a present-day worker would have to be over forty-five years of age, and few white-collar employees are. Only 28 per cent of clerks, 14 per cent of draughtsmen, and 15 per cent of laboratory technicians were over this age in 1961.[2] Most of today's white-collar workers were born during and after the Second World War, and they, like younger manual workers, do not share the fear of unemployment which overshadowed their parents.[3] Nor is there any reason why they should. For while the average annual unemployment rate between 1921 and 1939 was 13·9 per cent, that between 1946 and 1963 was only 1·9 per cent.[4] And most of this unemployment has been confined to manual workers.[5]

Given the relatively full employment conditions of the post-war period, white-collar workers can find new jobs fairly quickly even after a large-scale and geographically concentrated redundancy. Dorothy Wedderburn found that almost 50 per cent of the 1,000 white-collar workers made redundant by an aviation company in 1962 had secured other jobs even before their period of notice expired. Within two weeks of leaving, 70 per cent were working again, and 90 per cent had new jobs within six weeks of leaving.[6] The redundancy caused some hardship: the earnings of almost all the dismissed employees were lower in their new jobs, and many lost 'fringe benefits' such as pension rights. But the impact of the redundancy was softened by a system of *ex gratia* payments which averaged $3\frac{1}{2}$ to $4\frac{1}{2}$ weeks' pay for weekly staff and $3\frac{1}{2}$ to $4\frac{1}{2}$ months' pay for monthly staff. While it would be unwise to generalize too widely on the basis of a single case study, it at least indicates that the duration of white-collar unemployment may not always be as long and the hardship resulting from it may not be as great as is commonly believed.

[1] For a contrary view see W. D. Wood, 'An Analysis of Office Unionism in Canadian Manufacturing Industries' (unpublished Ph.D. thesis, Princeton University, 1959), p. 196.

[2] See *supra*, p. 46, table 4.2.

[3] For manual workers see N. Dennis, F. Henriques, and E. Slaughter, *Coal Is Our Life* (London: Eyre & Spottiswoode, 1956). For white-collar workers see Dale, op. cit., p. 20, table 10, and p. 84.

[4] *The British Economy: Key Statistics 1900–1966* (London: London & Cambridge Economic Service, 1967), p. 8, table E.

[5] Between September 1961, when the Ministry of Labour first began to segregate white-collar workers in the quarterly unemployment figures, and December 1963 the number of white-collar workers unemployed averaged only 19 per cent of total employment, although they composed roughly 36 per cent of the insured population. To look at it another way, the white-collar unemployment rate was roughly half the manual. See, for example, 'Occupational Analysis: Wholly Unemployed Adults and Unfilled Vacancies for Adults, December 1963', *Ministry of Labour Gazette*, lxxii (February 1964), pp. 66–7. The following occupational groups were counted as white-collar: clerical workers; shop assistants; and administrative, professional, and technical workers.

[6] Dorothy Wedderburn, *White-Collar Redundancy: A Case Study* (Cambridge: Cambridge University Press, 1964).

It has been suggested that white-collar workers are likely to be more affected by unemployment in the future than they have been in the past as a result of office automation, and that this will encourage them to seek job security through trade union membership.[1] But current research indicates that very few employees are discharged as a result of a computer installation. The most comprehensive survey of office automation undertaken in this country found that only 13 out of 331 organizations which had installed a computer discharged any staff as a result. In each case the employees discharged were few in number—averaging less than ten per installation—and were mostly married and part-time women workers.[2] This finding is supported by almost every detailed case study undertaken in this and several other industrialized countries.[3]

Nor is office automation likely to reduce greatly the over-all demand for office workers in the foreseeable future. The Ministry of Labour estimates that the net effect of office automation up to 1 January 1965 has been to reduce the total number of office jobs which otherwise would have been available by about three-quarters of 1 per cent. It goes on to predict that even by 1975 the number of additional office jobs being created will still outnumber those being eliminated, and concludes that the most probable effect of computers over the next decade will be to 'offer some relief to a growing shortage of office workers'.[4]

It might be argued that whatever the realities of the situation, white-collar workers will nevertheless fear that automation will make them redundant and join trade unions as a result. Unfortunately, the evidence available on white-collar workers' attitudes to automation is rather inconclusive. Some researchers report a considerable degree of anxiety and unrest while others are impressed by the equanimity with which office workers contemplate the change.[5]

[1] See, for example, the sources cited in *Effects of Mechanisation and Automation in Offices* (Geneva: ILO, 1959), p. 106; J. C. McDonald, *Impact and Implications of Office Automation* (Ottawa: Department of Labour, Economic and Research Branch, 1964), chap. 8; and G. M. Smith, *Office Automation and White-Collar Employment* (New Brunswick, N.J.: Rutgers University, Institute of Management and Labor Relations, Bulletin No. 6, 1959), pp. 19–20.

[2] Ministry of Labour, *Computers in Offices* (London: HMSO, 1965), p. 18.

[3] The literature on office automation is vast, but for a sample of the studies supporting this point see the following: R. B. Thomas, *Computers in Business* (Wednesbury: Staffordshire College of Commerce, 1964); Enid Mumford and Olive Banks, *The Computer and the Clerk* (London: Routledge & Kegan Paul, 1967); W. H. Scott, *Office Automation and the Non-Manual Worker* (Paris: OECD, 1962); Roy B. Helfgott, 'EDP and the Office Work Force', *Industrial and Labor Relations Review*, xix (July 1966), pp. 503–16; Ida R. Hoos, *Automation in the Office* (Washington: Public Affairs Press, 1961); Ida R. Hoos and B. L. Jones, 'Office Automation in Japan', *International Labour Review*, lxxxvii (June 1963), pp. 3–24; *Automation and Non-Manual Workers* (Geneva: ILO, 1967), pp. 99–100.

[4] Op. cit., pp. 7 and 42.

[5] See, for example, Thomas, op. cit., p. 36; Mumford and Banks, op. cit., pp. 196–9; and F. C. Mann and L. K. Williams, 'Some Effects of the Changing Work Environment in the Office', *Journal of Social Issues*, xviii (July 1962), pp. 90–101.

In any case, the purpose here is not to predict the future, but to explain the present. It may be uncertain whether office automation will cause unemployment and insecurity among white-collar employees in the future,[1] but it is clear that since the Second World War unemployment or insecurity among white-collar workers from this or any other cause has been negligible. Therefore, unemployment and the fear of unemployment cannot be considered important determinants of the existing pattern of white-collar unionism.

CONCLUSIONS

As white-collar workers have become more plentiful, they have played an increasingly important role in the social, political, and economic life of the nation. But, paradoxically, at the same time they have also become worth less in economic terms; their economic position relative to that of manual workers has become less favourable. White-collar workers have lost something, but only in the sense that the terms and conditions of employment to which they previously had an exclusive right are now being shared by other workers.

If this has left white-collar workers unhappy, it does not seem to have encouraged them to unionize. At least, it is not possible to demonstrate any connection between the economic position of various white-collar groups and the degree to which they are unionized. This does not mean that white-collar workers who join unions are not interested in higher salaries, better fringe benefits, and greater security. They obviously are. But then, so is almost everyone else including those white-collar workers who do not join trade unions. In fact, it may be just because these objectives are so widely appreciated and pursued, that the workers' economic position is not a differentiating factor with respect to their propensity to unionize. Whatever the reason, economic factors do not seem to offer an explanation for the occupational and industrial variations in the density of white-collar unionism in Britain. For this, attention must be directed to another aspect of the white-collar workers' environment—their work situation.

---

[1] The impact of office automation upon work techniques and conditions and the effect this might have on white-collar unionism is considered *infra*, pp. 82–4.

VI

THE WORK SITUATION

IN spite of the increased amount of leisure which has become available in modern society, for the majority of people work remains the most important aspect of adult life. Some sociologists have gone so far as to claim that 'work is not a part of life, it is literally life itself', and its impact 'is found in almost every aspect of living and even in the world of dreams and unconscious fantasies'.[1] Other sociologists have been more moderate in their statements, but nevertheless have seen experience at work as a pervasive influence on a person's life. In Lockwood's view, 'the most important social conditions shaping the psychology of the individual are those arising out of the organization of production, administration and distribution. In other words, the "work situation".'[2] Even economists have noted the importance of this factor. Alfred Marshall wrote that 'the business by which a person earns his livelihood generally fills his thoughts during by far the greater part of those hours in which his mind is at its best' and that 'during them his character is being formed by the way in which he uses his faculties in his work, by the thoughts and the feelings which it suggests, and by his relations to his associates in work, his employers or his employees'.[3]

Perhaps not surprisingly, therefore, social scientists have relied heavily upon this factor in analysing white-collar unionism. In explaining its growth they have isolated four aspects of the work situation as being particularly significant: the degree of employment concentration, opportunities for promotion, the degree of mechanization and automation, and proximity to unionized manual workers. Hence the following analysis will concentrate upon them.

EMPLOYMENT CONCENTRATION

There are several reasons why density of unionization is likely to be higher among larger rather than smaller groups of employees. To begin with, the larger the number of employees in a group the more necessary it becomes to administer them in a 'bureaucratic' fashion. In the present context the essential feature of bureaucratic administration 'is its emphasis on the office

---

[1] D. C. Miller and W. H. Form, *Industrial Sociology: An Introduction to the Sociology of Work Relations* (New York: Harper, 1951), p. 115.

[2] David Lockwood, *The Blackcoated Worker* (London: Allen & Unwin, 1958), p. 205.

[3] *Principles of Economics* (8th ed.; London: Macmillan, 1920), p. 1.

rather than upon the individual office-holder'.[1] This means that employees are treated not as individuals but as members of categories or groups. Their terms and conditions of employment as well as their promotion prospects are determined not by the personal considerations and sentiments of their managers but by formal rules which apply impersonally to all members of the group to which they belong. The result is that the group's working conditions tend to become standardized.[2]

'Bureaucratization' is the administrative answer to the problem of governing large numbers of employees. It makes for administrative efficiency in a business. But it is also likely to assist the growth of unionism. For, as Dubin has argued, making the rule for the work group rather than for the individual worker is likely to affect him in the following ways:

He becomes aware of his personal inability to make an individual 'deal' for himself outside the company rules and procedures, except under the circumstances of a 'lucky break'. He tends also to view himself as part of a group of similarly situated fellow-employees who are defined by the rules as being like each other. In addition, uniform rule-making and administration of the rules make unionism easier and, in a sense, inevitable. It should be reasonably clear that collective bargaining is joint rule-making. It is no great step to the joint determination by union and management of rules governing employment from the determination of them by management alone. Both proceed from the basic assumption that generally applicable rules are necessary to govern the relations between men in the plant. Once a worker accepts the need for general rules covering his own conduct, he is equally likely to consider the possibility of modifying the existing ones in his favor rather than to seek their total abolishment.[3]

Since the rules apply to him as a member of a group rather than as an individual, the most effective way of modifying them in his favour is by collective rather than individual bargaining.

But the greater degree of bureaucratization associated with larger groups of employees is not the only reason they are likely to be more highly unionized. Another reason is that trade unions tend to concentrate their recruiting efforts on such groups.[4] It is fairly obvious why they should do this. Larger groups of employees are probably more favourably disposed towards trade

---

[1] R. M. Blackburn and K. Prandy, 'White-Collar Unionization: A Conceptual Framework', *British Journal of Sociology*, xvi (June 1965), p. 117.

[2] The terms 'bureaucratic administration' and 'bureaucratization' have acquired a number of meanings in sociological writings. For a discussion of these see Richard H. Hall, 'The Concept of Bureaucracy: An Empirical Assessment', *American Journal of Sociology*, lxix (July 1963), pp. 32–40, and C. R. Hinings, *et al.*, 'An Approach to the Study of Bureaucracy', *Sociology*, i (January 1967), pp. 61–72. It is important to note that these terms are used in a very restricted sense throughout this study to refer simply to a method of administering the labour force.

[3] Robert Dubin, 'Decision-Making by Management in Industrial Relations', *Reader in Bureaucracy*, Robert K. Merton, *et al.*, editors (Glencoe, Ill.: The Free Press, 1952), p. 234.

[4] See *infra*, pp. 90–100 for a discussion of the policies followed by unions in recruiting members.

unionism because of the bureaucratic manner in which they are governed on the job, and they are therefore likely to be easier to recruit. They are also likely to be less expensive to recruit. Trade union recruiting is characterized by economies of scale: in general, the larger the group recruited the lower the *per capita* costs. Similarly, larger groups of members are less expensive for unions to administer: the larger the group the greater the probability that one or two of its members will possess the qualities required for leadership at the rank and file level, and the easier it is to police the collective agreement and ensure that its provisions are observed.[1] Moreover, collective agreements covering large groups of employees have a greater impact on the general level of salaries and conditions than a whole series of agreements covering small groups. Finally, the more members a union recruits the more power it is able to wield in negotiations with employers as well as within the labour movement.

This *a priori* reasoning tends to be supported by empirical evidence. In the United States, studies by the Bureau of Labor Statistics,[2] Cleland,[3] Meyers,[4] and Steele and McIntyre[5] have found a strong positive relationship between the size of establishments and the extent to which they are unionized. Studies in Norway,[6] Sweden,[7] Austria,[8] and Japan[9] indicate that the level of unionism is higher in larger than in smaller offices. In reviewing the extent and nature of white-collar unionism in eight countries, Sturmthal finds that its density is generally higher in the public than in the private sector of the economy, and concludes that this is primarily because public employees tend to be concentrated in large groups which are administered in a bureaucratic fashion.[10]

The argument is also supported by a wealth of evidence which Lockwood

[1] For example, when the Wages Council system was established, it was expected that it would promote the growth of voluntary collective bargaining. But this expectation has not been fulfilled largely because 'neither employers nor workers want to lose the services of the wages inspectorate in enforcing wage rates' in industries characterized by a large number of small firms. See Ministry of Labour, *Written Evidence*, p. 117.

[2] 'Extent of Collective Agreements in 17 Labor Markets, 1953–54', *Monthly Labor Review*, lxxviii (January 1955), p. 67.

[3] Sherrill Cleland, *The Influence of Plant Size on Industrial Relations* (Princeton, N.J.: Industrial Relations Section, Princeton University, 1955), pp. 14–21.

[4] Frederic Meyers, 'The Growth of Collective Bargaining in Texas—A Newly Industrialized Area', *Proceedings of the Industrial Relations Research Association*, ix (December 1956), p. 286.

[5] H. Ellsworth Steele and Sherwood C. McIntyre, 'Company Structure and Unionization', *The Journal of the Alabama Academy of Science* (January 1959), p. 38.

[6] Egil Fivelsdal, 'White-Collar Unions and the Norwegian Labor Movement', *Industrial Relations*, v (October 1965), p. 85 n. 7.

[7] Arne H. Nilstein, 'White-Collar Unionism in Sweden', *White-Collar Trade Unions*, Adolf Sturmthal, editor (Urbana: University of Illinois Press, 1966), pp. 275–6.

[8] Ernst Lakenbacher, 'White-Collar Unions in Austria', ibid., p. 53.

[9] Solomon B. Levine, 'Unionization of White-Collar Employees in Japan' ibid., pp. 222–3.

[10] Ibid., pp. 379–80.

has collected for Great Britain. He demonstrates that in the civil service, especially after the reorganization of 1920, 'a clear-cut classification of functions, qualifications, remuneration and criteria of advancement permitted a high degree of standardization of conditions throughout government departments . . . [and] . . . the resulting isolation of a clerical class, common to the service and made up of individuals whose chances of promotion were relatively small, provided the basis for the Civil Service Clerical Association'.[1] Similarly, he shows that in local government 'the growth of NALGO . . . has gone hand in hand with the subordination of local particularism in working conditions to a set of national standards common to the service'.[2] Conversely, he argues that the 'relatively small size of the office, the internal social fragmentation of the office staff through occupational, departmental and informal status distinctions, and the absence of any institutionalized blockage of mobility' characteristic of the private sector largely accounts for its low degree of clerical unionization.[3] He also establishes that the administration of banking is neither as bureaucratic as that of the public sector nor as paternalistic as that of private industry, and suggests that this explains why banking has an intermediate degree of clerical unionization. Other researchers have found that even the extent to which individual banks are unionized is partly accounted for by the extent to which their size forces them to administer their employees in a bureaucratic manner.[4]

If this line of reasoning is generally valid, then it should also help to account for the pattern of white-collar unionism in manufacturing industries in Britain. Fortunately, sufficient data is available to determine whether or not it does. Table 6.1 gives the average number of employees per establishment in each white-collar group for each manufacturing industry. These figures are the only readily available index of white-collar employment concentration in manufacturing industries, and hence they are extremely useful. But certain objections might be raised against using them to assess the validity of the argument being advanced here.

The degree of employment concentration would seem to be an accurate enough measure of the economies of scale characteristic of trade union recruitment and administration, but it might not adequately reflect the degree of bureaucratization which actually exists. An establishment may have relatively few white-collar employees yet administer them in a bureaucratic fashion, for it may be part of a much larger company with a centralized personnel policy. If this is the case, then regardless of how few employees there may be at a single establishment, they will probably be subject to company-wide grading schemes and salary structures, and, in general, their terms and conditions of employment will be determined by formal rules

[1] Op. cit., pp. 142–3. [2] Ibid., p. 145. [3] Ibid., p. 207.
[4] Blackburn and Prandy, op. cit., p. 118 n. 17.

TABLE 6.1

The Average Number of White-Collar Employees per Establishment in Great Britain by Industry and by Occupation

Occupational group	Food, drink, tobacco	Chemical	Metal manuf.	Metal N.E.S.	Eng. elect.	Ship. M.E.	Vehicles	Textiles	Leather, fur	Clothing, footwear	Bricks etc.	Timber, furn., etc.	Paper, print, pub.	Other manuf.	All manuf.
1. Foremen	3·12	6·27	8·48	1·68	4·52	7·80	12·86	4·02	1·12	1·39	2·29	1·01	1·77	3·65	3·25
2. All scientists, technologists, technicians	0·81	12·06	6·82	0·95	10·45	5·98	22·62	1·38	0·19	0·26	1·26	0·22	0·49	2·24	3·57
(a) Scientists, technologists	0·30	4·79	1·72	0·15	2·21	0·45	3·05	0·35	0·07	0·02	0·34	0·01	0·16	0·61	0·82
(b) All technicians	0·51	7·27	5·10	0·80	8·23	5·53	19·57	1·02	0·11	0·24	0·92	0·21	0·33	1·63	2·76
(i) Draughtsmen	0·12	0·84	1·54	0·51	4·34	4·07	8·65	0·18	0·01	0·03	0·44	0·14	0·05	0·53	1·21
(ii) Other technicians	0·39	6·42	3·56	0·28	3·89	1·45	10·91	0·83	0·10	0·21	0·48	0·07	0·27	1·10	1·55
3. Clerks	9·13	20·75	20·52	5·27	18·34	7·97	44·46	6·00	2·87	3·39	5·47	2·77	8·63	10·11	9·92
4. Other white-collar workers	4·13	8·56	5·78	1·41	6·03	3·43	12·08	1·32	0·60	0·62	1·64	0·59	3·40	4·56	3·31
5. All white-collar workers	17·21	47·66	41·61	9·33	23·61	25·20	92·04	12·73	4·78	6·12	10·68	4·60	14·32	20·59	20·07

which apply impersonally throughout the company. But although these employment concentration figures may therefore understate the degree of bureaucratization, there is no evidence to suggest that the extent to which they do so varies significantly from one manufacturing industry to another. Thus this limitation of the figures does not greatly reduce their usefulness for the purposes of the following analysis.

Rather than understating the degree of bureaucratization, the employment concentration figures may actually overstate it. Blackburn and Prandy have suggested that private employers generally try to resist the pressures towards greater bureaucratization because of an 'ideological need' for loyalty from their employees. Bureaucratic administration is often tempered with a measure of 'administrative particularism' whereby each employee is treated as much as possible as an individual and not just as one of a group.[1] But the more workers a firm employs the more difficult it becomes to administer them in a 'particularistic' manner. For as Ingham has argued:

In the absence of bureaucratic rules, effective coordination and control within an industrial organization requires a body of norms which are shared by both management and men . . . . small organizations, with a high degree of vertical inter-action, favour the development of these norms . . . large organizations inhibit such vertical interaction and therefore favour the use of bureaucratic rules in the problem of the administration of the labour force and its work. Such rules are, in this case, important in the 'remote control' of the organization which, by virtue of its size, is difficult to deal with in any other way.[2]

Thus although employers may be able to slow down the trend towards bureaucratization, the problems posed by governing large numbers of employees nevertheless tend to lead to bureaucratization. Hence it does not seem unreasonable to take the degree of employment concentration as a measure of bureaucratization.

Even granted this, the employment concentration figures still possess at least one other major limitation: they are expressed as averages and give no information about the dispersion of employment. Different occupational groups are likely to be distributed in different ways across the administrative hierarchy of a firm as well as across the various firms within an industry. For example, draughtsmen are generally concentrated in one or two departments within a firm, while clerks are usually dispersed among a number of different departments. Similarly, clerks are likely to be more evenly distributed across the various firms within an industry than are draughtsmen. Virtually all firms require at least a few clerks, but not all firms require draughtsmen. Thus the distribution of draughtsmen is probably skewed in the direction of larger work groups and larger firms, while that of clerks is

---

[1] Blackburn and Prandy, op. cit., pp. 117–18.
[2] Geoffrey K. Ingham, 'Organizational Size, Orientation to Work and Industrial Behaviour', *Sociology*, i (September 1967), pp. 243–4.

probably skewed in the direction of smaller work groups and smaller firms. But simply because the techniques of production dictate that in all industries clerks are more numerous than draughtsmen, the average number of clerks per firm is generally larger than the average number of draughtsmen. It would be wrong, however, to conclude that the employment of clerks is therefore more concentrated than that of draughtsmen. The reverse is undoubtedly true.

Fortunately, this limitation of the data can be overcome. If Table 6.1 is used for inter-industry as opposed to intra-industry or occupational comparisons, then the problem of different occupational dispersions does not arise. That is, it cannot be concluded that because the average number of clerks per establishment is greater than the average number of draughtsmen in, say, the vehicles industry, the employment of clerks in this industry is more concentrated than that of draughtsmen. But it does not seem unreasonable to conclude that because the average number of clerks per establishment is higher in vehicles than in metal manufacture, the employment of clerks is more concentrated in the former industry than in the latter.[1] Nor does it seem unreasonable, given the above argument, to expect to find for each white-collar group some relationship between the degree of employment concentration and the degree of white-collar unionism.

That there is at least some connection between employment concentration and the density of white-collar unionism in manufacturing industries can be seen simply by comparing each row of Table 3.8 with each row of Table 6.1. Many of the industries in which white-collar employment is most highly concentrated—vehicles, metal manufacture, shipbuilding and marine engineering, and engineering and electrical goods—are also those which have the highest density of white-collar unionism, while many of the industries in which white-collar employment is most diffused—timber and furniture, leather and fur, clothing and footwear, metal goods n.e.s., and bricks, pottery, glass, cement, etc.—are those with the lowest degree of white-collar unionism.[2]

It is also noticeable that the two most highly organized white-collar

---

[1] This assumes that for each occupational group the distribution of employment is roughly the same in each industry. This assumption is later relaxed. See *infra*, p. 80.

[2] Two exceptions to this generalization are chemicals which ranks second in terms of employment concentration but ninth in terms of membership density, and paper, printing, and publishing which ranks eighth in terms of employment concentration but first in terms of membership density. Low membership density in chemicals, in spite of high employment concentration, is explained by employer policies which are perhaps more unfavourable to the growth of white-collar unionism here than in any other manufacturing industry. (See Chapter VIII for a discussion of the importance of this factor.) In a sense paper, printing, and publishing is not really an exception for almost all the white-collar unionism in this area is in newspaper publishing which has a relatively high degree of employment concentration. It is also noticeable that employer policies towards the growth of white-collar unionism have been more favourable over a longer period in newspaper publishing than in any other manufacturing industry.

occupational groups in private industry—draughtsmen and journalists—are those whose employment is most highly concentrated. Draughtsmen are employed primarily in the engineering and shipbuilding industries, and, in addition, as the official historian of the draughtsmen's union has pointed out:

A particularly important factor for trade unionism among draughtsmen is that many drawing office workers in engineering and shipbuilding are employees of fairly large firms, and are brought together in relatively large offices. They work in circumstances where collective bargaining provides the only satisfactory method for determining certain common conditions of employment.[1]

Journalists are even more highly concentrated than draughtsmen. Although they tend to work in small groups, almost all of them are employed within one industry, and a large proportion of them are employees of the large national dailies, most of which are located on one London street.[2]

The conclusion suggested by these impressionistic observations was confirmed by regressing each row of Table 3.8 on each row of Table 6.1. With the exception of other technicians and other white-collar workers,[3] the association between density of unionization and the degree of employment concentration is quite strong for all the occupational groups. The correlation coefficients for draughtsmen; all technicians; clerks; all scientists, technologists, and technicians; foremen; and all white-collar workers are all significant at the 5 per cent level, while that for scientists and technologists is just below this level of significance.[4]

Scatter diagrams revealed that these correlation coefficients generally reflected quite accurately the strength of the relationship between employment concentration and union density. But, given the small number of industry observations, it is possible that in some cases the correlation

[1] J. E. Mortimer, *A History of the Association of Engineering and Shipbuilding Draughtsmen* (London: The Association, 1960), p. 417.

[2] Approximately 90 per cent of all journalists in the United Kingdom are employed in publishing (the remainder are employed in radio and television and public relations) and roughly 25 per cent of these are employed by the national dailies and Sundays. See George Viner, 'Basic Statistics on Journalism in the British Isles' (London: The NUJ, 1965), p. 2. (Mimeographed.)

[3] No explanation readily suggests itself for the very low correlation coefficient for other technicians, but, in the case of other white-collar workers, the low correlation coefficient probably is explained by the almost complete absence of unionism from this heterogeneous category except among journalists in printing and publishing.

[4] The correlation coefficient was 0·597 for foremen; 0·639 for all scientists, technologists, and technicians; 0·518 for scientists and technologists; 0·689 for all technicians; 0·911 for draughtsmen; 0·111 for other technicians; 0·756 for clerks; −0·065 for other white-collar workers; and 0·567 for all white-collar workers.

The textile observation was omitted for foremen because of classification difficulties which give it a greatly exaggerated density figure (see *supra*, p. 32 n. 3). Consequently, the equation for foremen has only 11 degrees of freedom, while those for all the other occupational groups have 12 degrees of freedom. With 11 degrees of freedom the correlation coefficient must exceed 0·553 and with 12 degrees of freedom it must exceed 0·532 to be significant at the 5 per cent level.

coefficients were unduly influenced by one or two outliers.[1] Consequently, the observations for foremen, scientists and technicians, draughtsmen, other technicians, and clerks were pooled and a further regression analysis was undertaken on these seventy observations. To overcome the difficulty of different occupational employment dispersions referred to above,[2] the same regression technique as was employed in Chapter IV was used here: explicit occupational variables were introduced into the density-employment concentration equation which allow for the fact that the level as well as the slope of the equation may shift with the type of occupation.[3] When the effect of occupation on the distribution of employment was allowed for in this way, employment concentration was found to be very significantly associated with the density of white-collar unionism.[4]

The above analysis has assumed that for each occupational group the distribution of employment is roughly the same in each industry. But this may not be the case. For example, it is conceivable that clerks are distributed both within and across firms in a different way in metal manufacture than in vehicles and that, therefore, averages do not adequately reflect differences in the actual degree of employment concentration between industries. Consequently, using the procedure adopted above in respect of different occupational dispersions, different industrial dispersions were allowed for and a further regression analysis was undertaken. But the results were not significant at the 5 per cent level,[5] and this suggests it is reasonable to assume that for each occupational group the distribution of employment is roughly the same in each industry.

Inasmuch as bureaucratization, as opposed to trade union recruitment and administration, is the reason why the density of unionization is higher among larger groups of employees, there is at least one major objection which might be raised against the above analysis. Several writers have suggested that bureaucratization and unionization are 'mutually cumulative' in their

---

[1] That is, a very large proportion of the variance is caused by one or two points, and hence any regression line which can pass near them will have a very high correlation.

[2] *Supra*, pp. 77–8.

[3] This is a standard econometric procedure and is explained fully in Appendix B.

[4] Allowing simply for the level of the equation to shift with occupation, the employment concentration coefficient had a t value of $2 \cdot 96$ and produced a $\bar{R}^2$ of $0 \cdot 177$. Allowing simply for the slope of the equation to shift with occupation, the employment concentration coefficient had a t value of $2 \cdot 14$ and produced a $\bar{R}^2$ of $0 \cdot 641$. Allowing for both the level and the slope of the equation to shift with occupation, the employment concentration coefficient had a t value of $2 \cdot 28$ and produced a $\bar{R}^2$ of $0 \cdot 669$. All of these are significant at the 5 per cent level.

[5] Allowing simply for the level of the equation to shift with industry, the employment concentration coefficient had a t value of $-0 \cdot 493$ and produced a $\bar{R}^2$ of $0 \cdot 139$. Allowing simply for the slope of the equation to shift with industry, the employment concentration coefficient had a t value of $0 \cdot 003$ and produced a $\bar{R}^2$ of $-0 \cdot 106$. Allowing for both the level and the slope of the equation to shift with industry, the employment concentration coefficient had a t value of $-0 \cdot 094$ and produced a $\bar{R}^2$ of $-0 \cdot 053$. None of these are significant at the 5 per cent level.

effects.[1] That is, the two variables are interdependent: not only does bureau-cratization encourage the growth of trade unions, but trade unions, by demanding the standardization of working conditions, further bureaucratization. But at least in British manufacturing industries the degree of interdependence between these two variables would seem to be slight. The over-all degree of white-collar unionism in this area is so low, roughly 12 per cent, that it is unlikely to have greatly furthered the process of bureaucratization. In any case, the unions catering for white-collar employees in these industries bargain minimum not maximum terms and conditions of employment, and, in addition, they generally do not prevent employers from administering large doses of 'administrative particularism' to their employees in, for example, the form of merit pay.[2] Even assuming that unionization has somewhat furthered the process of bureaucratization, there can be little doubt that the latter preceded the former.

OPPORTUNITIES FOR PROMOTION

Several researchers have found that blockage of promotion opportunities is favourable to the development of trade unionism among white-collar workers. C. Wright Mills claims to have found 'a close association between the feeling that one *cannot* get ahead, regardless of the reason, and a pro-union attitude' among a group of 128 white-collar workers in the United States.[3] British researchers have noticed that the process of bureaucratization in teaching and the civil service 'has been accompanied by a policy of recruitment from outside at two or more levels, with little or no opportunity for those recruited at low level to surmount the internal barriers blocking their promotion'. They claim that as a result:

'Elementary' school teachers, and civil servants without a university training who entered the clerical or executive classes, have had such poor chances of upward job and social mobility that their efforts to improve their lot have inevitably taken the form of creating powerful interest groups restricted to those whose promotion was virtually barred in this way. As the lower salariat often attracted socially aspiring individuals for whom the blockage of their upward mobility was especially frustrating, they often became the leading spirits in the formation and running of such organisations.[4]

Sykes interviewed ninety-six, or one-third, of the male clerks in the sales office of a Scottish steel company and found that almost all of them wanted promotion and felt they had a reasonable chance of obtaining it, but only

[1] See, for example, Lockwood, op. cit., p. 142, and Blackburn and Prandy, op. cit., p. 117.

[2] On this point see the author's *Trade Union Growth and Recognition* (London: HMSO, Royal Commission on Trade Unions and Employers' Associations, Research Paper No. 6, 1967).

[3] C. Wright Mills, *White Collar* (New York: Oxford University Press, 1956), p. 307.

[4] R. K. Kelsall, D. Lockwood, and A. Tropp, 'The New Middle Class in the Power Structure of Great Britain', *Transactions of the Third World Congress of Sociology*, iii (1956), p. 322.

four approved of trade unionism for clerks. Later, most of these clerks joined a trade union. Sykes then interviewed them again and found that as a result of the company introducing a management trainee scheme many of the clerks no longer felt that promotion to management level was a real possibility. He has suggested that the changed attitude of these clerks towards their promotion prospects explained their changed attitude towards trade unionism.[1]

These studies are too limited in scope to allow any firm conclusions to be drawn regarding the effect of restricted promotion opportunities on union growth.[2] But blockage of promotion opportunities is very often associated with bureaucratization,[3] and it is highly probable that the effects of the latter upon unionization are reinforced by the effects of the former. The fewer opportunities the members of a work group have to rise out of it the more likely they are to become aware of their common situation. But although this awareness may be reinforced by blocked promotion channels, it is, as Lockwood has argued, 'first and foremost a product of standard working conditions' produced by bureaucratization.[4]

While blockage of promotion opportunities may be favourable to the development of white-collar unionism, it is clear that this is not a necessary condition for its growth. Promotion prospects were not blocked in banking, but a considerable degree of unionization was nevertheless possible simply on the basis of large-scale bureaucratic organization.[5] Draughtsmen are a highly unionized occupational group, yet their promotion prospects are quite good and highly valued.[6] While there is no quantitative evidence available, it is also fairly obvious that another very highly unionized group in private industry, the journalists, have considerably better promotion prospects than poorly organized groups such as clerks.

MECHANIZATION AND AUTOMATION

The office is being increasingly mechanized.[7] Several authorities claim that this is altering both the techniques by which the work is performed and the

---

[1] A. J. M. Sykes, 'Some Differences in the Attitudes of Clerical and of Manual Workers', *Sociological Review*, xiii (November 1965), especially tables i and iv and pp. 307–10.

[2] Such as Sykes's that 'there is evidence for an association between opportunities for promotion and trade unionism among clerical workers generally' (ibid., pp. 308–9).

[3] The reasons why this may happen are given by Lockwood (op. cit., p. 142). First, with the bureaucratic emphasis on technical competency and formal qualifications, there may be direct recruiting to managerial positions from outside the organization. Second, the economies of administrative rationalization may lead to a reduction in the ratio of managerial to clerical functions.

[4] Ibid., p. 149.   [5] Ibid., pp. 149–50.

[6] A survey of 941 draughtsmen in 1960 showed that 'prospects of advancement' was the third most frequent advantage of the occupation (with 350 mentions), compared with 188 giving 'poor prospects of promotion' as a disadvantage. See Guy Routh, 'The Social Co-ordinates of Design Technicians', *The Draughtsman* (September 1961), p. 9.

[7] For a detailed account and history of office mechanization see A. A. Murdoch and J. R. Dale, *The Clerical Function* (London: Pitman, 1961), and *Effects of Mechanisation and Automation in Offices* (Geneva: ILO, 1959), chaps. 2 and 3.

conditions in which it is performed in such a way as to encourage office workers to unionize.[1] They argue that office mechanization reduces the average level of skill, increases the danger of 'dead-end' specialization thereby decreasing promotion prospects, facilitates 'assembly line' work flows, requires shift work, demands a faster work pace and more continuous attention to work, makes possible the introduction of work study and piecework methods thereby emphasizing the size rather than the quality of output, obliterates any inherent interest or pride of work in the job, and generally lowers the status of office work by erasing the line of distinction between the techniques of manual and non-manual work and creating a factory-like atmosphere in the office. The ostensible result is greater anxiety and nervous tension, increased physical and mental fatigue, and reduced morale and job satisfaction, all of which supposedly encourage white-collar workers to unionize.

The advocates of this argument have tended to exaggerate both the extent and the effects of office mechanization and automation. Their main impact has been on the jobs of clerical workers, and even relatively few of these have been affected. Less than 20 per cent of female clerks and 13 per cent of total clerks in manufacturing industries were engaged as typists or office-machine operators in 1951.[2] The electronic computer was first used for office work in Britain towards the end of 1953.[3] But their rate of introduction did not gather much momentum until around 1959, and by mid 1963 there were only about 300 computers operating in Britain. The majority of these were employed in sectors of the economy other than manufacturing. Even in those organizations which have introduced computers, only about 20 per cent of the office workers have been affected by them. The size of the office and the nature of office work itself have set fairly narrow limits to the application of technology. There is 'a vast amount of office business which must be conducted personally—salesmanship, the discussion of business, the writing and typing of individual letters, secretarial work—and these as far as can be foreseen at present are unlikely to become amenable to automatic methods to any great extent'.[4]

Even those office workers who have felt the impact of mechanization and automation have not been affected as adversely as is commonly imagined. It is difficult to generalize because the impact which mechanization and

[1] For example, see the sources cited in ibid., p. 106; *Automation and Non-Manual Workers* (Geneva: ILO, 1967), pp. 36–8; and by Albert A. Blum, *Management and the White-Collar Union* (New York: American Management Association, 1964), pp. 59–62; as well as F. D. Klingender, *The Condition of Clerical Labour in Britain* (London: Martin Lawrence, 1935), pp. 61–4.

[2] *Census of Population 1951*, England and Wales, Industry Table 7. The figures for the economy as a whole were 14·5 per cent for female clerks and 8·8 per cent for total clerks. The 1961 Census did not distinguish office-machine operators separately.

[3] All the information regarding computers in this paragraph is from *Computers in Offices* (London: HMSO, 1965). [4] Ibid., p. 6.

automation have upon office workers depends upon such factors as the nature of the functions taken over by the machine and the demographic characteristics of the clerks involved. But there has been a large quantity of research in this area, and, while there is not room here to report the findings in any detail, it is clear that so far office mechanization and automation have not had the dire results which some people have predicted.[1]

It is still probably too early to evaluate the effects of technical change upon white-collar workers with any certainty. But even if mechanization and automation should have an unfavourable impact upon white-collar workers in the future, it does not follow that this will necessarily encourage them to join trade unions. It has not been the least-skilled manual workers who have been most ready to join trade unions, but the most skilled. Similarly, it has not been those white-collar workers such as clerks and office-machine operators who have the most routine and monotonous work who are the most highly unionized, but those such as draughtsmen and journalists who have the more creative and interesting work.

This does not mean that office mechanization and automation will not have any effect on white-collar unionism in the future. The capacity of machines for handling large quantities of routine clerical work and facilitating managerial decision-making, as well as the economic necessity for keeping them fully occupied, will probably encourage the centralization of administrative functions within enterprises. If this occurs to any extent it will further the process of bureaucratization.[2] Hence by promoting bureaucratization, mechanization and automation may indirectly foster white-collar unionism. But this is mere speculation. The only thing which can now be stated with certainty is that office mechanization and automation are not yet sufficiently advanced to have had any appreciable impact on the pattern of white-collar unionism in Britain.

PROXIMITY TO UNIONIZED MANUAL WORKERS

Some writers[3] have argued that white-collar employees are more likely to join trade unions if they work in close proximity to unionized manual workers

---

[1] In particular, see the summary of the literature given in *Automation and Non-Manual Workers*, op. cit., chap. 1. See also *Computers in Offices*, op. cit., *passim*, and Enid Mumford and Olive Banks, *The Computer and the Clerk* (London: Routledge & Kegan Paul, 1967), especially pp. 183–94.

[2] This has already occurred to some extent. See *Effects of Mechanisation and Automation in Offices*, op. cit., pp. 31–4, and *Computers in Offices*, op. cit., p. 36.

[3] See, for example, Lipset, op. cit., pp. 19c–24; *Non-Manual Workers and Collective Bargaining* (Geneva: ILO, 1956), pp. 31–4; C. Wright Mills, op. cit., p. 306; and Joseph Shister, 'The Logic of Union Growth', *Journal of Political Economy*, lxi (October 1953), pp. 422–4. Lockwood's position on this point is somewhat vague. He claims that the degree of contact with manual workers and their unions 'is one of those factors which affect the distribution of membership within clerical unions, but are not generally decisive in determining the differences in the degree of concerted action between one union and another' (op. cit., p. 154). It is difficult to see how a factor can affect the industrial distribution of membership

or have trade unionists among their friends or relatives.[1] Two reasons are advanced in support of this argument. First, there are the so-called 'demonstration' and 'learning' effects. It is alleged that white-collar employees who work in close physical proximity to unionized manual workers are provided with a demonstration of the benefits of trade unionism while those who have trade unionists among their friends are given an opportunity to learn about these benefits, and this makes them anxious to obtain some of these for themselves. Second, it is suggested that the organization of white-collar workers may be stimulated by manual unions directly recruiting white-collar workers into their ranks or at least lending white-collar unions moral, financial, and strategic support in their organizing drives.[2]

There is some empirical evidence which can be used to support this argument. Draughtsmen are one of the most highly unionized groups in private industry and their work brings them into close contact with skilled manual workers.[3] Several of the union organizers interviewed for this study claimed that works clerks were easier to recruit than office clerks, and that clerks in offices attached to production plants were easier to recruit than those in the downtown head offices of the same companies. Some of the most successful white-collar unionism has arisen in those industries where manual workers are represented by strong unions; railways, printing, coal-mining, and, to a lesser extent, the metal industries provide examples in both this and other countries. Research studies have indicated that white-collar workers who have union-member friends are more likely to be trade unionists themselves than those without such friends.[4]

within a union without at the same time affecting the density of union membership within an industry.

[1] That part of the argument which concerns relatives has already been considered *supra*, pp. 43–6 and will not be further analysed here.

[2] This aspect of the argument will be considered more fully in the following chapter.

[3] Routh (op. cit., p. 8) found that 43 per cent of his sample of 941 DATA members entered the profession by serving a craft apprenticeship. Even those who enter the profession directly by serving a drawing-office apprenticeship generally spend two to three years of it in the workshop learning such manual skills as fitting, turning, and machining. According to the official historian of the draughtsmen's union (Mortimer, op. cit., pp. 415–16):

> Many draughtsmen remain fairly closely associated with the workshops throughout their career, long after they have finished their apprenticeship. This is particularly true of jig and tool draughtsmen, whose work usually brings them into close contact with toolmakers, foremen and machine setters in the workshops. Many draughtsmen on other kinds of work also often find it necessary to visit the workshops to consult foremen and other workpeople. These experiences help to familiarise draughtsmen with workshop life and with the issues which confront the shop workers.

[4] See C. M. Phillipson, 'A Study of the Attitudes Towards and Participation in Trade Union Activities of Selected Groups of Non-Manual Workers' (unpublished M.A. thesis, University of Nottingham, 1964), pp. 283–4. In America a study by the Opinion Research Corporation indicated that 44 per cent of those who said they had union-member friends or relatives favoured unions for white-collar workers, as contrasted with 21 per cent in favour among those who did not have such acquaintances or relatives (cited by Blum, op. cit., p. 66).

But much of the force of this argument is dispelled by a closer examination of the reasoning and the evidence by which it is supported. It is true that white-collar workers are highly unionized in industries where manual workers are also highly unionized. But the opposite is equally true. Manual workers are relatively highly unionized in the footwear and cotton industries, but white-collar workers are not. Moreover, those areas in which the degree of white-collar unionism is highest—national and local government, education, and banking—are those in which there is the least contact between white-collar and manual workers, while the area in which white-collar workers are most poorly organized—manufacturing industries—is that in which there is the closest proximity between the two groups.

Even in those areas where manual and white-collar unionism are both strong, it does not necessarily follow that the latter results from the former. It is at least equally plausible that both result from factors common to both groups of workers within the given area such as the size of the firm and the attitude of the employer towards trade unionism. Nor does it necessarily follow that because white-collar unionists have a higher proportion of union members among their friends than non-unionists, these friends persuaded them to join trade unions. It may simply indicate that white-collar workers, like most other people, choose friends with similar attitudes and values to their own.

Social and physical proximity to unionized manual workers is obviously not a necessary condition for the growth of white-collar unionism, although in some instances it may be a favourable condition. But in other instances it may be unfavourable. White-collar workers who are in close proximity to unionized manual workers may learn not only of the advantages of trade unionism but also of what may appear to them as its disadvantages—strikes, lack of democratic procedures, and so on. Proximity and familiarity may breed contempt as easily as they breed understanding and support.

CONCLUSIONS

This chapter has shown that some but not all of the variation in the pattern of white-collar unionism in Britain can be accounted for by variations in the degree of employment concentration. The greater the degree of employment concentration the greater the density of white-collar unionism. The explanation for this would seem to be that employees are more likely to realize the need for trade unionism and trade unions are more likely to be interested in recruiting them, the more concentrated their employment.

But needs do not inevitably create ways of meeting them; many social needs often persist without being met. Even if white-collar employees feel the need for trade unionism and trade unions are interested in recruiting them, there are at least two reasons why they may still not be organized. First, trade unions may have poorly designed structures and use inappropriate techniques for recruiting white-collar employees. Second, employers may be opposed to their 'staff' joining trade unions and pursue policies designed to discourage them from doing so. Hence it is also necessary to look at the trade unions which recruit white-collar workers and the employers of these workers to obtain a complete picture of the factors which may promote or hinder the growth of white-collar unionism.

VII

THE TRADE UNIONS

IN an attempt to isolate the factors which influence the growth of trade unionism among white-collar workers this study has so far examined their socio-demographic characteristics, economic position, and work situation. Many people have claimed that the trade unions themselves must also be taken into account. The public image of the trade union movement as well as the recruitment policies and structures of unions have all been alleged to affect profoundly their growth among white-collar workers.[1] The purpose of this chapter is to determine whether this is, in fact, the case.

THE PUBLIC IMAGE OF THE TRADE UNION MOVEMENT

'Image' is a greatly overworked term. But the public image presented by the trade union movement may nevertheless be of considerable importance in determining its growth. At least one American writer has been willing 'to crawl out on a limb and assert that the most basic thing affecting the possibility of unionization in new fields is . . . the general community attitude towards unions'.[2] A British writer feels that the prominence given by the mass-communication media to such things as unofficial strikes, demarcation disputes, undemocratic procedures, restrictive practices, charges of Communist infiltration, and internal union squabbles between 'left' and 'right' may have caused many people who would otherwise have joined unions not to do so.[3] Many of the employers interviewed for this study argued that the 'inability of the trade union movement to bring itself up to date and to present a better image of itself' was a major factor hindering its expansion among white-collar workers.

Unfortunately, the evidence necessary to test this argument is rather thin. But it is probably correct to suggest that during the first half of the twentieth

---

[1] Writers such as Shister have also claimed that the leadership of unions affects their growth. Union leadership as such is not analysed in this chapter. But this does not mean that its possible impact upon union growth is not considered. If leaders do have any effect upon union growth, it is, as Shister himself has pointed out, primarily by devising better recruiting policies and structures. The impact of these factors upon union growth is considered here and hence indirectly so is the influence of union leadership. See Joseph Shister, 'The Logic Of Union Growth', *Journal of Political Economy*, lxi (October 1953), pp. 429–30.

[2] S. M. Miller, 'Discussion of "The Occupational Frontiers of Union Growth"', *Proceedings of the Industrial Relations Research Association*, xiii (December 1961), pp. 214–15.

[3] Keith Hindell, *Trade Union Membership* (London: Political and Economic Planning, 1962), p. 183.

century the public attitude towards trade unions was least favourable just before the First World War and just after the General Strike of 1926 and most favourable during and immediately after both world wars.[1] Since 1950 the public attitude towards trade unions can be traced more systematically as a result of information provided by successive Gallup Polls. The question— 'Generally speaking, and thinking of Britain as a whole, do you think trade unions are a good thing or a bad thing?'—has been put to a national sample of the population at fairly regular intervals. The proportion of the sample which thought that trade unions were a 'good thing' is given for the period 1952–64 by Fig. 7.1.

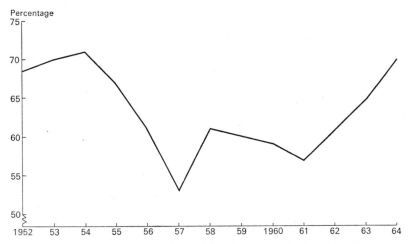

FIG. 7.1. Public standing of the trade unions, 1952–64. Percentage saying 'Trade Unions are a good thing for Britain'.

The major weakness of this argument is that the public image of the trade union movement is a general factor. It cannot, therefore, explain the occupational and industrial variations in the density of unionism which are of primary interest in this study. At best, it could only help to account for the country's over-all level of trade unionism. But it is doubtful if it even does this, for at least since 1950 the trends in over-all union membership have not followed very closely the trends in the public's attitude towards trade unions. Public opinion has been very fickle: the percentage of people who thought that trade unions were generally a 'good thing' has varied from a high of roughly 70 per cent in 1954 and 1964 to a low of 53 per cent in 1957. Yet, as Chapter III demonstrated, total union membership has remained on a gently rising plateau over the post-war period while its density has fairly steadily

[1] See W. G. Runciman, *Relative Deprivation and Social Justice* (London: Routledge & Kegan Paul, 1966), chap. 6, for references to the union movement's public image during the first half of the twentieth century.

declined. At the same time, white-collar union membership has increased very markedly while its density has remained more or less constant.

Prior to 1950 the periods in which the public attitude towards trade unions was most favourable—during and immediately after the two world wars—were also the periods of their greatest growth. But, as Chapter IX will demonstrate, the unions' membership increased primarily because government policies necessitated by war made it easier for them to obtain recognition from employers and their associations. The improvement in the unions' public image helped in this regard by making it harder for employers to deny unions recognition without at the same time damaging their own public image. But this improvement was itself largely a result of government policies which involved the unions in the administration of government and industry in order to help ensure maximum and uninterrupted production throughout the wars. Hence while the improvement in the unions' public image during the two world wars obviously assisted their growth, it was very much a secondary factor.

UNION RECRUITMENT POLICIES[1]

Recruitment strategy

The concept 'strategy' is generally applied to military affairs, but it may also be used to analyse trade union behaviour. Unions, like armies, are faced with solving the problem of strategic priorities. Just as armies must determine which fronts should get primary attention, so unions must decide, consciously or unconsciously, whether or not to concentrate their recruiting in certain geographical regions, occupations, industries, or firms.

White-collar unions in private industry generally do not focus their recruiting geographically, except inasmuch as certain industries tend to be concentrated in particular regions. Of course, the officials in charge of a particular region of a union may from time to time decide to concentrate on certain towns or cities within it, but these campaigns are generally very *ad hoc* and very short-run. Nor is there very much concentration of recruitment by occupation. Although most white-collar unions restrict their membership to fairly clearly defined occupational groups or 'crafts', this is more a reflection of craft consciousness, the Bridlington rules,[2] and union structures, than of occupational recruitment priorities. In other words, within the limits set

---

[1] Much of the information contained in the rest of this chapter was obtained from interviews with union officials. During 1965–6, the author interviewed sixty of the full-time officials of the major white-collar unions in private industry. The officials agreed to co-operate on the understanding that their replies would not be attributed to them as individuals. Consequently, material obtained from interviews has been cited in such a manner that it is not possible to identify the individual concerned.

[2] These were established at the 1939 TUC and place certain restrictions on inter-union competition for membership.

by the unions themselves and by external agreements, there is very little emphasis on occupational recruitment priorities.

The question of industrial recruitment priorities does not really arise for some white-collar unions. By virtue of its industrial structure, BISAKTA has tended to restrict its activities to the iron and steel industry. Similarly, NATSOPA tended to restrict itself to the printing and publishing industry. Given that the NUJ and DATA are relatively 'closed' unions in the sense that the former has tended to restrict its membership to journalists and the latter to technicians 'whose functions are related to design', it follows that they have generally confined their recruiting to those industries in which these two groups of workers are almost exclusively located, printing and publishing, and engineering and shipbuilding, respectively.

But all the other unions which cater for white-collar workers in private industry have a fairly 'open' structure, and they therefore must decide whether or not their recruiting should be concentrated industrially. Most of them have decided that at least to some extent it should be. In general, they have tended to concentrate on those industries in which their potential membership is greatest and in which they are generally recognized by employers. Thus the AScW concentrated on engineering,[1] chemicals, the National Health Service, and universities and technical colleges, and ASSET focused on engineering, civil air transport, and, to a lesser extent, chemicals and rubber. The NACSS has concentrated on engineering and road and water transport as well as those industries such as chemicals, rubber, and brewing in which their parent organization, the TGWU, is well established among manual workers. Since 1950 the CAWU has increasingly been concentrating its recruiting in the engineering industry, although to some extent the dispersed nature of clerical employment has forced it to recruit clerks wherever most of them were not already members of other unions.

Some unions have refined their recruitment priorities to the extent that they concentrate on certain types of firms. The NACSS has concentrated on those firms in which the TGWU is particularly well established among the manual workers. Within the engineering industry, the CAWU has tended to concentrate on the largest companies. ASSET went even further in this respect. After 1949 it increasingly concentrated its recruiting and collective bargaining on combines or large multi-plant firms and industries with a large proportion of such firms. ASSET believed that by concentrating on the pattern setters of British industry it would build up its financial and membership strength more rapidly and efficiently and have a greater impact on national salary and employment policies. In 1958 this strategy was formalized as a 'combines policy' and a directive was issued to all the union's full-time officers instructing them to:

[1] In this context the engineering industry is defined to include those areas covered by the Engineering Employers' Federation. See *infra*, p. 126 n. 3.

. . . above all else, develop mass membership in the largest manufacturing enterprises. This means a concentration of activity in the engineering, transport and chemicals industries. In all these sectors there is an unmistakable trend towards fusion or alliance and the growth of larger and larger units. It is, therefore, essential that planned organising and bargaining work should be undertaken as a *first* priority in these dominating firms in the selected industries.[1]

The union's executive then drew up a list of approximately eighteen key companies—including such firms as Associated Electrical Industries, the British Aircraft Corporation, Standard Telephone and Cables, International Computers and Tabulators, Hawker Siddeley, Plessey, and Rolls Royce—and allocated responsibility firm-by-firm to the union's staff.

Recruitment tactics

Whatever the nature of union recruitment strategies, they have generally not been pursued with much tactical aggressiveness. The recruitment process tends to be more passive than active in the sense that potential members generally have to contact a union before there is much attempt to recruit them. The comment of one AScW official is typical: 'Generally we wait for people to approach us before we try to recruit them. You might say that organization occurs by accident.' Even ASSET with its sophisticated recruitment strategy generally waited for individuals to contact it before trying to recruit them. But if the individuals happened to be from a combine, then they received special attention. As one ASSET official put it, 'I will drop everything to get them in. If they weren't from a combine I might not do this unless it was an extremely large group or I wasn't very busy.' But inasmuch as the union normally depended upon the initiative of potential members, its approach to recruitment was still more passive than active.

Reasons for this passivity are not hard to find. None of the unions recruiting white-collar workers in private industry are particularly wealthy organizations and all of them are understaffed relative to the size of the recruitment task they face. In 1964 the major unions catering for white-collar workers in private industry had a total of approximately 282,000 white-collar members and 98 full-time officers recruiting and servicing them, roughly 1 officer to every 3,000 members.[2] This relatively high ratio of members to officers[3] means that most officers spend the bulk of their time servicing

[1] Appendix to Organization Committee Minutes, 8 March 1958.

[2] The unions which were used to compute these figures include all those listed in Table 3.7 except BISAKTA. The TGWU has a few officers who handle manual as well as white-collar affairs. Each of them was counted as a half in deriving the figure for the total number of officers.

[3] This is relatively high by the standards of other countries. See S. M. Lipset, 'Trade Unions and Social Structure', *Industrial Relations*, i (February 1962), p. 93. But Lipset's calculation in this article of the ratio of members to full-time officers in Britain is incorrect. A survey in 1959 found that there were approximately 3,000 full-time trade union officers servicing a membership of approximately 9,600,000 roughly one to every 3,300 members. See H. A. Clegg, A. J. Killick, and Rex Adams, *Trade Union Officers* (Oxford: Blackwell, 1961), pp. 39 and 94.

existing members rather than recruiting new members. Only 3 of the 60 union officials interviewed for this study claimed they spent more than 10 per cent of their time recruiting.[1] Even unions such as the CAWU which have experimented with specialist recruitment officers in one or two regions have found that these officers eventually got drawn into servicing the members they recruited and increasingly had less time available for recruiting new members. 'It is very difficult', as one of these officers pointed out, 'for an official to do nothing but recruit. When people join a union they bring problems with them and it is only natural that the official who has been dealing with them up to the time they joined should continue to do so.'

Even if trade union officers were not so busy servicing existing members, the recruitment process probably would not become much less passive. For almost all of them believe, as one of them put it, that 'organizing in a vacuum is a waste of time and energy'. Most of the union officials who were interviewed had participated at one time or another in special recruitment campaigns complete with such techniques as blanket leafletting, loudspeaker vans, and open meetings advertised in the press, but none had found that these ever produced a significant number of new members. The following experiences are fairly typical. In 1964 an AScW official sent general recruiting literature to the laboratories of sixty companies and got no response whatsoever. In the early 1960s ASSET conducted a special recruitment drive along the Great West Road in London where several major companies have plants and offices, but, according to the official in charge of the campaign, it did not produce a single member. Almost every official who was interviewed spoke of several occasions on which he had advertised an open meeting for the staffs of a particular company and not one person had turned up. In view of these experiences it is not surprising that most union officials are convinced that 'a contact is an absolute prerequisite for successful recruiting' and refuse to engage in speculative recruiting unless they are 'desperate'.

But the recruitment process is not completely passive. Once a person approaches a union with a view to becoming a member, a number of steps are taken to ensure that he and, even more important, his colleagues actually do join. To begin with, an official usually writes to the contact enclosing some general recruiting literature and suggesting an informal meeting at his home or some other convenient location. When they meet, the official tries to get the facts regarding the company concerned and the employees' grievances, encourages the contact to try and get as many of his colleagues as possible to join the union, and suggests a time and place for a meeting of all interested employees. At this meeting the official generally gives an opening talk covering such topics as the union's history, structure, and policies, as well as the

---

[1] The 1959 survey also revealed that on average full-time officers felt that the recruitment of new members was only their fifth most important function in terms of the amount of time they spent on it. See ibid., p. 44.

case for trade unionism in general and his union in particular, and then discusses a variety of questions raised from the floor. Assuming a sufficient number of people are interested in joining the union, a committee is elected, and the employer is eventually approached to concede recognition.

The recruitment process is not entirely passive even prior to a contact being obtained. Although unions generally do not attempt to recruit until they have a contact, most of them take steps to increase the number of such contacts or at least to ensure that they come to their union rather than another. All the unions catering for white-collar workers in private industry lay great stress upon recruitment in their journals.[1] By such means as competitions, prizes, and just sheer exhortation they try to encourage existing members and lay officials to 'go out and talk unionism, and our union in particular, in community associations, churches, clubs, and pubs'; to distribute recruiting literature at their place of work and at day release classes; to pass their copy of the union journal which often contains recruitment propaganda and membership application forms to non-members; to convince their friends and relatives to join; to put their local officer in touch with any contacts they may have among groups of potential members; and, if transferred to a non-union area, to try on their own initiative to spark off organization there. Union officials sometimes make a tour of the trade councils in their regions and ask them to help in supplying recruiting contacts. They also approach manual unions for the same purpose. Officials of manual unions which also recruit white-collar workers are very favourably placed in this regard and often find that their manual members can put them in touch with friends or relatives on the white-collar side of a firm.

Some unions go much further than this and use the mass media to advertise their services. The AScW advertised on buses which followed routes servicing plants in which it had a special interest, as well as in certain technical and scientific journals.[2] The CAWU has also advertised on buses as well as on hoardings and the Waterloo and City underground railway which carries thousands of clerks into the City everyday. Although the NACSS has not used the mass media to advertise itself, its parent organization, the TGWU, has advertised on hoardings and buses, and, in addition, has experimented with give-away booklets of matches, drip mats, and car stickers.[3] Individual branches in most of these unions also advertise themselves in the local press from time to time.

The union which made the most extensive use of the mass media to publicize itself was ASSET. In fact this was the main way in which the union tried to implement its recruitment strategy. In words of one ASSET official,

---

[1] See, for example, 'Notes on Recruiting', *AScW Journal* (October 1949), pp. 13–15; 'A Call to Every Member', *AScW Journal* (January 1954), p. 3; and 'Organise! In Union There Is Strength', *TGWU Record* (December 1963), pp. 32–4.

[2] *Report of Executive Committee*, 1957, p. 16.

[3] See 'Union Publicity', *TGWU Record* (June 1964), pp. 28–32.

'Our major tactic is to get lots of publicity, to make employees aware of us so that when they have a grievance they know where to come to get it resolved.' The union advertised on buses, on hoardings, in the local press, in football league programmes, and even on commercial television.[1] But it placed even greater emphasis on getting news coverage of its activities and successes. It took every opportunity to issue press releases, and to have its officials, in particular its general secretary, write articles for newspapers and magazines, speak on the radio, and appear on television.[2]

Recruitment appeals[3]

Although the message which unions try to put across to potential members by such means as the mass media, leaflets, broadsheets, and speeches tends to vary from one union to another and from one situation to another, there are certain common elements. The basis of the appeals of all the white-collar unions in private industry is economic—improved salaries and conditions, greater job security, plus attractive services and benefits.

The theme 'you can get it too' is central to all the appeals. The recruitment literature is filled with examples of the improvements in salaries and conditions which the unions claim to have gained for their members and suggests that these are 'typical of the progress which can be made where staff workers join and support' trade unions. BISAKTA notes that 'the manual worker, through his trade union, has achieved higher wages, shorter working hours, better working conditions, holidays with pay and many other improvements' and points out that 'if staff workers would join a Trade Union they would enjoy the same protection as the manual workers'. In a special broadsheet addressed to white-collar workers the NACSS announces that it has launched 'a big campaign to stop the drift of staff earnings to well below those of many manual workers'. In one AScW leaflet a scientific worker asks the question— 'Why should I be organised in a Union?'—to which the answer is:

All kinds of professional people like doctors, teachers, civil servants, have associations registered as Trade Unions. These represent them and act as pressure groups for them. In this day and age the isolated individual cannot hope

---

[1] The union became convinced that the results from advertising on television did not justify the cost. Several white-collar unions have tried to get the TUC to underwrite an advertising campaign in the national press and on television, but it has argued that the cost of using such media make this impossible. See TUC's Non-Manual Workers' Advisory Committee, *Proceedings* of the Annual Conference for 1956, pp. 11–12; 1957, p. 6; and 1961, pp. 20–6; as well as the *Report* to the Annual Conference for 1957, pp. 7–9 and for 1961, pp. 6–9.

[2] See Hindell, op. cit., pp. 174–82, and R. M. Blackburn, *Union Character and Social Class* (London: Batsford, 1967), pp. 91–3, for a discussion of the recruitment methods used by other unions.

[3] Except where otherwise noted, all the quotations in the following section are taken from recruiting literature published by the major white-collar unions in private industry. The various pamphlets, leaflets, and broadsheets have not been specifically cited because in most cases they do not have a title and it is therefore difficult to identify them uniquely.

successfully to press his interests and claims by himself. If salaries of scientific workers are inadequate today this is a consequence of their lack of organization.

A DATA leaflet argues that 'in recent years there has been a steady decline in the position of design technicians relative to other sections of the community' and declares 'this is a process that D.A.T.A. is determined to reverse'. The CAWU lists a number of advantages which a clerk obtains by joining the union including 'above all the opportunity of working with your fellow clerks to improve your salaries and conditions of service'.

Job security is also emphasized. The NACSS observes that 'as firms (and offices) grow bigger, and take-overs and technical changes threaten the security of staff employment, more and more staff workers are realizing that they, too, need a strong Union behind them, to protect them and raise their incomes—so they join the N.A.C.S.S.'. The CAWU claims that electronic computers are having a greater and greater impact upon office work causing 'many thousands of office workers to wonder whether their future is in jeopardy' and adds that 'trade union membership is an insurance against the uncertainty of the future'. A piece of general recruiting literature published by the TUC's Non-Manual Workers' Advisory Committee warns that 'more mechanisation means new worries for staff—fear of losing a job or the chance of promotion' but suggests that 'the individual's anxieties are not nearly so great when there is a union to tackle the problems of redundancy, compensation for loss of employment, proper provision for those who are to retire early and retraining for those who seek another post'.

But improved salaries and greater security are not the only wares the unions offer to potential members. All the white-collar unions in private industry provide most of the usual 'friendly' benefits—accident pay, educational grants and facilities, funeral benefits, convalescent facilities, superannuation benefits, free legal advice, unemployment benefits, and benevolent grants—and stress them in their recruiting literature. In addition DATA offers, as did ASSET and the AScW, a discount trading scheme covering a wide range of goods and services including most types of insurance; ASSET and the AScW gave free advice on income tax and house purchase; DATA provides a continental holiday scheme, technical publications, and a technical circulating library; the CAWU gives actuarial advice on superannuation schemes; and the NACSS has access to the TGWU's Advice and Service Bureau which helps members with individual problems.

The purely white-collar unions embellish this general economic appeal by claiming that they cater exclusively for staff workers and as such can offer them a better service than an industrial or general union. ASSET declared itself to be the fastest growing trade union in Britain because supervisors and technicians 'are becoming increasingly aware of the need to join a specialist trade union catering for their particular requirements'. The CAWU points out that it 'employs specialist officials who know how to represent clerical and administrative staff' and adds that 'the CAWU is not a satellite of any other body. It is a completely free organisation, owned, managed, and controlled by clerical staff.' In speeches to potential members this point is made even more strongly. A CAWU official stated that in speaking to clerical workers he stresses that they 'will always play second fiddle in a manual workers' union'.

The partially white-collar unions try to counter these claims by arguing that in unity there is strength. The NACSS bills itself as 'the Union that puts real strength behind the staff' and declares that 'employers have extra respect for the N.A.C.S.S., for though it determines its own policies, and maintains its identity as an organisation of staff workers, it is also one of the key sections of the country's strongest Union—the Transport & General Workers' Union'. A BISAKTA official tells white-collar workers in the steel industry that 'we know more about the industry than specialist unions which have members in several industries and we can discuss your problems much better and give more effective representation'.

Finally, just in case the white-collar worker is convinced by these arguments but is reluctant to join a union for fear of what his employer may think, all the unions emphasize those areas in which they are recognized by employers and where relations with them are good. The NACSS informs potential members that 'it has won the respect of management by hard bargaining and fair dealing' and 'has full recognition rights to represent members of the Union in many industries and undertakings'. It also adds that it is possible to 'establish firm, but cordial relationships with employers, benefitting both the staff and the industry'. BISAKTA claims that 'most employers, recognising the need for organisation amongst themselves, are fair-minded enough to allow their staffs the same right' and states that it is 'the recognised union for clerical and supervisory staffs in the steel industry'. NATSOPA published a list of all the firms and employers' associations with which it had agreements and pointed out that 'relations with employers are good, and as long as print workers remain organised this should continue'. The NUJ mentions that it 'had to fight hard for its first newspaper agreement, but that phase of its work is now over' and today 'proprietors' organisations turn naturally to the Union when any question affecting the journalist's working life arises'. In a pamphlet addressed to planning engineers, DATA informs them that it is 'the only trade union which by national agreement with the Engineering Employers' Federation has the right to negotiate' on their behalf. A CAWU official tells potential members in the engineering industry that 'we don't want martyrs; your employers have already recognised us; negotiating machinery is sitting there waiting to be used'.

Impact upon union growth

Much more detail could be given on union recruitment policies. But the purpose here is not to provide a manual on recruiting practices. Rather, it is to determine what effect, if any, these have upon union growth, and sufficient information has now been provided for this to be done.

It is very doubtful if the pattern of aggregate white-collar unionism can be explained to any great extent by the nature of union recruitment policies. There are, as Chapter III indicated, considerable occupational, geographical, and industrial variations in the pattern of white-collar unionism. There are not, as this chapter has indicated, corresponding variations in union recruitment policies. The unions' general approach to recruiting white-collar workers differs little from one occupation to another or one region to another. It is true that most of the unions tend to concentrate their recruiting in certain industries, and, in the main, these are the industries in which the density of white-collar unionism is relatively high. While the latter no doubt helps to account for the former, it is very much a secondary factor. The generally higher levels of white-collar unionism in these industries are mainly explained by other factors—partly, as the previous chapter demonstrated, by the structure of these industries, and partly, as the following chapter will demonstrate, by the attitudes and behaviour of employers with regard to union recognition. In fact, it is precisely because industrial structure and employer attitudes are more favourable to unions in these industries that they have tended to concentrate their recruiting in them.

Although union recruitment policies are not of much significance in explaining variations in the pattern of white-collar unionism, they may nevertheless be important in accounting for its over-all level. The methods by which unions try to recruit white-collar workers have been shown to be more passive than active. It might be argued, therefore, that this is why the general level of unionization among these workers is relatively so low.

But no evidence has been produced which supports this view. In fact, when unions have adopted more aggressive recruiting tactics, they do not seem to have had much impact upon their recruiting performance. The experiences of the union officials who were interviewed for this study indicate that special recruitment drives generally do not produce an appreciable number of new members.[1] The AEU undertook a special recruitment campaign between 1957 and 1967 but its membership grew by only 15·5 per cent during this period as compared with 21·5 per cent during the period 1947–57 when there was no special campaign.[2] The NUGMW launched a special recruitment drive in 1960 and subsequently established several mobile teams of full-time recruitment officials and appointed a national recruitment officer. Yet the union's membership shrank by 0·3 per cent during the period 1960–6 as compared with a 1·1 per cent increase during the period

---

[1] See *supra*, p. 93.
[2] See Hindell, op. cit., p. 176. The AEU's membership in Great Britain was 721,902 in 1947, 877,641 in 1957, and 1,041,428 in 1967.

1954–60.[1] Even ASSET's 'combines policy' does not seem to have had all that much impact on its recruitment. Its growth from the time this policy was introduced in 1958 up to 1964 was greater than during the period 1951–8, but so was that of most of the other major white-collar unions in private industry.[2]

None of this proves that the recruitment policies of these unions did not have any impact upon their growth. It may be that their growth would have been even less if these policies had not been instituted. But the failure of these policies to make much of a difference in the rate of recruitment suggests that, if they are a factor in union growth, at least they are not very important and are easily outweighed by other factors. Until such time as better evidence becomes available, the most balanced conclusion would seem to be that unions are not much more than catalysts in the recruitment process. Before a group of workers can be successfully organized, there must be some irritating condition resulting in a widespread feeling of dissatisfaction. Unions cannot create this condition. They can only discover where it exists, emphasize it, and try to convince the workers that it can be removed by unionization. Union recruitment is by its very nature largely a passive process.[3]

Whether or not unions have been sufficiently energetic in carrying their appeals to white-collar workers, it has been argued, particularly by American writers, that more white-collar workers would have been recruited if unions had been more skilful in formulating their appeals to them. In Blum's view, American unions have failed to recruit large numbers of white-collar workers at least partly because

instead of fostering a feeling among white-collar employees that management had treated them unfairly and that a union would right their wrongs, labor organizations often unwittingly brought about a contrary opinion (or, at least, did nothing to refute it): that unions had forced management to give too much of industry's earnings to manual workers.

. . . . .

To use effectively whatever bitter feelings exist among white-collar employees concerning salaries, labor organizations must be able to convince these employees that the blame rests squarely on the corporation's shoulders.[4]

[1] See the NUGMW's *Report of the Forty-Fifth Congress*, 1960, pp. 43–4, and the *Report of the Fifty-First Congress*, 1966, pp. 38 and 209–10. The union's membership was 787,228 in 1954, 796,121 in 1960, and 792,995 in 1966.

[2] ASSET's membership increased by 69·9 per cent during the period 1958–64 as compared with 56·4 per cent during the period 1951–8; the AScW's by 72·5 per cent as compared with −11·9 per cent; the CAWU's by 57·4 per cent as compared with 25·4 per cent; the NUJ's by 22·3 per cent as compared with 19·1 per cent; and the NACSS's by 52·6 per cent as compared with 12·4 per cent. For the actual membership figures of these unions see Fig. 7·2 and the notes to this figure in Appendix A.

[3] For a contrary view, see Bernard Karsh, Joel Seidman, and Daisy M. Lilienthal, 'The Union Organizer and His Tactics', *American Journal of Sociology*, lix (September 1953), pp. 113–22.

[4] Albert A. Blum, *Management and the White-Collar Union* (New York: American Management Association, 1964), pp. 29 and 45.

It is Burns's judgement that the unions' approach to the white-collar worker has generally been wrong.

It is an approach that is primarily class-conscious instead of career-conscious, one dedicated to fighting the company and opposing the management rather than improving relations and co-operating with management to improve conditions of work for the worker. Too often the philosophy has been to proletarianize the white collar worker; to ask and induce him to accept the methods and mental outlook of workers in the mass production, maintenance and craft occupations, even though his problems, interests, and goals are not common and coterminus with those of shop workers. The slogan is standardization for the office worker.[1]

Strauss believes that unions would be more successful in recruiting white-collar workers if they used a 'middle-class approach' rather than a 'factory approach'. Instead of trying to convince white-collar workers to give up their 'vain delusions' that they are members of the middle class and hence above trade unionism, the unions should present themselves as the best means by which these workers can achieve their middle-class goals.[2]

It is difficult to prove or disprove this argument. But several observations can be made which cast serious doubt upon its validity. Firstly, the argument is largely based upon the impressions and speculations of its advocates; they have produced little, if any, reliable evidence which conclusively demonstrates that it holds even for America, let alone for Britain and other countries. Secondly, it is based on certain assumptions—that white-collar workers join unions primarily because of their concern for higher salaries, and are prevented from joining primarily because of their concern for their middle-class status—which this study has already shown to be inadequate.[3] Finally, the argument implies that if only the unions would employ good copy writers, they would be rewarded with a large influx of new members. To say the least, this is improbable. The evidence assembled in this study suggests that white-collar unionism is primarily a product of very powerful structural forces. If this is correct, then the density of white-collar unionism is unlikely to be significantly increased simply because a few sentences are changed in the unions' recruiting literature.

UNION STRUCTURES

Although the various recruitment techniques employed by unions do not seem to affect their growth to any great extent, the structures they use to recruit and service members may. Even if the question of union structure is confined to unions recruiting white-collar workers in private industry, it is

[1] Robert K. Burns, 'Unionization of the White Collar Worker', *Readings In Labor Economics and Industrial Relations*, Joseph Shister, editor (2nd ed.; Chicago; Lippincott, 1956), p. 71.
[2] George Strauss, 'White-Collar Unions Are Different!' *Harvard Business Review*, xxxii (September–October 1954), pp. 75–6.
[3] See *supra*, pp. 51–63 and 48–50.

still too broad to be fully discussed in this study. It is intended to examine here only those aspects of union structure which have been claimed to affect union growth. In considering union structure, it is convenient to distinguish between those aspects which concern the internal operation of unions and those which are external to particular unions and concern their relationships with each other and the wider labour movement.

Internal structures

Partially white-collar unions. The aspect of the partially white-collar unions' structure which has been most often examined in relation to their growth is their willingness and ability to represent the special interests of white-collar workers.[1] There can be no doubt that some of the ideals, aspirations, interests, and problems of white-collar workers are different from those of manual workers.[2] Nor can there be much doubt that manual workers often fail to appreciate this. In fact, a strong antipathy has traditionally existed between manual and white-collar workers in British industry. Manual workers often have a strong contempt for white-collar workers, regarding them as 'unproductive', as 'having its cushy', as snobs, as 'bosses' men', as one of 'them' and not one of 'us', and generally as poor material upon which to build trade union organization.[3]

Zweig has quoted a typical workman's view:

I start at 7.30 in the morning, an 'office-wallah' starts at 9. He works in a collar and tie and has clean hands, and I have to dirty my hands. What he does can be rubbed out with a rubber, while what I do stays. He keeps in with the boss class.[4]

Similar sentiments were expressed by *The Bradford News*, an organ of the Independent Labour Party, on the occasion of the decision of the National Union of Mineworkers to press for a special investigation into administrative efficiency in the National Coal Board in 1951.

All miners, and some other men and women who earn their living by hard work, under strict rules, feel suspicious, resentful and jealous of those chaps who sit on their you-know-whats in offices, and push pens. The administrators, technicians, and clerks recognise this, and feel a little nervous, ashamed, or defiant about it, according to how their minds work. . . . Actually, the clerk is an unskilled labourer under modern conditions, and there is no reason for treating him any differently from any unskilled labourer. Usually the qualification on which he gets his first job is good

---

[1] One sociologist has claimed that manual workers' unions have 'a kind of trained incapacity to deal with non-manual people' which he attributes to the fact that these unions 'have developed habits of work and behaviour which reflect working-class culture'. See Seymour M. Lipset, 'The Future of Non-Manual Unionism' (an unpublished paper, Institute of Industrial Relations, University of California, Berkeley, 1961), p. 60.

[2] If documentation is thought necessary, see Runciman, op. cit., especially chaps. 8–11.

[3] On this point see David Lockwood, *The Blackcoated Worker* (London: Allen & Unwin, 1958), pp. 100–5 and 125–32, and Ferdynand Zweig, *The British Worker* (Harmondsworth, Middlesex: Penguin Books, 1952), chap. 21.

[4] Ibid., p. 203.

appearance and address, and manners—the qualification appropriate to a footman or flunkey.[1]

The question of whether or not a clerk is 'productive' is evidently raised so frequently by the manual members of BISAKTA that one of the union's white-collar members was moved to write in the union journal that

The questioner is usually an ill-informed member of the manual labour force, and he is in fact not asking a question at all, but politely suggesting that a clerk is a social parasite living on the 'productive' worker and that if the parasite happens to be a male, then this unmanly scribbling should be left to the women while he gets himself 'a man's job'. Exactly what constitutes 'a man's job' is usually unspecified but presumably it requires a combination of 'brute force and ignorance'.[2]

Sometimes even prominent national officials of manual unions which try to recruit white-collar workers have a very negative attitude towards them. A full-time officer of the AEU, speaking in the prices and incomes debate at the 1966 TUC and commenting on the speeches made earlier by the general secretaries of NALGO and ASSET, had this to say:

Is it not the case that, as distinct from the manual workers in national and local government employment who are decidedly lower paid . . . that the third of a million people that Nalgo represent are enjoying salaries which range from as low as £1,000 per year to as high as £10,000 per year? . . . Is it not true that their lives are cushioned by security, by longer holidays, better superannuation and better sick schemes than the manual workers in these industries? I want to say to Nalgo that if they are really interested in the good of the nation, and bearing in mind that in the main, you know, (and somebody has got to say this) you represent people who produce nothing . . . is it asking too much of you to make a little contribution to solve our nation's problems under our Labour Government. I do not think so.

I now deal with the contribution made by that anarchistic anachronism, Clive Jenkins . . . 'Anarchistic' because he is against everything that is progressive and challenging, and 'anachronism' because he represents people who have run away from the struggle of the workshop floor, who do not want to be associated with manual workers' unions so they join this whatever-you-call-it. In fact, the vast majority of his people are people who have betrayed the manual workers' union . . . and, in my opinion, they are 40,000 Conservatives run by half-a-dozen Clive Jenkinses.[3]

But in spite of these attitudes, and, in some cases, perhaps because of them, most manual unions which also represent white-collar workers have made certain modifications in their structures to help ensure that the special needs and problems of these workers can be more easily articulated and catered for.

Even as traditional a manual union as BISAKTA has made a few structural

[1] Cited by David Rhydderch, ' "Fools Rush In" ', *The Clerk* (September–October 1951), pp. 246–7. The author of this article concludes that 'the lesson clerks must learn and learn very quickly is that Unions purporting to represent all and every type of manual workers cannot meet the special needs of clerks'.

[2] See *Man and Metal* (May 1950), p. 73, and (June 1950), p. 89.

[3] *TUC Annual Report*, 1966, p. 476.

concessions to its white-collar members. They are generally placed in separate branches from manual members,[1] and these are brought together in a special section of the union, Section J. This section, unlike the union's other ten sections, has been entitled since 1944 to hold an annual national conference, but this conference has no power to give effect to its recommendations. Like all the other sections, it is not autonomous on any matter and can only offer advice to the union's 'supreme governing and administrative authority', the Executive Council. Section J is allotted two representatives on the twenty-one member Executive Council.[2] Candidates for these positions must be nominated from and by the white-collar membership, but, like the executive candidates from other sections, they are voted upon by all the members of the union. Since the manual members outnumber the white-collar by roughly fifteen to one, they decide, in effect, which of several competing nominees will represent Section J on the Executive Council. In spite of being requested to do so by Section J,[3] BISAKTA has refused to appoint a full-time officer to be specifically responsible for white-collar affairs. In fact, since the union was formed in 1917 only one white-collar member has ever been appointed to a full-time position with the union.[4]

NATSOPA went much further than BISAKTA in providing separate facilities for its white-collar members.[5] The bulk of NATSOPA's membership was concentrated in London and Manchester.[6] In London it was divided into four branches, one of which was exclusively reserved for white-collar members.[7] The union's membership in Manchester was only large enough to justify one branch, but the branch committee was divided into three sub-committees, one of which exclusively represented clerical and administrative members. There was also only a single composite branch in all the other provincial areas of the union. The white-collar membership was not large

[1] This is partly the result of employer insistence. See *infra*, pp. 165–6.

[2] Prior to 1966 Section J was only entitled to one representative on the Executive Council.

[3] See Iron and Steel Trades Confederation, *Quarterly Report* (31 March 1960), p. 11, and (31 March 1964), p. 41.

[4] He was appointed as a Divisional Organiser in 1960 but left the service of the union in 1964. In 1950 the Vice-President of the union, who is elected from and by the Executive Council, was a Section J man, and the Parliamentary Representative in 1964 was also a Section J man.

While this book was being printed, BISAKTA announced a dramatic change in its structure: the appointment of a National Staff Officer and three Staff Organisers who would be specifically and exclusively responsible for white-collar affairs.

[5] The structure of NATSOPA which is described here is virtually identical to that of its successor, SOGAT, Division I.

[6] Roughly 57 per cent of the union's total membership was in London in 1964, while 3 per cent of it was in Manchester. Approximately 80 per cent of its white-collar membership was in London, while 7 per cent of it was in Manchester.

[7] This branch used to be called the London Clerical Branch. But in 1964 its name was changed to the London Branch of Clerical, Administrative and Executive Personnel. See NATSOPA, *Annual Report*, 1964, p. 19. For ease of expression it is still referred to as the London Clerical Branch in the following discussion. The other London branches were as follows: Machine, R & GA, and Ink & Roller.

enough in any of these to warrant a separate section on the branch committee, but white-collar members were generally placed in separate chapels and every attempt was made to see that they were represented on the branch committee in proportion to their numbers in the branch. Finally, in 1964 NATSOPA formed a National Association of Advertising Representatives to function as a separate branch within the union.

Of the twenty full-time officers which the union employed in 1964–5, three were exclusively concerned with white-collar affairs. There was a full-time National Clerical Officer at the union's Head Office in London.[1] The London Clerical Branch had a full-time secretary and assistant secretary and these had always been appointed from among the branch's members.[2] The Manchester Branch did not have a full-time official specially responsible for the white-collar section, but by a coincidence the branch secretary officiating in 1964–5 (who also acted as secretary of the union's North Western District) had been appointed from among the clerical members of the branch.[3] One of the other District Secretaries employed by the union in 1964–5 also had been a clerical worker prior to his appointment. Although District Secretaries had no special responsibility for their clerical members, many of them nevertheless spent a considerable amount of time servicing them.

The white-collar members of the union elected all their own representatives. The London Clerical Branch elected three of its members to sit on the union's Executive Council, while the provincial clerical members elected one of their members to do likewise. Representatives of white-collar members at the union's annual conference, the Governing Council, were also elected from and by these members.[4] For the purpose of electing representatives to attend the annual conferences of the PKTF, the TUC, and the Labour Party, the union's membership was divided into five groups, each entitled to elect one representative to attend each conference. One of these groups exclusively covered the London Clerical Branch.

[1] The position was founded in 1937. Its first incumbent was drawn from among the union's white-collar membership. In 1952 the union decided that strict departmentalization among officers should be discontinued and the National Clerical Officer was reclassified as one of three National Assistant Secretaries. It was understood, however, that while this officer no longer had an exclusive responsibility for the white-collar membership, he still had a special responsibility for it. This officer retired in 1963 and although his replacement was drawn from the London Machine Branch, it was still understood that he had a special but not exclusive responsibility for the white-collar membership. In 1966 SOGAT, Division I, decided once again to appoint a National Clerical Officer. The person who currently fills this position used to be a manual worker in the industry.

[2] For a history of this branch see G. D. Hill, 'Your Branch in Focus: London Clerical', *Natsopa Journal* (January 1959), pp. 8–9 and 21.

[3] The white-collar members in Manchester used to have their own full-time officer. But this position was dispensed with during the last war. See J. Collier, 'Your Branch in Focus: Manchester', *Natsopa Journal* (January 1958), pp. 12–13.

[4] In 1963, 21 out of 81 or 25·9 per cent of the delegates at the Governing Council were representatives of white-collar members, while 11,802 out of 45,832 or 25·7 per cent of the membership was white-collar.

Favourable as the white-collar members' position within the organizational structure of NATSOPA was, in one or two respects it was less favourable than that of the manual members. Although the London Clerical Branch was the largest in the union, it only had three seats on the nineteen-member Executive Council, while the next largest branch, London Machine, had four.[1] Similarly, the London Clerical Branch was only entitled to three representatives on the fourteen-member London Joint Branches Committee which co-ordinated the activities and business of the four London branches, while the London Machine Branch was entitled to five.[2] This under-representation was a sore point with the London Clerical Branch, and on various occasions it tried to get its representation on the Executive Council made equal to that of the London Machine Branch.[3] But the union opposed these moves on the grounds that, because of higher labour turnover, white-collar workers do not have the same length or continuity of service in the industry.[4]

The manual union in which the white-collar members' position has been most favourable is the TGWU.[5] This union has a dual structure. Its membership is divided into thirteen trade groups plus two trade sections, as well as thirteen geographical regions each of which is subdivided into several districts. In general, trade groups deal with questions concerning the wages and working conditions of their members and have primary responsibility for the recruitment of new members, while general administrative matters are dealt with by the regional organization. The lowest level of union government, the branches, also reflect the dual structure of the union. Not only is their coverage restricted to a specific geographical area, it is also generally restricted to the members of a single trade group. Even in those districts where some of the trade groups have so few members that they are placed in composite rather than separate branches, separate sectional meetings are generally held within these branches to deal with specific trade matters.

One of the TGWU's trade groups is occupational rather than industrial in scope. It caters exclusively for white-collar employees and is known as the

---

[1] In March 1965 the membership of the London Clerical Branch was 9,590 while the membership of the London Machine Branch was 9,272.

[2] It was also laid down that the chairman and secretary of this committee be from the London Machine Branch. The chairman could only vote in the event of a tie. See the union's *Rules*, 1962, p. 172.

[3] See the union's *Annual Report* for 1958, p. 102 and 1960, p. 12.

[4] But the union pointed out to the author that 'Over the years clerical representation at all levels, including the Executive Council and the London Joint Branches Committee, has continued to increase, and since this section has the greater organizational potential both in London and the provinces it is a fair assumption that this increased representation will be a continuing process.'

[5] The TGWU is the largest union in Great Britain and its structure is rather complicated and subject to vary slightly from one region, district, or trade group to another. The following description is of a very general nature. For a more detailed account see the union's *Rules* as well as the *Transport & General Worker's Union Home Study Course* (London: The Union, 1962), especially pts. i and ii.

National Association of Clerical and Supervisory Staffs. To overstress the point slightly, the NACSS is a union within a union. Although the NACSS, like all the other trade groups, is subject to the authority of the General Executive Council of the TGWU, it nevertheless enjoys a large measure of autonomy. In practice, this means that it must get the approval of the General Executive Council before it can call members out on strike or before it implements any policy which might affect the members of other trade groups, but is virtually a free agent with regard to all other trade matters. It is even affiliated in its own right to the CSEU.

The NACSS is governed at national level by a National NACSS Trade Group Committee. The form which the NACSS's government takes at regional level varies somewhat from one region to another. Most regions have a Regional NACSS Trade Group Committee.[1] In some regions, generally those in which the white-collar membership is very small, the NACSS representatives are linked with another Regional Trade Group Committee. There are also a few regions in which a district committee system rather than a trade group system exists. District Committees have the same function as Regional Trade Group Committees except that they generally represent the members of more than one trade group.[2] Where this is the case, NACSS members are represented on District Committees in proportion to their numbers within the district. Regardless of the form which the NACSS's government takes in the regions, the principle is retained of only allowing members of the NACSS to deal with its trade-group business including the election of its representatives.

Although the NACSS has a considerable measure of autonomy, it is also a part of the TGWU and as such is represented on all the administrative and co-ordinating bodies which make up its structure. Approximately 4 per cent of the TGWU's total membership is accounted for by the NACSS,[3] but its representation on these bodies is generally more favourable than this. Each of the regions has a Regional Committee to administer its affairs. Of the 283 people sitting on them in 1964-5, 14 or 4·9 per cent were chosen from and by the NACSS membership.[4] The supreme policy-making authority of the union is the Biennial Delegate Conference. In 1965, 38 or 4·6 per cent of the

[1] Although the NACSS is primarily an occupational rather than an industrial trade group, in regions where the NACSS membership is very large, steps are taken to ensure that different industrial interests are represented. For example, in Region 1, the Regional NACSS Trade Group Committee has two advisory committees—one for the Ford Motor Company and another for the road haulage industry. Nationally, *ad hoc* meetings of representatives from specific industries are held from time to time.

[2] Some district committees, especially in the Midlands, cover the members of only a single trade group. They are thus identical to Regional Trade Group Committees except that their scope is restricted to a district rather than a region.

[3] See *supra*, Table 3.6.

[4] All the information contained in this and the following paragraphs concerning the number and backgrounds of representatives and full-time officers was obtained from

820 delegates were chosen from and by the NACSS membership. The main governing body of the union between Biennial Delegate Conferences is the General Executive Council, and each of the trade groups, including the NACSS, is allowed one representative on it. In addition, the General Executive Council has a number of territorial representatives elected by the membership of each region irrespective of trade. In 1964–5 there were twenty-six of these, and three of them happened to be NACSS members. The General Executive Council elects eight of its members to sit on an executive committee, the Finance and General Purposes Committee. In 1964 one of these happened to be an NACSS member.

The NACSS has a full-time national secretary[1] as well as a number of full-time officers at regional level who help to administer its affairs. In 1964–5 there were thirteen regional officers who were exclusively responsible for NACSS affairs,[2] while six others had a special but not exclusive responsibility for them.[3] In addition, although the full-time regional and district secretaries of the TGWU do not have any special responsibility for NACSS members, they nevertheless spend a considerable amount of time servicing them. In fact, some of these officers have been drawn from the NACSS membership. Several of the present district secretaries and three of the thirteen regional secretaries were NACSS members prior to their appointment.[4]

Purely white-collar unions. The aspect of the purely white-collar unions' structure which has been most frequently examined in relation to their growth is their willingness and ability to represent the industrial or sectional interests of their members. All the purely white-collar unions in private industry are 'occupational unions' in the sense that their members are drawn from a particular white-collar occupation or related group of occupations.[5] While the members of any occupational group obviously have many common interests, they also have many that are dissimilar and which are primarily related to the industry or section of the industry in which they work. The major purely white-collar unions in private industry have recognized this and have tried to design their structures in such a way as to make it easier for the different industrial or sectional interests of their members to be articulated and catered for.

DATA places its members in branches according to where they work

correspondence and interviews with the national and regional officers of the NACSS and the TGWU.

[1] The present incumbent was drawn from among the manual members of the TGWU; the previous two were drawn from among its white-collar membership.

[2] All but one of these was appointed from among the members of the NACSS.

[3] All of these had been manual members of the TGWU prior to their appointment.

[4] The full-time secretary of one of the Regional General Workers Trade Group Committees was also a NACSS member prior to his appointment.

[5] See John Hughes, *Trade Union Structure and Government* (London: HMSO, Royal Commission on Trade Unions and Employers' Associations, Research Paper No. 5, 1967), pp. 2–6.

rather than where they live.[1] Branches are therefore more plant than geographically orientated. The Executive Committee of the union is advised of the particular interests of members employed in the shipbuilding industry, the aircraft industry, and the nationalized industries, as well as in the largest chemical employer, Imperial Chemical Industries, by a system of Advisory Panels which are elected by and from the members in these areas. DATA is also increasingly co-ordinating the activities of members in large combines or consortiums. In addition to being responsible for a particular geographical area, the union's full-time officers have recently been made responsible for liaison between the representatives of members employed in different factories of particular combines and for bringing them together in conferences as circumstances demand.[2]

The NUJ's branches generally cover a geographical area. But some of them, mainly those in the London area, are organized on a functional basis. They cover journalists working in periodical and book publishing, on trade and technical magazines, on evening papers, in public relations, in the High Court of Justice, and in government service, as well as parliamentary journalists. Freelance journalists are organized in separate branches in areas where their numbers are sufficient to warrant this, and in separate regional sections elsewhere. Freelance members are also entitled to elect representatives to a National Freelance Council which advises the union's National Executive Council on freelance matters. Radio and television journalists are members of their geographical branches except in London and Dublin where they have their own branches. There is also a Radio Journalists Council which advises the National Executive Council on radio and television affairs.[3]

Unlike draughtsmen or journalists, clerical workers are employed in virtually every section of the economy. Since the CAWU has generally tried to recruit them wherever they were not already members of other unions, it not surprisingly has found the problem of representing the different industrial and sectional interests of its members more difficult to solve than has DATA or the NUJ. Efforts to bring bank clerks and law clerks into the union in 1908 failed, according to the union's official historian, because special branches could not be formed for them. The membership refused to admit that clerkdom was divided into sections and that these should be reflected in union organization and policy.[4] The issue of occupational versus industrial organization has also resulted in a number of secessions from the CAWU.

[1] For a more complete account of DATA's structure see its *Rules* as well as *Draughtsmen's and Allied Technicians' Association: Its Structure and Work* (Richmond, Surrey: The Association, 1964).

[2] See DATA's *Report of Proceedings at Representative Council Conference 1966*, pp. 71–3.

[3] For a more complete account of the NUJ's structure see the union's *Rules* as well as *National Union of Journalists Officers' Guide* (London: The Union, n.d.).

[4] Fred Hughes, *By Hand and Brain: The Story of the Clerical and Administrative Workers Union* (London: Lawrence & Wishart, 1953), p. 23.

In 1920 most of the union's membership in London newspaper offices seceded to form first the London Press Clerks' Association, and then, after a few months, the clerical section of NATSOPA's London Branch.[1] In the same year, the CAWU's organizer for Wales resigned taking most of the Welsh membership with him to found a rival organization, and in 1921 a large proportion of the members employed in trade union offices seceded to join the Welsh breakaway as its London Guild.[2] Finally, in 1937 most of the union's membership in the iron and steel industry broke away or were taken away to become members of BISAKTA.[3]

The CAWU's first major attempt to solve the problem of occupational versus industrial organization took place just after the First World War. Largely as a result of the influence of Guild Socialist ideas upon some of the leading members of the union and the large number of clerks being organized outside the union along industrial lines, the CAWU decided to reorganize itself on an industrial basis in 1920. It established eight national guilds each of which was autonomous on matters concerning only the industries or services it covered.[4] There was also a General Council, composed of delegates from each of the guilds plus the union's national officers, which co-ordinated the guild's activities and ensured their mutual support. But the guild system was doomed almost from the start. Just before it was established, the post-war depression set in and the CAWU's membership not merely declined but slipped away 'in what began to look like a landslide'.[5] The guild system did not prove to be the panacea its advocates had claimed, and it soon came under heavy attack. In 1932 the union finally decided to abolish the guild system and revert to a geographical form of organization.

But as the depression eased during the 1930s and the CAWU's membership began to expand, it found that it was not sufficient simply to place members in branches,[6] group these into geographical areas governed by Area Councils, and let each of these elect a representative to the union's Executive Council. In 1941 it established an Advisory Council for the engineering industry and this became the prototype for several others. Today there are Advisory Councils for the union's membership in engineering, coal, co-operatives, electricity supply, and civil air transport.[7] These function at both regional

[1] Ibid., pp. 67–8 and *infra*, p. 149.
[2] Ibid., pp. 60–2 and 67. In the long run these defections failed to create a rival organization.
[3] See *infra*, pp. 164–5.
[4] A more detailed description of the guild system is scattered throughout Hughes, op. cit., chaps. 5–7.
[5] Ibid., p. 68. The CAWU's membership was 43,000 in 1919, 14,000 in 1921, and 7,000 in 1924. See *infra*, p. 214.
[6] Most of the union's branches are based on a single firm or industry.
[7] At one time there was also an Advisory Council for the membership employed in trade union offices, but this was disbanded because almost all of its membership was concentrated in London and it was not covered by any form of national negotiating machinery.

and national level. There are Area Advisory Councils which give advice to the Area Councils on matters within their own industry and appoint representatives to National Advisory Councils which offer similar advice to the Executive Council.[1] In addition, the CAWU convenes *ad hoc* 'multiple group meetings' where staff representatives working in different branches of large engineering establishments are brought together to discuss common problems. The union also tries to ensure that the sectional interests of its members are looked after by assigning a particular responsibility for certain industries or sections of industries to all its full-time national officers and many of its area officials.

The AScW's structure was similar in broad outline to that of the CAWU. It was organized primarily on a geographical basis: members were placed in branches based on a single establishment, or on a geographical area, or on both; these were grouped into geographical areas governed by Area Committees; and each of these elected a representative to the union's Executive Committee. But the union also made some provisions for ensuring that the different industrial interests of its members were represented. Advisory Committees were established at national and, in some cases, at area level covering members employed in medical research, universities, the chemical industry, the engineering industry, the National Health Service, and the National Coal Board. The function of these committees was 'to give expert advice to the Executive Committee, rather than to make final decisions, although the Executive Committee would seldom reject such advice and will often mandate an Advisory Committee to take a decision'.[2] The AScW also held 'interlocation meetings' to bring together the staff representatives in large multi-plant companies to discuss common problems.

The purely white-collar union in private industry which placed the greatest emphasis on representing the sectional as well as the occupational interests of its members during the post-war period was ASSET. For general administrative purposes, members were placed in branches generally based on a geographical area and these were grouped into districts governed by District Councils. The branches were entirely administrative units, and, in practice, their major function was the collection of subscriptions. They were also responsible for electing representatives to the union's National Executive Council and to its Annual Delegate Conference. The District Councils were responsible for seeing that the branches were properly administered. But neither the branches nor the districts had any direct control over industrial matters.

[1] Recently the union has strengthened representation on the Advisory Councils and, in this and other ways, made them more effective. See the union's *Annual Report and Balance Sheet* for 1961, Appendix B; 1962, pp. 8–10; 1963, pp. 8–10; and 1965, pp. 10–11.

[2] 'Democracy in the AScW', *AScW Journal* (May 1959), p. 11. See also 'AScW Organization', *AScW Journal* (March 1962), pp. 13–14 and 24, for a more detailed discussion of the AScW's structure.

For negotiating purposes, the 'group' was the operative unit in ASSET's structure. A group comprised all the union's membership in a particular factory or establishment and had its own grade representatives, chairman, and secretary who functioned as a shop steward. Although subject to the overriding authority of the National Executive Council, the group was, in practice, autonomous on those industrial matters which only concerned its members and was responsible for negotiating with local management. Representatives from all the groups within a single combine or multi-plant company were brought together at least once a year to enable them to exchange information and experiences and to formulate and co-ordinate bargaining objectives. In addition, in such areas as civil air transport, London Transport, railways, and Remploy where its membership was covered by national negotiating machinery, the union established National Industrial Councils. These were the union's 'principal medium for dealing with matters affecting the industry concerned'.[1]

ASSET's full-time officials had a dual responsibility. Each of them was responsible not only for the membership in a particular district, but also for that covered by one or more of the combines or National Industrial Councils, and was expected to study these intimately so as to be aware of their national policies as well as their local problems. The union even went so far as to purchase token shareholdings in the combines in which it had membership so that the specialist officer could attend their annual meeting and, if necessary, speak as a shareholder 'about issues which might be difficult to express as a trade-union officer in factory collective bargaining'.[2]

Impact upon union growth. When the different internal union structures are compared, there seems to be a prima facie case for assuming that some of them would be more attractive than others to white-collar workers. But even granted this, it is very difficult to assess what effect the relative attractions of these unions' internal structures have had upon their growth. For in making such an assessment little, if any, reliance can be placed on differences in the growth of their actual memberships. Such differences are very much influenced by differences in the growth of their potential memberships. Before the growth of one union can be meaningfully compared with that of another, the influence of labour force growth upon union growth must be controlled for. That is, growth-rates must be calculated on the basis of 'real' rather than actual memberships. But the membership densities of most of the unions being considered here cannot be calculated even for a single year, let alone over time. Either the area of potential membership cannot be delimited with any degree of exactness, or the government's

---

[1] See the union's *Rules*, 1964, p. 21.
[2] Clive Jenkins, 'Tiger in a White-Collar?', *Penguin Survey of Business and Industry 1965* (Harmondsworth, Middlesex: Penguin Books, 1965), pp. 58-9.

manpower statistics are not given in sufficient detail to allow this area
to be quantified.

Thus no firm conclusions can be drawn about the influence which these
unions' internal structures have had upon their growth. But one or two observations can be made which suggest that their influence has been minimal.
The potential membership areas of the CAWU and the NACSS are very
broadly similar.[1] Although one is a purely white-collar union and the other
is partially white-collar, their growth during the post-war period has been
almost identical. The CAWU's membership grew by 105·7 per cent during
the period 1948–64, while the NACSS's grew by 104·7 per cent.[2] Until very
recently BISAKTA would seem to have had the least attractive structure from
the white-collar workers' point of view and probably showed the least
aggressiveness in recruiting them, but this did not prevent it from establishing a relatively large membership among white-collar workers in the iron and
steel industry. Even if it is granted that this occurred in spite of its internal
structure,[3] it nevertheless indicates that any negative influence which its
structure may have had upon its growth was easily outweighed by other
factors.

Whatever the influence of internal structure upon individual union growth,
it does not seem rash to conclude that its influence upon aggregate union
growth has been negligible. At least it is extremely difficult to see how this
factor in any way helps to account for the variations in the pattern of white-
collar unionism in private industry. It might nevertheless be argued that the
over-all level of unionism among white-collar workers in private industry
would have been much higher if they had had more attractive union structures to choose from. But this seems most improbable. The unions catering
for white-collar workers in private industry show the same structural diversity as do British unions in general. With the exception of journalists and
draughtsmen who are already highly unionized, most white-collar workers
in private industry have had the opportunity to join either a purely or a partially white-collar union, and, in many instances, more than one variety of
each type. Either these white-collar workers are extremely pernickety regarding internal union structures, or it must be assumed that their reluctance
to join trade unions is explained by other factors.

External affiliations

All the major white-collar unions in private industry are affiliated to a large
number of other organizations. These range from educational and research

---

[1] But they are not identical. In addition to clerical workers, the NACSS also tries to
recruit supervisors, and, to a lesser extent, certain technical grades.

[2] See *supra*, Table 3.7.

[3] It is argued later in this study that BISAKTA has been relatively successful in recruiting white-collar workers because of the relatively favourable attitude which most iron and
steel employers have toward it. See *infra*, pp. 164–7.

bodies such as the Workers' Educational Association, the Labour Research Department, and Ruskin College, through such pressure groups as the National Council for Civil Liberties, the Consumers' Association Limited, and the Movement for Colonial Freedom, to federations of trade unions such as the National Federation of Professional Workers, the CSEU, the PKTF, the TUC, and the International Federation of Commercial, Clerical and Technical Employees, and to such political organizations as the Fabian Society and the Labour Party.[1] It is not possible or necessary to consider here the relationship between all these organizations and the various white-collar unions. Only their relationships with the TUC and the Labour Party will be examined here as these are the ones which have been most frequently mentioned as having an influence on union growth. In addition, the relationships between the various unions representing white-collar workers in private industry will be examined because such relationships have also been claimed to affect union growth.

Affiliation to the Trades Union Congress. All the major unions catering for white-collar workers in private industry are affiliated to the TUC. The CAWU (then the National Union of Clerks) affiliated in 1903; NATSOPA in 1901 before it began to recruit white-collar workers; BISAKTA and the TGWU in the very year they were founded, 1917 and 1922 respectively; DATA (then the Association of Engineering and Shipbuilding Draughtsmen) in 1918; and the AScW in 1942. When ASSET (then the National Foremen's Association) first applied to join the TUC in 1919, its application was turned down because a large proportion of its membership was already represented at the TUC by virtue of holding dual membership in the various craft unions. But ASSET reapplied in 1920, and in 1921 the TUC accepted it into membership.[2] The NUJ affiliated to the TUC in 1920, but it disaffiliated in 1923 after the TUC raised the affiliation fees of member-unions to help support the *Daily Herald.* The NUJ felt it was 'inimical' to its interests that journalists should be identified with the interests of any particular paper. It was not until 1940 that the membership of the NUJ once again decided that the union should affiliate to the TUC.

The question of TUC affiliation raised feelings of disquiet and sometimes outright opposition among certain sections of the memberships of some of these unions. When the motion to affiliate to the TUC was moved at DATA's annual conference in 1918 there was some opposition on the grounds that affiliation was 'premature' and a few delegates wished to be assured that affiliation did not mean that the union was committed to support the Labour Party.[3] The AScW (then the National Union of Scientific Workers) rejected

[1] This is by no means an exhaustive list.
[2] See Alan E. Williams, 'The Foreman's Story: An ASSET History', *ASSET* (January 1954), p. 5.
[3] J. E. Mortimer, *A History of the Association of Engineering and Shipbuilding Draughtsmen* (London: The Association, 1960), p. 61.

a proposal to affiliate to the TUC in the early 1920s,[1] and it was another two decades before this position was reversed by a ballot vote of the membership. Prior to the NUJ's affiliation to the TUC in 1920, the matter had been raised unsuccessfully at five of the union's annual conferences. As a result of affiliation a few members of the union's Parliamentary Branch resigned. There were also attempts at the union's annual conferences in 1921 and 1922 to get it to disaffiliate. After the union did disaffiliate over the *Daily Herald* issue, in 1923, the membership had to be balloted on six occasions before it finally agreed to reaffiliate.[2]

This opposition has led to suggestions that a union's affiliation to the TUC may hinder it from recruiting white-collar workers. Some writers have even gone so far as to suggest that the establishment of a separate central congress for white-collar unions such as exists in Sweden 'might be a valuable aid to the spread of organisation among non-manual employees, since it would demonstrate to many white-collar employees who are reluctant to join unions affiliated with the T.U.C., that there were associations which catered exclusively for their interests'.[3]

But the evidence does not lend much support to such views. The growth rates of the major white-collar unions in private industry as well as of trade union membership in general for the period 1900 to 1964 are shown by Fig. 7.2.[4] The points at which the various unions affiliated to the TUC are marked on the graph. The growth rates of DATA and the CAWU were roughly the same after affiliation as they were before which suggests that it had little impact upon their growth. The growth rates of ASSET and the AScW were slightly less after affiliation than they were before, while that of the NUJ was slightly less after its initial affiliation in 1920 and slightly more after its reaffiliation in 1940. The growth rates of all these unions immediately before and after affiliation tend to follow that of trade union membership in general which suggests that some factor common to all unions rather than something unique to the unions in question such as TUC affiliation was responsible for the change in growth rates. Nor have the growth rates of white-collar unions in other sections of the economy suffered as a result of affiliation to the TUC.[5] As David Lockwood has noted, 'It has yet to be shown

[1] Reinet Fremlin, 'Scientists and the T.U. Movement', *AScW Journal* (May 1952), p. 17.

[2] See the references to the TUC in the index of Clement J. Bundock, *The National Union of Journalists: A Jubilee History 1907–1957* (Oxford: Oxford University Press, 1957) for the history of the NUJ–TUC relationship.

[3] B. C. Roberts, 'Introduction', *Industrial Relations: Contemporary Problems and Perspectives*, B. C. Roberts, editor (London: Methuen, 1962), p. 5.

[4] This is a semilogarithmic graph and the slope of any series plotted on such a graph indicates its rate of growth.

BISAKTA is omitted from Fig. 7.2 because it was not possible to obtain its white-collar membership figures for most of the period 1917 to 1964, and, in any case, it affiliated to the TUC and the Labour Party in the very year it was founded.

[5] See D. Volker, 'NALGO's Affiliation to the T.U.C.', *British Journal of Industrial Relations*, iv (March 1966), pp. 66–7.

that TUC affiliation by any non-manual union has led to a decrease in membership.'[1]

Affiliation to the Labour Party. Most, but not all, of the major unions catering for white-collar workers in private industry are affiliated to the Labour Party. The CAWU affiliated in 1907, NATSOPA in 1916 before it began to recruit white-collar workers, BISAKTA and the TGWU immediately after they were formed in 1917 and 1922 respectively, DATA in 1944, and ASSET in 1947. All the attempts which have been made by certain sections of the AScW's membership to get it to affiliate to the Labour Party have been defeated. But in 1945 it established a political fund which was mainly employed for general lobbying activities to further the use of science and the interests of scientific workers. The Labour Party affiliation issue has never been formally raised within the NUJ.

The question of Labour Party affiliation has generally generated more heated feelings among white-collar trade unionists than that of TUC affiliation. More white-collar unions are affiliated to the TUC than are affiliated to the Labour Party.[2] When the CAWU decided to affiliate to the Labour Party in 1907, there was some debate as to whether this should be delayed in case it deterred more conservative clerks from joining.[3] DATA had to ballot its members in 1920, 1923, 1940, and 1944 before a majority of them agreed to allow a political fund to be established and the union to affiliate to the Labour Party.[4] After DATA and ASSET affiliated to the Labour Party, there were attempts by certain sections of their memberships to get them to disaffiliate.[5] Although the question of Labour Party affiliation has never formally arisen in the NUJ, it underlay the debate over affiliation to the TUC. The anti-affiliationists argued that the relationship between the TUC and the Labour Party was so close that TUC affiliation represented an inferential political alignment which journalists could not accept by the very nature of their profession.

Even when white-collar unionists do agree to establish a political fund and affiliate to the Labour Party, their contracting-out rates are generally higher than those of manual unionists.[6] In 1964 the contracting-out rate was 13 per cent in the London Clerical Branch of NATSOPA, 21 per cent in the CAWU,

---

[1] *The Blackcoated Worker* (London: Allen & Unwin, 1958), p. 182 n. 1.

[2] There are also more manual unions affiliated to the TUC than to the Labour Party. In fact, some of these unions are just as opposed to political action as are white-collar unions. See Martin Harrison, *Trade Unions and the Labour Party Since 1945* (London: Allen & Unwin, 1960), p. 325, and A. J. M. Sykes, 'Attitudes to Political Affiliation in a Printing Trade Union', *Scottish Journal of Political Economy*, xii (June, 1965), pp. 161–79.

[3] Hughes, op. cit., p. 21.

[4] Mortimer, op. cit., pp. 447–8, table 15. Harrison, op. cit., p. 23, claims DATA also went to the polls in 1921 over the question of establishing a political fund, but this is incorrect.

[5] Ibid., p. 25 n. 1.

[6] The following contracting-out figures have been obtained from either the union concerned or the Registrar of Friendly Societies.

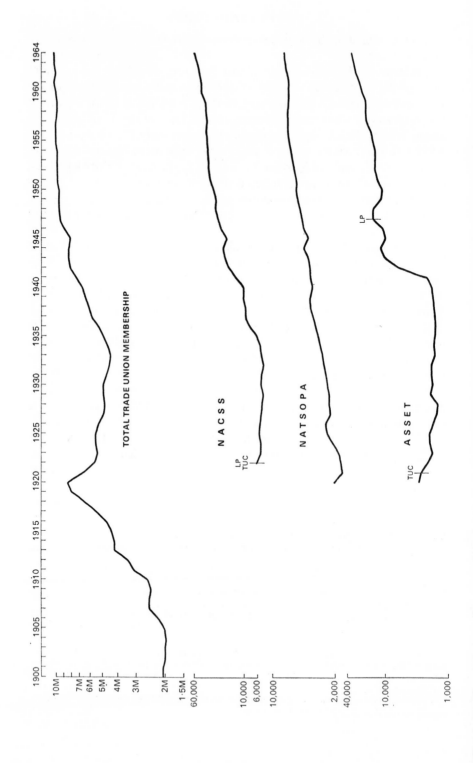

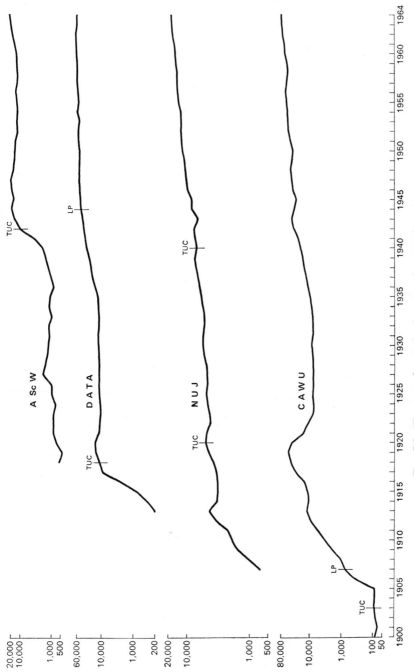

FIG. 7.2. The rate of growth of trade union membership.

48 per cent in ASSET,[1] and 58 per cent in DATA,[2] as compared to 3 to 4 per cent among the total memberships of NATSOPA, BISAKTA, and the TGWU.[3] The higher contracting-out rates among white-collar workers are hardly surprising since it is well known that they are more prone to support the Conservative or Liberal Parties than the Labour Party.[4]

Those writers who have suggested that TUC affiliation may restrict a union's ability to recruit white-collar workers have generally suggested the same, but even more strongly, regarding Labour Party affiliation.[5] Even some of the writers who have argued that TUC affiliation is not an obstacle to recruiting are not so sure about political affiliation. Lockwood notes that the CAWU was very active in the political sphere during the 1930s and suggests that 'it may very well be that clerks outside the union were deterred from joining because of its left-wing policy'.[6] But this is unlikely. The 1930s, as Fig. 7.2 demonstrates, was a period in which the CAWU's membership grew fairly rapidly. It might be argued that it would have grown even more rapidly if the union had not been so politically active, but there is no evidence to support this view. Nor does the evidence lend much support to the view that Labour Party affiliation

---

[1] This union's contracting-out rate had previously been much higher. In 1960 it was approximately 96 per cent. The reason for this seems to be that prior to 1960 most of its membership existed in a state of limbo in regard to the political fund. That is, they did not pay the political levy, nor did they contract out. The contracting-out form was included on the membership application form, and after 1960 the union refused to accept the application unless the political levy was included with it or the contracting-out section of the form was completed.

[2] DATA's contracting-out rate has been increasing over most of the post-war period. In 1948 it was only 35 per cent. The union's official historian (Mortimer, op. cit., pp. 421–2) claims that this high contracting-out rate is partly explained by the way in which the Association has collected the political levy.

> The contribution is levied monthly and a separate stamp is issued for it. On the membership card the monthly political contribution stamp is as prominent as the industrial contribution stamp. It thus serves as a constant reminder to the member that he is paying a monthly levy. If for some reason or other he feels momentarily dissatisfied with the labour movement he may be inclined to 'contract out' of paying the political contribution when the time comes for him to pay his monthly subscription. The method of collecting the political contribution in the A.E.S.D. almost requires from the individual member a conscious reaffirmation, month by month, of his support for the political labour movement.

This assessment seems to be at least partly correct. Towards the end of 1963, the union installed a computer, centralized its financial system, and issued a new membership card which did not distinguish between the political and industrial contributions. By 1966 the union's contracting-out rate had dropped to 47 per cent.

[3] The vast majority of the members who contract out in these unions are white-collar workers. No figures are available for Section J of BISAKTA or the NACSS, but in NATSOPA 86 per cent of the members who contracted out in 1964 were from the London Clerical Branch alone.

[4] See John Bonham, *The Middle Class Vote* (London: Faber, 1954), and D. E. Butler and Anthony King, *The British General Election of 1964* (London: Macmillan, 1965), p. 296.

[5] See, for example, Bernard Donoughue, *Trade Unions in a Changing Society* (London: Political and Economic Planning, 1963), pp. 195–6.

[6] Op. cit., p. 167.

has a detrimental effect on unions' growth rates. DATA's growth rate was roughly the same before and after affiliation which suggests that it had little impact upon the union's growth. The CAWU's growth rate was slightly less after affiliation, while that of ASSET was considerably less. But the CAWU's growth rate immediately before and after affiliation and, to a lesser extent, that of ASSET tend to follow that of trade union membership in general which suggests that some factor common to all unions rather than something unique to the unions in question such as Labour Party affiliation was responsible for the change in growth rates. Finally, it is worth noting that although many draughtsmen are opposed, or at least do not support, the political activities of DATA, they nevertheless join it.

Inter-union relations. In many countries the most bitter jurisdictional disputes have occurred among unions catering for white-collar workers, and this has frequently been offered as at least a partial explanation for why the density of white-collar unionism in these countries is not higher than it is.[1] Whatever the effect of jurisdictional disputes in other countries, their impact upon the growth of white-collar unionism in Britain has been negligible. The reason for this is very simple: there have been few serious jurisdictional disputes among British white-collar unions. Of course, the recruiting interests of these unions overlap at various points and they come into competition for members. There has been competition among the CAWU, the NACSS, and the NUGMW in most industries for clerical workers, as well as between them and BISAKTA in the iron and steel industry and between the CAWU and the National Union of Mineworkers in the coal industry. There has been competition between ASSET and DATA for such grades as planning, process, and methods engineers; between ASSET and the CAWU for such grades as progress chasers, schedulers, and programmers; between ASSET and the AScW for certain scientific and technical grades; and, to a much lesser extent, between the NUJ and the IOJ for journalists, and NATSOPA and the NUPBPW for certain white-collar grades, in the paper, printing, and publishing industry. There has also been competition between ASSET and such manual unions as the AEU, the ETU, the Boilermakers' Society, and BISAKTA for foremen and certain technical grades. But this competition has created relatively little friction and rarely broken into open conflict.[2] Almost all the disputes have been settled peacefully either by the unions themselves or by the Disputes Committee of the TUC.

In fact recent mergers between some of these unions have reduced what

---

[1] See, for example, *Non-Manual Workers and Collective Bargaining* (Geneva: ILO, 1956), p. 45, and Blum, op. cit., pp. 89–95.
[2] The recognition-jurisdiction dispute which occurred in the steel industry after it was nationalized is an outstanding exception to this generalization, but it is nevertheless an exception. See also G. S. Bain, *Trade Union Growth and Recognition* (London: HMSO, Royal Commission on Trade Unions and Employers' Associations, Research Paper No. 6, 1967), p. 54 n. 76.

little friction exists. NATSOPA and the NUPBPW merged in 1966 to form SOGAT, and the AScW and ASSET merged in 1968 to form the ASTMS. Although the NUJ and the IOJ have not merged, they have agreed on arrangements for dual membership and closer working. In addition, there is a Joint Consultative Committee of Staff Unions in the engineering industry which serves as a forum for the discussion of common problems by the ASTMS, the CAWU, DATA, the NACSS, and the NUGMW. But at least one writer has argued that there is room for further mergers among white-collar unions in private industry. Hindell feels that a merger of the clerical membership of the TGWU and the NUGMW with that of the CAWU would eliminate some needless competition and pool resources of staff and money so that both could be more efficiently deployed, and that this would have a beneficial effect upon membership growth.[1] Even granting that these mergers could be brought about, it is doubtful if they would greatly affect the density of white-collar unionism in private industry. At least, the existence of a single union for the distributive trades does not seem to have greatly boosted its level of unionism. Nor does the existence of a large number of unions for employees in national government seem to have retarded the growth of unionism in this area.

CONCLUSIONS

A great many people believe that the growth of unionism among white-collar workers is largely in the hands of the unions themselves. If only they would make themselves more popular with the general public, devote more time and energy to recruiting, and design structures more attractive to white-collar workers, they would be successful in recruiting these workers. But this belief is not supported by the evidence in Britain. While it is not reliable enough to permit the conclusion that the public image of the trade union movement, union recruitment policies, and union structures have had no effect upon the growth of white-collar unionism, it certainly does not permit the conclusion that their effect has been of great importance. What evidence there is suggests that these factors have been at most of very secondary importance in explaining the growth of white-collar unionism.

If the growth of white-collar unionism is not greatly influenced by the unions, perhaps it is by the employer. An American labour lawyer has noted that:

In an organizing drive, the advantages of communication are markedly with management. Labor has to rely pretty much on appeals via established media of communication, circulars distributed at the place of the employer, home solicitations, and meetings at a hired hall. Management has available to it a complete and accurate mailing list of employees, together with the plant itself, wherein employees can be addressed as a captive audience, or on their own time. Furthermore, during

[1] Op. cit., pp. 197–8.

working hours other than for relief and rest periods, management contacts with employees are more frequent and more sustained. This is a matter about which one could comment at length.[1]

This is also true in Britain, and the following chapter does comment on it at some length.

[1] Robert J. Doolan, 'Attitudes of White-Collar Workers Toward Unionization', *Addresses on Industrial Relations* (Ann Arbor: University of Michigan, Bureau of Industrial Relations, 1959), p. 11.

VIII

THE EMPLOYERS

THE employers' role in union growth has generally been neglected. Even when its importance has been noted, it has not received much systematic or detailed attention. This neglect is not justified. For even the most superficial reflection should indicate that employer policies and practices may profoundly affect the growth and development of trade unionism. Where employers disapprove of trade unions and pursue policies designed to discourage their employees from joining them, trade union growth is likely to be retarded. Conversely, where employers recognize and negotiate with trade unions and encourage their employees to belong to them, trade union growth is likely to be stimulated.

In order to obtain some information on employer attitudes and behaviour with regard to white-collar unionism, a survey of individual employers and employers' associations in private industry was carried out during May–July 1966.[1] The author interviewed twenty-five major firms and fourteen employers' associations, and, in addition, a questionnaire was sent to 142 employers' associations and fifteen firms. All of the organizations approached agreed to be interviewed; the response rate to the questionnaire was 86 per cent. The employers' associations were chosen, with the help of the Confederation of British Industry, as being the most important ones in the industry concerned. The firms were chosen partly because of their size and partly because they were known to have refused or to have granted recognition to white-collar unions.[2] This survey of employers was supplemented by information obtained from interviews during 1965–6 with sixty full-time officials of the major unions catering for white-collar employees in private industry and from documentary sources.

TRADE UNION GROWTH AND RECOGNITION

There are several reasons why the growth of unions is likely to be greater the more willing employers are to recognize them and the greater the degree of

[1] This survey was undertaken on behalf of and under the direction of the Royal Commission on Trade Unions and Employers' Associations, and the findings were published in G. S. Bain, *Trade Union Growth and Recognition* (London: HMSO, Royal Commission on Trade Unions and Employers' Associations, Research Paper No. 6, 1967).

[2] The various firms and employers' associations agreed to co-operate with the Royal Commission and the author on the understanding that their replies would not be attributed to them as organizations. Consequently, material obtained from interviews and questionnaires has been cited in such a manner that the organizations concerned cannot be identified.

recognition[1] which they are prepared to confer upon them. Firstly, workers, especially white-collar workers, tend to identify with management, and they are, therefore, less likely to join trade unions the more strongly management disapproves of them. A NACSS official claimed that before it is possible to recruit many white-collar employees 'you have to be able to show them that their employer is not really opposed, that they won't be disloyal by joining, and that all in all, there is not going to be much of a battle'.

Secondly, the more strongly management disapproves of trade unions, the less likely workers are to join them in case they jeopardize their career prospects. This point is brought out most clearly by the following letter to a union organizer from a technician in a drug-manufacturing company:

As I explained . . . all the members of the technical staff are fearful to have any-thing to do with our association or any other similar organisation. Apparently the —— Company will have nothing to do with trade unions and most of my colleagues appear to be surprised that I got a job at —— in spite of being a union member. To be quite honest I wasn't asked at the interview if I was a trade union member and to my knowledge they are still unaware of my affiliation.

The evidence which will be presented later in this chapter indicates that the fears of many white-collar workers in this regard are well founded.

Finally and most important, unions are usually accepted on instrumental rather than ideological grounds, 'as something to be used rather than as something in which to believe'.[2] Many employees want to see 'the proof of the pudding' before they join a union but 'the proof of the pudding comes once the union has been recognised'.[3] The less recognition an employer is prepared to give a union the more difficult it is for the union to participate in the process of job regulation and thereby demonstrate to employees that it can provide a service for them. In such circumstances not only are a large number of employees not likely to join the union, but many of those who have already done so are likely to let their membership lapse because the return they are getting on it is insufficient.

There is a considerable body of evidence which supports the argument that recognition is important in fostering union growth. A scholar who studied managerial unionism in the coal industry concluded that it is the 'fact of

---

[1] The assumption is often made that a union either possesses recognition or it does not. But, in reality, union recognition is a matter of degree. On the one extreme, the employer may oppose the union by force or by 'peaceful competition' and there is little or no recogni-tion. On the other extreme, the employer may bargain with the union on any matter it may wish to raise; meet any representatives that the union may appoint; accord the union the necessary facilities to collect subscriptions, hold meetings, and publicize its activities; en-courage his employees to join the union; and provide it with essential information for collective bargaining. Between these two extremes, there are many intermediate positions. The variety of employer policies with regard to union recognition should become clearer after reading this chapter.

[2] C. Wright Mills, *White Collar* (New York: Oxford University Press, 1956), p. 308.

[3] Allan Flanders, *Minutes of Evidence 62*, Q. 10,005.

recognition that explains the success of B.A.C.M. [the British Association of Colliery Management]' and the failure of rival organizations.[1]

A sociologist who studied unionization among bank clerks found that about one-fifth of the clerks in his sample who were not in NUBE said they were not members because it was not recognized. 'For one this was an ideological statement, but for the others it was a practical reason, expressing the belief that N.U.B.E. is powerless, or at least less effective than the staff association, because it is not recognised.'[2] He also found that 'the more hostility a bank has shown to the union, the lower has been the union's completeness [membership density]'.[3] This and other evidence he analysed led him to conclude that:

It does seem quite clear that the banks' attitudes, and their relationships with the two organisations, have played an important part in determining the completeness of N.U.B.E. and the staff associations, with recognition or non-recognition being a major factor in the situation. It is hard to doubt that if N.U.B.E. were generally recognised its completeness would be appreciably higher, or that if recognition were withdrawn from the staff associations their completeness would decline.[4]

The best illustration of the importance of employer policies and practices as a factor in union growth is provided in Great Britain by contrasting the public and private sectors of the economy. The density of white-collar unionism in the civil service, local government, and the nationalized industries is extremely high, even among managerial and executive grades.[5] Lockwood has suggested that this is explained by the large-scale bureaucratic administration characteristic of public employment.[6] But while 'bureaucratization' has obviously been very important in encouraging public employees to join trade unions, it has not been the only factor.

An equally and perhaps even more important reason for the high degree of unionism among white-collar employees in the public sector is that their employers have agreed to recognize and negotiate with their unions. In fact, most public employers have gone much further than this and have actively encouraged their employees to join trade unions. Each new entrant to the Civil Service is informed by the Treasury that he is

not only allowed but encouraged to belong to a staff association. Besides being a good thing for the individual civil servant to belong to an association, which can support him in his reasonable claims and put his point of view before the authorities on all kinds of questions affecting his conditions of service, it is also a good thing for Departments and for the Civil Service as a whole that civil servants should be strongly organised in representative bodies. It is only common sense to meet the

[1] Brian McCormick, 'Managerial Unionism in the Coal Industry', *British Journal of Sociology*, xi (December 1960), p. 367.
[2] R. M. Blackburn, *Union Character and Social Class* (London: Batsford, 1967), p. 250.
[3] Ibid., p. 249.   [4] Ibid., p. 251.   [5] See Table 3.2.
[6] David Lockwood, *The Blackcoated Worker* (London: Allen & Unwin, 1958), pp. 141–9, and *supra*, pp. 74–5.

wishes of the civil servant about his conditions of service as far as possible, for a contented staff will work much more efficiently than a staff which feels that its interests are being completely ignored by the 'management'. But it is hopeless to try to find out the wishes of a scattered unorganised body of individual civil servants each of whom may express a different view. When they get together in representative associations, their collective wish can be democratically determined and passed on to the 'management' with real force and agreement behind it; the 'management' know where they stand and can act accordingly.[1]

Most local authorities and nationalized industries have made similar pronouncements.

By contrast with the public sector, the density of unionization among white-collar employees in private industry is very low.[2] Lockwood has suggested that this is accounted for by the large number of small firms and the resulting low degree of 'bureaucratization' characteristic of private employment.[3] The present study has already shown that there is a great deal of truth in Lockwood's contention.[4] But again, it is by no means the whole story. Employer policies and practices are also important.

Most private employers, unlike those in the public sector, refuse to recognize and negotiate with white-collar unions, and many of them even pursue policies designed to discourage their white-collar employees from joining unions.[5] A survey undertaken by the British Employers' Confederation (one of the organizations which merged to form the present Confederation of British Industry) in 1963–4 revealed that:

Of 23 Member Organisations which replied to the questionnaire only one was able to state, without reservations, that trade union representation of staff workers is recognised, that this recognition is on a formal basis and that agreements covering the rates of pay, etc. of staff workers are made at national or company level. A few other Member Organisations indicated that staff unions are recognised to a limited extent (e.g. they are recognised only at certain firms or only in respect of certain types of staff workers) but generally such recognition is limited to procedure and does not cover the making of agreements.[6]

While this survey is somewhat limited in scope, its general conclusion is supported by an unpublished Ministry of Labour estimate that some 85 per cent of white-collar employees in manufacturing industries are not covered by collective agreements. In other words, probably not more than 15 per cent of staff workers in manufacturing industries have had their right to union representation recognized by employers.[7]

[1] Cited by Richard Hayward, *Whitley Councils in the United Kingdom Civil Service* (London: Civil Service National Whitley Council, Staff Side, 1963), p. 2.
[2] See Table 3.5.   [3] Loc. cit.   [4] *Supra*, pp. 72–81.
[5] These policies are described *infra*, pp. 131–5.
[6] 'Exchange of Views and Information: Staff Workers' (London: The Confederation, Wages and Conditions Committee, 1964), p. 1. (Mimeographed.)
[7] See also Bain, op. cit., especially pp. 68–72, and table 18.

There can be little doubt that a major reason for the difference in the density of white-collar unionism between the public and private sectors of the economy is the difference in employer policies and practices with regard to union recognition. Variations in such policies and practices also help to account for variations in the density of white-collar unionism within the private sector itself. Chapter III demonstrated[1] that union density among most white-collar workers in engineering and electrical goods, shipbuilding and marine engineering, vehicles, iron and steel, and newspaper publishing, as well as among foremen and overlookers in textiles[2] is relatively quite high. These are also the areas in which at least some form of white-collar union recognition has generally existed for several years.

The Engineering Employers' Federation[3] conceded recognition to the CAWU, the NACSS, and the NUGMW for clerical workers in 1920, 1940, and 1953 respectively, to DATA for draughtsmen and certain allied technical grades in 1924, and to the AScW for certain scientific and technical grades in 1944.[4] It also agreed to recognize ASSET in 1944, but only where the union had majority membership in a particular grade in a particular establishment of a member-firm. The Shipbuilding Employers' Federation recognized DATA in 1941.[5] In 1943 the Iron and Steel Trades Employers' Association adopted a policy of encouraging, but not compelling, member-firms to recognize BISAKTA for clerical workers, laboratory staffs, and departmental foremen, and, to a lesser extent, DATA for drawing-office staffs and the various craft unions for craft foremen.[6] The major proprietorial associations in the newspaper industry recognized the NUJ during and immediately after the First World War. The Newspaper Proprietors Association recognized the NUPBPW and NATSOPA for clerical and certain other administrative workers in 1919 and 1920 respectively, while the Newspaper Society recognized NATSOPA for clerical workers in 1938. Both the Newspaper Proprietors Association and the Newspaper Society recognized the Institute of Journalists in 1943. In the cotton-spinning industry, the British Spinners' and Doublers' Association granted a limited form of recognition to the Textile Officials' Association for supervisory grades in 1950.

Clearly, those areas of private industry in which some form of white-collar

[1] *Supra*, especially pp. 33–7 and Table 3.8.

[2] The density of unionization among foremen in textiles is relatively high partly, but not entirely, because weaving overlookers who are really manual workers had to be included in this category for reasons of comparability. See *supra*, p. 32 n. 3.

[3] This Federation covers the majority of firms in engineering and electrical goods, vehicles, certain sections of metal manufacture, and marine engineering (but not shipbuilding).

[4] The concessions of recognition which are mentioned in this chapter are described in much greater detail in Chapter IX.

[5] Since 1964 there have been further concessions of recognition to white-collar unions in the shipbuilding industry. See *infra*, p. 176 n. 3.

[6] Since 1964 the white-collar union recognition situation in the iron and steel industry has been considerably changed by the nationalization of the industry. See *infra*, p. 166 n. 1.

union recognition has generally existed for several years are also those in which the density of white-collar unionism is highest. Unfortunately, the relationship between union membership and recognition cannot be expressed more precisely. Recognition, unlike membership, does not lend itself very well to quantification. There are considerable differences in the degree or quality of the recognition which has been conceded to the various white-collar unions. The Engineering Employers' Federation compelled its member-firms to recognize ASSET only where it had majority membership; this condition still applies for its successor, Division I of the ASTMS. But all the other staff unions in engineering are recognized regardless of the extent of their membership. The British Spinners' and Doublers' Association has recognized the Textile Officials' Association but has refused to enter into formal joint agreements with it. Employers in the public sector have not only recognized the various staff unions, they have also encouraged their employees to join them. It is difficult, if not impossible, to know what weight should be assigned to these and numerous other qualitative differences in the form which white-collar union recognition has taken.

Although recognition cannot be quantified and subjected to statistical analysis, sufficient evidence has been presented here to establish that recognition is very closely associated with membership. But this does not prove that recognition produces membership. It may be that membership is the major factor influencing the employers' decision regarding recognition. Employers may withhold recognition from unions where they do not have sufficient membership to justify its concession and concede it where they do.

'Representativeness' is certainly the criterion which most employers claim to use in deciding whether or not to recognize a union for white-collar employees. Before the Treasury will recognize any staff union, it 'must show that it is representative of the category of staff concerned'.[1] The Engineering Employers' Federation insisted on majority membership as the criterion for recognizing ASSET, and still does for recognizing its successor, Division I of the ASTMS. Similarly, Unilever has claimed that its companies will

recognise a union's right to represent employees and negotiate on their behalf wherever it has established membership amongst a majority of the group which a company is prepared to accept as a 'negotiating group'. At the same time, our companies regard the building up of union membership to the level required for negotiating rights as primarily the responsibility of the unions themselves. Where a union fails to satisfy a company that it has a genuine majority amongst the 'negotiating group', it cannot be said to have earned the right to represent the group.[2]

Lloyds Bank claimed in its evidence to the Donovan Commission that it refused recognition to the National Union of Bank Employees because the bank's 'Staff Association . . . has (by a considerable margin) the greater

[1] H.M. Treasury, *Staff Relations in the Civil Service* (London: HMSO, 1965), para. 12.
[2] Unilever Limited, *Minutes of Evidence 46*, p. 1972.

membership'.[1] Another firm 'could only accept trade union representation in the knowledge that at least 80 per cent of our staff are members of the union'. Many of the other employers who were surveyed indicated that they had refused to recognize a staff union because it had not represented a 'substantial proportion' of the employees concerned.

But it is very doubtful if lack of representativeness is the major reason why most private employers refuse to recognize white-collar unions. In applying the criterion of representativeness it is necessary to define three concepts— 'representativeness', 'area of representation', and 'recognition'.[2] In Britain these concepts are usually defined by the employer and in such a way that either it is extremely difficult for the union to 'earn' recognition, or the form of recognition 'earned' is hardly worth having. When this is the case the criterion of representativeness merely becomes a 'respectable' device by which an employer can deny a union recognition.

There is no general agreement as to what 'representativeness' means. Some employers are prepared, and in engineering are obliged, to recognize a union if only one of their staff employees is a member. Other employers will not grant recognition unless a union has recruited 33 per cent of the employees concerned; others expect 51 per cent; and still others interpret 'substantial proportion' as substantial majority and expect unions to have 75 per cent or better. For a few employers the expected percentage tends to be a variable which moves upward as the union's actual membership in the firm increases. Employers who demand an extremely high degree of 'representativeness' would seem to be motivated not so much by their concern that the union should actually be representative of their employees as by their concern that it should not represent them at all.

Employers who claim that they will recognize a representative union, often choose the area over which it will be most difficult for the union to demonstrate its representativeness. Sometimes the firm may have an organizational reason for the choice of this area. But generally it is designed merely to keep the union out. One large company employing approximately 7,000 foremen in forty different establishments claimed it would only recognize ASSET if it could organize a majority of all the foremen and refused to recognize it on an establishment-by-establishment basis. In an interview the company admitted that there was no 'positive' reason for choosing this national area of representation, and that its primary purpose was to put 'a major hurdle in the way of the union'.

Another large company employing roughly 40,000 white-collar employees in seventy-five establishments has agreed to recognize a union for any particular occupational grade, but only if it has majority membership among this grade across the company as a whole. The company's justification for

[1] Lloyds Bank Limited, *Written Evidence*, p. 1.
[2] This is treated in greater detail in Bain, op. cit., paras. 202–4 and 211–12.

this policy was that it had a national salary structure. But it could advance no logical reason in support of such a salary structure or show that the efficiency of the company's operations would be impaired if local salary structures were adopted. In fact, the national salary structure was only introduced fairly recently, about the same time as various unions began recruiting the firm's staff. Moreover, as the company itself pointed out, this salary structure often resulted in overpaying in areas of labour surplus and underpaying in areas of labour shortage. The company's claim that it is not opposed to staff unionism is therefore not altogether convincing.

Some employers will recognize a union for staff employees only if it is representative across the whole industry concerned. Until 1967[1] the Shipbuilding Employers' Federation argued that the recognition of staff unions was a matter of interest to all employers in the industry, for what one firm did in this regard might set a precedent which other firms would be forced to follow. The Federation therefore advised all member-firms to agree only to informal discussions with a staff union over the grievances of an individual member and to refer all claims for recognition on behalf of staff grades to the Federation. For its part, the Federation would only recognize those unions which were representative of a 'substantial proportion' of a particular staff grade throughout the industry as a whole. As a result of this policy the CAWU was refused recognition both nationally and domestically, for while it claimed to have 'a high degree of organization at certain establishments' it did not represent a 'substantial proportion' of the clerical workers employed in the industry as a whole.

When employers say they are prepared to recognize any union which is representative of a particular area of staff employment, this does not necessarily mean that they are prepared to grant 'full' recognition. Many of them will not negotiate with the union; they are only prepared to enter into informal discussions regarding the grievances of individual members.

A leading employers' association covering a key area for future white-collar union growth gives the following advice to member-firms when they find that a substantial proportion of their white-collar employees are organized and 'contact with a staff union seems unavoidable':

It is desirable to confine recognition of a staff union to representations on behalf of members and to exclude *negotiations* on staff salary scales, etc. The difference between representation and negotiation is important. The first do not compel managements to act in the manner desired, while the latter may result in a binding agreement.

Similarly, the Wages and Conditions Committee of the British Employers' Confederation (one of the organizations which merged to form the Confederation of British Industry) took the view in June 1964 that:

Even where membership of staff unions is increasing employers are under no

[1] See *infra*, p. 176 n. 3.

obligation to recognise union representation . . . Even if 'recognition' were granted to a staff union, this need not include the negotiation of wages and conditions of employment, but might be limited to informal discussions or to the laying down of procedure for dealing with requests and complaints.[1]

But the Committee also took the view that:

There is a danger . . . that once a staff union has been recognised, for any purpose at all, as representing the interests of staff workers, it will be encouraged to press for the full rights of negotiation. Those members who have already granted full 'recognition' . . . confirmed that official representation of staff workers could be a source of much difficulty to employers. In the circumstances the Committee agreed that when discussions were held with staff unions it would be best if possible to avoid the use of the word 'recognition'.[2]

The possibility that partial recognition might lead to demands for full recognition has not worried some employers' associations, however, for when such demands have arisen, they have been refused and even the partial recognition has been revoked. One employers' association had a procedure agreement with DATA, but the union 'eventually chose to attempt to use the Agreement to further a claim for wages which was not the purpose of the Procedure Agreement. In the circumstances, and in accordance with the final paragraph of the Agreement, the —— terminated the Agreement.'

These examples indicate that even after a union has demonstrated its 'representativeness', many employers are only prepared to recognize it for purposes of discussion not negotiation. In fact, discussion rights are often granted in the hope that they will lessen the demands for negotiating rights. While it is useful for a union to be able to discuss with management the grievances of individual members, it is by no means wholly satisfactory. As the Ministry of Labour's definition of a trade union emphasizes, a major union function is the negotiation of wages and working conditions. Employers who refuse to allow a union to negotiate on these matters are thereby preventing it from exercising one of its major functions. Such employers can hardly claim to be speaking in good faith when they say they will 'recognize' any union which can demonstrate its representativeness.

Even if employers apply the criterion of representativeness in the fairest possible manner, there are still grounds for doubting that lack of representativeness is their major reason for refusing recognition. If an employer has no principled objection to white-collar unionism and his only desire is to ensure that a substantial proportion of his staff wish to be represented by a union

---

[1] Op. cit., p. 2. This document evidently recorded the views of the Wages and Conditions Committee only. In giving oral evidence to the Royal Commission, representatives of the CBI pointed out that it had not been considered by the BEC Council nor approved by its President or Director. The CBI's present policy was described as 'waiting and seeing', and neither to stimulate nor to prevent the development of white-collar unions. See *Minutes of Evidence 22*, p. 822.

[2] Loc. cit.

before he recognizes it, then he should be prepared to give his staff every opportunity to join the union and the union every opportunity to recruit them. If an employer who advances the criterion of representativeness is not prepared to do this, then his claim that he will recognize any representative union is rather hollow, for he is denying the union the means of obtaining recognition.

But few employers are prepared to allow an unrecognized union freedom of access to its potential members by giving it facilities to hold meetings during the lunch break or outside office hours, distribute literature or display notices, collect subscriptions, or process grievances on behalf of individual members. Few employers are even prepared to play a neutral role in the recruiting process, but pursue policies designed to discourage their staff from joining unions.

EMPLOYER POLICIES FOR DISCOURAGING WHITE-COLLAR EMPLOYEES FROM JOINING TRADE UNIONS

If union organizers are to be believed:

The employers will stop at nothing. Every case has to be fought through. They utilise staff associations, give general salary increases while a recruitment campaign is underway, and intimidate and victimise leading members. They also conduct a campaign of demoralisation through the supervisors—'You'll never get recognition', 'You won't be able to do anything, etc.'

Broadly speaking, there are two basic strategies which managers use to discourage their staff employees from joining trade unions—peaceful competition and forcible opposition.[1] These are not mutually exclusive strategies, and often both of them are pursued simultaneously. But it is useful to treat them separately for purposes of analysis.

The strategy of peaceful competition

The strategy of peaceful competition includes a variety of tactics: paying salaries equal to or better than those in unionized firms; granting salary increases during a union recruitment campaign; establishing welfare, profit sharing, and other benefit schemes; offering various types of rewards to 'loyal' employees; giving speeches and interviews designed to convince employees that their interests can be better cared for by management than by a union; granting monthly staff status; and establishing 'company unions'.

It is difficult to document adequately the use of these tactics. Salary increases during recruitment campaigns can be 'explained' by reference to the cost of living or increasing productivity; rewards to 'loyal' employees can be 'justified' on grounds of merit; and the thoughts expressed in speeches and

[1] These terms are used by Lloyd G. Reynolds, *Labor Economics and Labor Relations* (Englewood Cliffs, N.J.: Prentice-Hall, 1956), pp. 169–77.

interviews are rarely written down. Moreover, opposition to staff unionism may not be the only reason for following many of the tactics of peaceful competition. While they may be inspired primarily by a strong desire to keep the unions out, they may also be motivated by an employer's sincere wish to treat his staff employees properly and as a 'part of management'. Nevertheless, there are some examples of these tactics being deliberately used to discourage the growth of staff unionism.

An employers' association which 'has refused requests from staff unions for a national procedure agreement or negotiations on several occasions' offers the following service to member-firms:

... as a condition of avoiding negotiations with a staff union it is essential that member firms should be offering salaries and conditions which are not open to criticism and the Association is prepared to collect information, national and local, and give advice on such matters. This has been done on a limited and voluntary basis for a number of years and it is intended to extend and improve this service on behalf of members.[1]

An executive in the motor industry frankly admitted that:

ASSET had a fairly substantial membership among our chargehands and as a result we decided to grant some of them staff status and reduce the remainder to operative status.

Other employers were less specific, and simply stated that:

Every possible step is taken to ensure that staff do not reach the frame of mind which will make them think that membership of a union is appropriate.

Although many employers sponsor staff associations or committees to slow down 'the current somewhat sinister trend towards subversive influence from outside',[2] it is not possible to estimate how many of these exist in private industry. But several observations can be made regarding the nature of those which came to light as a result of the survey of employers. Only one was a staff association in the sense that all employees of the company were not automatically members but had to be recruited. The rest were staff committees in the sense that a group of employees was simply elected to represent the views of the staff to management. All were established on the initiative of the employer, in many cases after an approach from a staff union for recognition. Most had only 'consultative and advisory powers' and did not negotiate over salaries and other conditions of employment. Of those companies which allowed their staff committees to negotiate, only one permitted disputes to be submitted to independent arbitration. In all other cases the top executive officer of the company was the final court of appeal.

The best-known device for discouraging staff unionism in private industry

---

[1] Circular Letter from the Association to all Member-Firms, 1964.
[2] M. J. Ruddock, 'Some Problems of Clerical Staff Employment', *Factory Manager*, xxiii (April 1954), p. 90.

is the Foremen and Staff Mutual Benefit Society, a friendly society established in 1899 to provide pensions, life assurance, and sickness benefits to foremen and similar grades of staff in the engineering and shipbuilding industries.[1] The Society has two kinds of members: contributory members or employers; and ordinary members, the eligible employees of contributory members. Contributory members pay at least half of the total contributions in respect of each ordinary member. The FSMBS has over 2,600 contributory members and over 61,000 ordinary members, of whom 8,700 are in receipt of a pension.

Under the Society's rules an ordinary member may not belong to a trade union. In other words, to become a member of the Society an employee must resign any trade union membership which he may hold, or, if he wishes to join a trade union after he has become a Society member, he must resign from the Society. This means he must forfeit his claim to the contributions which the employer has made on his behalf, and that he receives only the surrender value (about 90 per cent) of his own contributions. In short, he must sacrifice 55 per cent of all moneys paid to the FSMBS on his behalf. For many staff employees, especially older ones, this is an obvious deterrent to becoming or remaining a union member.

The Engineering Employers' Federation and the Shipbuilding Employers' Federation advise member-firms to join the FSMBS and to encourage their eligible staff employees to do likewise. Although both Federations are opposed to firms coercing employees into the Society, a few of the employers who were interviewed admitted making an employee's promotion to foreman conditional upon his becoming a member of the Society. No doubt most contributory members do not adopt such measures, but it is clear from the interviews that most of them encourage foremen to join the FSMBS and give it facilities for recruiting and retaining members which are denied to staff unions.

The strategy of forcible opposition

The second strategy which employers use to oppose staff unionism is forcible opposition. This strategy is implemented by such tactics as: overlooking union members for promotion and pay rises, transferring active unionists from department to department, threatening to discontinue any 'extras' presently being paid above the union rate, sending management officials to union recruitment meetings to note the names of those employees attending, and dismissing leading union members. It is also difficult to document the use of these tactics, for threats are generally communicated verbally and victimization of leading members can often be 'justified' on grounds

[1] For a more complete account of this organization see FSMBS, *Written Evidence*; ASSET, *Minutes of Evidence 53*, pp. 2248–51; and *The FSMBS: British Industry's Friendliest Most Sincere Union-Hating Outfit* (London: ASSET, 1966).

such as inefficiency or insubordination. Only cases which trade unionists claim to be examples of forcible opposition can be cited here; the reader must judge for himself whether the facts substantiate these claims.

After recruiting approximately 40 per cent of the weekly paid clerical staff of a private bus company, the CAWU approached the firm for recognition late in 1965. But the company refused to recognize the union, even for discussion purposes, until it had recruited at least 75 per cent of the staff involved. Moreover, the company's general manager sent the following memorandum to each of his staff employees:

On my return to the office today I was very surprised to read a letter from the above Union seeking formal recognition of this Union as a negotiating body for my weekly paid staff.

Up to this time I have always assumed that the existing direct methods of negotiation between members of my staff and the management have been perfectly adequate and that the general relationship between management and staff has been a very happy one within this Company.

For the moment, the letter has been formally acknowledged as I feel it is only right that I should know the general opinion of all my staff in this matter, because any step which I would take must have far reaching consequences. I am asking you therefore to complete the enclosed form so that I shall know the numbers of my staff involved. This particular matter has been brought up at a rather unfortunate time, because recent discussions I have had with the Company Chairman have proved to be quite fruitful and there was to be a general review within the next two weeks and considerably wider in application than ever previously, embracing as it would have done the junior staff under 21 years of age. For the moment this will have to be suspended until I am aware of your views with regard to this Union.

Your replies will be handed to myself personally and will be treated in strict confidence.

The union claimed that this was an attempt to intimidate the staff. The company denied the charge and said it was entitled to know which members of the staff were union members. In spite of a boycott of the firm by the local Trades and Labour Council, a publicity campaign waged by the union in the local press, and attempts by the Ministry of Labour to obtain a settlement of the dispute, the company still refused to recognize the union.

During 1963 ASSET organized all of the thirty foremen employed at a Scottish factory of an optical manufacturing company. The firm refused to recognize the union and the foremen consequently staged a half-day protest stoppage. The company retaliated by dismissing all the foremen. Three months later, and after the intervention of the Scottish TUC, the firm rehired all but four of the foremen. The four men who were not rehired consisted of the union's local group secretary, treasurer, and chairman, plus the person who had first requested ASSET to recruit the company's foremen. Three of the men had sixteen years' service and the other had thirteen years'. ASSET claimed this was a case of victimization. The company denied this charge. It claimed that a reorganization had resulted in fewer employees being required and that it had the right to rehire 'the best people to serve the company'.[1]

It is not possible, of course, to give even a rough estimate of the extent to which the methods of peaceful competition and forcible opposition are used. But it is obvious from the above examples and many more which could be given that such methods are used by a considerable number of employers to discourage their white-collar employees from joining trade unions.

EMPLOYER BELIEFS REGARDING WHITE-COLLAR UNIONISM

The fact that most employers who claim not to recognize white-collar unions because of their lack of representativeness also try to ensure by one means or another that they do not become representative, suggests that this is not their major reason for withholding recognition. This conclusion is supported by the findings of the survey of employers. It indicates that there are more fundamental reasons why most private employers refuse to recognize white-collar unions. Some employers withhold recognition from certain unions because they consider them inappropriate for the occupation, firm, or industry concerned; other employers refuse to recognize any form of white-collar unionism because they believe it is unnecessary or because they fear it will have certain 'dire consequences'.

The 'appropriateness' of a union

Certain unions have been refused recognition because employers did not consider them 'appropriate' to represent a particular industry or category of staff. Before nationalization most iron and steel employers felt that BISAKTA was the appropriate union to represent the industry's clerks, laboratory staffs, and foremen, and they refused to recognize ASSET, the CAWU, or the AScW for these grades.[2] The British Spinners' and Doublers' Association granted recognition to the Textile Officials' Association because it was considered more appropriate than ASSET. The British Federation of Master Printers refused to recognize NATSOPA for clerical workers partly because it felt that a purely white-collar union with a wider industrial scope such as the CAWU would be more appropriate to represent these workers. Some companies have recognized the TGWU or the NUGMW for all their employees, both manual and non-manual, and have refused recognition to all other applicants because they did not wish to have a multiplicity of unions.

---

[1] ASSET also claimed that the company, as a government contractor, was in breach of clauses 4 and 5 of the Fair Wages Resolution, and it referred the firm, through the Minister of Labour, to the Industrial Court. The Industrial Court found that the firm was in breach of clause 5 (which concerns the posting up of copies of the Fair Wages Resolution) but not of clause 4 (which obliges contractors to recognize the freedom of their workpeople to be members of trade unions).

[2] See *supra*, p. 126 n. 6.

Still other employers have refused to recognize DATA or ASSET because they felt these unions were 'so politically motivated, that they would only be a disruptive force'.

Employers have applied the criterion of 'appropriateness' most often in deciding whether or not to recognize manual workers' unions on behalf of staff grades. Before most employers' organizations considered any of the manual workers' unions 'appropriate' to represent white-collar employees, these unions were required to make certain structural modifications: the Newspaper Society required NATSOPA to establish an autonomous clerical section; the Iron and Steel Trades Employers' Association demanded that BISAKTA establish separate branches for staff workers; the Engineering Employers' Federation insisted that the TGWU establish a separate organization, the NACSS, and that the NUGMW establish a separate clerical department with its own national officer and notepaper.

In particular, employers have used the criterion of 'appropriateness' in refusing recognition to manual unions on behalf of supervisory staffs. Most employers have traditionally argued that while they have no objection to supervisory workers retaining their membership in a central branch of a manual workers' union in order to maintain their rights to accumulated benefits or in case they are demoted to the tools, it is not desirable to have these unions actively representing and negotiating on behalf of foremen. This argument has been well stated by the Overseas Employers' Federation:

> Employers . . . argue against supervisory staff being members of a rank and file union, on the grounds that supervisory staff must obviously be in a position to exercise discipline, which must inevitably result in situations in which there is a conflict between their duty to the employer and their loyalty to their union and fellow members. Management would rightly regard it as prejudicial if action taken by the supervisors in the course of their duty could be called into question by the union, which can give directions to a supervisor by virtue of his membership in the union.

> · · · · ·

> This does not mean that he cannot join a union, but because of his relation to management, it is inappropriate for him to join the same bargaining unit as those whom he supervises and, in effect, sit on both sides of the bargaining table at once: he remains free to join a union whose objectives are not in conflict with his responsibilities.[1]

This question has been most fully debated in the shipbuilding and engineering industries. The major manual workers' unions, particularly the Boilermakers' Society and the AEU[2], have for several years been demanding the right to represent any supervisory grades which they have in membership. They claim it is up to the supervisor to choose which union he wishes to represent him, and, in any case, they are anxious 'to offset the infiltration of

---

[1] *Supervisory Staff and Trade Union Membership* (London: The Federation, 1963), pp. 2–3.
[2] Now the AEF. See p. xv.

less desirable forms of organisation' among supervisory grades (presumably a reference to ASSET).

The Boilermakers' Society first sought negotiating rights from the Shipbuilding Employers' Federation on behalf of foremen in 1950. In 1958 the union established a supervisory branch in each district, removed these branches from the jurisdiction of the rank and file District Committees, and placed them under the control of the full-time District Delegate who was made responsible to the Executive Council on all matters concerning foremen. The Shipbuilding Employers' Federation argued that in practice this structural modification did not mean a great deal for since the District Delegate was under the control of the District Committee on all other matters, it was still possible for the District Committee to bring pressure to bear on him and thereby indirectly control the supervisory branches. The employers pointed to several instances where in spite of this modification in the union's structure, it had been possible for foremen to be taken before District Committees and fined for 'acting contrary to the interests of the Society'. The Federation therefore felt that the Boilermakers' Society was not an appropriate organisation to represent supervisors and on several occasions refused the union recognition for supervisory grades. In 1967, the Federation revised this policy, but only after the Boilermakers' Society agreed to place all their supervisory membership in a central branch under the control of national rather than district officials.[1]

In 1963 the Executive Committee of the AEU decided to open a supervisory branch in each district, to remove these branches from the jurisdiction of the District Committees, and to make them directly responsible to the full-time Divisional Organizer. For some time the AEU pressed its claim for negotiating rights for foremen in individual engineering firms, and in November 1965 it requested such rights from the Federation itself. But the Federation refused the request because it did not consider that a manual union was appropriate for representing foremen.[2]

Staff unionism is unnecessary

Some employers refuse to accept that any union is appropriate to represent staff employees because they believe themselves better able than any 'outside organization' to look after the interests of these employees. These employers argue that their 'door is always open' and that they are prepared to give their personal consideration to any grievances which their staff employees may have. In any case, they claim that the terms and conditions of employment being observed by the firm for its staff employees are equal to, if not superior

---

[1] See *infra*, p. 176 n. 3.

[2] The NUGMW and the NACSS (Tgwu) have also requested negotiating rights for foremen from the Engineering Employers' Federation. The Federation has refused both these requests because it felt that the concession of such rights could not be confined to these two unions alone, but would have to be extended to other manual unions as well.

to, those negotiated by trade unions. Thus trade union representation for these workers, if not undesirable, is at least superfluous.

This general attitude was expressed by a medium-sized chemical firm which refused recognition to both the AScW and the NACSS because:

> At the present time our staff receive treatment over and above a standard likely to be negotiated by any staff union.

It was expressed even more fully by the personnel manager of a large department store chain who refused to recognize USDAW because:

> It is our company's policy . . . that all employees have the right to approach the management direct with any complaint or query which may arise, and which might affect the cordial relationship which exists between us. I cannot see any purpose, therefore, in our meeting to discuss the introduction of a third party into the relationship.

Another firm gave an even more intimate relationship with its staff as the reason for refusing to recognize the NACSS:

> We regard our Staff as being members of 'The Family'. There is no need for them to be represented by a Union.

While still another company refused to recognize the NACSS because:

> We already have a Staff Committee which provides the necessary facilities for staff to make known their views.

The 'dire consequences' of staff unionism

Employers' fear of the results of staff unionism is perhaps their most fundamental reason for refusing recognition. The 'dire consequences' which employers believe will result from staff unionism include restrictions on management's decision-making freedom, dissension and conflict among the members of the staff and between the staff and management leading to a decrease in morale and to divided loyalties, and practices which tend to promote mediocrity.

Restrictions on management's decision-making freedom. The most general fear is that ultimately staff unionism will restrict managements freedom to make decisions. A food manufacturer argued that after unions are recognized:

> Management's freedom to act without consultation in areas previously regarded as management's prerogative is restricted.

A brewers' association refused to recognize the NACSS because:

> The members of the Association feel that recognition of a white-collar union would seriously impair their responsibility for managing their own businesses.

Similarly, a wholesaler refused to recognize USDAW on the grounds that:

> A few of my staff may be members of a trade union and I do not interfere with their right to do so. Conversely I do not expect them to interfere with my right to

decide to run my business as I think fit, consistent with any duties I may have as a good employer.

Dissension and conflict leading to a decrease in morale and to divided loyalties. Some employers feel that white-collar unionism will result in dissension and conflict not only between themselves and their staffs who are regarded as 'part of management', but also among individual members of the staff. It is claimed that this will lead to a decrease in morale and divided loyalties and thereby reduce the over-all efficiency of the firm's operations. This view is particularly common with regard to supervisors and employees engaged on 'confidential' work, but many employers hold it in respect of all staff employees.

In its most general form this view is expressed by statements such as these:

The introduction of white-collar unions is a disturbing influence; some employees wish to be members, others do not, with the result that there is a divided camp with consequent jealousies, lack of team spirit, and a drop in morale.

If staff members are encouraged to join trade unions they must inevitably be liable to a conflict between their loyalty to the company and their responsibility to the union.

With particular reference to foremen, the argument takes the form:

I do not see how management or a part of management such as foremen can be part of an anti-management organisation such as a trade union, and still be an effective member of the management team.

We are reluctant to recognise a union for supervisors. We cannot give an entirely rational answer to why we are reluctant, but on the whole we feel that you are putting key people into a position whereby they are going to have a conflict of loyalty.

Practices which tend to promote mediocrity. Other employers believe that white-collar unions will introduce practices which will promote mediocrity and stifle ambition. Some employers think that staff unions, like manual unions, will introduce 'restrictive practices' such as limitations on the interchangeability of personnel and the closed shop, and that these will 'tend to restrict the ability of the employer to engage the best staff available for the job'.

But a more common belief is that white-collar unions will introduce practices which will prevent the employer from treating his staff employees on their merits as individuals. This fear is expressed in such statements as the following:

The real disadvantage is that any minimum scale of salaries and conditions tends to destroy incentive, initiative and ability and can result in a levelling down.

Unionisation, with consequent 'rate for the job', tends to restrict reward for individual merit, and to discourage interchangeability of work. Employees therefore become less versatile and possibly less ambitious.

Membership of a white-collar union would militate strongly against the prospects of a brighter individual in making his way up the ladder of success and would reduce all staff members to mediocrity. It would be difficult indeed to promote or to increase the remuneration of the 'better man or woman'.

We feel strongly that although we would recognise a staff union if it is the wish of the majority of the staff—we would not negotiate over salaries of staff as such negotiations might well lead to the underpayment of the efficient and the over-payment of the inefficient.

It has been argued elsewhere that most of these employer beliefs regarding white-collar unionism have little basis in fact.[1] But in the present context it matters little whether they do or not. What is important is that employers think they do and as a result refuse to recognize white-collar unions.

CONCLUSIONS

This chapter began by showing that union recognition is very closely associated with union membership. But this was not sufficient to prove that recognition produces membership. It was also necessary to demonstrate that white-collar unions are generally refused recognition not because they lack representativeness but because employers believe they are inappropriate, unnecessary, or will have certain 'dire consequences'.

But this is still not sufficient to prove that recognition produces membership. White-collar unions have been able to obtain recognition in a few areas in spite of the generally unfavourable attitudes and behaviour of employers. This may have been because they had sufficient membership strength to force employers to concede recognition. Hence before it can be said that recognition produces membership it is also necessary to demonstrate that these concessions of recognition were not made primarily because of the membership strength of these unions.

It is obviously impossible to examine every concession of recognition which individual employers have made to white-collar unions. Fortunately, it is also unnecessary to do this in order to explain why white-collar union recognition has generally come about. The vast majority of white-collar union recognition has resulted from employers' associations granting recognition on behalf of their member-firms or at least providing a framework within which recognition could fairly easily be obtained from individual firms. Employers' associations have many functions but, from the trade unionists' point of view, one of their most important has been to act as relay stations for transmitting minimum terms and conditions of employment, including union recognition, throughout an entire industry.

Employers' associations in private industry have granted recognition to

---

[1] Bain, op. cit., chap. 5.

white-collar unions in newspaper publishing, engineering, shipbuilding, iron and steel, and cotton spinning. Almost all these concessions of recognition were made during and immediately after the First World War, 1917–24, and immediately prior to and during the Second World War, 1938–45. In contrast, since 1945 only two employers' associations have conceded recognition to white-collar unions. The clustering of these concessions during two fairly short periods suggests that certain environmental factors may have been conducive to promoting union recognition. Consequently, in the following chapter every major concession of recognition to a white-collar union in private industry is considered in relation to the climate of its period.

IX

GOVERNMENT ACTION AND THE
SOCIAL CLIMATE

THE PERIOD 1917 TO 1924

The climate of the period

THE First World War enormously enhanced the power and prestige of the trade union movement. Between 1913 and 1920 the number of trade unionists more than doubled, giving the movement a total membership of over 8 million. More important, the density of union membership also more than doubled—from around 20 per cent in 1913 to almost 48 per cent in 1920, a figure which has yet to be surpassed.[1] Nor were all those who were joining trade unions manual workers; many white-collar employees were also taking collective action. Unions were formed among bank and insurance clerks; many white-collar unions affiliated to the TUC; various white-collar groups, including teachers, bank clerks, and insurance agents, threatened or undertook strike action; and civil servants demonstrated against the government. By 1920 almost three-quarters of a million white-collar employees belonged to trade unions,[2] and these unions were adopting an increasingly militant outlook.

The effect of the war on the status of the unions was of even greater importance than its effect on union membership. With the outbreak of war the government was compelled to seek the co-operation of the trade union movement to ensure that production would not be retarded by strikes or restrictive practices. On the whole, the movement was willing to co-operate, and it was soon acting as adviser to the government on labour matters. Trade unionists were given a share in the management of industry by being appointed to control boards which were responsible for allocating raw materials, deploying labour, and similar duties. A trade unionist was made a member of the War Cabinet, a Ministry of Pensions and a Ministry of Labour were formed with trade unionists as their heads, and altogether eight Labour men were given office in the wartime government. As the trade unions were brought into the

---

[1] See *supra*, Table 3.1, and A. G. Hines, 'Trade Unions and Wage Inflation in the United Kingdom, 1893–1961', *Review of Economic Studies*, xxxi (October 1964), pp. 250–1. Hines's density figures are understated because in computing them he uses the total occupied labour force which includes groups outside the scope of trade unionism such as employers, self-employed, and members of the armed forces.

[2] Sidney and Beatrice Webb, *The History of Trade Unionism* (London: Longmans Green, 1920), p. 503.

administration of government and industry, their position advanced to one of responsibility and respect and they were increasingly accepted as an essential element in the country's industrial and social organization.

The years immediately preceding the war had been marked by intense labour unrest. As the war went into its later stages and people began to think about reconstruction problems, they realized that industrial relations would have to be reconstituted on a new basis or massive labour unrest would re-occur in the post-war period. In 1916 the government established a committee under the chairmanship of J. H. Whitley to make recommendations 'for securing a permanent improvement in the relations between employers and workmen'. The Whitley Committee concluded that an essential condition for securing such an improvement was 'adequate organization on the part of both employers and workpeople'. It therefore recommended that in well-organized industries joint industrial councils representing both trade unions and employers' associations should be set up to give regular consideration to workers' problems, while in industries in which organization was inadequate to sustain collective bargaining, minimum wage-fixing machinery should be established. The government accepted these recommendations and instructed the Ministry of Labour to encourage employers and unions to establish machinery which would facilitate consultation and negotiation between them on industrial and labour problems.

The principle of trade union recognition which the Whitley Committee asserted was reaffirmed by the National Industrial Conference, the House of Commons, and the Mond-Turner talks. The National Industrial Conference of 1919 was held at the suggestion of the government to consider post-war labour problems and represented both employers and unions. It reported that 'the basis of negotiation between employers and workpeople should . . . be the full and frank acceptance of the employers' organisations on the one hand and the trade unions on the other as the recognised organisation to speak and act on behalf of their members'.[1] In 1923 the House of Commons debated a resolution stating '. . . that local authorities, banks, insurance and shipping companies, and other employers of professional and clerical workers should follow the example of the Government in recognising the organisations of these workers'. It received widespread support from members representing all shades of political opinion and was passed on a free vote without a division.[2] The Mond-Turner talks of 1928-9 resulted in a frank admission by the participating employers that it was 'definitely in the interests of all concerned in industry that full recognition should be given to . . . unions . . . as the appropriate and established machinery for the discussion and negotiation of all questions of working conditions'.[3]

[1] Cited in Henry Clay, *The Problem of Industrial Relations* (London: Macmillan, 1929), p. 155.
[2] See 162 *H.C. Deb.* 5 s., 18 April 1923, cols. 2160-2200.   [3] Clay, loc. cit.

Whitleyism created an atmosphere which was more favourable to the recognition of unions than that of any previous era. As one astute observer of this period of industrial relations has remarked, Whitleyism

constituted a public and official recognition of trade unionism and collective bargaining as the basis of industrial relations, that is perhaps surprising, when it is recollected that large groups of employers were still refusing to recognise the unions when war broke out.

.   .   .   .   .

Collective bargaining . . . was authoritatively pronounced normal and necessary, and was extended, potentially if not actually, over the whole field of wage-employment for the market.[1]

Finally, the war by causing a labour shortage, a rise in the cost of living, and the introduction of compulsory arbitration, accelerated a shift from local to national bargaining.[2] The labour shortage greatly enhanced the bargaining power of the unions and they were able with advantage to play off one employer or regional association against another. In order to counter these tactics, many employers began to negotiate on a national basis, a policy which happened to coincide with a traditional objective of most unions— the standardization of conditions of employment throughout an industry. The trend to national bargaining was also promoted by the rising cost of living and the existence of compulsory arbitration. As the war progressed, the arbitration tribunals soon became swamped with numerous local applications for wage increases, most of them supported by precisely the same argument— the rising cost of living. In order to reduce the number of claims and thereby simplify the work of the tribunals, the government encouraged employers and unions to settle wage claims on a national basis.

It was in this favourable environment of Whitleyism, and of increasing union membership, militancy, status, and employer willingness to engage in national bargaining, that many of the unions representing white-collar employees were recognized. The NUJ was recognized by the Newspaper Proprietors Association in 1917, by the Newspaper Society in 1918, and by the Scottish Daily Newspaper Society in 1921; the Newspaper Proprietors Association recognized the NUPBPW for circulation representatives in 1919, and NATSOPA for clerical workers in 1920; and the Engineering Employers' Federation recognized the CAWU in 1920 and DATA in 1924.

Recognition of these unions came just in time, for this favourable environment was soon to change. The two decades following 1920 were, on the whole, difficult ones for the unions and were characterized by severe depression, falling wages, massive unemployment, and drastically declining trade union funds and membership. The unsuccessful General Strike of 1926

---

[1] Clay, op. cit., pp. 154 and 177.

[2] See Allan Flanders, 'Collective Bargaining', *The System of Industrial Relations in Great Britain*, Allan Flanders and H. A. Clegg, editors (Oxford: Basil Blackwell, 1954), pp. 272–8.

resulted in a loss of prestige for the unions and in the passage of the Trade Disputes and Trade Union Act which introduced certain restrictions on strike action and political action by trade unions. After the recognition of DATA by the Engineering Employers' Federation in 1924, it was fifteen years before another major private employer or employers' association was to concede recognition to a union representing white-collar workers.

Recognition of white-collar unionism in the printing industry

Recognition of the NUJ. The NUJ was founded in 1907. For the first decade of its existence it followed a not very successful policy of trying to secure recognition and improvements in wages and working conditions on a firm-by-firm basis, but the outbreak of war in 1914 changed all this. As newspaper proprietors reduced wages and staffs causing real wages to fall and unemployment to rise, the union realized that 'the method of local action and individual approach had achieved its maximum possibilities'.[1]

During August 1917 the NUJ submitted an application for a general increase in salaries to the several associations of newspaper proprietors covering the whole of Great Britain. The Newspaper Federation, representing the daily papers of the north and the midlands and a rather indeterminate body of weekly papers, conceded this 'war bonus' to the NUJ in October; the Newspaper Proprietors Association, covering the national dailies of Fleet Street, granted the war bonus plus the principle of minimum salary scales in December; the Newspaper Society, acting for most of the English and Welsh weeklies, conceded the war bonus in April 1918; and the Southern Federation, negotiating for the dailies of the south, followed suit almost immediately. During 1919, the Northern Federation, the Newspaper Society, and the Southern Federation followed the lead of the Newspaper Proprietors Association by granting the principle of minimum salary scales. Before long, most of the individual Scottish proprietors began to follow these minimum rates and in 1921 the Scottish Daily Newspaper Society, covering the Scottish dailies, signed an agreement with the NUJ. The Scottish Newspaper Proprietors Association, formed in 1923 to act for the Scottish weeklies, agreed to follow the minimum rates in 1943.

No formal recognition or procedure agreement was ever signed by the NUJ and the employers' associations, although clauses concerning the procedure to be followed in settling disputes were inserted in later agreements. Recognition was established simply by the various federations of newspaper proprietors meeting the NUJ and granting its demand for a war bonus.

The strength of a union, as determined by the size of its membership and its willingness and ability to engage in industrial warfare, is one of the most obvious factors which might induce an employer to grant recognition. In 1917 the membership of the NUJ was almost 3,500 and it was estimated that

[1] F. J. Mansfield, *'Gentlemen, The Press!'* (London: W. H. Allen & Co., 1943), p. 218.

the union then represented almost 40 per cent of all working journalists.[1] Although the union was not anxious to engage in strike action, it had called a few successful strikes against individual firms in the years prior to 1917. In 1917–18 the Welsh journalists were threatening to strike if the employers did not grant recognition, and the union executive was threatening 'to ally itself with other trade union forces in the newspaper industry, with the object of securing by other means the justice which is denied through friendly nego-tiation'.[2] Moreover, the union's bargaining power was somewhat enhanced by the diversity of employers' associations which enabled the union to play off one proprietorial group against another. The granting of the war bonus by the Northern Federation was useful to the union in convincing the News-paper Proprietors Association to do likewise, while both these precedents were of great value in persuading the Newspaper Society and the Southern Federation to concede.

Despite this, the union was not yet well enough organized, nor was its membership sufficiently large, militant, or strategic to force the employers to grant recognition. By 1917 the union had been in existence only ten years, had no full-time officers, possessed negligible reserve funds with which to finance industrial action, and prior to its affiliation to the Printing and Kindred Trades Federation in July 1919, could not have expected any extensive help from the manual workers in the industry. Almost 1,500 of its 3,500 members were serving in the armed forces, the majority of journalists were far too genteel to participate in a strike and, even if they did, striking journalists were easily replaceable since the job could be undertaken by 'any one who has passed the sixth standard'.[3] As one officer of the union wisely remarked, 'The employers have not the slightest reason to fear a general, or even a district, strike.'[4]

A more significant factor than the strength of the union was the sympa-thetic attitude of a few of the key proprietors. An early president of the union claimed that it 'had many friends among the "executives" who sat in the Council of the N.P.A. to decide our economic fate'.[5] A few of the employers had risen from the ranks of journalism and had great affection for their origins. The greatest friend of the union was Lord Northcliffe, the owner of *The Times* and the *Daily Mail* and the most dominant figure of his day in British journalism. He backed the NUJ from the start: he argued that 'news-paper workers should adopt the methods of other professions, and form a society for mutual protection and encouragement';[6] he placed a column of the leader page of the *Daily Mail* at the disposal of the union for an article on

[1] F. J. Mansfield, '*Gentlemen, The Press!*' (London: W. H. Allen & Co., 1943), p. 298.

[2] Letter from NUJ to Newspaper Society, December 1917, cited ibid., p. 268.

[3] Speech in 1913 of H. M. Richardson, General Secretary of the NUJ, 1918–36, cited ibid., p. 297.

[4] Ibid., p. 204.   [5] Ibid., p. 224.

[6] Letter from Lord Northcliffe to Horace Sanders, Secretary, Central London Branch, NUJ, 2 October 1912, cited ibid., pp. 170–1.

its objectives; and when asked by the NUJ to use his influence to encourage the Newspaper Proprietors Association to recognize the union, he replied:

You know that the union will have my support in any reasonable negotiations with the newspaper proprietors. I am one of the few newspaper owners who have been through the mill of reporting, sub-editing and editing, and I have very vivid and resentful recollections of underpaid work for overpaid millionaires.[1]

It is difficult to explain fully the granting of recognition to the NUJ without reference to the sympathetic attitudes of a few of the leading newspaper proprietors of the day. No doubt these attitudes were conditioned by the environment of the era—the wartime trend to national bargaining and Whitleyism—but the fact nevertheless remains that they were favourable and almost certainly hastened the recognition of the NUJ.

The most important factors promoting the recognition of the journalists' union were the related developments of the consolidation among employers' associations and Whitleyism. In the early stages of the war wage negotiations were undertaken on an association-by-association basis. Many of the unions, particularly the Typographical Association (now the National Graphical Association) had improved their positions by obtaining concessions from one employers' association and then forcing these upon other associations. Partly to ensure that they would not be forced to follow agreements which they had not helped to make and partly to obtain the greater degree of policy co-ordination demanded by the wartime conditions, the proprietors began to unify their forces. A Newspaper Conference was established in 1914 and met weekly throughout the war to co-ordinate the actions of the proprietors' associations in their relationship with the government and also in their dealings with the unions. In 1916 the Newspaper Society reconstituted itself to act as a central co-ordinating body for the various regional associations. This process continued until by 1921 all the newspaper proprietors of England and Wales were represented by only two federations: the Newspaper Proprietors Association, covering the national dailies, and the Newspaper Society, covering all the provincial dailies and weeklies. By 1917 the proprietors were as anxious to bargain nationally, at least with the manual printing unions, as the unions themselves.

This trend toward national co-ordination of policy-making and bargaining by employers was reinforced by the Whitley spirit. Beginning in 1916, several conferences were held in the printing industry to discuss ways of reducing friction between employers and unions. While these conferences were in progress the Whitley Report was published. Since most of the newspaper proprietors eulogized its recommendations in their editorials, they could hardly reject them in practice. As the NUJ was quick to inform them, 'the whole

---

[1] Letter from Lord Northcliffe to F. J. Mansfield, President of NUJ in 1918, 15 December 1917, cited ibid., p. 226.

trend of things is in the direction of settling wages and conditions by bodies representing employers and workers, and what nearly every newspaper now advocates (take the articles on the Whitley Committee proposals as an example) cannot logically be refused to newspaper staffs'.[1] As a result of the above conferences and the stimulation given by the Whitley Reports, a Joint Industrial Council was formed, covering the various printing unions and all the provincial dailies and weeklies. The first objective of this JIC was 'to secure complete organisation of employers and employees throughout the trade'. In an industry as highly and extensively organized as printing, it was only logical that this meant all employees, including journalists. Logical or not, the manual workers through the offices of the PKTF, insisted that the NUJ should participate in the JIC.

Clearly, the recognition of the NUJ by the newspaper proprietors was very much a product of the First World War era. The union could claim to be fairly representative of journalists and thus had a good claim to recognition. But it was not bargaining from a position of strength, and, if it had been forced to rely strictly upon its own resources and undertake industrial action to try and force employers to grant recognition, it is highly unlikely that the union would have been successful. The liberal attitude of a few key employers and the increasing willingness of proprietors to engage in national bargaining because of the war, Whitleyism, and the bargaining tactics of some manual unions were the really crucial factors bringing about the recognition of the NUJ.

Recognition of the IOJ. The history of the recognition of unionism among journalists would not be complete without some mention of the Institute of Journalists. The IOJ was founded in 1884 primarily as a professional body to look after the interests of both working journalists and newspaper proprietors. During the war it became interested in negotiating with the employers over the wages and conditions of journalists, and in 1920 it was certified as a trade union. During the 1917–19 negotiations the employers suggested that a joint approach should be made to them by the NUJ and the IOJ, but the former refused. When the JIC was established, the employers also tried to have the IOJ included on the trade union side of the Council, but the PKTF gave solid backing to the NUJ and refused to accept the IOJ into membership. The employers then refused to recognize the IOJ on the grounds that they already bargained with a union affiliated to the PKTF for the grades covered by the Institute.

A twenty-year campaign by the IOJ for recognition ensued, and finally, in 1943, both the Newspaper Proprietors Association and the Newspaper Society gave way and recognized the Institute. Recognition of the IOJ came at the same moment as the NUJ was causing consternation in the employers' ranks by campaigning for the closed shop. Bundock has argued that the employers recognized the IOJ in the hope of being able to play off one

---

[1] Statement of NUJ to Newspaper Federation, 15 October 1917, cited F. J. Mansfield, *'Gentlemen, The Press!'* (London: W. H. Allen & Co., 1943), p. 264.

organization against the other and thereby undermine the NUJ's strength.[1] Strick has suggested that the IOJ was recognized in order to facilitate the merger talks of the NUJ and the IOJ which were then current and were being encouraged by both the employers and the PKTF.[2] While there is not sufficient evidence available to determine which of these interpretations is correct, it is at least clear that the employers were not forced to recognize the IOJ because of its industrial strength.

NPA recognition of the NUPBPW for circulation representatives and NATSOPA for clerks. In 1919 the Printing Trades Guild of the CAWU (then the National Union of Clerks) began to recruit clerical workers in the printing industry. By 1920 about 1,500 printing clerks belonged to the CAWU, and in March of that year the Newspaper Proprietors Association agreed to pay the union's minimum salary scales thereby recognizing the union. Almost simultaneously, the printing clerks became dissatisfied with the degree of autonomy they were allowed under the CAWU's rules and the vast majority of them seceded and formed the London Press Clerks' Association. In July 1920 the Association merged with NATSOPA, and soon afterwards the NPA recognized NATSOPA for clerical grades.[3] A few months earlier, in September 1919, the NPA and the NUPBPW (then the National Union of Printing and Paper Workers) had signed an agreement covering circulation representative and clerical grades employed in circulation, distribution, and competition departments. Although the CAWU has only a handful of printing clerks in membership, it is still recognized by the NPA and is a signatory to the National Clerical Workers (London) Agreement. But it does not take any part in negotiating this agreement.

The factors explaining the NPA's recognition of the NUPBPW and NATSOPA for white-collar workers are, in general terms, the same as those explaining the granting of recognition to the NUJ and they need not be repeated. It might be thought that the manual strength of NATSOPA and the NUPBPW was used to encourage employers to recognize these unions for white-collar workers. Although the device of manual leverage was to become important in later negotiations, there is no evidence to suggest that it was used in 1919–20. It should be remembered that the Newspaper Proprietors Association recognized a union with no manual connection, the CAWU, when it had roughly the same clerical membership as NATSOPA.

It only remains to explain why the NPA should have recognized the NUPBPW and NATSOPA for white-collar workers in 1919–20, while the Newspaper Society and the Scottish proprietors did not do so at this

[1] Clement J. Bundock, *The National Union of Journalists: A Jubilee History, 1907–1957* (Oxford: Oxford University Press, 1957), chap. 22.

[2] H. C. Strick, 'British Newspaper Journalism 1900 to 1956: A Study in Industrial Relations' (unpublished Ph.D. thesis, University of London, 1957), p. 284 and section 3.

[3] During 1964 NATSOPA established a National Association of Advertising Representatives, but so far the NPA has refused to recognize it.

time.[1] The explanation is twofold. First, the history of industrial relations in printing indicates that the NPA has always had a more favourable outlook towards trade unionism than the Newspaper Society or the Scottish proprietors' associations. This is due partly to the impact on the NPA of such personalities as Lord Northcliffe and partly to differences in the membership composition of these organizations. The NPA is composed of large London firms while the Newspaper Society and the Scottish associations are composed of much smaller provincial employers who tend to have a more restrictive outlook towards industrial relations. Second and more important, in 1920 virtually all the white-collar membership of the two unions was concentrated in London, that is, among the firms of NPA members. Even today most white-collar union membership in the printing industry is situated in the capital.

Recognition of the CAWU and DATA in the engineering industry

The other major development in white-collar union recognition during this period occurred in the engineering industry. The Engineering Employers' Federation[2] recognized the CAWU (then the National Union of Clerks) in 1920 and DATA (then the Association of Engineering and Shipbuilding Draughtsmen) in 1924.

The CAWU was founded in 1890. The union made its first systematic attempt to organize engineering office staffs in 1913, and in the following year it called its first strike. Throughout the war the union continued to call strikes against individual employers and, in addition, referred numerous cases to arbitration. Between 1914 and 1919 the union's membership increased from 10,000 to 43,000, and approximately 10,000 of these were employed in the engineering industry.[3] In December 1919 the Federation met the union to ascertain its objectives and policies, and on 9 August 1920 a memorandum of agreement was arrived at regarding the terms on which the Federation agreed to recognize the CAWU. After being discussed by the members of both organizations, it was signed and put into operation on 15 December 1921.[4]

[1] The Newspaper Society recognized NATSOPA for clerical workers in 1938 (*infra*, pp. 158–61) but still has not granted recognition for circulation representatives. Neither NATSOPA nor the NUPBPW ever formally requested recognition for white-collar workers from the Scottish Daily Newspaper Society, the Scottish Newspaper Proprietors Association, or the Society of Master Printers of Scotland.

[2] In 1920 the Engineering Employers' Federation was known as the Engineering and National Employers' Federations, and in 1924 as the Engineering and Allied Employers' National Federation.

[3] The total membership figures were supplied by the CAWU. The membership figure for the engineering industry is from Roy Grantham, 'How to Negotiate', *The Clerk* (January 1965), p. 4.

[4] The procedure agreement was amended in October 1946 to make it mandatory for employers to exhaust the procedure before introducing any downward alteration in wages or working conditions. See p. 151 n. 1; p. 169 n. 3; p. 172 n. 1; p. 174 n. 2; and p. 179 n. 3.

The draughtsmen's union was founded in 1913. By 1917 it had around 9,000 members, held its first national conference which elected a full-time General Secretary, and tried without success to meet the Federation for the purpose of negotiating a war bonus. In 1918 the union became a registered trade union affiliated to the TUC, and, although the Federation once again refused to meet the union for a national discussion on wages, two of the Federation's local associations recognized the union. Early in 1919 the union called its first strike. In December the Federation met the union to determine its aims and policies. Additional discussions were held the following year, but nothing came of them. By 1920 the union had close to 15,000 members on its books, and it decided to establish minimum rates below which no member of the Association would accept new employment. Finally in 1923 the union called several more strikes, the largest being at the English Electric plant at Rugby. After the English Electric dispute had lasted almost six weeks the Federation agreed to recognize the Association if it would call off the strike. This was agreed to by DATA in June 1923, and after fairly lengthy negotiations a procedure agreement was made operative as of 17 March 1924.[1]

The rapid build-up of membership in the CAWU and DATA during the war and the willingness of these unions to harass individual employers by strikes and by references to arbitration has led some people to argue that union strength was the most crucial factor forcing the Federation to grant recognition. Thus the General Secretary of the CAWU during this period claimed 'the union, started a guerilla campaign . . . [which] . . . resulted in a gradual general recognition, first by individual firms, then District Associations, and lastly by their National body'.[2] Similarly, the editor of the draughtsmen's journal at this time argued that DATA's experience showed that the Federation 'only recognise a union when compelled to do so by a demonstration of its industrial strength'.[3] While the industrial strength of these unions undoubtedly was a factor in getting the engineering employers to concede recognition, this strength was not as great nor as significant as the above remarks would suggest.

By the time the CAWU was recognized in 1921, the union was, in the words

---

[1] The 1924 procedure agreement covered draughtsmen employed in drawing-offices or in designing, calculating, estimating, and filing departments, and tracers, but excluded chiefs, principal staff assistants outside the jurisdiction of the chiefs, and apprentices. The scope of the agreement was enlarged to cover apprentices and young persons in 1938, planning engineers who previously had been employed as draughtsmen in 1944, all planning engineers in 1949, and draughtswomen in 1956.

The draughtsmen's procedure agreement also contained the 'general alterations clause' which was later to be inserted in the procedure agreements of most of the other staff unions in engineering. See p. 150 n. 4; p. 169 n. 3; p. 172 n. 1; p. 174 n. 2, and p. 179 n. 3.

[2] H. H. Elvin, 'Engineering for Engineering Clerks', The Clerk (February 1925), p. 19.

[3] Cited by J. E. Mortimer, A History of the Association of Engineering and Shipbuilding Draughtsmen (London: The Association, 1960), p. 106.

of its President, 'travelling through fog and storm'.[1] With the contraction of war industries and the onslaught of the post-war depression, its membership plummeted from 43,000 in 1919 to 14,000 in 1921, and eventually bottomed at 7,000 in 1924. By 1921 the union was virtually bankrupt; it was racked with factional conflict; and a majority of the branches in Wales and in the printing trades had seceded.

DATA was much better organized than the CAWU. This at least partly explains the inclusion of the 'general alterations clause' in DATA's procedure agreement and its omission until 1946 from that of the CAWU. But DATA's strength relative to that of the Federation was not impressive. Between 1920 and 1924 membership of the Association declined by close to a third, and the large expenditure occasioned by the unemployment among draughtsmen in 1921–2 seriously weakened the union's financial position. Moreover, in the depressed years of the early twenties the employers began to take the offensive. In 1921 they forced both manual and staff unions in the industry to accept wage cuts, and, in the following year, the Federation was strong enough to lock-out all the manual unions and force them to sign the Managerial Functions Agreement which clearly established management's right to exercise its 'prerogatives'.[2] The Kilmarnock strike, which occurred within a few months of the signing of the procedure agreement, best illustrates the draughtsmen's vulnerable position. After this strike had lasted almost four months the Federation threatened DATA with a national lock-out of its members and demanded the unconditional return to work of the strikers. In the face of this threat the union gave way and even had to accept the victimization of some of the strikers. Clearly, DATA was in no position in 1924 to force recognition upon an unwilling Federation.

Whatever may have been the industrial power of these two unions in the early twenties, it was not sufficient to convince the individual employers and local associations, against whom the brunt of the 'guerilla campaign' was directed, that the Federation should be allowed to concede recognition. In June 1917 and February 1919 the Federation inquired of local associations whether they were in favour of recognizing staff unions. On both occasions the vote was overwhelmingly in the negative. In August 1919 the Federation sent out a third letter of inquiry which actually recommended the recognition of staff unions and drew the attention of the local associations to the National Industrial Conference's favourable recommendation on union recognition.[3] In spite of this urging, the local associations agreed by only a bare majority to give the Federation the authority to grant recognition to the CAWU and DATA. After the Federation negotiated the memorandum of agreement

[1] Cited by Fred Hughes, *By Hand and Brain: The Story of the Clerical and Administrative Workers' Union* (London: Lawrence & Wishart, 1953), p. 68.

[2] For a more complete discussion see Arthur Marsh, *Industrial Relations in Engineering* (Oxford: Pergamon, 1965), pp. 74–9.

[3] Circular Letter from EEF to Local Associations, 2 August 1919.

with the CAWU in August 1920, a number of individual employers and local associations felt that the Federation had committed a 'great tactical blunder' by recognizing the union at a time when its strength was declining, and they threatened to secede unless recognition was withdrawn.[1] The Federation refused to revoke the CAWU's procedure agreement, but it decided to withhold recognition from DATA 'until circumstances make this course necessary'.[2] In order to pacify its dissident members, the Federation decided not to recognize DATA at the same time as the CAWU, but to wait until a tighter labour market and further industrial action by DATA would make it easier to justify recognition of this union to the local associations.

In view of the opposition of many local associations, the CAWU and DATA probably would not have been recognized at this time except for the leadership exercised by the Federation. If the war had not greatly increased the Federation's role in industrial relations, the Federation would probably have been unable to exercise successfully such leadership. Before the war all negotiations in engineering were carried out on a domestic or district basis, and the Federation only became involved if matters were referred to it through procedure. During the war a system of national bargaining for manual workers began to develop as a result of industry-wide arbitration awards, the emphasis which wartime governments placed on common action by employers and unions, and employers' desire to discourage competitive local wage settlements. These developments greatly increased the power and influence of the central body over the local associations and made it much easier for the Federation to get them to follow its lead on recognition.

The increased power and influence of the Federation as a result of wartime development suggests why it was able to lead a reluctant membership to grant recognition, but it does not explain why the Federation should wish to take this lead. The explanation of the latter is composed of two elements: the existence of a public opinion favourable to trade union recognition, and a belief by the Engineering Employers' Federation in the wisdom of a policy of union containment by conciliation.

Although Whitleyism itself had little impact on the engineering industry, the environment of opinion created by Whitleyism was most important in increasing the willingness of engineering employers to recognize staff unions. The Whitley Reports urged employers to recognize unions. This sentiment was reaffirmed by a unanimous recommendation of the National Industrial Conference.[3] At the second meeting of the Conference in the spring of 1919, the trade union side called upon the Federation to conform with this unanimous recommendation by recognizing the staff unions in the engineering industry. The Federation's representative at the Conference replied that lack

[1] Various letters of Local Associations to EEF.
[2] Circular Letter from EEF to Local Associations, 15 June 1923.
[3] *Supra*, p. 143.

of union recognition in engineering was a thing of the past.[1] During the summer the Federation sent a copy of the Conference's recommendation to all local associations and asked them for authority to recognize staff unions.[2] In the autumn the trade union side of the Conference informed the Federation that the question of staff union recognition in engineering would be raised again at the Conference's next meeting. In preparation for this the Federation decided to meet the staff unions in December to determine the 'conditions under which it might be possible to recognise the clerks' and draughtsmen's unions'.[3] In the period of idealism following the publication of the Whitley Reports, the Federation found it increasingly difficult to maintain a good public image while at the same time denying recognition to staff unions.

The pressure of public opinion upon the engineering employers was reinforced by the logic of the Federation's own industrial relations policy. The Federation's experience with manual workers' unions had strengthened its belief in the policy of trying 'to contain union claims, not by dramatic conflict, but by conciliation procedure—by the Provisions for the Avoidance of Disputes'.[4] The employers realized that the strength of the staff unions was on the decline due to the post-war depression and that the unions were in no position to force the Federation to grant recognition. The Federation felt, however, that it would be a most 'short-sighted policy' to withhold recognition from these unions until the economic situation improved and their strength increased.[5] Far better to recognize the unions now while the Federation was in a good bargaining position and was still able to get them to agree to procedural machinery which would tend to minimize strikes and contain union demands.

The results of the recognition negotiations justify the Federation's strategy. The Federation merely agreed to recognize formally the union's right to negotiate with individual firms and local associations, a right which many of the firms and local associations had already acknowledged. It refused to allow the staff unions to bargain nationally over wages[6] or to recognize the role of white-collar shop stewards or staff representatives in collective bargaining. In return the unions agreed to respect the workers' freedom to join or not to join a union, not to undertake joint action with manual workers'

[1] Mortimer, op. cit., p. 66.

[2] Circular Letter from EEF to Local Associations, 2 August 1919.

[3] From EEF documents.

[4] Marsh, op. cit., p. 43. The Engineering Employers' Federation recognized the manual unions in 1898 after completely defeating them by a lock-out lasting thirty weeks. In a sense, national recognition was merely a by-product of a policy of containment designed to allow managements maximum freedom to run their establishments as they thought best, while at the same time minimizing disruptions of production due to employee discontent.

[5] Report of a Federation meetings, 27 May 1921.

[6] National negotiations for white-collar workers in engineering did not occur until the Second World War.

unions, and to submit all grievances to a system of 'employer-conciliation'[1] before striking. Although the granting of formal recognition was to prove of great importance to the staff unions, it appeared at the time that the unions had paid a very high price for very little. It is not surprising that the procedure agreements were 'viewed with great suspicion' by many members of the CAWU and DATA.[2]

Thus in engineering, as in printing, the recognition of unions catering for white-collar workers was very much a product of the times. The ability of the CAWU and DATA to harass individual employers and to disrupt production obviously encouraged the employers to grant recognition. But if the war had not increased the power and influence of the Federation over its constituent associations, and if the pressure of public opinion and the logic of the strategy of union containment by conciliation had not convinced the Federation of the wisdom of granting recognition, it is highly unlikely that the CAWU and DATA would have obtained recognition in the engineering industry at this time.

THE PERIOD 1938 TO 1945

The climate of the period

The trade union movement, which had reached its peak membership of 8·3 million in 1920 and had fallen by 1933 to little more than half this level, began to recover as the depression eased in the later thirties. By 1938 the trade union movement had over 6 million members and by 1946 over 8·8 million. During the Second World War the unions gained not only in membership, but also in prestige just as they had done during the First World War. The Labour Party was strongly repesented in the Churchill Coalition, supplying two of the five members of the War Cabinet plus four other Ministers. As in the First World War, a trade unionist, Ernest Bevin, was appointed as Minister of Labour and National Service. In addition, trade union representatives were on a host of official bodies including: the National Joint Advisory Committee, the Joint Consultative Committee, the National Production Advisory Council, the Regional Boards for Industry, and the Local Joint Production Committees. The trade union movement was 'no longer prepared to be treated as a poor relation' but demanded and received 'an honourable recognition of its willingness and suitability for participation in the planning and execution of the production programme'.[3]

[1] 'Grievances arising are taken by trade unions to panels of employers at Local and Central Conference level. These panels have the dual responsibility, both of representing the member firms involved, and of conciliating between the firm and the trade union or unions concerned. Engineering carries with it the air of an industry in which employers act as judge and jury in their own cause on matters in procedure' (Marsh, op. cit., p. 13).

[2] Circular Letter from the Secretary of the EEF to members of the Management Committee, 2 May 1923.

[3] H. M. D. Parker, *Manpower* (London: HMSO, 1957), p. 67.

In view of the contribution the trade unions were making to the war effort, many trade unionists felt that the government should give more positive support to collective bargaining. The comments of one General Secretary are fairly representative of trade union feelings at this time:

It should not be necessary at this time of day, for any trade union to have to struggle for recognition for we have it on the authority of the Prime Minister himself that without the co-operation of the trade union movement during this war, our country might well face disaster.

It is my opinion that whilst we are prepared if necessary to sacrifice everything to preserve our democracy against Nazi and Fascist aggression, we must also, in the words of Sir Walter Citrine at the last Trades Union Congress, 'Keep an eye on the would-be tinpot Hitlers here'.

. . . If, indeed, with the enemy on the doorsteps and with our cities being blasted into ruins by the bombs of the enemy, there can still be found reactionary employers so feudal minded as to consider a bona fide trade union as one of their first enemies, there is something radically wrong.[1]

Such attitudes resulted in questions regarding union recognition in particular firms and industries being raised in the House of Commons, and eventually in discussions between the TUC and the Minister of Labour as to the possibility of introducing legislation to compel employers to recognize trade unions.

As a result of pressure from several unions,[2] the TUC began to examine the whole question of union recognition. They felt that there were three possible ways of dealing with the problem: an amendment to the Essential Works Orders making union recognition by a firm a condition of its being scheduled as essential under these Orders,[3] an amendment to the Conditions of Employment and National Arbitration Order (Order 1305)[4] making it possible to refer the question of union recognition to the National Arbitration Tribunal, and the introduction of legislation along the lines of the American Wagner Act.[5] After a fairly lengthy correspondence on this topic with the Minister of Labour during 1942, the TUC discussed the matter with him in person in February 1943. They agreed that the problems of defining what was meant by union recognition, a bona fide trade union, and representative membership were so great as to make it impossible to deal with the recognition problem by legislation. Although the war did not produce legislation on union recognition, the possibility of its introduction helped to create an environment in which it became more difficult for employers to refuse to recognize trade unions.

The Minister of Labour was reluctant to deal with the recognition problem by legislation, but he was not unwilling to handle the problem by other

[1] T. W. Agar, General Secretary of ASSET, 'Editorial', *The Foreman* (April 1941), pp. 1–2.

[2] See references to trade union recognition in the *TUC Annual Report* for 1942, 1943, and 1944.

[3] *Infra*, p. 157.   [4] *Infra*, p. 158.

[5] Most of this information is drawn from 'Trade Union Recognition' (an unpublished paper circulated to Regional Industrial Relations Officers by the Ministry of Labour, 1943).

means. He encouraged firms to recognize unions by using the Ministry's conciliation services and by employing 'the good services of the Supply Departments concerned with the firms' contracts ... to regularise the position'.[1] He also fostered union recognition by establishing Courts of Inquiry to investigate the causes and circumstances of a number of major recognition disputes. Since the Industrial Courts Act was passed in 1919, there have been nine Courts of Inquiry into recognition disputes. Four of them were set up during the Second World War, and all of these found in favour of the unions concerned.[2] In recommending that the employers should concede recognition, these Courts were influenced not only by the strength of the unions' arguments but also by the exigencies of the wartime situation. As the chairman of three of these Courts of Inquiry stated in one of his Reports:

We appreciate that Mr. King [the Managing Director of the firm concerned] is quite sincere in his belief that he is entitled to refuse to have any dealings with a Trade Union. . . . In peace time if he chooses to try to exercise this right and a trade dispute occurs in consequence the National Interest may not be gravely involved. In war time we think that however strongly individuals may desire to run their works in their own way, it is their duty to their country to fall into line with the vast majority of other good employers and assist the Government in the accepted methods of conciliation.[3]

Trade union recognition was also encouraged during the war by the passage of the various Essential Works Orders. These Orders enabled the Minister of Labour to prohibit workers from leaving employment in essential firms, as long as he was satisfied that their terms and conditions of employment were not less favourable than those generally recognized by trade unions and employers' organizations in the appropriate industry. Since the Essential Works Orders required firms to have 'recognized' conditions of employment before they could be scheduled as essential, the Orders encouraged the establishment of voluntary negotiating machinery. Altogether fifty-six JICs or similar bodies were revived or newly established in the years 1939 to 1946. In addition, the system of statutory wage regulation was extended by the passage of the Catering Wages Act in 1943 and the Wages Council Act in 1945. By 1946 the Ministry of Labour estimated that almost 90 per cent of the labour force was covered either by joint voluntary negotiating machinery or by statutory machinery.[4]

[1] *TUC Annual Report*, 1943, p. 118.
[2] *Report by a Court of Inquiry in the Matter of a Trade Dispute Apprehended at Briggs Motor Bodies Ltd., Dagenham*, Cmd. 6284, 1941; *Report by a Court of Inquiry into a Dispute Between Richard Thomas and Company Limited and the National Association of Clerical and Supervisory Staffs*, 30 June 1941 (*infra*, pp. 163–4); *Report by a Court of Inquiry into a Dispute Between Trent Guns and Cartridges, Limited, Grimsby and the National Union of General and Municipal Workers*, Cmd. 6300, 1941; and a *Report by a Court of Inquiry Concerning a Dispute Between the Clerical and Administrative Workers' Union and Certain Colliery Companies in South Wales and Monmouthshire*, Cmd. 6493, 1943.
[3] Cmd. 6300, p. 10.
[4] Cited in Allan Flanders, 'Collective Bargaining', op. cit., p. 285.

Another institution which promoted union recognition during the war was the National Arbitration Tribunal established in 1940 by Order 1305. This stipulated that there should be no stoppages of work and that, in the event of voluntary negotiation failing to settle a dispute, it should be submitted by either side, through the Ministry of Labour, to the National Arbitration Tribunal for a binding award. Although a recognition dispute could not be dealt with by the Tribunal under Order 1305, it was possible for any union, recognized or unrecognized, to request the Minister to refer a firm to the Tribunal for an award on wages and working conditions.[1] Simply by being involved in the determination of these conditions, the unions were given a form of implicit recognition. More important, when faced with the possibility of having a wage structure imposed upon them by a third party, many employers felt it was better to recognize the union and determine the firm's wage structure by collective bargaining. The National Arbitration Tribunal had many functions, but, from the trade unionists' point of view, its major one was 'to compel people who would not be reasonable enough to negotiate to face up to the question in the light of publicity, fact and reason before a body which would give a decision after hearing the case'.[2]

In this wartime atmosphere of growing union power and prestige, of demands for legislation compelling employers to recognize unions, and of increasing state intervention in industrial affairs, several more white-collar unions were recognized by employers: NATSOPA by the Newspaper Society in 1938, the NACSS by the Engineering Employers' Federation in 1940 and Richard Thomas in 1941, BISAKTA by the Iron and Steel Trades Employers' Association in 1943–5, and ASSET and the AScW by the Engineering Employers' Federation in 1944.

Recognition of NATSOPA by the Newspaper Society[3]

After receiving recognition for clerks from the Newspaper Proprietors Association in 1920, NATSOPA began to concentrate its clerical organizing in general printing and provincial newspapers. Agreements covering clerical workers in these areas were signed with Allied Newspapers of Manchester in 1924 and the *Newcastle Chronicle* in 1925. During these negotiations officials of the Newspaper Society acted on behalf of the firms concerned, but the Society itself was not a party to the agreements. Around this time NATSOPA also entered into discussions with the London Master Printers

---

[1] See Moshe Reisse, 'Compulsory Arbitration as a Method of Settling Industrial Disputes, with Special Reference to British Experience Since 1940' (an unpublished B.Litt. thesis, University of Oxford, 1963).

[2] A trade union official speaking at the Trades Union Congress of 1946. Cited by Allan Flanders, 'Collective Bargaining', op. cit., p. 284.

[3] Most of the information on this topic has been obtained from historical documents supplied by the Newspaper Society, relevant issues of the *NATSOPA Journal*, the union's Executive Council Minutes and Annual Reports, and James Moran, *NATSOPA: Seventy-Five Years* (Oxford: Oxford University Press, 1964), especially pp. 99–103.

Association, an affiliate of the British Federation of Master Printers, in an attempt to secure an agreement for all clerical workers in London weekly and periodical houses. The union's participation in the General Strike brought these talks to an end.

After the General Strike the matter of clerical recognition remained fairly dormant until 1932 when the union tried to extend the Manchester and Newcastle agreements to several other provincial newspaper houses. Most of these employers refused to grant recognition, and NATSOPA raised the whole question of clerical recognition with the Newspaper Society. Since its membership overlaps in some areas with that of the British Federation of Master Printers, the Newspaper Society referred the matter to the Joint Labour Committee of both organizations. But the employers' associations were no more willing to grant recognition than their individual members had been.

During the next four years the question of clerical recognition was frequently referred to the industry's Joint Industrial Council. In April 1937 the JIC expressed 'the hope that no employer will place an embargo on any employee who desires to do so joining a trade union'.[1] In October the Joint Labour Committee of the Newspaper Society and the British Federation of Master Printers replied that while they could not deny the right of clerical workers to join a trade union, they felt it was 'not desirable that clerical workers be members of an operatives' trade union', and they threatened to withdraw from the JIC if NATSOPA took any aggressive action to enforce its demands on this question.[2] NATSOPA felt the employers did not have the right to dictate to their employees which union they should join, and they threatened to strike. The dispute had now reached crisis proportions and the JIC appointed a Special Committee to go into the whole question of clerical recognition. The Special Committee was able to bring the parties together in March 1938, but the employers still refused to grant recognition.

NATSOPA then decided to ignore the British Federation of Master Printers[3] and to concentrate its attention on the Newspaper Society. This decision was probably prompted by two considerations. Most of the union's provincial clerical membership was situated among member-firms of the Newspaper Society. More important, the member-firms of the Newspaper

---

[1] Cited by George Issacs, 'Harmony or Discord in the Printing Industry', *NATSOPA Journal* (January 1938), p. 17.

[2] 'Organization of Clerical Workers in the Printing Industry', *NATSOPA Journal* (May 1938), p. 1.

[3] After the war, NATSOPA once again turned its attention to the British Federation of Master Printers. But in spite of a number of threatened stoppages, the Federation has yet to concede the union's request for clerical recognition. See NATSOPA's *Annual Report*, 1946, pp. 47–8; 1947, p. 43; 1948, p. 56; 1949, pp. 51–2; 1950, pp. 66–7; and 1951, pp. 58–9. See also Moran, loc. cit., and the *Report of a Committee of Investigation into a Dispute Between Waterlow & Sons Ltd. and the National Society of Operative Printers and Assistants*, 17 November 1948.

Society are more vulnerable to strikes than are those of the British Federation of Master Printers. General printers can stockpile their product and make up lost production to some extent; newspaper proprietors can do neither, and they therefore view strikes with greater apprehension than almost any other group of employers.

During the summer of 1938 NATSOPA tendered strike notices in several provincial newspaper houses. The Special Committee once again brought the two sides together. In July the Newspaper Society stated that its weekly members were still opposed to granting recognition, but it would be prepared to negotiate an agreement covering its daily members. NATSOPA accepted this offer and, after rather lengthy negotiations, an agreement was signed in December with effect from 17 November 1938.

The agreement was very restrictive in scope; it applied only to provincial daily newspaper houses and not to weekly houses; it excluded shorthand-typist-telephonists employed in editorial departments and employees engaged in a confidential capacity; and it covered only wages, specifically omitting hours and overtime provisions.[1] In addition, the union agreed to establish an autonomous clerical section and not to allow any joint action between the manual and clerical sections to be taken until the entire conciliation procedure of the JIC had been exhausted. The union felt that the agreement was most unsatisfactory, but they accepted it because 'it gave recognition of the right of clerical workers to join a trade union of their choice'.[2]

The union's industrial strength was the major factor persuading the Newspaper Society to recognize NATSOPA for clerical workers. NATSOPA participated in the general expansion and strengthening of unionism which began in the mid-thirties as the depression eased. The total membership of the union increased from 22,000 in 1933 to 28,000 by 1938, and in the latter year the London Clerical Branch had over 4,700 members. Clerical membership in the provinces was also increasing, and in 1937 a full-time national clerical officer was appointed. But it was not NATSOPA's strength among provincial clerical workers which forced the Newspaper Society to grant recognition. In 1944, the first year for which such figures are available, provincial clerical membership was only a little over 900 and in 1938 it was probably much less. What was significant in promoting clerical recognition was the printing clerks' strategic alliance with the printer. The union was prepared to call out not only its clerical membership in provincial newspaper houses but, more important, its manual membership. Moreover, NATSOPA would undoubtedly

---

[1] The agreement between NATSOPA and the Newspaper Society operative in 1964 covered substantive matters in addition to wages, but the other restrictions still applied. NATSOPA tried to extend the agreement to cover advertising representatives, but the Newspaper Society refused to do this on the grounds that the union did not have sufficient members among this occupational category to be representative of them. They also refused to recognize the NUPBPW for circulation representatives on similar grounds.

[2] George Issacs, 'Must We Fight Again?' *NATSOPA Journal* (December 1938), p. 21.

have called for and received the support of the other manual workers' unions in the industry through the PKTF. Faced with such pressure, the Newspaper Society could do little but give way. Thus the granting of clerical recognition to NATSOPA by the Newspaper Society resulted from the strength not of clerical but of manual unionism.

Recognition of DATA by the Shipbuilding Employers' Federation

DATA approached the Shipbuilding Employers' Federation[1] for recognition in 1918 and again in 1937, but on both occasions the Federation refused even to meet the Association in order to discuss the matter.[2] Towards the end of 1940 the union made another approach to the Federation for recognition. In considering this application the Federation felt that the recognition of the draughtsmen's union would be undesirable for several reasons:

There is obviously considerable objection on the part of shipbuilding and ship-repairing firms to any outside control of draughtsmen, who are regarded by firms as key men and constitute an important part of their responsible staff, being remunerated purely according to ability and qualifications. In these circumstances it was felt that the status of draughtsmen as staff employees would be adversely affected by recognition of their trade union, particularly bearing in mind that executive posts are generally filled from the drawing office. Quite apart from this, however, it was thought that any scale of minimum rates laid down by a union would tend to become standard rates, to stifle incentive for advancement, and militate against the best interests of the draughtsmen.[3]

In view of this the employers felt that the interests of draughtsmen would best be served by their joining the Foremen and Staff Mutual Benefit Society,[4] and the Federation therefore refused to grant recognition to DATA.

The union then approached the Minister of Labour, Ernest Bevin, and asked for his help in securing recognition from the Shipbuilding Employers' Federation. Bevin agreed to help, and in December 1940 the Ministry wrote to the Federation requesting their reasons for refusing to grant recognition to DATA, and pointing out that such a refusal was 'likely to create some trouble'.[5] The Federation furnished the Ministry with a statement of its attitude on the recognition of DATA and requested a meeting with the Ministry before it undertook any further action. Such a meeting took place and shortly afterwards the employers agreed to meet DATA for an informal discussion on the question of recognition. This meeting occurred in March

[1] During 1966 the Shipbuilding Employers' Federation merged with the Shipbuilding Conference (which was mainly a trading association for shipbuilders) and the Repairers' Central Council (which was mainly a trading association for repairers) to form the Shipbuilders' and Repairers' National Association.

[2] Mortimer, op. cit., pp. 45 and 181.

[3] Circular Letter No. 303/40 from the Shipbuilding Employers' Federation to Local Associations, 26 November 1940.   [4] Supra, pp. 132–3.

[5] Letter from the Ministry of Labour to Shipbuilding Employers' Federation, 31 December 1940.

1941, and in July the President of the Federation was able to report that the Executive Committee 'was practically unanimous after their meeting with the Draughtsmen in feeling that there was no reason for delaying any longer the recognition of their right to look after the interests of draughtsmen and tracers in the industry'.[1] It was agreed not to enter into any formal procedure agreement, but merely to circulate a letter to union branches and local associations indicating the general lines upon which disputes should be handled. In December 1941 DATA negotiated its first national wage agreement with the Shipbuilding Employers' Federation.[2]

In explaining the granting of recognition to DATA, the Federation claimed it was taking into account the extent to which the union was organized in the shipbuilding industry, and the existence for many years of a recognition agreement covering draughtsmen in the engineering industry to which many of the Federation's members with marine engineering departments were subject. No doubt these factors did condition the decision of the Federation to grant recognition to DATA. But the Federation was aware of both these factors in 1937 and 1940 when it refused even to meet the Association for discussions, and, therefore, they do not explain why the Federation should have changed its mind between 1940 and 1941. Nor does the industrial strength of the draughtsmen's union explain the Federation's change of outlook. In spite of claiming to have 70 per cent of the qualified draughtsmen in shipbuilding in membership, there was no talk in the union of taking strike action to bring pressure upon the Federation, and indeed there had been relatively few strikes among draughtsmen in shipbuilding since the union was founded in 1913.

The only element which was present in the spring of 1941 when the Federation agreed to meet DATA, that was not present in the autumn of 1940 when they refused to do so, was the influence of the Ministry of Labour. Although there is no record of what was discussed at the meeting between the Federation and the Ministry, it is fairly clear that the Ministry's influence was crucial in getting the Federation to agree to meet the union. Moreover, the very possibility that the Ministry might take further action on the matter was a factor which the Federation now had to take into consideration. Even if the employers' attitude on this question had evolved to a point where the Federation would have recognized the union in any case—and such a sudden unprompted change of attitude within the space of six months is most unlikely—then at the very least the Ministry's influence was a catalyst in the recognition process.

[1] Minutes of the meeting of the Central Board of the Shipbuilding Employers' Federation held at Carlisle, 25 July 1941.

[2] Although the Ship and Boat Builders National Federation, an association covering employers in the small ship (up to 150 feet in length) and boat-building industry, does not recognize DATA, it follows the agreements which DATA and the Shipbuilding Employers' Federation negotiate.

Recognition of white-collar unionism in the iron and steel industry

There were two major developments in the recognition of white-collar unionism in the iron and steel industry during this period: Richard Thomas and Company Limited[1] recognized the NACSS in 1942, and the Iron and Steel Trades Employers' Association recognized the Iron and Steel Trades Confederation (better known as the British Iron, Steel and Kindred Trades Association or BISAKTA) for staff grades in 1943–5. One of the most difficult struggles for recognition during the war occurred at Richard Thomas. It illustrates the lengths to which the Minister of Labour was willing to go in order to promote union recognition during the war, and it is worth considering in some detail.

Recognition of the NACSS by Richard Thomas.[2] During the late thirties, the NACSS began recruiting the white-collar employees of Richard Thomas. By the autumn of 1939 the union had approximately 50 per cent of the staff employees in the company's West Wales works in membership, and it requested recognition. In January 1940 the company replied that it did not feel that the union 'could improve the pleasant relationship which existed between it and the clerical staffs', and it refused to grant recognition.[3] The Ministry of Labour then tried to get the company to meet the union, but the Ministry's efforts were unsuccessful. In May the local branch requested permission to strike, but the union refused because of the general agreement not to undertake such action during the war. Order 1305 was passed in July, and almost immediately the NACSS submitted a claim to Richard Thomas for graduated salary scales and improved working conditions. The company refused these demands and the union then referred the claim, through the Minister of Labour, to the National Arbitration Tribunal.

At the first hearing in November, the Tribunal suggested to the employers that since the NACSS represented a majority of their staffs, they should seriously consider recognizing the union and negotiating the points at issue. After being allowed to consider this matter for a week, the employers still refused to grant recognition. Consequently the Tribunal held a further hearing in December and ordered that graduated minimum salary scales should be established for the company's staff. The award was a victory for the NACSS, but its value to the union was limited. The Tribunal only awarded the principle of minimum salary scales, it did not, and could not, force the company to negotiate with the union regarding the level of these scales.

---

[1] Later part of Richard Thomas and Baldwin Limited and now part of the British Steel Corporation.

[2] Most of the material for this section has been drawn from the *Report by a Court of Inquiry into a Dispute Between Richard Thomas and Company Limited and the National Association of Clerical and Supervisory Staffs, 30 June 1941.* This was supplemented by an interview on 7 November 1966, in Cardiff, with the late Mr. R. C. Mathias, who was then a Regional Secretary of the TGWU, and at the time of this dispute was secretary of the staff branch at Richard Thomas.   [3] Ibid., p. 1.

The company was thus able to implement the award without consulting the union. The union then claimed that the salary scales were unsatisfactory and, when in February 1941 the company refused to meet them to discuss the matter, the NACSS requested that the Minister of Labour refer the questions of the level of the salary scales to the National Arbitration Tribunal.

Meanwhile, a member of the union was suspended by the company for alleged insubordination. As a result the branch struck and refused to return to work until the company recognized the union. The strike quickly spread to thousands of manual workers in the tinplate works and in the steel works, and soon the scale of the strike made a return to work imperative in the national interest. The Ministry of Labour conciliated, and the strikers agreed to return to work on the understanding that the company would give favourable consideration to the question of recognition. But such consideration was not forthcoming and the staff were invited to join a company-sponsored staff association. The union once again referred the whole matter to the Minister of Labour and as a result a Court of Inquiry was established. In June 1941 it reported strongly in favour of the union's case for recognition, but the company still refused to recognize the NACSS. Finally, Bevin threatened to use the government's emergency powers to take over the firm for the duration of the war if it did not recognize the union. Faced with this prospect, Richard Thomas gave way and recognized the NACSS in January 1942.

The struggle of the NACSS for recognition at Richard Thomas shows very clearly how unions were able during the war to use the Ministry of Labour and the National Arbitration Tribunal to bring pressure upon employers to grant recognition. Most employers were not prepared to go to the lengths which Richard Thomas were to avoid granting recognition, and they gave way either as a result of conciliation or a threatened referral to the National Arbitration Tribunal. If they did not, however, and a situation arose which might disrupt production and thereby affect the national interest, the Minister had shown that he was prepared to establish a Court of Inquiry[1] and, as a last resort, even compel the employer to grant recognition. It was a lesson which no doubt was not lost on other employers.

Recognition of white-collar unionism by the Iron and Steel Trades Employers' Association.[2] The growth and recognition of staff unionism at Richard Thomas in 1939–42 was the first major breakthrough for unions catering for staff employees in the iron and steel industry, but it did not represent the first attempt to organize staff employees in this industry. The CAWU (then the National Union of Clerks) began recruiting iron and steel clerks before the First World War, and, after several manual unions merged in 1917 to form the Iron and Steel Trades Confederation or BISAKTA, this organization also

[1] *Supra*, p. 157.
[2] A great deal of the following material was obtained in interviews with various employers and trade unionists who, for fairly obvious reasons, asked not to be identified.

became interested in recruiting clerical workers. In 1920 these two unions came to an agreement whereby all clerical employees in the industry would join the CAWU, and it would affiliate to BISAKTA in respect of this part of its membership with the right to representation on BISAKTA's Executive Council. In a sense, clerical trade unionists in the iron and steel industry held membership in both organizations. The CAWU was responsible for recruitment, day-to-day administration, and negotiations; BISAKTA was responsible for certain administrative details and expenses, and had the right to be consulted on general policy matters regarding iron and steel clerks.[1]

The new alliance was not particularly successful in recruiting clerical workers. In 1919 there were four to five thousand clerical trade unionists in the industry, but by 1927 this number had dwindled to 338, and by the end of 1936 it had only risen to 399.[2] During 1936 the CAWU submitted a claim for improved salaries and working conditions to Colvilles Limited, but the company refused to recognize the union. With BISAKTA's approval the CAWU struck. The company then agreed to the establishment of a committee consisting of representatives of the Iron and Steel Trades Employers' Association, BISAKTA (but not the CAWU), and the employees concerned. The committee was to examine the whole situation created by the strike and to evolve a procedure for dealing with future staff disputes at Colvilles. It was agreed that any conclusions arrived at by the committee would 'not prejudice the right of the N.U.C. . . . to be recognized as acting on behalf of the staff'.[3] Very little documentation on the committee's proceedings is available, but it is clear that as a result of its deliberations BISAKTA was recognized for clerical workers at Colvilles, and the union became anxious to amend the 1920 agreement between itself and the CAWU. It seems likely, especially in view of later developments in the industry, that BISAKTA's change of attitude was brought about by a realization that the employers were more prepared to recognize it for staff workers than the CAWU. Whatever the reason, BISAKTA terminated the 1920 agreement on 30 June 1937 and took over most of the clerical membership in the iron and steel industry.

As white-collar unionism became more prevalent in the industry, the Iron and Steel Trades Employers' Association began seriously to consider the matter, and late in 1943 they laid down the following policy. The question of whether or not to recognize staff unionism was to be left to each individual firm to decide. If a firm decided in favour of granting such recognition, however, then it was to be guided by the following principles: recognition should not be granted for confidential employees or department heads; negotiations affecting staff grades should be kept separate from those affecting manual

[1] *CAWU Annual Report and Balance Sheet*, 1919, p. 19.

[2] Sir Arthur Pugh, *Men of Steel* (London: Iron and Steel Trades Confederation, 1951), p. 416, and the Iron and Steel Trades Confederation, *Quarterly Report* (31 March 1937), p. 48.

[3] Ibid., p. 49.

workers; staff employees should be left free to join or not to join the union as they saw fit; and recognition should be confined to those unions already established in the industry for manual workers, with the possible addition of DATA. Finally, it was decided that there would be no national negotiations for staff grades, but, if a firm wished, the Association would act on its behalf in negotiations with a union for any specified grade of staff.

In February 1945 this policy was formalized in three identical procedure agreements which the Iron and Steel Trades Employers' Association signed with BISAKTA for clerical workers, laboratory staffs, and departmental foremen.[1] These procedure agreements were entirely permissive and in no way forced an employer to recognize the union. Each agreement only applied to a firm after it had decided to grant recognition to BISAKTA for that specific staff grade. In spite of the permissive nature of the procedure agreements, most of the large firms in the industry adopted those for clerical workers and laboratory staffs, but not so many subscribed to the agreement for foremen. Among the firms which had signed the national procedure agreements, however, few actually entered into negotiations with BISAKTA for these grades; they merely tended to listen to what the union had to say regarding staff salaries and conditions and then made a unilateral announcement on these matters. Although there were no national procedure agreements for the other unions catering for staff employees in the industry, most firms negotiated with DATA and some negotiated with the various craft unions for craft foremen.

At the time of obtaining recognition, BISAKTA had only about 500 to 1,000 staff workers in membership. But other white-collar unions were beginning to make considerable headway. The CAWU still had a foothold in the industry which it was using as a base upon which to expand, and the NACSS, with the help of the government, had broken through in south Wales. Most important, ASSET, which the employers regarded as communist-controlled, had mounted an intensive recruitment campaign among foremen

[1] The white-collar union recognition situation in the iron and steel industry has been completely changed by the Iron and Steel Act of 1967 which nationalized a major section of the industry and obliged the British Steel Corporation to negotiate with any workers' organizations appearing to them to be appropriate. After consulting the TUC, the Corporation conceded recognition to BISAKTA, the National Union of Blastfurnacemen, the NUGMW, the TGWU, the Amalgamated Union of Building Trade Workers, and the National Craftsmen's Co-ordinating Committee (DATA is a member of this body) for 'staff, foremen, and supervising/technical grades' but excluding management grades above the level of foremen. The Corporation announced that the unionization of managerial grades up to but excluding departmental heads would be actively encouraged and that it would be prepared to recognize any organization on behalf of these grades which could demonstrate it was representative of them and free of influence from higher management. The ASTMS and the CAWU also claimed recognition from the Corporation and undertook industrial action to lend weight to their claim. In July 1968 a Court of Inquiry was set up to consider this matter, and it recommended that both these unions should be recognized. This recommendation is now under consideration by the Corporation. See the *Report of a Court of Inquiry Under Lord Pearson Into The Dispute Between the British Steel Corporation And Certain Of Their Employees*, Cmnd. 3754.

in south Wales and the north-east coast. In 1943 it called a strike at Dorman Long on Tees-side, and referred a claim against the company for increased wages to the National Arbitration Tribunal.[1] Shortly afterwards BISAKTA challenged ASSET's right to recruit in the iron and steel industry and the whole matter was referred to the Disputes Committee of the TUC.[2]

In view of these developments, the employers began to regard the growth of staff unionism in the industry as inevitable and felt that in such circumstances it was better to recognize, as one employer put it, 'the devil you know rather than the one you don't'. They argued that since all of BISAKTA's membership was in the iron and steel industry, the union's future progress was entirely dependent upon the prosperity of this industry and it was therefore likely to pursue 'reasonable' and 'statesmanlike' policies. 'Outside' unions such as ASSET, the CAWU, and the NACSS had only a small proportion of their membership in the iron and steel industry and were therefore in a position to follow 'militant' and 'irresponsible' policies in this industry without affecting the over-all progress of their organization. In short, the employers felt that over the years BISAKTA had been 'welded and educated into responsibility', and it was thus the best union to recognize for staff grades. Consequently, while the question of which union had the right to recruit staff workers in the iron and steel industry was being considered by the TUC, the Iron and Steel Trades Employers' Association tried to bolster BISAKTA's claim by granting them procedure agreements to cover these grades.

The employers granted these procedure agreements to BISAKTA not because of the union's strength among staff grades, but because of its weakness in this area. The procedure agreements were designed to serve a twofold purpose. In the first place, they provided a shield behind which an employer could take refuge if approached for recognition for staff grades by an 'outside' union. Without having to recognize BISAKTA, an employer could inform such a union that his Association's policy as formalized in the national procedure agreements only allowed him to recognize BISAKTA or other 'internal' unions. In addition, the procedure agreements ensured that if a firm decided to recognize BISAKTA for staff grades, it did so in a manner consistent with all the other firms in the industry which had also granted such recognition. The agreements thus prevented the union from playing off one employer against another.

Recognition of the NACSS, ASSET, and the AScW by the Engineering Employers' Federation

During and immediately after the First World War, when the Engineering Employers' Federation was granting recognition to the CAWU and DATA,[3] several more staff unions were being formed. The Association of Supervisory

[1] *Report of National Executive Council of ASSET*, 1945–6, p. 6.
[2] See *TUC Annual Report*, 1945, pp. 27–9.   [3] *Supra*, pp. 150–5.

Staffs, Executives and Technicians (then the National Foremen's Association) was founded in 1917 and in the following year the Association of Scientific Workers (then the National Union of Scientific Workers). In 1922, when several unions amalgamated to form the Transport and General Workers' Union, a white-collar section was also established within this organization. In the depressed years of the twenties and early thirties, the white-collar membership of these unions did not grow appreciably, and what little membership they had was concentrated in public employment. With the economic recovery in the later thirties and the outbreak of war, the membership of these unions began to grow both in the public and private sectors of the economy, and it was not long before they began to demand recognition from private employers, especially those in engineering.

Recognition of the NACSS. Of the fourteen unions which amalgamated to form the TGWU in 1922, at least five had white-collar workers in their ranks. The structure of the new union thus allowed for an administrative, clerical, and supervisory trade group to service and recruit this type of worker. At the time of the amalgamation almost all of the union's 6,000 white-collar members were concentrated in the Port of London Authority and in London Transport. This situation remained unchanged until 1935 when the union's white-collar membership began to grow, especially amongst the clerks of a few of the larger engineering firms. At the beginning of 1938 the TGWU approached the Engineering Employers' Federation for a procedure agreement similar to the one signed with the CAWU, but the Federation refused. The employers argued that a manual union 'was not an appropriate union for organizing clerical and supervisory staffs', and, in any case, if they recognized the TGWU for white-collar workers, they would soon be forced to recognize all the other manual unions for these grades as well.[1]

The TGWU then decided to reorganize the structure of the union in such a way that the Federation would not be involved in recognizing the right of a manual union to represent staff workers. Early in 1939 the administrative, clerical, and supervisory group of the union was reconstituted as the National Association of Clerical and Supervisory Staffs. The NACSS was still a part of the TGWU, but it was claimed that the relationship was similar to that between an individual union and the TUC or to that between a subsidiary company and a holding company. The new structure was devised not only because the union felt there was some substance to the employers' claim that manual and staff affairs should be kept separate, but also because it felt that the new structure 'would overcome the difficulties that appear to prevent certain classes of non-manual workers from joining our ranks'.[2]

In May 1939 the NACSS entered into discussions with the Federation over

---

[1] *Proceedings of a Conference Between the EEF and the NACSS*, London, 28 September 1939, p. 2.
[2] Secretary of the NACSS in the *TGWU Record* (May 1939), p. 299.

recognition. The union argued that the new structure met the employers' demands regarding the separation of staff and manual matters, and that by recognizing the NACSS they would not be recognizing a manual union for staff grades and thereby setting a precedent for manual unions to follow.[1] The NACSS gave an understanding not to be associated with the manual side of the TGWU and not to be a party to any dispute in which manual workers were engaged. It also assured the employers that the word 'supervisory' in its title referred to inspectors in passenger transport firms and not to engineering.[2] As a result of these discussions the Federation decided it would enter into a procedure agreement with the NACSS similar to the one negotiated with the CAWU in 1920, subject to two conditions: the union would act completely independently of the TGWU, and supervisors would be specifically excluded from the scope of the agreement. An agreement along these lines was signed on 23 April 1940.[3]

In granting recognition to the NACSS the Federation was aware that the union's membership in engineering was very small and confined to a few large firms. It was also not altogether convinced that the NACSS was in practice a separate and distinct entity from the TGWU since it shared the same offices and had a representative on the latter's executive. Yet it nevertheless recognized the NACSS for clerical workers.

Its reasons for doing so were twofold. The Federation still thought it was wise to continue the policy of union containment by granting recognition to a union in return for its agreement to submit all disputes to a system of 'employer conciliation'.[4] If a procedure agreement was not granted, the Federation argued to its local associations, firms would be 'in a somewhat vulnerable position should they become involved in a dispute with their clerical workers because there will be no official point of contact between the Association and the Union or between the Federation and the Union, and in the absence of any agreed procedure the clerical workers may look for support to the industrial side of their Organization'.[5] More important, the Federation felt that

to recognise one staff union as entitled to negotiate on behalf of clerical staff workers, and to refuse such recognition to other staff unions, is to place the employers in a position of appearing to dictate to their staff clerks as to which trade union they

[1] Actually, the NACSS did not have a fundamentally different structure from the old administrative, clerical, and supervisory trade group. The difference between the two organizations was simply one of name. The NACSS's constitutional position in relation to the TGWU is the same as that of the old white-collar trade group and that of the other trade groups within the TGWU.
[2] *Proceedings*, op. cit., p. 6.
[3] This agreement was amended on 14 May 1947 to make it mandatory for employers to exhaust the procedure before introducing any downward alteration in wages or working conditions. See p. 150 n. 4; p. 151 n. 1; p. 172 n. 1; p. 174 n. 2; and p. 179 n. 3.
[4] *Supra*, pp. 154–5.
[5] Circular Letter No. 64 from EEF to Local Associations, 28 March 1940.

must belong if they desire to raise a question through a union. It would also have the effect of creating a monopoly for the National Union of Clerks, thus leaving the Federation open to a charge of discriminating in favour of that particular union.[1]

Thus the NACSS's battle for recognition was actually won in 1920 when the Federation recognized the CAWU. For as soon as the NACSS could demonstrate that it had some membership in the engineering industry and that at least in a formal sense it was a separate organization from the TGWU, the precedent of 1920 forced the Federation to recognize the NACSS as well.

Recognition of ASSET.[2] While the recognition of the NACSS did not establish any major new principles and was therefore readily conceded by the Engineering Employers' Federation, the same was not true of the recognition of ASSET. The union was formed among engineering foremen in 1917 and had a membership which, for the next twenty-two years, fluctuated between 1,000 and 3,000. With the outbreak of the Second World War the union's membership began to grow, especially in engineering, and in May 1940 the General Secretary of ASSET wrote to the Engineering Employers' Federation 'to offer the loyal co-operation of our entire membership' and to request recognition for foremen.[3] The offer of loyalty was no doubt appreciated, but the Federation was not so overwhelmed that it responded by granting the union's request. In fact, it even sent a circular to member-firms instructing them to withold recognition of this union.

ASSET's first major breakthrough in the engineering industry came in 1941. By the beginning of that year it had recruited a large majority of the foremen at the two member-firms—Harland and Wolff Limited and Short and Harland Limited—of the Belfast Marine Engineering Employers' Association, and it approached the Association for recognition and the right to negotiate over wages. In accordance with the instructions issued by the national Federation, the Belfast Association refused to recognize or negotiate with the union. ASSET, with the support of the manual unions, then threatened a stoppage of work, got the Ministry of Labour and the Admiralty to bring pressure to bear upon the employers, and requested the Government of Northern Ireland to refer the matter to arbitration. As a result of these developments the Belfast Association turned the whole matter over to the national Federation. In view of the situation facing the Association, the Engineering Employers' Federation decided in August 1941 that the Association should maintain its refusal to recognize ASSET but the two member-firms should be allowed to recognize this union if they so desired. The firms then conducted a referendum among their foremen and since this resulted in a vote in both firms of about 70 per cent in favour of recognition, recognition was granted.

[1] Circular Letter No. 64 from EEF to Local Associations, 28 March 1940.
[2] Unless otherwise noted, the material for this section was drawn from EEF sources.
[3] Letter from ASSET to EEF, 27 May 1940.

Following this victory, the union once again approached the Federation for recognition on behalf of foremen and technical staffs,[1] and during the spring of 1942 informal discussions were held between the two organizations. But the Federation argued 'that as foremen are fundamentally part of the management, their conditions of employment cannot properly be the subject of negotiations by a trade union organization', and they therefore still refused to recognize ASSET.[2]

Not having sufficient strength to obtain recognition by means of industrial activity, ASSET transferred the struggle for recognition to the social and political arenas. It requested the help of the TUC whose General Secretary, Sir Walter Citrine, informed the Federation in October 1942 that the union had 'the fullest support of the Trade Union Congress' in its fight for recognition and hinted that the TUC was prepared to make this matter a major political issue.[3] Meanwhile ASSET submitted a claim for salary increases for foremen to the Witton, Birmingham, plant of the General Electric Company. Since the firm would not agree to negotiate a general claim for foremen, the union submitted a separate claim for each of its 280 members in the firm to the National Arbitration Tribunal at the beginning of December. This number of cases was sufficient to keep the Tribunal busy for well over a year, and by completely clogging the arbitration machinery would endanger production. It was rumoured in the press that arms output was threatened and that Bevin might intervene, and on 17 December the Federation discussed the possibility of it being adversely criticized in the Press because of its attitude towards the recognition of ASSET.[4]

On 28 December Citrine again wrote to the Federation, this time suggesting that they should recognize ASSET in those firms where the union had majority membership among the grades for which it catered. On 23 January 1943 ASSET held a mass meeting in London at which the main speaker was Sir Stafford Cripps, a socialist and the Minister of Aircraft Production. During the meeting it was announced that Cripps had invited ASSET to set up a committee to advise the Ministry of Aircraft Production on technical production matters. The Engineering Employers' Federation had an observer at this meeting who reported to them that the General Secretary of ASSET 'pointed out that this was tantamount to "recognition" and now they had this much from so important a Ministry they need not worry unduly about other quarters'. Five days later, on 28 January, the Federation decided to explore

[1] During 1942 the union changed its name from the National Foremen's Association to the Association of Supervisory Staffs and Engineering Technicians and began to recruit both foremen and technicians. In 1946 the union changed its name to the Association of Supervisory Staffs, Executives and Technicians.

[2] Letter from EEF to ASSET, 29 May 1942.

[3] Letter from Sir Walter Citrine to EEF, 7 October 1942.

[4] See *News Chronicle*, 1 December 1942; *Daily Telegraph*, 30 November 1942; and Minutes of the EEF Management Board, 17 December 1942.

the possibility of recognizing ASSET along the lines suggested by Citrine in his letter of 28 December.

ASSET then withdrew the 280 cases from the National Arbitration Tribunal, and negotiations to draft a procedure agreement began. There were numerous differences of opinion between the employers and the union over the details of the agreement, but it was finally signed on 1 January 1944. It was unlike any other procedure agreement in the engineering industry. It applied only where the union had a majority membership in a particular grade in a particular establishment operated by a member-firm of the Federation. It was agreed that every effort would be made to settle disputes without reference to ASSET. To ensure that no officials of the union would interfere with the employer–foremen relationship within an individual establishment, no provision was made for a works conference.[1]

During this period ASSET's membership, although increasing, was not impressive. At the beginning of 1942 the union had 2,500 members, and by the end of 1943 it had almost 10,000, but only about 4,000 were in engineering and most of these were in aircraft production.[2] The union could hardly claim to be representative, yet the Federation recognized it. The reasons for doing so were clearly pointed out by the Director of the Federation in a speech to his General Council on 28 January 1943:

> On the approach to the ASSET question the inclination of every man in this room is to say that no foremen ought to be made a member of a trades union and that no trades union should be recognised as having the right to speak for them. On the other hand we have to recognise that the war has brought in to management circles a whole host of people who are never likely to find a permanent home there. They will go back to the tools, and in the circumstances it is not reasonable to expect that during this interim period they should forego all their trades union inclinations.
>
> Again there is the question of public policy. There is very little doubt with the Trades Union Congress in the field, Government sentiment against us, the Minister of Aircraft Production in principle openly espousing the cause of ASSET [that] it only remains for ASSET to present one case where public opinion can be effectively challenged, and the verdict will be against the Federation. Public policy in war-time has a much stronger appeal to the popular mind than industrial reservations.

Given ASSET's lack of membership in the engineering industry, and the intense feeling of management on the question of the unionization of foremen, ASSET might not yet be recognized by the Engineering Employers' Federation if it had not been for the leverage the union was able to exercise because of the wartime situation.[3]

---

[1] On 1 February 1966 the 1944 agreement was amended to provide for a works conference stage in the procedure. The general alterations clause which is contained in the procedure agreements of all the other staff unions in engineering was never included in ASSET's agreement. See p. 150 n. 4; p. 151 n. 1; p. 169 n. 3; p. 174 n. 2; and p. 179 n. 3.

[2] *Reports of National Executive Committee of ASSET*, 1942–3, 1945–6.

[3] It became increasingly difficult after the war for unions to use statutory institutions to

Recognition of the AScW. The AScW was founded in 1918 and for the next two decades it restricted its membership to qualified scientists in universities and research establishments and acted in most respects simply as a professional association. Beginning in 1938 several industrial branches were formed and an increasing emphasis was placed on the trade union side of the Association's work. In 1940 the Association re-registered itself as a trade union[1] and amended its rules to enable unqualified assistants working in scientific and research departments to join the Association. The war resulted in a vast increase in the number of scientists and ancillary grades employed by industry and government, and the AScW's membership began to grow at a phenomenal rate. In 1939 there were fewer than 1,500 members in the Association; by 1946 there were over 17,000.

During 1941 the AScW submitted a claim for a salary increase to Napier Motors Limited and the resulting negotiations led to an informal conference being held between the union and the London Association of the Engineering Employers' Federation. At this conference the London Association suggested to the AScW that they should approach the national Federation with a view to obtaining a procedure agreement to cover their engineering membership. The union was active in several of the larger London electrical firms and the London Association was anxious that a procedure agreement be established to contain the AScW's demands.[2] Following the London Association's suggestion, the AScW wrote to the Engineering Employers' Federation in December requesting a procedure agreement. In February 1942 the Federation refused the union's request on the grounds that 'a close personal relationship exists between managements and their technical staffs and conditions of employment could be determined only through this personal relationship'.[3]

During the next few months several disputes arose between the AScW and

exert pressure upon employers for wage increases or recognition (*infra*, pp. 175–6). For example, in July 1952 ASSET submitted a claim to the Engineering Employers' Federation for a substantial increase in its members' salaries. The Federation refused to grant the claim because ASSET'S procedure agreement did not provide for national negotiations on wages, but was operative only where the union had majority membership in an individual grade in an individual establishment. In January 1953 ASSET requested the Minister of Labour to refer the dispute to the Industrial Disputes Tribunal for a binding award. Since there was no national joint negotiating machinery in the engineering industry for settling the wages of the workers covered by the claim, it was necessary, under Order 1376, for ASSET to demonstrate that it represented a 'substantial proportion' of these workers before the Minister could refer the dispute to the Tribunal. ASSET could not do this, and the Minister was unable to refer the dispute to compulsory arbitration.

[1] The AScW first registered as a trade union in 1918, but in 1926 it deregistered 'on the grounds of decreasing membership and the difficulties of recruiting'. See Reinet Fremlin, 'Scientists and the T.U. Movement', *AScW Journal* (May 1952), pp. 15–18; (October 1952), pp. 15–17.

[2] Letter from London Association to EEF, 18 July 1942.

[3] Cited in *Negotiations with the Engineering and Allied Employers' National Federation or a Procedure Agreement and Recognition Agreement*, an AScW pamphlet, 1943.

individual engineering firms, and in September the union reported these to the Ministry of Labour. The Ministry suggested that voluntary negotiating machinery should be established and they promoted a meeting between the union and the Engineering Employers' Federation to discuss this matter. At this meeting the Federation made it clear that its major objection to recognizing the AScW was that many of the union's members occupied executive or managerial posts in industry and the employers could not consent to such employees' salaries and working conditions being determined by negotiations with a trade union. The AScW pointed out that they were only interested in negotiating on behalf of the 'ordinary non-executive grade of technical or scientific worker'. They assured the Federation that they were prepared to exclude managerial and supervisory grades from the scope of any procedure agreement which might be signed, in the same way that such grades had been excluded from the DATA procedure agreement of 1924.[1] In view of this the Federation decided on 29 October that they were prepared to meet the AScW to negotiate an agreement.

The negotiations began in February 1943, and lasted on and off for the next fifteen months. The main points of difference centred around two issues: the specific grades to be excluded from the agreement, and whether a 'general alterations clause' similar to that in the DATA agreement of 1924, preventing employers from altering salaries and working conditions before they had exhausted the procedure, would be included in the agreement.[2] Negotiations temporarily broke down in December, but in March 1944 the AScW requested that negotiations be resumed. Finally on 11 May 1944 a procedure agreement was signed by the two organizations.

The Federation recognized the AScW not as a result of any great activity on the part of the Association, but largely as a by-product of recognizing DATA and ASSET. By the end of 1942 the AScW had only 3,400 members in the engineering industry, had never called a strike, and had yet to refer a case to the National Arbitration Tribunal. In fact, the union possessed so little strength that during the lengthy negotiations of 1943–4, it was forced to accept every major amendment to the draft procedure agreement which the Federation put forward. As in the case of the NACSS[3] the AScW obtained recognition largely as a result of precedents which had been set or were about to be set by other unions. As soon as the AScW made it clear that they were not interested in negotiating for managerial grades and were prepared to accept recognition on the same qualified basis as DATA, it became most

[1] *Supra*, p. 151 n. 1.

[2] The AScW was forced to accept the procedure agreement in 1944 without the inclusion of the general alterations clause, but in March 1947 the clause was added to the agreement. Moreover, the 1944 agreement did not provide for a works conference stage in the procedure. The agreement was amended to provide for such a stage on 17 December 1964. See p. 150 n. 4; p. 151 n. 1; p. 169 n. 3; p. 172 n. 1; and p. 179 n. 3.

[3] *Supra*, pp. 168–70.

difficult for the Federation to refuse the AScW recognition. As one local association pointed out to the Federation, the AScW 'represents a fairly compact and homogeneous body of employees in much the same way as the Draughtsmen do, and therefore it is difficult to see on what grounds they are excluded from a form of recognition which is granted to the Draughtsmen's Union'.[1] Even more important, at the same time as the AScW was trying to obtain a procedure agreement, ASSET was mounting its intensive public campaign for recognition. By the end of 1942 the Federation was on the brink of granting recognition to ASSET not only for technical grades but also for supervisory workers. Not to have granted recognition to the AScW as well would have been curiously illogical and would not have withstood public scrutiny.

THE PERIOD 1946 TO 1964

The climate of the period

The government replaced Order 1305 by Order 1376 in 1951 and Order 1376 by the Terms and Conditions of Employment Act in 1959. Each successive measure made it more difficult for unions to use statutory provisions to exert pressure upon employers for recognition. Under Order 1305 a union only had to be a party to a dispute over the terms and conditions of employment which should exist in a firm or industry for the dispute to be referred to compulsory arbitration. But under Order 1376 a union also had to be a recognized party to the joint voluntary negotiating machinery existing in the firm or industry concerned, or, in firms and industries where such machinery and recognition did not exist, the union had to represent a 'substantial proportion' of the employees concerned. Finally, under the Terms and Conditions of Employment Act a union cannot refer a dispute to compulsory arbitration. It can only request that the Industrial Court order an employer to observe terms and conditions of employment not less favourable than those contained in an agreement negotiated by organizations representing a substantial proportion of the workers and employers in the industry concerned. Even to do this, a union must be a party to the agreement; that is, the union must be generally recognized throughout the industry.

It is true that the government has taken some steps over the post-war period to encourage trade union recognition and collective bargaining, but none of them gives more than moral support to the principle of trade union recognition. In 1946 it strengthened the House of Commons Fair Wages Resolution by inserting a clause requiring all government contractors to recognize the freedom of their workpeople 'to be members of trade unions'. There is nothing in the Fair Wages Resolution, however, which forces an employer to recognize and bargain with a union. In 1949 and 1950 the

[1] Letter from London Association to EEF, 18 July 1942.

government ratified ILO Conventions No. 87 and 98 which state that workers shall have the right to 'join organisations of their own choosing' and enjoy 'adquate protection against acts of anti-union discrimination'.[1] But ratification does not make these conventions the law of the land. 'An Act would be required for that', as Professor Wedderburn has pointed out, 'and none has been passed.'[2]

Perhaps it is not surprising that in a period in which the legislative framework has become increasingly unfavourable to union recognition, no significant concessions of recognition have been made to unions catering for white-collar workers in manufacturing. While some individual firms have granted recognition to such unions during this period, only two employers' associations have done so: the British Spinners' and Doublers' Association to the Textile Officials' Association in 1950 and the Engineering Employers' Federation to the NUGMW for clerical workers in 1953.[3] In the first case, the nature of the recognition was so restricted as to make it unimportant; in the second, the outcome was determined by events which occurred during the Second World War.

Recognition of the Textile Officials' Association by the British Spinners' and Doublers' Association[4]

During the Second World War the wages and working conditions of textile operatives were improved on several occasions, but very few of these improvements were passed on to mill officials and supervisors. The relative economic position of textile officials therefore seriously deteriorated and they formed local associations to represent their views to the employers. These local associations began to band together, and in 1947 they merged with ASSET. Following the merger, ASSET established a National Sectional Council for the Textile Industry, appointed a full-time official to administer this Council,

---

[1] See C. Wilfred Jenks, *The International Protection of Trade Union Freedom* (London: Stevens, 1957).

[2] K. W. Wedderburn, *The Worker and the Law* (Harmondsworth, Middlesex: Penguin Books, 1965), p. 16.

[3] Since 1964 there has been another major concession of recognition to white-collar unions in manufacturing, and it supports the general argument of this chapter. For several years the Shipbuilders' and Repairers' National Association (see *supra*, p. 161 n. 1) refused to recognize unions representing clerical or supervisory grades (see *supra*, p. 129) but in 1967 it gave way and granted recognition. There is not room here to give all the details, but it is clear that the major factor bringing about this change of policy by the Association was a recommendation of the Geddes Committee. This Committee was established by the government in 1965 to examine the competitive position of the shipbuilding industry. It reported in 1966 and recommended, among other things, that there was a 'need for new national negotiating and consultative machinery for shipbuilding' which would 'be comprehensive embracing the shipbuilding employers on the one hand and *all* the unions operating in the industry on the other' (my italics). See *Shipbuilding Inquiry Committee 1965–1966 Report*, Cmnd. 2937, 1966, especially paras. 404–5. See also *supra*, p. 166 n. 1.

[4] The information for this section has been obtained mainly by correspondence and interviews with the parties concerned.

and provided him with an office in Manchester.[1] In 1948 the union approached the British Spinners' and Doublers' Association (then the Federation of Master Cotton Spinners' Association) for recognition.

But since the employers regarded ASSET as a communist-dominated union, they not only refused to recognize it but also encouraged their employees to resign from it and establish a separate organization. Many of the officials responded in 1949 by forming the Textile Officials' Association, a registered trade union unaffiliated to the TUC.

In February 1950 the British Spinners' and Doublers' Association recognized the Textile Officials' Association as 'the body which should be entitled to represent to the Federation the views and interests of its members', but stipulated 'that such recognition shall not extend to the entering into of formal joint agreements between the two bodies'.[2] In practice this has meant that the employers' association listens to what the union has to say, and sometimes responds by issuing recommendations to its member-firms regarding the 'appropriate' terms and conditions of employment for officials. But the British Spinners' and Doublers' Association refuses even to make recommendations regarding officials' salaries.[3] The employers argue that it is not possible to discuss meaningfully national minimum salary scales because of the wide variations in the responsibilities and qualifications of officials due to differences in the size and organizational structure of firms. But it is doubtful if this is more true of textiles than of other industries.

The British Spinners' and Doublers' Association granted recognition to the Textile Officials' Association to encourage its growth at the expense of ASSET's. At the same time, the form of recognition was designed to enable employers to keep themselves informed of the views of their officials without actually negotiating with them.

Recognition of the NUGMW for Clerical Workers by the Engineering Employers' Federation

In April 1940 the Engineering Employers' Federation recognized the NACSS (Tgwu).[4] This encouraged the NUGMW to approach the Federation for recognition on behalf of staff grades, and in September 1941 the two organizations met to discuss the matter.

The NUGMW argued that their 1922 Procedure Agreement with the

---

[1] *ASSET* (September 1947), p. 114.

[2] Cited in Allan Flanders, 'Collective Bargaining', op. cit., p. 255 n. 1.

[3] The British Spinners' and Doublers' Association has been willing to recommend that 'Assistant Overlookers who are engaged on the supervision of operatives as well as on maintenance and setting of processing machinery . . . should receive a wage not less than the average full staff wage earned by the highest paid grade of operative, under their supervision, on the standard number of machines', but it has not been willing to recommend specific minimum salary scales for the various staff positions. (See Circular Letter from British Spinners' and Doublers' Association to Chairman of Member-Firms, 12 March 1952.)

[4] *Supra*, pp. 168–70.

Federation[1] did not explicitly state that its scope was restricted to manual workers and that therefore the union was entitled under this agreement to raise questions concerning their engineering members whether they were manual, clerical, or supervisory employees. The Federation refused to accept this argument, but agreed to recognize the NUGMW for clerical workers if it would establish a separate organization similar to the NACSS. It was not possible to recognize the NUGMW itself for clerical workers for 'to admit the principle of recognition to one industrial union to negotiate on behalf of clerical staffs would at once open the door to similar claims by other industrial unions and this would create a position of chaos so far as negotiations covering clerical staffs were concerned'.[2] The NUGMW was not prepared to establish a separate organization, however, and claimed that union officials dealing with manual questions were also quite competent to handle staff affairs. Neither side would change its position and the meeting ended in a stalemate.

The matter lay more or less dormant until early in 1945 when the NUGMW began to press various London engineering firms for white-collar negotiating rights. The union threatened to refer certain of these firms to the National Arbitration Tribunal, and this resulted in an informal meeting between the union and the Federation and eventually, in October, to a formal conference between the two organizations. The NUGMW was now prepared to sign a separate agreement for clerical workers, establish separate branches for staff workers which would comprise a separate department within the union, appoint a national officer to administer the affairs of this department who would not participate in manual negotiations in engineering, assure the employers that there would be no joint action between the clerical and manual sections of the union, and even design distinctive notepaper for the use of the union's clerical department. In short, the NUGMW was prepared to conform to the same arrangements that had been made with the TGWU, except that it would not agree to establish a completely separate organization with its own name, constitution, and executive.

In spite of the unwillingness of the NUGMW to establish a separate organization, the Federation decided to recognize the union's 'clerical department', and in November 1945 the Management Board approved a draft procedure agreement for the NUGMW similar to the one signed with the NACSS. The draft agreement was also approved by the NUGMW. Before the agreement was signed by either organization, however, the National Engineering Joint Trades Movement (which was composed of the Confederation of Shipbuilding and Engineering Unions, the AEU, and a few other unions) requested the Federation to recognize the right of either itself or its individual member-unions to negotiate with federated firms, local associations, or the Federa-

[1] The Managerial Functions Agreement of 1922 which applies to all the manual workers' unions in the industry.

[2] Minutes of Meeting of the Management Board, 25 September 1941.

tion itself on behalf of all foremen and staff grades. The NEJTM's claim was prompted by the increased encouragement which the Federation and its member-firms were giving at this time to staff grades to join the Foremen and Staff Mutual Benefit Society,[1] the rules of which made union membership incompatible with membership of the Society. The Federation considered that since the NUGMW was a member of the CSEU, its claim was covered by that of the NEJTM. The Federation therefore postponed the ratification of the NUGMW's draft agreement pending consideration of the 'wider' claim. Negotiations on the wider claim took place with the NEJTM during 1946, but no agreement was reached and the unions let the matter drop.

After these negotiations ended, the NUGMW did not press the Federation to ratify the draft agreement, for the union was no longer willing to accept the 'restrictive' clause in the agreement prohibiting joint action by manual and staff workers.[2] Questions raised locally by the NUGMW in respect of clerical workers therefore continued to be dealt with on an informal basis until October 1952, when the NUGMW wrote to the Federation requesting that the 1945 draft agreement be made operative. After once again receiving assurances that clerical members would not be represented by an officer negotiating for engineering manual workers and that distinctive notepaper would be used by the clerical section, the Federation signed the procedure agreement on 17 June 1953.[3]

At the time of obtaining recognition, the NUGMW had fewer than 1,000 white-collar members in the engineering industry. In fact, the small number of white-collar workers in the union—less than 5,000 out of a total membership of almost 800,000—probably explains why it was so reluctant to establish a separate organization for them.[4] In spite of the union's small white-collar membership, the Federation felt 'it would be illogical . . . to refuse to extend to the National Union of General and Municipal Workers the same facilities as had been extended to the Transport and General Workers' Union'.[5] However, the two cases were not strictly comparable. In 1940 the Federation recognized an organization for clerical workers which, at least in a formal sense, was separate from any manual union, while in 1953 it recognized a department within a manual union for these grades. The Federation may well argue that in practice there is little to choose between the respective white-collar structures of the TGWU and the NUGMW. Nevertheless, the TGWU's structure formally respects the Federation's principle that manual unions should not represent staff workers, while the NUGMW's does not. The recognition of the NUGMW for clerical workers did more than just duplicate an existing precedent. It also created a precedent—one which in principle

[1] *Supra*, pp. 132–3.   [2] EEF documentary sources.
[3] The agreement was amended on 6 December 1962, to include the general alterations clause. See p. 150 n. 4; p. 151 n. 1; p. 169 n. 3; p. 172 n. 1; and p. 174 n. 2.
[4] From NUGMW documentary sources.
[5] Circular Letter No. 130 from EEF to Local Associations, 8 July 1953.

weakens the Federation's case against recognizing other manual unions for staff grades.

CONCLUSIONS

The industrial strength of trade unions, as determined by the size of their membership and their willingness and ability to engage in industrial warfare, is commonly believed to be the major, if not the only, factor encouraging employers to recognize these organizations. One student of trade union growth argues that

It is an axiom of trade unionism that 'employers recognise strength'. Non-manual unions have always found it difficult to demonstrate their strength by conducting a strike, yet in the face of a determined employer few other tactics are effective.[1]

The Labour Correspondent of *The Times* claims 'the history of trade unionism has shown that the only way to secure recognition from a reluctant employer is to strike for it'.[2] Similarly, the TUC maintains that unions have generally obtained recognition by overtly exercising their strength in the form of strike action and that this strength 'has been developed without the help of any external agency'.[3]

Some sociologists have argued that white-collar unions are recognized not so much as a result of their industrial strength as of the process of bureaucratization which makes recognition in the employer-managers' own self-interest. For 'if the bureaucratic rules are to be acceptable and friction in their operation is to be reduced to a minimum, they should clearly have been formulated in consultation with organised groups representative of all the main interests involved'.[4]

The evidence presented in this chapter supports neither of these arguments. None of the major concessions of recognition to white-collar unions in private industry came about because the employer-managers of bureaucratic organizations felt that their administrative burdens would thereby be lessened. In fact, the employer-managers of many of the larger and more bureaucratic firms have often been those who are most opposed to recognizing white-collar unions.[5] If employer-managers do realize the help which white-collar unionism can be to them, this would seem to occur only after it has been recognized and functioning in their organizations for some time and its effectiveness in this regard has been demonstrated.

[1] Keith Hindell, *Trade Union Membership* (London: Political and Economic Planning, 1962), p. 170.
[2] 'Recognition the Real Aim Behind Bank Unions Dispute With Employers', *The Times* (2 December 1963), p. 5.
[3] *Selected Written Evidence Submitted to the Royal Commission* (London: HMSO, 1968), p. 172.
[4] R. K. Kelsall, D. Lockwood, A. Tropp, 'The New Middle Class in the Power Structure of Great Britain', *Transactions of the Third World Congress of Sociology*, iii (1956), p. 322. See also Kenneth Prandy, *Professional Employees* (London: Faber, 1965), p. 147.
[5] See, for example, *supra*, pp. 128–9.

Nor did the major concessions of recognition to white-collar unions in private industry generally come about wholly or even primarily because of their industrial strength. The industrial strength of these unions was generally a factor in getting employers' associations to concede recognition, but it was rarely the most important factor. In fact, in only one instance—the Newspaper Society's recognition of NATSOPA for clerical workers in 1938—was the industrial strength of a union the major reason for its obtaining recognition, and even in this instance it was the strength of the manual membership rather than the white-collar membership which was crucial. In all other instances, employers' associations granted recognition to these unions long before they had sufficient strength to force the employers to do so, and generally even before the unions represented a substantial proportion of the employees concerned.

The Iron and Steel Trades Employers' Association recognized BISAKTA for staff grades and the British Spinners' and Doublers' Association recognized the Textile Officials' Association largely to encourage the growth of these unions at the expense of others which the employers considered to be less desirable. The NUJ, NATSOPA, and the NUPBPW obtained recognition from the newspaper proprietors as did DATA from the Shipbuilding Employers' Federation, and the CAWU, DATA, and ASSET from the Engineering Employers' Federation, largely as a result of government policies necessitated by war. Both world wars resulted in government policies which made it easier for unions to exert pressure for recognition and harder for employers to resist it. Finally, the NACSS, the AScW, and the NUGMW were recognized by the Engineering Employers' Federation almost solely because of the precedents it had established by recognizing the CAWU, DATA, and ASSET. In short, most white-collar union recognition in private industry has come about, directly or indirectly, as a result of government policies and the favourable climate they created for trade unionism.

There is not room here to give all the details, but it is clear that government action and the favourable climate which it produced were also the major factors bringing about the recognition of white-collar unions in the public sector of the economy. The bureaucratization of the civil service may have made unionization 'a virtual necessity with or without the accompaniment of Whitleyism'.[1] But there can be little doubt that the change in the government's negative attitude towards the unionization of its own employees was brought about after the First World War not because it suddenly became aware of the administrative advantages of trade unionism but because of the recommendations of the Whitley Reports. In Professor Clay's words, the government 'could hardly now refuse to adopt for itself the treatment it prescribed for other employers'.[2]

[1] Kelsall, *et al.*, loc. cit.
[2] Op. cit., p. 162. See also B. V. Humphreys, *Clerical Unions in the Civil Service* (Oxford: Blackwell, 1958), especially chaps. 5–8.

The government has also placed the industries which it has nationalized under a duty to recognize and bargain with appropriate trade unions, and this largely explains the recognition of white-collar unions in the coal-mining, road transport, civil air transport, electricity, and gas industries.[1] Whitleyism also helped NALGO to obtain limited recognition and negotiating rights from a few local authorities immediately after the First World War. But it was not until the Second World War and the passage of Order 1305 that NALGO was able to persuade all the local authorities of the wisdom of granting recognition and negotiating rights by referring several of them to the NAT.[2]

[1] See O. Kahn-Freund, 'Legal Framework', *The System of Industrial Relations in Great Britain*, op. cit., p. 54.

[2] See Alec Spoor, *White-Collar Union. Sixty Years of NALGO* (London: Heinemann, 1967), especially chaps. 9, 16, and 17.

X

CONCLUSIONS

OST of this study has been taken up with examining the relationship between the industrial and occupational pattern of white-collar
unionism and a range of factors which might conceivably affect this
pattern. Its findings can be briefly summarized. No significant relationship
was found between the growth of aggregate white-collar unionism and any
of the following factors: (*a*) such socio-demographic characteristics of white-
collar workers as their sex, social origins, age, and status; (*b*) such aspects of
their economic position as earnings, other terms and conditions of employment, and employment security; (*c*) such aspects of their work situation as the
opportunities for promotion, the extent of mechanization and automation,
and the degree of proximity to unionized manual workers; and (*d*) such
aspects of trade unions as their public image, recruitment policies, and
structures. While the evidence regarding some of these factors was not
sufficiently reliable to permit them to be discounted completely, it was
satisfactory enough to reveal that at most they have been of negligible
importance.

But the findings of this study are by no means entirely negative. It also
found that the growth of aggregate white-collar unionism was significantly
related to the following factors: employment concentration, union recognition, and government action. The relationship between these key independent
variables and between them and the dependent variable can be usefully summarized in a two-equation descriptive model.[1]

$$D = f(C, R) \tag{1}$$
$$R = g(D, G) \tag{2}$$

where D = the density of white-collar unionism;
$\quad\quad\;\; C$ = the degree of employment concentration;
$\quad\quad\;\; R$ = the degree to which employers are prepared to recognize unions
$\quad\quad\quad\quad$ representing white-collar employees; and
$\quad\quad\;\; G$ = the extent of government action which promotes union recog-
$\quad\quad\quad\quad$ nition.

The first equation specifies that the density of white-collar unionism is a

---

[1] The term 'model' is used here simply to mean 'a number of co-ordinated working
hypotheses which give a simplified and schematized picture of reality'. See Maurice Duverger,
Introduction to the Social Sciences (London: Allen & Unwin, 1964), pp. 243-4.

function of the degree of employment concentration and the degree to which employers are prepared to recognize unions representing white-collar employees. The more concentrated their employment the more likely employees are to feel the need to join trade unions because of 'bureaucratization', and the more easily trade unions can meet this need because of the economies of scale characteristic of union recruitment and administration. While employment concentration is a favourable condition for the growth of white-collar unions, it is not by itself sufficient. Employers must also be prepared to recognize these unions. The greater the degree to which employers are willing to do this the more likely white-collar employees are to join unions. This is because they are less likely to jeopardize their career prospects by joining, they can more easily reconcile union membership with their 'loyalty' to the company, and they will obtain a better service as their unions will be more effective in the process of job regulation.

But the degree to which employers are prepared to recognize unions representing white-collar employees is to some extent dependent upon the membership density of these unions. This is why the second equation is necessary. It specifies that the degree of recognition is a function of the density of white-collar unionism and the extent of government action which promotes union recognition. Employers generally do not concede recognition to a union before it has at least some membership in their establishments. The only exception to this is when employers recognize a union prior to it having obtained any membership in order to encourage its growth at the expense of other 'less desirable' unions. Even in these cases, recognition is at least partly a function of membership density—that of the 'less desirable' unions. But while a certain density of membership is a necessary condition for any degree of recognition to be granted, the findings of this study suggest that it is generally not a sufficient condition. The industrial strength of white-collar unions, as determined by the size of their membership and their willingness and ability to engage in industrial warfare, has generally not been sufficient in itself to force employers to concede recognition. This has also required the introduction of government policies which have made it easier for unions to exert pressure for recognition and harder for employers to resist it.

There are several respects in which this model might be thought to be incomplete. Bureaucratization and the density of white-collar unionism have been claimed to be interdependent; not only does bureaucratization encourage the growth of trade unions, but trade unions by demanding the standardization of working conditions are alleged to further bureaucratization. Inasmuch as bureaucratization is associated with employment concentration, this argument implies that employment concentration and the density of union membership are also interdependent. But the findings of this study suggest that the degree of interdependence between bureaucratization and unionization is very slight.[1] Employment concentration and the bureaucratization associated with it are primarily a function of the techniques of production and as such are exogenous to the industrial relations system. Even if the degree of interdependence between these variables is stronger than the findings of this study suggest, this could easily be allowed for by simply adding a third equation to the model:

$$C = h\,(D, T) \qquad\qquad (3)$$

where $T =$ the techniques of production.

Some writers have argued that white-collar unions are recognized as a result of the process of bureaucratization which makes recognition in the employer-managers' own self-interest. Inasmuch as bureaucratization is associated with employment concentration, this argument implies that white-collar union recognition is determined by employment concentration, and that the second equation should be rewritten as

$$R = g\,(D, G, C). \qquad\qquad (4)$$

But if employer-managers do realize the help white-collar unionism can be to them, this would seem to occur only after it has been recognized and functioning in their organizations for some time and its effectiveness in this regard has been demonstrated. For the findings of this study suggest that the degree to which employers are prepared to recognize white-collar unionism can be adequately explained by its density and the extent of government action which promotes recognition.[2]

Some of these same writers have also argued that the more opposed employers are to recognizing white-collar unionism, the more they will try to resist the administrative pressures which lead to bureaucratization and eventually to the unionization of their staff employees.[3] Inasmuch as bureaucratization is associated with employment concentration, this argument implies that the degree of employment concentration is determined by the extent to which employers are prepared to recognize unions representing white-collar employees, and that the third equation should be rewritten as

$$C = h\,(D, T, R). \qquad\qquad (5)$$

But while employers may be able to slow down the trend towards bureaucratization, the problems posed by governing large numbers of employees prevent them from reversing or even stopping this process. It thus seems safe to conclude that this refinement to equation three, like the equation itself, is superfluous.

Finally, it might be argued that the extent of government action which promotes union recognition is not an exogenous variable as the model suggests, but is determined by the industrial and political strength of the trade union movement, of which the density of union membership is a rough

---

[1] *Supra*, pp. 80–1.   [2] *Supra*, pp. 180–2.   [3] *Supra*, pp. 75–7.

quantitative index. Even granting this, the extent of such action by the government is still exogenous from the point of view of this model. For if the extent of government action is determined by the density of union membership, then it is determined by the density of *total* union membership and not by the density of *white-collar* union membership, the variable of concern to this model. The density of white-collar union membership comprises only a small part of the density of total union membership, especially in the period when the government action which promoted union recognition occurred, and it therefore seems safe to treat the latter as an exogenous variable from the point of view of this model.

It may even be that the extent of government action is also largely an exogenous variable from the point of view of the industrial relations system. This cannot be established very firmly, but the evidence gathered for this study suggests that at least the government policies which have promoted union recognition in Britain were not introduced because of pressure from the trade union movement.[1] In fact, the primary purpose of these policies was not to promote union recognition, but to deal with the social and economic exigencies created by world wars. In some instances these policies infringed the right to strike and the free movement of labour and were only most reluctantly agreed to by the trade union movement. In a sense, their favourable effect on union recognition was simply a by-product, and, in many cases, an unexpected by-product. In short, if the world wars had not occurred, it is most improbable that these policies would have been introduced, and, given the wars, some of them would have had to be introduced even if there had not been a trade union movement.

This two-equation descriptive model of the growth of aggregate white-collar unionism in Britain is therefore claimed to be complete: the number of equations is just enough to determine all the endogenous variables, given the exogenous variables. This does not mean that the model gives a 'complete' explanation of the growth of aggregate white-collar unionism. 'To attempt to account for the unique or even the rare event', as Moore has noted, 'is to set an impossibly high standard for theory'.[2] This study has therefore concentrated on the systematic and repetitive features of union growth rather than its exceptional or deviant aspects. All that is claimed on behalf of the model is that the variables it includes are those which have a systematic influence on aggregate union growth, while those it excludes behave in a random manner. If the equations in this model were to be estimated, they would both have to contain an error term which would represent not only the errors of measurement in the variables, but also the influence of the omitted variables which have a sporadic and unsystematic influence on union growth.

[1] *Supra*, Chap. 9.
[2] W. E. Moore, 'Notes for a General Theory of Labor Organization', *Industrial and Labor Relations Review*, xiii (April 1960), p. 387.

A model is generally constructed for the ultimate purpose of solving the equations simultaneously to obtain the values of the variables that are contained in them, thereby making a prediction. Unfortunately, given the nature of available statistical data and technique, not all the variables in this model can be satisfactorily quantified and the system of equations cannot be solved. While this may make the model less useful, it does not make it less valid. Although not practically quantifiable, the model is nevertheless conceptually quantifiable and operational in nature. It gives an adequate explanation of the growth of aggregate white-collar unionism in Britain, and, in addition, has some important implications for research on this subject as well as for the function of unions in modern industrial society, and for the future growth of white-collar unionism.

The model claims that the growth of aggregate white-collar unionism in Britain can be adequately explained by three strategic variables—employment concentration, union recognition, and government action. This in no way implies that other factors, including some of those discounted in this study, are not of importance in accounting for less aggregative patterns of union growth. For example, while the strategic variables may explain the existence of unionism *per se* among a given group of workers, which *particular* union is successful in organizing the group may be determined by union structures and recruitment policies. Similarly, the explanation of why one worker in a given environment joins a union while another worker in the same environment does not, may well be found in the different personality or attitude structures of the two individuals. But the strategic variables predominate, and unless they are held constant any explanation of these less aggregative patterns of union growth is likely to be obscured or distorted. Regrettably, most of the studies which have tried to ascertain the determinants of the individuals' propensity to unionize by means of attitude surveys have not controlled for these strategic variables.[1]

It is becoming increasingly fashionable to argue that with industrial progress, greater affluence, and more enlightened management, unions are losing their function. No less a social critic than John Kenneth Galbraith sees unions as having 'a drastically reduced function' and as being 'much less essential for the worker' in the modern industrial system.[2] Much of this argument assumes that the major, if not the only, function of trade unions is their

---

[1] See J. R. Dale, *The Clerk in Industry* (Liverpool: Liverpool University Press, 1962), chap. 4; Kenneth Prandy, *Professional Employees* (London: Faber, 1965), especially chaps. 5, 6, and 8; R. M. Blackburn, *Union Character Social Class* (London: Batsford, 1967), especially chap. 4; E. W. Bakke, 'Why Workers Join Unions', *Readings in Labor Economics and Industrial Relations*, Joseph Shister, editor (New York: J. B. Lippincott & Co., 1956), pp. 30–7; Joel Seidman, Jack London, and Bernard Karsh, 'Why Workers Join Unions', *Annals of the American Academy of Political and Social Science* (March 1951), pp. 75–84; and K. N. Vaid, 'Why Workers Join Unions', *Indian Journal of Industrial Relations*, i (October 1965), pp. 208–30.

[2] *The New Industrial State* (London: Hamish Hamilton, 1967), chap. 23.

ability to achieve economic benefits for their members. Even granting the highly controversial contention that unions possess such ability, the model seriously challenges this assumption. It suggests that white-collar workers value trade unions and join them not so much to obtain economic benefits as to be able to control more effectively their work situation.[1] As their employment becomes more concentrated and bureaucratized, individual white-collar workers find that they have less and less ability to influence the making and the administration of the rules by which they are governed on the job. In order to rectify this situation, they join trade unions and engage in collective bargaining. Given that employment concentration and bureaucratization will continue, trade unions will be just as necessary and useful to the white-collar workers of the twentieth century as they were to the 'sweated' manual workers of the nineteenth century.

The final implication of the model concerns the future growth of white-collar unionism. The model suggests that white-collar unions will continue to grow in the future as a result of increasing employment concentration, but that their growth will not be very great unless their recognition by employers is extended. The model also suggests that the strength of these unions will generally not be sufficient in itself to persuade employers to concede recognition; this will also require the help of the government. In short, the future growth of white-collar unionism in Britain is largely dependent upon government action to encourage union recognition.

The Government, following a recommendation of the Donovan Commission,[2] established a Commission on Industrial Relations in February 1969 which, among other things, is empowered to hear recognition disputes and to make recommendations for their settlement. In order to ensure that employers respect the Commission's rulings and bargain with unions in 'good faith', the Government also intends to give the Commission the power to recommend that a union should have the right of unilateral arbitration.[3] The CIR has got off to a somewhat uncertain start. But given that it creates an environment which makes it easier for unions to exert pressure for recognition and harder for employers to resist it, then the argument advanced in this study suggests that there will be a significant increase in the degree of white-collar unionism in Britain during the 1970s.

[1] On this point see Allan Flanders, 'Collective Bargaining: A Theoretical Analysis', *British Journal of Industrial Relations*, vi (March 1968), pp. 24–6.

[2] Report of the *Royal Commission on Trade Unions and Employers' Associations* (London: HMSO, 1968), chap. 5, and the White Paper, *In Place of Strife*, Cmnd. 3888, 1969.

[3] In order to allow the CIR to begin work without delay, it was established as a Royal Commission. Provisions to put it on a statutory basis and to give it the power described here will be included in an Industrial Relations Bill which the Government intends to present to Parliament as soon as possible.

APPENDIX A

NOTES TO TABLES AND FIGURES

TABLE 2.1

Table 2.1 is largely based on Guy Routh, *Occupation and Pay in Great Britain* (Cambridge: Cambridge University Press, 1965), pp. 4–5, table 1. His work, in turn, is based upon the *Census of Population* of Scotland and England and Wales for 1911, 1921, 1931, and 1951. For a detailed account of the methods Routh used to compile his table, see p. 6, n. 1, and Appendix A of his book.

The following modifications were made to Routh's figures to derive Table 2.1:

(*a*) In order to obtain a separate category for salesmen and shop assistants the following occupations were abstracted from the 1951 Census: 715, 730–741, 749, and 755. For 1921 and 1931 the comparable occupations were abstracted. In 1911 a distinction between employers and proprietors in the distributive sector of the economy was not always drawn. Consequently, they were subdivided on the basis of the 1921 ratio.

(*b*) The new salesmen and shop assistant category (with the exception of occupation number 755) was subtracted from Routh's semi-skilled group. Then this new semi-skilled category, the unskilled, and the skilled categories were added together to give the 'all manual workers' group.

(*c*) The 'insurance agent and canvassers' category (code number 755 in the 1951 Census) was subtracted from Routh's clerical workers to give the 'clerks' in Table 2.1. As mentioned in (*a*) above, occupation 755 was also included in the salesmen and shop assistant group.

(*d*) The figures in Table 2.1 do not always agree exactly with Routh's, due to rounding.

The occupations in the 1961 Census were classified as follows:

(*a*) *All manual workers.* The members of the following occupations were included in this group after subtracting all persons designated as 'employers and managers' and 'foremen and supervisors': 000–007, 010–015, 020–021, 030–034, 040–045, 050–056, 060–078, 080–085, 090–093, 100–108, 110–113, 120–124, 130–135, 140–143, 150–154, 160–161, 170–174, 180–188, 191, 193–211, 235, 250, 251 (officers–men divided according to 1951 ratio and men assigned to this group), 252, 254–264, 266–267, and 320–321 (officers–men divided on the 1951 ratio and men assigned to this group).

(*b*) *Salesmen and shop assistants.* The members of the following occupations were included in this group after subtracting all persons designated as 'employers and managers': 232, 233, 234, 237, and 239.

(*c*) *Clerks.* The members of the following occupations were included in this group after subtracting all persons designated as 'employers and managers': 220 and 221.

(*d*) *Foremen and inspectors.* The members of all manual occupations listed in (*a*) who were designated as 'foremen and supervisors' were included in this group. The members of all white-collar occupations designated as 'foremen and supervisors' were included with the occupations they supervised.

(*e*) *Lower professionals and technicians.* The members of the following occupations (including those designated as 'employers and managers' and 'foremen and supervisors') were included in this group: 190, 192, 265, 282–287, 294, 295, 310, and 312–314.

(*f*) *Higher professionals.* The members of the following occupations (including those designated as 'employers and managers' and 'foremen and supervisors') were included in this group: 280, 281, 288–293, 296–299, 311, and 320–321 (officers–men divided on 1951 ratio and officers assigned to this group).

(*g*) *Managers and administrators, employers and proprietors.* In 1961, as in 1931, both these groups were combined. The members of all occupations designated as 'employers and managers' (except those following an occupation in the higher or lower professional groups) were included in this category, plus all the members of the following occupations: 222–223, 230–231, 236, 238, 251 (officers–men divided according to 1951 ratio and officers assigned to this group), 253, and 270–278. This total group was then divided according to the 1951 ratio.

(*h*) As in previous years, the 'totally economically inactive' and the 'inadequately described occupations' were excluded from all occupational groups.

Thus all occupations which are generally considered to be non-manual have been included in the white-collar group except for the following marginal groups: fire-brigade officers, photographers, storekeepers, radio operators, and telephone and telegraph operators. Thus, if anything, the white-collar totals are slightly understated.

Table 2A.1 is the master table derived as described above and on which Table 2.1 and Table 2.3 are based.

TABLE 2.2

Table 2.2 was abstracted from the *Census of Population* of Scotland and England and Wales for 1921, 1931, 1951, and 1961. The 'scientists and engineers' are composed of the following 1961 occupations: 288, 289, 290, 291, 292, 297, 311; 'draughtsmen' are occupation 312 (and exclude industrial designers); 'laboratory technicians' are occupation 313. Comparable occupations were used for 1921, 1931, and 1951.

TABLE 2.3

Table 2.3 was derived from Table 2A.1.

TABLE 2.4

The source of Table 2.4 is 'The World's Working Population: Its Industrial Distribution', *International Labour Review*, lxxiii (May 1956), p. 508, table 3.

The primary sector comprises agriculture. The secondary sector comprises: mining and quarrying; manufacturing; building; gas, electricity, and water. The tertiary sector comprises transport and communications, distributive trades, public administration and defence, professional services, and miscellaneous services.

TABLE 2.5

The figures in Table 2.5 are from the following sources: (*a*) For 1907, *Final Report of the First Census of Production of the United Kingdom, 1907*, p. 12. To obtain the figures for manufacturing industries the following industries were subtracted from the total given in the above *Report*: mining and quarrying; clay, stone, building, and contracting trades; public utility services; and factory owners—power only. (*b*) For 1924 and 1930, *Final Summary Tables of the Fifth Census of Production,*

TABLE 2A.1

The Occupied Population of Great Britain by Major Occupational Group by Sex, showing the Number in Each Group as a Percentage of the Total Occupied Population, 1911–61

Occupational group	Males					Females					Total				
	1911	1921	1931	1951	1961	1911	1921	1931	1951	1961	1911	1921	1931	1951	1961
1. Employers and proprietors	1,000 (7.7)	1,048 (7.7)	1,129 (7.6)	894 (5.7)	907 (5.7)	232 (4.3)	270 (4.7)	278 (4.4)	223 (3.2)	232 (3.0)	1,232 (6.7)	1,318 (6.8)	1,407 (6.7)	1,117 (5.0)	1,139 (4.7)
2. All white-collar workers	2,409 (18.6)	2,556 (18.7)	3,109 (21.1)	4,006 (25.7)	4,705 (29.4)	1,024 (18.9)	1,538 (27.0)	1,732 (27.7)	2,942 (42.5)	3,775 (49.4)	3,433 (18.7)	4,094 (21.2)	4,841 (23.0)	6,948 (30.9)	8,480 (35.9)
(a) Managers and administrators	506 (3.9)	584 (4.3)	670 (4.5)	1,056 (6.8)	1,072 (6.7)	125 (2.3)	120 (2.1)	100 (1.6)	189 (2.7)	196 (2.6)	631 (3.4)	704 (3.6)	770 (3.7)	1,245 (5.5)	1,268 (5.4)
(b) Higher professionals	173 (1.3)	186 (1.4)	222 (1.5)	399 (2.6)	648 (4.1)	11 (0.2)	10 (0.2)	18 (0.3)	36 (0.5)	70 (0.9)	184 (1.0)	196 (1.0)	240 (1.1)	435 (1.9)	718 (3.0)
(c) Lower professionals and technicians	208 (1.6)	276 (2.0)	300 (2.0)	492 (3.2)	697 (4.4)	352 (6.5)	403 (7.1)	428 (6.8)	567 (8.2)	721 (9.4)	560 (3.1)	679 (3.5)	728 (3.5)	1,059 (4.7)	1,418 (6.0)
(d) Foremen and inspectors	227 (1.8)	261 (1.9)	295 (2.0)	511 (3.3)	612 (3.8)	10 (0.2)	18 (0.3)	28 (0.4)	79 (1.1)	70 (0.9)	237 (1.3)	279 (1.4)	323 (1.5)	590 (2.6)	682 (2.9)
(e) Clerks	654 (5.1)	696 (5.1)	758 (5.1)	932 (6.0)	1,045 (6.5)	178 (3.3)	560 (9.8)	646 (10.3)	1,409 (20.3)	1,951 (25.5)	832 (4.5)	1,256 (6.5)	1,404 (6.7)	2,341 (10.4)	2,996 (12.7)
(f) Salesmen and shop assistants	641 (5.0)	553 (4.1)	864 (5.9)	616 (4.0)	631 (3.9)	348 (6.4)	427 (7.5)	512 (8.2)	662 (9.6)	767 (10.0)	989 (5.4)	980 (5.1)	1,376 (6.5)	1,278 (5.7)	1,398 (5.9)
3. All manual workers	9,516 (73.6)	10,031 (73.6)	10,522 (71.3)	10,685 (68.6)	10,378 (64.9)	4,169 (76.8)	3,889 (68.3)	4,254 (67.9)	3,765 (54.3)	3,642 (47.6)	13,685 (74.6)	13,920 (72.0)	14,776 (70.3)	14,450 (64.2)	14,020 (59.3)
4. Total occupied population	12,925 (100.0)	13,635 (100.0)	14,760 (100.0)	15,585 (100.0)	15,990 (100.0)	5,425 (100.0)	5,697 (100.0)	6,264 (100.0)	6,930 (100.0)	7,649 (100.0)	18,350 (100.0)	19,332 (100.0)	21,024 (100.0)	22,515 (100.0)	23,639 (100.0)

Note: Numbers in brackets are percentages. All other numbers are in thousands.

1935, p. 11, table 4*b*. (*c*) For 1935 and 1949, *Censuses of Production for 1950, 1949, and 1948: Summary Tables*, part I, table 1. (*d*) For 1958, *The Report on the Census of Production for 1958*, part cxxxiii, table 3. (*e*) For 1963, 'Census of Production Results for 1963', *Board of Trade Journal* (24 December 1965), pp. 2–4.

All figures are for the United Kingdom (England, Wales, Scotland, and N. Ireland) except for 1907 when the whole of Ireland was included.

The value of these figures is enhanced by the rather consistent definition given to the white-collar and manual groups in the successive Censuses of Production. In 1958 'administrative, technical and clerical employees' (the white-collar group) were defined to include:

> managers, superintendents, and works foremen; research, experimental, development, technical and design employees (other than operatives); draughtsmen and tracers; travellers; and office (including works office) employees. For Great Britain, but not for N. Ireland, they include directors, other than those paid by fee only.

Working proprietors are also included in the white-collar group except for 1949. 'Operatives', or the manual group, was defined to include:

> all other classes of employees, that is, broadly speaking, all manual wage earners. They include those employed in and about the factory or works; operatives employed in power houses, transport works, stores, warehouses and, for 1958, canteens; inspectors, viewers and similar workers; maintenance workers; and cleaners. Operatives engaged in outside work of erection, fitting etc., are also included, but outworkers [i.e., persons employed by the firm who worked on materials supplied by the firm in their own homes, etc.] are excluded.[1]

The definitions used by the Ministry of Labour are the same as those used in the Censuses of Production.

From the viewpoint of the present study, the main difficulty with the Ministry of Labour-Census of Production definition of a white-collar employee is that it excludes all foremen except works foremen and includes all grades of managerial personnel. Nevertheless, the figures give a relatively good idea of the increasing importance of the white-collar group over time.

The figures in Table 2.5 differ slightly from those given in Seymour Melman, *Dynamic Factors in Industrial Productivity* (Oxford: Blackwell, 1956), p. 73, table 10. Periodically, the Board of Trade (Census of Production) issues revised figures for former years and these have been used here. Melman used the figures which appeared in the Census of Production for the year in question. The differences in percentage terms, however, are negligible.

Table 2A.2 gives the number of white-collar employees in manufacturing industries in Great Britain from 1948 to 1964. The figures in Table 2.5 are for the United Kingdom and are from unpublished Ministry of Labour data.

TABLE 2.6

The figures for 1948 and 1959 are from 'Administrative, Technical and Clerical Workers in Manufacturing Industries', *Ministry of Labour Gazette*, lxix (January 1961), p. 9. All the figures for 1948 are not given because the differences between

---

[1] 'Introductory Notes', *The Report on the Census of Production 1958*, part i, p. 11.

the 1948 and 1958 Standard Industrial Classification were so great in respect of these industries that all comparability was destroyed. Even for those industries for which a 1948 figure is given, there are small differences in classification between 1948 and 1959. These differences, however, did not affect the percentages to any significant extent in 1959 when the figures were given according to both classification systems, and it has been assumed that this was also true for 1948.

TABLE 2A.2

Employment[a] in Manufacturing Industries in Great Britain, 1948–64

Year	White-collar (000s)	Manual (000s)	Total (000s)	White-collar as a percentage of total (%)
1948	1,286	6,749	8,035	16·0
1949	1,351	6,887	8,238	16·4
1950	1,397	7,067	8,464	16·5
1951[b]	1,462	7,141	8,603	17·0
1952	1,538	6,957	8,495	18·1
1953	1,586	7,126	8,712	18·2
1954	1,647	7,303	8,950	18·4
1955	1,741	7,422	9,163	19·0
1956	1,811	7,334	9,145	19·8
1957	1,852	7,318	9,170	20·2
1958	1,898	7,052	8,950	21·2
1959[c]	1,950	7,205	9,155	21·3
1959	1,801	6,709	8,510	21·1
1960	1,878	6,932	8,810	21·3
1961	1,957	6,892	8,849	22·1
1962	1,979	6,763	8,742	22·6
1963	1,976	6,702	8,678	22·8
1964	2,031	6,774	8,805	23·1

[a] These figures exclude the unemployed.

[b] The figures for 1948–51 inclusive are for December of each year; the figures for 1952–64 inclusive are for October of each year.

[c] All figures prior to 1959 are classified according to the Standard Industrial Classification 1948, while all those after 1959 are classified according to the Standard Industrial Classification 1958. For 1959 the figures are given on both bases.

The figures for 1964 are from 'Administrative, Technical and Clerical Workers in Manufacturing Industries', *Ministry of Labour Gazette*, lxxii (July 1964), p. 291.

The figures for 1948 and 1959 are for the end of October; those for 1964 are for April.

The Ministry of Labour uses the same definition of white-collar employees as the Board of Trade in the Censuses of Production (see notes to Table 2.5).

TABLE 2.7

The figures for foremen in Table 2.7 were abstracted from the *Census of Population 1961*, Industry Table 5, for both Scotland, and England and Wales. Foremen are defined as 'employees (other than managers) who formally and immediately supervise others engaged in manual occupations, whether or not themselves engaged in such operations'.[1] There is a three-year difference between the figures for foremen

[1] *Classification of Occupations 1960* (London: HMSO, 1960), p. xii.

and all other occupations in the table. This is unfortunate, but the 1961 figures for foremen are the only ones available. While the category has undoubtedly expanded since 1961, it is assumed that the relative importance of foremen between industries has remained the same.

The figures for all other occupations were derived from 'Occupations of Employees in Manufacturing Industries', *Ministry of Labour Gazette* (December 1964), pp. 492–502; (January 1965), pp. 11–19. These figures are for 16 May 1964 and relate to all firms with eleven or more employees in manufacturing in Great Britain. The details of the sampling techniques are given on p. 492. The Ministry of Labour's 'managers, work superintendents, departmental managers' group was excluded from Table 2.7.

All figures exclude the self-employed and the unemployed.

The occupational groups are defined as follows (all definitions taken from instructions to employers who were requested to complete the questionnaire forms):

(*a*) Scientists and technologists 'include persons engaged on, or being trained for, technical work for which the normal qualification is a university degree in science or technology and/or membership of an appropriate professional institute (e.g., A.M.I.Mech.E.)'. Managers and technical directors possessing such qualifications are excluded.

(*b*) All technicians 'include persons carrying out functions of a grade intermediate between scientists and technologists on the one hand and skilled craftsmen and operatives on the other, whether in research or development, production, testing, or maintenance'. Some of the main job titles are: draughtsmen, laboratory technicians, service engineers, production planners, testers and inspectors, technical writers, work-study specialists, and others.

(*c*) The draughtsmen category includes all people so designated. It does not include tracers who are included in the 'other white-collar worker' group.

(*d*) Other technicians include all the occupations specified in (*b*) except for draughtsmen.

(*e*) Clerks 'include shorthand typists, typists, office-machine operators, automatic data programmers, telephone operators, etc.'.

(*f*) Other white-collar workers 'include all other administrators, technical and commercial staff not included above, e.g., personnel and welfare assistants, occupational health nurses, safety officers, management trainees, tracers, salesmen and representatives, etc.'. In the paper, printing, and publishing industry this category also includes, in addition to the above, salaried reporters and journalists (not those working on a freelance basis), press photographers, circulation travellers, and advertising representatives.

The figures for shipbuilding and for marine engineering were given separately by the Ministry of Labour and no occupational breakdown of the administrative, technical, and clerical workers' category was given for shipbuilding (see table 6). Consequently, this over-all figure for shipbuilding was distributed on the same basis as the occupational distribution in marine engineering, and then the two sets of figures were added together to give Order VII, Shipbuilding and Marine Engineering.

For the printing and publishing sub-industry (see table 17) the 'designers and typographers' category was omitted as it was considered that they were manual employees and not covered by the scope of any of the unions under consideration.

The labour force figures relate to private employment except that Royal Ordnance Factories and railway workshops are classified under Engineering and Electrical Goods. The figures for foremen in Shipbuilding and Marine Engineering include those in Royal Navy Dockyards; employees in all other white-collar occupations in this industry are excluded if in Royal Naval Dockyards.

The actual numbers for Table 2.7 are given in Table 2A.3.

TABLE 2.8

Table 2.8 is based on the same sources as Table 2.7. The actual number of females in each white-collar occupation is given by Table 2A.4. The percentages in Table 2.8 were obtained by merely taking a figure in Table 2A.4 as a percentage of the corresponding figure in Table 2A.3.

The number of females in each white-collar occupation in Shipbuilding and Marine Engineering was determined in the same manner as total employment in each white-collar occupation in this industry (see *supra*, p. 194).

Obviously, the number of male white-collar workers can be obtained by merely subtracting Table 2A.4 from Table 2A.3.

TABLE 3.1

The trade union membership figures are those published annually by the Ministry of Labour. See, for example, 'Membership of Trade Unions in 1964', *Ministry of Labour Gazette*, lxxiii (November 1965), pp. 480–1. The figures from 1955–64 are only provisional and are subject to revision as additional information becomes available. For each year the latest revised figure was used.

These membership figures 'relate to all organisations of employees . . . which are known to include in their objects that of negotiating with employers with a view to regulating the wages and working conditions of their members'.[1] They thus include all trade unions and staff associations, whether they be registered or unregistered, affiliated or unaffiliated to the TUC, whose headquarters are situated in the United Kingdom. More specifically, they include all unions listed in the *Directory of Employers' Associations, Trade Unions, Joint Organisations, Etc., 1960*. Unfortunately, the figures also include the membership of British unions located in branches in the Irish Republic and overseas as well as members serving with H.M. Forces. Total union membership at the end of 1964 included 49,000 members in the Irish Republic and 89,000 in other branches outside the United Kingdom. It is not possible to adjust the revised total membership figures because the non-United Kingdom membership figures are not published on the revised basis. In any case, the numbers involved are relatively small and do not significantly affect the density figures.

The labour force figures for 1959–64 are from the 'Number of Employees (Employed and Unemployed) June 1964', *Ministry of Labour Gazette*, lxxiii (February 1965), pp. 61 and 64. The figures for 1948–58 were supplied from unpublished data and are comparable with the 1959–64 series. The figures exclude employers, self-employed, and members of the armed forces, but include the unemployed.

The labour force figures for 1891, 1901, 1911, 1921, 1931, 1933, and 1938 were derived as follows. The total occupied population of the United Kingdom was estimated by Professor Bowley for the above years as being 15,783,000, 17,648,000,

---

[1] *Ministry of Labour Gazette*, lxxiii (November 1965), p. 480.

TABLE 2A.3

Total White-Collar Employment in Manufacturing Industries in Great Britain, 1964

Occupational group	Food, drink, tobacco	Chemical	Metal manuf.	Metal N.E.S.	Eng. elect.	Ship, M.E.	Vehicles	Textiles	Leather, fur	Clothing, footwear	Bricks, etc.	Timber, furn, etc.	Paper, print, pub.	Other manuf.	All manuf.
1. Foremen	28,850	22,380	24,400	17,870	67,910	9,800	29,450	34,030	2,180	13,410	12,060	10,130	16,650	12,390	301,510
2. All scientists, technologists, technicians	7,560	43,030	19,630	10,110	156,700	7,516	51,800	11,680	370	2,540	6,660	2,230	4,670	7,620	332,116
(a) Scientists, technologists	2,770	17,100	4,950	1,610	33,220	570	7,000	3,000	150	200	1,800	50	1,560	2,080	76,060
(b) All technicians	4,790	25,930	14,680	8,500	123,480	6,946	44,800	8,680	220	2,340	4,860	2,180	3,110	5,540	256,056
(i) Draughtsmen	1,150	3,010	4,440	5,430	65,160	5,114	19,820	1,580	10	310	2,320	1,420	490	1,800	112,054
(ii) Other technicians	3,640	22,920	10,240	3,070	58,320	1,832	24,980	7,100	210	2,030	2,540	760	2,620	3,740	144,002
3. Clerks	84,380	74,000	59,020	55,830	275,050	10,004	101,770	50,820	5,590	32,610	28,770	27,660	80,940	34,290	920,734
4. Other white-collar workers	38,190	30,550	16,640	14,990	90,440	4,306	27,670	11,230	1,170	10,190	8,630	5,940	31,950	15,480	307,376
5. All white-collar workers	158,980	169,960	119,690	98,800	590,100	31,626	210,690	107,760	9,310	58,750	56,120	45,960	134,210	69,780	1,861,736

TABLE 2A.4

Total Female White-Collar Employment in Manufacturing Industries in Great Britain, 1964

Occupational group	Food, drink, tobacco	Chemical	Metal manuf.	Metal N.E.S.	Eng. elect.	Ship M.E.	Vehicles	Textiles	Leather, fur	Clothing, footwear	Bricks, etc.	Timber, furn., etc.	Paper, print, pub.	Other manuf.	All manuf.
1. Foremen	5,520	2,010	360	1,510	4,330	20	390	4,340	360	7,370	850	400	2,740	1,660	31,860
2. All scientists, technologists, technicians	1,240	4,560	680	320	3,940	180	1,150	2,150	20	900	310	140	330	640	16,560
(a) Scientists, technologists	330	1,190	140	80	720	..	80	250	10	70	100	..	90	50	3,110
(b) All technicians	910	3,370	540	240	3,220	180	1,070	1,900	10	830	210	140	240	590	13,450
(i) Draughtsmen	..	10	40	70	1,310	128	280	10	..	110	30	30	20	20	2,058
(ii) Other technicians	910	3,360	500	170	1,910	52	790	1,890	10	720	180	110	220	570	11,392
3. Clerks	57,810	50,720	31,370	37,740	170,240	4,398	49,430	33,610	4,690	26,950	17,190	18,850	51,590	23,190	577,778
4. Other white-collar workers	5,130	4,450	2,810	3,680	18,240	1,137	5,290	2,900	40	4,730	1,390	760	4,800	3,330	58,687
5. All white-collar workers	69,700	61,740	35,220	43,250	196,750	5,735	56,260	43,000	5,110	39,950	19,740	20,150	59,460	28,820	684,885

19,615,000, 19,840,000, 21,620,000, 21,810,000, and 22,660,000, respectively.[1] To obtain the union potential, it was necessary to exclude the employers, self-employed, and the armed forces from the above figures. The number of employers and self-employed in Great Britain were obtained from an analysis of census data undertaken by Dr. Routh,[2] and the number in the armed forces was obtained directly from the Censuses of Population. The total of such occupational groups was 1,921,000 in 1911, 2,163,000 in 1921, and 2,238,000 in 1931. Out of the total gainfully occupied population of 18,347,000 in 1911, 19,333,000 in 1921, and 21,029,000 in 1931, the employers, self-employed, and armed forces represented 10·5 per cent, 11·2 per cent, and 10·6 per cent in the respective years. Professor Bowley's estimates of the total occupied population of the United Kingdom were then reduced by the same proportions. The 1911 ratio was used to reduce the 1891, 1901, and 1911 totals;[3] the 1921 ratio was used to reduce the 1921 total; and the 1931 ratio was used to reduce the 1931, 1933, and 1938 totals to give the figures shown in Table 3.1. These figures, like the 1948–64 series, include the number of unemployed.

TABLE 3.2

The employment figures for 1948 are based upon those appearing in 'The Employed Population, 1948–1952', *Ministry of Labour Gazette*, lxi (February 1953), pp. 39–47, and those for 1964 are from the 'Number of Employees (Employed and Unemployed) June 1964', *Ministry of Labour Gazette*, lxxiii (February 1965), pp. 59–64. The Standard Industrial Classification was changed between 1948 and 1964, and some of the 1948 figures had to be adjusted to make them even broadly comparable to the 1964 figures.[4]

The figures for the various industries were obtained in the following manner (the industries which are not mentioned below were considered to be broadly comparable and were simply abstracted unaltered from the above sources):

(a) *Professional and business services.* In 1948 this category was derived by subtracting 'Education' from 'Professional Services', and in 1964 by subtracting 'Educational Services' from 'Professional and Scientific Services'.

(b) *Distribution.* The 1948 figure was obtained by adding 'Wholesale Bottling' to 'Distributive Trades'.

(c) *Metals and engineering.* The 1948 figure was obtained by adding together 'Metal Manufacture', 'Engineering, Shipbuilding, and Electrical Goods', 'Vehicles', 'Metal Goods N.E.S.', 'Jewellery, Plate and refining of precious Metals', 'Scientific, Surgical and Photographic Instruments, etc.', 'Manufacture and Repair of Watches and Clocks', and then subtracting 'Motor Repairers and Garages'. The 1964 figure was derived by adding together the following: 'Metal Manufacture', 'Engineering

---

[1] The figures for 1891, 1901, and 1911 are from A. L. Bowley, *Wages and Income in the United Kingdom Since 1860* (Cambridge: Cambridge University Press, 1937), pp. 134–5. The figures for the other years are from his *Studies in the National Income 1924–1938* (Cambridge: Cambridge University Press, 1944), p. 56.

[2] Guy Routh, op. cit., pp. 4–5, table 1.

[3] The total number of employers and self-employed are not conveniently available in the 1891 and 1901 Censuses and hence the 1911 ratio was used for these years. Moreover, union membership figures only became available in 1892. No labour force figure is available for that year so the 1891 figure had to be used.

[4] For a brief summary of the major changes in the system of industrial classification see 'Standard Industrial Classification', *Ministry of Labour Gazette*, lxvii (February 1959), p. 55.

and Electrical Goods', 'Shipbuilding and Marine Engineering', 'Vehicles', and 'Metal Goods N.E.S.'.

(d) *Food, drink and tobacco.* The 1948 figure was obtained by subtracting 'Wholesale Bottling' from 'Food, Drink and Tobacco'.

(e) *Other transport and communication.* For both years this is simply 'Transport and Communication' minus 'Railways'.

(f) *Theatres, cinemas, sport, etc.* The 1948 figure was derived by adding together 'Theatres, Cinemas, Music Halls, Concerts, etc.', and 'Sport, Other Recreations, and Betting'. The 1964 figure was obtained by adding together 'Cinemas, Theatres, Radio, etc.', 'Sports and Other Recreations', and 'Betting'.

(g) *Furniture, timber, etc.* The 1948 figure is the one which appears for 'Manufactures of Wood and Cork'.

(h) *Footwear.* The 1948 figure is the one which appears for the 'Manufacture of Boots, Shoes, Slippers and Clogs (exc. rubber)'.

(i) *Clothing.* The 1948 figure was obtained by subtracting the 'Manufacture of Boots, Shoes, Slippers and Clogs (exc. rubber)' and 'Repair of Boots and Shoes' from 'Clothing'. The 1964 figure was derived by subtracting 'Footwear' from 'Clothing and Footwear'.

(j) *Textiles other than cotton.* The 1948 figure was obtained by subtracting 'Cotton Spinning, Doubling, etc.' and 'Cotton Weaving, etc.' from 'Textiles'. The 1964 figure was derived by subtracting 'Spinning and Doubling of Cotton, Flax and Man-Made Fibres' and 'Weaving of Cotton, Linen and Man-Made Fibres' from 'Textiles'.

(k) *Cotton.* The 1948 figure was obtained by adding together 'Cotton Spinning, Doubling, etc.', and 'Cotton Weaving, etc.'. The 1964 figure was derived by adding together 'Spinning and Doubling of Cotton, Flax and Man-Made Fibres' and 'Weaving of Cotton, Linen and Man-Made Fibres'.

In spite of the adjustments which have been made to the 1948 figures, there are still some differences in coverage between the figures of 1948 and 1964. Consequently, the percentage change in employment between 1948 and 1964 is subject to some error and should be used cautiously.

Most of the industry headings are self-explanatory regarding their scope. The ones which require special mention are:

(a) *Professional and business services.* This category includes accounting, legal, religious, and other services, as well as medical and dental services, i.e., personnel employed by the National Health Services.

(b) *Other transport and communication.* This category includes road, air, sea, and inland water transport, as well as the postal services and telecommunications.

The union density figures for 1960 were obtained from Keith Hindell, *Trade Union Membership* (London: Political and Economic Planning, 1962), p. 191, table 7. Hindell also gives (p. 156, table 1) the density figures for 1958 and the change in density between 1948 and 1958. It is therefore possible to determine what the actual density figure was in 1948 and then calculate the change in density between 1948 and 1960.

The method used by Hindell to calculate these density figures is given in the notes to table 1, p. 156, and also in the Appendix, pp. 199–200, of the above publication. These density figures are little more than rough approximations. In preparing their industrial analysis of union membership, the Ministry of Labour allocates the total membership of a union to the industry in which the majority of its members are employed. In the case of industrial unions such as **BISAKTA** and the National Union of Mineworkers, for example, this procedure involves little error. In the case of other unions such as the Amalgamated Engineering Union and the Electrical Trades Union, this procedure involves considerable error. The Ministry of Labour allocates the total membership of these unions to the Metals and Engineering industries, but thousands of their members are employed outside this group of industries. Although Hindell has been able to remove some of the inaccuracies inherent in this procedure, it is not possible in most cases to correct the figures as the unions do not prepare detailed industrial classifications of their membership. The percentage change in density between 1948 and 1960 is subject to additional error because for 1948 Hindell used the labour force figures published in 1949 rather than the very much revised figures published in 1953.[1] Nevertheless, Hindell's figures are the most detailed and reliable ones published.[2]

TABLE 3.3

The membership figures of non-TUC unions were obtained from the files of the Ministry of Labour.

The membership figures of the purely white-collar unions affiliated to the TUC were abstracted from the TUC's *Annual Reports*. In general, these figures give a fairly reliable idea of the size of the various unions. Some unions, however, do not always affiliate their actual membership to the TUC; for various reasons they affiliate more or less than the true total.

The white-collar membership of the partially white-collar unions was obtained from the the unions concerned. Only those partially white-collar unions which keep a record of their white-collar membership or could give a fairly reliable estimate of its size were included in this list. Nevertheless, the list includes all the major partially white-collar unions. In general, these figures slightly understate the actual totals. In some areas where there are very few white-collar members, these members will be placed in manual branches and no separate record kept of them. In addition, although many unions have separate supervisory branches, some newly promoted supervisors prefer to retain their membership in their old branch on grounds of class loyalty, tradition, etc. The totals of the Amalgamated Engineering Union and the Electrical Trades Union are likely to be particularly understated for the above reasons.

The membership figures for the purely and partially white-collar unions which are affiliated to the TUC are given in Table 3A.1, Parts A and B. The figures in Part C of Table 3A.1 were derived as follows:

(*a*) *Line 1*. This line was obtained by summing all the figures, except those in brackets, in each column of Part A of the table.

(*b*) *Line 2*. This line was obtained by summing all the figures, including the ones in brackets, in each column of Part A of the table.

(*c*) *Line 3*. The 1964 figure was obtained by summing all the figures in the 1964 column of Part B of the table. The 1955 and 1948 figures were obtained by assuming

---

[1] Cf. 'Estimated Number of Employees Insured Under the National Insurance Schemes at Mid-1948', *Ministry of Labour Gazette*, lvii (February 1949), p. 44, and 'The Employed Population, 1948–1952', *Ministry of Labour Gazette*, lxi (February 1953), pp. 44–7.

[2] Cf. Guy Routh, 'Future Trade Union Membership', *Industrial Relations: Contemporary Problems and Perspectives*, B. C. Roberts, editor (London: Methuen, 1962), p. 72, table 4.

that the white-collar membership of the partially white-collar unions grew at the same rate as the adjusted membership of the purely white-collar unions. In other words, the 1955 figure is 21·4 per cent less than the 1964 figure, and the 1948 figure is 36·2 per cent less than the 1964 figure. If anything, these figures understate the growth of the partially white-collar unions. It can be seen from Part B of the table that the white-collar membership of most of the partially white-collar unions for which growth figures are available grew by amounts substantially in excess of 36·2 per cent.

(*d*) *Line 4*. Since the number of partially white-collar unions affiliated to the TUC between 1948 and 1964 remained unaltered, line 4 is the same as line 3.

(*e*) *Line 5*. These figures were abstracted from the TUC's *Annual Report* for each year.

(*f*) *Line 6*. This line was obtained by adding line 2 to line 8.

(*g*) *Line 7*. This line was obtained by subtracting line 1 from line 5.

(*h*) *Line 8*. This line was obtained by adding to line 7 the membership figures of those manual unions which affiliated to the TUC since 1948 for the years they were unaffiliated; plus the membership figures of those unions which were temporarily expelled during the period (except the ETU) for the years they were expelled; minus the membership figures of those unions which left or were expelled during the period and have not reaffiliated. The membership figures of the following manual unions were added to line 7 for the years indicated (all dates inclusive): Card Setting Machine Tenters' Society, 1948; Amalgamated Union of Sailmakers, 1948–51; Nottingham and District Hosiery Finishers' Association, 1948–53; Leicester and Leicestershire Hosiery Trimmers and Auxiliary Association, 1948–54; Screw, Nut, Bolt and Rivet Trade Society, 1948–54; Watermen, Lightermen, Tugmen and Bargemen's Union, 1948–55; Northern Carpet Trades Union, 1948–57; National Union of Waterworks Employees, 1948–62; and the National Engineers' Association, 1955–7. The membership figures of the following manual unions were subtracted from the above total for the years indicated: National Amalgamated Association of Nut and Boltmakers, 1948–53; Nelson and District Preparatory Workers' Association, 1948–53; National Amalgamated Stevedores and Dockers, 1948–57. The figures of all these unions were obtained either from the union concerned or from the Registrar of Friendly Societies.

(*i*) *Line 9*. This line was obtained by adding line 1 and line 3.

(*j*) *Line 10*. This line was obtained by adding line 2 and line 4.

(*k*) *Line 11*. This line was obtained by subtracting line 9 from line 5.

(*l*) *Line 12*. This line was obtained by subtracting line 10 from line 6.

(*m*) *Line 13*. This line was obtained by taking line 1 as a percentage of line 5.

(*n*) *Line 14*. This line was obtained by taking line 10 as a percentage of line 5.

Since Table 3A.1 was prepared, five more purely white-collar unions have affiliated to the TUC: the Society of Telecommunication Engineers (7,154), the Customs and Excise Preventive Staff Association (2,600), the County Court Officers Association (4,725), the Association of Teachers in Technical Institutions (25,000), and the Prison Officers' Association (10,313).

TABLE 3.4

The union membership figures used in calculating the densities in this table came from the same sources as those used in Table 3.3.

The labour force figures for 1948 and 1964 are from the same source as those used in Table 3.1 and have been broken down into manual and white-collar categories using the proportion of white-collar and manual employees in the total occupied population (excluding employers) as given by the 1951 and 1961 *Census of Population*.

TABLE 3.5

Table 3.5 is largely an abridged version of Table 3.2. The derivation of the manufacturing figure is explained in the notes to Table 3.8.

Although the labour force of the 'Professional and Business Services' sector is predominantly white-collar, it was omitted from Table 3.5 because the Ministry of Labour classified both the AScW and the CAWU under this heading. As is shown in Table 3.6, however, a substantial proportion of the membership of both these unions was situated in manufacturing industries. Thus to have included this industry group along with the manufacturing group would have resulted in a considerable amount of double counting.

TABLE 3.6

All membership figures in this table are for the year 1963. The total membership figures of the AScW, ASSET, the CAWU, DATA, and the NUJ, and the total white-collar membership figures of BISAKTA, NATSOPA, and the TGWU are from Table 3.7. The total membership figure of BISAKTA is from its *Quarterly Report*, 31 March 1964, p. 7, and is its total contributing membership minus its retired members; NATSOPA's is from its *Annual Report*, 1964, p. 17; and the TGWU's is from its *Report and Balance Sheet*, 1963, p. 7.

The occupational and industrial composition of the unions' white-collar memberships was determined by the survey described in the notes to Table 3.8.

TABLE 3.7

All the membership figures in Table 3.7, with the exception of BISAKTA's, were supplied by the unions concerned. In all cases, they are the contributing or paying membership figure and generally exclude non-contributory and retired members. The figures in this table are more accurate than the ones given in Table 3A.1. As was explained in the notes to the latter table, unions do not always affiliate their true membership to the TUC.

The white-collar membership of BISAKTA was derived by analysing its quarterly list of individual branch membership. The list of the branch audits for December 1964 appear in the *Quarterly Report*, 31 March 1965, pp. 23–5. There are similar lists in earlier Reports. For 1964 it was possible to obtain a list of all the white-collar branches and to abstract their membership totals from the quarterly branch audits. This list was also used to help identify white-collar branches in earlier years. In any case, virtually all BISAKTA's white-collar membership is in separate branches and most of the branches are designated in such a way that they indicate the type of personnel covered, e.g., 'Britannia Clerical', 'Tees-side Chemists', etc. Not all the branches are audited every quarter. For each branch the audited figure closest to December of each year was used. Sample passers were included as white-collar, ambulance drivers were not. Although the annual totals are obviously subject to some error, they should nevertheless give a reliable idea of the growth of BISAKTA's white-collar membership between 1948 and 1964.

The figures of all the unions, except ASSET, are for December of each year. ASSET's figures are for March of each year and are therefore slightly understated relative to the others. This does not affect the relative size of its increase over the period.

TABLES 3.8 AND 3.9

The labour force statistics used to calculate the density figures of Table 3.8 are the ones which appear in Table 2A.3, and they refer to May 1964. A geographical analysis of these figures is not available.

The labour force figures used to calculate the density figures of Table 3.9 are from unpublished Ministry of Labour data sheets (H.Q.W. 474-200 4/64 EC). These data sheets give a geographical breakdown of the industrial and occupational analysis of the labour force which was published as: 'Occupations of Employees in Metal Manufacture, Engineering and Electrical Goods, Vehicles and Metal Goods', *Ministry of Labour Gazette*, lxxi (December 1963), pp. 474–80; and 'Occupations of Employees in Manufacturing Industries (Other than the Metal Group of Industries)', *Ministry of Labour Gazette*, lxxii (April 1964), pp. 132–42. The figures in these articles refer to May 1963 and exclude all employers, self-employed, and unemployed. They conform to the standard Ministry of Labour definition of a white-collar employee and include managers but exclude all foremen (except works foremen).

Unfortunately, the May 1963 survey does not include a regional analysis of the Shipbuilding labour force and, in addition, it includes the figures for Marine Engineering with those of Engineering and Electrical Goods. Thus in order to obtain comparable manpower statistics, the white-collar labour force of Marine Engineering had to be subtracted from that of Engineering and Electrical Goods (including Marine Engineering), and a geographical breakdown of the white-collar labour force of Shipbuilding and Marine Engineering had to be obtained. This was done in the following manner. In the May 1964 survey[1] the white-collar labour force of Marine Engineering was given separately as 10,240. This figure was distributed according to the regional distribution of *total* employment in the Shipbuilding and Marine Engineering industry in the United Kingdom as given by the Census of Production.[2] Then the regional white-collar labour force totals obtained by this procedure were subtracted from the Engineering and Electrical Goods (including Marine Engineering) regional white-collar labour force totals to leave the category Engineering and Electrical Goods. The May 1964 survey also gives the total white-collar labour force for the combined industrial order, Shipbuilding and Marine Engineering, as 26,890. This figure was distributed regionally using the same geographical distribution of total employment described above. The figure for Shipbuilding and Marine Engineering in Wales is not given because the Census of Production does not give the percentage of total employment in this industry in Wales as this would disclose information about individual establishments.

The scope of the regions is given by 'Definition of Standard Regions', *Ministry of Labour Gazette*, lxxiii (January 1965), p. 5. The Ministry of Labour's 'Midlands' and 'Yorkshire and Lincolnshire' regions are combined in Table 3.9. The union membership figures were originally classified according to the Standard Regions of the United Kingdom used by the Census of Production and the Census of Population.[3] At a later stage in the research it became necessary and desirable to use the manpower

[1] 'Occupations of Employees in Manufacturing Industries', *Ministry of Labour Gazette* (December 1964), pp. 492–502; (January 1965), pp. 11–19.
[2] *Report on the Census of Production 1958* (London: HMSO, 1963), table 8, pp. 134/8–134/9.
[3] Ibid., part 134, pp. 129–31; and *Census of P nulation 1961* Industry Tables, pp. x–xiii.

statistics of the Ministry of Labour. Unfortunately, the Ministry uses a different system of regional classification and the only way the two systems could be rationalized was to combine the union membership figures in the manner shown in Table 3.9. If the two systems of classification are compared, the need for the regional groupings used in Table 3.9 will become obvious.

The Ministry of Labour's system of regional classification and the manner in which it has had to be modified obscures much of the regional variation in union density. By grouping low-density areas with high-density areas (e.g., east midlands with west midlands and north Wales with south Wales) and by not giving separate labour force figures for the major conurbations, extremely broad and general average density figures result. The geographical analysis of May 1963 gives an occupational breakdown (except for Shipbuilding and Marine Engineering) but because of sampling difficulties the Ministry of Labour advised against using it.

The actual geographical labour force figures used in Table 3.9 are given by Table 3A.2.

In general, the union membership figures are for 1 January 1964 and exclude all retired and non-contributory members. The figures relate to Great Britain.

Unions keep their membership records in the most administratively convenient manner rather than in accordance with the official systems of industrial and regional classification. Consequently, in order to get membership figures which would be comparable with the government's labour force statistics, it was necessary to undertake a detailed analysis of the membership records of the major unions catering for white-collar workers in manufacturing industries. In general, this analysis was carried out in the following manner: (a) A list of the branches in the union and the names and addresses of the firm or firms covered by each branch were obtained from the union. (b) The membership in each firm or in each branch (if the branch covered only one firm) was determined. (c) The firm was then checked in *Kelly's Directory of Merchants, Manufacturers and Shippers* (Kingston-upon-Thames, Surrey: Kelly's Directories Ltd., 1963) to determine the type of product manufactured. This was checked against information supplied by the union. (d) The type of product manufactured was then located in the *Standard Industrial Classification: Alphabetical List of Industries* (London: HMSO, 1959) and the industry to which it was assigned was noted. (e) Finally, the membership of the firm was assigned to the industry as determined in (d) and the region as determined by its address in (a).

The specific details of the manner in which the membership of each of the major unions was regionally, industrially, and occupationally classified are given below:

(a) *AScW*. The AScW's membership records were decentralized. Each of the regional offices was visited and the membership files were analysed according to the general procedure described above. The AScW catered for two broad categories of personnel: qualified scientists and engineers, and laboratory technicians. The former were classified as Section I members and the latter as Section II members. The distinction between Section I and Section II members conforms very closely to the distinction the Ministry of Labour draws in its manpower statistics between 'scientists and technologists' and 'other technicians'.[1] The union stopped keeping an exact count of the two categories of membership in 1955 when Section I members comprised 31·4 per cent of the total. At that time the proportion of Section I members in the union was declining, and the AScW estimated that by 1964 only 25 per cent of

---

[1] Cf. notes to Table 2.7 and *AScW Rules*, 1964, p. 2, rule 5.

TABLE 3A.2

Total White-Collar Employment in Manufacturing Industries in Great Britain by Industry and by Region, 1963

Region	Food, drink, tobacco	Chemical	Metal manuf.	Metal N.E.S.	Eng. elect.	Ship. M.E.	Vehicles	Textiles	Leather, fur	Clothing, footwear	Bricks, etc.	Timber, furn., etc.	Paper, print, pub.	Other manuf.	All manuf.
1. Scotland	15,560	10,320	11,110	4,450	41,731	6,561	9,490	11,360	220	2,970	3,510	3,940	12,340	5,310	138,872
2. Northern	5,140	15,070	8,290	2,260	25,497	5,916	2,220	2,150	480	2,970	2,850	2,010	2,490	2,660	80,003
3. North-west	23,160	39,720	6,860	11,810	77,437	3,657	27,870	28,600	2,390	11,030	9,570	4,220	14,970	12,000	273,294
4. East and West Ridings of Yorkshire and East and West Midlands	33,890	26,940	59,550	53,380	152,119	1,076	74,410	47,640	3,530	19,590	22,100	10,990	20,020	16,210	541,445
5. Wales	4,320	5,300	17,720	3,470	11,200	n/a	2,700	4,320	380	1,220	1,040	1,360	2,180	2,170	57,380
6. Eastern and Southern	18,120	20,430	4,830	6,590	79,823	3,011	38,640	3,850	330	8,650	6,470	7,810	16,450	5,970	220,974
7. London and South-east	38,270	51,190	9,750	22,640	193,730	2,205	34,150	7,740	2,760	18,570	13,590	18,270	67,730	21,260	501,855
8. South-western	12,790	3,560	1,500	2,030	23,353	1,882	23,870	2,580	1,050	3,440	2,390	3,400	7,510	3,730	93,085
9. ALL REGIONS	151,250	172,530	119,610	106,630	604,890	24,308	213,350	108,240	11,140	68,440	61,520	52,000	143,690	69,310	1,906,908

its total membership was in Section I. Consequently, after the industrial and geographical analysis of its membership was completed, 25 per cent of the AScW's membership was assigned to the 'scientists and technologists' category and the remainder to the 'other technicians' group of Table 3A.3.

(b) *ASSET*. ASSET supplied a list of firms in each of its branches and the membership in each firm. The industrial and regional classification of the union's membership was then carried out according to the general procedure described above. Generally speaking, ASSET catered for two groups of personnel: 'foremen' and 'other technicians'. It also claimed to cater for executive grades but the number of such personnel in the union was negligible. Unfortunately, ASSET did not know the exact number of members in each occupational group. Consequently, the Industrial Officer in charge of each District of the union was interviewed and asked to estimate the proportion of his membership which was in the two groups. The national weighted average of all the estimates indicated that approximately 50 per cent of the union's membership was among foremen and 50 per cent among 'other technicians'. Thus after the industrial and geographical classification of ASSET's membership was completed, one-half of its membership was assigned to the foremen category and the other half to 'other technicians'.

(c) *BISAKTA*. As was explained in the notes to Table 3.7, a list of the white-collar branches in BISAKTA was obtained and then the membership figures for these branches were abstracted from the quarterly list of branch audits. Each branch's membership is not audited every quarter; for each branch the figure closest to 31 December 1963 was used (for the December 1963 audit see *Quarterly Report*, 31 December 1963, pp. 251–8). All of BISAKTA's membership is in the iron and steel industry, so industrial classification was no problem. Most of the branches have a geographical or company name and, in addition, are arranged according to the union's geographical divisions, so regional classification was relatively easy. Occupational classification was made possible by the union's practice, largely at management insistence, of keeping different types of white-collar personnel in different branches. The names of most of the white-collar branches indicate the type of personnel covered, e.g., 'Britannia Clerical', 'Tees-side Foremen', etc. The occupational composition of BISAKTA's membership as determined by analysing the branch audits was: clerks, 63 per cent; other technicians, 19 per cent; foremen, 16 per cent; other white-collar workers, 2 per cent. The membership of the few mixed white-collar branches was distributed on the same basis.

(d) *CAWU*. From the master membership file in the CAWU's head office it was possible to determine the names and addresses of the firms in each branch and the membership in each firm. Then the general procedure described above was carried out. The membership of the union is entirely composed of clerical and administrative workers so all of it was assigned to the 'clerks' category of Table 3A.3.

(e) *DATA*. DATA's membership was classified by analysing the returns the union received from its annual statistical survey. Between October and March the union sends out a questionnaire to each of its members requesting information on salaries. It also asks for the name, address, and product of the firm in which the member works. In the October 1963–March 1964 survey, 41,046 of the 61,446 members or 67 per cent of the total membership returned these questionnaires. These questionnaires were then classified industrially and geographically according to the general procedure described above. It was assumed that the 67 per cent return was a representative sample of DATA's total membership, and the total draughtsmen membership

TABLE 3A.3

Total White-Collar Union Membership in Manufacturing Industries in Great Britain by Industry and by Occupation, 1964

Occupational group	Food, drink, tobacco	Chemical	Metal manuf.	Metal N.E.S.	Eng. elect.	Ship. M.E.	Vehicles	Textiles	Leather, fur	Clothing, footwear	Bricks, etc.	Timber, furn, etc.	Paper, print, pub.	Other manuf.	All manuf.
1. Foremen	..	364	2,025	567	9,232	285	3,192	9,590	24	..	242	390	n/a	588	26,499
2. All scientists, technologists, technicians	130	4,120	5,184	262	44,112	4,023	19,676	308	2	24	556	280	21	1,073	79,771
(a) Scientists, technologists	33	2,136	148	..	1,280	77	162	50	..	6	79	..	4	69	4,044
(b) All technicians	97	1,984	5,036	262	42,832	3,946	19,514	258	2	18	477	280	17	1,004	75,727
(i) Draughtsmen	..	176	2,234	..	32,694	3,434	15,848	..	..	..	..	..	..	217	54,603
(ii) Other technicians	97	1,808	2,802	262	10,138	512	3,666	258	2	18	477	280	17	787	21,124
3. Clerks	4,592	1,735	12,595	1,337	33,749	556	22,701	1,715	..	180	1,508	112	12,310	3,498	96,588
4. Other white-collar workers	..	252	..	..	1,655	160	803	..	..	..	..	..	19,022	12	21,913
5. All white-collar workers	4,722	6,228	20,056	2,166	88,748	5,024	46,372	11,613	26	204	2,306	782	31,353	5,171	224,771

of 58,485 and the total tracer membership of 2,961 in 1963 were distributed on the same basis. The draughtsmen were assigned to the 'draughtsmen' category of Table 3A.3, and the tracers to the 'other white-collar workers' group. DATA also has a few 'other technicians' in membership, but the numbers in this group are not yet significant. Nevertheless, their inclusion in the draughtsmen category slightly inflates the density figures for draughtsmen and understates the density figures for 'other technicians'.

(f) *NATSOPA*. Among white-collar workers NATSOPA recruited only clerical and administrative grades and restricted itself entirely to the paper, printing, and publishing industry. Thus the occupational and industrial classification of its white-collar membership was straightforward. The union supplied a geographical breakdown of its membership.

(g) *NUJ*. A list of all the branches in the NUJ and their membership at December 1963 is given in the union's *Annual Report*, 1964, pp. 53–6. All the membership in Freelance, Radio and T.V., and Public Relations branches was excluded (the Ministry of Labour classifies these activities to professional and scientific services, and miscellaneous services), and then the remaining membership was assigned to the 'other white-collar workers' category in the paper, printing, and publishing industry. Most of the branches cover a specific geographical area and hence it was relatively easy to classify the membership regionally.

(h) *TGWU*. Approximately 30 per cent of the TGWU's white-collar membership was situated in the union's Region 1 (London and Home Counties). From the white-collar membership file in the regional office it was possible to determine the names and addresses of the firms in each branch and the membership in each firm. At the request of head office the other twelve regions of the TGWU analysed their white-collar membership and supplied the author with the names and addresses of the firms covered, the membership in each firm, and the product manufactured. The membership was then distributed industrially and regionally according to the general procedure described above. Among white-collar workers, the TGWU caters in the main for clerks and supervisors. Almost all its supervisory members are employed in public transport; the number in private industry is negligible. Consequently, all the TGWU's white-collar membership in manufacturing industries was assigned to the 'clerks' category of Table 3A.3.

The white-collar membership of the above unions accounted for almost 95 per cent of total white-collar union membership in manufacturing industries. In order to determine what other unions catered for white-collar workers in manufacturing, a questionnaire was sent to all unions organizing in the manufacturing area of the economy plus most of the major purely white-collar unions. The questionnaire asked the union if it had any white-collar workers in manufacturing industries in membership and, if so, to give an occupational, industrial, and geographical analysis of it. During the summer of 1964, 210 questionnaires were sent out and almost 60 per cent of these were completed and returned. Often, additional correspondence was engaged in to clarify aspects of the union's initial reply. Many of these unions restricted their white-collar recruiting to a single occupational group and/or a single industry and thus classification was relatively easy.

The following unions had white-collar members in manufacturing industries in 1964 and were therefore included in Tables 3A.3 and 3A.4: Amalgamated Engineering Union; Electrical Trades Union; National Union of General and Municipal Workers; National Union of Stove Grate and General Metal Workers; Heating and Domestic Engineers' Union; United Patternmakers Association; Philanthropic

TABLE 3A.4

Total White-Collar Union Membership in Manufacturing Industries in Great Britain by Industry and by Region, 1964

Region	Food, drink, tobacco	Chemical	Metal manuf.	Metal N.E.S.	Eng. elect.	Ship M.E.	Vehicles	Textiles	Leather, fur	Clothing, footwear	Bricks, etc.	Timber, furn., etc.	Paper, print, pub.	Other manuf.	All manuf.
1. Scotland	187	312	1,586	53	8,734	2,148	2,642	266	4	..	139	22	1,928	493	18,514
2. Northern	87	525	3,330	176	6,231	1,428	184	179	..	..	..	34	875	245	13,294
3. North-west	299	1,617	1,357	223	17,047	551	7,193	6,181	..	24	637	43	2,944	1,011	39,127
4. East and West Ridings of Yorkshire and East and West Midlands	2,282	1,004	6,299	595	25,572	288	16,317	4,213	22	72	503	67	2,691	1,631	61,556
5. Wales	64	132	6,836	369	2,715	19	1,387	503	..	50	46	299	533	240	13,193
6. Eastern and Southern	386	267	308	605	8,865	457	7,683	3	..	..	211	64	1,756	346	20,951
7. London and South-East	835	2,128	328	145	16,569	88	4,705	55	..	58	346	235	19,824	465	45,781
8. South-western	582	243	12	..	3,015	45	6,261	213	..	..	424	18	802	740	12,355
9. ALL-REGIONS	4,722	6,228	20,056	2,166	88,748	5,024	46,372	11,613	26	204	2,306	782	31,353	5,171	224,771

Society of Journeymen Coopers of Sheffield and District; National Union of Lock and Metal Workers; British Association of Chemists; Chemical Workers Union; Tobacco Workers' Union; United French Polishers' Society; Supervisory Staffs Federation of the Glove Industry; London Foremen Tailors' Mutual Association; Institute of Journalists; National Union of Printing, Bookbinding and Paper Workers; Civil Service Clerical Association (a few members in Royal Ordnance Factories); General Union of Associations of Loom Overlookers; National Federation of Scribbling Overlookers Association; Managers and Overlookers' Society; Yorkshire Association of Power Loom Overlookers; Textile Officials' Association; Scottish Lace and Textile Workers Union; Halifax and District Carpet Power Loom Tuners' Association; Trade Society of Machine Calico Printers; Amalgamated Society of Textile Workers and Kindred Trades; and the National Woolsorters' Society.

The following unions reported that they had a few white-collar members in manufacturing industries but were unable to determine the number: National Union of Leather Workers; Amalgamated Society of Woodworkers; Amalgamated Society of Boilermakers, Shipwrights, Blacksmiths and Structural Workers; National Union of Commercial Travellers; Union of Shop, Distributive and Allied Workers; Amalgamated Society of Lithographic Printers; National Union of Furniture Trade Operatives; Birmingham and Midland Sheet Metal Workers' Society; National Union of Boot and Shoe Operatives; Electrical Power Engineers' Association; and the Association of Supervising Electrical Engineers. The following public service unions reported that they had membership among white-collar workers in railway workshops and Royal Ordnance Factories: National Union of Railwaymen; Transport Salaried Staffs' Association; Society of Technical Civil Servants; Society of Civil Servants; Association of Government Supervisors and Radio Officers; and the Institution of Professional Civil Servants.

The Ministry of Labour's definition of a trade union, given in the notes to Table 3.1 (*supra*, p. 195) was followed in constructing Tables 3A.3 and 3A.4. However, the membership of three organizations of white-collar personnel listed in the *Directory* were excluded from the present survey. It was clear from the rules of the British Pottery Managers' Association and the Printers' Managers and Overseers Association and from correspondence with the general secretaries of these two organizations that they were not trade unions within the scope of the above definition. Their main purpose was to provide professional education and superannuation benefits. The British Iron and Steel Management Association caters for all managerial grades in the iron and steel industry and has been excluded because managers are outside the scope of the present study. BISMA was founded when the iron and steel industry was nationalized in 1951 but made little headway after the industry was denationalized in 1953. In 1964 it had approximately 400 members and has expanded since the industry was renationalized. As such, BISMA is not an exception to the general rule that managerial personnel in private industry tend to advance their interests by individual rather than collective bargaining.

Whether or not press telegraphists, proof readers, and lithographic artists are white-collar employees is a very debatable question. Fortunately, this question did not have to be resolved in this study. The memberships of the National Union of Press Telegraphists, the Association of Correctors of the Press, and the Society of Lithographic Artists, Designers, Engravers and Process Workers had to be excluded from Tables 3A.3 and 3A.4 on grounds of comparability. The Ministry of Labour does not include the occupations covered by these unions in its white-collar labour force figures.

The density figures in Table 3.8 are obviously subject to some error. Undoubtedly, the author's industrial classification of some firms, particularly those with joint products, does not always agree with that of the government statisticians. (Since the Statistics of Trade Act, 1947, prohibits the disclosure of any statistical information about individual enterprises, Ministry of Labour statisticians are not able to inform individual researchers how various firms are industrially classified.) Moreover, because it was not possible to classify USDAW's membership on an industrial basis, the density figures for food, drink, and tobacco are probably understated. Due to the difficulties in obtaining membership data for foremen (see *supra*, p. 200) the density figures for foremen are probably also somewhat understated. But, in general, it is felt that Table 3.8 gives a reliable account of the differences in the density of white-collar unionism between various occupations and industries in manufacturing.

The labour force figures used to construct Table 3.9 are subject to a wide margin of error and are not strictly comparable with the union membership figures, so this table should be used with great caution.

TABLE 4.1

This table is taken from 'The Size of Manufacturing Establishments', *Ministry of Labour Gazette*, lxx (April 1962), p. 145.

The unit measured in this table is not the firm or the enterprise but the establishment. 'In most cases the establishment is a single factory engaged in one type of industrial activity. Establishments have been counted separately and firms with more than one establishment are represented more than once in the figures.' But there are some exceptions to this, for these see the above reference.

TABLE 4.2

The figures for foremen were derived from Occupation Table 9 and those for all the other occupational groups from Occupation Table 2 of the 1961 *Census of Population*.

TABLE 4.3

This table was abstracted from Industry Table 13 of the 1961 *Census of Population*. These figures not only include foremen supervising manual workers but also those supervising white-collar workers. But in manufacturing industries the former make up approximately 90 per cent of total foremen so the inclusion of the latter is unlikely to affect the totals to any great extent. (Compare 'foremen and supervisors—manual' as given by Industry Table 5 with 'foremen and supervisors' as given by Industry Table 13.)

TABLE 5.1

This table was derived from Guy Routh, *Occupation and Pay in Great Britain 1906–60* (Cambridge: Cambridge University Press, 1965), tables 30, 33, 37, and 47. Where Routh gives quartiles, medians, or deciles for clerks, averages were calculated by giving a weight equal to one-third of the number of clerks in the relevant sector. Routh's skilled, semi-skilled, and unskilled manual categories were averaged together (using the weights given for 1951 in his table 1) to obtain the 'all manual workers' category.

TABLE 5.2

This table was obtained by converting the average annual money earnings figures given in Table 5.1 into index numbers with 1922–4 as the base year, and then taking the white-collar indices for each year as a percentage of the manual index for that year.

TABLE 5.3

The white-collar earnings figures for 1959–63 are derived from the annual October survey of white-collar earnings begun in 1959. See, for example, 'Earnings of Administrative, Technical and Clerical Employees, October 1963', *Ministry of Labour Gazette*, lxxii (March 1964), pp. 92–3. The Ministry of Labour supplied unpublished information which made it possible to separate the leather, leather goods, and fur industry from other manufacturing industries. They also supplied unpublished data which made it possible to compute the average annual earnings of all white-collar workers (column 7).

The average annual earnings of manual workers for 1959–63 were supplied from unpublished sources by the Ministry of Labour. They are comparable with the white-collar earnings figures. The figures for males include the earnings of men (21 years and over) and the earnings of youths and boys (under 21). The figures for females include the earnings of women (18 years and over), both full-time and part-time, and the earnings of girls (under 18).

In computing the 1954–63 relatives, the white-collar and manual earnings figures used for 1963 were obtained from the sources indicated above. For 1954, both white-collar and manual earnings figures were obtained from the Board of Trade, *The Report on the Census of Production for 1958* (London: HMSO, 1962), part 133, table 3.

For a definition of the term 'white-collar' as used in this table see the notes to Table 2.5.

TABLE 5.4

The white-collar earnings figures for each manufacturing industry are the same as those used in Table 5.3 for the period 1959–63. The earnings indices for male and female white-collar employees in the economy as a whole were abstracted from 'Index of Average Salaries', *Ministry of Labour Gazette*, lxxii (May 1964), p. 195. A similar index for males and females combined was obtained from *Statistics on Incomes, Prices, Employment and Production*, No. 13 (June 1965), p. 48, table B. 17.

TABLE 5.5.

Derived from *Occupational Pension Schemes: A Survey by the Government Actuary* (London: HMSO, 1958), p. 4, and *Occupational Pension Schemes: A New Survey by the Government Actuary* (London: HMSO, 1966), p. 12.

TABLE 6.1

The labour force figures used in computing these averages are those appearing in Table 2A.3, *supra*, p. 196. The number of establishments in each industry were obtained from *The Report on the Census of Production for 1958*, part 133, table 4, and are given in the following table.

TABLE 6A.1

Number of Establishments in Manufacturing Industries in the United Kingdom, 1958

Industry	Number of establishments
Food, drink, and tobacco	9,233
Chemicals and allied	3,566
Metal manufacture	2,876
Metal goods N.E.S.	10,588
Engineering and electrical goods	14,992
Shipbuilding and marine engineering	1,255
Vehicles	2,289
Textiles	8,461
Leather, leather goods, and fur	1,945
Clothing and footwear	9,592
Bricks, pottery, glass, cement, etc.	5,252
Timber, furniture, etc.	9,976
Paper, printing, and publishing	9,371
Other manufacturing industries	3,389
All manufacturing industries	92,785

For a definition of the term 'establishment' as used in this table see the notes to Table 4.1.

FIGURE 7.1

Trade Unions and the Public in 1964 (London: The Gallup Poll, 1964).

FIGURE 7.2

The membership figures of the major white-collar unions in private industry are given in Table 7A.1. Those of ASSET, the AScW, the CAWU, DATA, and the NUJ were obtained from the Registrar of Friendly Societies. Those of the NACSS and NATSOPA were supplied by these unions. The membership figures of NATSOPA for the period 1920–43 are for the London Clerical Branch only. Those for 1944 onwards represent the union's total clerical membership. The manner in which BISAKTA's membership figures were obtained is described in the notes to Table 3.7. The membership figures given in Table 7A.1 do not always agree with those in Table 3.7 primarily because they are given for different months of the year.

APPENDIX A

TABLE 7A.I

The Membership of Major White-Collar Unions in Manufacturing Industries, 1890–1964

Year	AScW	ASSET	BISAKTA	CAWU	DATA	NACSS	NATSOPA	NUJ
1890								
1891				..				
1892				30				
1893				30				
1894				59				
1895				28				
1896				39				
1897				50				
1898				68				
1899				80				
1900				82				
1901				70				
1902				80				
1903				85				
1904				84				
1905				84				
1906				350				
1907				750				738
1908				1,000				1,004
1909				1,870				1,525
1910				3,166				1,925
1911				5,225				2,155
1912				8,840				3,338
1913				11,750	200			4,407
1914				10,206	350			3,232
1915				10,843	750			3,127
1916				12,738	2,500			3,095
1917				26,572	9,000			3,342
1918	545	..		36,302	10,911			3,629
1919	428	..		43,222	13,500			4,343
1920	681	2,905		33,949	14,570		2,000	4,888
1921	808	2,758		14,204	11,460		1,500	4,680
1922	809	2,159		11,044	10,764	6,129	1,590	4,190
1923	821	1,770		7,442	9,975	5,263	1,697	4,275
1924	725	1,938		7,056	10,176	5,272	2,134	4,484
1925	922	2,006		7,570	10,730	5,649	2,565	4,827
1926	930	1,749		7,303	10,785	5,047	2,729	4,579
1927	1,700	1,498		7,183	10,794	5,120	2,314	4,522
1928	1,500	1,443		7,397	10,690	5,056	2,335	4,638
1929	1,250	1,859		7,666	10,735	4,861	2,340	5,071
1930	1,216	1,762		8,146	11,670	5,270	2,572	5,486
1931	1,094	1,839		7,482	12,147	5,074	2,759	5,477
1932	1,154	1,803		7,362	11,188	4,891	2,886	5,226
1933	900	1,636		7,510	10,943	5,106	3,120	5,182
1934	1,109	1,654		8,100	11,400	5,335	3,412	5,441
1935	1,031	1,625		9,030	12,140	6,040	3,623	5,806
1936	800	1,585		10,335	15,147	7,960	3,976	6,090
1937	988	1,606	..	12,423	17,920	9,115	4,322	6,522
1938	1,177	1,674	..	14,217	20,179	9,214	4,738	6,978
1939	1,379	1,743	..	15,943	23,137	9,934	4,787	7,245
1940	1,764	1,829	..	18,478	27,350	10,038	4,513	6,984
1941	3,246	2,142	..	21,470	30,920	13,137	4,702	7,031
1942	9,474	5,833	..	29,422	34,383	17,020	4,888	7,357
1943	14,010	9,823	..	33,902	38,418	18,964	5,096	6,627
1944	16,275	11,495	..	30,093	40,752	20,746	5,977	7,897

TABLE 7A.1 (cont.)

Year	AScW	ASSET	BISAKTA	CAWU	DATA	NACSS	NATSOPA	NUJ
1945	15,632	9,661	..	25,247	43,466	18,341	5,257	8,216
1946	17,158	10,618	..	31,370	47,038	22,428	5,995	9,277
1947	18,387	15,090	..	34,127	47,668	24,957	6,506	9,711
1948	15,623	15,069	4,774	37,071	46,734	27,620	6,846	10,256
1949	14,133	11,592	4,609	33,429	46,792	27,147	7,446	10,559
1950	13,206	10,934	4,797	31,781	46,712	29,133	7,879	11,267
1951	13,014	12,738	6,002	38,266	49,039	32,936	7,873	11,684
1952	11,190	13,971	6,154	43,014	49,994	34,391	8,627	12,013
1953	11,592	13,975	6,163	43,337	48,642	34,615	9,153	11,815
1954	11,289	14,199	6,226	46,113	54,325	35,202	9,855	12,175
1955	11,740	15,422	6,306	50,375	50,438	36,525	10,462	12,874
1956	12,446	16,532	6,433	52,993	52,725	37,044	10,911	13,499
1957	11,886	18,970	6,237	50,517	54,449	36,373	10,981	13,910
1958	11,474	19,930	6,479	47,990	55,664	37,143	11,066	13,917
1959	11,720	19,942	6,601	49,306	56,242	38,150	10,808	14,371
1960	11,888	22,945	7,167	56,501	58,945	44,491	10,800	14,737
1961	14,119	25,270	7,662	60,739	61,368	44,655	10,615	15,053
1962	16,170	28,636	7,430	65,336	62,513	47,571	11,308	16,034
1963	18,195	30,159	7,520	71,527	59,679	51,337	11,802	16,478
1964	19,796	33,880	9,039	75,558	60,381	56,541	12,250	17,030

APPENDIX B

A NOTE ON THE REGRESSION TECHNIQUES
USED IN THIS STUDY

THIS is not the place to give a complete exposition of regression analysis. The reader who does not understand this method of analysis and wishes to have more information regarding it than was provided in Chapter I should consult the sources referred to there (*supra*, p. 8 n. 1) as well as more advanced texts such as C. E. V. Leser, *Econometric Techniques and Problems* (London: Griffin, 1966), chaps. 2 and 3, and J. Johnston, *Econometric Methods* (New York: McGraw-Hill, 1963), chaps. 1-4. Only a few comments regarding the specific way in which the technique was used in this study will be made here.

The technique of simple regression analysis is used in this study to analyse the relationship between: (*a*) the density of unionization among foremen and the age distribution of foremen, *supra*, p. 47; (*b*) the density of white-collar unionism and the earnings of white-collar workers, *supra*, pp. 55 and 61; and (*c*) the density of white-collar unionism and the degree of employment concentration, *supra*, p. 79. Except where otherwise noted, the equation for each occupational group is fitted to fourteen industry observations; since the 'all manufacturing' observation is simply the total of these fourteen industry observations, it was excluded from the analysis.

Because of the relatively small number of industry observations, data from the various occupational groups has been pooled where possible and subjected to a multiple regression analysis. That is, a single equation was fitted to the seventy observations which result from pooling the data for the following occupational groups: (*a*) foremen, (*b*) scientists and technologists, (*c*) draughtsmen, (*d*) other technicians, and (*e*) clerks. The observations for (*a*) all scientists, technologists, and technicians, (*b*) all technicians, and (*c*) all white-collar workers, were excluded as they are simply the total of the observations for various of the occupational groups referred to above. The observations for other white-collar workers were also excluded because of the extremely heterogeneous nature of this category and the almost complete absence of unionism from it except among journalists in the paper, printing, and publishing industry. It proved possible to pool the data in analysing the relationship between: (*a*) the density of white-collar unionism and the proportion of women in the labour force, *supra*, pp. 41-2, and (*b*) the density of white-collar unionism and the degree of employment concentration, *supra*, pp. 79-80.

Pooling the data in this way implies that the relationship between the density of white-collar unionism and the various independent variables is the same for all occupational groups and all industries. But this need not be the case. The relationship may change systematically from one occupation to another and from one industry to another as a result, for example, of variations in employment distributions between different occupations and different industries (see *supra*, pp. 77-8). To allow for the possibility that the level as well as the slope of the equation may shift by occupation and by industry, dummy variables were included in each equation. These variables take the value of zero or one according to whether the particlar observation is from that specific occupation or industry.[1]

Allowing for only the level of the equation[2] to shift with occupation, then, for a five-occupation model, the linear equation $y = a+bx+u$ becomes:

$$y = a_0+a_1\phi_1+a_2\phi_2+a_3\phi_3+a_4\phi_4+bx+u \tag{1}$$

where $\phi_i = 1$ in the ith occupation and 0 in all other occupations ($i = 1, 2, 3$, etc.). Allowing for only the slope of the equation to shift with occupation, then for a five-occupation model, the linear equation $y = a+bx+u$ becomes:

$$y = a_0+b_1\phi_1x+b_2\phi_2x+b_3\phi_3x+b_4\phi_4x+bx+u. \tag{2}$$

Allowing for both the level and the slope of the equation to shift with occupation, then for a five-occupation model, the linear equation $y = a+bx+u$ becomes:

$$y = a_0+a_1\phi_1 \ldots +a_4\phi_4+b_1\phi_1x \ldots +b_4\phi_4x+bx+u. \tag{3}$$

Allowing for only the level of the equation to shift with industry, then, for a fourteen-industry model, the linear equation $y = a+bx+u$ becomes:

$$y = a_0+a_1I_1 \ldots +a_{13}I_{13}+bx+u \tag{4}$$

where $I_i = 1$ in the ith industry and 0 in all other industries. Allowing for only the slope of the equation to shift with industry, then, for a fourteen-industry model, the linear equation $y = a+bx+u$ becomes:

$$y = a_0+b_1I_1x \ldots +b_{13}I_{13}x+bx+u. \tag{5}$$

Allowing for both the level and the slope of the equation to shift with industry, then, for a fourteen-industry model, the linear equation $y = a+bx+u$ becomes:

$$y = a_0+a_1I_1 \ldots +a_{13}I_{13}+b_1I_1x \ldots +b_{13}I_{13}x+bx+u. \tag{6}$$

Allowing for only the level of the equation to shift with both occupation and industry, then, for a five-occupation-fourteen-industry model, the linear equation $y = a+bx+u$ becomes:

$$y = a_0+a_1\phi_1 \ldots +a_4\phi_4+a_1I_1 \ldots +a_{13}I_{13}+bx+u. \tag{7}$$

Allowing for only the slope of the equation to shift with both occupation and industry, then, for a five-occupation-fourteen-industry model, the linear equation $y = a+bx+u$ becomes:

$$y = a_0+b_1\phi_1x \ldots +b_4\phi_4x+b_1I_1x \ldots +b_{13}I_{13}x+bx+u. \tag{8}$$

Allowing for both the level and the slope of the equation to shift with both occupation and industry, then, for a five-occupation-fourteen-industry model, the linear equation $y = a+bx+u$ becomes:

$$y = a_0+a_1\phi_1 \ldots +a_4\phi_4+a_1I_1 \ldots +a_{13}I_{13}+b_1\phi_1x$$
$$\ldots +b_4\phi_4x+b_1I_1x \ldots +b_{13}I_{13}x+bx+u. \tag{9}$$

[1] This is a standard econometric procedure and is used, for example, by L. R. Klein *et al.*, *An Econometric Model of the United Kingdom* (Oxford: Blackwell, 1961), pp. 42–4, in adjusting for seasonal variation. The account of the procedure which is given below relies very heavily on this source. This technique is also used by Keith Cowling and David Metcalf, 'Wage–Unemployment Relationships: A Regional Analysis for the U.K. 1960–65', *Bulletin of the Oxford University Institute of Economics and Statistics*, xxix (February 1967), pp. 30–9, in adjusting for regional variation. See also Johnston, op. cit., pp. 221–8.

[2] The reader who does not find the following equations very illuminating may find that his understanding of this technique will be increased by studying the graphical description given by Klein, op. cit., pp. 45–6.

In analysing the relationship between variables, all these equations were fitted to the seventy observations. But, in each case, the results are cited only for those equations which possessed the correct signs and were most significant.

The above exposition has been given only in terms of linear relationships. But the following curvilinear relationships were also tested for: semilog, double-log, inverse, and log inverse (see Johnston, op. cit., chap. 2). The linear relationships always proved to be the most significant, and all results cited in the text are for such relationships.

All the computations were carried out by the Oxford University Computing Laboratory using their multiple regression programme FAKAD 2300 designed by Emiel van Broekhoven. The measure of correlation calculated by this programme is the $\bar{R}^2$ and has been adjusted for degrees of freedom.

APPENDIX C

BIBLIOGRAPHICAL NOTE

T H E sources which were consulted most frequently in the research for this study have been cited in the footnotes, and there seems little point in listing them once again here. Nor does there seem much point in compiling here a comprehensive bibliography on British white-collar unionism as this has been done elsewhere. See G. S. Bain and Harold Pollins, 'The History of White-Collar Unions and Industrial Relations: A Bibliography', *Labour History*, No. 11 (Autumn 1965), pp. 20–65. In addition, many of the more important foreign sources are listed in *Bibliography on Non-Manual Workers* (Geneva: ILO, 1959). But it might be of some use to the student who wishes to pursue this subject further to list here the major works in Britain and a few other countries.

The most important works on the subject in Great Britain include the following: F. D. Klingender, *The Condition of Clerical Labour in Great Britain* (London: Martin Lawrence, 1935); David Lockwood, *The Blackcoated Worker* (London: Allen & Unwin, 1958); J. R. Dale, *The Clerk in Industry* (Liverpool: Liverpool University Press, 1962); Geoffrey Millerson, *The Qualifying Associations* (London: Routledge & Kegan Paul, 1964); Kenneth Prandy, *Professional Employees* (London: Faber, 1965); and R. M. Blackburn, *Union Character and Social Class* (London: Batsford, 1967).

In the United States, as in Britain, histories of some of the more important white-collar unions have been written, and most of these are listed in Maurice F. Neufeld, *A Representative Bibliography of American Labor History* (Ithaca, N.Y.: Cornell University Press, 1964), especially pp. 74, 94–8, 111, 124–9. The more analytical studies include C. Wright Mills, *White Collar* (New York: Oxford University Press, 1951); G. Strauss, 'White-Collar Unions Are Different!' *Harvard Business Review*, xxxii (September–October 1954), pp. 73–82; B. Goldstein, 'Some Aspects of the Nature of Unionism Among Salaried Professionals in Industry', *American Sociological Review*, xx (April 1955), pp. 199–205; J. W. Riegel, *Collective Bargaining as Viewed by Unorganized Engineers and Scientists* (Ann Arbor: University of Michigan, Bureau of Industrial Relations, 1959); S. M. Lipset, 'The Future of Non-Manual Unionism' (unpublished paper, University of California, Institute of Industrial Relations, Berkeley, 1961); Richard E. Walton, *The Impact of the Professional Engineering Union* (Boston, Mass.: Harvard University, Division of Research, 1961); Albert A. Blum, *Management and the White-Collar Union* (New York: American Management Association, 1964); Archie Kleingartner, *Professionalism and Salaried Worker Organization* (Madison, Wisconsin: University of Wisconsin, Industrial Relations Research Institute, 1967); and by the same author, 'The Organisation of White-Collar Workers', *British Journal of Industrial Relations*, vi (March 1968), pp. 79–93. There is also a symposium on 'Professional Workers in Industry' in *Industrial Relations*, ii (May 1963), pp. 7–65.

The writing on the subject in Australia has been small in volume but generally analytical in nature. See, in particular, D. W. Rawson, 'The Frontiers of Trade Unionism', *Australian Journal of Politics and History*, i (May 1956), pp. 196–209; N. F. Dufty, 'The White Collar Unionist', *Journal of Industrial Relations*, iii (October 1961), pp. 151–6; R. M. Martin, 'Class Identification and Trade Union Behaviour: The Case of Australian White Collar Unions', *Journal of Industrial Relations*, vii (July 1965), pp. 131–48; and by the same author, *White-Collar Unions in Australia* (Sydney: Australian Institute of Political Science, 1965).

Finally, there are two sources which contain essays on white-collar unionism in the above countries as well as in Austria, France, Germany, Israel, Japan, Norway, and Sweden. See the symposium on 'Professional and White-Collar Unionism: An International Comparison' in *Industrial Relations*, v (October 1965), pp. 37–150; and Adolf Sturmthal, editor, *White-Collar Trade Unions* (London: University of Illinois Press, 1966).

INDEX

[The author is grateful to Kate Buckley who skilfully compiled this index.]

PRINTED IN GREAT BRITAIN
AT THE UNIVERSITY PRESS, OXFORD
BY VIVIAN RIDLER
PRINTER TO THE UNIVERSITY